044291

Mill Woods

RC
565
.B52
v.4

The Biology of alcoholism.

SOCIAL ASPECTS
OF ALCOHOLISM

THE BIOLOGY OF ALCOHOLISM

Volume 1: Biochemistry

Volume 2: Physiology and Behavior

Volume 3: Clinical Pathology

Volume 4: Social Aspects of Alcoholism

Volume 5: The Treatment and Rehabilitation of the Chronic Alcoholic

SOCIAL ASPECTS OF ALCOHOLISM

Edited by
Benjamin Kissin and Henri Begleiter

Division of Alcoholism and Drug Dependence
Department of Psychiatry
State University of New York
Downstate Medical Center
Brooklyn, New York

PLENUM PRESS • NEW YORK-LONDON

Library of Congress Cataloging in Publication Data

Kissin, Benjamin, 1917-
　The biology of alcoholism.

　Includes bibliographies and index.
　CONTENTS: v. 1. Biochemistry.—v. 2. Physiology and behavior.—v. 3. Clinical pathology.—v. 4. Social aspects of alcoholism—v. 5. The treatment and rehabilitation of the chronic alcoholic.
　1. Alcoholism. 2. Alcoholism—Physiological effect. I. Begleiter, Henri, joint author. II. Title. [DNLM: 1. Alcoholism WM274 K61b]
RC565.K52　　　　　　　　　　616.8'6'1　　　　　　　　　　74-131883
ISBN 0-306-37114-6 (v. 4)

© 1976 Plenum Press, New York
A Division of Plenum Publishing Corporation
227 West 17th Street, New York, N.Y. 10011

All rights reserved

No part of this book may be reproduced, stored in a retrieval system, or transmitted, in any form or by any means, electronic, mechanical, photocopying, microfilming, recording, or otherwise, without written permission from the Publisher

Printed in the United States of America

Contributors

Joan Ablon, *Department of Psychiatry, University of California School of Medicine, San Francisco, California; and The Community Mental Health Training Program, Langley Porter Neuropsychiatric Institute, San Francisco, California*

Margaret K. Bacon, *Department of Anthropology, Livingston College, Rutgers, the State University, New Brunswick, New Jersey*

Howard T. Blane, *University of Pittsburgh, Pittsburgh, Pennsylvania*

Don Cahalan, *School of Public Health, University of California, Berkeley, California*

Ira H. Cisin, *The George Washington University, Washington, D.C.*

Harold M. Ginzburg, *The Johns Hopkins University, School of Hygiene and Public Health, Baltimore, Maryland*

Edith S. Gomberg, *School of Social Work, University of Michigan, Ann Arbor, Michigan; and Rutgers University Center of Alcohol Studies, New Brunswick, New Jersey*

Dwight B. Heath, *Department of Anthropology, Brown University, Providence, Rhode Island*

Jan de Lint, *Addiction Research Foundation, Toronto, Ontario, Canada*

Wallace Mandell, *The Johns Hopkins University, School of Hygiene and Public Health, Baltimore, Maryland*

Kai Pernanen, *Social Studies Department, Addiction Research Foundation, Toronto, Ontario*

Robert E. Popham, *Addiction Research Foundation, Toronto, Ontario, Canada*

Paul M. Roman, *Department of Sociology, Tulane University, New Orleans, Louisiana*

Wolfgang Schmidt, *Addiction Research Foundation, Toronto, Ontario, Canada*

Harrison M. Trice, *Department of Organizational Behavior, New York State School of Industrial and Labor Relations, Cornell University, Ithaca, New York*

Julian A. Waller, *Department of Epidemiology and Environmental Health, University of Vermont, Burlington, Vermont*

Allan F. Williams, *Research Department, Insurance Institute for Highway Safety, Washington, D.C.*

Preface

The first three volumes of this series have dealt with materials which generally justify the title, *The Biology of Alcoholism*. This is only remotely true of the present volume, *Social Aspects of Alcoholism*, or of the final volume to come, *Treatment and Rehabilitation*. Except for small portions of the treatment section which involve pharmacotherapy, much of these last two volumes deals with the psychological aspects of alcoholism and still more with the social. It is interesting to review the evolution of this new pattern over the past seven years, a pattern which, had it existed initially, would have resulted, if not in a different format, at least in a different title.

Our initial selection of areas to be covered was influenced by our desire to present as "hard" data as possible, in an attempt to lend a greater aura of scientific rigor to a field which was generally considered as "soft." When we completed our review of this material in volumes 1–3, we recognized that what we might have gained in rigor, we had more than lost in completeness. These volumes presented a picture of a biological disease syndrome for which the remedies and preventive measures were presumably also biological. And yet, most workers in the field readily accept the significant contributions of psychological and social factors to the pathogenesis and treatment of alcoholism. To ignore these factors only because sociology may not yet be as rigorous as biochemistry would be as unreasonable as to deny the presence of political influence in government because "politics" is not an exact science.

Accordingly, we have devoted the last two volumes of this series to psychological and social factors as they relate to the pathogenesis, treatment and social consequences of alcoholism. The general terminology and style of these volumes are necessarily "looser" than in the previous ones. By no means, however, does this mean that the data or the conclusions based upon them are necessarily less valid. Not a few of the biological conclusions in the earlier volumes, based on presumably solid data, are overtly contradictory. An example of this is the question of the rate of ethanol metabolism in American Indians as against Caucasians, where it has been variously reported as lower,[1] the same,[2] and higher.[3] Science, whether physical, biological, or social, is a trial and error pursuit of knowledge, where impressions are subject to investigation, theories require validation, and "facts" are open to reappraisal. Not to have added these last two volumes would have been, in our opinion, a serious omission, which would have resulted, not only in a skewed presentation of the field, but also in a failure on our part to complete our original goal, the development of a comprehensive overview of the disease, alcoholism.

Benjamin Kissin
Henri Begleiter

New York
1976

---

[1] Fenna, D., Mix, L., Schaefer, O., and Gilbert, J. A. L. (1971). Ethanol metabolism in various racial groups, *Canadian Medical Association Journal* 105:472–475.

[2] Bennion, L. J. and Li, T.-K. (1976). Alcohol metabolism in American Indians and whites—lack of racial differences in metabolic rate and liver alcohol dehydrogenase, *The New England Journal of Medicine,* 294(1):9–13.

[3] Reed, T. E., Kalant, H., Gibbins, R. J., Kapur, B. M., and Rankin, J. G. (1976). Alcohol and acetaldehyde metabolism in Caucasians, Chinese and Amerinds, *Canadian Medical Association Journal,* in press.

Contents of Volume 4

Contents of Earlier Volumes

Volume 1 xix
Volume 2 xxi
Volume 3 xxii

Chapter 1
Alcohol Use in Tribal Societies
by Margaret K. Bacon

Introduction 1
Descriptive Studies 3
Range of Cultural Variation 7
Crosscultural Investigations 11
    Transcultural Variables in Drinking Behavior 11
    Sex Differences in the Use of Alcohol 14
    Determinants of Drinking Behavior 17
Discussion and Summary 29
References 31

Chapter 2
Anthropological Perspectives on the Social Biology of Alcohol: An Introduction to the Literature
by Dwight B. Heath

Introduction	37
General Sources	39
Alcohol in Archaeology and History	40
Alcohol and Linguistics	42
Alcohol and Physical Anthropology	42
Alcohol and Sociocultural Anthropology	44
Drinking Patterns: Customs and Cultures	44
Change and Persistence	46
Crosscultural Correlations	47
Prospective	48
References	50

Chapter 3
Drinking Behavior and Drinking Problems in the United States
by Don Cahalan and Ira H. Cisin

American Values and Attitudes Concerning Alcohol	77
Historical Heritage and Value Conflict in America	78
Current Contradictory State of Values and Attitudes	80
Current Drinking Practices in the United States	83
Findings of the 1964–65 Drinking Practices Survey	85
Correlates of Drinking	86
Ethnocultural Groups	86
What People Drink	90
Behavioral Correlates of Drinking	90
Alcohol-Related Problems in the United States	91
Definitions of Problems	92
Drinking Problems among Men 21–59	96
Other Factors Influencing Problem Drinking	99
Patterns of Change in Drinking Behavior	103
Summary of Survey Findings	103
Epidemiologic and Behavioral Science Perspectives	104
Alternative Models for Studying "Alcoholism" or "Problem Drinking"	104
Contributions to Social Control	109
References	112

Chapter 4
Alcoholism in Women
by Edith S. Gomberg

Introduction	117
Sex Differences	118
Sex Differences in Psychopathology	120
The Social Drinking of American Women	124
Sanctions	125
Intoxication	126
Effects of Drinking	127
Alcoholism in Women	128
Early Life Experience	129
Patterns of Drinking and Alcoholism	133
Psychodynamics	139
Complications of Alcoholism	149
Differences: Social Class and Race, Religion, and Ethnic Background	151
Therapy and Prognosis	155
Discussion	158
Suggestions for Research	159
References	161

Chapter 5
Youthful Alcohol Use, Abuse, and Alcoholism
by Wallace Mandell and Harold M. Ginzburg

Introduction	167
Youth's Perception of Adult Use of Alcohol	169
Prevalence of Alcohol Use among Youth	169
Factors Influencing Youth Alcohol Use Pattern	174
Influence of Parents	174
Influence of Friends	176
Sociocultural Factors and Youthful Drinking	177
Summary	179
Alcohol Use and Socially Unacceptable Behavior among Youth	180
Personal Characteristics and Alcohol Abuse among Youth	184
Juvenile Alcoholism	189
Youthful Characteristics Predictive of Adult Alcoholism	191
Juvenile Delinquency and Teenage Drinking	195
Discussion	197
References	202

Chapter 6
Family Structure and Behavior in Alcoholism: A Review of the Literature
by Joan Ablon

Introduction	205
Theoretical Approaches to the Study of the Family	205
The American Family	207
The Problem Family	208
The Literature on the Alcoholic Family	210
"The Wife of the Alcoholic"	211
Husbands of Women Alcoholics	216
Interaction in the Alcoholic Marriage	217
The Children of Alcoholics	226
Factors Affecting Family Actions in Regard to the Alcoholic Problem	227
Treatment	229
Group Therapy	230
Family Casework	230
Al-Anon Family Groups	232
Sociocultural Variables	234
Overview	235
The Importance of Sampling in the Direction of Future Research	238
References	239

Chapter 7
The Alcoholic Personality
by Allan F. Williams

Views on the Role of Personality Factors in Alcoholism	243
Approaches to Studying Personality Factors in Alcoholism	244
Studying Alcoholics	245
Alternative Methods	246
Studying Emotional Effects of Alcohol	247
Studies of Prealcoholics	249
Studies of Young Heavy Drinkers and Problem Drinkers	250
The Power and Dependency Theories of the Alcoholic Personality	253
Description of the Power Theory	253
Description of the Dependency Theory	254
Evidence for the Power Theory	254
Evidence for the Dependency Theory	255

Power and Dependency Concerns in Alcoholics 256
Effects of Alcohol on Power and Dependency Concerns 260
Status of the Power and Dependency Theories 262
Other Theories of the Alcoholic Personality 264
Jessor's Social Psychological Theory 264
The Tension Reduction Theory 266
Alcoholism in Women 268
References 269

Chapter 8
Alcoholism and Mortality
by Jan de Lint and Wolfgang Schmidt

Introduction 275
The Definition of Alcoholism 276
Coincidence Studies 278
Retrospective Studies 280
Prospective Studies 282
Cause-Specific Mortality 285
    Tuberculosis 286
    Syphilis 286
    Cancers of the Upper Digestive and Respiratory Tracts 287
    Cancer of the Liver 288
    Other Neoplasms 288
    Diabetes 289
    Alcoholism 290
    Heart Diseases 290
    Apoplexy 290
    Diseases of the Respiratory Organs 291
    Gastroduodenal Ulcers 291
    Cirrhosis of the Liver 292
    Accidents, Poisonings, and Violence 292
    Suicide 293
Some Specific Issues 294
    Social Class 294
    Male–Female Mortality 294
    Abstinence 295
    Type of Beverage 295
Postscript 295
References 296

Chapter 9
Alcohol and Unintentional Injury
by Julian A. Waller

Injury as a Public Health Problem	307
Historical Perspective	307
A Model of Injury Events and Their Outcome	308
Highway Crashes	310
History, and Problems of Measurement	310
Fatal Crashes	312
Nonfatal Crashes	315
Alcohol in the Preinjury, Injury, and Postinjury Phases	315
Other Transportation Injuries	318
Nontransportation Injury	319
Deliberate Injury	322
Alcohol in Combination with Other Drugs	323
Characteristics of Alcohol Users in Injury Events	324
Historical Perspective	324
Highway Injury	325
Nonhighway Injury	330
Countermeasures to Alcohol-Related Injury	331
Overview of Options within the Systems Approach	331
Controlling Availability of Alcohol	332
Control of User Behavior to Reduce Impairment or Exposure to Risk	333
Early Identification of Alcohol Abusers and Their Removal from Hazardous Situations	337
Environmental Controls Aimed at the Preinjury Phase	340
Moderation of Energy Transfer During the Injury Phase	342
Improvement of Emergency and Definitive Care for the Injured	343
Summary	344
References	344

Chapter 10
Alcohol and Crimes of Violence
by Kai Pernanen

Introduction	351
Review of Findings	354
Association between Acute Alcohol Use and Interpersonal Noninstrumental Crimes of Violence	355

Association between Alcoholism and Interpersonal
        Noninstrumental Crimes of Violence 364
Explanatory Models 369
    Explanation of the Positive Relationship between Alcohol
        Use and Crimes of Violence 371
    Explanation of the Positive Relationship between
        Alcoholism and Violent Crime 428
References 436

Chapter 11
Alcohol Abuse and Work Organizations
by Paul M. Roman and Harrison M. Trice

Introduction 445
Rationale for Work-Based Programs 446
    Costs of Alcohol Abuse to Work Organizations 446
    Preventive Potential in the Workplace 452
    A Neglected Rationale: Etiological Factors in Work Roles 454
The Emergence of "Occupational Alcoholism" 459
    Ingredients of Occupational Alcoholism Policies 460
    Organizational Support for Work-Based Alcoholism and
        Assistance Programs 462
    Formation of the Occupational Programs Branch of the National
        Institute on Alcohol Abuse and Alcoholism 465
    Other Organizations Involved in Occupational Alcoholism 469
    Development of Occupational Alcoholism and Assistance
        Programs for Federal Employees 477
Prospects and Problems for Occupational Alcoholism Programs 481
    The "Broad-Brush" or "Employee Assistance" Approach: Its
        Nature and Problems 482
    Constituency Support: The Beneficiaries of Occupational
        Programs 489
    Issues in the Identification Process 496
    Issues in the Referral Process 499
    Role of Labor Unions in Occupational Policies and Programs ... 500
    Evaluation of Occupational Alcoholism Programs 505
    Conclusion: Needed Research 509
References 514

Chapter 12
Education and the Prevention of Alcoholism
by Howard T. Blane

Introduction	519
Contemporary Models of Prevention	520
Social Science	520
Distribution of Consumption	526
Proscriptive	531
Traditional Public Health	532
Techniques of Prevention	535
Public Information and Education	536
Manipulation of Substance, Person, and Environmental Factors	541
Secondary Prevention	544
Education in School Systems	547
Basic Assumptions	547
Philosophies of Alcohol Education	548
Materials	552
Target Groups	553
Curricular Placement	554
Methods of Instruction	555
Teacher Preparation	556
Issues in Implementation	558
Issues in Contemporary Approaches to Prevention	561
Untested Assumptions and Interpretations	561
Research and Evaluation	562
Development of Realistic Preventive Policy	565
Summary	571
References	572

Chapter 13
The Effects of Legal Restraint on Drinking
by Robert E. Popham, Wolfgang Schmidt, and Jan de Lint

Introduction	579
Control of Outlet Frequency	581
Regulation of Type and Location of Outlets	586
Control of Hours and Days of Sale	591
Limitation of Drinking Age	592
Price Control	595

Differential Taxation 601
The Monopoly System of Control 606
Models of Prevention 609
References 619

Index 627

Contents of Earlier Volumes

Volume 1: Biochemistry

Chapter 1
Absorption Diffusion, Distribution, and Elimination of Ethanol: Effects on Biological Membranes
by Harold Kalant

Chapter 2
The Metabolism of Alcohol in Normals and Alcoholics: Enzymes
by J. P. von Wartburg

Chapter 3
Effect of Ethanol on Intracellular Respiration and Cerebral Function
by Henrik Wallgren

Chapter 4
Effect of Ethanol on Neurohumoral Amine Metabolism
by Aaron Feldstein

Chapter 5
The Role of Acetaldehyde in the Actions of Ethanol
by Edward B. Truitt, Jr., and Michael J. Walsh

Chapter 6
The Effect of Alcohol on Carbohydrate Metabolism: Carbohydrate Metabolism in Alcoholics
by Ronald A. Arky

Chapter 7
Protein, Nucleotide, and Porphyrin Metabolism
by James M. Orten and Vishwanath M. Sardesai

Chapter 8
Effects of Ethanol on Lipid, Uric Acid, Intermediary, and Drug Metabolism, Including the Pathogenesis of the Alcoholic Fatty Liver
by Charles S. Lieber, Emanuel Rubin, and Leonore M. DeCarli

Chapter 9
Biochemistry for Gastrointestinal and Liver Diseases in Alcoholism
by Carroll M. Leevy, Abdul Kerim Tanribilir, and Francis Smith

Chapter 10
Alcohol and Vitamin Metabolism
by Joseph J. Vitale and Joanne Coffey

Chapter 11
The Effect of Alcohol on Fluid and Electrolyte Metabolism
by James D. Beard and David H. Knott

Chapter 12
Mineral Metabolism in Alcoholism
by Edmund B. Flink

Chapter 13
Alcohol-Endocrine Interrelationships
by Peter E. Stokes

Chapter 14
Acute and Chronic Toxicity of Alcohol
by Samuel W. French

Chapter 15
Biochemical Mechanisms of Alcohol Addiction
by Jack H. Mendelson

Chapter 16
Methods for the Determination of Ethanol and Acetaldehyde
by Irving Sunshine and Nicholas Hodnett

Chapter 17
The Chemistry of Alcoholic Beverages
by Chauncey D. Leake and Milton Silverman

Volume 2: Physiology and Behavior

Chapter 1
Effects of Alcohol on the Neuron
by Robert G. Grenell

Chapter 2
Peripheral Nerve and Muscle Disorders Associated with Alcoholism
by Richard F. Mayer and Ricardo Garcia-Mullin

Chapter 3
The Effects of Alcohol on Evoked Potentials of Various Parts of the Central Nervous System of the Cat
by Harold E. Himwich and David A. Callison

Chapter 4
Brain Centers of Reinforcement and Effects of Alcohol
by J. St.-Laurent

Chapter 5
Factors Underlying Differences in Alcohol Preference of Inbred Strains of Mice
by David A. Rogers

Chapter 6
The Determinants of Alcohol Preference in Animals
by R. D. Myers and W. L. Veale

Chapter 7
Voluntary Alcohol Consumption in Apes
by F. L. Fitz-Gerald

Chapter 8
State-Dependent Learning Produced by Alcohol and Its Relevance to Alcoholism
by Donald A. Overton

Chapter 9
Behavioral Studies of Alcoholism
by Nancy K. Mello

Chapter 10
The Effects of Alcohol on the Central Nervous System in Humans
by Henri Begleiter and Arthur Platz

Chapter 11
Changes in Cardiovascular Activity as a Function of Alcohol Intake
by David H. Knott and James D. Beard

Chapter 12
The Effect of Alcohol on the Autonomic Nervous System of Humans: Psychophysiological Approach
by Paul Naitoh

Chapter 13
Alcohol and Sleep
by Harold L. Williams and A. Salamy

Chapter 14
Alcoholism and Learning
by M. Vogel-Sprott

Chapter 15
Some Behavioral Effects of Alcohol on Man
by J. A. Carpenter and N. P. Armenti

Volume 3: Clinical Pathology

Chapter 1
The Pharmacodynamics and Natural History of Alcoholism
by Benjamin Kissin

Contents of Earlier Volumes: Volume 3

Chapter 2
Heredity and Alcoholism
by *Donald W. Goodwin and Samuel B. Guze*

Chapter 3
Psychological Factors in Alcoholism
by *Herbert Barry, III*

Chapter 4
Interactions of Ethyl Alcohol and Other Drugs
by *Benjamin Kissin*

Chapter 5
Acute Alcohol Intoxication, The Disulfiram Reaction, and Methyl Alcohol Intoxication
by *Robert Morgan and Edward J. Cagan*

Chapter 6
Acute Alcohol Withdrawal Syndrome
by *Milton M. Gross, Eastlyn Lewis and John Hastey*

Chapter 7
Diseases of the Nervous System in Chronic Alcoholics
by *Pierre M. Dreyfus*

Chapter 8
Metabolic and Endocrine Aberrations in Alcoholism
by *D. Robert Axelrod*

Chapter 9
Liver Disease in Alcoholism
by *Lawrence Feinman and Charles S. Lieber*

Chapter 10
Diseases of the Gastrointestinal Tract
by *Stanley H. Lorber, Vicente P. Dinoso, Jr., and William Y. Chey*

Chapter 11
Acute and Chronic Pancreatitis
by *R. C. Pirola and C. S. Lieber*

Chapter 12
Diseases of the Respiratory Tract in Alcoholics
by Harold A. Lyons and Alan Saltzman

Chapter 13
Alcoholic Cardiomyopathy
by George E. Burch and Thomas D. Giles

Chapter 14
Hematologic Effects of Alcohol
by John Lindenbaum

Chapter 15
Alcohol and Cancer
by Benjamin Kissin and Maureen M. Kaley

Chapter 16
Alcoholism and Malnutrition
by Robert W. Hillman

Chapter 17
Rehabilitation of the Chronic Alcoholic
by E. Mansell Pattison

SOCIAL ASPECTS
OF ALCOHOLISM

CHAPTER 1

Alcohol Use in Tribal Societies

Margaret K. Bacon

Department of Anthropology
Livingston College
Rutgers, the State University
New Brunswick, New Jersey

INTRODUCTION

The drinking of alcoholic beverages is both ancient in origin and widespread throughout the peoples of the world. Forbes (1954) reports that wine was in use in Mesopotamia some time prior to 3000 B.C. The oldest known code of laws, that of Hammurabi of Babylonia (c. 1700 B.C.), regulated the sale of wine and forbade riotous assembly in the house of the wine seller. The medicinal use of alcohol dates back some four thousand years. A clay tablet found at Nippur, dated about 2100 B.C., records in Sumerian cuneiform directions for making various remedies; beer was the usual solvent (Keller, 1958). The ancient Egyptians have depicted their alcoholic excesses in Theban wall paintings in which women are shown as drunk to the point of nausea. Drinking practices in ancient Greece and Rome and among other peoples of antiquity have been documented in detail (McKinlay, 1948a, 1948b, 1949a, 1949b, 1951).

Alcoholic drinks are also well-known to many of our preliterate contemporaries. A worldwide census of such societies is not yet available, but extrapolation from the sample now known suggests that the majority of the cultural groups of this planet make use of some kind of alcoholic beverage in one context or another. Tribal peoples in all major parts of the world (with the exception of Oceania and most of North American) knew alcoholic drink from their early beginnings onward (Mandelbaum, 1965). And those societies that did not develop a drinking custom early in their history did so readily on contact with Western civilization. The historical evidence also suggests that once accepted by a society, the drinking custom has rarely been relinquished except temporarily. An exception to this appears to be the Moslems, most of whom do not drink. This is in keeping with the religious proscription laid down by Muhammad, reportedly as a result of widespread excess (Bales, 1946).

The custom of drinking is thus persistent, easily diffused, and widely pervasive in known cultural groups. As such it may be viewed as a behavioral domain of very considerable theoretical interest. Since it is a complex but easily identified area of social activity that is culturally regulated, its occurrence in widely dispersed societies provides the basis for comparative study. The range of cultural variation in the complex of social behavior surrounding the drinking of alcoholic beverages is underlined by such study. Against this background, the drinking customs of any given group acquire added significance. At the same time, behavioral regularities within the context of the drinking custom may be delineated in a sample of social groups and may be systematically explored in relation to other possibly significant universal variables.

The study of customs of drinking in so-called primitive societies may thus serve a number of functions. From it we may learn something of the role, function, significance, regulation, etc., of the use of alcohol in social groups and of what and how and why and under what circumstance people drink or abstain, and perpetuate their custom in succeeding generations. As a widely varying but relatively discrete pattern of social behavior, developing in different ecological and cultural contexts, it may also provide us through comparative study with some clues as to significant transcultural variables in social behavior, beyond the drinking context.

Scientific interest in the study of drinking customs, per se, has been relatively slow in developing, despite the obvious need for such an approach. Bacon (1944) emphasized the importance of studying drinking as a social custom several decades ago and his analysis is still pertinent. However, research interest in drinking behavior has been largely dominated by a social-problem orientation and has focused mainly on deviant aspects of drinking. In recent decades the major concern has been with the subject of alcoholism, with methods of diagnosis, prevention, treatment, social control, and the attempt to define a dichotomy of "normal" and "problem" drinking. In short, interest has concentrated on drunkenness rather than drinking, and drinking as deviance rather than drinking as social behavior. The reasons for this research bias appear to be both historical and pragmatic. The temperance movement of the last century bequeathed to American society a profoundly negative image of drinking (Bacon, 1967). The later development of the disease concept of alcoholism helped modify the earlier conceptualization of drinking as evil but still emphasized its negative aspects. At the same time, the increasing mechanization of our society has underlined the danger of drug-induced incompetence and the urgent need for social controls.

It is clear, however, that drinking cannot be understood solely within the framework of disease and deviation. The patterns of alcohol use, function, and meaning are enormously influenced by the cultural context in which drinking occurs. Drinking may be, and is, perceived variously as a divine gift, a symbol of virility, a curse brought by the white man, an act of profound religious significance, etc. Drinking is a complex of learned social behavior (and possibly learned autonomic behavior) surrounding the oral ingestion of alcohol and its accompanying pharmacologic effects. It occurs as a persistent social custom across the societies of the world and can be best understood within this range of variation and uniformity.

Sociocultural studies of drinking customs have been confined largely to the past 30 years. In general they have approached the problem of understanding drinking as social behavior from two basically different points of view. The first has emphasized the accumulation of descriptive accounts of drinking customs in different societies. These have varied greatly in focus and breadth of coverage but have in general served to demonstrate the cultural relativity of all aspects of drinking behavior. The second approach has involved crosscultural studies of large samples of societies, and has sought to explore the interrelationship of universal variables in drinking behavior underlying a wide range of cultural variation.

This chapter will be concerned with the field of sociocultural research based on the study of tribal societies and relevant to these two basic approaches. The main body of the chapter will be directed to a consideration of the findings of large-scale crosscultural studies.

DESCRIPTIVE STUDIES

Early reports of drinking customs among tribal peoples appeared in the literature of the last century as a part of general descriptive accounts of the ways of life of such groups. Such reports were made by missionaries, travelers, and anthropologists and sometimes included a fair amount of information on drinking practices. The Reverend C. T. Wilson and R. W. Felkin, for example, include in their report on the Baganda several paragraphs on the manufacture and customary use of banana wine (1882). Similarly, Wilhelm Junker in his "Travels in Africa During the Years 1882–1886" detailed at some length the drinking customs of the Azande (1892). Following early accounts of this sort a voluminous ethnographic literature has developed that, while not primarily concerned with the study of drinking, nevertheless contains hundreds of descriptions of drinking customs in a wide range of societies. A series of examples of such reports may be found in Washburne (1961).

Descriptive studies, explicitly concerned with the drinking customs of a given social group, have been much fewer in number but have occurred with increasing frequency in the past 30 years. Beginning in the 1940s and increasing rapidly in the past decade, these studies seem to have received their impetus from two main sources. One of these may be found in the early interest in Jewish drinking and its trouble-free characteristics (Myerson, 1940) that led eventually to Snyder's classic study, "Alcohol and the Jews" (1958). During this same period a number of other studies of drinking in different subcultural groups appeared. These also emphasized the influence of the total cultural context on drinking customs and differing rates of problem drinking. Examples include, among others, Bales's (1946, 1962) comparisons of Irish and Jewish drinking, Glad's (1947) comparison of drinking in Irish and Jewish male youth, Barnett's (1955) study of drinking among the Cantonese of New York, Williams and Straus's (1950) description of the drinking patterns of Italians in New Haven, the investigation of drinking customs in the black ghetto by Sterne (1966), the report of drinking in Italian culture by Lolli *et al.* (1958), and "Drinking in French Culture" by Sadoun *et al.* (1965).

Studies such as these have clearly demonstrated the influence of cultural factors in the incidence of alcoholism and other problems associated with drinking. In all of these groups drinking is prevalent, <u>but the frequency of problems associated with the use of alcohol varies greatly</u>. Attempts to discover those features of culture that are most significantly associated with these differences have resulted in a number of suggestions. For example, the religious context of drinking among the Jews may operate as a generalized constraint on all types of drinking. The cultural disapproval of drunkenness found among both the Jews and certain Chinese groups may tend to prevent excessive indulgence. The association of drinking with food among the Italians may possibly be a factor in their relative immunity to problems. The more troublesome drinking of the Irish might be related to inadequate social controls, greater inclination to drink for purely convivial reasons, unusual early socialization experiences with eating, etc. From such suggestive analyses have come the current emphasis on concepts of social controls integrated into drinking customs and categories of "kinds" of drinking or "reasons" for drinking, i.e., drinking the purely "social" reasons and drinking for "personal" reasons, which are conceived to lie at opposite ends of a continuum of "safe" and "unsafe."

A second group of studies explicitly directed to the study of drinking customs in different cultural groups seems to have developed in conjunction with the culture and personality movement of anthropology. In 1940 Ruth Bunzel published her pioneer study in which she presented a detailed description and analysis of the sharply divergent drinking customs found in two distinct cultural groups in Central America. Bunzel's study was followed closely by that of Honigmann and Honigmann (1945), which contrasted drinking patterns among Indians and whites in a small trading post in northern Canada. Similar descriptions during this period of investigation of drinking practices in small preliterate groups included Devereaux's study of the function of alcohol in a Mohave community (1948) and Sandoval's analysis of motivations for drinking among the Indians of Ecuadorean Sierra (1945).

The next two decades saw increasing numbers of such anthropological case studies, detailed reports of drinking customs in relatively small and homogeneous cultural groups. These studies have ranged widely in geographic area, extent of analysis, focus of interest, and interpretation. Some, like Bunzel's and the Honigmanns', have compared two or three culturally distinct social groups with sharply differing customs of drinking and rates of inebriety and analyzed the role of drinking in the total cultural context. Thus, Berreman (1956) compared drinking practices among three Aleutian communities. Sayres (1956) described ritual and nonritual drinking patterns in three different ethnic status groups (Indian and mestizo) in a rural district in southern Colombia. Madsen and Madsen (1969) contrasted drinking patterns in a Nahuatl Indian village and the mestizo town of Tepepan near Mexico City. And Lemert contributed a series of comparative studies: the original study of drinking among Salish Indian tribes of the northwest coast (1954), followed by comparisons of drinking customs in several Hawaiian plantation communities (Lemert, 1964b) and in three Polynesian societies (Lemert, 1964a).

Other investigations have emphasized the interrelationships of acculturative influences and drinking practices: Gallagher (1965) for the tribal Bihar of India, Hamer (1965) for the Forest Potawatomi, Curley (1967) for the Mescalero Apache, and Robbins and Pollnac (1969) for rural Baganda. Heath (1964) has also considered the problems resulting from the influence of a dominant culture in his excellent series of studies on drinking among the Navaho. Maynard (1969) describes drinking among the Oglala Sioux as an adjustment to acculturation stress. Lubart (1969) discusses excessive drinking among the MacKenzie Delta Eskimos as an adaptation to social and economic change. Similarly, Lobban (1971) points to cultural problems and drunkenness in an Arctic population in Inuvik, Canada, and Honigmann and Honigmann (1965) describe the difficulties experienced by the Baffin Island Eskimo in learning to use alcohol.

Another group of studies emphasizes the degree to which drinking patterns are integrated into the social customs of the community. Netting (1964) discusses beer as a locus of value among the Kofyar who are subsistence farmers in northern Nigeria. Levy (1966) presents an extensive analysis of Ma'ohi drinking patterns in the Society Islands, where drinking is an accepted part of life and is not generally disruptive. Heath (1958) suggests that among the Bolivian Camba alcohol serves a socially integrative function in that it facilitates rapport between people who are normally isolated and introverted. Mangin (1957), in his description of drinking among the Andean Indians points out the long association of drinking with ceremonial activity among these people. Leacock (1964) describes ceremonial drinking among the members of an urban religious cult in northern Brazil. Simmons (1968), also gives an account of the sociocultural integration of alcohol use in a Peruvian coastal community. Lomnitz (1969a, 1969b, 1969c) describes the native drinking patterns of the Mapuche Indians of southern Chile as an institutionalized ritual of friendship. Wolcott (1974) reports on integrated drinking in the African beer gardens of Bulawayo.

Other studies explore possible psychodynamic factors in drinking as well as socially integrative functions. Kearney (1970) has studied drunkenness and religious conversion in a peasant farming community in Mexico. Whittaker (1962, 1963), in his study of drinking among the Standing Rock Sioux tribe, attributes drinking problems in this group to increased intrapsychic tension, attitudes toward alcohol that encourage its use to relieve tension, and absence of social sanctions against the heavy drinker. Savard (1968) in a recent study of alcoholism among Navaho men found excessive use of alcohol related to the suppression of anger (a prominent social proscription in Navaho culture). Dailey (1968) has related drinking among certain North American Indian tribes to the cultural value placed on dreaming and other psychic experiences resembling intoxication. Carpenter (1959) has traced historical changes in motivations for drinking among the Iroquois. In the seventeenth century the Iroquois used alcohol to stimulate dreams and visions and as a means of inducing a revelation of the will of the spirits. To them it was a positive spiritual experience. Later the emphasis changed, but the earliest interpretation was never entirely superseded and may persist still to a limited degree.

In the last few years a number of studies have appeared applying the newly developing methods of ethnoscience to drinking behavior. The most widely known example of this type of analysis is found in Spradley's (1970) description of "Skid Row" hobos in Seattle. Everett (1971) has used this approach in a series of studies of drinking among the White Mountain Apache. Topper (1972) has made similar studies of Navaho drinking patterns. This approach to ethnography, pioneered by Goodenough (1970, 1971) and others (Frake, 1964a, 1964b; Geoghegan, 1969, 1971; and Keesing, 1971, 1972), seeks in essence to elicit the cognitive categories and dimensions utilized by a given cultural group in structuring significant aspects of a domain of behavior (participants, contexts, time, etc.). With further knowledge of the decision-making process, it may be possible to predict behavior and thus test the validity of the analysis. Studies of cognitive structuring such as these are, without question, of considerable importance. The development of reliable and valid methods to study this significant set of variables is difficult but progress in this direction is being made.

Intensive case study analyses of drinking customs, such as the examples cited in the preceding paragraphs, are of unquestioned heuristic value. They have variously explored the social functions of alcohol, the contexts of drinking, the extent of patterning of inebriety, the motivations for drinking, the socially conditioned responses to drinking, and the forms and extent of social controls. And by this means they have illuminated the wide range of cultural variation in the dimensions of the drinking custom and the complexity of the behavioral variables involved. However, as units they are necessarily limited in their applicability, since no matter how comprehensive they may be, they permit no generalizations about drinking behavior. Statements about drinking in any given cultural group cannot, of course, be assumed to hold true in a different group. Similarly, no relationship of drinking to other features of culture that is observed in one group can be assumed to occur in the same way in all groups. Only through the study of drinking across a sample of societies will it be possible to discover significant variables in drinking behavior common to all peoples, and the relationship of these variables to other cultural variations.

RANGE OF CULTURAL VARIATION

The vast body of descriptive material that has been accumulated through the case-study method can be organized in terms of cultural variation in various features of drinking behavior that are common to all drinking customs. A sample of people, if studied, will be found to differ in various aspects of their use of alcohol. Similarly, societies, taken as units, will be found to vary over a wide range in commonly held features of the drinking custom. Examples of the variation found for some aspects of drinking behavior may serve to illustrate the range of such differences and at the same time exemplify the basic method of hologeistic or crosscultural analysis.

Societies in which drinking is an accepted custom differ in the type of beverage used and how it is made and stored. Volumes have been written illustrating the extent of man's inventiveness in this regard (Emerson, 1908). Societies also differ with respect to how often a typical person in a given society may have something to drink. In some cases drinking is very infrequent. The Papago, according to their aboriginal custom, drank only once a year. This tribal society of southwestern United States made a fermented liquor from the fruit of the giant cactus that was the central feature of their annual rainmaking ceremony. At this time it was the duty of all men to drink to the point of saturation. Many women joined in the drinking (Underhill, 1946).

In contrast to this are societies where the typical adult drinks daily. Among the Jivaro, for example, it is reported than men have traditionally drunk quantities of their native beer every day. They consume it in part as food, drinking with meals, and taking a quantity of fermented paste wrapped in a banana leaf with them on war and hunting expeditions, so that they may not find themselves deprived of their customary drink (Karsten, 1935).

Societies also vary as to the extent of drinking, i.e., the proportion of people in the society who drink at all. In a number of societies, drinking is confined to adult men. Among the Lepcha, on the other hand, everyone drinks, including even very small children, who are reported to drink at times to the point that they are unsteady on their feet (Gorer, 1938; Morris, 1938).

A wide range of variation among societies may also be found in the amount that they drink and the proportion of all cultural effort (measured by such indices as money, time, and labor) that is expended in preparing or procuring alcoholic beverages. In some societies people drink infrequently and in small amounts and spend little effort in procuring alcohol. Other societies would be ranked high on such a variable: the Lovedu of Africa are an example. Beer drinking is the principal source of recreation among the Lovedu. The following passage from Krige and Krige (1943) suggests the frequency of drinking among these people:

> No occasion, whether social or ritual or religious or economic, is complete without beer. During some seasons of the year, people are in a perpetual state of exhilaration: some are singing or gossiping near the beer pots; some are expectantly watching their concoctions mature; and the rest are recovering from the last beer drink.

An appreciable amount of the economic effort of this society is directed toward the procurement of beer. This is indicated in the descriptions of its preparation and in such statements as these (Krige and Krige, 1943, p. 25):

> Porridge of kaffir corn or millet is very seldom eaten, most of it being used for beer. A man will complain, not because he has no cattle, but because he cannot brew beer to maintain his prestige.

The degree to which the drinking of alcoholic beverages is surrounded by ritual is another way in which societies differ widely. In some societies, there is little or no ritual attached to the act of drinking. Most North American tribal peoples, who had no alcoholic drinks before contact with Western culture, now drink without the accompaniment of ritual behavior. In contrast to this are societies whose customs of drinking involve elaborate ritual such as the Ainu of northern Japan. The principal alcoholic beverage among these people is saké, the rice wine obtained from the Japanese, but other alcoholic drinks are made locally from millet and from the juice of a tree. By tradition most of the drinking among the Ainu occurs on ceremonial occasions such as the first tattooing of a young girl, housewarmings, funerals, anniversaries of funerals, etc. On many occasions libations are offered to spirits at a sacred spot within each family's hut. Drinking also plays an important part in ancestor worship. One ethnographer describes in some detail a ceremony of drinking to the gods: a saké cup with its stand is placed before the chief man who sits on the floor before the fire. He places a mustache stick across the top. Then, stroking his beard in a ceremonial gesture, he lifts the cup and stand with both hands and bows his head. When the saké is poured into the cup he begins to recite a long formula that continues during the succeeding preparations. He takes the stick in the right hand, dips the end into the saké, and gracefully moves the stick forward as though throwing some saké into the fire. Dipping the stick once more, a drop is thrown over the left shoulder. These ceremonial operations are repeated two, three, or more times. The stick is then replaced on the cup, the whole is again raised, and finally the stick is used to lift the mustache while drinking (Hitchcock, 1890). This ritual, which precedes the first drink, consumes at least five minutes, and before a man can have another drink himself he must fill the cup for someone else. Among the Ainu, carved mustache lifters are especially valued, since failure to keep the mustache out of the cup is considered disrespectful to the other people present and to the gods (Batchelor, 1892).

Ethnographic reports also yield a great deal of information on the contexts of drinking—the general setting in which drinking occurs, i.e., as a beverage with meals, as a part of religious ritual, as a mark of hospitality, as a part of any festivity. Studies of drinking in ethnic subcultures, such as Snyder's (1958) study of drinking among the Jews, Bales's (1962) of the Irish, and Lolli's et al. (1958) of the Italians, suggest that the context in which drinking customarily occurs is an important variable in the social control of drinking. Snyder, for example, suggested that the degree to which drinking is integrated with core social symbols, sentiments, and activities might be significant in the control of inebriety.

Societies differ greatly in the extent to which their drinking customarily occurs in some sort of ceremonial context. At one extreme are peoples like the Navaho, whose drinking practices have been reported in detail by Heath (1952, 1964). Drinking in this group is a social activity in the sense that people never drink alone but always with someone else, usually a relative. At the same time, it is an isolated activity that has little to do with the other customs of the society. It is usually not done at home, and so is not a part of the custom of eating and hospitality. Drinking does take place at religious festivals but only among small groups of onlookers on the outskirts of a crowd and never as an integral part of the ceremonial. For the most part, drinking is done in small isolated groups on the periphery of other activities.

The drinking practices of the Cuna (Stout, 1947) stand in sharp contrast to those of the Navaho with regard to contexts of drinking. Drinking among the Cuna is exclusively ceremonial in nature. It occurs only in connection with various rites and observances marking some special event. For example, the making of *chicha* (a fermented drink usually made of either corn or sugarcane) is a communal affair in which the whole village is involved. *Chicha* is usually brewed in large amounts but only at periodic intervals; it is then consumed within a few days and no more is available until the next occasion.

The ceremonial occasions on which *chicha* is drunk are numerous: births, weddings, religious festivals, etc. The most important occasion occurs at the time of female puberty rites, which involve elaborate ceremonies lasting for days. There are songs and intricate rituals and dances, and each step of the ceremony is marked by the drinking of *chicha*. There are a number of other ceremonial occasions when *chicha* is used: the ritual for making ceremonial flutes, for the making of hammocks, for the ceremonial cleaning of rattles and flutes, etc.

Like the Lovedu, the Cuna perform fairly elaborate rituals while drinking. The actual taking of a drink is always done in a certain manner. People do not serve themselves but serve others, who in turn serve them. On some occasions the drinker must empty a magical four cups. Cups or calabashes must be rinsed in certain specified ways. Special chanters and chanters' assistants officiate at the various *chicha* ceremonies. Clearly, the Cuna can be conceived to fall at the opposite extreme from the Navaho on a dimension of drinking in ceremonial context. Every act of drinking for these people is embedded in some ceremony and there is no casual drinking outside of such contexts. Between the times of celebration, the Cuna do not drink.

Other dimensions of variation in the drinking custom can be abstracted from the ethnographic reports available in the literature, e.g., frequency of inebriety, degree of problem, approval of drinking, behavior while drinking such as sociability, exhibitionism, hostility, rule-breaking, etc. Approval of drunkenness, for example, ranges widely, from societies where it is strongly disapproved to those where it is not only approved but sought for and highly valued. Again the Cuna provide an example. They greatly enjoy being drunk and look forward to such occasions with eager anticipation. During festivities men frequently drink to the point of stupor and must be carried home and cared for by their wives. They apparently suffer no loss of dignity by such behavior. In other societies drunkenness is sought as a means of experiencing visions and obtaining spiritual power (Carpenter, 1959).

CROSSCULTURAL INVESTIGATIONS

Crosscultural studies of drinking have sought to quantify on a comparative basis pancultural variables of the drinking custom such as those illustrated. By this means it becomes possible to study in a sample of societies the interrelationship of variables of drinking and their association with other features of culture.

The crosscultural method, in its present form, developed out of the early work of Horton (1943), Whiting and Child (1953), Barry, Bacon, and Child (1957), and others. In the past 30 years it has been used in a variety of investigations; Naroll (1970) has recently reviewed 150 such studies. The method has been described in detail elsewhere (Whiting, 1969). In essence each society of a worldwide sample is independently rated on a comparative basis by at least two raters with regard to the presence or strength of a given variable of drinking behavior. Each variable is defined as precisely as possible in operational terms. Ratings are made (usually on a seven-point scale) when the ethnographic material is judged sufficient to permit such quantification. Those societies, for which ratings cannot be reliably made by raters independently judging the same material, are discarded from the sample.

Comparative studies of drinking customs across a wide range of societies have been relatively few in number. Horton's (1943) pioneering study on a sample of 56 societies stood alone in this area until Field (1962) published further crosscultural findings based on a reanalysis of Horton's data. In 1965, Bacon, Barry, and Child (Bacon *et al.*, 1965a,b, Barry *et al.*, 1965; Child *et al.*, 1965a,b, hereinafter referred to as the BBC study) published the results of a crosscultural study based on 139 societies, including Horton's original sample. Klausner (1964) investigated the religious use of alcohol in a sample of societies. Further crosscultural studies in the past decade include the investigation of the relationship between alcohol variables from the BBC study and folktale content (McClelland *et al.*, 1966, 1972) and Schaefer's exploration of problem drinking (1973).

Within the limitations imposed by the ethnographic material available, crosscultural studies of drinking such as these have produced significant and suggestive findings in the field of sociocultural research. Such data can provide a testing ground for some of the constructs and hypotheses derived from more intuitive case-study approaches, and can thus serve an integrative function in sociocultural research on drinking behavior.

Transcultural Variables in Drinking Behavior

One of the major functions of crosscultural analysis may be found in the delineation of significant transcultural variables in drinking behavior. This is well illustrated by an examination of that dimension of drinking behavior that has come to be labeled "integrated drinking." In his classic study of Jewish drinking, Snyder elaborately detailed an aspect of the custom that he considered important in explaining the coexistence of widespread drinking and absence of alcohol-related problems among Jews. He pointed out that most Jewish drinking occurs in a ceremonial context, and that the act of drinking tends to be surrounded by ritual of religious significance. Customary patterns of drinking tend to be strongly reinforced by early socialization and by adult disapproval of drinking that deviates from the culturally prescribed pattern.

In the BBC crosscultural study, quantitative ratings were obtained on a large number of alcohol-related variables in each of the societies of the sample. These variables were concerned with the availability of alcoholic beverages; the extent, frequency, and quantity of drinking; frequency of drunkenness; attitudes toward drinking and drunkenness; contexts of drinking (religious, ceremonial, party, household, segmented, solitary); behavior associated with drinking (sociability, exhibitionism, hostility, rule-breaking), etc.* The selection of variables to be rated was determined in part by intuition, in part by the results of national surveys (Mulford and Miller, 1959; Straus and Bacon, 1953; Maxwell, 1952), in part by insightful suggestions from earlier case-study reports, and in part by knowledge of the kind of material likely to be found in ethnographic reports. Factor analysis was performed on the intercorrelations of those variables that were rated on 50 or more societies. Four relatively independent clusters of measurements were revealed by this analysis. The first (which accounted for 29 percent of the total variance) was labeled the "integrated drinking factor." Variables with a high loading on this factor were those related to the use of alcohol in a ceremonial and ritualized context. In societies high on this factor, drinking is an integral part of ceremonial and other social occasions. It occurs in a context of positive social meaning. The drinking practices of the Cuna, described earlier, provide a clear example of the alliance between drinking and social ceremonies. The Ainu of Japan with their extensive concern with ritual in all aspects of their drinking custom would also rank high on this factor.

The delineation of Integrated drinking as a dimension of alcohol-related behavior in a worldwide sample of societies provides crosscultural validation of Snyder's conceptualization based on his study of drinking in a single cultural group. The degree to which drinking occurs in a ceremonial context is a significant variable in the drinking custom (Child *et al.*, 1965a).

The crosscultural findings yield further information with regard to the interrelationship of Integrated drinking with other variables of the drinking

---

* Precise definitions for these variables are included in the original publication (Bacon *et al.*, 1965).

custom. Societies that rank high in Integrated drinking tend also to be those where drinking is generally approved and where most people take part in the drinking custom. This would be generally expected and would be predicted from Snyder's case study. However, the crosscultural finding that societies rated high in Integrated drinking tend also to show a high rate of consumption of alcohol would not have been predicted and is of considerable interest. This finding is contrary to the generally accepted belief in our society that heavy drinking is deviant and, almost by definition, socially threatening. It strongly suggests rather that a high rate of alcohol consumption is not necessarily disruptive, but may be entirely compatible with a pattern of drinking that is intimately linked with the positive social values of the group. The Lepcha exhibit drinking customs that exemplify this association.

Snyder's study indicated that Jewish drinking customs unite a strong tendency toward ceremonial drinking with strong disapproval of drunkenness. Crosscultural findings, however, indicate that Integrated drinking is significantly associated with approval of drunkenness rather than disapproval. Societies where drinking is highly integrated into the social customs of the group are generally able to tolerate drunkenness. Certainly the presence of Integrated drinking does not presuppose the existence of a disapproving attitude toward drunkenness.

Another implication derived from studies of Jewish drinking has been that the low rate of problems associated with the use of alcohol that is characteristic of this group is causally linked with their custom of drinking in ceremonial contexts. Here again, the crosscultural findings do not support such a view. In a sample of societies, the correlation between the presence of Integrated drinking and Frequency of Drunkenness is close to zero ($r = +.05, N = 52$) indicating that the two variables are unrelated. In other words, the presence of Integrated drinking does not predict the relative presence or absence of drunkenness.

Investigation of the relationships of Integrated drinking to other features of culture serve to illuminate further the characteristics of this drinking variable. One of the most striking relationships found in a sample of societies is that between Integrated drinking and the presence of drinking in aboriginal times. A large proportion (75 percent) of the societies in the BBC sample where drinking occurred aboriginally were found to have Integrated drinking. On the other hand, in societies reported as drinking after contact but not aboriginally, Integrated drinking was present in only 8.5 percent of the cases. Societies high in Integrated drinking were also found to have a more highly organized and stratified social structure ($r = +.51, N = 57, p < .01$), a more densely populated settlement pattern ($r = +.37, N = 58, p < .01$), and a subsistence economy that permits the accumulation of food ($r = +.52, N = 58, p < .01$). These features appear to be characteristic of a relatively settled and organized cultural group and contribute to the conception of Integrated drinking as a cultural adjustment to drinking, an adaptation and integration of the drinking of alcoholic beverages into the way of life of the group.

If Integrated drinking is examined in relation to transcultural variables of child-rearing, it is found that societies rated high in this type of drinking also tend to show cultural pressures toward responsibility and obedience (either during the childhood training period or in adult life) together with a low expectation of achievement. Thus, pressures toward Compliance (responsibility and obedience) rather than Assertion (self-reliance and achievement) are correlated with a high incidence of Integrated drinking. This pattern of socialization pressures has also been found to be characteristic of societies whose economy permits an accumulation of food (Barry et al., 1959).

Sex Differences in the Use of Alcohol

In the literature on alcohol, sex differences in drinking have been frequently noted but have not been systematically studied. In most Western societies women tend to drink less than men. The degree of difference between the sexes varies considerably from one group to another and in some cases is quite marked (Lisansky, 1957). It is usually dismissed with no effort at explanation other than the suggestion that there are greater sanctions against drinking for women than for men. The preponderance of drinking among men observed in Western societies has apparently led some researchers in this field to feel that this represents a pervasive sex difference in all cultural groups. Available crosscultural evidence, however, indicates that this is not the case.

Horton (1943) reported some information about drinking customs of men and women in 30 of the societies of his sample. In 14 of these he found no evidence of any sex difference and in the other 16, women drank less than men. In three societies (Abipone, Choroti, and Toba, all located in the Chaco region of South America) Horton's evidence indicated that alcohol consumption was limited entirely or almost entirely to men.

Horton's observations on sex differences in drinking were largely incidental to the main purpose of his study. However, Child et al. (1965b) made a systematic attempt to investigate sex differences in drinking in a crosscultural perspective. Their findings indicated that both men and women drank in 109 of the 113 societies on which there was sufficient information to make a judgment. Drinking was restricted to one sex in only four societies and in each of these (the Ifaluk of Micronesia, the Abipone and Choroti of South America, and the Creek of the southeastern United States) the men drank but the women did not. Comparisons of the ratings of alcohol-related variables for the two sexes indicated that the majority of societies showed no sex difference in quantitative ratings on any given variable. Whenever the judges agreed in making a different rating for the two sexes, it was almost always in the direction of a higher rating for men than for women. Two exceptions were observed: Among the Aleut, women were rated higher than men on Intensity of Hostility while drinking; among the Omaha, women were related higher in Intensity of Exhibitionism.

Based on the judgments of an additional rater, there were 53 societies with definite evidence of sex differences in the use of alcohol and 36 societies where there was either no evidence of a sex difference or both sexes were judged equal. The remaining 50 societies were excluded for a variety of reasons: conflicting evidence, insufficient information, or evidence that the members of the society did not drink at all. The overall evidence strongly suggests that while the tendency for men to drink more than women is a frequently observed pattern in a sample of societies, it is by no means universal.

When the two groups of societies (sex differences reported present versus sexes judged equal or no difference reported) were compared with regard to alcohol variables and other features of culture, the most conspicuous difference found was that related to the aboriginal use of alcohol. Alcohol was used aboriginally, rather than being introduced postcontact, in 81 percent of the societies with a definite sex difference, but in only 45 percent of those without evidence of a sex difference. This difference has a very low probability of being due to chance (chi square = 9.54, $df\, 1, p < .01$).

Societies with a definite sex difference were also preponderantly higher than those without evidence of a sex difference in Frequency of Ceremonial drinking ($t = 3.49, df = 55, p < .01$). Similarly, other alcohol-related variables tend to show the same direction of relationship with presence of aboriginal drinking as with a definite sex difference in the use of alcohol. It seems clear that the presence of sex differences in drinking is positively associated, not only with the presence of aboriginal drinking, but also with Integrated drinking. A positive relationship among these three variables seems logical. The integration of drinking customs into the total cultural pattern of a given society may be reasonably assumed to be a function of time; such integration would be more likely to have occurred where drinking had been a custom for several generations than in a society to which alcohol had been only recently introduced. Likewise, it seems not unreasonable that sex differences in drinking might be more likely to develop in societies where it had been a custom during the aboriginal period than in those were the experience with alcohol had been more brief. Possibly the universal preoccupation of women with child care is a factor in this relationship. Women's activities seem often limited by the degree to which they are compatible with the care of children (Brown, 1970). Also, the linkage between the presence of Integrated drinking and sex differences in the consumption of alcohol might be due in part to the fact that men usually take part in ceremonial observances to a greater extent than women. The use of alcohol in such rites could thus be an important factor in the development of sex differences in drinking.

Child's study showed further that drinking customs tend to conform to the adult sex-differentiation pattern. In societies where differences between male and female roles are strongly emphasized, sex differences in drinking are more apt to be present. The relationship of large sex differences in drinking to a nomadic or rural settlement pattern ($t = 3.10$, $df = 32$, $p < .01$), an economy based on hunting ($t = 2.50$, $df = 34$, $p < .02$), low accumulation of food resources ($t = 2.09$, $df = 41$, $p < .05$), and strong child-training pressures toward achievement ($t = 2.92$, $df = 27$, $p < .01$) are all consistent with the associations previously found with degree of sex difference in child training (Barry, et al., 1949).

Two additional findings are of interest. If the influence of aboriginal drinking is held constant by examining only those societies with aboriginal drinking present, then it is found that societies with a definite sex difference have significantly higher scores on occurrence of Extreme Hostility than do those without evidence of a sex difference ($t = 2.47$, $df = 29$, $p < .02$). The latter, in turn, tend to be higher in availability of alcoholic beverages ($t = 1.88$, $df = 46$, $p < .10$).

Extreme Hostility while drinking tends to be found in men rather than women. Its occurrence in societies with sex differences in drinking may reflect a more unrestrained expression of aggression in societies where sex differences are generally emphasized. In such societies women may occupy an inferior position and thus lack permission or opportunity to intervene when drunken brawls occur. Possibly the presence of women in the drinking situation, when it is permitted (i.e., sex differences are less), operates as an inhibiting factor with regard to the expression of hostility. It is interesting to view this finding as a contrast to the interpretation suggested by Field (1962) and McClelland et al. (1966) that male solidarity, masculine domination, etc., lead to increased control of drinking and lower frequency of intoxication.

A low score in availability of alcoholic beverages is usually related to seasonal restriction. The tendency for this situation to be found in societies with definite sex differences in drinking may indicate that, when the supply of alcohol is restricted or uncertain, men may be motivated to limit the amount that women may drink.

Further crosscultural studies of sex differences in drinking are obviously needed. That women do drink and that there is a widespread but not universal tendency for them to drink in different ways and often less than men is clearly evident. However, the gaps in crosscultural information on the drinking practices of women are extensive. In many ethnographic studies of drinking, information on the drinking of women is limited to the statement that women participate in the drinking custom. To what degree, in what contests, in what ways like or different from the drinking of men, etc., is not recorded. And yet questions such as these and others are important. It cannot be assumed that the determinants of drinking for women are totally different from those for men. Comparative studies of sex differences in crosscultural perspective may therefore contribute significantly to our understanding of the underlying variables involved in the development, perpetuation, and differentiation of drinking behavior.

Determinants of Drinking Behavior

Questions as to reasons why people drink have been asked many times and addressed in different ways. Many anthropologists feel that drinking customs in a given society can only be understood within the context of the social customs of that society. By this view any attempt to abstract and quantify variables across a sample of societies in order to make general statements about drinking behavior represents a distortion of meaning. Other researchers tend to avoid motivational constructs and may adopt the social learning approach that would see drinking behavior as acquired as a part of the learning of appropriate social behavior, relying only on the concept of norm acquisition to explain drinking (see Davis, 1972, for a summary of this approach). Such an approach, however, does not account for variations on either side of the socially accepted norm and would appear to necessitate different explanations for different "kinds" of drinking. Thus, Jessor's (1968) tri-ethnic study invokes motivational constructs to explain "heavy drinking," which he conceives as a type of deviance. One of the difficulties with this approach lies in the definition of deviance and also in the apparent need for separate conceptualizations of differing degrees of participation in the custom, e.g., drinking behavior that deviates in the direction of abstinence.

Other researchers have sought psychological explanations for drinking behavior. Most of the research in this area is concerned with causal factors in the development of alcoholism, which, in turn, is perceived as a disease entity. This has led to a heavy emphasis on a search for various symptoms of maladjustment or emotional difficulty that might be antecedent to or causally related to the development of alcoholism. While it is recognized that drinking is a necessary antecedent to alcoholism, it is also clear that alcoholism is not a necessary consequence of drinking. It is not synonymous with drinking, heavy drinking, or drunkenness. At the same time, in order to understand causal factors in the development of alcoholism, it seems important to understood not only why people drink to the point of drunkenness, but also why people drink in the first place.

Horton (1943, 1959) posed the question in crosscultural perspective. He noted the prevalence of the drinking custom throughout the cultures of the world, its ease of diffusion and persistent survival despite competing customs and organized attempts at prohibition. He emphasized the importance of understanding the nature of the universal human need that is apparently satisfied in some way by the drinking of alcoholic beverages. The need for greater understanding of motives for drinking, of the facilitating consequences associated with the consumption of alcohol has been repeatedly emphasized (Jellinek and McFarland, 1940; Lisansky, 1959; Coopersmith, 1964). A recent increase in publications in this field suggests a renewed interest (Williams, 1966, 1968; Kalin et al., 1965; Kastl, 1969; Hetherington and Wray, 1964; Mello, 1972).

Clearly, the factors influencing the consumption of alcohol are multiple, both at the individual and the societal level. Crosscultural studies have provided a testing ground for hypotheses concerned with the determinants of drinking as they relate to two main questions: (1) What are the psychological states or needs that seem to be associated with drinking? (Subsidiary to this main query is that referring to the facilitating or reinforcing effects that accompany or follow drinking.) (2) What is the nature of the cultural influences that either facilitate or inhibit the drinking response? Included here are all factors related to the social control of drinking.

One of the most persistent ideas concerning motivations for drinking is that relating alcohol consumption to the reduction of anxiety. This idea has been long supported by clinical evidence as well as folk knowledge and recurs regularly in the literature as an explanatory concept. A number of experiments with animals have clearly illustrated the efficacy of alcohol in reducing fear (Barry and Miller, 1962; Conger, 1951; Kaplan, 1956; Masserman et al., 1945; Masserman and Yum, 1946; Scarborough, 1957; Smart, 1965). Greenberg and Carpenter's work (1957) with the galvanic skin response in human subjects strongly suggests that moderate amounts of alcohol may reduce emotional tension. Williams (1966), investigating drinking in a social context among male college students, also found that anxiety and depression decreased significantly at moderate levels of alcohol consumption, increasing again to nearly the pre-party level with higher levels. On the other hand, psychiatric observations of alcoholic subjects during prolonged experimental drinking sessions have suggested that, under these conditions, increases in anxiety may occur at higher levels of consumption (Mendelson et al., 1964; McNamee et al., 1968). These results have been interpreted as indicating that anxiety reduction does not act as a motivation for drinking among alcoholics. (See Mello's (1972) discussion of the effect of alcohol on the affective state of the alcoholic.) Kissin and Platz (1968), however, found that not only were alcoholics more anxious and depressed than the "normal" population, but also that alcohol ingestion significantly decreased their anxiety and tension. These authors interpret Mendelson's findings as not inconsistent, but as indicating that the immediate effect of alcohol for the alcoholic is tension reduction followed by an increase in anxiety with prolonged drinking. They suggest that the alcoholic's impulsivity and intolerance of tension may lead him to seek short-term relief even though the after-effects may leave him more anxious than ever. For a review of the literature dealing with anxiety reduction and alcohol, see Wallgren and Barry (1970), also a recent review article by Cappell and Herman (1972).

The first attempt to test the transcultural validity of this relationship was made by Horton (1943). Beginning with the basic assumption that the primary function of alcohol in all societies is the reduction of anxiety, Horton developed a set of hypotheses relating variations in frequency of drunkenness in a sample of societies to variations in anxiety level as measured by degree of subsistence insecurity and acculturation.

As a test of these hypotheses, Horton made judgments on a three-point scale of the degree of insobriety among men in a sample of 56 societies of wide geographic distribution. Since no direct measure of the anxiety level of a society existed, he chose as a possible indirect measure, the level of "subsistence insecurity" that might be conceived to accompany a more or less insecure food supply. It was assumed that societies whose subsistence depended primarily on fairly well-developed agricultural techniques might have a food supply that was more reliable than societies whose subsistence depended primarily on hunting. Other indices of subsistence insecurity were based on threats to the food supply. It was assumed, for example, that such insecurity would be relatively greater where there were constant threats of food shortage (as a consequence of drought, insect plagues, crop failures, etc.) than where such dangers did not exist. In addition to these two measures of insecurity, the degree of acculturation of the society was also used as an indirect measure of generalized anxiety.

Horton's data showed a significant positive association between his measures of insobriety and subsistence insecurity, indicating that in societies where various threats to the food supply were present, the men of the society tended to drink to more prolonged and higher levels of intoxication. This association became even stronger when the degree of acculturation was included in the measure of anxiety. Indeed, all cases of acculturation in Horton's sample were accompanied by extreme drunkenness, further supporting the general association between level of anxiety and insobriety in men, as measured by this study.

Field (1962) reanalyzed Horton's data and offered alternative interpretations for his findings. He correlated Horton's measures of insobriety with various measures of fear from the Whiting and Child study (1953) on the 27 overlapping societies from the two samples. On the basis of a lack of association between these two sets of measures, he concluded that level of fear is not related to extent of drunkenness in primitive societies. This finding does not seem to represent a crucial test of Horton's hypothesis, especially in view of the fact that the measures of fear from the Whiting and Child study were actually indirect measures of level of anxiety associated with various explanations of illness. On the other hand, as Field suggests, it is possible to interpret the association Horton found between type of subsistence economy and degree of drunkenness in terms other than level of anxiety. Field points out that the economy measure used by Horton may represent a continuum of social organization as well as one of subsistence insecurity. By this argument, hunting societies would be conceived to be lower on a scale of social organization than societies where subsistence activities depend largely on herding and agriculture; Horton's findings would be interpreted to indicate an association between level of social organization and sobriety rather than between level of anxiety and drunkenness. Some additional evidence was presented by Field of a negative association between frequency of drunkenness and certain other measures of social structure from Murdock (1957) (i.e., unilineal kin groups, patrilocal residence, approach to an exogamous clan community, political integration). These associations led Field to conclude that more highly organized societies tend to have interpersonal relationships organized along hierarchical lines and conducive to the control of informal drinking bouts. By this argument, increased social organization would be associated with increased control of drinking behavior, i.e., decreased drunkenness. Field also reported relationships between degree of drunkenness and certain child-training variables that he felt supported this interpretation.

Field's argument relates to the second of the two areas of inquiry listed above, i.e., cultural factors that inhibit or control drinking and drunkenness. Thus, he views frequency of drunkenness in terms of influences that prevent or control its occurrence. This point of view is also held by others in the field of alcohol studies who see variations in drinking and drunkenness as primarily a function of variations in social control. By this view, facilitating influences are either not considered or assumed held constant. In essence the search appears to be for the determinants of sobriety rather than for the determinants of drinking and drunkenness.

The societies that comprised the Horton sample were included in the later BBC study (1965) to provide a check on similar measures from both sets of data, and on correlates found by both Horton and Field. In general it was found that the correlations between Horton's three alcohol-related variables and three similar measures from the BBC study were impressively high (+.68 to +.71). These findings mutually confirm the validity and reliability of the respective alcohol measures in both studies. Horton's findings relating frequency of drunkenness to measures of anxiety based on subsistence insecurity were supported in part by the BBC study. The relationship with type of economy was confirmed, but other measures of insecure food supply and acculturation did not show a statistically significant relationship. On the other hand, findings of a significant positive association between frequency of drunkenness and the degree to which there is unrealistic storing of food would offer additional support to an anxiety reduction hypothesis such as Horton's.

Field's findings with respect to kinship, economy, and child-training were also reanalyzed with a more sensitive measure of association (product moment rather than tetrachoric correlations) and with an equivalent measure of drunkenness in a larger sample of societies (Barry *et al.*, 1965). Under these conditions, the majority of the significant associations reported by Field were not confirmed. This was especially true for various kinship features and child-training variables. Thus, the positive correlation between Childhood Indulgence and Frequency of Drunkenness was not only not replicated, but, indeed, was in the opposite direction. However, there was confirmation of Field's finding of a significant positive association of frequency of drunkenness with a more primitive hunting economy as opposed to an economy based on agriculture, and with child-training pressures as found in earlier studies (Barry *et al.*, 1959) to be associated with this economy continuum (pressures toward Obedience and Responsibility in predominantly agricultural societies versus pressures toward Self-reliance and Achievement in hunting and gathering societies). Thus, the association of inebriety with the economy continuum seems clearly established. What remains in doubt is how this should be best interpreted: (a) in terms of subsistence insecurity operating to facilitate drinking, (b) as a social control measure operating to inhibit drinking, or (c) as a function of some other mediating influence (Bacon *et al.*, 1965; McClelland *et al.*, 1972).

In addition to the question of the effect of alcohol on anxiety, other recurring themes in the alcohol literature have been subjected to crosscultural study. One of these is the idea that drinking behavior is somehow related to the psychological state of dependence (variously defined).

Lolli (1956) described the alcoholic as characteristically longing for the dependent state of infancy and at the same time desiring self-respect and independence. Bailey (1961) noted the tendency of the alcoholic to assume a dependent role in marriage and Lemert (1962) has made similar observations. McCord and McCord (1960) offered a dependency-conflict interpretation of their data from a longitudinal study of a group of alcoholics. Witkin *et al.* (1959) found evidence of perceptual field dependence in alcoholics, and related this mode of perceiving to basic personality characteristics of passivity, lack of self-esteem and undifferentiated body image. Blane (1968) has described dependence–independence conflict in alcoholics. Barry (1974) has reviewed some of the literature related to this question. It should be emphasized that these studies all relate different aspects of dependence to alcoholism, which is by no means synonymous with drinking or even drunkenness. Nevertheless, they suggest a possible relationship between some aspect of dependence and drinking or drunkenness.

Crosscultural evidence on possible relationships between measures of drinking and drunkenness, on the one hand, and dependence–independence conflict, on the other, was presented in the BBC study (Bacon et al., 1965) and has been recently reexamined (Bacon, 1974). The rationale underlying the dependency-conflict hypothesis may be briefly reviewed. Drinking, especially drinking to drunkenness, is conceived to be associated with a type of dependence–independence conflict. Dependence, in this context, refers to a universally acquired set of behaviors related to seeking for help. The human infant is born helpless and cannot survive unless his physical needs are met by some nurturant agent. As a consequence of the long period of helplessness at the beginning of life, every child develops an elaborate set of help-seeking sequences of behavior that are responded to with help-giving or nurturant parallels by the people in the child's social environment. In all societies the dependent, help-seeking behavior of the infant and child is subject to socialization. The help-giving behavior of the nurturant agents is modified as the child develops, in conformance with the physical capabilities of the child and cultural expectations with respect to autonomy. (For a review of the literature on this subject, see Maccoby and Masters, 1970; Gewirtz, 1972.) It is assumed that adults in any society may show variations in habits, attitudes, feelings, and beliefs about asking for help, as a function of the nature of this socialization process. Thus, an individual in any society who is considered old enough to drink, however this may be defined by his cultural group, will have in his behavioral repertoire a set of help-seeking techniques including attitudes and feelings about asking for help that have been "shaped" by his experience as a member of his social group. Although the necessity of asking for help is less obvious in adulthood than in infancy, it is nevertheless universally present in one form or another.

Earlier crosscultural studies have demonstrated that societies vary significantly in different aspects of their customary treatment of infants and children (Whiting and Child, 1953; Barry et al., 1957). For example, societies show a wide range of variation in the degree to which they respond to the dependence needs of infants, i.e., give them affection, feed them adequately and consistently, protect them from discomfort, etc. They differ in the number of people who provide nurturant care for the typical infant. Societies also vary in their expectations for self-reliant and achieving behavior* during childhood and

---

* Detailed definitions for these variables as utilized in earlier studies may be found in Ford (1967). In general, self-reliant behavior refers to learning to take care of one's self and become independent of the assistance of other people in supplying one's needs and wishes. Pressures toward achievement refer to socialization pressures toward high standards of performance, successful competition, etc.

the manner in which they enforce these expectations. In some societies high standards for self-reliance and achievement are set at an early age and techniques may be punitive. In other societies socialization with regard to this kind of behavior is nonpunitive and standards set are lower and imposed later in childhood.

In terms of a dependence–independence conflict conceptualization, it might be reasonably assumed that the severity of such conflict would be related to the strength of opposing motivations generated by the socialization process. Thus, the motive for self-reliant and achieving behavior was conceived to be in opposition to previously learned help-seeking behavior. On the other hand, a low level of indulgence of dependence needs in infancy was viewed as a frustration of dependence needs and as such also related to the development of dependence–independence conflict.

A third set of variables that seemed logically related to the dependence–independence continuum concerned the extent to which typical adults in a sample of societies could ask for and expect help from each other without criticism or any other indication of disapproval. An uncritical acceptance of dependent behavior in adulthood would appear to be associated with a low degree of dependence–independence conflict.

It was hypothesized that adequate indulgence of the dependence needs of infants, mild and nonpunitive socialization pressures toward self-reliance and achievement, and the acceptance of help-seeking behavior in adulthood would be associated with a low level of conflict over dependence and with relative sobriety. Conversely, high levels of drinking and drunkenness would be associated with the reverse conditions. It was hypothesized further that the drinking situation might permit the temporary resolution of such conflict by permitting simultaneously these effects: the satisfaction of dependence or help-seeking motives, the enjoyment of fantasies of achievement and success, and the reduction of anxiety associated with the state of conflict. This resolution of conflict might then operate as a reinforcement to drinking behavior.

Substantial evidence in support of this hypothesis was presented in the 1965 study that was based on a crosscultural sample of 139 societies. Significant correlations were found upholding the predictions that crosscultural measures of alcohol consumption and drunkenness would be negatively associated with degree of indulgence of dependence in infancy and adulthood, and positively associated with socialization pressures toward self-reliance and achievement in childhood. Of the 28 correlations of Consumption and Drunkenness, reported in Tables I and II (Bacon *et al.*, 1965a, pp. 34 and 37), with variables related to these three areas, 25 were in the predicted direction and 11 were statistically significant. For the correlates of Frequency of Drunkenness alone, the association was even stronger. Out of 14 correlations, all were in the expected direction and half were statistically significant.

McClelland *et al.* (1966, 1972) has questioned the interpretation of these results and has presented crosscultural findings that he views as negative with regard to a dependency-conflict interpretation and supportive of an hypothesis relating drinking in men to motivations for power. His crosscultural evidence is based on measures of drinking taken from the BBC study (Overall Consumption plus Frequency of Drunkenness) and an analysis of the frequency of occurrence of what are considered to be pertinent words in folktales. His test of the dependency-conflict hypothesis rests on the determination of the degree of association between the alcohol measures and the frequency of occurrence of such words as *ask, carry, help*. This is not a very convincing test. The semantic problems involved in the interpretation of these words are very large, especially in view of the fact that the folktales utilized in the study are translations from the original. And even if the problems of translation were solved, the interpretation of the frequent or infrequent appearance of such words in folktales is by no means clear. Societies use folktales in different ways. The frequent appearance of the word *help* in the folktales of a society where such tales are used primarily to teach children proper behavior might, for example, be subject to different interpretation than its appearance in folktales typically used in different contexts.*

McClelland proposes an alternate hypothesis: that men are motivated to drink out of a frustrated need for power. Women, apparently, are assumed to drink for totally different reasons. The crosscultural findings cited in support of the power motive are again not very convincing. The association of Frequency of Drunkenness and high consumption of alcohol with frequency of occurrence in folktales of the words *hunt*, entering implements such as *arrow, knife, spear, eating (hunger, eat, meat, fat, fruit)* is interpreted as an expression in fantasy of a need for power. However, it seems more likely that this association is mediated by the crosscultural association of drunkenness with economy. As already noted (Horton, 1943; Field, 1962; and Barry *et al.*, 1965), societies whose subsistence economy depends largely on hunting and gathering show a higher frequency of drunkenness than societies who depend mainly on agriculture and herding. This being the case, it would seem that the increased frequency of occurrence of such words as *hunt, spear, knife, arrow, eat,* and *meat* might be more reasonably and parsimoniously interpreted as representing a practical interest in the manner of one's livelihood than as an expression of a psychological need for power. This is, of course, not inconsistent with the power motive, but neither is it inconsistent with the motive for success and achievement postulated by the dependency-conflict hypothesis.

---

* A more detailed consideration of the folktale data is not possible, since a search of the relevant listed publications (Kalin, *et al.*, 1966; McClelland, *et al.*, 1966, 1972) has not revealed a listing of the sample of societies from which the folktales were taken (a procedure usually followed in crosscultural studies).

As support for the masculine power hypothesis, Wanner (1972) makes use of findings from the BBC study with regard to behavior associated with drinking. In the BBC study ratings were made of the following kinds of behavior: sociability (friendly interaction), exhibitionistic behavior (all behavior by which individuals call attention to themselves), hostility and resentment (directly expressed in interpersonal relationships), rule-breaking (transgressions from the standard of "good" behavior), boisterousness. Three different aspects of each type of behavior were rated separately: (1) intensity or degree of the behavior, (2) extent of change in the behavior during drinking as compared with the sober state, and (3) occurrence of extreme behavior. Detailed definitions and directions for coding are described in the original study. Wanner reports a significant positive correlation between the drinking measure employed in his study (Consumption of Alcohol plus Frequency of Drunkenness) and each of these measures of behavior associated with drinking: exhibitionism, boisterousness, and occurrence of extreme hostility. Wanner interprets these findings as indicative of a loud, assertive, often aggressive display suggesting that drinking provides an immediate means of gratifying the need to feel powerful. On the surface this seems to be a possible interpretation. However, it should be noted that, when these measures of behavior while drinking are correlated separately with Consumption of Alcohol and Frequency of Drunkenness, the difference between the two sets of correlations indicates that it is Frequency of Drunkenness that is significantly correlated with these kinds of behavior. Further examination of these data reveals that Frequency of Drunkenness also shows a statistically significant correlation with Sociability ($r = +.49$, $N = 37$, $p < .01$) and Sociability Change ($r = +.47$, $N = 35$, $p < .01$). It seems difficult to rationalize a marked increase in friendly interaction as an expression of a need for power. Also, some note should be made of the fact that all of the above-listed types of behavior associated with drinking showed significant positive correlations with Frequency of Drunkenness. This inevitably raises the question whether the observations of the presence of these kinds of behavior did not in fact contribute materially to the definition of drunkenness.

A recent reconsideration of data from the BBC study has yielded even stronger confirmation of the significance of variables related to the dependence–independence continuum in accounting for variations in frequency of drunkenness (Bacon, 1974). In the BBC article the correlations reported between measures of drinking and variables related to dependence and achievement were computed for all societies in the larger sample for which measures were available for each successive pair of variables. The findings indicated that these variables were, as predicted, individually related to measures of drinking and drunkenness. However, the method of statistical analysis used provided no information as to how these variables might interact. The original hypothesis predicted that variations in frequency of drinking and drunkenness would be a function of the combined effect of three sets of variables in this area. Clearly, a more crucial test of the dependency-conflict hypothesis would be provided by a multiple regression analysis. This technique would make it possible to determine how much of the total variation in drunkenness could be accounted for by the three sets of dependence–independence variables acting together.

With Frequency of Drunkenness as a dependent variable, the following independent variables were chosen on the basis of the earlier findings as having some predictive relationship: Diffusion of Nurturance in infancy, Pressures toward Achievement in childhood, Instrumental Dependence and Emotional Dependence in adulthood.* A total of 38 societies from the original sample had ratings available on all five of these variables.

The multiple regression findings indicated that the combined interaction of the independent variables entered into the analysis produces a multiple correlation (R) of .67 with the dependent variable, Frequency of Drunkenness. In other words, 46 percent of the variance of the dependent variable (R^2) is explained by the four independent variables. This is a highly significant finding and provides strong support for the original hypothesis.

The multiple regression analysis also provides the necessary information for an equation stating the relationship between Frequency of Drunkenness and the four independent variables. The findings may be summarized as follows:

$$\text{Freq. Drunk.} = 13.8 - .45 \text{ Dif.Nur.} + .46 \text{ P.Ach.} - .05 \text{ I.Dep} - .33 \text{ E.Dep}$$
$$(5.57) \quad (2.17) \quad\quad (3.24) \quad\quad (.34) \quad\quad (2.30)$$

The numbers in the parentheses represent "t" values. The abbreviations refer to the independent variables.

The first coefficient (13.8) in this equation is a constant term. It represents the point at which the regression line intercepts the Y axis. The other coefficients of the equation indicate the direction and degree to which each independent variable contributes to the association with drunkenness when the influence of the other variables is held constant. In all cases, the sign of the relationship is as predicted, i.e., the variables theoretically related to indulgence of dependence are negatively associated with frequency of drunkenness, while the achievement variable is postively associated. The "t" values included in parentheses with each independent variable represent a measure of the statistical significance of the contribution of each variable. The "t" statistics given in the equation are all statistically significant except for the independent variable, Instrumental Dependence. Because this variable (I.Dep) is highly correlated with another independent variable (E.Dep) ($r = +.67$), the interpretation of

---

* Emotional dependence was defined generally as seeking support from other people in times of crisis, avoiding isolation, seeking security in group contact, etc. Instrumental dependence referred to seeking help of others in satisfying needs for food, clothing, shelter, transportation, etc.

these two variables in the regression is more complicated. In general the correlation between independent variables is called multicollinearity, and results in large standard errors (and hence in low "t" statistics) for the coefficients of these particular variables. The consequence of multicollinearity in this case is that the joint effect of I.Dep and E.Dep on the Frequency of Drunkenness is clearly significant. However, we can*not* conclude that I.Dep by itself has no effect on Frequency of Drunkenness. For this reason, it has not been dropped from the regression, even though it has a very low "t" statistic.*

The regression equation permits the precise statement of the direction and degree of change in Frequency of Drunkenness (in a crosscultural sample) as a consequence of a change of one unit in each of the independent variables taken one at a time with the remaining variables held constant. Thus, a change of one unit in Diffusion of Nurturance in infancy results in a reduction in Frequency of Drunkenness of .45 units (with pressures toward Achievement and Indulgence of Dependence in adulthood held constant). Similarly, a change of one unit in Pressures toward Achievement results in an increase in Frequency of Drunkenness of .46 units (with Diffusion of Nurturance and Indulgence of Dependence in adulthood held constant), etc. These variables can, in fact, be collapsed even further into two combined variables that, when entered into the multiple regression analysis, explain 44 percent of the variance of the dependent variable. The regression equation in this instance would be:

$$\text{Freq. Drunk.} = 14.01 + .42 \text{ Comb.} - .20 \text{ O.Dep.}$$
$$(6.36) \quad (3.74) \quad (3.46)$$

In this case *Comb.* = Pressure toward Achievement minus Diffusion of Nurturance (for each society) and *O.Dep* refers to Instrumental Dependence plus Emotional Dependence in adulthood. The numbers within the parentheses again refer to "t" statistics all of which are significant at the .001 level.

The results of the multiple regression analysis provide strong confirmation of the significance of these variables in predicting drunkenness in a crosscultural sample of societies. When nearly half the variance in frequency of drunkenness can be accounted for by the interaction of four variables (or two), then these variables cannot be dismissed as inconsequential. Within what set of explanatory constructs they can best be rationalized is another question. They may be interpreted as providing powerful support for a dependency-conflict hypothesis relating frequency of drunkenness to societal customs that (1) limit the indulgence of dependence in infancy, especially the diffusion of nurturance among many caretakers, (2) emphasize demands for achievement in childhood, and (3) in some way limit dependent behavior in adulthood. Conversely, this

---

* For a clear discussion of multicollinearity, and the problems it causes in regression analysis, see Wonnacott and Wonnacott (1972).

evidence indicates that societies whose members drink but tend in general toward sobriety are those where the dependence needs of infants and small children are indulged by many people, where the demands for achievement in childhood are not stressed and where adults freely behave in a dependent manner. These findings are in essential agreement with the predictions of the original hypothesis (Bacon et al., 1965) with regard to the nature of the significant variables, the direction of their influence, and the effect of their interaction.

The possibility of alternate emphases also exists. The variable, Diffusion of Nurturance, as noted, provides a measure of the number of adults in the environment who provide nurturant care for the young child. Two dimensions are clearly included in this measure. One involves the degree of nurturance given to the child and the other refers to the degree of dispersion of this nurturance among many or few adults. The interpretation of the associations found will vary somewhat according to which aspect of this variable is assumed to be operative in this interaction. If the degree of nurturance is emphasized, then the interpretation would be of a negative association between frequency of drunken and the degree of indulgence of dependence needs. On the other hand, if the significant factor is assumed to lie in the degree to which this nurturance is dispersed among adults, the interpretation would shift. The suggestion in this case would be that in societies with high diffusion of nurturance, children are not only indulged with respect to their dependence needs but are taught to expect and ask for help from a large number of nurturant people, rather than primarily one (the mother or mother-surrogate). Possibly this kind of childhood experience leads to adults who ask freely for help when they feel the need and experience no conflict or hesitance about doing so.

In considering the essential nature of this variable as it interacts in this context, it may be of significance to note that Diffusion of Nurturance in early childhood is found to be significantly correlated with the following: Communal eating ($r = +.28$, $N = 71$, $p < .01$), Hospitality ($r = +.21$, $N = 69$, $p < .05$), Adult Nurturance ($r = +.22$, $N = 64$, $p < .05$), and Adult Responsibility ($r = +.36$, $N = 61$, $p < .002$). This cluster of associations might be tentatively viewed as cultural expressions of affiliative motives that may be seen as another aspect of dependence.

The second child-training variable found to be a significant predictor of frequency of drunkenness was Pressures toward Achievement. This measure tended to emphasize ethnographic evidence of encouragement of competitiveness and expectations of individual rather than group attainment of high standards of performance. Perhaps it is the individualistic aspect of this variable that is crucial in this association. It is also possible that pressures toward achievement (individualistic) in childhood involve anxieties not characteristic of socialization with regard to other kinds of behavior. Measures of Anxiety over

Achieving were found to be significantly associated with Frequency of Drunkenness ($r = +.35$, $N = 54$, $p < .01$). This variable was defined as the amount of anxiety inherent in the achievement situation, such as fear of arousing hostility in others by winning the competition. Perhaps the significant factor involved here is anxiety. Possibly an important dimension is one of positive and negative affect. Tomkin's (1966, 1973) theoretical model of addictive smoking might be relevant here. By such a formulation societal tendencies toward diffusion of nurturance, absence of pressure toward achievement and permission of adult dependence might represent conditions of positive affect associated, in turn, with drinking with a low frequency of drunkenness.

Certainly the possibility exists that the independent variables included in the regression equation reported above can be conceptualized as representing an affiliative–individualistic dimension rather than or in addition to a dependence–independence conflict dimension. In this connection it might be noted that other studies relating drunkenness to factors of social deviance and alienation seem not unrelated to conceptualizations involving absence of affiliation or, at the extreme, social isolation (e.g., Jessor *et al.*, 1968). Similarly, it is not difficult to compare McClelland's "search for power" with a motivation toward egoistic dominance.

The significant dimension is, of course, that which best rationalizes the existing findings. At this time, the independent variables found here to be significant predictors of frequency of drunkenness still seem best conceived as representing aspects of a dependency-conflict dimension. Such a conceptualization seems to offer some explanation for more of the observed sociocultural variation than any other yet stated. Here might be included variations in drinking among women as well as variations between men and women, variations between ethnic groups and in economic achievement categories. Also rationalized within this framework is the success of Alcoholics Anonymous and other types of group treatment of drinking problems that emphasize the surrender of autonomy and the development of new social networks providing freely expressed interdependence among members.

DISCUSSION AND SUMMARY

The importance of crosscultural studies of drinking customs in the field of research on alcohol seems clearly established. Although the methodological problems involved in the collection of data are numerous, they do not appear to be insurmountable and, even at present, operate as a limiting rather than invalidating condition. There is an obvious need for more extensive and more uniformly organized ethnographic data on drinking practices as a basis for coding.

The development of a field guide for the study of drinking customs that would facilitate the collection of such material is long overdue. Also, a crosscultural file of classified materials on drinking practices would assist in extending the body of coded materials available to researchers.

The interpretation of crosscultural findings also presents problems that seem to have received less emphasis than those involved in the collection of data. At the risk of seeming gratuitous, attention might again be called to the insidious effect of one's own cultural bias on the development of hypotheses and theory in this field. The strongly negative image of drinking and drunkenness implicitly held by many people in American society has already been noted, and drinking as a symbol of masculinity is a familiar theme. At the same time many cultural subgroups place a high value on achievement and independence training, especially for male children. The same set of values tends to view dependence as associated with weakness and femininity, and "indulgence" as a kind of "giving in" to children that causes them to be "spoiled." Perhaps it is this combination of connotations that leads some writers to associate indulgence of dependence and frequency of drunkenness in a positive direction (Naroll, 1970) rather than a negative one as the evidence indicates. On an intuitive basis, it also seems likely that most American males who like to drink would prefer to think that heavy drinking represented a frustrated need for power than a "hang-up" about being dependent. Data relevant to such questions must, of course, await research on the cognitive categories and dimensions utilized by various groups in organizing their perceptions of drinkers and drinking behavior.

Another problem of interpretation which is perhaps of exceptional significance in crosscultural research concerns the interpretation of the variables measured. The need, when rationalizing findings, to return to the operational definition of the variables cannot be overemphasized. Names for these variables have been arbitrarily chosen as reasonably descriptive cues, but should not be interpreted on the basis of their apparent meaning in terms of common usage. For example, the term, "infant Indulgence," has been applied to a quantified variable of child-rearing, first described in research on the socialization of dependence (Barry *et al.*, 1957). It has been used frequently in crosscultural studies and has sometimes been interpreted rather loosely in terms of its apparent surface meaning. Actually, the term represents a summary label based on seven subvariables, each of which has been defined in some detail and separately labeled. Ratings listed under this heading should not be utilized as measures of an aspect of child-rearing without review of the procedures on which the ratings were based. A considerable number of such quantified variables representing different dimensions of cultural variation in samples of societies are now available in the literature. They are frequently used to explore associations between variables and to test hypotheses. The definitions of these variables and the operations whereby they were quantified are in nearly all cases clearly stated in the original publications (See Textor's "A Cross-Cultural Summary," 1967). The interpretation of such correlational findings is necessarily dependent on a clear understanding of the operations whereby the interacting variables were delineated.

This chapter has attempted an overview of the contributions of studies of drinking in tribal societies to the field of alcohol research. Ethnographic data on drinking customs now available in the literature derive from general descriptions of the ways of life of preliterate societies or from specific studies of drinking in such groups. As case studies, such descriptive units provide dramatic illustration of the range of variation in all dimensions of the drinking custom to be found in cultural groups throughout the world. This descriptive material may also be organized in categories and roughly quantified to form the basis for hologeistic studies of variables of the drinking custom. Crosscultural findings from such studies that are pertinent to three main areas of interest in the study of alcohol have been considered here: integrated drinking, sex differences in the use of alcohol, and motivations for drinking. In each of these areas the crosscultural perspective has provided heuristic insights as well as a testing ground for existing hypotheses. In some cases, hypotheses have been supported in a crosscultural sample. In others, associations found to exist within a single case were not replicated across societies. The need to test all intracultural observations against a broader framework of cultural variation is clear. With an improved data base, crosscultural studies should prove increasingly useful in the pursuit of this task.

ACKNOWLEDGMENTS

The author wishes to thank Professors Mark Keller and Sylvan Tomkins for their generous assistance in reading and commenting on this manuscript.

REFERENCES

Bacon, M. K., 1974, The dependency-conflict hypothesis and the frequency of drunkenness: Further evidence from a cross-cultural study, *Quart. J. Stud. Alc.* 35:863.
Bacon, M. K., Barry, H. III, and Child, I. L., 1965a, A cross-cultural study of drinking: II. Relations to other features of culture. *Quart. J. Stud. Alc.* Suppl. No. 3:29.
Bacon, M. K., Barry, H. III, Child, I. L., and Snyder, C., 1965b, A cross-cultural study of drinking: V. Detailed definitions and data, *Quart. J. Stud. Alc.* Suppl. No. 3:78.
Bacon, S. D., 1944, "Sociology and the Problems of Alcohol: Foundations for a Sociological Study of Drinking Behavior," Hillhouse Press, New Haven.
Bacon, S. D., 1967, The Classic Temperance Movement of the U.S.A.: impact today on attitudes, action and research, *Brit. J. Addict.* 62:5.

Bailey, M., 1961, Alcoholism and marriage; a review of research and professional literature, *Quart. J. Stud. Alc.* 22:81.
Bales, R. F., 1946, Cultural differences in rates of alcoholism, *Quart. J. Stud. Alc.*, 6:480.
Bales, R. F., 1962, Attitudes toward drinking in the Irish culture, *in* "Society, Culture and Drinking Patterns" (D. J. Pittman and C. R. Snyder, eds.), pp. 157–187, Wiley, New York.
Barnett, M. L., 1955, Alcoholism in the Cantonese of New York City: An anthropological survey, *in* "Etiology of Chronic Alcoholism" (O. Diethelm, ed.), pp. 179–227, Thomas, Springfield, Ill.
Barry, H. III, 1974, Psychological factors in alcoholism, *in* "The Biology of Alcoholism, Vol. 3, Clinical Pathology" (B. Kissin and H. Begleiter, eds.), pp. 53–107, Plenum Press, New York.
Barry, H. III, and Miller, N. E., 1962, Effects of drugs on approach-avoidance conflict tested repeatedly by means of a "telescope alley," *J. Comp. Physiol. Psychol.* 55:201.
Barry, H. III, Bacon, M. K., and Child, I. L., 1957, A cross-cultural survey of some sex differences in socialization, *J. Abnorm. Soc. Psychol.* 55:327.
Barry, H. III, Child, I. L., and Bacon, M. K., 1959, Relation of child training to subsistence economy. *Amer. Anthrop.* 61:51.
Barry, H. III, Buchwald, C., Child, I. L., and Bacon, M. K., 1965, A cross-cultural study of drinking: IV comparisons with Horton ratings, *Quart. J. Stud. Alc.* Suppl. No. 3:62.
Batchelor, J., 1892, "The Ainu of Japan," Revell, New York.
Berreman, G. D., 1956, Drinking patterns of the Aleuts, *Quart. J. Stud. Alc.* 17:503.
Blane, H. T., 1968, "The Personality of the Alcoholic," Harper and Row, New York.
Brown, J., 1970, A note on the division of labor by sex, *Amer. Anthrop.* 72:1073.
Bunzel, R., 1940, The role of alcoholism in two Central American cultures, *Psychiatry* 3:361.
Cappell, H., and Herman, C. P., 1972, Alcohol and tension reduction: A review, *Quart. J. Stud. Alc.* 33:33.
Carpenter, E. S., 1959, Alcohol in the Iroquois dream quest, *Amer. J. Psychiat.* 116:148.
Child, I. L., Bacon, M. K., and Barry, H. III, 1965a, A crosscultural study of drinking: I Descriptive measurements of drinking customs, *Quart J. Stud. Alc.* Suppl. No. 3:1.
Child, I. L., Barry, H. III, and Bacon, M. K., 1965b, A cross-cultural study of drinking: III Sex differences, *Quart. J. Stud. Alc.* Suppl. No. 3:49.
Conger, J. J., 1951, The effects of alcohol on conflict behavior in the albino rat, *Quart. J. Stud. Alc.* 12:1.
Coopersmith, S., 1964, The effects of alcohol on reactions to affective stimuli, *Quart. J. Stud. Alc.* 25:459.
Curley, R. T., 1967, Drinking patterns of the Mescalero Apache, *Quart. J. Stud. Alc.* 28:116.
Dailey, R. C., 1968, The role of alcohol among North American Indian tribes as reported in Jesuit Relations. *Anthropologica* 10:45.
Davis, K. E., 1972, Drug effects and drug use, *in* "Social Psychology in the Seventies" (L. S. Wrightsman, ed.), pp. 517–546, Wadsworth, Belmont, Calif.
Devereaux, G., 1948, The function of alcohol in Mohave society, *Quart. J. Stud. Alc.* 9:207.
Emerson, E. F., 1908, "Beverages, Past and Present: An Historical Sketch of their Production, Together with a Study of the Customs Connected with their Use," Putnam Sons, New York.
Everett, M. W., 1971, Drinking, talking and fighting: An Apache dilemma, Paper, American Anthropological Association Meeting, New York.
Field, P. B., 1962, A new cross-cultural study of drunkenness, *in* "Society, Culture and Drinking Patterns" (D. J. Pittman and C. R. Snyder, eds.), pp. 48–74, Wiley, New York.
Forbes, R. J., 1954, Chemical, culinary and cosmetic arts, *in* "A History of Technology" (C. Singer, E. J. Holmyard, and A. R. Hall, eds.), pp. 238–298, Oxford Univ. Press, London.
Ford, C. S., 1967, "Cross-Cultural Approaches: Readings in Comparative Research," HRAF Press, New Haven.

Frake, C. O., 1964a, Notes on queries in ethnography, *Amer. Anthrop.* 66:132.
Frake, C. O., 1964b, A structural description of Subanum "Religious Behavior," in "Explorations in Cultural Anthropology" (W. Goodenough, ed.), pp. 111–130, McGraw-Hill, New York.
Gallagher, O. R., 1965, Drinking problems among tribal Bihar, *Quart. J. Stud. Alc.* 26:617.
Geoghegan, W., 1969, Decision making and residence on Tagtabon Island *Univ. Calif. Language Behavior Research Lab.,* Working Paper No. 17, Berkeley.
Geoghegan, W., 1971, Information processing systems in culture, in "Explorations in mathematical anthropology" (P. Kay, ed.), pp. 4–35, MIT Press, Cambridge, Mass.
Gewirtz, J. L. (ed.), 1972, "Attachment and Dependency," Wiley, New York.
Glad, D. D., 1947, Attitudes and experiences of American-Jewish and American-Irish male youth as related to differences in adult rates of inebriety, *Quart. J. Stud. Alc.* 8:406.
Goodenough, W., 1970, "Description and Comparison in Cultural Anthropology," Aldine, Chicago.
Goodenough, W., 1971, "Culture, Language, and Society," Addison-Wesley McCaleb Module.
Gorer, G., 1938, "Himalayan Village: An Account of the Lepchas of Sikkim," Michael Joseph, London.
Greenberg, L. A., and Carpenter, J. A., 1957, The effect of alcoholic beverages on skin conductance and emotional tension. I. Wine, Whisky and alcohol., *Quart. J. Stud. Alc.* 18:190.
Hamer, J. H., 1965, Acculturation stress and the function of alcohol among the Forest Potawatomi, *Quart. J. Stud. Alc.* 26:285.
Heath, D. B., 1952, "Alcohol in a Navaho Community," A. B. thesis, Harvard College.
Heath, D. B., 1958, Drinking Patterns of Bolivian Camba, *Quart. J. Stud. Alc.* 19:491.
Heath, D. B., 1964, Prohibition and post-repeal drinking patterns among the Navajo. *Quart. J. Stud. Alc.* 25:119.
Hetherington, E. M., and Wray, N. P., 1964, Aggression, need for social approval, and humor preferences, *J. Abnorm. Psychol.* 68:685.
Hitchcock, R., 1890, "The Ainus of Yezo Japan," Report of U.S. Nat'l. Museum under direction of Smithsonian Inst., Washington.
Honigmann, J. J., and Honigmann, I., 1945, Drinking in an Indian-white community, *Quart. J. Stud. Alc.* 5:575.
Honigmann, J. J., and Honigmann, I., 1965, How the Baffin Island Eskimo have learned to use alcohol, *Social Forces* 44:73.
Horton, D., 1943, The functions of alcohol in primitive societies: A cross-cultural study, *Quart. J. Stud. Alc.* 4:199.
Horton, D., 1959, Primitive societies, in "Drinking and Intoxication" (R. McCarthy, ed.), pp. 251–262, Free Press, Glencoe, Ill.
Jellinek, E. M., and McFarland, R. I., 1940, Analysis of psychological experiments on the effects of alcohol, *Quart. J. Stud. Alc.* 1:272.
Jessor, R., Graves, T. D., Hanson, R. C., and Jessor, S. L., 1968, "Society, Personality, and Deviant Behavior," Holt, Rinehart and Winston, New York.
Junker, W., 1892, "Travels in Africa During the Years 1882–1886," Chapman and Hall, Ltd., London.
Kalin, R., McClelland, D. C., and Kahn, M., 1965, The effects of male social drinking on fantasy, *J. Pers. Soc. Psychol.* 1:441.
Kalin, R., Davis, W. N., and McClelland, D. C., 1966, The relationship between use of alcohol and thematic content of folktales in primitive societies, in "The General Inquirer: A Computer Approach to Content Analysis" (P. J. Stone, D. C. Dunphy, M. S. Smith, and D. M. Ogilvie, eds.), pp. 569–88, MIT Press, Cambridge, Mass.
Kaplan, H. S., 1956, Effects of alcohol on fear extinction, *Diss. Abst.*, 55:571.

Karsten, R., 1935, "The Head Hunters of Western Amazonas: The Life and Culture of the Jivaro Indians of Eastern Ecuador and Peru," Societas Humanarum Litterarum, Vol. VII, No. 1, Centraltryckeriet, Helsingfors.
Kastl, A. J., 1969, Changes in ego functioning under alcohol, *Quart. J. Stud. Alc.* 30:371.
Kearney, M., 1970, Drunkenness and religious conversion in a Mexican village, *Quart. J. Stud. Alc.* 31:132.
Keesing, R., 1971, Kwaio fosterage, *Amer. Anthrop.* 71:991.
Keesing, R., 1972, Formalization and the construction of ethnographies, *in* "Explorations in Mathematical Anthropology" (P. Kay, ed.), pp. 36–50, MIT Press, Cambridge, Mass.
Keller, M., 1958, Beer and wine in ancient medicine, *Quart. J. Stud. Alc.* 19:153.
Kissin, B., and Platz, A., 1968, The use of drugs in the long-term rehabilitation of chronic alcoholics, *in* "Psychopharmacology: Review of Progress, 1957–1967" (D. H. Efron, ed.), Public Health Service Publication No. 1836, pp. 835–851.
Klausner, S. Z., 1964, Sacred and profane meanings of blood and alcohol, *J. Soc. Psychol.* 64:27.
Krige, E. J., and Krige, J. D., 1943, "The Realm of a Rain-Queen", Oxford, Univ. Press, New York.
Leacock, S., 1964, Ceremonial drinking in an Afro-Brazilian cult, *Amer. Anthrop.* 66:344.
Lemert, E. M., 1954, Alcohol and the Northwest Coast Indians, U. Calif. Publications Cult. Soc., 2:303.
Lemert, E. M., 1962, Dependency in married alcoholics, *Quart. J. Stud. Alc.* 23:590.
Lemert, E. M., 1964a, Forms and pathology of drinking in three Polynesian societies, *Amer. Anthrop.* 66:361.
Lemert, E. M., 1964b, Drinking in Hawaiian plantation society, *Quart. J. Stud. Alc.,* 25:689.
Levy, R. I., 1966, Ma'ohi drinking patterns in the Society Islands, *J. Polynes. Soc.* 75:304.
Lisansky, E. S., 1957, Alcoholism in women: Social and psychological concomitants. I. Social history data, *Quart. J. Stud. Alc.* 18:588.
Lisansky, E. S., 1959, Psychological effects, *in* "Drinking and Intoxication" (R. G. McCarthy, ed.), Publications Division Yale Center of Alcohol Studies, New Haven.
Lobban, M. D., 1971, Cultural problems and drunkenness in an Arctic population, *Brit. Med. J.* 1:344.
Lolli, G., 1956, Alcoholism as a disorder of the love disposition, *Quart. J. Stud. Alc.* 17:96.
Lolli, G., Serianni, E., Golder, G., and Luzzatto-Fegis, P., 1958, "Alcohol in Italian Culture," Free Press, Glencoe, Ill.
Lomnitz, L., 1969a, Patterns of alcohol consumption among the Mapuche, *Hum. Organiz.* 28:287.
Lomnitz, L., 1969b, Patrones de ingestion de alcohol entre migrantes mapuches en Santiago, *Amer. indig.* 29:43.
Lomnitz, L., 1969c, Funcion del alcohol en la sociedad Mapuche, *Actapsiquiat. psicol. Amer lat.* 15:157.
Lubart, J. M., 1969, Field study of adaptation of MacKenzie Delta Eskimos to social and economic change, *Psychiatry* 32:447.
Maccoby, E. E., and Masters, J. C., 1970, Attachment and dependency, *in* "Carmichael's Manual of Child Psychology" (P. H. Mussen, ed.), Vol. II., pp. 73–157, Wiley, New York.
Madsen, W., and Madsen, C., 1969, The cultural structure of Mexican drinking behavior, *Quart. J. Stud. Alc.* 30:701.
Mandelbaum, D. G., 1965, Alcohol and culture, *Curr. Anthrop.* 6:281.
Mangin, W. P., 1957, Drinking among Andean Indians, *Quart. J. Stud. Alc.* 18:55.
Masserman, J. H., and Yum, K. S., 1946, An analysis of the influence of alcohol on experimental neurosis in cats, *Psychosom. Med.* 8: 36.
Masserman, J. H., Jacques, M. G., and Nicholson, M. R., 1945, Alcohol as a preventive of experimental neuroses, *Quart. J. Stud. Alc.* 6:281.

Maxwell, M. A., 1952, Drinking behavior in the state of Washington, *Quart. J. Stud. Alc.* 13:219.

Maynard, E., 1969, Drinking as part of an adjustment syndrome among the Oglala Sioux, *Pine Ridge Res. Bull., S. D.* 9:35.

McClelland, D. C., Davis, W., Wanner, E., and Kalin, R., 1966, A cross-cultural study of folktale content and drinking. *Sociometry* 29:333.

McClelland, D. C., Davis, W. N., Kalin, R., and Wanner, E., 1972, "The Drinking Man," Free Press, New York.

McCord, W., and McCord, J., 1960, "Origins of Alcoholism," Stanford Univ. Press, Stanford.

McKinlay, A. P., 1948a, Early Roman Sobriety, *Classic. Bull.* 24:52.

McKinlay, A. P., 1948b, Ancient experience with intoxicating drinks: non-classical peoples, *Quart. J. Stud. Alc.* 9:388.

McKinlay, A. P., 1949a, Roman sobriety in the later Republic, *Classic Bull.* 25:27.

McKinlay, A. P., 1949b, Ancient experience with intoxicating drinks: Non-Attic Greek states, *Quart. J. Stud. Alc.* 10:289.

McKinlay, A. P., 1951, Attic temperature, *Quart. J. Stud. Alc.,* 6:102.

McNamee, H. B., Mello, N. K., and Mendelson, J. H., 1968, Experimental analysis of drinking patterns of alcoholics: Concurrent psychiatric observations, *Amer. J. Psychiat.* 124:1063.

Mello, N. K., 1972, Behavioral studies of alcoholism, *in* "The Biology of Alcoholism, Vol. 3, Clinical Pathology" (B. Kissin and H. Begleiter, eds.), pp. 219–291, Plenum Press, New York.

Mendelson, J. H. (ed.), 1964, Experimentally induced chronic intoxication and withdrawal in alcoholics (Part 3, Psychiatric Findings by Mendelson, J. H., LaDou, J., and Solomon, P.) *Quart. J. Stud. Alc.* Suppl. No. 2:40.

Morris, J., 1938, "Living with Lepchas," Heinemann, London.

Mulford, H. A., and Miller, D. E., 1959, Drinking in Iowa: Sociocultural distribution of drinkers, with a methodological model for sampling evaluation and interpretation of findings, *Quart. J. Stud. Alc.,* 20:704.

Murdock, G. P., 1957, World ethnographic sample, *Amer. Anthrop.* 59:664.

Myerson, A., 1940, Alcohol: A study of social ambivalence, *Quart. J. Stud. Alc.,* 1:13.

Naroll, R., 1970, What have we learned from cross-cultural surveys? *Amer. Anthrop.* 72:1227.

Netting, R. M., 1964, Beer as a locus of value among the West African Kofyar, *Amer. Anthrop.* 66:375.

Robbins, M. D., and Pollnac, R. B., 1969, Drinking patterns and acculturation in rural Baganda, *Amer. Anthrop.* 71:276.

Sadoun, R., Lolli, G., and Silverman, M., 1965, "Drinking in French Culture," Rutgers Center of Alcohol Studies, New Brunswick, N.J.

Sandoval, R. L., 1945, Drinking motivations among the Indians of the Ecuadorean Sierra, *Primitive Man* 18:39.

Savard, R. J., 1968, Cultural stress and alcoholism: a study of their relationship between Navaho alcoholic men. PhD dissertation, University of Minnesota (University Microfilm No. 69-1532).

Sayres, W. C., 1956, Ritual drinking, ethnic status, and inebriety in rural Colombia, *Quart. J. Stud. Alc.* 17:53.

Scarborough, B. B., 1957, Lasting effects of alcohol on the reduction of anxiety in rats, *J. Genet. Psychol.* 91:173.

Schaefer, J. M., 1973, A hologeistic study of family structure and sentiment, supernatural beliefs and drunkenness, PhD Thesis, SUNY Univ. Microfilms.

Simmons, O. G., 1968, The sociocultural integration of alcohol use. A Peruvian study, *Quart. J. Stud. Alc.,* 29:152.

Smart, R. C., 1965, Effects of alcohol on conflict and avoidance behavior, *Quart. J. Stud. Alc.* 26:187.

Snyder, C. R., 1958, "Alcohol and the Jews: A Cultural Study of Drinking and Sobriety," Free Press, Glencoe, Ill.
Spradley, J., 1970, "You Owe Yourself a Drunk: An Ethnography of Urban Nomads," Little Brown, Boston.
Sterne, M. E., 1966, "Drinking Patterns and Alcoholism among American Negroes," Social Sci. Inst., Washington Univ., St. Louis, Missouri.
Stout, D. B., 1947, "San Blas Cuna Acculturation: An Introduction," Viking Fund Publications in Anthropology, No. 9, New York.
Straus, R., and Bacon, S. D., 1953, "Drinking in College," New Haven, Yale Univ. Press.
Textor, R. B., 1967, "A Cross-Cultural Summary," HRAF Press, New Haven.
Tomkins, S., 1966, Psychological model for smoking behavior, *Amer. J. Publ. Hlth.* 56:17.
Tomkins, S., 1973, The experience of affect as a determinant of smoking behavior, *J. Abnorm. Psychol.* 81:172.
Topper, M., 1972, Alcohol and the young Navaho male: A study in drinking and culture change, MS., Northwestern Univ.
Underhill, R. M., 1946, Papago Indian religion, Columbia Univ. Contributions to Anthropology, No. 30, Columbia Univ. Press, N.Y.
Wallgren, H., and Barry, H. III, 1970, "Actions of Alcohol," Vol. 1. "Biochemical, Physiological, and Psychological Aspects," Vol. 2. "Chronic and Clinical Aspects," Elsevier, Amsterdam.
Wanner, E., 1972, Power and inhibition: A revision of the magical potency theory, *in* "The Drinking Man" (McClelland, D. C., Davis, W. N., Kalin, R., and Wanner, E.), pp. 73–98, Free Press, New York.
Washburne, C., 1961, "Primitive Drinking: A Study of the Uses and Functions of Alcohol in Preliterate Societies," College and Univ. Press, New Haven.
Whiting, J. W. M., and Child, I. L., 1953, "Child Training and Personality: A Cross-Cultural Study," Yale Univ. Press: New Haven.
Whiting, J. W. M., 1969, Methods and problems in cross-cultural research, *in* "The Handbook of Social Psychology" (G. Lindzey and E. Aronson, eds.), pp. 693–728, Addison Wesley, Reading, Mass.
Whittaker, J. O., 1962, Alcohol and the Standing Rock Sioux Tribe I: The pattern of drinking, *Quart. J. Stud. Alc.* 23:468.
Whittaker, J. O., 1963, Alcohol and the Standing Rock Sioux Tribe II: Psychodynamic and cultural factors in drinking, *Quart. J. Stud. Alc.,* 24:80.
Williams, A. F., 1966, Social drinking, anxiety and depression, *J. Pers. Soc. Psychol.* 3:689.
Williams, A. F., 1968, Psychological needs and social drinking among college students, *Quart. J. Stud. Alc.,* 29:355.
Williams, P. H., and Straus, R., 1950, Drinking patterns of Italians in New Haven, *Quart. J. Stud. Alc.,* 11:51, 247, 452, 586.
Wilson, C. T., and Felkin, R. W., 1882, "Uganda and the Egyptian Soudan," Sampson Low, Marston, London.
Witkin, H. A., Karp, S. A., and Goodenough, D. R., 1959, Dependence in alcoholics, *Quart. J. Stud. Alc.* 20:493.
Wolcott, H. F., 1974, "The African Beer Gardens of Bulawayo: Integrated Drinking in a Segregated Society," Rutgers Center of Alcohol Studies, New Brunswick, N.J.
Wonnacott, T. H., and Wonnacott, R. J., 1972, "Introductory Statistics, 2nd ed.", Wiley, New York.

CHAPTER 2

Anthropological Perspectives on the Social Biology of Alcohol: An Introduction to the Literature

Dwight B. Heath
Department of Anthropology
Brown University
Providence, Rhode Island

INTRODUCTION

It is apparent that throughout human history the drug most widely used for altering human consciousness has been alcohol.* It is probably the oldest, and certainly now the most widespread mind-altering drug, but it is also distinctive in its versatility. In various times and in various contexts, human beings have used alcohol as an energizer or tranquilizer, as a "super-ego solvent" or sacred symbol, as a medicine or food, as a social leveler or source of subjective feelings of power, and for a host of other social or symbolic functions.

From this range of examples, it is clear that the chemistry of the substance and the physiology of its effects on the human body, however important they

---

* Throughout this chapter, I will use the convenient and familiar convention of using the word "alcohol" to stand for "ethanol" and/or "alcoholic beverages."

may be, do not fully determine the ways it will be used or the meanings that will be attached to it. With this in mind, it should be obvious too that alcohol can fruitfully be studied from a number of different points of view, even when we limit our attention to the meanings and uses of alcohol for *homo sapiens*. Presumably this is what the editors had in mind when they chose to include in this series a volume on its social aspects. This is certainly the principal reason why, although virtually no anthropologists have studied alcohol, there is a considerable corpus of anthropological literature on human beliefs and behaviors as they relate to alcohol.

My main purpose in this chapter is to provide a brief introduction to the range of such studies for the benefit of both anthropologists who are not familiar with alcohol research and colleagues in other disciplines who are not familiar with the relevant work of anthropologists. The literature is extremely diverse and widely scattered. In part, this is normal with respect to the full range of writings about alcohol by scholars in whatever discipline. With respect to anthropology, however, the problem may be aggravated by the breadth of the discipline and the publishing habits of anthropologists.

In general, as the term is used in the United States, anthropology comprises the following major subfields: archaeology and ethnohistory (the use of prehistoric and/or historical data in order to infer patterns of belief and behavior among populations who can no longer be observed in action); linguistics (which focuses on the symbolizing and communicative parts of human behavior); physical anthropology (human biology); and sociocultural anthropology (which deals with the range of social and cultural patterns extant today). The diversity of interests within anthropology is both a strength and a weakness. One of the problems is the proliferation of specialized journals in which anthropologists publish; such fractionation is aggravated by the fact that most sociocultural anthropologists also work in societies other than their own, so that area-oriented journals also carry many anthropological contributions. Books and specialized monographs often include rich, descriptive data on alcohol, embedded in a context and under a title that would not signal that fact to a reader unfamiliar with the material.

One of the most useful ways of characterizing anthropological perspectives may be to briefly contrast some of the emphases of the material discussed in this chapter, with emphases that predominate in other social sciences. In archaeology, for example, many of the same data and methods are used as in classics, Egyptology, art history, history, and related disciplines that also focus on the past, but the preoccupation with attempting to understand the workaday life of all segments of the population is unusual. Anthropological linguistics, in like manner, shares much with comparative philology, historical linguistics, semiotics, and other studies of symbols and communication; but the readiness to work with unwritten languages, to relate the forms of language to the cosmography of a people, and, in general, to attempt to deal with language in relation to other aspects of social action is distinctive. Among physical anthropologists there are comparative anatomists, geneticists, biophysiologists, primatologists, and others, who may share little in common but who tend to focus on certain topics that are not regularly studied by other biologists; among these are race, primate evolution, somatology, population genetics, and so forth.

Sociocultural anthropology shares much with comparative sociology, except that anthropologists have tended to pay more attention to unfamiliar ways of life among peoples whose history is not in the mainstream of Western civilization, and have usually attempted, with more or less success, to describe such systems in holistic terms. In dealing with "primitive"* groups, or with peasantries and other components in complex and pluralistic societies, anthropologists have also tended to emphasize different methods; participant-observation, undirected interviewing, and other means of collecting information sacrifice the standardization, quantification, and other kinds of comparability that questionnaires, for example, would provide. This apparent disadvantage may be outweighed, however, by the better opportunity for understanding things from the point of view of the people being studied, and nonrigorous methods of research often reveal unexpected and valuable insights. Sociocultural anthropology also shares many concerns with psychology, but in dealing with values, sentiments, attitudes, emotions, needs, and other motivational data, anthropologists tend to focus on modal or preferred patterns and on the range of variation within a population, rather than on specific individuals.

It should be obvious that these gross characterizations are intended merely to suggest the range of concerns of anthropologists, and some of the ways in which the literature under discussion here may differ from that which predominates in other discussions of social and cultural aspects of alcohol. Fortunately, the division of labor among academic disciplines is not strict; one of the most telling demonstrations of that is the fact that fewer than two thirds of the authors cited in this review consider themselves anthropologists.

GENERAL SOURCES

The fact that an anthropologist has been invited to contribute a chapter to a work entitled *The Biology of Alcoholism* is in itself one sign of the impact that anthropological perspectives have had in recent years on the study of alcohol. Another indication that anthropological approaches to alcohol are recognized as important is the recent sponsorship, by the National Institute on

---

* The word "primitive" is set in quotation marks to signal its specific and distinctive meaning in anthropological usage, i.e., connoting the absence of a written language, with *no* implication that such societies are "simple," "backward," "archaic," or otherwise conform to the lay imagery of "primitive man."

Alcohol Abuse and Alcoholism, of an international and interdisciplinary conference on the subject, the proceedings of which are in press (Everett et al., 1976). In that volume, some of the major controversies are reviewed, and some new directions of study are exemplified. One chapter (Heath, 1976) provides a historical review of the anthropological literature. Another recent review article deals primarily with ethnographic studies, and is organized in terms of categories that are prevalent in the literature on alcohol, as described, analyzed, and interpreted by members of other disciplines (Heath, 1975).

For a single volume that provides summaries of the roles that drinking* plays in a wide range of societies throughout the world, Washburne (1961) is convenient. Excellent anthologies that bring together different kinds of studies of alcohol in sociocultural context have been edited by McCarthy (1959), and by Pittman and Snyder (1962).

The key bibliographic sources are the Classified Abstracts Archive of the Alcohol Literature (see Keller, 1976), and the monumental bibliographies compiled by the Rutgers Center for Alcohol Studies (Keller, 1966–   ). A valuable early contribution (Popham and Yawney, 1967) deserves to be updated.

A number of salient theoretic questions about alcohol and culture were raised by Mandelbaum et al. (1969), and several volumes of broad scope include substantial reference to anthropological findings (e.g., Chafetz and Demone, 1962; Lucia, 1963; MacAndrew and Edgerton, 1969; McClelland et al., 1972; Patrick, 1952; Pittman, 1967; Rouechè, 1960; and others).

Because anthropologists so often conduct their research among groups which are not closely identified with nation-states, it is fortunate that a few review articles are available to provide guides to the scattered and diverse sources that deal with drinking in various regions of the world. For North America, Leland (1976) and Everett and Waddell (forthcoming) will probably soon serve as such guides. For Latin America, Hartmann (1958) surveys the tribal data, and Heath (1974) attempts broader coverage. Marshall's article on Oceania (1976) is up-to-date, but Seekirchner's early synthesis of the African material (1931) has not been supplanted. Ahlström-Laakso's survey of recent work in Scandinavia (1976) is the nearest we have to a review of the European literature, and no one has yet ventured to survey the material on Asia.

ALCOHOL IN ARCHAEOLOGY AND HISTORY

Fermentation is a simple natural process that produces alcohol without human intervention in many foods. Several contemporary peoples use simple

---
* As another familiar and convenient convention, I will use the word "drinking" throughout this chapter not only in its generic sense but also occasionally to stand for "the drinking of alcoholic beverages"; the latter meaning is often clearly implied by context, as in this instance.

beers and wines made from wild plants even when they do not cultivate anything; presumably the same was true of Paleolithic societies prior to the beginnings of agriculture.

Some of the earliest writings we know contain references to alcohol and its effects, and as early as ca. 1700 B.C., the Code of Hammurabi specified in detail restrictions on the sale of wine. The sources that deal with alcohol in archaeology and ancient history are uneven in quality. Among the broad surveys, some offer little more than catalogs of exotica (e.g., Peeke, 1917; Gilder, 1921); others provide interesting detail but lack the scholarly apparatus of documentation (e.g., Emerson, 1908; Morewood, 1838); and some are remarkably encyclopedic (e.g., Cherrington, 1925-30; Crawley, 1929). Biblical and ancient Jewish patterns are discussed, often in considerable detail, by Danielow (1949), Fenasse (1964), Goodenough (1956), Jastrow (1913), Keller (1970), and others. Various aspects of drinking among other peoples in ancient Egypt and the Near East are in Cornwall (1939), Crothers (1903), Drower (1966), Lutz (1922), Modi (1888), Piga (1942a), and others. The roles of alcohol in classical cultures of the Mediterranean Basin are highlighted by Brown (1898), Buckland (1878), Gremek (1950), Hirvonen (1969), Jellinek (1961), McKinlay (1939, 1944, 1945, 1948a,b, 1948-1950a,b, 1951, 1953), Rolleston (1927), Suolahti (1955), and many others. Alcohol in ancient India is the theme of Aalto (1955), Bose (1922); Mitra (1873), Prakash (1961), and Ravi (1950), to mention only a few.

Historians who have dealt with more limited themes during the last millenium have tended to produce more detailed accounts of drinking behaviors and how they related to the sociocultural context. For example, Lender (1973) looked at drunkenness in relation to Puritan ethics; Bruman (1940) was able to determine the areal distribution of native alcoholic beverages in early colonial Mexico, and Taylor (in preparation) is surveying Indian criminality and drinking in the same period. Alcohol was important for both Indians and whites, on the North American frontier also, as noted (e.g., by Belmont, 1840; Brown, 1966; Dailey, 1968; Gray, 1972; Jacobs, 1950; MacLeod, 1928). For similar problems in colonial South America, see Bejarano (1950), Rojas (1960), and Ruiz (1939). Strict limitations surrounded alcoholic beverages in pre-Columbian Mexico (see, e.g., Barrera, 1941; Beals, 1932; Calderon, 1968; Gonçalves [ca. 1956]; Martin, 1938). Most of Oceania, however, had not used alcohol before contact with Europeans, so their drinking history is short but generally violent (see, e.g., Gunson, 1966; Wheeler, 1839).

Historical accounts that highlight little-known aspects of drinking in Europe and Asia include Connell (1968), Glatt (1969), Harrison and Trinder (1969), Jacobsen (1951). On Africa, see Dumett (1974), Seekirchner (1931), and Leis (1964), who have made pioneering attempts at providing historical depth to the rich ethnographic literature.

Surprising as it may seem, in only a few instances has there been a systematic attempt to compare drinking patterns over any significant length of time among a non-Western people (e.g., Levy and Kunitz, 1974; Lomnitz, 1976; Dailey, 1964), or to evaluate the impact of particular social changes (e.g., Heath, 1964, 1965, 1971; Ogan, 1966; Toit, 1964).

A number of other kinds of historical studies deal with specific aspects of alcohol rather than with regional drinking patterns. Baird (1944-48) traced the evolution of legal controls; Rolleston (1941) compiled the folklore of drinking. Jellinek (1965) paid special attention to the symbolism of alcoholic beverages, as did Klausner (1964). Whereas some authors have focused on medical uses of alcohol in history (e.g., Keller, 1958; Wright-St. Clair, 1962), others have focused on the history of medical problems caused by misuse of alcohol (e.g., Brown, 1898; Hirsh, 1953; Medical Practitioner, 1830; Mesa, 1959).

ALCOHOL AND LINGUISTICS

Applications of linguistics to alcohol studies have not been numerous, but they do reflect the ways in which recent trends in anthropology have diverged from more traditional structural and historical approaches to language. Illustrative examples are attempts to "translate" the argot of English-speaking urban gangs of chronic inebriates (Rubington, 1968, 1971; Spradley, 1971), Frake's (1964) pioneering attempt to analyze drinking behavior in terms of some Philippine native categories, Hage's (1972) compotential analysis in Munich, and Topper's (1976) dogged delineation of "verbal plans" among the Navajo.

ALCOHOL AND PHYSICAL ANTHROPOLOGY

Insofar as they deal with human biology, anthropologists often focus on health, nutrition, and race. Beliefs and attitudes of various peoples with respect to these subjects are varied, and sometimes even contradictory, just as is the case with respect to many other realms of concern. Although many African peoples consider alcoholic beverages to be healthful (e.g., Jeffreys, 1937; Krige, 1932; Leyburn, 1944; Platt, 1955; Tadesse, 1958), others consider them harmful (see Gelfand, 1966; Graviere, 1957; McGlashen, 1969; Mears, 1942). Similarly, some Asiatic populations favor drinking (Hasan, 1964; Kerketta, 1960; Ravi, 1950), whereas others view it as damaging (Feliciano, 1926). In many Latin-American societies, drinking is seen as wholesome (R. Anderson *et al.*, 1946; Busch, 1952; Heath, 1958; Simmons, 1960; Vazquez, 1967), but some emphasize medical problems it can cause (International Labor Organization, 1953; Sariola, 1961). In Oceania, Lemert has studied groups that treat alcohol as salubrious (1964b) and others that treat it as harmful (1962, 1964a). The same holds in Europe: some populations consider drinking heathful (Strubing, 1960; Wright-St. Clair, 1962), whereas others believe the opposite (Blum and Blum, 1964; Bruun, 1959).

An interesting sidelight on the discussion of alcohol and human physiology, is the apparent absence of such features as blackout, hangover, or addiction among many populations, even where drunkenness is commonplace (e.g., Heath, 1958; Kennedy, 1963; Lemert, 1954, 1962, 1964a,b; Mangin, 1957); unfortunately, these data are all anecdotal. Among the Navaho, another unusual feature is "negative tolerance," with long-term heavy drinkers reporting that they require progressively *less* in order to get drunk (Levy and Kunitz, 1974).

A theme of more widespread interest, and also of great emotional and political sensitivity, is the question of differential "racial" tolerance to alcohol. For the past 30 years, most of the writings by anthropologists about alcohol have stressed sociocultural factors, emphasizing how beliefs, values, rituals, and similarly learned patterns affect drinking, drunkenness, and drunken behavior. This cultural bias, linked with rejection of popular simplistic racism, led to dismissal of the so-called "firewater myth" as a supposed stereotypical expression of prejudice; see Leland (1976) for a historical review of the subject. Reaction against the idea that different populations might be affected differently by alcohol even became widely accepted by many physicians (*Journal of the American Medical Association,* 1954), although some, recognizing other kinds of physiological variation among populations, predicted that some such differences might be identified with respect to the metabolism of alcohol (e.g., Collard, 1962; B. Vallee, 1966).

It is unfortunate that political and philosophical views have so strongly colored both sides of this controversy. The North American image of "the drunken Indian" has sometimes been reinforced by scholars (e.g., Dailey, 1964; Hentig, 1945; Kunitz *et al.*, 1969; Lemert, 1958; Stewart, 1964), and sometimes contradicted (e.g., Brody, 1971; Honigmann and Honigmann, 1945). During the 1970s, a few experiments that seem to vindicate the folk-wisdom are having a considerable impact (e.g., Ewing and Rouse, 1973; Fenna *et al.,* 1971; Wolff, 1972, 1973), although some of the sampling and methods in those studies have been criticized (Lieber, 1972; Hanna, 1976). MacAndrew and Edgerton mustered considerable historical and ethnographic evidence to the effect that "... drunken comportment is an essentially *learned* affair"(1969: p. 88; italics in original). *Medical Tribune and Medical News* (1972) reported a distinctive physiological feature that created alcohol from rice during the digestive process in one population. The recent work by Korsten *et al.* (1975) on acetaldehyde levels between alcoholic patients and "nonalcoholics" may bear on the differences that are noted in terms of subjective as well as objective responses to alcohol among individuals in different ethnic categories. The subject will undoubtedly be hotly debated for some time, and it is to be hoped that controlled empirical data will play a greater role than has been the case until recently.

ALCOHOL AND SOCIOCULTURAL ANTHROPOLOGY

It is in the realm of social and cultural aspects that anthropologists have heretofore made the largest contribution to alcohol studies.

Drinking Patterns: Customs and Cultures

One of the most obvious values of anthropological perspectives is the ethnographic and comparative view, documenting in detail the enormous range of variation that occurs throughout the world in terms of who drinks, what beverages people drink, where and when they drink, in what manners, for what ostensible purposes, with what effects, and so forth. What is now common knowledge was striking news to some investigators a few decades ago, and it is still noteworthy when an occasional ethnographic report reveals drinking patterns that expand our view of that range of variation. The volume of such studies is enormous, and it will probably increase geometrically as does the general concern with focal "problems" in social research, and as more students become interested in alcohol studies specifically.

In a recent review of the literature, Heath (1976) has traced the historical evolution of this corpus. In another recent review, M. Bacon (1976b) summarized much of the range of variation of alcohol uses in tribal societies; a number of good articles have combined a few such examples with more on the variety of drinking to be found among subgroups within pluralistic Western society (e.g., E. Blacker, 1966; Heath, 1972; Lemert, 1969; Pittman, 1965; Riley and Marden, 1947; Wechsler et al., 1970); and Reader (1967) did the same, with an emphasis on excessive drinking. The proceedings of a major conference on Alcohol and Culture reflect not only what has been learned about human beliefs and practices with respect to alcohol, but also indicate new directions that are being explored in this connection (Everett et al., 1976).

The richness and ready accessibility of sources on drinking customs and cultures is such that it would be superfluous to repeat many of the themes in this context. Comparisons of rates of alcoholism among Jews, Irish, and ascetic Protestants in the United States would nowadays be more controversial with respect to the dubious delineation of the "ethnic groups" in question than with respect to accounting for such markedly different rates (for historical purposes, however, see Bales, 1946; Glad, 1947; Knupfer and Room, 1967; Snyder, 1958; Skolinik, 1958). Comparable studies have also been made in Canada (Room, 1968; Negrete, 1970) and Australia (Sargent, 1971). The pattern of frequent drinking with rare alcoholism among Chinese is also well-known (e.g., Barnett, 1955; Chu, 1972; LaBarre, 1946; Lee and Mizruchi [1960?]; Wang, 1968).

The confusion of national, religious, and other kinds of criteria for classifying "ethnic groups" creates many conceptual problems, but, for purposes of introducing the literature, we can use the simple expedient of following the usage of the respective authors. Gypsies have been studied in Sweden (Tillhagen, 1957), and in Yugoslavia (Vasev and Milosavčević, 1970). Drinking among Muslims, for example, is enormously varied, as reflected in the works of Chafetz (1964), Drower (1966), Quichaud (1955). The ambivalence of East Indians toward alcohol is amply documented by Carstairs (1954), Chopra *et al.* (1952), Mitra (1873), Patnaik (1960), Pertold (1931), Prakash (1961), Rao (n.d.), and Ravi (1950). On Japan, see Ando and Hasegawa (1970), and the works of Yamamuro (1954–1969). A few anthologies offer a convenient sampling of the range of cultural variation in drinking beliefs and practices, such as McCarthy (1959) and Pittman and Snyder (1962); the bibliography compiled by Popham and Yawney (1966) is good and has an index.

Rather than attempt to characterize any significant number of these studies individually, it may be more useful for this readership simply to underscore a few of the significant propositions that do emerge from the extremely fragmented and uneven literature on the ethnography of alcohol. At the outset, it is worth mentioning that there has been little attempt to study drinking systematically even in most of the populations mentioned here; the majority of the studies cited were tangential or serendipitous by-products of research efforts that were focused in other directions. Furthermore, in terms of alcohol studies there has been little attempt at systematic sampling, or at long-term investigation in any non-Western society. Despite these limitations, the following general statements may be noteworthy.

Drinking is normally a social act, and it is embedded in a context of values, attitudes, and conceptions of reality. These values, attitudes, and conceptions of reality are often implicit, and they often vary from one population to another. To a significant extent, the outcomes of drinking are affected by those values, attitudes, and conceptions of reality, as well as by the social context in which the drinking takes place.

In many societies, it is important to distinguish between drinking, drunkenness, drunken behavior, and the inebriate as a person, since attitudes and degrees of tolerance toward each of those may be very different. Drunkenness not only has different meanings and values in different societies, but it is also expressed in very different forms of behavior; drunken behavior is patterned to such a degree that it appears to be, in large part, the resultant of a learning process.

Alcoholism—even in the very general sense of problems associated with drinking (World Health Organization, 1952)—is rare in most societies of the world, although it is fast becoming more commonplace, particularly in areas where modern Western culture has made a close and sustained impact.

Change and Persistence

In a very real sense, cultural change is the stuff of history; similarly, cultural change is crucial whenever we talk about "acculturation" as a societal process, or "alcoholic" as a label put on an individual. Insofar as there has been any dominant theme throughout the diversely burgeoning anthropological literature of the past twenty-five years, it has been a recognition of the importance of change in patterns of human thought and action, and this is, to some extent, reflected in contributions to alcohol studies.

Most of the sources already cited in connection with alcohol in archaeology and history are descriptive rather than analytic. Often they are based on fragmentary evidence in ways that leave us wondering how representative they may be. Anthropological perspectives can more clearly be discerned in studies that emphasize the impact that some critical event may have had on the cultural patterns of a people, or attempt to factor out the significant covariants when it is recognized that the drinking patterns of a group have been significantly altered.

Among the most obvious situations in which temporal variation in alcohol use can fruitfully be studied are those of the imposition or repeal of prohibition. The importance of prohibition in United States history is signaled by the fact that, in two hundred years, that is the only constitutional amendment that was ever repealed. But other prohibitions have affected other populations (e.g., Heath, 1964; Ogan, 1966; Varma, 1959). Similarly, economic shifts can be identified as having an impact on drinking customs (see, e.g., Coffey, 1966; Harwood, 1964; R. Robbins, 1973).

Political and other forces often help to shape the roles of alcohol in a society, as in Bolivia, where interethnic drinking changed in diametrically opposing ways as part of a revolution (Heath, 1971); and Keller (1970) has directed our attention to some organizational concomitants of "the great Jewish drink mystery." MacAndrew and Edgerton (1969) have convincingly reconstructed evidence that "the firewater myth" gained currency and some validity only when some North American Indian tribes began to evince violent and debauched patterns of drunkenness, often after years of controlled and sometimes even sedate drinking; the case of the Iroquois is especially illuminating in terms of successive "stages" in terms of drinking patterns (Dailey, 1964). A curious case of diffusion occurred in Polynesia, where alcoholic beverages had not been known prior to contact with Europeans, but were accepted in roughly *inverse* relationship to the duration of such contact (Defer, 1969).

Much more common is the pattern by which popular acceptance of drinking, and increasing problems associated with alcohol, progress as the pressures of acculturation are increasingly felt by native peoples. Throughout the world, most groups who did not have alcoholic beverages prior to contact with Western civilizations have adopted them, often with deleterious consequences. Furthermore, those peoples who had been using traditional beers or wines before adopting new beverages from alien societies, often encountered "social problems" that had previously been unknown. It would be misleading to infer a simple causal relationship in such situations, since altered drinking patterns constitute only a small part of a complex combination of changes that occur at the same time in such situations.

Interest in the social welfare of non-Western peoples has combined with interest in the impact of modernization, however, to yield a rich literature on the impact that relations with a dominant sociocultural system have on the lives of members of subordinate populations. Most of the authors who write about this in connection with alcohol have tended to emphasize the adoption of drunkeness as an "escape" for people who are suffering ideological or social dislocation or anomie (e.g., Berreman, 1956; Dozier, 1966; Ervin, 1971; Jessor *et al.*, 1968; Norick, 1970; F. Vallee, 1968). Opposing views usually stress either continuity with precontact patterns (e.g., Hurt and Brown, 1965; Levy and Kunitz, 1971, 1974), protest, in a symbolic interactionist sense (Lurie, 1971; Everett and Waddell, forthcoming), or the larger cultural context as a whole, explicitly denying the relevance of ethnicity (e.g., Graves, 1970, 1971).

A few of the many other topics that have been analyzed in relation to alcohol and change are values (Leis, 1964), peasant organizations (Heath, 1965), native political organization (Hutchinson, 1961), trade (Dumett, 1974), prestige (Defer, 1969), urbanization (Hellmann, 1934), symbolism (Washburne, 1968), and epidemiology (Lint and Schmidt, 1970; Whitehead, 1972).

Crosscultural Correlations

Parallel and concurrent with the anthropological concern for seeking out and describing the range of variation in terms of human beliefs and practices, there is also a concern for comparing diverse cultures, and for identifying and analyzing those features that are common or covariant in the human experience.

Excellent review articles offer up-to-date summaries of the literature on crosscultural studies of a clinical nature (Westermeyer, 1974, 1976), and those of a hologeistic nature (i.e., statistical examination of the degrees of correlation between selected variables as they occur in a broad sample of sociocultural systems throughout time and space; see M. Bacon, 1976a; Schaefer, 1976, forthcoming). There have also been recent critiques of the methods used in crosscultural studies (Cahalan, 1976; Room, 1972; Stull, 1975), so it would be superfluous to do any of those here. By way of introducing social aspects of the biology of alcohol, however, it may be worth simply underscoring the remarkable impact that such studies have made on theories about alcoholism.

Probably the most widely quoted general statement about human motivations for using alcohol is Horton's conclusion that "... *the primary function of alcoholic beverages in all societies is the reduction of anxiety*" (1943, p. 223; italics in original). His pioneering methods have been criticized, but with minor refinements and adaptations, continue to reveal striking correlations among aspects of culture in large samples. Field (1962) reworked the same data, and found that kin groupings and other features of social organization were correlated, in more statistically significant degree, with patterns of sobriety. An expanded sample and more rigorous methodology were used by M. Bacon *et al.* (1965), H. Barry *et al.* (1965), and Child *et al.* (1965), demonstrating the association between conflict over dependence (resulting from discontinuities in the socialization of individuals), and drunkenness. A similarly global theory about motivations for drinking emphasizes that an individual finds in alcohol "... a means of giving himself at least a temporary boost in his feelings of potency" (McClelland *et al.*, 1972, p. 197). Much of this "power theory" is based on crosscultural analyses (see, e.g., Boyatzis, 1976; Kalin *et al.*, 1966; McClelland *et al.*, 1966; Wanner, 1972); but contrast the conclusions of Cutter *et al.* (1973), and Wilsnack (1972).

A few smaller crosscultural studies have pointed to interesting associations between drinking and symbolism (Klausner, 1964), between drinking and ambivalence (Whitehead and Harvey, 1974), and between drinking and anomie (Hanson, 1975). Advances in the scope of the sample and advances in the methods of analysis hold promise for broader and more convincing use of the hologeistic approach. This is particularly promising because, at the same time, the proliferation of detailed studies of drinking patterns from a variety of other points of view will increase our sample, which is still not large in terms of societies where the roles of alcohol have been described with any degree of comprehensiveness.

PROSPECTIVE

Not even the most socially oriented anthropologist would contend that culture *determines* biology. But, to an extent that is not popularly recognized, cultural factors do *condition* biology. Some of the most familiar examples of this are the universal human experience of early training in sphincter control, the customary timing of hunger pangs, views about what is edible and what is not, and patterns of where, when, and how people sleep. Many other less frequent behaviors could be cited as illustrations of ways in which cultural conditioning affects the human organism, such as the practice of colonic irrigation in yoga, visual or olfactory stimuli that produce vomiting in members of some groups but not in others, and so forth.

One of the reasons why the field of alcohol studies has flourished in recent years is that virtually every realm of investigation can be brought to bear in a fruitful manner on the complex interrelations of humanity and beverage alcohol. Within this, anthropological perspectives have been useful—and will probably become increasingly so—in pointing out the occurrence and prevalence of different patterns of thinking and acting with respect to alcohol among various populations, and the ways in which beliefs and behaviors interact to produce strikingly different kinds of drunkenness, drunken behavior, and alcoholism among individuals.

On the basis of increasing familiarity with the reality of cultural variation, it has been increasingly practicable to develop or to offer limited tests of theories about drinking in relation to such varied themes as aggression (Bunzel, 1940; Hamer, 1965; Jay, 1966; Rohrmann, 1972), tension reduction (Curley, 1967; Sayres, 1956; Boalt, 1961; Cappell, 1975; Cappell and Herman, 1972), ambivalence (Yawney, 1969; Lemert, 1954; Simmons, 1960; Kearney, 1970), and functional integration of society (e.g., Clairmont, 1962; Gregson, 1969; Holmberg, 1971; Honigmann, 1963).

Perhaps more striking is the fragmentary evidence that alcohol causes strikingly different reactions, "physiological" (such as facial flushing, headache, accelerated heartbeat, etc.) as well as in terms of gross physical actions. This should not be surprising, in view of widely recognized culturally patterned differences in blushing, fainting, crying, belching, and so forth, but virtually all such observations have been impressionistic until recent years. It is unfortunate that few biologists have worked on alcohol studies among non-Western populations, and equally unfortunate that anthropologists who often do so are not more competent in biology.

With respect to the social biology of alcohol, anthropological perspectives have much to contribute. At the same time, anthropologists have begun to recognize that they also have much to learn. This is an area in which collaboration of an interdisciplinary nature is not merely desirable but virtually indispensable.

ACKNOWLEDGMENTS

The Addiction Research Foundation of Ontario and the Brown University Libraries provided settings in which it was possible to track down many of the sources that are not widely available.

Phil Leis has, against all odds, managed to maintain our department in such a way that teaching and research are complementary activities.

A.C. is the best friend and colleague one could have.

REFERENCES

The diverse and scattered nature of the literature that deals with alcohol in terms of anthropological perspectives is at the same time a strength and a problem. It reflects the virtual absence of orthodoxy, but makes it difficult for the beginner to gain any degree of mastery of the source materials on a given problem or area.

For these reasons, it seems appropriate in a review of this scope to list not merely those works that were specifically cited in the text, but also to list the broad range of relevant material that is available. Furthermore, a few unpublished sources are listed because they deal with peoples, topics, or methods that are not yet well represented in the published literature.*

Aalto, P., 1955, Alkoholens Ställning i Indiens Klassiska Kultur, *Alkoholpolitik* 18 (2):32–46.
Ablon, J., 1976, Family behavior and alcoholism, *in* "Cross-Cultural Approaches to the Study of Alcohol: An Interdisciplinary Perspective" (M. W. Everett, J. O. Waddell, and D. B. Heath, eds.), Mouton, The Hague.
Ablon, J., 1976, Family structure and behavior in alcoholism: A review of the literature," *in* "The Biology of Alcoholism, Vol. 4, Social Aspects" B. Kissin and H. Begleiter, eds.), Plenum Press, New York.
Adandé, A., 1954, Le vin de palme chez les Diola de la Casamance, *Notes Africaines* (Dakar) 61:4–7.
Adler, N., and Goleman, D., 1969, Gambling and alcoholism: Symptom substitution and functional equivalents, *Quarterly Journal of Studies on Alcohol* 30:733–736.
Adrianes, S. L., and Lozet, F., 1951, Contributions à l'étude des boissons fermentées indigénes au Ruanda, *Bulletin Agricole du Congo Belge* 42:933–950.
Aguilar, G. Z., 1964, Suspension of control: A sociocultural study on specific drinking habits and their psychiatric consequences, *Journal of Existential Psychiatry* 4:245–252.
Ahlfors, U. G., 1969, "Alcohol and Conflict: A Qualitative and Quantitative Study on the Relationship between Alcohol Consumption and an Experimentally Induced Conflict Situation in Albino Rats," Alcohol Research in Northern Countries, 16, Finnish Foundation for Alcohol Studies, Helsinki.
Ahlström-Laakso, S., 1976, European drinking habits: A review of research and some suggestions for conceptual integration of findings, *in* "Cross-Cultural Approaches to the Study of Alcohol: An Interdisciplinary Perspective" (M. W. Everett, J. O. Waddell, and D. B. Heath, eds.), Mouton, The Hague.
Aiyappan, A., (in preparation), [Alcohol and anxiety in Orissa, India].
Alba, M. de, 1926, The Maguey and pulque, *Mexican Folkways* 2 (4):12–15.

---

* For historical purposes, books and articles are cited as of their first publication; in cases where significant revision have been made, the most recent edition is also indicated, in parentheses. Titles are in the original language, except where an English translation of the complete source is available. Place of publication is mentioned for those journals that might be confused with others of the same name, or that might otherwise be unusually difficult to locate. Anonymously authored articles are alphabetized under the name of the responsible source. This bibliography was closed in April, 1975. The author would welcome additional pertinent references.

Alcohol, Health, and Research World, 1974, Self-help programs: Indians and native Alaskans, *Alcohol Health and Research World* (experimental issue, summer, 1974):11–16.
"Alcohol, Science, and Society," 1945, Journal of Studies on Alcohol, New Haven.
Alhava, A., 1949, Väkijuomaolojen erikoisluonne Lapissa, *Alkoholiliikkeen Aikakauskirja* 12:35–37.
Allardt, E., 1956, Alkoholvanorna på Landsbygden i Finland, *Alkoholpolitik* 19:73–77.
Allardt, E., 1957, Drinking norms and drinking habits, *in* "Drinking and Drinkers" (E. Allardt, et al.), Finnish Foundation for Alcohol Studies 6, Helsinki.
Almeida V., M. 1962, Investigación clinica sobre la evolución de alcoholismo, *Revista de Neuropsiquiatría* (Lima) 25:97–122.
América Indigena, 1954, El alcohol y el indio, *América Indígena,* 14:283–285.
American Medical Association, 1956, "Manual on Alcoholism," American Medical Association, Chicago (revised edition, 1967).
Anderson, B. G., 1969, How French children learn to drink, *Trans-action* 5:20–22.
Anderson, R. K., Calvo, J., Serrano, G., and Payne, G., 1946, A study of the nutritional status and food habits of Otomi Indians in the Mezquital Valley of Mexico, *American Journal of Public Health* 36:883–903.
Ando, H., and Hasegawa, E., 1970, Drinking patterns and attitudes of alcoholics and nonalcoholics in Japan, *Quarterly Journal of Studies on Alcohol* 31:153–161.
Angrosino, M. V., (1974), "Outside is Death: Alcoholism, Ideology and Community Organization among the East Indians of Trinidad" Wake Forest University, Overseas Research Center, Medical Behavioral Science Monograph 2, Winston-Salem, N.C.
Auersperg, A. P., and Derwort, A., 1962, Beitrag zur vergleichenden Psychiatrie exogener Psychosen vom soziokulturellen Standpunkt, *Nervenarzt* 33:22–27.
Bacon, M. K., 1976a, Cross-cultural studies of drinking: integrated drinking and sex differences in the use of alcoholic beverages, *in* "Cross-Cultural Approaches to the Study of Alcohol: An Interdisciplinary Perspective" (M. W. Everett, J. O. Waddell, and D. B. Heath, eds.), Mouton, The Hague.
Bacon, M. K., 1976b, Alcohol use in primitive societies, *in* "The Biology of Alcoholism, Vol. 4: Social Aspects" (B. Kissin and H. Begleiter, eds.), Plenum Press, New York.
Bacon, M. K., Barry, H., and Child, I. L., 1965, A cross-cultural study of drinking: II. Relations to other features of culture, *Quarterly Journal of Studies on Alcohol Supplement* 3:29–48.
Bacon, M. K., Barry, H., Child, I. L., and Snyder, C. R., 1965, A cross-cultural study of drinking: V. Detailed definitions and data, *Quarterly Journal of Studies on Alcohol Supplement* 3:78–111.
Bacon, S. D., 1943, Sociology and the problems of alcohol: Foundations for a sociological study of drinking behavior, *Quarterly Journal of Studies on Alcohol* 4:399–445.
Bacon, S. D., 1945, Alcohol and complex society, *in* "Alcohol, Science and Society," Journal of Studies on Alcohol, New Haven; revised, *in* Pittman and Snyder, 1962.
Bacon, S. D., 1955, Current research on alcoholism: V. Report of the Section on Sociological Research, *Quarterly Journal of Studies on Alcohol* 16:551–564.
Bacon, S. D. (ed.), 1958, "Understanding Alcoholism," Annals of the Academy of Political and Social Science 315, Philadelphia.
Bacon, S. D., 1973, The process of addiction to alcohol: Social aspects, *Quarterly Journal of Studies on Alcohol* 34:1–27.
Baddeley, F. J., 1966, African Beerhalls, thesis (School of Architecture), University of Cape Town, Cape Town.
Baird, E. G., 1944–48, The alcohol problem and the law, *Quarterly Journal of Studies on Alcohol* 4:535–556; 5:126–161; 6:335–383, 7:110–162, 271–296; 9:80–118.
Baker, J. L., 1959, Indians, alcohol and homicide, *Journal of Social Therapy* 5:270–275.

Baldus, H., 1950, Bebidas e narćoticos dos indios do Brasil, *Sociologia* (São Paulo) 12:161–169.
Bales, R. F., 1946, Cultural differences in rates of alcoholism, *Quarterly Journal of Studies on Alcohol* 6:480–499.
Banay, R. S., 1945, Cultural influences in alcoholism, *Journal of Nervous and Mental Diseases* 102:265–275.
Banks, E., 1937, Native drink in Sarawak, *Sarawak Museum Journal* 4:439–447.
Bard, J., Mare, C., Williams, C., and Wolpaw, I., [ca. 1955], Effect of intragroup competition on alcohol consumption in primitive cultures, manuscript.
Barnett, M. L., 1955, Alcoholism in the Cantonese of New York City: An anthropological study, *in* "Etiology of Chronic Alcoholism" (O. Diethelm, ed.), Charles C Thomas, Springfield, Ill.
Barrera V., A., 1941, "El pulque entre los Mayas," Cuadernos Mayas 3, Mérida, Mexico.
Barry, E., 1775, "Observations Historical, Critical, and Medical, on the Wines of the Ancients . . . ," T. Cadell, London.
Barry, H., 1968, Sociocultural aspects of addiction, *The Addictive States* 46:455–471.
Barry, H., 1976, Cross-cultural evidence that dependency conflict motivates drunkenness, *in* "Cross-Cultural Approaches to the Study of Alcohol: An Interdisciplinary Perspective" (M. W. Everett, J. O. Waddell, and D. B. Heath, eds.), Mouton, The Hague.
Barry, H., Buchwald, C., Child, I. L., and Bacon, M., 1965, A cross-cultural study of drinking: IV. Comparisons with Horton ratings, *Quarterly Journal of Studies on Alcohol Supplement* 3:62–77.
Beals, R. L., 1932, "The Comparative Ethnology of Northern Mexico before 1750," University of California Press, Berkeley.
Beaubrun, M. H., 1967, Treatment of alcoholism in Trinidad and Tobago, 1956–1965, *British Journal of Psychiatry* 113:643–658.
Beaubrun, M. H., 1968, Alcoholism and drinking practices in a Jamaican suburb, *Alcoholism* 4:21–37.
Beaubrun, M. H., 1971, The influence of socio-cultural factors in the treatment of alcoholism in the West Indies, *in* "29th International Congress on Alcoholism and Drug Dependence" (L. G. Kiloh and D. S. Bell, eds.), Butterworths, Sydney, Australia.
Beaubrun, M. H., and Firth, H., 1969, A transcultural analysis of Alcoholics Anonymous: Trinidad/London, Paper read at American Psychiatric Association Meeting, Ocho Rios, Jamaica.
Beidelman, T. O., 1961, Beer drinking and cattle theft in Ukaguru: Intertribal relations in a Tanganyika chiefdom, *American Anthropologist* 63:534–549.
Bejarano, J., 1950, "La derrota de un vicio: Orígen e historia de la chicha," Editorial Iqueima, Bogotá.
Bellmann, H., 1954, Destillation bei den Naturvölkern, *Wissenschaftliche Zeitschrift der Friedrich Schiller Universität* 3:179–185.
Belmont, F. V. de, 1840, "Histoire de l'Eau-de-Vie en Canada," Société Litteraire de Quebec, Quebec.
Bernier, G., and Lambrecht, A., 1960, Étude sur les boissons fermentées indigènes du Katanga, *Problèmes Sociaux Congolais* 48:5–41.
Berreman, G. D., 1956, Drinking patterns of the Aleuts, *Quarterly Journal of Studies on Alcohol* 17:503–514.
Bett, W. R., [and others], 1946, Alcohol and crime in Ceylon: A preliminary communication [and discussion], *British Journal of Inebriety* 43:57–60.
Bismuth, H., and Menage, C., 1960, Alcoolisation du Niger; . . . du Senegal; . . . d'Haut Volta; . . . des états de langue française de l'Afrique Occidentale; Aspects de l'alcoolisation du Dahomey; Aperçu de l'alcoolisation de la Guinée, Haut, Comité d'Étude et d'Information sur l'Alcoolisme (multigraphed), Paris.

Blacker, E., 1966, Sociocultural factors in alcoholism, *International Psychiatry Clinics* 3, (2):51–80.
Blacker, H., 1971, Drinking practices and problems abroad: The Isle of Reunion; Tahiti, *Journal of Alcoholism* 6, (2):61–63.
Bleichsteiner, R., 1952, Zeremoniale Trinksitten und Raumordnung bei Turko-Mongolischen Nomader, *Archiv für Völkerkunde* 6–7:181–208.
Block, M. A., 1965, "Alcoholism: Its Facets and Phases," John Day, New York.
Blom, F., 1956, On Slotkin's "Fermented drinks in Mexico," *American Anthropologist* 58:185–186.
Bloom, J. D., 1970, Socio-cultural aspects of alcoholism, *Alaska Medicine* 12:65–67.
Blum, R. H., and Blum, E. M., 1964, Drinking practices and controls in rural Greece, *British Journal of Addiction* 60:93–108.
Blyth, W., 1972, Transcultural studies in alcoholism in rural catchment areas as pertaining to three cultures: White Americans, American Negroes, American Indians (multigraphed).
Boalt, G., 1961, "A Sociological Theory of Alcoholism," International Bureau Against Alcoholism, Selected Articles 4, Lausanne.
Bose, D. K., 1922, "Wine in Ancient India," K. M. Connor, Calcutta.
Bourguignon, E. E., 1964, Comment on Leacock's "Ceremonial drinking in an Afro-Brazilian cult," *American Anthropologist* 66:1393–1394.
Bourke, J. G., 1893, Primitive distillation among the Tarascoes, *American Anthropologist* (OS) 6:65–69.
Bourke, J. G., 1894, Distillation by early American Indians, *American Anthropologist* (OS) 7:297–299.
Boyatzis, R. E., 1976, Drinking as a manifestation of power concerns, *in* "Cross-Cultural Approaches to the Study of Alcohol: An Interdisciplinary Perspective" (M. W. Everett, J. O. Waddell, and D. B. Heath, eds.), Mouton, The Hague.
Boyer, L. B., 1964, Psychological problems of a group of Apaches: Alcoholic hallucinosis and latent homosexuality among typical men, *in* "The Psychoanalytic Study of Society" (W. Muensterberger and S. Axelrad, eds.), Vol. 3, International Universities Press, New York.
Braidwood, R. J., Sayer, J. D., Helback, H., Mangelsdorf, P. C., Coon, C. S., Linton, R., Steward, J., and Oppenheim, A. L., 1953, Symposium: Did man once live by beer alone? *American Anthropologist* 55:515–526.
Brigham Young University (Department of Audio-Visual Communication), 1963, "Bitter Wind," 16 mm., color/sound, 30 min. film.
Brody, H., 1971, "Indians on Skid Row," Northern Science Research Group, Department of Indian Affairs and Northern Development Publication 70-2, Ottawa.
Brown, D. N., (forthcoming), Pueblo Indian drinking patterns, *in* "Alcohol, Drinking, and Drunkenness among North American Indians" (M. W. Everett and J. O. Waddell, eds.)
Brown, J. H., 1966, "Early American Beverages," Charles E. Tuttle, Rutland, Vt.
Brown, J. S., and Crowell, C. R., 1974, Alcohol and conflict resolution: A theoretical analysis, *Quarterly Journal of Studies on Alcohol* 35:66–85.
Brown, W. L., 1898, Inebriety and its "cures" among the ancients, *Proceedings of the Society for the Study of Inebriety* 55:1–15.
Brownlee, F., 1933, Native beer in South Africa, *Man* (OS) 33:75–76.
Bruman, H. J., 1940, Aboriginal drink areas in New Spain, PhD dissertation (Geography), University of California.
Bruman, H. J., 1944, Asiatic origin of the Huichol still, *Geographical Review* 34:418–427.
Bruun, K., 1959, Den Sociokulturella Backgrunden till Alkoholismen, *Alkoholpolitik* 22:54–58.
Bruun, K., 1959, Significance of role and norms in the small group for individual behavioral changes while drinking, *Quarterly Journal of Studies on Alcohol* 20:53–64.

Buckland, A. W., 1878, Ethnological hints afforded by the stimulants in use among savages and among the ancients, *Journal of the Royal Anthropological Institute* 8:239–254.
Bunzel, R., 1940, The role of alcoholism in two Central American cultures, *Psychiatry* 3:361–387
Bunzel, R., 1976, Chamula and Chichicastenango: A reexamination, *in* "Cross-Cultural Approaches to the Study of Alcohol: An Interdisciplinary Perspective" (M. W. Everett, J. O. Waddell, and D. B. Heath, eds.), Mouton, The Hague.
Busch, C. E., 1952, Consideraciones médico-sociales sobre la chicha, *Excelsior* (Lima) 217:25–26.
Cagol, A., 1936, A note on Bapedi beverages, *Primitive Man* 9:32.
Cahalan, D., 1976, Observations on methodological considerations for cross-cultural alcohol studies, *in* "Cross-Cultural Approaches to the Study of Alcohol: An Interdisciplinary Perspective" (M. W. Everett, J. O. Waddell, and D. B. Heath, eds.), Mouton, The Hague.
Cahalan, D., Cisin, I. H., and Crossley, H. M., 1969, "American Drinking Practices: A National Study of Drinking Behavior and Attitudes," Rutgers Center of Alcohol Studies Monograph 6, New Brunswick, N.J.
Calderón N., G., 1968, Consideraciones acerca del alcoholismo entre los pueblos pre-hispánicos de México, *Revista del Instituto Nacional de Neurologia* 2 (3):5–13.
Cappell, H., 1975, An evaluation of tension models of alcohol consumption *in* "Research Advances in Alcohol and Drug Problems" (R. J. Gibbins, Y. Israel, H. Kalant, R. E. Popham, W. Schmidt, and R. Smart, eds.), Vol. 2, John Wiley and Sons, New York.
Cappell, H., and Hermàn, C. P., 1972, Alcohol and tension reduction: A review, *Quarterly Journal of Studies on Alcohol* 33:33–64.
Carpenter, E. S., 1959, Alcohol in the Iroquois dream quest, *American Journal of Psychiatry* 116:148–151.
Carpenter, J. A., and Armenti, N. P., 1972, Some behavioral effects of alcohol on man, *in* "The Biology of Alcoholism, Vol. 2: Physiology and Behavior" (B. Kissin and H. Begleiter, eds.), Plenum Press, New York.
Carstairs, G. M., 1954, Daru and bhang: Cultural factors in the choice of intoxicant, *Quarterly Journal of Studies on Alcohol* 15:220–237.
Cavan, S., 1966, "Liquor License: An Ethnography of Bar Behavior," Aldine, Chicago.
Central African Journal of Medicine, 1958, Native liquors in southern Rhodesia, *Central African Journal of Medicine* 4:558–559.
Chafetz, M. E., 1964, Consumption of alcohol in the Far and Middle East, *New England Journal of Medicine* 271:297–301.
Chafetz, M. E., and Demone, H. W., Jr., 1962, "Alcoholism and Society," Oxford University Press, New York (revised edition, 1965).
Chaiaramonte, J., 1969, Mumming in Deep Harbour: Aspects of social organization in mumming and drinking *in* "Christmas Mumming in Newfoundland: Essays in Anthropology, Folklore, and History" (H. Halpert and G. M. Storey, eds.), University of Toronto Press, Toronto.
Cheinisse, L., 1908, La race juive, jouit-elle d'une immunité a l'égard de l'alcoolisme? *Semaine Médicale* 28:613–615.
Cherrington, E. H. (ed.), 1925–30, "Standard Encyclopedia of the Alcohol Problem" (6 vols.), American Issue Publishing Co., Westerville, Ohio.
Child, I. L., Bacon, M. K., and Barry, H., 1965, A cross-cultural study of drinking: I. Descriptive measurements of drinking customs, *Quarterly Journal of Studies on Alcohol Supplement* 3:1–28.
Child, I. L., Barry H., and Bacon, M. K., 1965, A cross-cultural study of drinking: III. Sex differences, *Quarterly Journal of Studies on Alcohol Supplement* 3:49–61.
Chopra, R. N., Chopra, G. S., and Chopra, J. C., 1942, Alcoholic beverages in India, *Indian Medical Gazette* 77:224–232, 290–296, 361–367.
Chu, G., 1972, Drinking patterns and attitudes of rooming-house Chinese in San Francisco, *Quarterly Journal of Studies on Alcohol Supplement* 6:58–68.

Cinquemani, D. K., (in preparation), [Research on drinking and alcoholism in Middle America].
Clairmont, D. H., 1962, "Notes on the Drinking Behavior of the Eskimos and Indians in the Aklavik Area: A Preliminary Report," Northern Coordination and Research Centre, Department of Northern Affairs and National Resources, Ottawa.
Clairmont, D. H., 1963, "Deviance among Indians and Eskimos in Aklavik," Northern Coordination and Research Centre, Department of Northern Affairs and National Resources, Ottawa.
Claudian, J., 1970, History of the usage of alcohol, in "International Encyclopedia of Pharmacology and Therapeutics" (J. Trémolières, ed.), Sec. 20, Vol. 2, Pergamon Press, Oxford.
Clemmesen, C., 1958, Oversigt over alkoholproblemet på Grønland, *Ugeskrift for Laeger* 120:1374–1379.
Clinard, M. B. (ed.), 1964, "Anomie and Deviant Behavior: A Discussion and Critique," Free Press/Macmillan, New York.
Collard, J., 1962, Drug responses in different ethnic groups, *Journal of Neuropsychiatry* 3:5114–5121.
Collins, T., 1970, Economic change and the use of alcohol among American Indians, Paper read at American Anthropological Association meeting, San Diego.
Collins, T. (forthcoming), Variance in Northern Ute drinking, in "Alcohol, Drinking, and Drunkenness among North American Indians." (M. W. Everett, and J. O. Waddell, eds.).
Collins, T., and Dodson, J., 1972, Arapahoe, Shoshone and Ute drinking behavior: A comparative analysis, Paper read at American Anthropological Association meeting, Toronto.
Collis, C. H., Cook, P. J., Foreman, J. K., and Palframan, J. F., 1971, A search for nitrosamines in East African spirit samples from areas of varying oesophageal cancer frequency, *Gut* (London) 12:1015–1018.
Collis, C. H., Cook, P. J., Foreman, J. K., and Palframan, J. F., 1972, Cancer of the oesophagus and alcoholic drinks in East Africa, *Lancet* 1972, 1:441.
Collocott, E. V., 1927, Kava ceremonial in Tonga, *Polynesian Society Journal* 36:21–47.
Commission to Study Alcoholism among Indians, 1956, Report [to U.S. Department of Interior, Bureau of Indian Affairs] (multigraphed), Washington.
Conger, J. J., 1951, The effects of alcohol on conflict behavior in the albino rat, *Quarterly Journal of Studies on Alcohol* 12:1–29.
Conger, J. J., 1956, Reinforcement theory and dynamics of alcoholism, *Quarterly Journal of Studies on Alcohol* 17:296–305.
Connell, K. H., 1968, Illicit distillation in "Irish Peasant Society: Four Historical Essays" (K. H. Connell), Clarendon Press, Oxford.
Cooley, R. (forthcoming), Indian alcohol program training needs, in "Alcohol, Drinking and Drunkenness among North American Indians" (M. W. Everett and J. O. Waddell, eds.).
Cooper, J. M., 1949, Stimulants and narcotics, in "Handbook of South American Indians: Vol. 5, The Comparative Ethnology of South American Indians," (J. H. Steward, ed.), Bureau of American Ethnology Bulletin 143, Washington.
Cornwall, E. E., 1939, Notes on the use of alcohol in ancient times, *Medical Times* 67:379–380.
Crahan, M. E., 1964, "Early American Inebrietatis," The Zamorano Club, Los Angeles.
Crawley, A. E., 1912, Drinks, Drinking, in "Encyclopaedia of Religion and Ethics" (J. Hastings, ed.), Vol. 5, Charles Scribner's Sons, New York.
Crawley, E., 1931, "Dress, Drink and Drums," Methuen, London.
Crothers, T. D., 1903, Inebriety in ancient Egypt and Chaldea, *Quarterly Journal of Inebriety* 25:142–150.
Csikszentmihalyi, M., 1968, A cross-cultural comparison of some structural characteristics of group drinking, *Human Development* (Basel) 11:201–216.
Curley, R. T., 1967, Drinking patterns of the Mescalero Apache, *Quarterly Journal of Studies on Alcohol* 28:116–131.

Cutler, H. C., and Cadenas, M., 1947, Chicha: A native South American beer, *Harvard University Botanical Museum Association Leaflet* 13:33–60.
Cutter, H. S. G., 1964, Conflict models, games, and drinking patterns, *Journal of Psychology* 58:361–367.
Cutter, H. S. G., 1969, Alcohol, drinking patterns, and the psychological probability of success, *Behavioral Science* 14:19–27.
Cutter, H. S. G., Key, J. C., Rothstein, E., and Jones, W. C., 1973, Alcohol, power and inhibition, *Quarterly Journal of Studies on Alcohol* 34:381–389.
Dailey, R. C., [1964], "Alcohol and the Indians of Ontario: Past and Present," Addiction Research Foundation Substudy 1-20-64, Toronto.
Dailey, R. C., [1966], "Alcohol and the North American Indian: Implications for the Management of Problems," Addiction Research Foundation Substudy 2-20-66, Toronto.
Dailey, R. C., 1968, The role of alcohol among North American Indian tribes as reported in the Jesuit Relations, *Anthropologica* 10:45–59.
Danielou, J., 1949, "Les repas de la Bible el leur signification," La Maison Dieu, Paris.
Davis, W. N., 1972, Drinking: A search for power or nurturance? *in* "The Drinking Man" (D. McClelland, W. Davis, R. Kalin, and E. Wanner), Free Press, New York.
Defer, B., 1969, Variations épidémiologiques de toxicomanies associées à des contacts de culture, *Toxicomanies* (Quebec) 2:9–18.
Deihl, J. R., 1922, Kava and kava drinking, *Primitive Man* 5:61–68.
Desai, A. V., 1965, An exploratory survey of drinking in Suraf and Bulsar Community, Department of Psychology, S. B. Garda College, Navsari, India, (manuscript).
Devenyi, P., 1967, "Sociocultural Factors in Drinking and Alcoholism," Alcoholism and Drug Addiction Research Foundation of Ontario, Clinical Division Substudy 17-1967, Toronto.
Devereux, G., 1948, The function of alcohol in Mohave Society, *Quarterly Journal of Studies on Alcohol* 9:207–251.
Dobyns, H. F., 1965, Drinking patterns in Latin America: A review, Paper read at American Association for the Advancement of Science meeting, Berkeley.
Doughty, P. L., 1971, The social uses of alcoholic beverages in a Peruvian community, *Human Organization* 30:187–197.
Douyon, E., 1969, Alcoolisme et toxicomanie en Haiti, *Toxicomanies* (Quebec) 2:31–38.
Doxat, J., 1971, "Drinks and Drinking: An International Distillation," Ward Lock, London.
Dozier, E. P., 1966, Problem drinking among American Indians: The role of sociocultural deprivation, *Quarterly Journal of Studies on Alcohol* 27:72–87.
Driver, H. E., 1961, "Indians of North America," University of Chicago Press, Chicago (2nd ed., 1969).
Drower, E. S., 1966, "Water into Wine: A Study of Ritual Idiom in the Middle East," John Murray, London.
Dumett, R. E., 1974, The social impact on the European liquor trade on the Akan of Ghana (Gold Coast and Asante), 1875–1910, *Journal of Inter-Disciplinary History* 5:69–101.
Eddy, R., 1887, "Alcohol in History, An Account of Intemperance in All Ages: Together with a History of the Various Methods Employed for its Removal," National Temperance Society and Publication House, New York.
Efron, V., 1970, Sociological and cultural factors in alcohol abuse, *in* "Alcohol and Alcoholism" (R. E. Popham, ed.), University of Toronto Press, Toronto.
Eis, G., 1961, Altdeutsche Hausmittel gegen Trunkenheit und Trunksucht, *Medizinische Monatsschrift* (Stuttgart) 15:269–271.
Elwin, V., 1943, "Maria Murder and Suicide," Oxford Unviersity Press, Bombay.
Emerson, E. R., 1908, "Beverages Past and Present: An Historical Sketch of their Production, together with a Study of the Customs connected with their Use" (2 vols.), G. Putnam's Sons, New York.

Eriksson, K. and Kärkkäinen, K., 1971, Pullo-ja tölkkijatteen Kasaantriminen luontoon Suomessa uvonna 1970, *Alkoholipolitiika* 36:175-186.
Erlich, V. S., 1965, Comment on D. Mandelbaum's "Alcohol and Culture" *Current Anthropology* 6:288-289.
Ervin, A. M., [ca. 1971], "New Northern Townsmen in Inuvik," Department of Indian Affairs and Northern Development, Mackenzie Delta River Project 5, Ottawa.
Escalante, F., (forthcoming), Yaqui drinking groups, *in* "Alcohol, Drinking and Drunkenness among North American Indians" (M. W. Everett and J. O. Waddell, eds.).
Everett, M. W., (forthcoming), "Drinking" and "trouble": The Apachean experience, *in* "Alcohol, Drinking and Drunkenness among North American Indians" (M. W. Everett and J. O. Waddell, eds.).
Everett, M. W., Baha, C. J., Declay, E., Endfield, M. R., and Selby, K., 1973, Anthropological expertise and the "realities" of White Mountain Apache adolescent drinking, Paper read at Society for Applied Anthropology, Tucson.
Everett, M. W., and Waddell, J. O. (eds), (forthcoming), "Alcohol, Drinking, and Drunkenness among North American Indians: An Anthropological Perspective."
Everett, M. W., Waddell, J. O., and Heath, D. B., (eds.), 1976, "Cross-Cultural Approaches to the Study of Alcohol: An Interdisciplinary Perspective," Mouton, The Hague.
Ewing, J. A., and Rouse, B. A., 1973, Alcohol sensitivity and ethnic backgrounds, *in* "Scientific Proceedings in Summary Form: 126th Annual Meeting of the American Psychiatric Association," Washington.
Ezell, P. H., 1965, A comparison of drinking patterns in three Hispanic cities, Paper read at American Association for the Advancement of Science meeting, Berkeley.
F., H., 1933, Chikaranga cocktails, *Nada* (Salisbury) 11:116-117.
Fairbanks, R. A., 1973, The Cheyenne-Arapaho and alcoholism: Does the tribe have a legal right to a medical remedy? *American Indian Law Review* 1 (1):55-77.
Fallding, H., 1969, The source and burden of civilization, illustrated in the use of alcohol, *Quarterly Journal of Studies on Alcohol* 25:714-724.
Feldman, W. M., 1923, Racial aspects of alcoholism, *British Journal of Inebriety* 21:1-15.
Felice, Ph.de, 1936, "Poisons Sacrés, Ivresses Divines," Albin Michel, Paris.
Feliciano, R. T., 1926, Illicit beverages, *Philippine Journal of Science* 29:465-474.
Fenasse, J. M., 1964, La Bible et l'usage du vin, *Alcool ou Santé* 63:17-28.
Fenna, D., Mix, L., Schaefer, O., and Gilbert, J. A. L., 1971, Ethanol metabolism in various racial groups, *Canadian Medical Association Journal* 105:472-475.
Ferguson, F. N., 1968, Navaho drinking: Some tentative hypotheses, *Human Organization* 27:159-167.
Ferguson, F. N., 1970, A treatment program for Navaho alcoholics: Results after four years, *Quarterly Journal of Studies on Alcohol* 31:898-919.
Ferguson, F. N., 1976, Similarities and differences among a heavily arrested group of Navajo Indian drinkers in a southwestern American town, *in* "Cross-Cultural Approaches to the Study of Alcohol: An Interdisciplinary Perspective," (M. W. Everett, J. O. Waddell, and D. B. Heath, eds.) Mouton, The Hague.
Field, P. B., 1962, A new cross-cultural study of drunkenness, *in* "Society, Culture and Drinking Patterns" (D. J. Pittman and C. R. Snyder, eds.), John Wiley and Sons, New York.
Fort, J., 1965, Cultural aspects of alcohol (and drug) problems, *in* "Selected Papers presented at the 27th International Congress on Alcohol and Alcoholism," Vol. 1, International Bureau against Alcoholism, Lausanne.
Fouquet, P., 1965, Alcool et religions, *Revue d'Alcoolisme* 11:81-92.
Frake, C. O., 1964, How to ask for a drink in Subanun, *American Anthropologist* 66:(6,2):127-132.

Frederikson, O. F., 1932, "The Liquor Question among the Indian Tribes in Kansas, 1804–1881," Bulletin of the University of Kansas 33, 8, Lawrence.

Frølund, B., 1965, Drinking patterns in Zambiza (Pichincha), in "Drinking Patterns in Highland Ecuador" (Eileen Maynard, ed.), Ithaca, N.Y.

Fukui, K., 1970, Alcoholic drinks of the Iraqu, *Kyoto University African Studies* 5:125–148.

Galang, R. C., 1934, Pangasi: The Bukidnon wine, *Philippine Magazine* 31:540.

Gallagher, O. R. 1965, Drinking problems of the tribal Bihar, *Quarterly Journal of Studies on Alcohol* 26:617–628.

García, A., A., 1972, El maguey y el pulque en Tepetlaoxtoc, *Comunidad* (México) 7, (38):461–474.

Gearing, F., 1960, Toward an adequate therapy for alcoholism in non-Western cultures: An exploratory study of American Indian drinking, Department of Anthropology, University of Washington, (manuscript).

Geertz, C., 1951, Drought, death and alcohol in five Southwestern cultures, Department of Social Relations, Harvard University, (manuscript).

Gelfand, M., 1966, Alcoholism in contemporary African society, *Central African Journal of Medicine* 12:12–13.

Gelfand, M., 1971, The extent of alcohol consumption by Africans: The signficance of the weapons at beer drinks, *Journal of Forensic Medicine* 18:53–64.

Genin, A. M. A., 1924, "La cerveza entre los antiguos mexicanos y en la actualidad" [no publisher], México.

Ghosh, S. K., 1973, Alcohol and alcoholism in the North-East Frontier Area, Paper read at International Congress of Anthropological and Ethnological Sciences, Chicago.

Gilder, D. D., 1921, Drink in the scriptures of the nations, *Anthropological Society of Bombay* 12:172–189.

Glad, D. D., 1947, Attitudes and experiences of American-Jewish and American-Irish male youth as related to differences in adult rates of inebriety, *Quarterly Journal on Alcohol* 8:406–472.

Glatt, M. M., 1969, Hashish and alcohol "Scenes" in France and Great Britain 120 years ago, *British Journal of Addiction* 64:99–108.

Glover, E., 1932, Common problems in psycho-analysis and anthropology: Drug ritual and addiction, *British Journal of Medical Psychology* 12:109–131.

Gómez H., N., 1966, Importancia social de la chicha como bebida popular en Huamanga, *Wamani* 1 (1):33–57.

Gómez, J., 1914–15, Chichismo: Estudio general, clínico y anatomopatológico de los efectos de la chicha en la clase obrera de Bogotá, *Repertorio Médicina y Cirugía* 5:302–320; 366–379; 424–440; 483–497; 540–559; 588; 652–667; 6:179.

Gonçalves de Lima, O. [ca. 1956], "El maguey y el pulque en los códices mexicanos," Fondo de Cultura Económica, México.

Goodenough, E. R., 1956, "Jewish Symbols in the Greco-Roman Period, Vols. 5–6: Fish, Bread, and Wine," Bollingen Series 37, Pantheon Books, New York.

Górski, J., 1969, Alkohol u kulturze i obyczaju, *Problemy Alkoholizu* 17 (7–8):10–11.

Grace, V., 1957, Wine jars, *Classical Journal* 42:443–452.

Gracia, M. F., 1973, Analysis of incidence of alcoholic intake by Indian population in one state of U.S.A. (Montana), Paper read at International Congress of Anthropological and Ethnological Sciences meeting, Chicago.

Grant, A. P., 1963, Some observations on alcohol consumption and its results in northern Ireland, *Ulster Medical Journal* 32:186–191.

Graves, T. D., 1967, Acculturation, access, and alcohol in a tri-ethnic community, *American Anthropologist* 69:306–321.

Graves, T. D., 1970, The personal adjustment of Navajo Indian migrants to Denver, Colorado, *American Anthropologist* 72:35–54.

Graves, T. D., 1971, Drinking and drunkenness among urban Indians, *in* "The American Indian in Urban Society" (J. O. Waddell and O. M. Watson, eds.), Little, Brown, Boston.
Gravière, E.la, 1957, The problem of alcoholism in the countries and territories south of the Sahara, *International Review of Missions* 46 (183):290-298.
Gray, J. H., 1972, "Booze: The Impact of Whiskey on the Prairie West," Macmillan of Canada, Toronto.
Gregson, R. E., 1969, Beer, leadership, and the efficiency of communal labor, Paper read at American Anthropological Association meeting, New Orleans.
Gremek, M. D., 1950, Opojna píca i otrovi antiknih Ilira, *Farmaceutski Glasnik* (Zagreb) 6:33-38.
Guiart, J., 1956, "Un siècle et demi de contacts culturels à Tanna, Nouvelles Hébrides," Publications de la Société des Océanistes 5, Paris.
Gunson, N., 1966, On the incidence of alcoholism and intemperance in early Pacific missions, *Journal of Pacific History* 1:43-62.
Haas, S. J. de, and Jonker, C., 1965, Horton's hypothese getoetst (manuscript).
Haavio-Mannila, E., 1959, Alkohelens Roll vid Byslagsmålen i Finland, *Alkoholpolitik* 22:16-18, 44.
Hage, P. 1972, A structural analysis of Munchnerian beer categories and beer drinking, *in* "Culture and Cognition: Rules, Maps and Plans" (J. P. Spradley, ed.), Chandler, San Francisco.
Hamer, J. H., 1965, Acculturation stress and the functions of alcohol among the forest Potawatomi, *Quarterly Journal of Studies on Alcohol* 26:285-302.
Hamer, J. H., 1969, Guardian spirits, alcohol, and cultural defense mechanisms, *Anthropologica* 11:215-241.
Hamer, J. H., and Steinbring, J., (eds.), (in preparation), "Alcohol and the North American Indian."
Hanna, J. M., 1976, Ethnic groups, human variation, and alcohol use, *in* "Cross-Cultural Approaches to the Study of Alcohol: An Interdisciplinary Perspective" (M. W. Everett, J. O. Waddell, and D. B. Heath, eds.), Mouton, The Hague.
Hansen, E. C., [ca. 1971], From political association to public tavern: Two phases of urbanization in rural Catalonia (manuscript).
Hanson, D. J., 1975, Anomie theory and drinking problems: A test, *Drinking and Drug Practices Surveyor* 10:23-24.
Harford, C. F., 1905, Drinking habits of uncivilized and semi-civilized races, *British Journal of Inebriety* 2:92-103.
Hartmann, G., 1958, "Alkoholische Getränke bei den Naturvölkern Südamerikas," Free University of West Berlin, Berlin.
Hartmann, G., 1968, Destillieranlagen bei südamerikanischen Naturvölkern, *Zeitschrift für Ethnologie* 93:225-232.
Hartocollis, P., 1966, Alcoholism in contemporary Greece, *Quarterly Journal of Studies on Alcohol* 27:721-727.
Harwood, A., 1964, Beer drinking and famine in a Safwa village: A case of adaptation in a time of crisis, Paper read at East African Institute of Social Research conference, Kampala.
Hasan, K. A., 1964, Drinks, drugs and disease in a north Indian village, *Eastern Anthropologist* 17:1-9.
Hasan, K. A., 1965, Comment on D. Mandelbaum's "Alcohol and Culture," *Current Anthropology* 6:289.
Havard, V., 1896, Drink plants of the North American Indians, *Bulletin of the Torrey Botanical Club* 23:33-46.
Hawthorn, H. B., Belshaw, C. S., and Jamieson, S. M., 1957, The Indians of British Columbia and alcohol, *Alcoholism Review* 2 (3):10-14.

Hays, T. E., 1968, San Carlos Apache drinking groups: Institutional deviance as a factor in community disorganization, Paper read at American Anthropological Association meeting, Seattle.
Heath, D. B., 1952, Alcohol in a Navajo community, A. B. thesis (Social Relations), Harvard University.
Heath, D. B., 1958, Drinking patterns of the Bolivian Camba, *Quarterly Journal of Studies on Alcohol* 19:491-508 (revised, in Pittman and Snyder, 1962).
Heath, D. B., 1964, Prohibition and post-repeal drinking patterns among the Navaho, *Quarterly Journal of Studies on Alcohol* 25:119-135.
Heath, D. B., 1965, Comment on D. Mandelbaum's "Alcohol and Culture," *Current Anthropology* 6:289-290.
Heath, D. B., 1971, Peasants, revolution, and drinking: Interethnic drinking patterns in two Bolivian communities, *Human Organization* 30:179-186.
Heath, D. B., 1974, Perspectivas socioculturales del alcohol en América latina, *Acta Psiquiátrica y Psicológica de América Latina*, 20 (2):99-111.
Heath, D. B., 1975, A critical review of ethnographic studies of alcohol use, in "Research Advances in Alcohol and Drug Problems", (R. J. Gibbins, Y. Israel, H. Kalant, R. E. Popham, W. Schmidt, and R. Smart, eds.), Vol. 2, John Wiley and Sons, New York.
Heath, D. B., 1976, Anthropological perspectives on alcohol: An historical review, in "Cross-Cultural Approaches to the Study of Alcohol: An Interdisciplinary Perspective" (M. W. Everett, J. O. Waddell, and D. B. Heath, eds.), Mouton, The Hague.
Heilizer, F., 1964, Conflict models, alcohol, and drinking patterns, *Journal of Psychology* 57:457-473.
Helgason, T., 1968, Rapport fràn Island: Alkoholismens Epidèmiologi, *Alkoholfràgan* 62:219-230.
Hellmann, E., 1934, The importance of beer-brewing in an urban native yard, *Bantu Studies* 8:38-60.
Henderson, N. B., 1967, Cross-cultural action research: Some limitations, advantages and problems, *Journal of Social Psychology* 73:61-70.
Henderson, N. B., 1972, Indian problem drinking: Stereotype or reality? A study of Navajo problem drinking, Paper read at American Psychological Association meeting, Honolulu.
Hentig, H. von, 1945, The delinquency of the American Indian, *Journal of Criminal Law and Criminology* 36:75-84.
Herrero, M., 1940, Las viñas y lose vinos del Perú, *Revista de Indias* 1, (2):111-116.
Hes, J. P., 1970, Drinking in a Yemenite rural settlement in Israel, *British Journal of Addiction* 65:293-296.
Hirsch, J., 1953, Historical perspectives on the problem of alcoholism, *Bulletin of the New York Academy of Medicine* 29:961-971.
Hirvonen, K., 1969, Antiikin alkoholijuomat, *Alkoholipolitiikka* 34:138-142, 191-194, 244-248, 300-305.
Hocking, R. B., 1970, Problems arising from alcohol in the New Hebrides, *Medical Journal of Australia* 2:908-910.
Hoff, E. C., 1958, "Cultural Aspects of the Use of Alcoholic Beverages," New Hampshire State Department of Health, Division on Alcoholism Publication 22, Concord.
Hoffman, M., 1956, "5000 Jahre Bier," Alfred Metzner, Berlin.
Holloway, R., 1966, "Drinking among Indian Youth: A Study of the Drinking Behaviour, Attitudes and Beliefs of Indian and Metis Young People in Manitoba," Alcohol Education Service, Winnipeg.
Holmberg, A. R., 1971, The rhythms of drinking in a Peruvian coastal mestizo community, *Human Organization*, 30:198-202.
Honigmann, J. J., 1963, Dynamics of drinking in an Austrian village, *Ethnology* 2:157-169.

Honigmann, J. J., 1965, Comment on D. Mandelbaum's "Alcohol and Culture," *Current Anthropology* 6:290-291.

Honigmann, J. J., 1971, Alcohol in its cultural context, Paper read at Interdisciplinary Symposium on Alcoholism, Washington.

Honigmann, J. J., and Honigmann, I., 1945, Drinking in an Indian-White community, *Quarterly Journal of Studies on Alcohol* 5:575-619.

Honigmann, J. J., and Honigmann, I., 1965, How Baffin Island Eskimo have learned to drink, *Social Forces* 44:73-83.

Honigmann, J., and Honigmann, I., 1968, Alcohol in a Canadian northern town, Institute for Research in Social Science, University of North Carolina, Chapel Hill (multigraphed).

Horton, D. J., 1943, The functions of alcohol in primitive societies: A cross-cultural study, *Quarterly Journal of Studies on Alcohol* 4:199-320.

Horwitz, J., Marconi, J., and Adis C., G., (eds.), 1967, "Bases para una epidemiología del alcoholismo en América latina," Fondo para la Salud Mental, Buenos Aires.

Howay, F. W., 1942, The introduction of intoxicating liquors amongst the Indians of the Northwest Coast, *British Columbia Historical Quarterly* 6:157-167.

Hrdlička, A., 1904, Method of preparing tesvino among the White River Apaches, *American Anthropologist* 6:190-191.

Hunt, G. M., and Azrin, N. H., 1973, A community-reinforcement approach to alcoholism, *Behavior Research and Therapy* 14:91-104.

Hurt, W. R., and Brown, R. M., 1965, Social drinking patterns of the Yankton Sioux, *Human Organization* 24:222-230.

Hutchinson, B., 1961, Alcohol as a contributing factor in social disorganization: The South African Bantu in the nineteenth century, *Revista de Antropologia* (São Paulo) 9:1-13.

International Labor Office, 1953, Alcoholism and the mastication of coca in South America, *in* "Indigenous Populations," I.L.O., Geneva.

Irgens-Jensen, O., 1970, The use of alcohol in an isolated area of northern Norway, *British Journal of Addiction* 65:181-185.

Jackson, C., 1970, Some situational and psychological correlates of drinking behavior in Dominica, W.I., paper read at American Anthropological Association meeting, San Diego.

Jacobs, W. R., 1950, "Diplomacy and Indian Gifts: Anglo and French Rivalry along the Ohio and Northwest Frontiers, 1748-63," Stanford University Press, Stanford.

Jacobsen, E., 1951, Alkohol als soziales Problem, *in* "Rauschgifte und Genussmittel" (K. O. Møller, ed.), Benno Schwabe, Basel.

Jarvis, D. H., 1899, "Report of the Cruise of the U.S. Revenue Cutter *Bear*, and the Overland Expedition for the Relief of the Whalers in the Arctic Ocean," House Document 511 (56th Cong., 2nd Sess., Vol. 93), U.S. Government Printing Office, Washington.

Jastrow, M. Jr., 1913, Wine in the Pentateuchal Codes, *Journal of the American Oriental Society* 33:180-192.

Jay, E. J., 1966, Religious and convivial uses of alcohol in a Gond village of middle India, *Quarterly Journal of Studies on Alcohol* 27:88-96.

Jay, M., 1971, L'Evolution de l'alcoolisme à la Reunion, *Alcool ou Santé* 104:32-38.

Jeffreys, M. D. W., 1937, Palm wine among the Ibibio, *Nigerian Field* 22:40-45.

Jellinek, E. M., 1952a, Alkoholbruket såsom en Folksed, *Alkoholpolitik* 15:36-40.

Jellinek, E. M., 1952b, Phases of alcohol addiction, *Quarterly Journal of Studies on Alcohol* 13:673-684; revised, *in* Pittman and Snyder, 1962.

Jellinek, E. M., 1957, The world and its bottle, *World Health* 10 (4):4-6.

Jellinek, E. M., 1960, "The Disease Concept of Alcoholism," Hillhouse Press, New Haven.

Jellinek, E. M., 1961, "Drinkers and Alcoholics in Ancient Rome," Addiction Research Foundation Substudy 2-J-61, Toronto.

Jellinek, E. M., [1965], "The Symbolism of Drinking: A Culture-Historical Approach," Addiction Research Foundation Substudy 3-2 & Y-65, Toronto.

Jessor, R., Graves, T. D., Hanson, R. C., and Jessor, S. L., 1968, "Society, Personality and Deviant Behavior: A Study of a Tri-Ethnic Community," Holt, Rinehart and Winston, New York.

Jilek-Aal, L., 1972, Alcohol and the Indian-White relationship: The function of Alcoholics Anonymous in coast Salish society, MA thesis, University of British Columbia, Vancouver.

Jochelson, W., 1906, Kumiss festivals of the Yakut and the decoration of kumiss vessels, *in* "Boas Anniversary Volume" (B. Laufer, ed.), Stechert, New York.

Journal of American Medical Association, 1954, Alcohol intoxication in Indians, *Journal of American Medical Association* 156:1375.

Jupp, G. A., 1971, Socio-cultural influences on drinking practices, *Brewers Digest* 46:76 *et seq.*

Kalant, H., 1969, Problems of alcohol and drugs: Relationships and non-relationships from the point of view of research, *in* "Proceedings of 28th International Congress on Alcohol and Alcoholism" (M. Keller and T. G. Coffey, eds.), Vol. 2, Hillhouse Press, Highland Park, N.J.

Kalin, R., Davis, W. N., and McClelland, D. C., 1966, The relationship between use of alcohol and thematic content of folktales in primitive societies, *in* "The General Inquirer" (P. J. Stone, D. C. Dunphy, M. A. Smith, and D. M. Ogilvie, eds.), M.I.T. Press, Cambridge, Mass.; revised, *in* D. C. McClelland, W. N. Davis, R. Kalin, and E. Wanner, 1972.

Kant, I., 1798, "Anthropologie in pragmatischer Hinsicht," F. Nicolovius, Koenigsberg.

Kaplan, B., 1962, The social functions of Navaho "heavy drinking," Paper read at Society for Applied Anthropology meeting, Kansas City.

Kearney, M., 1970, Drunkenness and religious conversion in a Mexican village, *Quarterly Journal of Studies on Alcohol* 31:132–152.

Keehn, J. D., 1969, Translating behavioral research into practical terms for alcoholism, *Canadian Psychologist* 10:438–446.

Keehn, J. D., 1970, Reinforcement of alcoholism: Schedule control of solitary drinking, *Quarterly Journal of Studies on Alcohol* 31:28–39.

Kelbert, M., and Hale, L., [1965] "The Introduction of Alcohol into Iroquois Society," Addiction Research Foundation Substudy 1-K & H-65, Toronto.

Keller, M., 1958, Beer and wine in ancient medicine, *Quarterly Journal of Studies on Alcohol* 19:153–154.

Keller, M., 1960, Definition of alcoholism, *Quarterly Journal of Studies on Alcohol* 21:125–134 (revised, *in* Pittman and Snyder, 1962).

Keller, M., 1966, Alcohol in health and disease: Some historical perspectives, *Annals of the New York Academy of Sciences* 113:820–827.

Keller, M. (ed.), 1966– , "International Bibliography of Studies on Alcohol" (3+ vols.), Rutgers Center of Alcohol Studies, New Brunswick.

Keller, M., 1976, A documentation resource for cross-cultural studies on alcohol, *in* "Cross-Cultural Approaches to the Study of Alcohol: An Interdisciplinary Perspective," (M. W. Everett, J. O. Waddell, and D. B. Heath, eds.), Mouton, The Hague.

Kennedy, J. G., 1963, Tesguino complex: The role of beer in Tarahumara culture, *American Anthropologist* 65:620–640.

Kerketta, K., 1960, Rice beer and the Oraon culture: A preliminary observation, *Journal of Social Research* (Ranchi) 3:62–67.

Kermorgant, A., 1909, L'alcoolisme dans les colonies françaises, *Bulletin de la Société de Pathologie Exotique et de Ses Filiales* 2:330–340.

Kim, Y. C., 1972, "A Study of Alcohol Consumption and Alcoholism among Saskatchewan Indians: Social and Cultural Viewpoints," The Research Division, Alcoholism Commission of Saskatchewan, Regina.

Kircher, K., 1910, Die sakrale Bedeutung des Weines im Altertum, *Religionsgeschichtliche Versuche und Vorarbeiten* 9:2.
Klausner, S. Z., 1964, Sacred and profane meanings of blood and alcohol, *Journal of Social Psychology* 64:27–43.
Knupfer, G., 1960, Use of alcoholic beverages by society and its cultural implications, *California's Health* 18:9–13.
Knupfer, G., and Room, R., 1967, Drinking patterns and attitudes of Irish, Jewish and white Protestant American men, *Quarterly Journal of Studies on Alcohol* 28:676–699.
Koplowitz, I., 1923, "Midrash Yayin Veshechor: Talmudic and Midrashic Exegetics on Wine and Strong Drink" [no publisher], Detroit.
Korsten, M. A., Matsuzaki, S., Feinman, L., and Lieber, C. S., 1975, High blood acetaldehyde levels after ethanol administration in alcoholics, *New England Journal of Medicine* 292 (8):386–389.
Krige, E. J., 1932, The social significance of beer among the Balobedu, *Bantu Studies* 6:343–357.
Kubodera, I., 1935, Ainu no Kozoku, Sake no Jōzō Oyobi sono saigi, *Minozokugaku Kenkyū* 1:501–532.
Kunitz, S. J., and Levy, J. E., 1974, Changing ideas of alcohol use among Navajo Indians, *Quarterly Journal of Studies on Alcohol* 35:243–259.
Kunitz, S. J., Levy, J. E., and Everett, M. W., 1969, Alcoholic cirrhosis among the Navaho, *Quarterly Journal of Studies on Alcohol* 30:672–685.
Kunitz, S. J., Levy, J. E., Odoroff, C. L., and Bollinger, J., 1971, The epidemiology of alcoholic cirrhosis in two southwestern Indian tribes, *Quarterly Journal of Studies on Alcohol* 32:706–720.
Kuttner, R. E., and Lorincz, A. B., 1967, Alcoholism and addiction in urbanized Sioux Indians, *Mental Hygiene* 51:530–542.
La Barre, W., 1938, Native American beers, *American Anthropologist* 40:224–234.
La Barre, W., 1946, Some observations on character structure in the Orient: I. The Chinese, Part 2, *Psychiatry* 9:375–395.
La Barre, W., 1956, Professor Widjojo goes to a koktel parti, *New York Times Magazine* (Dec. 9, 1956): 17 *et seq*.
Lane, E. W., 1883, "Arabian Society in the Middle Ages," Chatto and Windus, London.
Langness, L. L., and Hennigh, L., 1964, American Indian drinking: Alcoholism or insobriety, Paper read at Mental Health Research meeting, Fort Steilacoom, Washington.
Lanu, K. E., 1956, "Control of Deviating Behavior: An Experimental Study on the Effect of Formal Control over Drinking Behavior," Finnish Foundation for Alcohol Studies Publication 2, Helsinki.
Larni, M., 1960, Kinesiska Dryckesseder, *Alkoholpolitik* 23:116–118.
Leacock, S., 1964, Ceremonial drinking in an Afro-Brazilian cult, *American Anthropologist* 66:344–354.
Leake, C., and Silverman, M., 1966, "Alcoholic Beverages in Clinical Medicine," World, Cleveland.
Ledermann, S. (ed.), 1956–64, "Alcool, Alcoolisme—Alcoolisation" (2 vols.), Institut National d'Etudes Demographique, Travaux et Documents Cahiers 29 and 41, Presses Universitaires de France, [Paris].
Lee, R. H., and Mizruchi, E. H., [ca. 1960], A study of drinking behavior and attitudes toward alcohol of the Chinese in the United States (manuscript).
Leibowitz, J. O., 1967, Acute alcoholism in ancient Greek and Roman medicine, *British Journal of Addiction* 62:83–86.
Leis, P. E., 1964, Palm oil, illicit gin, and the moral order of the Ijaw, *American Anthropologist* 66:828–838.

Lejarza, F. de, 1941, Las borracheras y el problema de las conversiones en Indias, *Archivo Ibero-Americano* 1–2:111–142; 3:229–269.
Leland, J., 1976, "Firewater Myths: North American Indian Drinking and Alcohol Addiction," Rutgers Center of Alcohol Studies Monograph 11, New Brunswick, N.J.
Lemert, E. M., 1954, Alcohol and the Northwest Coast Indians, *University of California Publications in Culture and Society* 2:303–406.
Lemert, E. M., 1956, Alcoholism and the sociocultural situation, *Quarterly Journal of Studies on Alcohol* 17:306–317.
Lemert, E. M., 1958, The use of alcohol in three Salish tribes, *Quarterly Journal of Studies on Alcohol* 19:90–107.
Lemert, E. M., 1962, Alcohol use in Polynesia, *Tropical and Geographical Medicine* 14:183–191.
Lemert, E. M., 1964a, Forms and pathology of drinking in three Polynesian societies, *American Anthropologist* 66:361–374.
Lemert, E. M., 1964b, Drinking in Hawaiian plantation society, *Quarterly Journal of Studies on Alcohol* 25:689–713.
Lemert, E. M., 1965, Comment on D. Mandelbaum's "Alcohol and Culture," *Current Anthropology* 6:291.
Lemert, E. M., 1967, Secular use of kava in Tonga, *Quarterly Journal of Studies on Alcohol* 28:328–341.
Lemert, E. M., 1969, Socio-cultural research on drinking, *in* "28th International Congress on Alcohol and Alcoholism" (M. Keller and T. Coffey, eds.), Vol. 2, Hillhouse Press, Highland Park, N.J.
Lender, M., 1973, Drunkenness as an offense in early New England: A study of "Puritan" attitudes, *Quarterly Journal of Studies on Alcohol* 34:353–366.
Lenoir, R., 1925, Les fêtes de boisson, *in* "Compte-Rendue de la 21 Congrès Internationale des Américanistes" (pt. 2), Museum Göteborg.
Levy, J. E., and Kunitz, S. J., 1971, Indian reservations, anomie, and social pathologies, *Southwestern Journal of Anthropology* 27:97–128.
Levy, J. E., and Kunitz, S. J., 1974, "Indian Drinking: Navajo Practices and Anglo-American Theories," Wiley-Interscience, New York.
Levy, J. E., Kunitz, S. J., and Everett, M. W., 1969, Navajo criminal homicide, *Southwestern Journal of Anthropology* 25:124–152.
Levy, R. I., 1966, Ma'ohi drinking patterns in the Society Islands, *Journal of the Polynesian Society* 75:304–320.
Leyburn, J. G., 1944, Native farm labor in South Africa, *Social Forces* 23:133–140.
Lickiss, J. N., 1971, Alcohol and Aborigines in cross-cultural situations, *Australian Journal of Social Issues* 6:210–216.
Lieber, C. S., 1972, Metabolism of ethanol and alcoholism: Racial and acquired factors, *Annals of Internal Medicine,* 76:326–327.
Lindner, P., 1933, El secreto del "Soma," bebida de los antiguos indios y persas, *Investigación y Progreso* 7:272–274.
Lint, J. de, 1976, The epidemiology of alcoholism, with special reference to sociocultural factors, *in* "Cross-Cultural Approaches to the Study of Alcohol: An Inter-Disciplinary Perspective" (M. W. Everett, J. O. Waddell, and D. B. Heath, eds.), Mouton, The Hague.
Lint, J. de, and Schmidt, W., [1970], "The Epidemiology of Alcoholism," Addiction Research Foundation Substudy 12-10 & 4-70, Toronto.
Lint, J. de, and Schmidt, W., 1971, Consumption averages and alcoholism prevalence: A brief review of epidemiological investigations, *British Journal of Addiction* 66:97–107.
Little, M. A., 1970, Effects of alcohol and coca on foot temperature responses of highland Peruvians during a localized cold exposure, *American Journal of Physical Anthropology* 32:233–242.

Littmann, G., 1965, Some observations on drinking among American Indians in Chicago, *in* "Selected Papers Presented at 27th International Congress on Alcohol and Alcoholism," Vol. 1, International Bureau against Alcoholism, Lausanne.

Littmann, G., 1970, Alcoholism, illness and social pathology among American Indians in transition, *American Journal of Public Health* 60:1769–1787.

Lobban, M. C., 1971, Cultural problems and drunkenness in an Arctic population, *British Medical Journal* 1:344.

Loeb, E. M., 1948, Primitive intoxicants, *Quarterly Journal of Studies on Alcohol* 4:387–398.

Loeb, E. M., 1960, Wine, women and song: Root planting and head-hunting in southeast Asia, *in* "Culture and History" (S. Diamond, ed.), Columbia University Press, New York.

Lolli, G., 1955, Alcoholism as a medical problem, *Bulletin of the New York Academy of Medicine*, 31:876–885.

Lolli, G., Serriani, E., Golder, G. M., and Luzzatto-Fegiz, P., 1958, "Alcohol in Italian Culture: Food and Wine in Relation to Sobriety among Italians and Italian Americans," Yale Center of Alcohol Studies Monograph 3, New Haven.

Lomnitz, L., 1969a, Patrones de ingestión de alcohol entre migrantes mapuches en Santiago, *América Indígena* 29:43–71.

Lomnitz, L., 1969b, Función del alcohol en la sociedad mapuche, *Acta Psiquiátrica y Psicológica de América Latina* 15:157–167.

Lomnitz, L., 1969c, Patterns of alcohol consumption among the Mapuche, *Human Organization* 28:287–296.

Lomnitz, L., 1973, Influencia de los cambios políticos y económicos en la ingestión del alcohol: el caso mapuche, *América Indígena* 33:133–150.

Lomnitz, L., 1976, Alcohol and culture: The historical evolution of drinking patterns among the Mapuche, *in* "Cross-Cultural Approaches to the Study of Alcohol: An InterDisciplinary Perspective" (M. W. Everett, J. O. Waddell, and D. B. Heath, eds.), Mouton, The Hague.

Long, J. K., (in preparation), [Drinking and witchcraft as indices of tension in Latin America].

Lubart, J. M., 1969, Field study of the problems of adaptation of Mackenzie Delta Eskimos to social and economic change, *Psychiatry* 32:447–458.

Lucia, S. P. (ed.), 1963, "Alcohol and Civilization," McGraw-Hill, New York.

Lundberg, G., (in preparation), Sociocultural change and drinking patterns in British Honduras.

Lurie, N. O., 1971, The world's oldest on-going protest demonstration: North American Indian drinking patterns, *Pacific Historical Review* 40:311–332.

Lutes, S. V., (in preparation), [Drinking patterns in a rural Peruvian community].

Lutz, H. F., 1922, "Viticulture and Brewing in the Ancient Orient," J. C. Heinrichs, Leipzig.

MacAndrew, C., and Edgerton, R. B., 1969, "Drunken Comportment: A Social Explanation," Aldine, Chicago.

McCall, G., (in preparation), [Drinking patterns of Basques in Europe].

McCarthy, R. G. (ed.), 1959, "Drinking and Intoxication: Selected Readings in Social Attitudes and Controls," Free Press, Glencoe, Ill.

McClelland, D. C., Davis, W. N., Kalin, R., and Wanner, E., 1972, "The Drinking Man," Free Press, New York.

McClelland, D. C., Davis, W., Wanner, E., and Kalin, R., 1966, A cross-cultural study of folktale content and drinking, *Sociometry* 29:308–333; revised, *in* D. C. McClelland, W. N. Davis, R. Kalin, and E. Wanner, 1972.

McCloy, S. G., (in preparation), [Drinking patterns and religion in the Outer Hebrides].

Maccoby, M., 1965, El alcoholismo en una comunidad campesina, *Revista de Psicoanálisis, Psiquiatria y Psicología* 1:38–64.

Maccoby, M., 1972, Alcoholism in a Mexican village, *in* "The Drinking Man" (D. McClelland, W. N. Davis, R. Kalin, and E. Wanner), Free Press, New York.

McCord, W., and McCord, J., with Gudeman, J., 1960, "Origins of Alcoholism," Stanford University Press, Stanford.
McFarland, R. A., and Forbes, W. H., 1936, The metabolism of alcohol in man at high altitudes, *Human Biology* 8:387–398.
McGlashan, N. D., 1969, Oesophageal cancer and alcoholic spirits in Central Africa, *Gut* 10:643–650.
McGuire, M. T., Stein, S., and Mendelson, J. H., 1966, Comparative psychosocial studies of alcoholic and nonalcoholic subjects undergoing experimentally induced ethanol intoxication, *Psychosomatic Medicine* 28:13–25.
McKinlay, A. P., 1939, The "indulgent" Dionysius, *Transactions of the American Philosophical Association* 70:51–61.
McKinlay, A. P., 1944, How the Athenians handled the drink problem among their slaves, *Classical Weekly* 37:127–128.
McKinlay, A. P., 1945, The Roman attitude toward women's drinking, *Classical Bulletin* 22:14–15.
McKinlay, A. P., 1948a, Temperate Romans, *Classical Weekly* 41:146–149.
McKinlay, A. P., 1948b, Early Roman sobriety, *Classical Bulletin* 24:52.
McKinlay, A. P., 1948–49, Ancient experience with intoxicating drinks: Non-classical peoples; Non-Attic Greek states, *Quarterly Journal of Studies on Alcohol* 9:388–414; 10:289–315.
McKinlay, A. P., 1950a, Bacchus as health-giver, *Quarterly Journal of Studies on Alcohol* 11:230–246.
McKinlay, A. P., 1950b, Roman sobriety in the Early Empire, *Classical Bulletin* 26:31–36.
McKinlay, A. P., 1951, Attic temperance, *Quarterly Journal of Studies on Alcohol* 12:61–102.
McKinlay, A. P., 1953, New light on the question of Homeric temperance, *Quarterly Journal of Studies on Alcohol* 14:78–93.
MacLeod, W. C., 1928, "The American Indian Frontier," Alfred A. Knopf, New York.
MacLeod, W. C., 1930, Alcohol: Historical aspects, *in* "Encyclopaedia of the Social Sciences" (E. R. A. Seligman and A. Johnson, eds.), Vol. 1, Macmillan, New York.
McNair, C. N., [1969], "Drinking Patterns and Deviance in a Multi-Racial Community in Northern Canada," Addiction Research Foundation Clinical Division Substudy 32-1969, Toronto.
Madsen, W., 1964, The alcoholic agringado, *American Anthropologist* 66:355–361.
Madsen, W., 1965, Comment on D. Mandelbaum's "Alcohol and Culture," *Current Anthropology* 6:291–292.
Madsen, W., 1973, "The American Alcoholic: The Nature-Nurture Controversy in Alcoholic Research and Therapy," Charles C Thomas, Springfield, Ill.
Madsen, W., 1976, Body, mind and booze, *in* "Cross-Cultural Approaches to the Study of Alcohol: An Interdisciplinary Perspective" (M. W. Everett, J. O. Waddell, and D. B. Heath, eds.), Mouton, The Hague.
Madsen, W., and Madsen, C., 1969, The cultural structure of Mexican drinking behavior, *Quarterly Journal of Studies on Alcohol* 30:701–718.
Maha Patra, S. K., (in preparation), [Alcohol and alcoholism among the tribal communities in Orissa, India].
Mail, P. D., 1967, The Prevalence of Problem Drinking in the San Carlos Apache, MPH thesis, Yale University Medical School, New Haven.
Malen, V. D., (in preparation) [Value orientation and alcoholism on the Pine Ridge Sioux Reservation].
Mandelbaum, D. G., Erlich, V. S., Hasan, K. A., Heath, D. B., Honigmann, J. J., Lemert, E. M., and Madsen, W., 1965, Alcohol and culture [with comments], *Current Anthropology* 6:281–294.

Mangin, W., 1957, Drinking among Andean Indians, *Quarterly Journal of Studies on Alcohol* 18:55-66.
Marconi, J., 1969, Barreras culturales en la comunicación que afectan el desarrollo de programas de control y prevención del alcoholismo, *Acta Psiquiátrica y Psicológica de América Latina* 15:351-355.
Marroquin, J., 1943, Alcoholismo entre los aborígenes peruanos, *Crónica Médica* (Lima) 60:226-231.
Marshall, M., 1974, Research bibliography of alcohol and kava studies in Oceania, *Micronesia,* 10:299-306.
Marshall, M., 1976, A review and appraisal of alcohol and kava studies in Oceania, *in* "Cross-Cultural Approaches to the Study of Alcohol: An Interdisciplinary Perspective" (M. W. Everett, J. O. Waddell, and D. B. Heath, eds.), Mouton, The Hague.
Martín del Campo, R., 1938, El pulque en México precortesiano, *Universidad Nacional Autónoma de México, Anales del Instituto de Biología* 9:5-23.
Martindale, D. and Martindale, E., 1971, "The Social Dimensions of Mental Illness, Alcoholism, and Drug Dependence," Greenwood, Westport, Conn.
Masserman, J. H., and Yum, K. S., 1946, An analysis of the influence of alcohol on experimental neurosis in cats, *Psychosomatic Medicine* 8:36-52.
Maynard, E., 1965, Drinking patterns in the Colta Lake Zone (Chimborazo), *in* "Drinking Patterns in Highland Ecuador" (E. Maynard, B. Frøland, and C. Rasmussen), Cornell University, Ithaca.
Maynard, E., 1969, Drinking as part of an adjustment syndrome among the Oglala Sioux, *Pine Ridge Reservation Bulletin* 9:35-51.
Maynard, E., Frøland, B., and Rasmussen, C., 1965, Drinking patterns in highland Ecuador, Andean Indian Community Research and Development Program, Department of Anthropology, Cornell University (multigraphed), Ithaca.
Mears, A. R. R., 1942, Pellagra in Tsolo District, *South African Medical Journal* 16:385-387.
Medical Practitioner, A, 1830, "Notices respecting Drunkenness, and of the Various Means which have been Employed in Different Countries for restraining the Progress of that Evil," William Collins, Glasgow.
Medical Tribune and Medical News, 1972, Unusual intoxication laid to GI fermentation, *Medical Tribune and Medical News* 13, (43):23.
Medina C., E., and Marconi, J., 1970, Prevalencia de distintos tipos de bebedores en adultos mapuches de zona rural en Cautín, *Acta Psiquiátrica y Psicológica de América Latina* 16:273-285.
Merry, J., 1966, The "loss of control" myth, *Lancet* 1966 (1); 7449:1257-1258.
Mesa y P., S. A., 1959, Historia del alcohol y el alcoholismo en Europe en America, *Orientaciones Médicas* 8 (107):5-6.
Metzger, D. G., 1964, Interpretations of Drinking Performances in Aguacatenango, PhD dissertation (anthropology), University of Chicago, Chicago.
Metzger, D. G., and Williams, G., [ca. 1963], Drinking patterns in Aguacatenango: Code and content (manuscript).
Midgley, J., 1971, Drinking and attitude toward drink in a Muslim community, *Quarterly Journal of Studies on Alcohol* 32:148-158.
Miles, J. D., 1965, "The Drinking Patterns of Bantu in South Africa," National Bureau of Educational and Social Research Series 18, Department of Education, Arts and Sciences, [Johannesburg].
Missions des Iles, 1952, "Alcool en Océanie," Paris.
Mitra, B. R., 1873, Spirituous drinks in ancient India, *Journal of the Asiatic Society* (Bengal) 43:1-23.

Mizruchi, E. H., and Perucci, R., 1962, Norm qualities and differential effects of deviant behavior: An exploratory analysis, *American Sociological Review* 27:391-399.
Modi, J. J., 1888, "Wine among the Ancient Persians," Bombay Gazette Steam Press, Bombay.
Mohatt, G., 1972, The Sacred Water: The quest for personal power through drinking among the Teton Sioux, *in* "The Drinking Man" (D. McClelland, W. Davis, R. Kalin, and E. Wanner), Free Press, New York.
Montell, G., 1937, Distilling in Mongolia, *Ethnos* 2:321-332.
Montoya y F., J. B., 1903, El alcoholismo entre los aborígenes de Antioquia, *Anales de la Academia de Medicina* (Medellín) 12:132.
Moore, M., 1948, Chinese wine: Some notes on its social use, *Quarterly Journal of Studies on Alcohol* 9:270-279.
Morewood, S., 1838, "A Philosophical and Statistical History of the Invention and Customs of Ancient and Modern Nations in the Manufacture and Use of Inebriating Liquors," William Curry, Jun. and William Carson, Dublin.
Morote B., E., 1952, Chicha, *Impulso* 1 (3):1-6.
Mossman, B. M., and Zamora, M. D., 1973, Culture-specific treatment for alcoholism, Paper read at International Congress of Anthropological and Ethnological Sciences, Chicago.
Muelle, J. C., 1945, La chicha en el distrito de San Sebastián, *Revista del Museo Nacional* (Lima) 14:144-152.
Myerson, A., 1940a, Alcohol: A study of social ambivalence, *Quarterly Journal of Studies on Alcohol* 1:13-20.
Myerson, A., 1940b, The social psychology of alcoholism, *Diseases of the Nervous System* 1:43-50.
Nagler, M., 1970, "Indians in the City: A Study of the Urbanization of Indians in Toronto," Canadian Research Centre for Anthropology, Saint Paul University, Ottawa.
Negrete, J. C., 1974, Factores culturales en estudios epidemiológicos sobre alcoholismo, *Acta Psiquiátrica y Psicológica de América Latina* 20:112-120.
Nelson, G. K., Novellie, L., Reader, D. H., Reuning, H., and Sachs, H., 1964, Psychological, nutritional, and sociological studies of Kaffir beer, Johannesburg Kaffir Beer Research Project, South African Council for Scientific and Industrial Research (multigraphed), Pretoria.
Netting, R. Mc., 1964, Beer as a locus of value among the West African Kofyar, *American Anthropologist* 66:375-384.
New Mexico Association on Indian Affairs Newsletter, July, 1956, The liquor problem among Indians of the Southwest, *New Mexico Association on Indian Affairs Newsletter.*
Newsletter of Southwestern Association on Indian Affairs, January, 1959, Drinking and Indian problems, *Newsletter of Southwestern Association on Indian Affairs.*
Nida, E. A., 1959, Drunkenness in indigenous religious rites, *Practical Anthropology* 6:20-23.
Nissly, C. M., (in preparation), [Chicha in Peru].
Nolan, R. W., (in preparation), [Beer among the Bassari of Senegal, West Africa].
Norick, F. A., 1970, Acculturation and drinking in Alaska, *Rehabilitation Record* 11 (5):13-17.
Obayemi, A. M. U., 1976, Alcohol usage in an African society, *in* "Cross-Cultural Approaches to the Study of Alcohol: An Interdisciplinary Perspective," (M. W. Everett, J. O. Waddell, and D. B. Heath, eds.), Mouton, The Hague.
Ogan, E., 1966, Drinking behavior and race relations, *American Anthropologist* 68:181-187.
Otélé, A., 1959, Les boissons fermentées de l'Oubangui-Chari, *Liaison* (Brazzaville) 67:34-42.
Owen, R. C., (in preparation), [Alcohol consumption and use in several Brazilian population segments: an evolutionary approach].
Park, P., 1962, Problem drinking and role deviation: A study of incipient alcoholism, *in* "Society, Culture and Drinking Patterns" (D. J. Pittman and C. R. Snyder, eds.), John Wiley and Sons, New York.
Park, P., 1973, Developmental ordering of experiences in alcoholism, *Quarterly Journal of Studies on Alcohol* 34:473-488.

Parkin, D. J., 1972, "Palms, Wine, and Witnesses: Public Spirit and Private Gain in an African Farming Community," Chandler, San Francisco.
Pascal, G. R., and Jenkins, W. O., 1966, On the relationship between alcoholism and environmental satisfactions, *Southern Medical Journal* 59:698–702.
Patnaik, N., 1960, Outcasting among oilmen for drinking wine, *Man in India* 40:1–7.
Patrick, C. H., 1952, "Alcohol, Culture, and Society," Duke University Sociological Series 8, Durham, N.C.
Peeke, H. L., 1917, "America Ebrietatis: The Favorite Tipple of our Forefathers and the Laws and Customs Relating Thereto" [no publisher], New York.
Pelto, P. J., 1960, Alcohol use in Skolt Lapp society, Paper read at American Ethnological Society meeting, Stanford.
Pelto, P. J., 1963, Alcohol use and dyadic interaction, Paper read at Northeastern Anthropological Association meeting, Ithaca, N.Y.
Pendered, A., 1931, Kubika Wawa: Beer making, *Nada* (Salisbury) 9:30.
Perisse, J., Adrian, J., Rerat, A., and Le Berre, S., 1959, Bilan nutritif de la transformation du sorgho en bière: Preparation, composition, consommation d'une bière du Togo, *Annales de la Nutrition et de l'Alimentation* 13:1–15.
Pertold, O., 1931, The liturgical base of mahuda liquor by Bhils, *Archiv Orientální* (Prague) 3:400–407.
Piga P., A., 1942a, Influencia del uso de las bebidas fermentadas en la primitiva civilización egipcia, *Actas y Memorias de la Sociedad Española de Antropología, Etnografía, y Prehistoria* 17:61–86.
Piga P., A., 1942b, La lucha antialcohólica de los españoles en la época colonial, *Revista de Indias* 3:711–742.
Pitt, P., 1971, Alcoholism in Nepal, *Journal of Alcoholism* (London) 6:15–19.
Pittman, D. J., 1965, Social and cultural factors in drinking patterns, pathological and nonpathological, *in* "Selected Papers Presented at 27th International Congress on Alcohol and Alcoholism," Vol. 1, International Bureau against Alcoholism, Lausanne.
Pittman, D. J. (ed.), 1967, "Alcoholism," Harper and Row, New York.
Pittman, D. J., 1971, Transcultural aspects of drinking and drug usage, *in* "29th International Congress on Alcoholism and Drug Dependence" (L. G. Kiloh and D. S. Bell, eds.), Butterworths, Australia.
Pittman, D. J., and Snyder, C. R. (eds.), 1962, "Society, Culture, and Drinking Patterns," John Wiley and Sons, New York.
Platt, B. S., 1955, Some traditional alcoholic beverages and their importance in indigenous African communities, *Proceedings of the Nutrition Society* 14:115–124.
Plaut, T. F., 1967, "Alcohol Problems: A Report to the Nation by the Cooperative Commission on the Study of Alcoholism," Oxford University Press, New York.
Podlewski, H., and Catanzaro, R. J., 1968, Treatment of alcoholism in the Bahama Islands, *in* "Alcoholism: The Total Treatment Approach," (R. J. Catanzaro, ed.), Charles C Thomas, Springfield, Ill.
Poirier, J., 1966, Ethnologie et sociologie des alcoolismes a Madagascar, Centre International d'Alcoologie (multigraphed), Lausanne.
Polacsek, E., Barnes, T., Turner, N., Hall, R., and Weise, C. (comps.), 1972, "Interaction of Alcohol and Other Drugs," Addiction Research Foundation Bibliography Series 3, Toronto.
Poot, A., 1954, Le "munkoyo" boisson des indigenes Bapende (Katanga), *Bulletin des Séances de l'Institut Royal Colonial Belge* 25:386–389.
Popham, R. E., 1959a, Some problems of alcohol research from a social anthropologist's point of view, *Alcoholism* 6 (2):19–24.
Popham, R. E., 1959b, Some social and cultural aspects of alcoholism, *Canadian Psychiatric Association Journal* 4:222–229.

Popham, R. E., [1968], "The Practical Relevance of Transcultural Studies," Addiction Research Foundation Substudy 9-2-70, Toronto.
Popham, R. E. (ed.), 1970, "Alcohol and Alcoholism," University of Toronto Press, Toronto.
Popham, R. E., and Yawney, C. D. (comps.), 1966, "Culture and Alcohol Use: A Bibliography of Anthropological Studies," Addiction Research Foundation, Toronto; 2nd ed., 1967.
Pozas A., R., 1957, El alcoholismo y la organización social, *La Palabra y el Hombre* 1:19–26.
Poznanski, A., 1956, Our drinking heritage, *McGill Medical Journal* 25:35–41.
Prakash, O., 1961, "Food and Drinks in Ancient India: From Earliest Times to c. 1200 A.D.," Munshi Ram Manohar Lal, Delhi.
Prestán, A., (in preparation), [Chicha among the Cuna of Panama].
Preuss, K. T., 1910, Das Fest des Erwachens (Weinfest) bei den Cora-Indianern, *in* "16th International Congress of Americanists" (pt. 2), Vienna.
Quarcoo, A. K. (in preparation), [Alcohol and traditional religions in Ghana].
Quichaud, J., 1955, Problèmes médico-sociaux d'outre-mer: l'alcoolisme en Guinée, *Semaine Médicale Professionelle et Médico-Sociale* 31:574–575.
Radović, B., 1937, Pechenje Rakije ve Nashem Narodu, *Glasnik Etnograficheskog Muzea* 55:69–112.
Raman, A. C., 1968, Cultural factors in alcoholism, Paper read at International Congress of Mental Health, London.
Rao, M. S. A. (in preparation), A Religious Temperance Movement and its Impact on a Toddy-Tapping Case in Kerala.
Rasmussen, C., 1965, Drinking patterns in Peguche (Imbabura), *in* "Drinking Patterns in Highland Ecuador" (E. Maynard, B. Frøland, and C. Rasmussen), Department of Anthropology, Cornell University, Ithaca.
Ravi Varma, L. A., 1950, Alcoholism in Ayurveda, *Quarterly Journal of Studies on Alcohol* 11:484–491.
Ray, R. B. J.-C., 1906, Hindu method of manufacturing spirit from rice, *Journal of the Asiatic Society of Bengal* (NS) 2 (4):16–28.
Raymond, I. W., 1927, "The Teaching of the Early Church on the Use of Wine and Strong Drink," Columbia University Studies in History, Economics, and Public Law 286, New York.
Reader, D. H., 1967, Alcoholism and excessive drinking: A sociological review, *Psychologia Africana Monograph Supplement* 3, National Institute for Personnel Research, [Johannesburg].
Reader, D. H., and May, J., 1971, "Drinking Patterns in Rhodesia: Highfield African Township, Salisbury," University of Rhodesia, Department of Sociology Occasional Paper 5, Salisbury.
Redding, C., 1860, "A History and Description of Modern Wines" (3d ed.), Henry G. Bohn, London.
Reiche, C. E., 1970, Estudio sobre el patrón de embriaguez en la región rural altaverapacense, *Guatemala Indígena* 5:103–127.
Ribstein, M., Certhoux, A., and Lavenaire, A., 1967, Alcoolisme au rhum: étude de la symptomatologie et analyse de la personalité de l'homme martiniquais alcoolique au rhum, *Annales Médico-Psychologiques* 125:537–548.
Riffenberg, A. S., 1956, Cultural influences and crime among Indian-Americans of the Southwest, *Federal Probation* 10:38–41.
Riley, J. W., Jr., 1946, Sociological factors in the alcohol problem, *Scientific Temperance Journal* 54:67–74.
Riley, J. W., Jr., and Marden, C. F., 1947, The social pattern of alcoholic drinking, *Quarterly Journal of Studies on Alcohol* 8:265–273.
Robbins, M. C., and Pollnac, R. B., 1969, Drinking patterns and acculturation in rural Buganda, *American Anthropologist* 71:276–284.

Robbins, R. H., 1969, Role reinforcement and ritual deprivation: Drinking behavior in a Naskapi village, *Papers on the Social Sciences* 1:1-7.
Robbins, R. H., 1973, Alcohol and the identity struggle: Some effects of economic change on interpersonal relations, *American Anthropologist* 75:99-122.
Roca W., D., 1953, Apuntes sobre la chicha, *La Verdad* 42 (2004): 3.
Rodríguez S., L., 1945, Drinking motivations among the Indians of the Ecuadorean sierra, *Primitive Man* 18:39-46.
Roebuck, J. B., and Kessler, R. G., 1972, "The Etiology of Alcoholism: Constitutional, Psychological and Sociological Approaches," Charles C Thomas, Springfield, Ill.
Rohrmann, C. A. [ca. 1972], Drinking and violence: A cross-cultural survey (manuscript).
Rojas G., F., 1942, Estudio histórico-etnográfico del alcoholismo entre los indios de México, *Revista Mexicana de Sociología*, 4:111-125.
Rojas, U., 1960, La lucha contra las bebidas alcohólicas en la época de la colonia, *Repertorio Boyacense* 46:877-883.
Rolleston, J. D., 1927, Alcoholism in classical antiquity, *British Journal of Inebriety* 24:101-120.
Rolleston, J. D., 1933, Alcoholism in mediaeval England, *British Journal of Inebriety* 31:33-49.
Rolleston, J. D., 1941, The folklore of alcoholism, *British Journal of Inebriety* 39:30-36.
Room, R., 1968, Cultural contingencies of alcoholism: Variations between and within nineteenth-century urban ethnic groups in alcohol-related death rates, *Journal of Health and Social Behavior* 9:99-113.
Room, R., 1972, Some propositions on the analysis of cross-cultural data on alcohol, *Drinking and Drug Practices Surveyor* 6:1 seq.
Roth, W. E., 1912, On the native drinks of the Guianese Indian, *Timehri Demerara* (ser. 3) 2:128-134.
Rotter, H., 1957, Die Bedeutung des alkoholischen Milieus für den Alkoholismus, *Wiener Medizinische Wochenschrift* 107:236-239.
Roueché, B., 1960, "The Neutral Spirit: A Portrait of Alcohol," Little, Brown, Boston.
Roufs, T. G., and Bregenzer, J. M., 1968, Some aspects of the production of pulque, *in* "Social and Cultural Aspects of Modernization in Mexico" (F. Miller and P. Pelto, eds.), Department of Anthropology, University of Minnesota, Minneapolis (multigraphed).
Rubington, E., 1968, The bottle gang, *Quarterly Journal of Studies on Alcohol* 29:943-955.
Rubington, E., 1971, The language of "drunks," *Quarterly Journal of Studies on Alcohol* 32:721-740.
Rüden, E., 1903, Der Alkohol in Lebensprozess der Rasse, *Internationale Monatsschrift zur Erforschung des Alkoholismus und Bekampfung der Trinksitten* 13:374-379.
Ruiz M., A., La lucha antialcohólica de los Jesuitas en la época colonial, *Estudios* (Buenos Aires) 62:339-352, 423-446.
S., H., 1963, Non-drinking societies, [Institute for the Study of Human Problems?], ms. 46 (multigraphed), Stanford, Cal.
Sadoun, R., Lolli, G., and Silverman, M., 1965, "Drinking in French Culture," Rutgers Center of Alcohol Studies Monograph 5, New Brunswick, N.J.
Salone, E., 1907, Les sauvages du Canada et les maladies importeés de France au xviie et au xviiie siècle: la picote et l'alcoolisme, *Journal de la Société des Américanistes de Paris* (NS) 4:7-20.
Salonen, A., 1957-58, Dryckesseder före och efter Muhammed, *Alkoholpolitik* 20:50-52; 81-83; 107-109.
Sangree, W. H., 1962, The social functions of beer drinking in Bantu Tiriki, *in* "Society, Culture, and Drinking Patterns" (D. J. Pittman and C. R. Snyder, eds.), John Wiley and Sons, New York.
Sargent, M. J., 1967, Changes in Japanese drinking patterns, *Quarterly Journal of Studies on Alcohol* 28:709-722.

Sargent, M. J., 1971, A cross-cultural study of attitudes and behavior towards alcohol and drugs, *British Journal of Sociology* 22:83–96.

Sargent, M. J. 1976, Theory in alcohol studies, *in* "Cross-Cultural Approaches to the Study of Alcohol: An Interdisciplinary Perspective" (M. W. Everett, J. O. Waddell, and D. B. Heath, eds.), Mouton, The Hague.

Sariola, S., 1954, "Lappi ja Väkijuomat," Vakijuomaksmyksen Tutkimussäätiö, Helsingfors (English translation, 1956).

Sariola, S., 1956, Indianer och Alkohol, *Alkoholpolitik* 19:39–43.

Sariola, S., 1961, Drinking customs in rural Colombia, *Alkoholpolitik* 24:127–131.

Savard, R. J., 1968a, Cultural stress and alcoholism: A study of their relationship among Navaho alcoholic men, PhD dissertation (Sociology), University of Minnesota, Minneapolis.

Savard, R. J., 1968b, Effects of Disulfiram therapy on relationships within the Navaho drinking group, *Quarterly Journal of Studies on Alcohol* 29:909–916.

Sayres, W. C., 1956, Ritual drinking, ethnic status and inebriety in rural Colombia, *Quarterly Journal of Studies on Alcohol* 17:53–62.

Schaefer, J. M., 1976, Drunkenness and culture stress: A holocultural test, *in* "Cross-Cultural Approaches to the Study of Alcohol: An Interdisciplinary Perspective" (M. W. Everett, J. O. Waddell, and D. B. Heath, eds.), Mouton, The Hague.

Schaefer, J. M., (forthcoming), "Drunkenness: A Hologeistic Treatise," H.R.A.F. Press, New Haven.

Schmidt, W., and Popham, R. E., 1961, "Some Hypotheses and Preliminary Observations Concerning Alcoholism among Jews," Addiction Research Foundation Substudy 1-4 & 2-61, Toronto.

Schreiber, G., 1958, Der Wein und die Volkstumsforschung: Zur Sakralkultur und zum Genossenrecht, *Rheinisches Jahrbuch für Volkskunde* 9:207–243.

Schwartz, T., and Romanucci-Ross, L, 1974, Drinking and inebriate behavior in the Admiralty Islands, Melanesia, *Ethos* 2:213–231.

Seekirchner, A., 1931, Der Alkohol in Afrika, *in* "Atlas Africanus" (L. Frobenius and R. von Wilm, eds.), Vol. 8, W. de Gruyter, Berlin.

Seltman, C., 1957, "Wine in the Ancient World," Routledge and Kegan Paul, London.

Serebro, B., 1972, Total alcohol consumption as an index of anxiety among urbanized Africans, *British Journal of Addiction* 67:251–254.

Shalloo, J. P., 1941, Some cultural factors in the etiology of alcoholism, *Quarterly Journal of Studies on Alcohol* 2:464–478.

Shore, J. H., and Fumett, B. V., 1972, Three alcohol programs for American Indians, *American Journal of Psychiatry* 128:1450–1454.

Sievers, M. L., 1968, Cigarette and alcohol usage by southwestern American Indians, *American Journal of Public Health* 58:71–82.

Siliceo P., P., 1920, El pulque, *Ethnos* 2:60–63.

Simmons, O. G., 1959, Drinking patterns and interpersonal performance in a Peruvian mestizo community, *Quarterly Journal of Studies on Alcohol* 20:103–111.

Simmons, O. G., 1960, Ambivalence and the learning of drinking behavior in a Peruvian community, *American Anthropologist* 62:1018–1027.

Simmons, O. G., 1968, The sociocultural integration of alcohol use: A Peruvian study, *Quarterly Journal of Studies on Alcohol* 29:152–171.

Singer, K., 1972, Drinking patterns and alcoholism in the Chinese, *British Journal of Addiction* 67:3–14.

Singer, K., 1974, The choice of intoxicant among the Chinese, *British Journal of Addiction* 69 (3):257–268.

Singh, S., 1937, Preparation of beer by the Loi-Manipuris of Sekami, *Man in India* 17:80.

Skolnik, J. H., 1958, Religious affiliation and drinking behavior, *Quarterly Journal of Studies on Alcohol* 19:452-470.
Skorzyski, L., 1889, L'alcool chez les peuples primitifs de Russie, *in Congrès Internationale contre l'Abus des Boissons Alcooliques,* Paris.
Slotkin, J. S., 1954, Fermented drinks in Mexico, *American Anthropologist* 56:1089-1090.
Smythe, D. W., 1966, Alcohol as a symptom of social disorder: An ecological view, *Social Psychiatry* (Berlin) 1:144-151.
Snyder, C. R., 1958a, "Alcohol and the Jews: A Cultural Study of Drinking and Sobriety," Yale Center of Alcohol Studies Monograph 1, Free Press, Glencoe, Ill.
Snyder, C. R., 1958b, Culture and Jewish sobriety: The ingroup-outgroup factor, *in* "The Jews: Social patterns of an American Group," (M. Sklare, ed.), Free Press, Glencoe, Ill.
Synder, C. R., and Pittman, D. J., 1968, Drinking and alcoholism: Social aspects, *in* "International Encyclopedia of the Social Sciences," Vol. 4, Macmillan and Free Press, New York.
Sølling, L., 1974, Alcohol and the subjective experience of power, Paper read at the International Symposium on Circumpolar Health, Yellowknife.
Solms, H., 1966, Sozio-kulturelle und wirtschaftliche Bedingungen der Giftsuchten, des Medikamentenmissbrauches und des chronischen Alkoholismus, *Hippokrates* 37:184-192.
Spaulding, P., 1966, The social integration of a northern community: White mythology and metis reality, *in* "A Northern Dilemma: Reference Papers" (A. K. Davis, ed.), Western Washington State College, Bellingham.
Spindler, G. D., 1964, Alcohol symposium: Editorial preview, *American Anthropologist* 66:341-343.
Spradley, J. P., 1970, "You Owe Yourself a Drunk: Ethnography of Urban Nomads," Little, Brown, Boston.
Spradley, J. P., 1971, Beating the drunk charge, *in* "Conformity and Conflict: Readings in Cultural Anthropology" (J. P. Spradley and D. W. McCurdy, eds.), Little, Brown, Boston.
Steiner, C., 1971, "Games Alcoholics Play: The Analysis of Life Scripts," Grove Press, New York.
Stewart, O. C., 1960, Theory for understanding the use of alcoholic beverages, Paper read at American Anthropological Association meeting, Minneapolis.
Stewart, O. C., 1964, Questions regarding American Indian criminality, *Human Organization* 23:61-66.
Straus, R., and McCarthy, R. G., 1951, Nonaddictive pathological drinking patterns of homeless men, *Quarterly Journal of Studies on Alcohol* 12:601-611.
Strübing, E., 1960, Vom Wein als Genuss und Heilmittel im Alterum mit Plinius und Asklepiades, *Ernährungsforschung* 5:572-594.
Stull, D., 1975, Hologeistic studies on drinking: A critique, *Drinking and Drug Practices Surveyor,* 10:4-10.
Suolahti, J., 1955, Alkoholmissbruket under Antiken, *Alkoholpolitik* 20:77-78.
Suolahti, J., 1956, Statlig Alkoholpolitik i Rom under Kejsartidens Slutskede, *Alkoholpolitik* 22:5-12.
Swanson, D. W., Bratrude, A. P., and Brown, E. M., 1971, Alcohol abuse in a population of Indian children, *Diseases of the Nervous System* 32:835-842.
Szwed, J. F., 1966, Gossip, drinking and social control: Consensus and communication in a Newfoundland parish, *Ethnology* 5:343-441.
Tadesse, E., 1958, Preparation of täg among the Amhara of Šäwa, *Bulletin of Addis Ababa University College in Ethnology and Sociology* 8:101-109.
Tapia P., I., Gaete A., J., Muñoz L., C., Sescovitch, S., Miranda, I., Minguell I., J., Perez P., G., and Orellana A., G., 1966, Patrones socioculturales de la ingestión de alcohol en Chiloé: Informe preliminar, algunos problemas metodológicos, *Acta Psiquiátrica y Psicológica de América Latina* 12:232-240.

Taylor, W. B., (in preparation), Reply to conquest: Indians and alcohol in colonial Mexico.
Thorner, J., 1953, Ascetic Protestantism and alcoholism, *Psychiatry* 16:167–176.
Tillhagen, C.- H., 1957, Food and drink among the Swedish Kalderaša Gypsies, *Journal of the Gypsy Lore Society* 36:25–52.
Titcomb, M., 1948, Kava in Hawaii, *Journal of the Polynesian Society* 57:105–169.
Toit, B. M. du, 1964, Substitution: A process in culture change, *Human Organization* 23:16–23.
Topper, M., 1976, The cultural approach, verbal plans, and alcohol research, *in* "Cross-Cultural Approaches to the Study of Alcohol: An Interdisciplinary Perspective" (M. W. Everett, J. O. Waddell, and D. B. Heath, eds.), Mouton, The Hague.
Toulouse, J. H., 1970, High on the hawg, or how the western miner lived, as told by the bottles he left behind, *Great Plains Journal* 4:59–69.
Trice, H. M., and Pittman, D. J., 1958, Social organization and alcoholism: A review of significant research since 1940, *Social Problems* 5:294–306.
Udvalget for Samfundsforskning i Grønland, 1961, "Alkoholsituationen i Vestgrønland," Dansk Bibliografisk Kontor, København.
Ullman, A. D., 1958, Sociocultural backgrounds of alcoholism, *Annals of the American Academy of Political and Social Sciences* 315:48–54.
Umunna, I., 1967, The drinking culture of a Nigerian community: Onitsha, *Quarterly Journal of Studies on Alcohol* 28:529–537.
United States, Department of Health, Education and Welfare, Indian Health Service, 1970, "Alcoholism—A High Priority Health Problem: A Report of the Indian Health Service Task Force on Alcoholism," U.S. Government Printing Office, Washington.
United States, Department of the Interior, Bureau of Indian Affairs, 1956, Report [of the Commission to Study Alcoholism among Indians], (multigraphed), Washington.
Vachon, A., 1960, L'eau-de-vie dans la société indienne, *Canadian Historical Association Annual Report:* 22–32.
Valenzuela R., B., 1957, Apuntes breves de comidas y bebidas de la región de Carahue, *Archivo Folklórico* 8:90–105.
Vallee, B. L. 1966, Alcohol metabolism and metalloenzymes, *Therapeutic Notes* 14:71–74.
Vallee, F. G., 1968, Stresses of change and mental health among the Canadian Eskimos, *Archives of Environmental Health* 17:565–570.
Vanderyst, H., 1920, Le vin de palm ou malafu, *Bulletin Agricole du Congo Belge* 8 (11):219–224.
Varlet, F., 1956, Fabrication et composition de l'alcool de Bangui, *Notes Africaines* 71:74–75.
Varma, S. C., 1959, Problem of drinking in the primitive tribes, *Eastern Anthropologist* 12:252–256.
Vasev, C., and Milosavčević, V., 1970, Alkoholizam kod Cigana, *Alkoholizam* (Beograd) 10:47–57.
Vatuk, V. P., and Vatuk, S., 1967, Chatorpan: A culturally defined form of addiction (manuscript).
Vázquez, M. C., 1967, La chicha en los paises andinos, *América Indígena* 27:265–282.
Vedder, H., 1951, Notes on the brewing of kaffir beer in South West Africa, *South West Africa Scientific Society Journal* 8:41–43.
Velapatiño O., A., 1976, Summary of an alcoholism study in the Apurímac-Ayacucho River Valley, Peru, *in* "Cross-Cultural Approaches to the Study of Alcohol: An Interdisciplinary Perspective" (M. W. Everett, J. O. Waddell, and D. B. Heath, eds.), Mouton, The Hague.
Viñas, T., E., 1951, La composición química de las diferentes variedades de chicha que se consumen en el Perú, Ministerio de Salud Pública y Asistencia Social, Departamento de Nutrición (multigraphed), Lima.
Viqueira, C., and Palerm, A., 1954, Alcoholismo, brujería y homicidio en dos comunidades rurales de México, *América Indígena* 14:7–36.

Vogel-Sprott, M.D., 1967, Alcoholism as learned behavior: Some hypotheses and research, *in* "Alcoholism: Behavioral Research, Therapeutic Approaches" (R. Fox, ed.), Springer, New York.

Voss, H. L., (in preparation), [Drinking in Oahu, Hawaii].

Waddell, J. O., 1973, For individual power and social credit: The use of alcohol among Tucson Papagos, Paper read at Society for Applied Anthropology meeting, Tucson.

Waddell, J. O., 1974, "Drink, friend!" Social contexts of convivial drinking and drunkenness among Papago Indians in an urban setting, *in* "Proceedings of 1st Annual Institute on Alcohol Abuse and Alcoholism," National Institute on Alcohol Abuse and Alcoholism, Washington.

Waddell, J. O., (forthcoming), Social and pathological dimensions of Papago drinking behavior, *in* "Drinking, Drunkenness, and Alcoholism among North American Indians" (M. W. Everett and J. O. Waddell, eds.).

Wallgren, H., 1960, The alcoholism problem as a cultural question, *Alkoholiliike* 7086.

Wang, R. P., 1968, A study of alcoholism in Chinatown, *International Journal of Social Psychiatry* 14:260–267.

Wanner, E., 1972, Power and inhibition: A revision of the magical potency theory, *in* "*The Drinking Man*" (D. McClelland, W. Davis, R. Kalin, and E. Wanner), Free Press, New York.

Washburne, C., 1956, Alcohol, self and the group, *Quarterly Journal of Studies on Alcohol* 17:108–123.

Washburne, C., 1961, "Primitive Drinking: A Study of the Uses and Functions of Alcohol in Preliterate Societies," College and University Press, New York.

Washburne, C., 1968, Primitive religion and alcohol, *International Journal of Comparative Sociology* 9:97–105.

Webe, G., (in preparation), [A distributional study of drinking among South American Indians].

Wechsler, H., Demone, H. W., Jr., Thum, D., and Kasey, E. H., 1970, Religious-ethnic differences in alcohol consumption, *Journal of Health and Social Behavior* 11:21–29.

West, L. J., 1972, A cross-cultural approach to alcoholism, *Annals of the New York Academy of Sciences* 197:214–216.

Westermeyer, J. J., 1971, Use of alcohol and opium by the Meo of Laos, *American Journal of Psychiatry* 127:1019–1023.

Westermeyer, J. J., 1972, Options regarding alcohol use among the Chippewa, *American Journal of Orthopsychiatry* 42:398–403.

Westermeyer, J. J., 1974, Alcoholism from the cross-cultural perspective: A review and critique of clinical studies, *American Journal of Drug and Alcohol Abuse* 1:27–38.

Westermeyer, J. J., 1976, Cross-cultural studies of alcoholism in the clinical setting: A review and evaluation, *in* "Cross-Cultural Approaches to the Study of Alcohol: An Interdisciplinary Perspective" (M. W. Everett, J. O. Waddell, and D. B. Heath, eds.), Mouton, The Hague.

Westermeyer, J., and Brantner, J., 1972, Violent death and alcohol use among the Chippewa in Minnesota, *Minnesota Medicine* 55:749–752.

Wheeler, D., 1839, "Effects of the Introduction of Ardent Spirits and Implements of War among the Natives of the South Sea Islands and New South Wales," Harvey and Darton, London.

White, M. F., 1971, Drinking Behavior as Symbolic Interaction, PhD dissertation (Sociology), University of Kentucky.

Whitehead, P. C., 1972, Toward a new programmatic approach to the prevention of alcoholism: A reconciliation of the socio-cultural and distribution of consumption approaches, *in* "30th International Congress on Alcoholism and Drug Dependence" (E. Tongue and Z. Adler, eds.), Lausanne.

Whitehead, P. C., Grindstaff, C. F., and Boydell, C. L. (eds.), 1973, "Alcohol and Other Drugs: Perspectives on Use, Abuse, Treatment and Prevention. Holt, Rinehart and Winston, New York.

Whitehead, P. C., and Harvey, C., 1974, Explaining alcoholism: An empirical test and reformulation, *Journal of Health and Social Behavior* 15:57–65.

Whittaker, J. O., 1962–63, Alcohol and the Standing Rock Sioux tribe: I. The pattern of drinking; II. Psychodynamic and cultural factors in drinking, *Quarterly Journal of Studies on Alcohol* 23:468–479; 24:80–90.

Whittet, M. M., 1970, An approach to the epidemiology of alcoholism: Studies in the highlands and islands of Scotland, *British Journal of Addiction* 65:325–339.

Wilkinson, R., 1970, "The Prevention of Drinking Problems: Alcohol Control and Cultural Influences," Oxford University Press, New York.

Williams, A. F., 1966, Social drinking, anxiety, and depression, *Journal of Personality and Social Psychology* 3:689–693.

Wilson, G. C., 1963, Drinking and drinking customs in a Mayan community, Cornell-Columbia-Harvard-Illinois Summer Field Studies Program in Mexico, Harvard University, Cambridge (multigraphed).

Wolcott, H. F., 1974, "The African Beer Gardens of Bulawayo: Integrated Drinking in a Segregated Society," Rutgers Center of Alcohol Studies Monograph 10, New Brunswick, N.J.

Wolcott, H. F., 1975, Feedback influences on fieldwork, or, a funny thing happened on the way to the beer garden, *in* "Urban Man in Southern Africa" (C. Kileff and W. Pendleton, eds.), Mambo Press, Gwelo, Rhodesia.

Wolff, P. H., 1972, Ethnic differences in alcohol sensitivity, *Science* 175 (4020):449–450.

World Health Organization, Expert Committee on Mental Health, 1951, "Report of the First Session of the Alcoholism Subcommittee," W.H.O. Technical Report Series 42, Geneva.

World Health Organization, Expert Committee on Mental Health, 1952, "Second Report of the Alcoholism Subcommittee," W.H.O. Technical Report Series 48, Geneva.

Wright-St. Clair, R. E., 1962, Beer in therapeutics: An historical annotation, *New Zealand Medical Journal* 61:512–513.

Yamamuro, B., 1954, Notes on drinking in Japan, *Quarterly Journal of Studies on Alcohol* 15:491–498.

Yamamuro, B., 1958, Japanese drinking patterns: Alcoholic beverages in legend, history and contemporary religions, *Quarterly Journal of Studies on Alcohol* 19:482–490.

Yamamuro, B., 1964, Further notes on Japanese drinking, *Quarterly Journal of Studies on Alcohol* 25:150–153.

Yamamuro, B., 1968, Origins of some Japanese drinking customs, *Quarterly Journal of Studies on Alcohol* 29:979–982.

Yawney, C. D., [1967], "The Comparative Study of Drinking Patterns in Primitive Cultures," Addiction Research Foundation Substudy 1-Y-67, Toronto.

Yawney, C. D., 1969, Drinking patterns and alcoholism in Trinidad, *McGill Studies in Caribbean Anthropology Occasional Papers* 5:34–48.

Younger, W., 1966, "Gods, Men and Wine," World, New York.

Zentner, H., 1963, Factors in the social pathology of a North American Indian society, *Anthropologica* 5:119–130.

Zingg, R. M., 1942, The genuine and spurious values in Tarahumara culture, *American Anthropologist* 44:78–92.

CHAPTER 3

Drinking Behavior and Drinking Problems in the United States

Don Cahalan
School of Public Health
University of California
Berkeley, California

and

Ira H. Cisin
The George Washington University
Washington, D.C.

AMERICAN VALUES AND ATTITUDES CONCERNING ALCOHOL

Anyone interested in assessing drinking practices and the epidemiology of drinking problems must take into account the values and attitudes prevailing among major subgroups in America, for such values and attitudes play a very large role in determining the direction and persistence of drinking behavior. As just one of many possible illustrations, Jews in America have a very high proportion of persons who drink at least a little but a very low proportion who get into trouble over their drinking, while the Irish-Americans have a lower proportion of drinkers but a fairly high proportion who get into trouble (Glad, 1947). Such subgroup differences in drinking practices seem to be due not so much to stress as to deep-seated cultural and environmental influences. Thus, the history and present state of American values and attitudes about drinking are not merely of antiquarian or humanistic interest, but are central to understanding the dynamics of American drinking behavior and current social and health problems related to alcohol.

Historical Heritage and Value Conflict in America

Probably no other country has had so many clashes of values and beliefs concerning alcohol as America has had, beginning in Colonial times. Our conflictful heritage of practices and beliefs concerning alcohol has been well set forth by such writers as Keller (1971), Gusfield (1962), and Bacon (1962) and need not be repeated here in detail. Keller reminds us that the very early colonists (including even the Puritans) were anything but abstemious and that there was much sporadic drunkenness, even though the punishments were at times rather severe, such as being put in the stocks and exposed to public ridicule. As Keller summarizes it (1971, pp. 7–8), "Contemporary drinking ways are a mix imported by successive waves of immigrants from different geographic areas and ethnocultural backgrounds. In time, some indigenous customs developed in the particular American situation, notably the reckless frontier drinking in the eighteenth and nineteenth centuries and, later, the cocktail hour." The beginning of the Industrial Revolution had its repercussions in the form of social disorganization and the growth of extremes of wealth and poverty (and of drunkenness) in the nineteenth century.

Differences in drinking behavior and values were intensified by the westward thrust of the frontier, with the development of virile, hard-drinking, cowboy, mining, Indian-fighting, riverboating ways of life with many dangers and few social controls. Meanwhile, back in the effete East and spreading westward, there developed strong social movements (led by middle-class and managerial elements) that attempted to stem the tide of alcohol. These movements first preached moderation and later attempted (successfully for a brief period) to impose teetotalism in America. The Temperance Movement was for many decades a most effective political force in middle America because as Gusfield (1962) says, the Women's Christian Temperance Union and allied groups managed to link the Temperance (Prohibition) Movement to status-enhancing middle-class morality and self-improvement.

Ultimately, the dry and wet power struggle heightened, particularly after the Temperance Movement culminated in national Prohibition in 1920. On the one side were aligned the forces of middle-class and feminist morality led by conservative religious forces, white Anglo-Saxon Protestants, and women's groups. On the other, the wet forces consisted largely of a masculine element with many Irish-Catholics and other non-English Europeans who had poured into the U.S. to build the railroads and to man the factories. The polarization and politicization of the wet-dry controversy, perhaps best exemplified in the 1928 campaign between a morally righteous small-town Protestant, Herbert Hoover, and a permissive big-city Irish Catholic, Al Smith, have left us with a legacy of tension and anxiety and guilt surrounding the topic of alcohol, even though Prohibition lasted only 13 years and came to an official end about 40 years ago.

These historical waxings and wanings of attempts to control alcohol consumption are reflected in the alcohol consumption trends compiled by Keller (1971, pp. 10–12). In addition, statistics show a change in the relative popularity (and/or availability) of spirits and beer during the latter part of the nineteenth century: At midcentury, 85 percent of the alcohol consumed was in the form of distilled spirits; by the end of the century, about 50 percent of the alcohol consumption occurred in the form of beer drinking. Keller attributes this shift not to the temperance movement, but to the immigration after 1850 of large blocks of people from traditionally beer-drinking countries. The Keller reports survey data show that in recent years the proportion of drinkers in the United States population has increased somewhat; and it also appears from his alcohol statistics that there has been a significant increase in per capita consumption of alcoholic beverages within the last few years. However, no conclusive evidence has been marshaled to show whether there has been any significant increase in the number of "alcoholics" or "problem drinkers" within the last few years. It is certain that more people are now drinking than in the 1960s, but it is less certain whether the increased aggregate consumption may be accounted for simply by a large increase in the number of moderate drinkers or by much more heavy consumption on the part of the addictive drinker, or by some combination of both factors. However, the stresses of modern American society and the weakening of social controls demonstrably are sufficient to make the possibility of increases in alcoholism something to be guarded against.

National survey findings (reported later) show that problem drinking is heavily dependent upon the individual's social supports and restraints: Thus, the highest rates of problem drinking are to be found among those who live within highly permissive or indifferent social environments, who live in large cities, are psychologically alienated from middle-class values, and who are economically insecure. This is congruent with Bacon's observations (1962) that a society that has many individuals who are highly self-centered, ignorant of others' interests and activities, and prone to aggressive and competitive behavior is more likely to resort to excessive use of alcohol; concurrently, the complexity of modern society demands great levels of regularity, precision, individual responsibility, and integration of work and social activities through self-control and cooperation with others. Accordingly, the complexity of modern society makes alcohol more accessible and more highly prized by man, and also increases its dangers. Consistent with this, in relation to drugs other than alcohol, the First Report of the National Commission on Marihuana and Drug Abuse (March, 1972) discusses the implications of the growth of drug use in an era that is characterized by rapid change in leisure time and affluence; an increase in the inability to forecast the future; growing alienation and anomie associated with growth of urbanism and the generation gap; and a rise in cynicism and disillusion over our society's failure to solve urgent problems of war, poverty, and pollution of our environment.

Even if the use of alcohol is increasing, however, it is most unlikely that the United States will ever return to Prohibition, not only because of the bad taste Prohibition left in most people's mouths, but because alcohol has so many utilities as a relaxant and social lubricant that many would regard the cure to be worse than the disease. As Chafetz says, "Man's desire to alter the unpleasant realities in life may well be one of the most ancient, most persistent, and most comprehensible of all human needs. Certainly the use of the drug—alcohol—to effect a change in the perception of these realities has been a characteristic of many people in virtually every culture" (Oct. 1, 1971, p. 2).

As Terris (1968) and other epidemiologists have pointed out, Prohibition actually *worked*—and worked very well—in bringing down the rate of liver cirrhosis temporarily. (The rate resumed its climb after repeal, and is now as high as rates before Prohibition.) However, since two thirds of the adults in the United States drink more than once a year, the "noble experiment" is not likely to be repeated. Therefore, if alcoholism and problem drinking are to be kept within acceptable bounds, it would appear that treatment and preventive efforts must be increased.

Current Contradictory State of Values and Attitudes

Much confusion and ambivalence is evident in American attitudes and values related to alcohol, presumably because of the crosspressures between Calvinistic middle-class reformism and lower-class and youthful and frontier hedonism and machismo. This ambivalence is evident in our frequent observation of wry jokes and cartoons about the "funny drunk," coupled with our antipathy and sorrow about alcoholics when they impinge on our lives. On the one hand, many persons feel that the "drunken driver" should be run off the road and jailed indefinitely or worse; and on the other, judges and juries are often reluctant to convict first offenders for drunken driving, for "there but for the grace of God, go I." The ambivalence and confusion concerning what should be our appropriate stance vis-à-vis drunkenness is further evident in many surveys of values and attitudes concerning alcohol and alcoholics, as noted below.

Haberman and Scheinberg (1969) in a recent survey found about two thirds of a New York City sample considered alcoholism to be an "illness," but that fewer than half agreed with the illness concept in connection with various symptoms indicating addictiveness of drinking; and fewer than half thought a physician was the one who was best qualified to help to correct the manifestations of alcoholism. Mulford and Miller (1961), in another survey in Iowa, found their sample of adults apparently saw no inconsistency between feeling that the alcoholic would be best described as "sick" (65 percent) and "morally weak" (75 percent), with many persons agreeing to both propositions. The first nationwide survey of drinking practices based on a probability sample of adults found that while 68 percent drank at least once a year (Cahalan, Cisin, and Crossley, 1969, p. 14), 35 percent said that "nothing good" can be said about drinking (p. 133). In the same survey three fourths of the total sample (and even a majority of male heavy drinkers) said they thought drinking "does more harm than good"; and three fourths rated alcoholism as a serious public health problem in the United States (p. 29).

A recent national survey by Louis Harris and Associates, Inc., for the National Institute of Alcohol Abuse and Alcoholism covered a wide range of issues concerning American attitudes and values about drinking and alcoholism. A summary of these findings (from the report of the Harris organization to the NIAAA, December, 1971) is as follows:

> Only 17 percent said they believed it should be against the law to drink. This represented a drop of five percentage points from the last previous survey conducted on this topic two years before. (On the other hand, 67 percent of this national sample of adults aged eighteen and over believed marihuana should remain illegal.) However, when asked what age they believed to be the safe and proper one for young men to begin drinking, 39 percent answered "never," indicating that while few would have us return to Prohibition, a considerable minority regard drinking as "never" safe or proper (Harris, 1971, p. 46).
>
> On the pros and cons of drinking, a majority (51 percent) agreed with the proposition, "Liquor is destructive to people's health and morals; people should not drink at all" (p. 48). On the other hand, 73 percent agreed with the statement, "A mature and healthy person is his own best judge of what and where to drink"; and 70 percent agreed that "A little social drinking makes people friendlier, and often releases their inhibitions so that they feel more relaxed and open." While a majority of this national sample tended to agree with the statement, "It is all right to drink moderately on a regular basis as long as a person eats properly," and 65 percent agreed that "Beer or wine with meals is not harmful," a majority of those with opinions disagreed that "With the pressures of life today, a drink or two in the evening is a real benefit." The report infers from these findings that "What underlies much of

this unwillingness to endorse regular drinking is a very real fear of its long-term consequences . . ." (p. 48). As many as 48 percent concurred that "So-called 'social drinking' sooner or later leads to excessive drinking or alcoholism"; and 56 percent agreed that "Drinking often leads to a breakdown in people's sexual morals"; 60 percent agreed that "Even if a person drinks only moderately, over a long period of time it will damage his health." In addition, 52 percent agreed that "Drinking causes more absenteeism from the job than any other factor," and 70 percent concurred with the statement that "Drink is still the 'curse of the working classes'; it breaks up homes and causes unemployment."

This Harris survey also revealed that a large proportion of this national sample was poorly informed on the effects of alcohol, even though they were in general uneasy about its long-term effects. Substantial numbers agreed with the inaccurate statements that "Drinking black coffee and dousing your head with cold water will help you sober up quickly" (42 percent), "Alcohol is a stimulant that peps up a person and makes him feel sharper" (31 percent), and "There's no real risk to driving a car if a person has had only one or two drinks" (21 percent) (p. 69). And the estimates of what constituted "heavy" drinking were rather liberal in that 51 percent said a person would have to drink more than six ounces of hard liquor *each day* before they would consider him to be a heavy drinker; and 44 percent said a person would have to drink *more than six drinks* per occasion before they would consider him to be drinking heavily (p. 173). The report concludes: "Clearly, the drinking public wishes to separate itself from the somewhat pejorative label of 'heavy drinker.'" What is one to make of these seeming inconsistencies? It appears that what is often labeled as ambivalence about alcohol might also be described as a stance of circumstances alter cases: that what is too much for one person or in one setting may be acceptable for another person or circumstance. In any event, the Harris study reveals a general state of misunderstanding, misinformation, mythology, and misgiving: "The thrust of much of the foregoing data is that many people have trouble evaluating how much is too much, largely because they are quite uncertain as to what constitutes the difference between social drinking and problem drinking."

To sum up the present state of American attitudes and values concerning alcohol, events in our history, reinforced by the findings of recent surveys of the general public, lead to the conclusion that a large proportion of American people are rather uneasy and misinformed about the subject of drinking and its consequences. Even the term "drinker" tends to have pejorative connotations, although the vast majority drink at least occasionally; and many people (particularly those of medium to lower socioeconomic status, and women) tend to see alcohol as a real threat to their security (Cahalan et al., 1969, p. 196). The prevalence of uneasy and contradictory attitudes about drinking would appear to be a handicap in campaigns for moderation in drinking; for in order to position lighter beverages (such as wine, beer, or relatively weak mixed drinks) as adjuncts to gracious living, to be sipped with enjoyment and not be gulped in pursuit of intoxication, the general public will need to be well informed and to have confidence that they *can* (and should) keep their drinking within moderate limits. Where the climate is permissive about heavy drinking (as in the core areas of large cities), those who are subject to heavy stresses will tend to slip, if unchallenged, from "social" drinking to problem drinking. And where drinking *at all* is forbidden or frowned upon (such as in small towns or rural areas in the Southeast or Plains regions), young men of drinking age, being well aware that drinking exists in other parts of the country, have ritualized the art of "spree" drinking through the common practice of getting a bottle at the ABC store in another town, or a jug of moonshine, and driving out of their hometown to make a night or weekend of it. As will be noted later in this chapter, drinking practices in the United States vary by region of the country, so public health measures need to take into account the very heavy drinking in central cities in the Northeast and on the West Coast, and also the binge or explosive drinking (accompanied by midnight or weekend auto accidents and high accidental alcohol poisoning rates) in the Southeast and some other Bible Belt areas. In a country where even the term "temperance" for many years has meant Prohibition or complete abstinence, it may take many years of education and retraining to bring about a norm of socially controlled drinking that permits one or two drinks at a time, with clearly predictable sanctions for frequent exceeding of prescribed limits.

CURRENT DRINKING PRACTICES IN THE UNITED STATES

Until recent years, most of the research on drinking behavior was focused on anecdotal or anthropological studies of primitive peoples or special subgroups, or upon institutionalized alcoholics, rather than upon a broad cross section of the general public. Case studies and histories of nonindustrial peoples are interesting and illustrative of the fact that subcultures vary tremendously in their drinking behavior; but they can tell us little about the drinking practices and the incidence of alcohol problems for our country as a whole. And as regards the study of institutionalized alcoholics, obviously such information is not necessarily representative of the cultural and personality characteristics of problem drinkers in the noninstitutionalized population, as has been discussed by many writers (Room, 1970, p. 8; Trice and Wahl, 1958; Jellinek, 1960, p. 38; Wolf, Chafetz, Blane, and Hill, 1965; Blane, Overton, and Chafetz, 1963). The process of becoming an institutionalized alcoholic ordinarily takes many years. While later in this chapter it will be shown that the highest incidence of all types of drinking problems actually occurs among men in their early twenties, the average institutionalized alcoholic is in his forties; and therefore the analysis of the characteristics that predict the later development of alcoholism obviously needs to be conducted relatively early in life and among the general population, rather than trying to reconstruct the process of alcoholic deterioration from the questionable recollections of the institutionalized alcoholic.

To meet the need for objective descriptive studies of drinking practices throughout the United States, a long-term series of studies was set up at the beginning of the 1960s under grants from the National Institute on Alcohol Abuse and its predecessors within the National Institute of Mental Health. These studies, a portion of which are summarized below, were begun in California by Ira H. Cisin and Wendell Lipscomb in 1959 and were continued by Cisin and Cahalan at The George Washington University (Cisin, 1963; Cahalan, Cisin, Kirsch, and Newcomb, 1965; and Cahalan *et al.*, 1969; and by Knupfer, Fink, Clark, and Goffman in the San Francisco area, 1963). These studies in the 1960s drew in part upon the experiences of others in earlier state and local surveys (Mulford and Miller, 1959, 1960a, 1960b, 1960c; and Maxwell, 1952), and upon the Riley and Marden national quota sample reported in 1947.

Most of the findings reported below are drawn from three national probability sample surveys of U.S. adults twenty-one and over, conducted under the direction of the authors at the Social Research Group, The George Washington University, with further analysis conducted at the School of Public Health of the University of California, Berkeley. The three national surveys were as follows:

1. The 1964–65 national survey, published in "American Drinking Practices" (Cahalan *et al.*, 1969); hereafter referred to as the ADP survey). This survey measured drinking practices and related attitudes within a scientifically selected sample of 2,746 persons who were representative of the national adult population resident in households.

2. A 1967 followup of the ADP survey gathered information on drinking problems in some detail, from a subsample of 1,359 men and women ("Problem Drinkers," Cahalan, 1970).

3. A new national sample of drinking problems among men within the high-risk age group of twenty-one to fifty-nine was interviewed in 1969. Data in this 1969 survey have been combined with those from the 1967 follow-up and presented in a new monograph by Cahalan and Room (1974). This two-survey pool of combined data has permitted a rather detailed analysis of the correlates of drinking problems of subgroups. While drinking problems among American women are summarized below from the "Problem Drinkers" book, the primary emphasis in the discussion of problem drinking will be upon the drinking problem characteristics of the men aged twenty-one to fifty-nine as reported in the Cahalan and Room monograph (1974).

Since it will be seen that the correlates of merely *drinking* and of *having drinking problems* are often quite different, drinking practices and drinking problems will be discussed separately below.

Findings of the 1964–65 Drinking Practices Survey*

The first national (ADP) survey found that 68 percent of the adult population drink at least once a year: 77 percent of the men and 60 percent of the women. It is also found that 10 percent of the adult population used to drink but have stopped, and that 22 percent never have drunk alcoholic beverages. Of the sample studied 12 percent were described as heavy drinkers, including 21 percent of the men and 5 percent of the women. Gallup (American Institute of Public Opinion, 1966) and Mulford (1964) reported similar proportions of drinkers and nondrinkers.

It should not be concluded, however, that most American adults are regular drinkers. The adult population was found to be rather evenly divided between the 47 percent who do not drink at all or drink less than once a month, and the 53 percent who drink once a month or more.

Findings of this 1964–65 national survey are consistent with Gallup Poll trend data that indicate that the proportion of persons who drink has increased somewhat since World War II, particularly among women (American Institute of Public Opinion, 1966). The ADP and Gallup findings are congruent in that the ADP survey found a much larger proportion of younger women drank than was true for women over fifty (ADP, p. 22). However, a single study cannot establish whether the proportion of heavy drinkers in America is increasing or decreasing. That the per capita consumption of alcohol has increased rather modestly since World War II (Efron and Keller, 1970), in the face of an apparent increase in the proportion of persons drinking (especially among women), suggests that the increase in the number of drinkers may be balanced to some extent by a decrease in the proportion of heavy drinkers. This could occur, however, through the usual cutting-down or quitting by older people and the likelihood that the newer drinkers may drink relatively little. The findings do not necessarily imply any reduction in heavy drinking by the continuing drinking population. The current series of longitudinal studies should clarify the ways in which social pressures operate to bring about increases and decreases in the amount and kind of drinking.

---

* This section is largely drawn from the ADP monograph by Cahalan, Cisin, and Crossley, and from summaries prepared subsequently by Cahalan for the report to Congress on "Alcohol and Health" (NIAAA, 1971).

Correlates of Drinking

Differential drinking patterns occur among various groups. The summary (Figure 1) shows those groups most likely to be drinkers; and among drinkers, those most likely to be heavy drinkers (ADP, p. 189):

It is seen that drinking and heavy drinking tend to be distributed somewhat differently as regards social and occupational status, education, urbanization, and ethnic background. It should be kept in mind, however, that some of the differences in drinking behavior are relatively small, and also that some of the variables are themselves intercorrelated (e.g., occupation and education, urbanization, and ethnic and religious backgrounds). The following discussions of major variables will clarify some of these differences and interrelationships. Fuller details appear in the detailed report of the study (ADP).

Ethnocultural Groups

The United States no longer is a melting pot of recent immigrants: Three fourths of the adults interviewed in the 1964 survey reported that their fathers

Most Likely to be Drinkers:	*Among Drinkers, Most Likely to be Heavy Drinkers:*
Men under 45 years	Men aged 45–49
Men and women of higher social status	Those of lower social status
Professional, business, and other white-collar workers	Operatives; service workers
College graduates	Men who completed high school but not college
Single men	Single, divorced, or separated men and women
Residents of the Middle Atlantic, New England, East North Central, and Pacific areas	Residents of Middle Atlantic, New England, and Pacific areas
Residents of suburban cities, towns	Residents of largest cities
Those whose fathers were born in Ireland or Italy (age-adjusted findings)	Fathers were born in Ireland, Latin America, or the Caribbean, or the United Kingdom (age-adjusted)
Jews, Episcopalians	Protestants of no specific denomination; Catholics; those without religious affiliation

FIGURE 1. Relatively heavy drinkers are here rather arbitrarily defined as those who drink nearly every day with five or more drinks per occasion at least once in a while, or about once weekly with usually five or more per occasion. While these include most of those with severe drinking problems, not all of these so-called heavy drinkers have drinking problems.

were born in this country (ADP, p. 48). Nevertheless, ethnic background still plays an important part in determining patterns of American life, including drinking habits. Survey findings are in general congruent with Ullman's hypothesis (1958) that the rate of alcoholism will tend to be low within groups in which the drinking customs, values, and sanctions are well established, known, and agreed upon by all, and congruent with the rest of the culture; but the rate of alcoholism will tend to be high in groups with marked ambivalence toward alcohol. Lolli, Serianni, Golder, and Luzzato-Fegiz (1958) found that Italians in Italy and first-generation Italians in the United States drink very frequently but have low rates of alcoholism or problem drinking, but that subsequent-generation American Italians have higher rates of heavier drinking (see also Jessor, Graves, Hanson, and Jessor, 1968). Sadoun, Lolli, and Silverman (1965) found that the rate of alcoholism among the Italians is substantially lower than among the French, even though both countries have a wine-drinking culture. Snyder (1958, 1962) has discussed the ritual drinking of Jews in relation to their low level of alcoholism and alcohol-related delinquency; and Glad (1947) and Bales (1962) have presented research and anecdotal evidence regarding the high rate of alcoholism among the Irish.

These studies imply that there is no necessary correlation between widespread *drinking* and a high incidence of *drinking problems*. Blum (1967, p. 31) summarizes the essence of past studies as follows:

> ... When drinking is part of an institutionalized set of behaviors which include important other people in roles of authority and when drinking is part of ritualized or ceremonial activities (e.g., family meals, festivals, religious occasions, etc.) as opposed to leisure time or private use, it is not likely to be associated with high individual variability (unpredictability, loss of control) in conduct nor with the growth of drug dependency nor with the judgment by observers of "abuse" or "alcoholism." Further, when parents themselves reflect safe or model drinking behavior (i.e., are not problem drinkers), when drinking occurs shortly before or with food taking, and when the drinks used are wine or beer, the risks of either long- or short-term adverse effects are quite slim. Adverse effects nevertheless can still occur.

As can be seen in Figure 1, the findings of the national survey are congruent with these earlier studies of ethnocultural differences in drinking behavior. Respondents born outside the United States included a materially higher percentage of drinkers than did native-born respondents: but of moderate, rather than heavy, drinkers. Of those whose fathers were born outside the United States, 80 percent were drinkers, compared to 64 percent among those whose fathers were native born.

As shown in Figure 2, when standardized on age level, those whose fathers were born in Ireland had the highest proportions of both drinkers and heavy drinkers. Those whose fathers came from Latin America or the Carib-

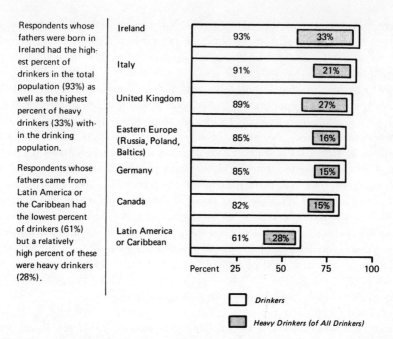

FIGURE 2. Percent of drinkers among adults (age 21+; standardized for age level) and heavy drinkers among all drinkers, by father's country of origin (selected countries) U.S.A. 1964-1965. (From NIAAA, "Alcohol and Health," 1971, p. 25.)

bean had the highest adjusted proportion of abstainers; but those of that group who drank had a high proportion of heavy drinkers.

Jews and Episcopalians had the lowest proportions of abstainers of any religious groups (less than 10 percent). Those who belonged to the more conservative or fundamentalist Protestant denominations had relatively high proportions of abstainers (48 percent), and relatively few heavy drinkers (7 percent).

Catholics had above-average proportions both of drinkers (83 percent) and heavy drinkers (23 percent of the drinkers).

Age. Among men, a majority in each age group up to sixty-five drank at least once a month. The highest proportions of heavy drinkers among the drinkers were found among men aged forty-five to forty-nine (40 percent). Among women, however, half or more of every age group either did not drink at all or drank less than once a month. The highest proportions of heavy drinkers among women drinkers were only about 15 percent, occurring at ages forty-five to forty-nine and twenty-one to twenty-four.

These apparent age trends are consistent with the findings of Glenn and

Zody (1970) in their analysis of Gallup Poll data from 1945 to 1960, in which the apparent maturing-out process of becoming an abstainer with advancing age was found to hold constant in separate measurements of different age cohorts over a fifteen-year period.

Social Status. The proportion of drinkers was lowest among women over sixty in the lowest social status group (34 percent) and was highest among young men aged twenty-one to thirty-nine in the highest status group (88 percent). The lower-status groups had smaller proportions of light and moderate drinkers than among upper-status persons, while the proportion of heavy drinkers was about the same at all social levels. Among drinkers, however, the proportion of heavy drinkers tended to be a little higher at the lower social levels, thus modifying perspectives on the "abstemious middle classes" (Dollard, 1945): relatively more of the well-to-do and middle-class people reported drinking at least occasionally; but fewer of those who did drink were heavy drinkers.

A relatively larger proportion of those of higher social status started drinking later in life, and also continued drinking to a more advanced age, than was true among those of lower status. This finding is consistent with the general differential in the phasing of various activities, remarked on by Kinsey, Pomeroy, and Martin (1948), whereby upper-status people tend to initiate certain activities associated with adulthood (e.g., sex, smoking, drinking) at a relatively later age than do those of lower status, but who tend also to continue to an older age.

Drinking seems to have different implications for people in lower-status groups than for those of upper status. Proportionately more of those of high than those of low social status drink, but drink moderately and see alcoholic beverages as a relatively harmless part of their life-style. On the other hand, a larger share of those of lower status see alcohol as harmful to themselves or to their families. A case could be made that alcohol is more likely to constitute a threat to those of lower status than those of higher status, on the grounds that the well-being of the lower-status person is more easily jeopardized by any untoward event. As Knupfer *et al.* have noted (1963), those of upper status have a much greater range of options in life, including less threat of being fired for showing the effects of drinking.

Regional Differences. There are considerable differences in regional drinking patterns. The highest proportions of both drinkers and heavy drinkers were found in the Middle Atlantic, New England, Pacific, and East North Central areas, all of which are heavily urban in population. The lowest percentage of drinkers occurred in the East South Central states, followed by other southern areas and the Mountain States. Factors that affect the high rate of abstention in the South include its less urban character and its higher proportion of members of conservative Protestant denominations that frown upon alcohol (ADP, pp. 37–38). Large cities had the highest proportion of drinkers, a finding also noted by Riley and Marden (1947) and Mulford (1964). Those drinking at least once a year ranged from only 43 percent among farm residents to 87 percent among those living in cities of fifty thousand to 1 million population that were not the central cities of their metropolitan areas (chiefly large suburbs). However, the highest rate of heavy drinkers among drinkers was found among residents of the very largest central cities.

The above findings are consistent with those of Mulford (1964) and with Malzberg's analysis of admission rates to hospitals for alcoholic psychosis (1960), which revealed geographic and urbanization patterns, marital status, occupational levels, and race similar to those shown for heavy drinking in the ADP survey.

What People Drink (ADP, Chap. 3)

Four out of ten respondents drank wine as often as once a year, but only 1 percent of the total sample were heavy drinkers of wine. Half of the respondents drank beer at least once a year; 7 percent were heavy drinkers of beer. Distilled spirits (either straight or in mixed drinks) were drunk by 57 percent at least once a year; 6 percent were heavy drinkers of spirits. One fourth said they drank all three beverages at least once a year.

Compared to the other two types of beverages, wine was drunk relatively more often by women than by men, by moderate than by heavy drinkers, and by persons of upper social status; by residents of the wine-producing Pacific and Middle Atlantic states, and by those living in larger suburbs. Beer was drunk by above-average proportions of the heavier drinkers, by men, and by younger persons. Spirits were drunk by relatively more of the heavier drinkers, those of upper status, men in their thirties and forties, and women in their twenties.

Behavioral Correlates of Drinking (ADP, Chap. 4)

Larger proportions of younger persons and those of higher social status had parents who drank frequently and who approved of drinking. Permissiveness of parents was generally correlated with the proportion of drinkers. Heavy drinking by women was closely correlated with heavy drinking by their husbands and vice versa.

In general, alcoholic beverages were served more often when people met socially with persons with whom they worked than even when they were with their close friends. Drinking occurred most often with friends (including those from work), next most often with family members, and least often alone. Heavy drinkers were more likely to say that they drank alone, but less than a quarter said they did so "fairly often." Thus, even heavy drinking tends to be social drinking, and the solitary heavy drinker is seen to be relatively rare.

Comparing the two extreme groups (abstainers and heavy drinkers) with the light drinkers found abstainers as more likely to be older people and lower than average in social status and income. Relatively more of them lived in the South and in rural areas, had native-born parents, belonged to conservative or fundamentalist Protestant denominations, and took part in religious activities frequently. Abstainers also tended to have a more gloomy perspective on life than others; but abstainers and heavy drinkers were not always poles apart, because both groups tended to be somewhat more alienated from society and more unhappy with their lot in life than were moderate drinkers.

ALCOHOL-RELATED PROBLEMS IN THE UNITED STATES

The George Washington University national surveys on problem drinking reported below were built upon the foundations established by a 1964 San Francisco survey of problem drinking (Knupfer, 1967; Clark, 1966), the Iowa surveys of Mulford and Miller (1960c), and a Washington Heights (New York City) survey (Bailey, Haberman, and Alksne, 1965). These national surveys represent an early stage in the investigation of problem drinking, focusing as they do primarily upon the correlations between certain types of drinking and specific types of problems. In order to establish more clearly whether the drinking caused the problem or the problem caused the drinking, two subsequent stages of research will be needed. The first will be the completion of the current longitudinal studies in which changes in people's lives between Time A and Time B are related to subsequent changes in drinking behavior or health or interpersonal relations. The final stage would be the conducting of controlled experimental studies in which alternative remedial measures are undertaken over a period of time, with ultimate results measured in terms of the effects of remedial measures upon problems related to drinking.

As mentioned earlier, two national surveys on drinking *problems* were conducted, the first being a 1967 follow-up of the 1964–5 survey of drinking practices reported upon earlier in this chapter, in which a subsample of 1,359 respondents (men and women of all ages twenty-one and over) were interviewed. The second survey on drinking problems was a special supplemental sample of men aged twenty-one to fifty-nine, interviewed in 1969 to provide additional interviews for analysis of this high-risk group. The 1967 findings are summarized from Cahalan (1970; designated as PD for "Problem Drinkers"), and the combined 1967 and 1969 findings for men aged twenty-one to fifty-nine are from Cahalan and Room (1972).

The studies here reported use the concept of *problems associated with use of alcohol* rather than that of "alcoholism." The concept utilized here subsumes Plaut's definition in the findings of the Cooperative Commission on the Study of Alcoholism: "Problem drinking is a *repetitive use of beverage alcohol causing physical, psychological, or social harm to the drinker or to others*. This definition stresses interference with functioning rather than any specific drinking behavior" (1967, pp. 37–38). This definition is consistent with the more general definition that "a problem—any problem—connected fairly closely with drinking constitutes a drinking problem" (Knupfer, 1967, p. 974).

These national surveys of drinking problems were designed to cover not only actual problems but *potential* problems as well, so as not to miss any significant problems. Thus, while fairly heavy drinking may not in itself constitute a problem, a high level of alcohol intake was included as a potential problem.

The national surveys of drinking problems have attempted to take into account the variables of severity of problems, certainty or reliability of measurement, the currency of the problem, and specific types of problems:

1. The severity of problems is assessed in terms of the frequency of the problem and the degree to which the problem exists (e.g., the number of times the respondent has had trouble with police over his drinking).

2. Certainty or reliability of measurement enters into the assessment of the drinker's problems in terms of the number of items or indicators of a problem (e.g., if five questions were asked concerning one's drinking problems vis-à-vis one's wife, the individual would be given a higher score on this problem than if the problem were reflected in responses on only two items).

3. Currency of problems was measured in terms of whether the problem had occurred during the last three years, or prior to that time.

4. A dozen types of specific problems were covered, which in turn may be combined into three general types: (a) potential problems related to high alcohol intake, (b) tangible consequences of either an interpersonal, a health/injury, or financial nature, and (c) miscellaneous potential problems and belligerence related to drinking).

Definitions of Problems

The types of problems or potential problems covered in this national survey were as follows:

Frequent Intoxication. A "high" score on this potential problem was attained by drinking a minimum of five or more drinks at least once a week; or eight or more drinks on one of the most recent two drinking occasions and twice in the last two months; or 12 or more drinks on one of the last two occasions and twice in the last year; or currently getting high or tight at least once a week. On this index of frequent intoxication 14 percent of the men, and only 2 percent of the women, had a "high" score.

Binge Drinking. This problem consisted of being intoxicated for at least several days at a time or for two days on more than one occasion. Binge drinking is one manifestation of Jellinek's "loss of control" (1960, p. 41), characterized as Epsilon alcoholism (p. 39). Relatively few (3 percent of men, and less than one half of 1 percent of women) qualified as binge drinkers.

Symptomatic drinking. This term refers to signs of Jellinek's Gamma alcoholism (1960, p. 37), the exhibition of signs of physical dependence and loss of control (e.g., drinking to get rid of a hangover, having difficulty in stopping drinking, blackouts or lapses of memory, skipping meals while drinking, tossing down drinks for quicker effect, or sneaking drinks). Positive responses on three or more of seven items were required to qualify respondents for a high score on this potential problem, which was attained by 8 percent of the men and 3 percent of the women.

Psychological Dependence. Drinking to alleviate depression or nervousness or to escape from the problems of everyday living constitutes psychological dependence. A high score (attained by 8 percent of the men and 3 percent of the women) could be acquired by rating at least one out of five psychological effects of alcohol as being very helpful or important, plus two others rated fairly helpful or important.

Problems with Spouse or Relatives. This category of problems included the spouse's leaving or threatening to leave the respondent, or becoming concerned over the respondent's drinking, or the spouse's or relative's asking the respondent to cut down on his drinking, or the respondent judging his drinking as having had a harmful effect on his home life. On this type of problem 8 percent of men, and 1 percent of the women, had a high score. Trouble with spouse or relative was usually accompanied by other drinking problems: Persons with a high score on spouse or relatives trouble had an average of 2.4 *additional* problems related to drinking.

Problems with Friends or Neighbors. These included the respondent's report that friends or neighbors had suggested he cut down, or that he himself felt his drinking had been harmful to his friendships and social life. Relatively few (only 2 percent of men, and less than one half of 1 percent of women) reported such problems in rather severe form, and only 7 percent of men and 3 percent of women reported *any* degree of this type of problem within the last three years. It is believed that such friends and neighbors problems are somewhat understated, to the extent that some heavy drinkers may gravitate toward friends or associates who do not disapprove of their drinking behavior.

Job Problems. The area of job problems includes having lost or nearly lost a job because of drinking, having co-workers suggest that one cut down on drinking, or rating oneself as having harmed one's work opportunities through drinking. On job problems 3 percent of the men and 1 percent of the women qualified as having a "high" score, through losing or nearly losing a job because of drinking, or having people at work suggest he should cut down, or if he felt drinking had harmed his career.

Problems with Law, Police, and Accidents. Such difficulties were reported by 1 percent.

Health. A high score in the category of health problems was based on reporting both that drinking had been harmful to health and that a physician had advised the respondent to cut down. Such health problems during the last three years were reported by 6 percent of the men and 4 percent of the women. That a health problem related to drinking often may be a relatively isolated problem is implied in the finding that half of those with a high score on the health problem did not achieve high scores on any other potential problems.

Financial Problems. A high score on the issue of whether drinking had had a harmful effect on their finances during the prior three years was reported by 3 percent of men and 1 percent of women.

Belligerence Associated with Drinking. Feeling aggressive or cross, or getting into a fight or heated argument after drinking, was reported in marked form by 4 percent of the women and 3 percent of the men. (It should be noted that this may be only a potential problem rather than necessarily an actual problem, for belligerence must be overt in order for it to have any immediate consequences.)

Figure 3 gives aggregate percentages, separately for men and women, for the 11 specific types of drinking problems plus a Combined Problems Score in the national survey.

The chief specific problems for men shown in Figure 3 were frequent intoxication, symptomatic drinking, psychological dependence, and problems with spouse or relatives. None of the specific problems showed a high score for women in excess of the 4 percent for health problems.

Each respondent, besides being scored on individual problems, was also given an overall combined score of moderate or high, taking into account all of the problems he reported having, and noting the severity. A surprising 31 percent of the total population had some degree of one or more problems connected with drinking within the last three years: 43 percent of the total male population and 21 percent of the women. Even when only more severe involvement is considered, 15 percent of the men and 4 percent of the women could be said to have drinking problems.

Many people had more than one drinking problem. Frequent intoxication

Drinking Behavior and Drinking Problems in the United States   95

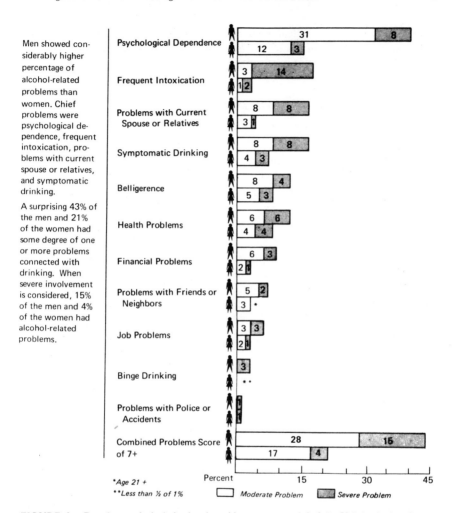

FIGURE 3.   Prevalence of alcohol-related problems among adults* in U.S.A. during 3 years prior to 1967. (From NIAAA, "Alcohol and Health," 1971, p. 31.)

was found most often to be accompanied by symptomatic drinking and psychological dependence, and binge drinking most often was associated with symptomatic drinking and problems with spouse or relatives. Problems with friends and neighbors were associated in most instances with frequent intoxication, symptomatic drinking, problems with spouse or friends, and health problems, which may be another way of saying that by the time friends and neighbors (who have no clear responsibility for the respondent) get around to remonstrating with him about his drinking, the individual usually will have accumulated a host of drinking problems in other areas of his life.

Drinking Problems among Men 21-59

The remainder of the discussion of drinking problems is concentrated upon survey results for men aged twenty-one to fifty-nine, where the rate of drinking problems is sufficiently high to make it easier to study subgroup differences. That this is a high-risk group is indicated by the finding that fully half reported at least a minimal problem, and one third had one or more fairly severe problems within the last three years (Cahalan and Room, 1974). Almost three fourths had had one or more problems with drinking at some time in their lives. Heavy intake and psychological dependence and wife problems were the most common, and police problems and loss of control least common.

Roughly twice as many men aged twenty-one to fifty-nine who report a *current* problem in any area will report having *ever* had a problem in that area. This implies a rate of remission of drinking problems far greater than could be accounted for by any formal agencies of intervention. This is consistent with the finding, in the national follow-up survey reported in 1970, that a substantial proportion of people go into—or out of—the problem-drinking population within a three-year period (Cahalan, 1970, pp. 114-115). These findings cast doubt on characterizations of drinking careers in terms of an irreversible progression or "snowball effect" (DeLint and Schmidt, 1968).

Table 1 shows that *all* types of drinking problems are most prevalent among men in the youngest age group, and that the percentage with a high current overall problems score among men aged twenty-one to twenty-four is almost twice as high (40 percent) as it is for any of the older groups. One infers from these findings that there is an apparent rapid decline in drinking problems after the age of twenty-five, and perhaps also that the seeds of longer-term serious problems with alcohol are usually sown by one's drinking habits in one's early twenties and not so much by habits not acquired until one's forties.

Socioeconomic Status

Most studies of drinking *problems* have found them more prevalent among the poor, even though *drinking at all* is more prevalent among those of higher status. Figure 4 bears out the higher prevalence of drinking problems among the poor, and particularly the *younger* men of lower social status (Hollingshead's Two-Factor Index of Social Position, 1957). Note also that while drinking problems appear to taper off rather regularly among upper-status men

TABLE 1. Higher Severity Level for Specific Current Problems (in Percent) by Eight Age Groups (Combined 1967–69 Surveys)

	21–24 (147)[a]	25–29 (204)	30–34 (186)	35–39 (216)	40–44 (226)	45–49 (201)	50–54 (199)	55–59 (182)	Total (1561)
1. Heavy intake	7	7	5	7	6	3	5	6	6
2. Binge drinking	10	3	3	3	1	2	4	2	3
3. Psychological dependence	5	4	4	4	4	5	4	3	4
4. Loss of control	12	5	4	5	7	5	4	4	6
5. Symptomatic drinking	26	11	8	7	6	10	9	3	9
6. Belligerence	15	12	10	8	7	8	6	2	8
7. Problems with wife	19	17	15	10	9	9	11	6	12
8. Problems with friends or neighbors	15	5	7	5	4	4	6	4	6
9. Problems on job	10	4	3	5	5	5	6	2	5
10. Police problems	10	4	2	2	2	2	4	1	3
11. Health or injuries from drinking	8	4	5	6	4	6	8	6	6
12. Finances	11	4	6	4	3	2	4	3	4
13. Current overall problems score 7+	40	22	20	21	17	17	17	11	20

[a] Number of respondents is shown in parentheses.

as they move into their forties, among lower-status men there actually is an apparent *rise* in problems at the age of forty-five to forty-nine.

The bottom line of Table 2 presents aggregate findings on a five-category *typology* of drinking problems, in which we distinguish between nondrinkers, drinkers without problems, those with potential problems only, heavy or binge drinkers, and those with tangible consequences of heavy drinking. Note that more than half either did not drink or had no drinking problems within the last three years; and that the heavy or binge drinkers who had no apparent consequences to their drinking (12 percent) were almost as numerous as the proportion with consequences (14 percent).

Table 2 bears out the higher prevalence of drinking problems among the *poor*. Note also that the ratio of consequences of drinking to heavy intake or binge drinking is highest among those of lower status and youngest age. In the words of the old music-hall song, "It's the poor wot gits the blyme."

The same typology is used for region, urbanization, and social position (see Table 3) and for ethnoreligious groupings (Table 4). Table 3 shows that abstention is almost uniformly higher for those living in rural areas, those living in "dryer" regions, and those of lower status, even after controlling for each

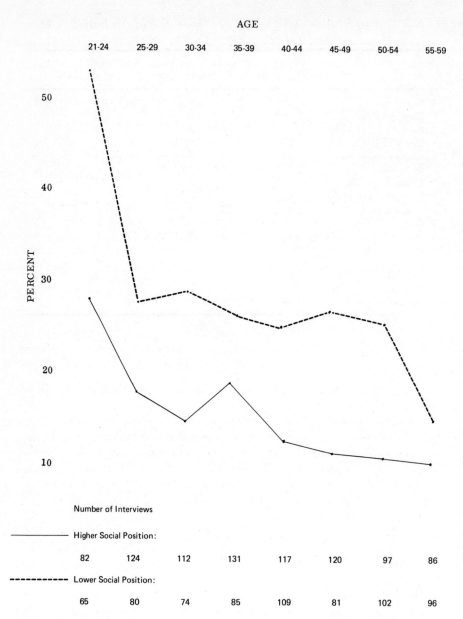

FIGURE 4. High current overall drinking problems score of 7+ by eight age groups within two socioeconomic groups, for men aged 21–59. Combined U.S. National Surveys of 1967 and 1969, by Social Research Group, The George Washington University and School of Public Health, UC Berkeley.

TABLE 2. Drinking Problems Typology (in Percent) for Four Social Position Groups,[a] Men Aged 21–59

	Total number	Non-drinker	Drank, no problems	Potential problems only	Heavy intake or binge, not consequences	High consequence score
Lowest social postion	281	19	22	19	14	26
Lower middle	411	17	38	19	10	17
Upper middle	401	14	39	24	15	8
Highest social position	468	11	47	22	11	9
Total	1561	15	38	21	12	14

[a] Key: High consequences score of 3+: tangible consequences, i.e., social consequences, health, or injury problems associated with drinking, or financial problems.
Intake or binge: not in above, but at least minimal severity intake or binge problems.
Potential problems only: not in the above groups, but a problem of at least minimal severity in any problem area.
Drank, no problems: has been a drinker within the last three years, but not in any of the above groups.
Nondrinkers: did not drink during last three years.

of the other variables. The historically dryer areas tend to have fewer drinkers; but people in the dryer areas tend more often to get into trouble when they *do* drink heavily. Table 4 bears out the historical connections between religious affiliation and use of alcohol found in earlier studies (Gusfield, 1962) and in our first national survey. Here it is shown that among Catholics and Liberal Protestants of most national origins there are relatively few abstainers and many heavy drinkers, that most Jewish men aged twenty-one to fifty-nine drink at least a little but that few drink heavily, and that conservative protestant denominations (those favoring complete abstinence) show a fairly high percentage of abstainers (ADP, pp. 55–61), but a relatively high ratio of high consequences score in relation to heavy intake or binge drinking. One infers that the conservative Protestants are subject to an above-average amount of social pressure against drinking and intemperate drinking, particularly if they are living in the dryer areas.

Other Factors Influencing Problem Drinking

In addition to the demographic factors discussed above, there are many environmental and personality variables found to be related to drinking behavior. Various writers have emphasized a wide variety of such intervening

TABLE 3. Current Problems Typology by Region, Urbanization and Social Position,[a] in Percent (Combined 1967–69 Surveys)

	Total number	Non-drinker	Drank, no problems	Potential problems only	Heavy intake or binge not consequences	High (3+) consequences score
Higher ISP						
Wetter regions:[b]						
Central cities	154	8	35	28	20	9
Other cities and towns	269	6	49	23	15	7
Rural areas	93	11	46	26	9	9
Total	516	7	44	25	15	8
Dryer regions:[c]						
Central cities	112	15	38	24	12	11
Other cities and towns	86	14	44	23	9	9
Rural areas	155	28	43	17	5	7
Total	353	21	42	21	8	9
Lower ISP						
Wetter regions:						
Central cities	189	11	25	24	14	27
Other cities and towns	118	13	34	18	20	16
Rural areas	86	17	42	19	11	12
Total	393	13	31	21	15	20
Dryer regions:						
Central cities	72	13	21	14	11	42
Other cities and towns	77	18	44	17	8	13
Rural areas	150	33	29	15	6	17
Total	299	24	31	15	8	22

[a] Key: See Table 2.
[b] Wetter regions: New England, Middle Atlantic, East North Central, and Pacific States.
[c] Dryer regions: South Atlantic, E. South Central, W. South Central, W. North Central, and Mountain States.

variables, as reviewed by Cahalan (1970, pp. 63–95). Plaut (1967, p. 49) summarizes the role of such variables as follows:

> A tentative model may be developed for understanding the causes of problem drinking, even though the precise roles of the various factors have not yet been determined. An individual who (1) responds to beverage alcohol in a certain way, perhaps physiologically determined, by experiencing intense relief and relaxation, and who (2) has certain personality characteristics, such as difficulty in dealing with and overcoming depression, frustration, and anxiety, and who (3) is a member of a culture in which there is both pressure

to drink and culturally induced guilt and confusion regarding what kinds of drinking behavior are appropriate, is more likely to develop trouble than will most other persons. An intermingling of certain factors may be necessary for the development of problem drinking, and the relative importance of the differential causal factors no doubt varies from one individual to another.

The concepts used in analyzing the correlates of drinking problems in the national surveys reported in this chapter are congruent with Plaut's model of problem drinking. They are adapted to a considerable extent from the model of Jessor and his associates in their tri-ethnic community survey of behavior deviancy including problem drinking (1968), which drew upon Merton's

TABLE 4. Current Problems Typology by Ethnoreligious Categories,[a] in Percent (Combined 1967–69 Surveys)

Ethnoreligious category[a]	Total number	Non-drinker	Drank, no problems	Potential problems only	Heavy intake or binge not consequences	High (3+) consequences score
British						
Catholic	34	12	35	27	18	9
Liberal Protestant	48	13	31	38	8	10
Conservative Prot.	204	20	39	22	3	16
Irish						
Catholic	77	4	33	27	16	21
Conservative Prot.	74	27	46	15	3	10
German						
Catholic	76	7	40	21	22	11
Liberal Protestant	86	12	47	24	9	8
Conservative Prot.	120	22	42	21	7	9
Italian						
Catholic	64	5	52	14	23	6
Latin-American						
Catholic	42	10	10	21	17	43
Jewish	40	8	60	25	5	3
Black						
Conservative Prot.	97	18	23	16	13	31
Eastern European						
Catholic	71	6	38	21	21	14
Other ethnicity						
Catholic	114	7	42	26	11	13
Liberal Protestant	56	5	43	18	23	11
Conservative Prot.	158	34	35	15	8	8

[a] Key: See Table 2.
[b] Categories with small n's are omitted from this table. Ethnicity is defined by religion for Jews, by race for blacks, by "country most ancestors come from" for the remainder.

theory of social structure and anomie (1957), Cloward and Ohlin's theories on delinquent gangs (1960), and Rotter's "social learning" theories (1954). Jessor and his associates summarize their intervening correlates of deviance as follows: "The likelihood of deviant behavior will vary directly with the degree of personal disjunction, alienation, belief in external control, tolerance of deviance, and tendencies toward short time perspective and immediate gratification characterizing an individual at a given moment in time" (op. cit., p. 111).

Table 5 presents a multiple correlation based on all of the intervening and demographic variables utilized in the combined national data for men twenty-one to fifty-nine, against overall problem drinking scores. The principal finding here is that the intervening variables involving drinking attitudes and drinking environments were understandably correlated most highly with problem drinking. It can be argued that the association between one's attitudes and the drinking behavior and attitudes of one's associates vis-à-vis one's own problem drinking may be as much the result of one's prior heavy drinking as the "cause" of it. However, support for the hypothesis that drinking attitudes may "cause" later drinking problems is found in the earlier two-stage study in the same series, in which attitudes toward drinking were found to precede drinking problems more often than drinking problems preceded the development of favorable attitudes toward drinking (PD, p. 125).

The fact that environmental factors such as the permissiveness of one's family and one's socioeconomic status and ethnic background are the leading correlates of problem drinking may suggest that preventive and remedial programs be tailored to take into account these background characteristics. The findings are also consistent with those of M. C. Jones (1968), Bailey et al., (1965), Zucker (1968), and Williams (1965), in which rebelliousness, undercontrol, and alienation and low self-evaluation were found to be related to prob-

TABLE 5. Multiple Correlation of 51 Variables Against Combined Problems Score of 7+ (1,561 Men Aged 21–59)

Step No.	Multiple correlation[a]	Partial correlation	Simple correlation (Pearson r)
1. Drinking by significant others	.26	.15	.26
2. Tolerance of deviance	.33	.13	.22
3. Own attitude toward drinking	.37	.18	.26
4. Index of social position	.41	.11	.16
5. Black	.42	.09	.15
6. Nonhelpfulness of others	.43	.06	.16

[a] Cumulative multiple correlation for 51 variables (including demographic and intervening variables) was .47, or 22 percent of the total variance in problem drinking.

lem drinking. However, the findings are also not inconsistent with the earlier findings of Syme (1957), Armstrong (1958), and Rosen (1960) to the effect that the problem drinker does not represent a unique personality or psychiatric type.

Patterns of Change in Drinking Behavior

The magnitude and character of change in drinking habits over time are especially crucial issues in the planning of preventive and remedial programs to control problem drinking. While this series of studies is yet to be completed, the available evidence clearly indicates that many people change their drinking behavior markedly over a span of time. Half reported they had changed their drinking habits since starting to drink, either by quitting or by drinking more or less (ADP, chap. 5). Even one third of the abstainers said they used to drink (ADP, p. 113). The 1967 national survey found that as many persons had fairly severe drinking problems only prior to three years ago as had drinking problems within the last three years (PD, p. 119). The retrospective data on past changes in drinking behavior are borne out by separate measurements in the 1964–65 and 1967 surveys: During that short period 15 percent of the total persons interviewed had moved into—or out of—the group reporting themselves as drinking five or more drinks per occasion at least some of the time (ibid., p. 145).

The finding that even those with severe drinking problems do show considerable fluctuation in their drinking implies that the average problem drinker has occasions when it is easier to bring his drinking under better control. The process of maturing out of problem drinking appears to be more successful among upper-status men in their forties and fifties and among people in smaller towns and rural areas (ibid., p. 143). An analysis of men aged twenty-one to fifty-nine in the combined national surveys also found that those who matured out of problem drinking fastest (on the basis of retrospective reports) were more secure in their jobs, less impulsive, and lived in an environment that was less permissive of heavy drinking than was true for those who were still problem drinkers (Cahalan and Room, 1974, p. 159). Recollections of changes in problem drinking in a 1964 San Francisco study found that those who stayed in the problem drinker class tended to show more signs of childhood stress, drinking problems in the home while growing up, anxiety, depression, maladjustment, guilt, and need for approval (Knupfer, 1971, p. 24).

Summary of Survey Findings

1. In the United States, in most areas drinking is typical behavior, and both abstinence and heavy and problem drinking are atypical.

2. The proportions of those who drink are highest among those of upper social status; but the proportions of heavy and problem drinkers *among drinkers* are highest among those of lower status.

3. Heavy drinking and all types of drinking problems are at their height among men in their early twenties, indicating that the drinking habits and attitudes that are followed by later drinking problems probably become established at a relatively early age.

4. The proportion of drinkers (especially among women) appears to have increased since World War II. However, there is little solid evidence as to whether problem drinking is on the increase or decrease.

5. Whether a person drinks at all is primarily a sociocultural variable, as shown in the great differences in drinker status by such variables as sex, age, social status, region, degree of urbanization, and religion. However, certain personality measures are found useful in explaining some of the variations in heavy and problem drinking; these include such measures as alienation, neurotic tendencies, and impulsivity.

6. There is a fairly high turnover in the problem drinking status of many individuals, in addition to a general tendency for older persons to drop out of the drinking and problem drinking classes. The implications are that the average problem drinker frequently presents opportunities to help him to mobilize his resources to bring his drinking under better control. In short, problem drinking may not be nearly as intractable as it is commonly supposed.

EPIDEMIOLOGIC AND BEHAVIORAL SCIENCE PERSPECTIVES

Alternative Models for Studying "Alcoholism" or "Problem Drinking"*

At least three models for the definition of "alcoholism" or "problem drinking" coexist at the present time; and these models have widely varying implications for the epidemiology and treatment of alcoholism. The three models are the "disease" model, the "vice" model, and the "social problems" model. Each model is intended to be highly functional. The disease model was developed with the primary motive of getting the derelict alcoholic more decent treatment than being dumped into the drunk tank in the city jail, only to repeat the same process until illness, malnutrition, and accidents have taken their toll. The vice model is the old-fashioned, puritanical assumption that "Any of these drunks could straighten out and become decent citizens if they only tried." Often this vice model becomes linked with the disease model in that the patient or client is

* Part of this discussion is adapted from chap. II of Cahalan and Room (1974).

given only meager and reluctant medical aid, under demeaning and punishing circumstances. The social problems model assumes that drinking problems arise primarily out of the stresses and disjunctures of society; and that if environmental arrangements and controls were just managed properly, there would be far fewer advanced alcoholics.

While adherents of each of these models tend to claim for their model a sweeping applicability to most of the field of whatever they define as alcoholism, the scope of each of the models is limited. The vice model deals primarily with the behavior of the alcoholic: heavy intake, binge drinking, and perhaps symptomatic drinking behaviors and belligerence. The disease or medical model emphasizes primarily the sufferer's mind (loss of control and psychological dependence) or body (health consequences of excessive drinking). The primary focus of the social problems model is upon such social consequences or correlates of heavy drinking as problems with one's wife, with friends, relatives, on the job, and with the police.

As background for the methodological discussion that follows, we are here enumerating some of the components of the three basic models. For a more detailed discussion of such models, the reader should consult Siegler, Osmond, and Newell (1968).

1. The vice model subsumes the following:

(a) The crime model, in which disapproved behavior is dealt with by the processes of criminal justice and other formal sanctions.

(b) The bad habits model, in which the person is to be retrained toward self-control. Two widely divergent philosophies may become intertwined in the use of the bad habits model, one being the classical, moral mind-over-matter or free-will process of improvement in which the individual is expected to take most of the initiative; and the other being the behaviorist or operant point of view, in which the determinism of the conditioning process is manipulated by the trainer with the active consent and understanding of the trainee. There are great variations within this habit model in the degree of voluntariness of treatment as well as in the types of treatment, which range from moral suasion through various types of positive reinforcements and such negative conditioning as the use of electric shock or emetine.

2. The disease model has two somewhat different facets:

(a) The classical disease concept, in which compulsive drinking is defined as a disease of the mind or will. The chief symptom utilized in defining the disease is loss of control. The primary method of treatment often has a moral tone almost indistinguishable from some behavior therapy methods, with emphasis upon group reinforcement and other social supports for sobriety. Thus, the classical precepts of Alcoholics Anonymous (and the AA's championing of the disease model) are tailor-made for promoting a close alliance between the physician and AA, with the AA referring patients to the physician for treatment of withdrawal and other medical symptoms, and the physician referring the patient to AA for further group therapy and aftercare of a sort. Because psychiatric therapy for individual patients is generally found to be too time-consuming or otherwise professionally unrewarding for the average physician (including the average psychiatrist), there is increasingly more recourse to group therapy methods after the client has regained sobriety and his withdrawal symptoms have diminished.

(b) The physiological-disease or medical-consequences submodel, which concentrates on treating such consequences of heavy drinking as cirrhosis and delirium tremens.

3. The social problems model may draw upon one of three alternative sociological perspectives:

(a) Labeling theory, which puts primary emphasis not upon the individual's *initial* behavior, but upon the social forces that single out and label the individual, or in which the individual may label himself as a deviant. The consequences of the labeling process are seen as a widening of the gap between the labeled individual and the mainstream of society, in which he either seeks out other deviants like himself for social support or becomes an isolate. The implications of labeling theory are that the process of identifying the alcoholic needs to be handled very carefully in order that the individual may not be permanently damaged.

(b) Anomie theory, which has at least two subtypes: that conflicting norms produce deviant behavior, and that an absence of norms produce deviance (Mizruchi and Perrucci, 1970). Anomie theory would put considerable emphasis on the role of cultural ambivalence and confusion of norms in encouraging the development of alcoholism.

(c) The "differential association" process hypothesized by Sutherland (1955) lays emphasis upon the role of the permissiveness of the individual's subculture in his drinking behavior. A historical illustration of the application of this hypothesis was the Temperance Movement's focus upon the saloon or tavern as the transmitter and reinforcer of a heavy-drinking subculture.

Each of the models implies both a different etiology of alcoholism and also a different type of prevention or treatment. Each of the models taken by itself is grossly inadequate to explain much of the variance in drinking problems among subgroups in American society, as will become clear upon re-examination of the implications of the survey evidence presented earlier.

The limitations of the disease model have been reviewed at some length previously by Cahalan (1970) and will not be repeated here in any detail. Suffice it to say that many authorities are of the opinion that while obviously alcoholism has medical implications and while obviously alcohol is one of the most dangerous drugs from a medical standpoint, alcoholism has more of the manifestations of a chronic illness or maladjustment than of an acute infectious or physiological-deficiency disease. Further, the classical nosology of Jellinek (1952), which divides alcoholism up into Alpha, Beta, Gamma, Delta, and Epsilon types has been questioned by Room (1970) and Hoff (1968), who show that the early Jellinekian progression of symptoms (e.g., from gulping drinks and morning drinking to blackouts and to helpless addiction) does not occur in the real world sufficiently often to have any practical significance for diagnosis or treatment. Also, many writers such as Reinert (1968), Roman and Trice (1967), Scott (1968), Wexberg (1951), Chafetz (1966), and Seeley (1962) point out that although the disease model of alcoholism was adopted primarily to get better treatment for the problem drinker, the disease model tends to be counterproductive to a considerable extent on the various grounds that do not account for much of the behavior and symptoms of the sufferer. It may be used as a crutch by the alcoholic who is unwilling to change his habit-patterns, and it implies an all-or-nothing division point between alcoholics and nonalcoholics that may be evaded by the problem drinker who stoutly maintains, "I'm not an alcoholic yet."

The unworkability of the vice model of alcoholism seems patently obvious: Our nation in modern times has decided, through its courts and social agencies, that we will no longer hold the drunkard legally responsible merely for being intoxicated in public (NIAAA, 1971, chap. VII). Further, it has dawned upon the American consciousness generally that mere punishment or public shame seldom reforms the alcoholic.

The social problems model of alcoholism in turn has its limitations, for the stresses of modern society certainly cannot be sufficient by themselves to explain alcoholism, as evidenced by the fact that Jews (who generally live in an urbanized and competitive business or professional world with a high level of anxiety-evoking stresses) seldom develop chronic alcohol problems despite the fact that most of them, according to our national surveys, do drink at least occasionally and thus are at potential risk.

Particularly in such an emotionally laden area as alcohol problems, epidemiologists and behavioral scientists must be particularly cautious lest they fall into ways of thinking that are influenced more by ideological commitments than by facts. In the field of alcoholism, one discerns a pressure from many social planners and practitioners in the area of treatment to resort to simplistic single-cause explanations for alcoholism, despite the fact that no single cause has been found to explain the development of alcoholism. Alcoholism in the form of deep-seated addictive drinking (accompanied by withdrawal symptoms when drinking is interrupted) is almost invariably the outcome of many years of very heavy drinking, under circumstances permissive of very heavy drinking, lacking in sufficient incentives for the individual to moderate his drinking, and perhaps accompanied by certain physiological liabilities (including sometimes the capability and hardihood to be able to ingest huge quantities of alcohol). In order to be able to explain most chronic problem drinking, it is necessary to adopt a "system" approach not unlike that used in understanding the mixed etiology of such chronic conditions as tuberculosis, where there is indeed a beginning "cause"—the bacillus—but a bacillus that is omnipresent and that cannot result in full-blown tuberculosis unless extreme stresses and deprivations are present for protracted periods of time.

It should be emphasized again that since there are a host of types of problems associated with drinking—such as health problems, risk of accidents, and deterioration of social relations and economic status—it is unlikely that any one etiology can explain all alcohol-related problems, nor that any one treatment modality is appropriate. Obviously a different approach to prevention of adverse drinking consequences is necessary in the rural South, where drinking tends to be sporadic and hectic when it occurs, and in the coastal metropolitan centers, where cirrhosis rates are high. And obviously a different approach will be needed in dealing with the youthful problem drinker (hopefully, even before his drinking habits become deeply ingrained) than in dealing with the chronic alcoholic derelict. However, many influences have teamed up to slow down progress in appropriate prevention and treatment methods.

One influence is the pervasiveness of fairly heavy drinking among middle-aged influentials, in which the difference between the politician and the statesman often lies in the latter's facility in holding his liquor with more aplomb. It is suspected that many physicians are reluctant to pry into a patient's drinking history not only for fear of giving offense, but because the patient is not drinking much more heavily than the physician, and thus the physician finds it difficult to see where the seat of problems lies in heavy drinking. Another influence is the legacy of laissez-faire individualism, in which one does not make other people's problems one's business until one is officially called upon to do so. Another related factor militating against early identification of drinking problems is an American aversion to being labeled a bluenose or an anti-civil libertarian, a legacy from our sad experiences with the excesses of Prohibition. Finally, the difficulty in formulating and applying appropriate preventive methods in relation to alcohol problems lies very much in the ordinarily very slow process of the evolution of severity of drinking problems over many years time. As we have seen earlier, the peak in incidence of drinking problems is in the early twenties; yet the average institutionalized labeled alcoholic is in his forties. The length of time it takes to become a labeled alcoholic is so long that it makes it difficult to motivate legislators and social planners to put into effect the really massive preventive educational and early case-finding programs that will be necessary if we are to hope to reduce the incidence of the more troublesome kinds of problems related to drinking.

Contributions to Social Control

What can the society do to establish controls over drinking problems? More specifically, how can the skills of the epidemiologists and the behavioral scientists be brought to bear in a cooperative effort to achieve the universally acknowledged objective of minimizing the burden placed by problem drinking on the individual and the society in terms of survival, economic cost, and the quality of life?

As a start, we could do with a bit less dogmatism. As indicated above, adherents of competing models for the definition and explanation of alcohol problems frequently fail to communicate with one another because they are indeed concentrating on different phenomena; and they are frequently more concerned with the exclusion of alternative models than they are with the expansion and accommodation of their models to fit the facts, or with the development of composite models that might work better.

Certainly the disease model has served a useful purpose in helping to achieve better treatment for the problem drinker and more understanding of the fact that excessive drinking is not merely a failure of will power. Its popularity has had another effect in the area of social control: By implication, the disease model delegates responsibility for the management of the problem to the medical subsystem of the society; it specifies the physician as the appropriate social agent for the handling of individual cases, and it allocates to the public health subsystem the initiation, execution, and evaluation of programs at the societal level. Delegation of authority and responsibility is comforting; once again we have entrusted our problem to a highly educated, socially motivated, trustworthy, and trusted professional group. Moreover, the allocation of praise and blame is simplified, since we now know whom to credit with success and whom to blame for failure.

The development of epidemiology as a discipline is closely linked to the development of medical science and practice; and, in fact, many epidemiologists hold medical degrees. As a consequence, epidemiologists tend to share many of the values of the medical profession. One of these values, linked to the phenomenal success and prestige of the medical profession in preventive medicine during the first half of the twentieth century, includes a reliance upon single-point intervention to disrupt a chain of infection or deterioration, such as drying up the source of infections by treating typhoid-laden water, spraying anopheles mosquitoes, destroying *aftosa*-infected cattle, and the like. Traditional epidemiologists have viewed the causal chain as a closed loop, consisting of an agent that can infect the host, but only in a suitable environment. Given this point of view, to the traditional epidemiologist the reduction of drinking problems is a relatively simple issue in which the "agent" in the host-agent-environment paradigm is subject to simple control. As Terris has pointed out

(1968), the most direct way to reduce liver cirrhosis is to reduce the consumption of alcohol sharply; and he notes that in this regard, Prohibition was indeed very much of a success.

That the disease concept tends to lead to simplistic and socially unacceptable solutions such as Prohibition certainly weakens its appeal. On the other hand, it seems patently obvious that a labeling theory of problem drinking drives relativity into a *reductio ad absurdum*. In its extreme form, labeling theory denies the reality of the phenomenon of problem drinking and argues that (given any drinking at all) the problem is essentially an aesthetic one in the mind of the beholder. While such a view is certainly consonant with the nonethical stance of scientific research, it seems to deny to the society any right to impose a social aesthetic—and no organized society can be expected to tolerate behavioral anarchy.

Are there any grounds for rapprochement? Certainly, most epidemiologists and most behavioral scientists in the field of drinking agree on the multiplicity of the problem. The question of whether the multiple problems are facets of a single entity worthy of a class-name or whether the multiple problems are indeed separate and perhaps even unrelated phenomena can be readily seen as a problem of semantics that need not interfere with a cooperative effort toward social control. Although there may be vigorous disputes about priorities, it would seem that most epidemiologists and most behavioral scientists in this field would have little trouble in agreeing that a program of social control of drinking problems must involve:

1. *Prevention in the form of early education.* School programs that merely identify abstinence with virtue are simply out of tune with reality. Similarly, parents who communicate unhealthy attitudes to their children (emphasizing abstinence until some magic turning point is reached)—especially parents who communicate to their children their own unhealthy guilt with respect to drinking—simply have to be taught (or their children have to be taught) to handle this subject in a more intelligent manner. As we have said many times, we are not advocating that children should be taught to drink. They should be taught, however, that most people in our society do drink and that the vast majority of those who drink do so without harmful consequences. Since this is true, perhaps we can teach about drinking the way we teach swimming and driving, not ignoring the dangerous aspects of the activity, but by emphasizing the precautions required to avoid danger.

2. *Early case finding through self-identification.* Attempts to define danger signs have usually ended up with lists of behavior that would describe the "far-gone" alcoholic rather than the person who (perhaps twenty years younger) is showing the first signs of potential problems. Widespread publicity concerning behaviors that seem to lead to trouble could encourage self-exami-

nation and early awareness of problems *and what to do about them,* without magnifying the latent guilt of the problem-free drinker.

3. *Guidance and therapy.* The success of a program of early case finding depends entirely on the adequacy of guidance and therapy programs for the found cases. This area is perhaps the weakest link in the chain. Whether the troubled person turns to his clergyman, his social worker, or his physician, what is the probability that he can be helped? Unless we are ready to throw up our hands in despair, the increase in the probability of rehabilitation must be the primary focus of our society's control system. Our efforts to legitimate problem drinking as a medical phenomenon and to shift the burden from the will of the drinker to the skill of the helping agent have been reasonably successful. However, until we are ready to instruct the helping agents with respect to helping techniques, we are wasting our time. Whatever other disagreements may exist in this field, no one disagrees with the thesis that research on rehabilitative procedures deserves primary emphasis.

The proposals enumerated above are so obvious that we can expect support for them from epidemiologists and from behavioral scientists who may still have serious disagreements on models, on methods, on definitions, on etiologies, and indeed on ideologies. Ultimately, the disagreements that remain may involve quite fundamental philosophic problems that tend to separate decision-makers from civil libertarians. These are not merely communications difficulties, nor are they merely reflections of the eternal conflict between value-laden policy research and value-free scientific research. The fact is that as a society we suffer from severe ambivalence with respect to the right of self-destruction; more specifically, as a society we seem to behave inconsistently with respect to intervention when the behavior in question is viewed less as a threat to social order and more as a threat to the individual perpetrator. An excellent example of this ambivalence is found in official acts with respect to cigarette smoking. On the basis of expert interpretation of the evidence, it seems clear that heavy cigarette smoking is dangerous at least to the smoker. The obvious (though perhaps repressive) move would be to outlaw the agent; but such a move is doomed to failure not only for political and economic reasons but also because many of us interpret our civil liberties to include the right to choose dangerous behavior. So, instead of outlawing the agent, we have pecked away at the problem, first by appealing to the rationality of the smoker in the form of a warning on the product; and second (in a marvelous confession of ambivalence), by outlawing the advertising of the product—but only in certain of the mass media.

However, if we are lucky, this kind of ambivalence will not interfere with mechanisms for the social control of problem drinking. As a society we are clear that outlawing the agent simply will not do: Alcohol serves socially desirable functions that we are not willing to sacrifice. At the same time, there is unquestioned consensus is government, in industry and in the public at large that social control of problem drinking is legitimate and necessary. So far at least, the ambivalence that is reflected in anti-smoking compaigns has not impeded campaigns against problem drinking and its social consequences.

We believe that it is essential for epidemiologists and other medically oriented personnel and behavioral scientists to learn to work together in harmony and mutual understanding as aides in the social planning process, in order to bring about effective preventive programs related to reducing alcohol-connected problems. In doing so, it is important that all investigative disciplines learn to recognize their own ideological, methodological, and subcultural hang-ups; and, particularly, that they learn to keep a clear distinction between the dual roles in which they are cast, first as scientists and second as makers of public policy.

REFERENCES

American Institute of Public Opinion, 1966, Gallup political index; political, social and economic trends, Report, Princeton, N.J.
Armstrong, J. D., 1958, The search for the alcoholic personality, *Annals of the American Academy of Political and Social Science* 315:40–47.
Bacon, S. D., 1962, Alcohol and complex society, *in* "Society, Culture and Drinking Patterns" (D. J. Pittman and C. R. Synder, eds.), pp. 78–94, Wiley, New York.
Bailey, B., Haberman, W., and Alksne, H., 1965, The epidemiology of alcoholism in an urban residential area, *Quart. J. Stud. Alc.* 26:19–40.
Bales, R. F., 1962, Attitudes toward drinking in the Irish culture, *in* "Society, Culture and Drinking Patterns" (D. J. Pittman and C. R. Synder, eds.), pp. 157–187, Wiley, New York.
Blane, H. T., Overton, W. F., and Chafetz, M. E., 1963, Social factors in the diagnosis of alcoholics. I. Characteristics of the patient, *Quart. J. Stud. Alc.* 24:640–663.
Blum, H. (assisted by Lauraine Braunstein), 1967, Mind-altering drugs and dangerous behavior: Alcohol, *in* "Task Force Report: Drunkenness" (Appendix B), President's Commission on Law Enforcement and Administration of Justice, pp. 29–49, U.S. Government Printing Office, Washington, D.C.
Cahalan, D., 1970, "Problem Drinkers," Jossey-Bass, San Francisco.
Cahalan, D., Cisin, I. H., and Crossley, H. M., 1969, "American Drinking Practices: A National Survey of Drinking Behavior and Attitudes," Monograph No. 6, Rutgers Center of Alcohol Studies, New Brunswick, N.J.
Cahalan, D., Cisin, I. H., Kirsch, A. D., and Newcomb, C. H., 1965, Behavior and Attitudes Related to Drinking in a Medium Sized Urban Community in New England, Social Research Group Report No. 2, George Washington University, Washington, D.C.
Cahalan, D., and Room, R., 1974, "Problem Drinking Among American Men," Monograph No. 7, Rutgers Center of Alcohol Studies, New Brunswick, N.J.
Chafetz, M., 1966, Alcohol excess, *Annals of the New York Academy of Sciences,* 133:808–813.
Chafetz, M. E., 1971, The Problem of Alcoholism in the United States, Paper presented at the International Symposium on Alcoholism and Drug Addiction, Zagreb, Yugoslavia, October 1.

Cisin, I. H., 1963, Community studies of drinking behavior, *Annals of New York Academy of Sciences* 107:607-612.
Clark, W. B., 1966, Operational definitions of drinking problems and associated prevalence rates, *Quart. J. Stud. Alc.* 27:648-688.
Cloward, R. A., and Ohlin, L. E., 1960, "Delinquency and Opportunity," Free Press, New York.
DeLint, J., and Schmidt, W., 1968, The distribution of alcohol consumption in Ontario, *Quart. J. Stud. Alc.* 29:968-973.
Dollard, J., 1945, Drinking mores of the social classes, *in* "Alcohol, Science and Society," pp. 95-101, *Quart. J. Stud. Alc.,* New Haven.
Efron, V., and Keller, M., 1970, Selected statistics on consumption of alcohol (1850-1968) and on alcoholism (1930-1968), Rutgers Center of Alcohol Studies, New Brunswick.
Glad, D. D., 1947, Attitudes and experiences of American-Jewish and American-Irish male youth as related to differences in adult rates of inebriety, *Quart. J. Stud. Alc.* 8:406-472.
Glenn, N. D., and Zody, R. E., 1970, Cohort analysis with national survey data, *The Gerontologist* 10:233-240.
Gusfield, J. R., 1962, Status conflict and the changing ideologies of the American temperance movement, *in* "Society, Culture and Drinking Patterns" (D. J. Pittman and C. R. Snyder, eds.), pp. 101-121, Wiley and Sons, New York.
Haberman, P. W., and Scheinberg, J., 1969, Public attitudes toward alcoholism as an illness, *American Journal of Public Health* 59:1209-1216.
Harris, L. and Associates, Inc., 1971, American attitudes toward alcohol and alcoholics, Prepared for the National Institute on Alcohol Abuse and Alcoholism.
Hoff, E., 1968, The Alcoholisms, Paper presented at the 28th International Congress on Alcohol and Alcoholism, September, Washington, D.C.
Hollingshead, A. B., 1957, Two-Factor Index of Social Position (mimeo), New Haven.
Jellinek, E., 1952, Phases of alcohol addiction, *Quart. J. Stud. Alc.* 13:673-684.
Jellinek, E. M., 1960, "The Disease Concept of Alcoholism," Hillhouse Press, Highland Park, N.J.
Jessor, R., Graves, T. D., Hanson, R. C., and Jessor, S. L., 1968, "Society, Personality, and Deviant Behavior: A Study of a Tri-Ethnic Community," Holt, Rinehart and Winston, New York.
Jones, M. C., 1968, Personality correlates and antecedents of drinking patterns in adult males, *Journal of Consulting and Clinical Psychology* 32:2-12.
Keller, M., 1971, Ethanol: The basic substance in alcoholic beverages, *in* "Alcohol and Health," First special report to the U.S. Congress on, National Institute on Alcohol Abuse and Alcoholism, from the Secretary of Health, Education and Welfare, Government Printing Office, Washington, D.C.
Kinsey, A. C., Pomeroy, W. B., and Martin, C. E., 1948, "Sexual Behavior in the Human Male," Saunders, Philadelphia.
Knupfer, G., 1967, The epidemiology of problem drinking, *American Journal of Public Health* 57:973-986.
Knupfer, G., 1971, Ex-Problem drinkers, Presented at the Fourth Conference on Life History and Psychopathology, November, St. Louis, Missouri.
Knupfer, G., Fink, R., Clark, W. B., and Goffman, A. S., 1963, Factors Related to Amount of Drinking in an Urban Community, California Drinking Practices Study, Report No. 6, Division of Alcoholic Rehabilitation, State Department of Public Health, Berkeley, California.
Lolli, G., Serianni, E., Golder, G. M., and Luzzato-Fegiz, P., 1958, "Alcohol in Italian Culture; Food and Wine in Relation to Sobriety Among Italians and Italian Americans" (Monographs of the Rutgers Center of Alcohol Studies, No. 3), Rutgers Center of Alcohol Studies, New Brunswick, N.J.

Malzberg, B., 1960, The alcoholic psychoses: Demographic aspects at midcentury in New York State, Yale Center of Alcohol Studies, New Haven.

Maxwell, M. A., 1952, Drinking behavior in the State of Washington, *Quart. J. Stud. Alc.* 13:219–239.

Merton, R. K., 1957, "Social Theory and Social Structure," Free Press, Glencoe, Ill.

Mizruchi, H. and Perrucci, R., 1970, Prescription, proscription and permissiveness: Aspects of norms and deviant drinking behavior, *in* "The Domesticated Drug: Drinking Among Collegians" (G. Maddox, ed.), pp. 234–253, College and University Press, New Haven.

Mulford, H. A., 1964, Drinking and deviant drinking, U.S.A., 1963, *Quart. J. Stud. Alc.* 25:634–650.

Mulford, H. A., and Miller, D. E., 1959, Drinking in Iowa. I. Sociocultural distribution of drinkers; with a methodological model for sampling evaluation and interpretation of findings, *Quart. J. Stud. Alc.* 20:704–726.

Mulford, H. A., and Miller, D. E., 1960a, Drinking in Iowa. II. The extent of drinking and selected sociocultural categories, *Quart. J. Stud. Alc.* 21:26–39.

Mulford, H. A., and Miller, D. E., 1960b, Drinking in Iowa. III. A scale of definitions of alcohol related to drinking behavior, *Quart. J. Stud. Alc.* 21:267–278.

Mulford, H. A., and Miller, D. E., 1960c, Drinking in Iowa. IV. Preoccupation with alcohol and definitions of alcohol, heavy drinking, and trouble due to drinking, *Quart. J. Stud. Alc.* 21:279–291.

Mulford, H. A., and Miller, D. E., 1961, Public definitions of the alcoholic, *Quart. J. Stud. Alc.* 22:312–320.

National Commission on Marihuana and Drug Abuse, 1972, Marihuana: A Signal of Misunderstanding, First Report, Government Printing Office, Washington, D.C.

National Institute on Alcohol Abuse and Alcoholism, 1971, "Alcohol and Health," First special report to the U.S. Congress on, from the Secretary of Health, Education and Welfare, Government Printing Office, Washington, D.C.

Plaut, T. F., 1967, "Alcohol Problems: A Report to the Nation by the Cooperative Commission on the Study of Alcoholism," Oxford University Press, New York.

Reinert, R., 1968, The concept of alcoholism as a bad habit, *Bulletin of the Menninger Clinic* 32:21–25.

Riley, J. W., and Marden, C. F., 1947, The social pattern of alcoholic drinking, *Quart. J. Stud. Alc.* 8:265–273.

Roman, P., and Trice, H., 1967, Alcoholism and Problem Drinking as Social Roles: The Effects of Constructive Coercion, Paper presented at the 17th annual meeting of the Society for the Study of Social Problems, August, San Francisco, California.

Room, R., 1970, Assumptions and Implications of Disease Concepts of Alcoholism, Paper delivered at the 29th International Congress on Alcoholism and Drug Dependence, February, Sydney, Australia.

Rosen, A. C., 1960, A comparative study of alcoholic and psychiatric patients with MMPI, *Quart. J. Stud. Alc.* 21:253–266.

Rotter, J. B., 1954, "Social Learning and Clinical Psychology," Prentice-Hall, New York.

Sadoun, R., Lolli, G., and Silverman, M., 1965, "Drinking in French Culture," Monograph No. 5, Rutgers Center of Alcohol Studies, New Brunswick, N.J.

Scott, P., 1968, Offenders, drunkenness, and murder, *The British Journal of Addiction* 63:221–226.

Seeley, J., 1962, Alcoholism is a disease: Implications for social policy, *in* "Society, Culture, and Drinking Patterns" (D. J. Pittman and C. R. Snyder, eds.), pp. 586–593, Wiley and Sons, New York.

Siegler, M., Osmond, H., and Newell, S., 1968, Models of alcoholism, *Quart. J. Stud. Alc.* 29:571-591.
Snyder, C. R., 1958, "Alcohol and the Jews; a Cultural Study of Drinking and Sobriety," Monograph No. 1, Rutgers Center of Alcohol Studies, New Brunswick, N.J.
Snyder, C. R., 1962, Culture and Jewish sobriety: the ingroup-outgroup factor, *in* "Society, Culture and Drinking Patterns" (D. J. Pittman and C. R. Snyder, eds.), pp. 188-225, Wiley, New York.
Sutherland, E., and Cressey, D., 1955, "Principles of Criminology" (5th ed.), Lippincott, New York.
Syme, L., 1957, Personality characteristics of the alcoholic, *Quart. J. Stud. Alc.* 18:228-301.
Terris, M. A., 1968, A social policy for health, *American Journal of Public Health* 58:5-12.
Trice, H. M., and Wahl, J. R., 1958, A rank order analysis of the symptoms of alcoholism, *Quart. J. Stud. Alc.* 19:636-648.
Ullman, A. D., 1958, Attitudes and drinking customs, *in* "Mental Health Aspects of Alcohol Education," U.S. Public Health Service, Government Printing Office, Washington, D.C.
Wexberg, L., 1951, Alcoholism as a sickness, *Quart. J. Stud. Alc.* 12:217-230.
Williams, A. F., 1965, Self-concepts of college problem drinkers. I. A comparison with alcoholics, *Quart. J. Stud. Alc.* 26:586-594.
Wolf, I., Chafetz, M. E., Blane, H. T., and Hill, M. J. 1965, Social factors in the diagnosis of alcoholism. II. Attitudes of physicians, *Quart. J. Stud. Alc.* 26:72-79.
Zucker, Robert A., 1968, Motivational Factors and Problem Drinking Among Adolescents, Paper presented at 28th International Congress on Alcohol and Alcoholism, September, Washington, D.C.

CHAPTER 4

Alcoholism in Women

Edith S. Gomberg

School of Social Work
University of Michigan
Ann Arbor, Michigan

and

Rutgers University Center of Alcohol Studies
New Brunswick, New Jersey

INTRODUCTION

The present women's liberation movement has an important historical link with alcohol problems and the Temperance Movement. From about 1880 on, there was a vigorous struggle for legal, educational and occupational rights of "the new woman." The Temperance Movement had been going for some time, but late in the nineteenth century, under the leadership of Frances Willard, the Women's Christian Temperance Union became an aggressive, militant organization with radical positions on a variety of social issues: It supported labor unions, prison reform, universal education, and women's rights. The Temperance Movement withered away, largely destroyed by its own success, the prohibition amendment, but like all movements—women's liberation included—it encompassed a variety of people with a variety of motives, ascetic, religious, or radical (Bacon, 1967).

If we are to think about alcohol problems in women, we must first look at the role and status of women in American society because, unless we accept moral or hereditary explanations of alcoholism, we have to look for causes in the alcoholic person's history and development within the society in which he or she lives. From the onset, then, we are making assumptions: that people do not become alcoholics because they are sinners or because they were born that way but because of a complex of etiological reasons, possibly biochemical, probably psychological, very likely social. What this means is that a person's body chemistry may make him or her more vulnerable because alcoholic beverages have quite special effects on him or her; more significant, we believe, is the role of life experience, character development, patterns of coping, family and other group experiences, and the frustrations and stresses that are part of the human condition.

Sex Differences

Since mankind began, there has been commentary on the differences between the sexes. The Bible, Shakespeare, poetry, grand opera, the writings of philosophers abound with comments on sex differences. Psychologists have displayed research interest in this difference only lately. Psychology is a young science but the lack of research interest may have been because the two major theories of psychology, psychoanalysis and learning theory, both centered on needs believed common to both sexes (Kagan, 1964). The differences between the sexes, of which psychologists must have been aware, were not thought to be a significant variable. This was a kind of equalitarianism, not really thought through, just as it seemed important to ignore or deny differences between ethnic, religious, and social class groups.

Recent work in the social sciences on sex differences and the role and status of women has a wide range. We will distinguish three types of literature although, admittedly, they may overlap.

First, there are *empirical-measurements-test-laboratory studies* of sex differences summarized in books on differential psychology (Anastasi, 1949) or in books on the psychology of women (Bardwick, 1971). Nor by any means are such empirical works confined to psychologists, there are sociological studies such as Komarovsky's study of blue-collar marriages (1962) and Mead's (1928, 1930) anthropological reports on growing up in Samoa and New Guinea. Generally, the studies reported are of normal, nondeviant women. Empirical research on sex differences has only occasionally dealt with aberrant behavior. Bardwick (1971) makes this explicit, stating that her work is not concerned with psychopathology.

Second, there are *theoretical-clinical writings* and these have centered largely around psychoanalytic ideas. Freud's description of the Oedipal crises differentiated male and female early childhood experience; to oversimplify, the girl's primary crisis was the discovery of the difference between brother and herself. The phrase "penis envy" emphasized both anatomical difference and her secondary status. This view of female psychology was born in the late nineteenth century and, considering the role and status of women at the time, was appropriate to the period. Freud wrote primarily a psychology of male development and although he treated and wrote about women patients, the structure of psychoanalysis is built around *masculine* experience.

The Freudian view of women is best presented in Deutsch's (1944, 1945) classical work, but by the time it appeared, rifts had developed among psychoanalysts. Horney (1939) and Thompson (1941, 1942) were writing of "cultural pressures" in the psychology of women as Adler (1927) had done earlier. Adler wrote from an equalitarian's point of view about the "alleged" inferiority of women, societal premiums placed on maleness, and the compensatory mechanisms used by women, which he summarized as "masculine protest." The main issue was whether sex roles were fixed, immutable, and universal because they were based on physical and physiological differences. Theorists such as Mead, Horney, and Thompson argued that sex role was learned and culturally determined. But it is not really an either-or choice. Women's sexual roles are based on anatomical and physiological differences, and sex roles are learned and do vary from culture to culture. The most useful psychology of women will have to encompass both "Anatomy is destiny" *and* "cultural pressures."

The third group of writings may be called *polemical*. In our time, from Mary Wollstonecraft in the eighteenth century, John Stuart Mill in the nineteenth, and people such as Margaret Sanger, Frances Willard, and Emma Goldman in the twentieth, this group became articulate with de Beauvoir's book on "the second sex" (1953), a scholarly work, and Friedan's (1963) highly popularized book about "the feminine mystique." These writers and those who came a little later, such as Millett (1970) and Greer (1971), are trained social scientists and their works are a combination of scholarship and political tract. Women's liberation theorists have not presented a single social-political position. Some writing takes an equalitarian point of view; equality of opportunity and equal pay for equal work implies equality of men and women in aptitude and performance. Other writers recognize that all societies define male and female roles as different and they attribute some validity to distinguishable sexual roles, but also urge women to break with the traditional sex role. Indeed, the collection of writings in "Women in Sexist Society" (Gornick and Moran, 1971) was described in a review as providing emotional support so that the women who read it could break out of traditional sex roles. The epithet, "male sexist pig," brassiere burning, and slogans such as, "Run for office, not for coffee," suggest that the oppressor and the enemy is the male. A discussion of women and insanity (Chesler, 1972) presents the thesis that definitions of mental illness, therapy, and cure are "patriarchal," and that there is a double standard of mental health.

The fact is that in our modern technological society the differences between male and female roles, based on differences in strength and biology, do indeed become less important. In an urbanized, technologically advanced society, jobs become interchangeable. This has, inevitably, tremendous impact on the traditional family structure and traditional sex role.* For women, it means a wider range of possibilities as work outside the home is ever more available; for men, it means a reorganization and redefinition of maleness and the male role.

The reexamination of women's role, its shifts, and its discontents has its counterpart in recently expanding interest in men's work and its discontents. Studies and reports about the problems of the blue-collar worker and his attitude toward his work follow earlier works on the discontents of middle class, white-collar, organization men. Quite relevant to our interest are discussions of the use of alcohol and other drugs in industry, on the assembly line, among executives, to lower the tensions or break up the monotony.

The percentage of women working outside the home continues to increase but this freedom, the spate of sex manuals and the right of orgasm, the larger freedoms permitted, including the right to swear and the right to drink publicly—does not seem to have produced better-adjusted women. One explanation is that women have low self-esteem because they are second-class citizens. Another explanation is that women in contemporary American society are faced with an unprecedented and terrifying amount of freedom and that they are burdened with too many choices (Decter, 1972).

Sex Differences in Psychopathology

The question can be put many different ways. If we agree on the criteria of mental illness, do we have the same incidence of schizophrenia or neurosis or alcoholism among women as among men? Do they look the same clinically? Are the causes, the course, and the therapies the same? Are we measuring deviation from the same definition of acceptable normality?

---

* The impact is even greater and seen more clearly in societies where industrialization and employment of women outside the home is suddenly and rapidly introduced. This has been a subject for extensive research, for instance, in Puerto Rico. The traditional family roles do not change with the same rapidity for *all* groups; it appears that those who benefit the least and who remain poor despite industrialization tend to maintain more traditional family patterns (Rodriguez, 1970). Interestingly enough, it was reported by several psychiatrists and internists in Puerto Rico that the incidence of drinking problems among women has increased and that these women are, for the most part, middle and upper class.

Differences tend to be reported more than nondifferences. If a medication is administered and 60 percent of the men and 60 percent of the women show improvement, it is reported that 60 percent of the people to whom it was administered show improvement. If, on the other hand, 80 percent of the men improve and 40 percent of the women, we tend to report such findings as research on sex differences. At the annual meeting of the American Psychological Association in 1972 (Proceedings, 1972), many papers on sex-difference research were reported and they included studies of liberal-conservative attitude, helping behavior, children's aggressive behavior, and so on. The question, it seems to us, is not whether differences exist. Let us assume that they do. If they are looked for, they are usually found but the question remains: Which are the *important* differences? It depends, of course, on the purposes of the questioner: That men tend to prefer active sports somewhat more than do women is important to manufacturers of recreational material, and that women tend to be more conscientious students may be important to the college dean. One of our biggest problems in studying sex differences in psychopathology is that we are not sure where to start first, i.e., which are the more important differences between men and women. Symptoms? Response to medication? Attitude of relatives? Marital status? Of the array, incidence, clinical features, etiology, course of illness, therapy, etc., where should we start?

To date, research on sex differences in psychopathology is meager. The most developed area of reporting so far is in the *epidemiology* of mental disorder. Statistics of mental hospital admissions, attendance at outpatient clinics, and psychiatric help-seeking are reported frequently. Perhaps because it is more acceptable for women to look for help or perhaps because they are an unhappier lot, women outnumber men in all these statistics. Even in those studies that are not directed toward a search for help but rather toward symptoms of psychological distress, women outnumber men. In a national survey of everyday, common symptoms of psychological distress (1970), people—not patients—were asked about their experience with 12 symptoms of psychological distress (they were asked as part of a medical history questionnaire). The symptoms were: reported nervous breakdown, feelings of an impending nervous breakdown, nervousness, psychological inertia, insomnia, hand-trembling, nightmares, perspiring hands, fainting, headaches, dizziness, and heart palpitations. Without exception, the women had higher rates than the men for all 12 symptoms. Research also documents a greater frequency of depression reported by women at all ages studied.

Of course, part of the answer is determined by *which* disorder you ask about. It is very important to distinguish between mental disorders such as schizophrenia, depression, and the psychoneuroses and another class of disorders called variously psychopathic, sociopathic, acting-out or social-deviance disorders. In this class, which includes crime, delinquency, and the addictions, *men outnumber women*. In alcohol problems, although there is variability in the ratio of men to women, depending on whether one gathers data in a detoxification center or in a private hospital or physician's office, the ratio of *more* men than women stays the same. There are, then, differences in the frequency with which different kinds of disorders appear among men and women, and this does seem to be tied to what is more acceptable and less acceptable behavior for men and women. To be depressed is somehow more consistent with the female role as we define it; to be an alcoholic is more consistent with the male role. Furthermore, we believe that when the same disorder appears, there are differences; thus, a male depressed patient is not the same as a female depressed patient, and a male alcoholic is not the same as a female alcoholic. But how are they different?

A recent report on drinking practices and problems in a suburb of London (Edwards, Chandler, Hensman, and Peto, 1972) compares alcohol problems with other disorders, and it would appear that the situation is approximately similar in Great Britain:

> Recent community mental health studies in Britain have ... consistently shown up a situation which is converse to that in alcoholism, crime and narcotic dependence: mental disorder seems generally to be 2 or 3 times more prevalent among women than men. (page 126)

Still looking for the more important and less important sex differences in the causes, course, and therapies of mental disorders, we turn to the literature on mental illness. There is a good deal written about the etiology, the psychodynamics, the clinical symptoms, and effective therapies, but is has been written almost entirely with the male patient in mind and the implicit assumption is that it is, of course, generalizable to the women patient. A generation ago, it was common to describe the parents of the mental patient as a weak or passive father and a domineering mother, but the question did not seem to arise if that was the common family constellation of the women patient.

There are some studies of sex differences in schizophrenia that are relevant. Cheek (1964), studying the family interaction of schizophrenic patients, came upon a curious difference. *Male* schizophrenic patients interacting with their families showed the traditionally described schizophrenic pattern of withdrawal, low activity, and passivity. But *women* schizophrenic patients turned out to be *more active and more dominating* than the women in the normal control group. Cheek described this as a "reversal of sex roles." McClelland and Watt (1968) reported on what they called "sex role alienation" among a sample of schizophrenic patients: The women patients tended to react on the tests administered in an "assertive" way, the men in a more "sensitive" way.

These would seem to relate to data reported by Opler (1959) not on sex differences but on ethnic background differences among schizophrenics. He compared Irish and Italian schizophrenic patients in a New York City hospital by methods psychiatric, anthropological, and psychological, and found striking differences in preoccupation with sin and guilt, control of impulses, fixity of delusions, somatic preoccupation, and attitude toward authority. It is a longstanding finding in alcohol studies that ethnic groups differ in their rates of alcoholism and Opler's schizophrenic patients show this difference: Of the Italian subjects 3 percent had a history of alcoholism accompanying the schizophrenic disorder whereas 63 percent of the Irish schizophrenic patients had such a history. Recently, Kelleher (1972) has compared Irish and English patients in London in their manifestations of obsessional symptoms.

The point is that the different subcultural groups to which we all belong shape us and determine the styles of problem-solving and coping and *the styles of breakdown,* too. A relationship between social class and mental illness exists in diagnosis and in selected agency of therapy (Hollingshead and Redlich, 1958). The clinical manifestations of schizophrenia vary by ethnic background (Opler, 1959). And it seems likely that the manifestations of mental disorder would vary with color, age, and sex, if these variables were studied. Is it not possible that many of the ambiguous and conflicting findings of psychological and psychiatric studies are related to a mixing together of subjects who should have been subgrouped and studied separately? Apart from our obvious methodological and measurement difficulties, perhaps one of our grossest errors has been in selection of subjects for study. Abnormal behavior has seemed sufficient to group together widely varying, heterogeneous groups. Certainly that has been true in alcoholism research where the subject is an "alcoholic" and that is all we need to know about him to put him in a research sample of alcoholics.

Related to the studies of sex differences in schizophrenia (Cheek, 1964; McClelland and Watt, 1968) are several studies that indicate that women's psychiatric wards are characterized by more noise, more excitement, more belligerence, and more interpersonally disruptive behavior than is true of men's psychiatric wards (Chesler, 1972). These findings suggest a basis for the often-repeated comment that women are "worse" patients than are men; from the point of view of the authority who needs to maintain an orderly ward, the more disruptive behavior of women patients does indeed make them less manageable, i.e., "worse" patients. Of great interest to us is the fact that this is reported from wards where the women confined were schizophrenic. When ward personnel report from alcoholism treatment facilities, prisons, and narcotic addiction treatment units that women are "worse"—i.e., less cooperative—patients, one of the interpretations has been that they behave that way because they are more deviant than their male counterparts. An interesting report of impression and data about women narcotic addicts follows:

> In a narcotics hospital the manner of female addict presentation appears quite different from her male counterpart. The females are considerably more attention-seeking, are more erratic and frequent medical clinics more often with severe complaints of non-organic origin. Not infrequently, females appear to be psychotic or borderline psychotic. These and other impressions prompted this study. Interestingly, however, the data revealed much more uniformity between male and female addicts than clinical impressions had led us to believe. (Ellinwood, Smith, and Vaillant 1966)

The research reports we have discussed indicate that women are noisier, more excitable, more belligerent patients in other disorders, even in those disorders where they are a majority, the nondeviant disorders. But the data are too sparse and the definition of good or patient patient behavior too ambiguous and judgmental to draw any conclusion.*

We believe that mental disorders and deviance disorders do differ in cause, course, and manifestation by age, sex, ethnic grouping, and social class—to name the most obvious subgroups.

THE SOCIAL DRINKING OF AMERICAN WOMEN

If drinking is forbidden behavior, then obviously those who observe the taboo cannot develop drinking problems. But there is no clear relationship between taboos and the development of drinking problems and those groups that are quite relaxed about drinking and where almost universal moderate drinking prevails, e.g., Italians and Jews, have low rates of alcoholism. There is another dimension: Sociologists and anthropologists suggest that where the rules are *clear and congruent with the culture,* the likelihood of alcoholism is low. And, logically, those societies that have clear rules about sexual roles tend also to have less ambiguous rules about the sexes' drinking norms. The definition of acceptable behavior will include proscriptions and rules about drinking:

> Societies with a definite sex difference in drinking tended significantly to have a large rather than a small sex difference in child-training practices and also tended to have characteristics . . . which appear to encourage or require greater differentiation between the sexes in their adult role. (Child, Barry, and Bacon 1965)

---

* There are a number of reports that suggest that women may make "better" patients than men. Garai (1970) has noted that women accept being mental patients more readily. Zigler and Phillips (1960) found that ". . . hospitalized women tend to be more effectively adaptive in society than hospitalized men," the men appearing more aggressive and nonconforming. And a report by Madeddu et al. (1969) of Italian women tubercular patients with drinking problems describes the behavior of the patients in the sanatorium as ". . . more satisfactory," the women patients showing more independence and cooperation. It seems to depend on the therapeutic setting and the definition of "a good patient."

Sanctions

How clear are the rules about women's social drinking in American society and how do the rules, particularly the sanctions relating to drinking, relate to sexual roles? What we can say with certainty is that the drift of change since World War I has been in the direction of more permissiveness. One evidence of this is in the advertising material used by liquor companies, a reflection of changing mores. As with cigarette advertisements, the first "breakthrough" was the mere presence of a woman while the man was drinking, and the second was having glasses clearly intended for both or even in their hands. Recently, for the first time, women have begun to appear in liquor advertising on their own, one company advertising Scotch and the single girl.

The hard evidence indicates that although more American men than women drink, the gap between the two sexes has become smaller. Riley and Marden (1947) compared their findings with an earlier study done only several years before (Ley, 1940) and reported that female drinking showed a "sharp rise" from 1940 to 1946, the year of their survey. Male drinking did not show so sharp a rise; therefore, ". . . the gap between men and women, particularly in the age brackets under 50, is smaller than it was six years ago" (page 267).

Twenty-two years later, the report by Cahalan, Cisin, and Crossley (1969) on American drinking practices indicated that 77 percent of the male population drank and 60 percent of the female population drank. The percentage of increase since the Riley and Marden survey of 1946 was 2 percent for the male population and 4 percent for the female population. Each percentage point represents approximately a million people. Percentagewise, the gap between men and women who drink has become smaller still.

Sex differences in the drinking patterns in the United States are discussed in other chapters, and for our purposes we need only note that more men than women drink and that women's drinking varies with age and social class. Just as the percentage of men who drink drops between the ages of forty and sixty, so does the percentage of women (Cahalan *et al.*, 1969). Among lower income groups, the percentage of men who drink is higher than among higher income groups but the opposite is true of women; by and large, the lower the income, the smaller the percentage of women who drink (Lawrence and Maxwell, 1962). According to the most recent survey by Cahalan *et al.* (1969), both men and women of lower income groups tend to have a smaller percentage of those who do any drinking at all than men and women of higher income groups. Cahalan and his associates also report the finding that the percentage of those who drink *heavily* peaks for both men and women in the age group forty-five to forty-nine. Interestingly enough, there is another age category in which heavy drinking peaks for women, i.e., age twenty-one to twenty-four. This is probably the dating, going-steady, and partying age before many of these women settle into wife-and-mother role.

Intoxication

Interestingly enough, although there is no question that more women in American society drink than ever before, and although no one will argue that the rules are more permissive than they were a couple of generations ago—e.g., about women's drinking in public places—*social attitudes of disapproval toward women's drunkenness are apparently unchanged over time.* Two reports indicate that tolerance toward drunkenness in women in very low. Lawrence and Maxwell (1962) reported that *both* sexes in *all* social classes reported more intolerance toward drunkenness in women than in men. Knupfer (1964) has also reported that both men and women maintain a double standard on this point. Asked, "What do you think of a man who has too much to drink? What do you think of a woman who has too much to drink?" Knupfer's men and women respondents judge drunkenness in women to be *much worse.* Asked why they believed drunkenness to be worse in a women, about two thirds made comments relating to the social role of women.

Knupfer believes the relationship between drinking and women's role involves two aspects, one relating to the division of labor between the sexes and impaired response to the needs of others, and the other relating to loss of customary sexual restraints and inhibitions. On the division of labor, Child et al. (1965) in a crosscultural study of sex differences in drinking customs note:

> It seems reasonable to expect that most societies would limit drinking and drunkenness in women more than in men. Under the generally prevailing conditions of human life, temporary incapacity of a women is more threatening than is temporary incapacity of a man. For example, care of a field can be postponed for a day, but care of a child cannot. The general social role of the sexes makes drunkenness more threatening in women than in men. (page 60)

Knupfer makes an excellent point in regard to women's work role and alcohol. Her alcohol intake may not markedly lower the efficiency of a housewife in using a vacuum cleaner or washing machine or in cooking but it does impair "precisely the quality of sensitivity to the needs of others" (Knupfer, 1964) that we value in her familial role.

Knupfer's second aspect, relating to the loss of customary sexual restraints and inhibitions, needs little comment. Even in our permissive contemporary scene, "sowing wild oats" is still more acceptable male than female behavior. While the relationship between alcohol and sex is more talked about than studied, it is a fact of life that women are able to participate or at least be available as sexual objects even when they are drinking heavily. Males, on the other hand, are likely to be rendered impotent by large amounts of alcohol. Perhaps more girls are seduced when under the influence than when sober, but it seems more likely that there is a high degree of relationship between drinking, sexual activity, and the site and group chosen for drinking. The more sexually available the girl or woman, the more likely she is to choose a place and time and companions for drinking that make sexual contact more probable. But one must also remember that rape has increased in recent years and that the woman drinking in public attracts notice and attention.

Despite the fact that the clear division of labor between the sexes becomes more obscured in a technologically advanced society such as ours, and despite the sexual revolution and new freedoms, there is no evidence of a more tolerant attitude toward women's drunkenness at this time. Curlee (1967), in fact, notes that even among alcoholic women, the attitude expressed is one of disgust, and Curlee comments, ". . . no one likes to think that the hand that rocks the cradle might be a shaky one."

Effects of Drinking

The question of sex differences in the effects of alcohol on various psychomotor and intellectual tasks has been raised (Carpenter, 1962) but not resolved. Currently, there is developing some new research questions revolving more around the function that alcohol serves, or its reinforcement value, for individuals. One such study used volunteer college students as subjects and investigated the students' expectations of having needs for affection and achievement satisfied (Jessor, Carman, and Grossman, 1968). The investigators found that the lower the expectation of having one's needs satisfied, the greater the recourse to alcohol, *especially among the women subjects.* It might be a good idea to pursue this further and to relate such behavior to the data of Jones (1971), which showed that women problem drinkers, as adolescents, were rated as pessimistic and self-defeating. One might speculate about the relationship of low expectations of satisfaction in the heavier alcohol users among the young women and the high incidence of depression among women problem drinkers.

Using college students in their twenties as subjects, Wilsnack (1972) has studied the effect of social drinking ("parties") on her subjects' fantasy productions (Thematic Apperception Test stories). She found that:

> . . . social drinking reduces women's personal power fantasies . . . Drinking appears to increase the female drinker's sense of "being orientation," the sense of spontaneous enjoyment of the present which has also been found to characterize the fantasies of mothers during breast feeding. These effects are interpreted as reflecting enhanced womanly feelings.

One wonders if the unprecedented amount of freedom and terrifyingly wide range of choices now available to women, as described by Decter (1972), do not produce, among other results, a longing and search for "womanliness." It would be ironic indeed if the increase in women's drinking were to be interpreted not as a manifestation of assertiveness, masculinity, and sexual equality, but as a search for "womanliness." It is an intriguing thought.

Wilsnack's point, that social drinking for women does not serve merely to reduce power or aggression fantasies but to increase "womanly feelings" relates well to Jones's (1971) description of the adolescent who will later become a problem drinker: She escapes into "ultrafemininity." It seems striking that studies of both men and women problem drinkers as adolescents (Jones, 1968, 1971) emphasize the attempt to play a hypermasculine or hyperfeminine role. How this will relate to later failures in marriage, dependency needs and the inability to accept dependency relationships, and to alcohol as coping substance and psychic anesthetic are matters about which we are still speculating.

ALCOHOLISM IN WOMEN

In 1957, the present writer reviewed the literature then available on the alcoholic woman and summarized it as dealing primarily with seven issues: degree of psychopathology, variability, specific precipitants, sexual promiscuity, rapidity of shift from controlled to uncontrolled drinking, relationship to physiological functions, and medical complications (Lisansky, 1957). Two recent literature reviews (Lindbeck, 1972; Schuskit, 1972) have appeared, the first emphasizing the theme of "concealed alcoholism" among women, the second stressing positive family history, particularly psychiatric illness among the relatives of alcoholic women.

Most research reports about the alcoholic woman compare her with her male counterpart; the question is that since more is known (or at least written) about the male alcoholic, can we simply generalize our knowledge of the male alcoholic to the woman alcoholic? Some (Berner and Kryspin-Exner, 1965) have believed the problem to be similar enough so that similar therapeutic institutions are recommended. Others, the present writer among them, believe the disorders to be reasonably different phenomena and that knowledge of the differences in etiology, manifestations, and therapeutic needs is of primary importance in trying to help the patient.

Many research reports make comparisons *within* the group of alcoholic women under study, e.g., remitted and unremitted patients, those diagnosed as primary alcoholics and those diagnosed as affective alcoholics, outpatient clinic patients, and women in penal institutions who have alcohol problems. These reports go a step behind the male versus female alcoholic comparisons because they assume not only that female alcoholism differs from male alcoholism, but that female alcoholism is a complex phenomenon and requires descriptions of different *subgroups* of alcoholic women.

Finally, there are a very few studies with "normal" control groups and of the three that have appeared recently, one uses as control group a sample of wives of alcoholic men, the other two a randomly selected group of nonalcoholic, nonpsychiatrically disturbed women. Comparisons with nonalcoholic women, inpatient, outpatient, and nonpatient, psychiatric and normal, will hopefully be the next direction of research.

We will summarize the research reports about alcoholic women under six major headings: (1) early life experience, (2) patterns of drinking and alcoholism, (3) psychodynamics, (4) complications of alcoholism, (5) differences: social class, race, religion, and ethnic background, and (6) treatment.

Early Life Experience

We assume that where a parent was missing, alcoholic, or psychiatrically ill during a person's childhood, a stress is added during the formative years that increases the likelihood of breakdown of one sort or another in later life.* Research reports are not strictly comparable because percentages of "disruption" in early life sometimes refer to loss of a parent by death or abandonment and sometimes refer to a wide variety of parental and familial difficulties.

Loss. If we consider the absence of a parent during childhood years, 40 percent of women patients seen by Rosenbaum (1958) reported such loss. In a study of parental loss among alcoholics, de Lint (1964) reported an overrepresentation of last-borns among female alcoholics† and related this to more frequent loss of a parent. As many as 37 percent of women alcoholics reported loss of a parent before the age of six as compared to 13 percent of men alcoholics.

In comparing results obtained by different investigators, it is well to keep in mind the *social class status* of the alcoholic women being studied. Kinsey

---

* The trend of epidemiological research indicates that a staggering percentage of the population show mild or marked psychiatric symptoms (Srole, Langner, Michael, Opter, and Rennie, 1962). If only about 20 percent of the population show no definable psychiatric stress, then problem-free parents are definitely in the minority. However, we are referring here to those families in which divorce or death has left one parent to raise a child, in which one or both parents were alcoholic, in which a parent was overtly psychiatrically ill. There is much evidence that such stresses appear more frequently—though not inevitably—in the histories of psychiatrically or behaviorally disturbed individuals.

† Both DeLint (1964) and Smart (1963) have reported an overrepresentation of later-borns among women alcoholics, i.e., they tend to be the younger members of a family more often than chance would predict. But another recent report presents data that indicate that overrepresentation of last-borns is characteristic of male alcoholics but *not* of female alcoholics (Blane, Barry, and Barry, 1971). Thus, the birth order issue, and its relation to female alcoholism, are still unresolved.

(1966), with a state hospital group of alcoholic women, relatively low in educational achievement, occupation, and other class indices, reported 72 percent of his subjects growing up with "the father absent from the home because of death, desertion, inebriety or some type of psychosis," while Curlee (1970), working in a private hospital, reported 43 percent who experienced "some type of disruption in . . . early family life."

Loss of a parent in early life does appear to occur more frequently in the histories of alcoholic women than of alcoholic men. Loss of a parent is a relatively objective index; one gets into more problems when asking about "unhappy childhoods" (Sclare, 1970) or even about alcoholism in parents. However, although these judgments are subjective and women apparently more sensitive to and more prone to report unhappy family situations in early life, the fact is that alcoholic women do indeed report such events to a greater extent than do alcoholic men.

Alcoholism in Relatives. Estimates of the percentage of women patients whose fathers were alcoholic range from 33 percent (Johnson, de Vries, and Houghton, 1966) to 51 percent (Wood and Duffy, 1964), and with alcoholism in either parent counted, the percentage goes to 60 percent for an English sample of women alcoholics (Rathod and Thomson, 1971). A high incidence of alcoholism among the siblings of research subject patients has been reported by a number of investigators (Lisansky, 1957; Johnson *et al.,* 1966; Jacob and Lavoie, 1971). The incidence of alcoholism does seem to be higher in the family history of women alcoholics than of men alcoholics (Winokur and Clayton, 1967; Winokur and Clayton, 1968) and this has been confirmed again quite recently (Jones, 1972).

Some caution seems in order. In one of the earlier studies of alcoholism in women (Wall, 1937), more men than women reported alcoholism in the family. The different percentages reported by men and women patients, while usually a larger percentage reported by women, are often not significantly different (Lisansky, 1957; Sclare, 1970).

Psychiatric Problems in Parents. Winokur and Clayton (1967, 1968) have reported that psychiatric problems are more common in the family background of women alcoholic patients than they are among men. Schuckit (1971, 1972) has reported that more mental illness appears among the relatives of women alcoholics; at least more affective disorders do.

Disruption in Early Life Experience. Now, what do all these findings suggest? They touch upon the old controversy of genetic versus learned components of behavior. Deviant behavior seems to occur more frequently among people who come from families where there is deviance or disruption manifested by the parents. It is not clear whether this is useful information for a therapist or a social planner. One thing does seem evident: There is more loss, more alcoholism, more depression, and other psychiatric problems in the families of people who become alcoholics than there is among those who do not and there appears to be more loss, more alcoholism, and more familial psychiatric problems among women alcoholics than among men alcoholics. However, this kind of family history, disruption and psychopathology, characterizes *many other psychodiagnostic groups*. Broken homes and emotional deprivation in childhood appear more frequently in the history of *any* psychiatrically ill population than they do among normal persons' histories. Robins (1966) reports strong similarities in the childhood histories of groups of hysterics, alcoholics, and sociopaths. She found this in her St. Louis follow-up study of child-guidance clinic patients and she raises the question "whether these are in fact all subtypes of one disease." This still leaves open the question of those factors that determine whether a person from a broken, unhappy, disrupted childhood home will become alcoholic or depressed or schizophrenic or hysterical or will manifest a combination of disorders. Accepting the relationship between disruptive early history and later psychopathology, the question still is why particular people manifest psychopathology in one form and not another.

One more point before we move on to the adolescence of women who become alcoholic. If we have established that disrupted home and psychiatrically ill or alcoholic parents occur more frequently among people who become alcoholics, we have just opened the problem. The absence or illness of a parent or sibling has a variable impact depending on the nature and quality of family life, whether the ill or missing person is the mother or the father, whether the affected person is a son or a daughter, and the group identity of the family (religion, ethnic background, social class, to mention only some of the relevant variables).

The Woman Alcoholic as Adolescent. Longitudinal studies of women who later became alcoholics are few. The McCords's (1960) study was of the origins of male alcoholism. Robins, Bates, and O'Neal (1962) included both males and females in their follow-up study of children who were referred to a child guidance clinic, and made the point that the rather high rate of alcoholism developed in later life among the female patients reflects the severity of their problems:

> The relatively high rate of alcoholism among the female patients probably reflects the fact that the clinic population was largely male—as is the case in most child guidance clinics—and that the girls referred to such a clinic, therefore, tend to be even more atypical of the total population of girls than the males are of boys in general. (page 400)

The only longitudinal study of women problem drinkers to date is that of Jones (1971). People who participated in the Oakland (California) Growth Study, now in their late forties, were surveyed and asked about their current drinking practices. Jones found that women, divided into problem drinkers, heavy drinkers, moderate drinkers, light drinkers, and abstainers showed rather different characteristic cores of personality traits "discernible to raters in the early adolescent period." Some of the adjectives used in describing women problem drinkers include vulnerable, pessimistic, withdrawn, and self-defeating; but more significantly, "Problem drinkers are judged to be submissive as youngsters, rebellious as adults." Interestingly enough, almost the identical description occurs in a clinical study of alcoholic women that appeared a few years earlier (not listed by Jones as a reference). Wood and Duffy (1964) note, "Our patients grew up submissive and passively resentful . . ."

Jones presents a case summary sketch, typical of those who later become problem drinkers:

> At 15, life is full of adolescent self-doubt and confusion. She fears and rejects life, is distrustful of people, follows a religion which accentuates judgment and punishment. She escapes into ultrafemininity. This protective coloration will keep her going through the mating season but very likely she will recognize the emptiness and impotence in later years.

The adolescent period seems to be a more severe crisis period for the girls who will later become problem drinkers; clinical appraisals refer to "marked changes" and to "an adolescent upheaval." Whatever the reason, adolescence is experienced more painfully than it was for the women in the other groups.

An interesting aspect of Jones's work is the similarity, on many points, between women problem drinkers and women abstainers. And the ways in which these two groups differ is suggestive. Speaking of abstainers, Jones writes:

> . . . judged to be responsible, conventional, consistent, ethical, and emotionally controlled. They are also able to accept a dependency relationship at an age when this may be a determining ingredient for mental health . . . Unlike the problem drinker, this girl's qualities are of long standing and not related to an adolescent upheaval.

Again, the clinical observations of Wood and Duffy (1964) of a sample of alcoholic women lend support to Jones's data. The difficulty in accepting dependency relationship appears in the growing-up years when the young woman "developed a powerful need for love, coupled with an inability to accept it."

What emerges then is a picture of a submissive, passively resentful girl, undergoing a stormy, conflict-ridden adolescent upheaval, already showing signs of difficulty in impulse control, afraid of dependency relationships, trying to solve her problems with superfemininity. As we shall see, this attempt to solve problems with superfemininity manifests itself in her adult life in a tremendous concern with the maternal role, her erotic-wife role not having been a particularly satisfactory one. But the isolation and the anger and the rebelliousness we see later is there before the problem drinking starts.

Patterns of Drinking and Alcoholism

This is not the place for a debate on the definition of alcoholism but we need to repeat that the etiology and the manifestations of alcoholism will undoubtedly differ for men and for women alcoholics. It would be remarkable if it were otherwise. When Sclare (1970) asked his Scottish alcoholics about "provocative situations leading to hospital referral," no one need be surprised that "domestic stress" is mentioned significantly more often by women, and "employment stress" occurs significantly more often as "provocative situation leading to hospital referral" among men. Taking cognizance of the fact that male : female alcoholism ratios will vary, depending on whether one bases the ratio on arrests or on deaths from cirrhosis of the liver, Edwards et al. (1972) raise the question of differences in "drinking troubles" for men and for women. That the ratio depends on the criteria of pathology is clear from the evidence that the ratio of men to women alcoholics *varies* in different facilities: in doctors' offices, social agencies, private and public hospitals, and prisons (Keller and Efron, 1955).

Whatever the definition or social agency or culture, alcoholism appears more frequently among men than among women. Edwards et al. (1972) raise the question whether this is because of taboos related to drinking or because of taboos related to "problematic behavior." We are inclined to agree with Knupfer's (1964) idea of "cultural protection, i.e., the inhibiting effects of constraints placed on heavy drinking by women. Certainly there is no indication that in other kinds of "problematic behaviors" (neuroses? psychoses?), the statistics show fewer women.

An intriguing exception to the male : female ratio is described by Rosin and Glatt (1971) in a report on drinking problems in an English geriatric population. In the age group sixty-five and above, women with drinking problems outnumbered men with drinking problems. Rosin and Glatt (1971) say:

> The excess of women in the series was partly due to the larger number of women in the elderly population in general; but even in the relatively younger psychiatric group (65 to 70) 55% were women. . . . the predominance of men alcoholics over women appears to recede in later years, and . . . there is a higher chance of encountering drinking problems among elderly women than among men.

While it is true that the British male : female alcoholism ratios in general show a higher proportion of women than is true in the United States, and while differences in life expectancy is a factor here, the report still raises questions about loneliness, reactive depression, aging, and the sex ratios for alcohol problems in geriatric populations.

Telescoping: Later Onset

One of the most frequently repeated statistics about alcoholic women is that their drinking and problem drinking begins later in life than among men and that alcoholism seems to develop more rapidly. There is a good deal of evidence to support this. Fort and Porterfield (1961) report later onset and shorter duration for women alcoholics and a tendency for the stages or phases in the development of alcoholism to blur and be less distinct than is true for male alcoholics. Women patients tend to appear at about the same age, but with a shorter duration of alcoholism, at hospitals (Winokur and Clayton, 1968), outpatient clinics (Lisansky, 1957), and mental health centers (Wanberg and Knapp, 1970). The pattern of later onset and shorter duration is reported from countries other than the United States, e.g., in England (Rathod and Thomson, 1971), Scotland (Sclare, 1970), France (Lengrand, 1964), and the Soviet Union (Averbukh, 1966).

There are occasional exceptions in the literature. Pemberton (1967) reports from a hospital serving Scotland and North England that men and women alcoholic patients were similar in age on admission, age of onset, and duration.

What may be more important is an analysis of different subgroups of women alcoholics *differentiated by age of onset.* In Kinsey's (1966) study of state hospital alcoholic women, he distinguishes three subgroups: those with early beginning and rapid development, those with early beginning and later development, and those with later beginning and rapid development. Cramer and Blacker (1963) have described two groups of problem drinkers among women prisoners, early and late starters, and they were able to distinguish the two in terms of family background and prognosis. Curlee (1969) had described a group of middle-aged women who develop alcoholism rather late in life and relate it to the "empty nest," i.e., the end of the maternal role. And the report by Rosin and Glatt (1971) raises the question of geriatric alcoholism and the vulnerability of older women.

On the whole, it appears to be true—taking the whole alcoholic population as we know it—that women alcoholics in a given facility or setting present themselves at approximately the same age but with a shorter duration of alcoholism than do male alcoholics. That means that onset does tend to come later and that the period of time from onset to presenting oneself at the facility is shorter. Having made that point, it is necessary now to examine the different temporal patterns of the development of alcoholism among women and see how these relate to therapy and prognosis, and also to explore the relationship between differences in temporal patterns of development of alcoholism among men and women and how these relate to sex roles, constraints, social class, and changes over time. A study of the stages or phases of developing alcoholism and how these differ for men and women remains to be done.

Precipitating Stresses

A recent summary of the literature by Schuckit (1972) says that the woman alcoholic ". . . reports her drinking to be in relation to stress." This has been reported consistently from an early paper by Wall (1937) to a recent report by Curlee (1970). In the latter study, significantly more women alcoholics than men reported a definite precipitating stress but Curlee adds a sensible note:

> In most cases, the woman had been drinking prior to the "trauma," but the particular event or situation seems to have tipped the scale toward uncontrolled, clearly alcoholic drinking.

Johnson (1965), reporting physicians' attitudes toward alcoholism in women, says that physicians believe that much more often than with men, women drink in response to "a crisis situation." But Johnson and her co-workers (1966) found, in interviews with women alcoholics (volunteers from Alcoholics Anonymous or physicians' practices), that few reported their alcoholism to be related to "a crisis situation such as the death of a loved one." Most of their volunteer subjects spoke of frustrated needs and feelings of loneliness and difficulty in close interpersonal relationships, i.e., those with husband, parents, and children.

There are two questions, one dealing with the *proneness* of women to justify or rationalize by reporting a precipitating event, and the other dealing with the *nature* of precipitating stresses. The nature of such stresses will obviously vary for men and women; society defines their sex roles and goals differently. But are women more sensitive and vulnerable to stress than men or more likely to report a precipitating stress defensively? As mentioned before, Sclare (1970) asked about the "provocative situation leading to hospital referral" and found, as common sense suggests, that "domestic stress" was significantly more often a precipitant for women, and "employment stress" was significantly more often a precipitant for men. It does not add much to say that women alcoholics relate their drinking to a concrete life situation more than men do unless we can relate this to etiology or therapy or prevention.

The nature of precipitating stresses are not only different for men and women but they vary very much among the women reporting. Sometimes a single event, a point in time, is reported, e.g., loss of a loved person. Sometimes the single event is psychophysiological in nature, e.g., childbirth, a miscarriage, a hysterectomy. But very often the precipitating stress is loneliness (Johnson et al., 1966) or marital conflict (Rosenbaum, 1958) or a middle-age identity crisis (Curlee, 1969), i.e., general life situations that extend over time and that involve the breakdown of interpersonal relationships. The mixture of specific traumas, general life situations, and personality variables that are included as precipitating stresses is listed by Lindbeck (1972) in her review:

> The reasons women give for their chronic, excessive use of alcohol are premenstrual tension, dysmenorrhea, menopause, hysterectomy, fertility, abortion, postpartum depression, miscarriage, frigidity, death, desertion, divorce, the demands of small children, marital troubles, and boredom ... insecurity in the feminine role and functioning, sensitization to loss, dependency, low self-esteem, fear of inadequacy.

The idea of precipitating stresses has become so all-inclusive as to be rendered meaningless. It might be a good idea to go back to Wall's (1937) idea of "a specific life event," or to evaluate "the provocation situation leading to hospital referral" (or any kind of referral) as Sclare (1970) had done. The former permits some specification of those experiences that may be particularly painful danger points for the alcoholism-prone woman. The latter points up the clear difference in the nature of stresses that act as precipitant, men reporting employment stress as primary and women reporting domestic stress.

Drinking in Public and "Closet Drinking"

One of the consistent findings is that women tend to drink at home and alone much more often than men do (Wood and Duffy, 1964; Johnson et al., 1966; Wanberg and Knapp, 1970; Jacob and Lavoie, 1971). Since women are at home much more than are men, this is hardly surprising. In a study of symptom clusters of men and women alcoholics in a mental health center, Horn and Wanberg (1971) report the finding that the women "... are more likely than men ... to be characterized by a solitary-drinker syndrome."

The descriptions of the solitary drinking of women alcoholics, however, go beyond drinking at home alone. They are described as "hidden drinkers" (Lindbeck 1972) or "secret drinkers" (Pemberton, 1967). Pemberton (1967), in his study of women alcoholics from Scotland and North England, contrasts them with the male alcoholics in the same hospital. The men tend to be quite free and open in their discussions of their drinking problems among themselves and with staff, while the women alcoholics, according to clinical observation, "... remain secretive about their reasons for being in the hospital." In Johnson's (1965) study of physicians' attitudes toward alcoholic women, the description of differences between men and women alcoholics on this point tells us a great deal about the advantages and disadvantages women have in going to their physicians with their alcohol problems:

> Sixty-eight per cent of the physicians believe that the behavior of women alcoholics differs from that of men alcoholics. The major difference is that women are more protected by their position in the home so it is easier for them to hide the problem ... they are able to hide it for a longer period of time. The women are more secretive about their drinking; they do their drinking at home and they drink alone more than men do. Another difference is that liquor is more accessible to women at home than it is for a man at his place of business. Physicians believe that women minimize the amount of liquor they drink more than men and that they are more sensitive about others knowing how much they have been drinking. Women drink gin or vodka more to keep from being detected. They are more sly, evasive, and sneaky about their drinking.

A study by Kinsey (1966) of alcoholic women in a state hospital setting suggests from a number of indices that this group falls socially between outpatient women and state farm women studied by Lisansky (1957). This may account for the fact that more than half of Kinsey's subjects did most of their drinking " ... in bars, taverns, or cocktail lounges." The younger the women, the more likely they were to do their drinking in public places. Neither Kinsey (1966) nor Fort (1949), whose subjects were members of Alcoholics Anonymous who volunteered for study, report any figures on arrests, but Fort, writing of the "social repercussions of alcoholism," says:

> The general impression (is) that members of the "higher social classes" are more protected against public disgrace....

It should be noted that there will be differences from place to place relating to cultural attitudes and to police attitudes. Sclare's (1970) study of men and women alcoholics in a Glasgow hospital shows the same pattern of women more likely to drink at home and men more likely to drink in the public house. However,

> The female patients had a remarkably high rate (32 percent) of law infringement, the pattern of which, however, displayed merely social fecklessness in comparison with the more violent patterns of the males.

The percentage of male subjects who had been convicted of infringements of the law was 40 percent. These convictions were largely for breach of the peace and theft; the women were convicted for being "... drunk and incapable, or drunk and disorderly." It is interesting to speculate why, if these women drink at home as do their American counterparts, the Scottish police seem readier to arrest them. It is surely a matter of cultural–national differences in alcohol-related behavior and attitudes.

When an alcoholic woman drinks at home, alone or with others, she is less likely to have an arrest record, but drinking at home and low arrest rates are both related to social class. The upper or middle-class woman alcoholic is indeed "more protected" in one sense, but Fort (1949) suggested that the "social repercussions" varied with the women's social status; within middle- and upper-class groups, there is more marked and severe punishment *within* the family, while women of lower social status are more likely to get into trouble *outside* the family.

There is a continuum from private practice and private hospitals, through voluntary agencies and outpatient clinics, through state hospitals, to prisons, and the public versus private drinking of the female alcoholic population in these different facilities will vary along with many other sources of variability such as social class. Differences in the male alcoholic populations of different facilities has been demonstrated by Pattison, Coe, and Doerr (1973). Lisansky's (1957) study demonstrated some of the differences between an outpatient and state farm population of women alcoholics; this should be carried further for women alcoholics as Pattison *et al.* have done for male alcoholics.

Drinking with the Spouse

Where the question is raised, it usually appears that the frequency of drinking problems among the husbands of alcoholic women is considerably greater than the frequency in the general population, and greater than the frequency of their opposite number, the wives of alcoholic men. Figures vary widely. Jacob and Lavoie (1971) report for a French-Canadian sample of alcoholic women that almost 56 percent of them view their husbands as alcoholics. Massot, Hamel, and Deliry (1956) note that more than half, and Basquin and Osouf (1965) note that one third of their French women alcoholic patients had alcoholic husbands. Lisansky (1957) reported 35 percent of outpatient women, and 56 percent of women alcoholics in a prison had husbands who were probably alcoholic. Rosenbaum's (1958) Boston sample reported that half the husbands studied were alcoholic. But Wood and Duffy's (1964) sample of middle-class outpatient women alcoholics "surprisingly" reported that only four out of 69 husbands, less than 6 percent, were alcoholic.

This, of course, does not really tell us whether alcoholic women tend to drink with their spouses; the figures only deal with reported alcoholism on the part of the spouse. There is a related observation made about women narcotic addicts by O'Donnell, Besteman, and Jones (1966):

> The data clearly and consistently indicate that the transmission of narcotics use in marriage was from husband to wife much more often than from wife to husband.

The Lisansky study (1957) found spouses' problem drinking occurred four times as often among outpatient women than among outpatient men alcoholics, and this statistically significant difference seems to be confirmed by subsequent research. It is logical to assume that greater frequency of husband problem drinkers means that husbands of alcoholic women are far more likely to be drinking with them than are the wives of alcoholic men to be drinking with their husbands. In a large sample of alcoholic patients attending a mental health center, men alcoholics report that they are apt *not* to drink with their spouses, while woman alcoholics significantly more often do drink with their husbands (Wanberg and Knapp, 1970). It is, of course, related to the issue of private versus public drinking and social attitudes toward women drinking and becoming intoxicated in public. Men, social-drinking or alcoholic, are more likely to drink in public, i.e., in bars, taverns, cars, at work, and with groups of fellow drinkers. Women alcoholics are more to be found drinking at home so thay they are more likely to be drinking alone or with a member of the family.

Psychodynamics

Variability

Thirty years ago, Van Amberg (1943) wrote of women alcoholics:

> Most of them felt a vague dissatisfaction with their patterns of life, and the causes of this dissatisfaction tended to be varied and highly individual.

The question of relative variability was raised in Lisansky's paper (1957) and we may raise it again. Horn and Wanberg (1971) conclude from factor analyses of drinking history questionnaires, "There is no one type of female alcoholic." This, however, may also be said of male alcoholics. Those who survey adolescent and adult drinking behavior and those who treat alcoholics report more heterogeneity among women. In countries other than the United States, women alcoholics are reported to be a very heterogeneous group (Zetterman, 1966). This may be a reflection of women's greater variability in all things: Her life is more differentiated by physiological shifts into stages than is a man's, and the choice of roles offered the contemporary woman as wife, mother, wage earner, careerist, community activist, and so on is wider. The question of relatively greater variability or heterogeneity among women alcoholics than among men is not resolved but the likelihood is that they vary more. This question of more variability does not seem relevant to some therapists (Blane, 1968), but it would suggest a need for a wider range of therapeutic approaches for women patients.

The greater variability makes it even more difficult to form generalizations about women patients than about male alcoholics. Women seem to vary more in psychological test results although that is not altogether clear. A check on Minnesota Multiphasic Personality Inventory reports on sex differences among alcoholics suggests that ". . . the female alcoholics were more variable (Zelen, Fox, Gould, and Olson, 1966). Another study showed that there were *no* common personality features among women alcoholics although there were several common features among men alcoholics (Mogar, Wilson, and Helm, 1970). On the other hand, both Rosen (1960) and Curlee (1970) found fewer sex differences in the MMPI test results than they expected.

Sexual Role

Because of the complexity of sexual role, various components will be discussed separately: biological phenomena, sex role identity, sexual adjustment, marriage and children.

Biological Phenomena. In his discussion of alcoholism in women, Lolli (1953) wrote:

> Women are more tied than men to their biological selves ... the link between these biological events (menstruation, menopause) and excessive, uncontrolled drinking is often clear-cut.

Three sets of biological events are involved and three relationships may be postulated: relationships between alcoholism in women and (1) premenstrual difficulties, (2) a high incidence of miscarriages, infertility problems, abortions and hysterectomies, and (3) menopausal depression.

The question whether *premenstrual tensions* may trigger off a drinking bout is not really a question about etiology but rather about precipitating stresses. A disappointment, a quarrel, a headache may also act as precipitant. Note that we talk of triggering off a *bout*; a relationship between premenstrual tension and alcoholism presupposes a cyclical pattern of drinking bouts.

The relationship between premenstrual tension and a drinking bout is usually reported in research as based on the patients' reports rather than on objective measures of estrual status. A recent review of literature dealing with the relationship between suicide and menstruation (Wetzel and McClure, 1972) turned up "... many confusing and even contradictory results." Most researchers relied on self-reports, and since many women have irregular cycles, the self-reports are not particularly reliable data. Furthermore, the review reported that two studies, well thought of, reported increased risk of suicide at two quite different points in the menstrual cycle.

There is the question whether premenstrual tension, as reported by women, should be considered a physiological or a psychophysiological response. The complicated relationship between physiology, sexual role, and attitude toward one's own psychosexual functioning is rather startingly illustrated in the not uncommon symptom of the Couvade syndrome; as many as one out of five expectant *fathers* report nausea and vomiting and many admit having morning sickness (Howells, 1972). The question might be put this way: If we could measure women alcoholics with objective measurements, would we find that she does indeed show signs of more "premenstrual tensions" than other women? Could it be one more source of stress to a more vulnerable woman with less tolerance for stress than other women? Is her report of premenstrual tension expressive of her attitude toward being a woman? Is it a rationalization, a defensive excuse for drinking?

Belfer, Shader, Carroll, and Harmatz (1971) reported that of 34 women alcoholics attending an outpatient clinic, more than half of them stated that their drinking began or increased in the premenstruum. However, with the other measures used in the study—tests of depression, tests of anxiety, tests of femininity—there were no differences between those women patients who related their drinking to premenstrual tension and those who did not. The position taken by Lisansky (1957), Podolsky (1963), and others that it is the emotional perception and acceptance–nonacceptance of feminine physiological function that are critical in correlating premenstrual tension and drinking seems to be supported by the data. Premenstrual discomfort is an added source of tension for those women who have problems about being women. Belfer *et al.* (1971) concluded:

> Subtle acceptance or nonacceptance of feminine role behavior, heightened by the perception of premenstrual physiologic changes, may serve as a significant stress. This interaction might be looked for in our observed correspondence between the menstrual cycle and alcohol intake.

The relationship between *a high incidence of various gynecological–obstetrical problems such as infertility, miscarriages, hysterectomy,* and alcoholism has been found by a number of investigators (Kinsey, 1966; Wilsnack, 1972). Kinsey qualifies:

> ... in a majority of the cases, patterns of heavy drinking had usually been established prior to the emergence of the specific physiological problem given by the respondent as the reason for excessive drinking.

This raises three possibilities: First, gynecological–obstetrical problems may *precede* the onset of alcoholism and play a contributing role as cause and/or precipitant; Curlee (1970) has pointed out that most of her women subjects had been drinking before the "trauma" but that the stressful event "... tipped the scale" toward alcoholism. A second possibility is that the prior-existing heavy drinking, i.e., the heavy alcohol intake, played a contributing role in the gynecological–obstetrical difficulties; there is an early paper from the Soviet Union (Vogel, 1919) but little else on this possibility.* The third, and it seems to the writer the most likely possibility, is that there may be a *common* pattern of personality characteristics that are shared by women who manifest obstetrical-gynecological difficulties in general and by alcoholic women. Research on women with psychosomatic menstrual and pregnancy symptoms suggest that these symptoms may be related to dependency, passivity, and subtle manifestations of aggression; and the descriptive terms used for women patients with such symptoms are strikingly similar to terms used to describe alcoholic women

---

* Reports on the fetal alcohol syndrome have appeared since this was written.

(Bardwick, 1971; Howells, 1972). One of the more important research questions for the future is the study of these two groups of women with gynecological–obstetrical problems: those who are alcoholic and those who are not.

There is no question but that a strong relationship exists between depression in the menopausal—i.e., middle age—years and female alcoholism. Curlee's (1969) report on alcoholism and the "empty nest" states it clearly. There is a very strong relationship between alcoholism and depression in general; depression is a concomitant, preceding or coexisting with female alcoholism to a very large extent, and women in general, alcoholic or not, manifest more depressive symptoms than men.

It is not a denial of biological phenomena, but the reports of patients do suggest that further investigation with objective measures and depth interviews may be in order. Women's physiological functions and their disorders are *psychosomatic behaviors,* tied to ". . . emotional adjustment to and acceptance of these feminine physiological functions" (Lisansky, 1957; Podolsky, 1963). Another subject for investigation might be the effectiveness of medication in minimizing anxiety and depression during times of female physiological stress (Belfer *et al.*, 1971) but there are caveats: the overreadiness of physicians to prescribe pills for women, and the tendency of these women patients to generalize drug dependencies.

Sex Role Identity. As Wilsnack (1972) has pointed out, such terms as "role confusion," "masculine identification," or "inadequate adjustment to the adult female role" occur frequently in the literature on female alcoholism. A study by Levine (1955) suggested that a major sexual problem for women alcoholics was frigidity. Divorce, marital disruption, and unhappy marital situations are far more prevalent among alcoholic women than in the general population of women. At this time, there is a good deal of interest in sex role identity studies of alcoholic women. One problem of course is that the measurement of "femininity" or sex role identity is limited by the methods available to researchers, usually a questionnaire. Three of the four studies we will describe involve comparison with nonalcoholic women but we do need to deal with the research question whether alcoholic women have more difficulty in sex role identity than neurotic women, schizophrenic women, and women who present obstetrical–gynecological difficulty. The trend of the evidence is that women with psychological problems do tend to have difficulty with sex role identity. Finally, one great difficulty in studying the sex role identity of alcoholic women is that normative standards of sex role identity are not quite clear. We know they are changing—the percentage of women working outside the home, legalized abortion, and related social phenomena indicate that change—and that these changes vary with age, class, and ethnic group.

Early family interaction reported frequently is a controlling or cold mother and a passive or absent or alcoholic father (Lolli, 1953; Rosenbaum, 1958; Wood and Duffy, 1964; Kinsey, 1966). Again, we need to know to what extent such interactions occur in families of normal women and women who present other kinds of problem behavior. Studies of the effects of different kinds of parental behaviors on sex typing and sex role identity have been reviewed (Sears, Maccoby, and Levin, 1957; Kagan, 1964).

The assumption has been that alcoholic women were, somehow, more *masculine* than other women. For one thing, their psychopathology, i.e., alcoholism, is more common among men. For another, since drinking is more acceptable as masculine than as feminine behavior, women who drink are acting in a masculine way (Curlee, 1967). What may also have influenced thinking was earlier psychoanalytic writing about alcoholism that related it to homosexuality.

In general, findings are that alcoholic women do not reject femininity and the female role but tend, if anything, to overemphasize and overvalue the feminine role. These were the results, for example, with Kinsey's (1966) state hospital subjects and he reasons, in fact, that the women alcoholic's failure as a woman, a "previously internalized" and valued role, becomes ". . . one important source of chronic tension and anxiety" that keeps her drinking.

Belfer *et al.* (1971) compared 34 alcoholic women with ten nonalcoholic wives of alcoholic men. The alcoholic women's scores on tests of anxiety and depression were significantly higher than for the control group. But femininity scores (the femininity scale of the California Psychological Inventory) were "normal" for alcoholic women, *not* significantly different from other women's scores.

Positive attitudes toward the female role, toward femininity and motherhood are confirmed from several sources. Jones (1971) reported from the longitudinal data, the follow-up of the Oakland Growth Study, that women who become problem drinkers in later life are typically described in adolescence as manifesting "ultrafemininity." It is a common clinical observation that women patients are often deeply concerned about their children (e.g., Fort, 1949), and that threat of removing the children evokes intense response. The alcoholic woman's failure in marriage and sexual relationship is frequent, and if she fails as "erotic woman" (Deutsch, 1944), perhaps she puts all the more energy and longing and expectation into her role as maternal woman.

That sex role identity is more complicated than expressed positive attitudes toward the female role is demonstrated in several recent research reports that suggest the conflicts and ambivalences that lie beneath the surface. Parker (1972) gave the Terman–Miles masculinity–femininity test to 56 women alcoholics and a matched group of moderate drinkers, and found that the femininity measured by "role-relevant preferences" was lower, while femininity measured by "emotionality" was higher among the women alcoholics. His findings confirm Kinsey's (1966) suggestion of *conflict* in sex role identity; i.e., Kinsey's alcoholic subjects had not succeeded in adjusting to feminine role requirements but still valued them highly. Parker speaks of "overidentification" as a kind of compensatory mechanism for women alcoholics' "rejection of the female role in other respects."

Wilsnack (1973) administered to 28 alcoholic women and 28 matched normal women several tests designed to gauge (1) gender identity or unconscious masculinity–femininity, (2) sex role style, and (3) attitudes and interests or conscious masculinity–femininity. On the latter measures of conscious femininity, the alcoholic women appeared to value the maternal role ". . . perhaps to an exaggerated, 'hyperfeminine' degree." However, on sex role style and unconscious masculinity–femininity measures, the alcoholic womens' responses were more masculine than the those of the controls. The alcoholic women also had a significantly higher incidence of obstetrical and gynecological disorders than the controls, possibly another indicator of their conflicted sex role identity.

What emerges in the work done to date is not a picture of sex role identity that tends toward the masculine. On the contrary, such words as "overidentification" (with female role) or "hyperfeminine" are used. What does emerge is a theme of frustration–aggression: The "ultrafemininity" (Jones, 1971) of the adolescent years is still manifest, but the frustration and anger are now acted out in the drinking.

A description of a group of women ambivalent about being female defines.

> . . . an infantilizing vicious cycle in which they accept responsibilities (especially those that are normal in the traditional female role), are frustrated by failure (and, based on a lifetime of experience, probably anticipate failure), feel anger, fear rejection if they act upon their anger, feel guilt about their hostility . . . (Bardwick, 1971)

These are women with psychosomatic gynecological–obstetrical symptoms, and the symptom ". . . obliquely expresses hostility and removes the sufferer from the reality of an unsolvable conflict" (Bardwick, 1971). So does alcohol. The next research question might be why some women express the ambivalence and resolve the conflict by psychosomatic symptoms and some by alcoholism.

Sexual Adjustment. Since the available information on sexual behavior of alcoholic women was last summarized (Lisansky, 1957), there is little to add. Levine's (1955) observation of the lack of sexual interest of many alcoholic women, and the description by Massot *et al.* (1956) of ". . . disturbed sexual adjustment, including repulsion of men, frigidity, guilt feelings, and masochism," just about sum up the state of our knowledge. A more recent study (Kinsey, 1966) of state hospital alcoholic women indicates that no less than 33 out of 46 subjects report frigidity in sexual relations. It is not clear but unsatisfying sexual relationships are implied in Rosenbaum's (1958) description of "marital conflict" as a major precipitant of drinking among her subjects; and Wood and Duffy (1964) comment, "The outstanding situation in the current lives of our married patients has been an emotionally unrewarding marriage." It is probably fair to say that most alcoholic women find their sexual lives unfulfilling, although the extent to which this appears as compared with a similar female population, psychiatric or general, we must guess.

The question of "promiscuity" has come up less often lately as sexual standards of behavior shift. There is also more awareness of the alcoholic women's pattern of drinking alone at home which is seen so frequently. If promiscuity is many casual affairs, it is undoubtedly true that some alcoholic women are promiscuous, and the chances are that the more public the drinking, the more the likelihood of promiscuity. Fairly casual sexual relations exist among individuals and groups who drink relatively little. Much is said about new sexual freedom, alternatives to marriage, suburban spouse-swapping; how this would apply to women with drinking problems, we don't know. In the absence of data, each one decides the issue for himself. Thus, the doctors asked by Johnson (1965) about their views of women alcoholics tend on the whole to agree among themselves, but:

> There was disagreement as to the woman alcoholic's moral code. Some felt that she had loose sexual morals, had more psychosexual conflict as homosexuality, and was more likely to get into social difficulties. Others felt that the married women kept good contact with their families. They stated that within the family, they seldom stray away as the male alcoholic has a tendency to do.

Marriage and Children. We do have information on the marital status and marital disruptions among alcoholic women. They are no different from the general population in the percentage who marry (remember the picture of adolescence and the wish for feminine satisfactions among these potential alcoholics), but the number of marital disruptions is significantly greater than among the general population. The single woman alcoholic, the woman who has never married, and the widowed woman alcoholic who loses control of her drinking after her husband dies are rather special cases, the kinds of failures that are different from those involved in broken marriages. But the one-third to two-thirds incidence of broken marriages reported in Schuckit's (1972) literature review is high.* It is noted by many investigators (Fort, 1949; Lisansky, 1957; Rosenbaum, 1958; Kinsey, 1966; Curlee, 1970), further, it is stated by Pemberton (1967): "Disturbed marital status in the females was related to a poor prognosis."

There are no research reports about the husbands of alcoholic women. We

---

* One would wonder as the divorce rate for all American marriages goes up whether the divorce rate for marriages in which one or both partners are alcoholic also goes up. Maybe, with the divorce rate for the general population rising, the gap is narrowing?

know that it is far more usual in a marriage in which the wife is alcoholic for the husband to be an alcoholic, a problem drinker, or at least a heavy drinker, and for the two of them to drink together; this occurs far less frequently in the marriages of alcoholic men. Lindbeck's (1972) review of the literature makes some clinical observations about the husbands of alcoholic women: She notes, for example, the extensive use of *denial* by the husband. In case conferences, the writer has heard therapists remark about the subtle ways in which a husband may encourage his wife's drinking because she is more sexually responsive, because it relieves his guilt, because it makes him an object of sympathy, and for a wide variety of reasons, constructive and destructive. Lindbeck (1972) comments:

> The nonalcoholic spouse of the alcoholic woman is described as reacting to his wife's drinking, once he ceases to deny its existence, either protectively, with bewilderment, unforgivingly, or sadistically, and sometimes in a variety of these ways.

In a paper entitled "Do wives drive husbands to drink?" (heavy and implicative drinking), the authors (Zucker and Barron, 1971) report that there is no relationship between the wife's personality and her husband's drinking behavior, but "... if anything, there is more evidence that the husband's personality shows a relationship to the wife's drinking."

Let us go back a moment to the picture of the woman alcoholic as adolescent. She has a great deal in common with the woman who later becomes an abstainer, but she differs in two important ways: less impulse control and less capacity to accept a dependency relationship. The potential for trouble lies in the intense need for love and reassurance, relating to some of the deprivations of early childhood, coupled with the block in accepting dependency relationships. Adolescence is a crisis period, and the potential woman alcoholic "... escapes into ultrafemininity" (Jones, 1971). Fundamentally distrustful, she looks to marriage and a man for ultimate fulfillment, and more often than not "... marriage is ... a painful, disappointing experience" (Lindbeck, 1972).

Blane (1968) has described the alcoholic woman as one who, frustrated in childhood, develops more than usual dependency needs and "... an aggressive insistence that her needs be fulfilled"; she derives her sense of worth vicariously, and responds to the personal crisis of loss by turning to alcohol. We would add to this her difficulty in accepting dependency relationships, based on lack of trust, and her difficulty in impulse control.

The absorption with the maternal role, the greater than average maternal manifestations reported by Wilsnack (1973) would appear to be reactive to failure as wife and sexual partner. Asked how many children they would have liked to have had, Wilsnack's women alcohol subjects averaged 4.15 and the controls 2.94, a significant difference.

One should not lose sight of the fact that from birth on, the two sexes are trained quite differently for parental roles. That the male alcoholic who is in many ways immature himself should often not be a conscientious, protective father is not so surprising. Women are trained for their parental roles and this is the aspect of her femininity that alcoholic women frequently cling to. Fort (1949) observed in her group of women from Alcoholics Anonymous:

> ... social discrimination against the children, especially against the daughter ... is to the woman alcoholic a great shock while it may not even be noticed by an alcoholic father.

Depression

The very significant amount of depression manifested by alcoholic women has been noted in all research studies. It is, of course, a common symptom among male alcoholics as well. Its importance in female alcoholism was highlighted in a study by Winokur and Clayton (1968), who reported significantly more secondary diagnoses of depressive reaction and significantly higher incidence of suicidal thoughts and delusions among female alcoholics compared with male alcoholics. Schuckit, Pitts, Reich, King, and Winokur (1969) have carried this further, distinguishing between two groups of women alcoholics, those with primary alcoholism and those for whom alcoholism is secondary to affective disorder. Schuckit (1972) has also presented data indicating that the latter group tends to be younger and have a shorter duration of alcoholism, and despite more suicide attempts, their prognosis is better than the primary alcoholics.

Observations of depression associated with female alcoholism have been made in a number of different places: in Quebec (Jacob and Lavoie, 1971), in Scotland (Sclare, 1970), in England (Rathod and Thompson, 1971), in France (Lengrand, 1964), as well as in the United States. While it is true that the alcoholic behavior produces all sorts of social and personal difficulties that compound the depression, there is strong evidence that in many alcoholic women, the depressive disorder may *precede* the loss of control of drinking. Curlee's (1970) private hospital population showed big differences between men and women patients on this score: More women had been psychiatric patients, had more frequent hospitalizations and for longer periods, virtually all of this related to depression. Rathod and Thompson (1971) report about their English patients that 30 percent of the women and none of the men had ". . . affective (depressive) disorders needing treatment before the onset of problem drinking." And Jones (1971) in the Oakland Growth Study follow-up, compares men and women problem drinkers and finds that although they have characteristics in common, e.g., dealing with difficulty in impulse control, the women uniquely show ". . . depressive, self-negating, distrustful tendencies."

Schuckit's point is well taken. If depression appears almost universally among alcoholic women, it becomes important that we learn about the symptoms and severity and their chronology. Depressive breakdown *before* the onset of alcoholism, and depression *reactive to* the rejections and feelings of failure intrinsic to the alcoholism, may orient the therapist to deal with the patient in different ways.

The point should be made that severe depression and manic-depressive psychosis is a much more frequent diagnosis for women psychiatric patients than it is for men (Chesler, 1972). Although females have no monopoly on depression, it is more characteristically a symptom of women in the United States than of men.

Other Characteristics and Behaviors—Speculations

We will speculate. Taking into account the great heterogeneity among women alcoholics, we will try to draw a typical history of a middle-class, white woman alcoholic. Our woman has had more parental deprivation than her peers, and like most other girls, she enters adolescence looking for satisfactions mainly drawn from interpersonal dealings. She may need reassurance more than other girls, but she carries attitudes and behaviors that make that reassurance less rather than more available. For one thing, she is *distrustful* and unable to form healthy dependency relationships. She also does not have sufficient social talents and interpersonal skills to satisfy her great need for closeness to others. One compensation is to become *ultrafeminine,* to be *submissive,* to look for a husband, to buy wholly and try and live the traditional female role. Later, after failure and the development of drinking problems, we see emergent some of the *difficulties in impulse control.* If there are two characteristics that describe the woman alcoholic, they are *social isolation and anger.* Jones (1971) wrote of her that she is "... judged to be submissive as youngster, rebellious as adult." Wood and Duffy (1964) say that their patients "... grew up submissive and passively resentful," and they describe very well the role of alcohol for their patients:

> Alcohol, initially the "enabling drug" which provided the missing feelings of adequacy and acceptability, became later the releaser of hostility, tool of revenge, and means of self-punishment. All patients exploded their anger when drunk.

Another group of women vulnerable to alcoholism are relatively empty women who are quite devoid of interests of their own but live with reasonable success vicariously through their husbands and/or children. Some of these women turn to alcohol when they are bored, i.e., when this pattern is not so successful. These women often turn to alcohol when the marriage ends, one way or the other.

The descriptions given by the physicians Johnson (1965) tell us how alcoholic women often appear to trained observers:

> ... (compared to male alcoholics) they are more hostile, angry, unhappy, self-centered, withdrawn, depressed, and more subject to mood swings; they are more emotional, lonely, nervous; they have less insight, and are not as likable as men alcoholics.

The alcoholism serves to express anger and punish the people around her, but it is self-defeating and deepens the isolation and loneliness that triggered it in the first place. She still longs for femininity; both Kinsey (1966) and Wilsnack (1972) emphasize their alcoholics' identification with and high valuation of the traditional female roles. Perhaps there is more emotional investment in the maternal role because the role of wife and lover has not been so successful. Although there are homosexual, Lesbian women alcoholics, they seem to be a special subgroup: The typical woman alcoholic is a woman who has not been very successful in the female role, although she has valued it highly.

Complications of Alcoholism

Medical Disorders

It was noted some years ago (Spain, 1945) that death from portal cirrhosis seemed to occur more frequently among women alcoholics than among men. There are varying reports: A Boston study (Summerskill, Davidson, Dible, Mallory, Sherlock, Turner, and Wolfe, 1960) showed a relatively high ratio of cirrhotic women alcoholics; and recent research in Australia (Wilkinson, Santamaria, and Rankin, 1969; Wilkinson, Santamaria, Rankin, and Martin, 1969) suggests that alcoholic cirrhosis occurs more commonly among women alcoholics. However, Sclare's (1970) women patients showed less hepatic and cerebral damage than did the men patients. In the United States, a recent report (Barcha, Stewart, and Guze, 1968) notes that 25 percent of alcoholic males and 39 percent of alcoholic women on general hospital wards presented one of these disorders: peptic ulcer, cirrhosis of the liver or fatty liver, and pancreatitis.

Spain (1945) suggested the possibility of more faulty fat metabolism or a more "severe degree" of alcoholism among the women as an explanation of differences. Wilkinson et al. (1969a) found the prevalence of cirrhosis to increase slightly with age among men but to increase quite markedly among women 40 to 60 and they hypothesize that the hormonal imbalances of menopause might predispose the liver to the toxic effects of alcohol. A paper from the Soviet Union (Averbukh, 1966) suggests that women are "... much more sensitive" to the toxic effects of alcohol and hence show more of the somatic disorders of alcoholism than do men.

Mortality and Suicide

In an earlier paper, Schmidt and de Lint (1969) found women alcoholics resembled men alcoholics in "mode of death," and suggested that these data contraindicated ". . . the clinical impressions concerning sex differences in alcohol use and mental health." However, a more recent paper by the same authors (1972) reports excess mortality to be highest among younger alcoholics and particularly among women alcoholics:

> . . . alcoholic women are as likely to die as alcoholic men, but since women in the general population have lower death rates than men, a sex difference obtains in excessive mortality.

Insofar as suicide goes, suicide rates for both men and women alcoholics are higher than suicide rates for men and women in general (Palola, Dorpat, and Larson, 1962). One finds the same sex differentials between *attempted* and *completed* suicide that seem to prevail in the general population. Suicide is much more common among men than among women in all the Western countries, but it appears that suicide *attempts* occur more frequently among women than among men (Clinard, 1968). Palola *et al.* (1962) report a study in which men outnumber women in *both* attempted and completed suicide; this may relate to the fact that the study was done in the state of Washington, where a suicide attempt is a criminal violation.

Winokur and Clayton (1968) report a higher frequency of suicidal *thoughts* among women patients than among men patients. Pemberton (1967), Curlee (1970), and Rathod and Thompson (1971) report a higher incidence of suicide *attempts* among alcoholic women subjects than among alcoholic men. And there is evidence that shows that women alcoholics actually commit suicide less often than do men (Rushing, 1969; Schmidt and de Lint, 1972).

But both men and women alcoholics commit suicide more frequently than the general population. This is a complex problem, involving as it does issues of sex differences in depression concurrent with alcoholism, the acting out of self-destructive impulses, the cry for help, manifestations of anger and hostility. A comparative study needs also to take into account alcohol-related "accidents" in cars and in the home.

Use of Drugs Other Than Alcohol

In the United States, there is evidence that women alcoholics tend to use drug substances other than alcohol to a greater extent than do men alcoholics. Curlee (1970), comparing the same number of male and female alcoholics admitted to a private treatment center, noted that more than twice as many women as men reported use of minor tranquilizers and sedatives. Of the number using, again more than twice as many women as men were considered to have problems relating to usage. Horn and Wanberg (1971) report that women with alcohol problems are more likely to be characterized ". . . by a solitary-drinker syndrome, in which use of drugs other than alcohol may be prominent."

Here we find some interesting national differences, because two papers from Great Britain report *no* significant differences between men and women alcoholics in drug abuse (Sclare, 1970; Rathod and Thomson, 1971). The British and United States histories of dealing with drug problems have been quite different and the male-to-female ratio of alcoholism is also different. The proportion of British alcoholics who are women is higher. It would seem that national differences in usage and attitude and law relate to these results, and this question merits further research.

Within the United States and Canada, the greater usage by women of tranquilizers and sedatives has appeared in a number of research reports (Differential Drug Use Within the New York State Labor Force, 1971; Smart, Laforest, and Whitehead, 1971; Cooperstock, 1971). Cooperstock (1971) found 69 percent of *all* prescriptions written for mood-modifying drugs to be for women, and this compares with a 67 percent finding in the United States, nor can this be accounted for simply by sex ratios of patients' visits.

We have, on the one hand, a possible greater degree of vulnerability to substance dependency among women alcoholics, and a greater tendency on the part of physicians to prescribe medication. Cooperstock (1971) writes.

> Much evidence exists to demonstrate that holding diagnosis constant, women patients receive more mood-modifying drugs than men. The study of psychiatric illness in general practice (Britain) found that the only types of treatment that distinguished the male and female neurotic patients were the proportions of tranquilizers, stimulants, and sedative drugs given them. . . . Among the alcoholic patients treated in the various clinics (Canada), women patients also received a disproportionate share of mood-modifying prescriptions.

If anything, this is probably a conservative picture of the prescription situation in the United States. It is hardly surprising, considering U.S. drug abuse problems and the easy availability of some drug substances, the vulnerability of the women involved, and the tendency for medication to be prescribed more readily and more often than for men, that women alcoholics use drugs other than alcohol more than do male alcoholics. This is a complex issue relating to the whole drug scene but one should be very cautious in interpreting the greater use of drug substances other than alcohol by women alcoholics as a sign of more profound pathology or poorer prognosis.

Differences: Social Class and Race, Religion, and Ethnic Background

Social Class

In a lecture delivered 30 years ago at the Yale Summer School of Alcohol Studies, Dollard (1945) described the "drinking mores of social classes," and contemporary surveys indicate that those descriptions were reasonably accurate—with one major exception. Dollard's description of "lower-lower" class drinking behavior included "socially unrestrained" weekend binges for both men and women, although he did make the point that the sexes usually did not drink together. But it turns out that, with age, regional, and rural-urban variations, a much smaller percentage of low-income women drink. Put another way, there are far more abstainers among poor women than among middle- or upper-income women (Lawrence and Maxwell, 1962; Knupfer and Room, 1964).

Virtually all the literature on the alcoholic woman, derived as it is from study of patients who come to doctors' offices, outpatient clinics, family agencies, and public and private hospitals, deals with the "respectable" woman alcoholic. Usually, she has not been arrested, and when we see her, she is not serving time in a penal institution. But there *is* a population of alcoholic women in these institutions, and when they were compared with an alcoholic group for an outpatient clinic (Lisansky, 1957), a great number of differences emerged: age, educational achievement, occupational level, disruption in early family life, marital conflict, public drinking, and sexual promiscuity. Since that time (1957), a number of other prison studies have appeared. Several of these are reports of clinical observations and problems of therapeutic work. Extreme early and present deprivation and extreme isolation are characteristic, and there is much promiscuity and acting out (Myerson, 1959; Mayer and Green, 1967). It has been noted several times that these women alcoholics have virtually no social resources; there are no helping or supportive persons in their social environments (Lisansky, 1957; Myerson, 1959).

There are two reports by Cramer and Blacker (1963, 1966). The earlier report compares women's reformatory inmates who have had drinking problems since adolescence with other inmates whose drinking problems started *later* in life. Disturbing childhood experiences were found to be more characteristic of "early" than of "late" drinkers. Those who were *later* problem drinkers seem to be less disorganized and more like the "respectable" women alcoholics. In their second report, Cramer and Blacker found that when they made comparisons within a group of female drunkenness offenders on the basis of ". . . small differences in status," there were social class differences even within this relatively homogeneous group:

> Female drunkenness offenders who originated in modest but respectable homes tended to be more isolated from social contacts, to display a different drinking style, to indicate a different developmental history, and to define themselves differently as drinkers.

These are the kinds of differences found in research when one stops thinking of the "economically disadvantaged" or "poor" or "low income groups" as an undifferentiated unity and, indeed, it turns out that all sorts of differences exist between the respectable, working-class poor, and the stormy, disrupted, multiproblem poor often in trouble with the law.

Cramer and Blacker's female drunkenness offenders from "modest but respectable homes" seem to have much in common with Kinsey's state hospital sample (1966). There is a social continuum within the research literature: from private facility, through outpatient clinics, through state and city hospitals to penal institutions. This is similar to what has been described by Pattison, Coe, and Rhodes (1969) in their comparison of different treatment facilities. As was true of Pattison, Coe, and Doerr (1973) subjects, there is every reason to believe that populations of women alcoholics at different facilities differ in social competence; and furthermore, the philosophy and methods of facilities need to be matched to the specific needs of women alcoholic subpopulations as well as to the needs of men alcoholics.

There is a study of drinking behavior of delinquent girls (Widseth, 1971; Widseth and Mayer, 1971) in which 40 percent of the 100 girls charged with delinquency in several Massachusetts detention centers were identified as "high risk" heavy drinkers. Several findings are significant. Although the girls were between the ages of thirteen and sixteen, . . . the incidence of excessive, solitary, and pathological drinking was high." While delinquent boys almost never report drinking alone, as many as 30 percent of the girls in Widseth's sample got high or drunk when alone; might this be how the more public drinking of the adult female alcoholic prisoner gets started? Heavy drinking in this group of delinquent girls is related to more intense feelings of rejection by mother and to hostility toward mother; the heavy-drinking girl feels rejected by and rejects her mother in turn, refusing to seek help from her. Once again, we have a pattern of experienced emotional deprivation, plus difficulty in dependency relationships. It would be illuminating to examine the data of Widseth's heavy-drinking, delinquent, young adolescent girls side by side with the data of Jones's (1971) women problem drinkers as adolescent girls.

Race, Religion, and Ethnic Background

The drinking practices of Americans as they vary by race, religion, and ethnic background are discussed in contemporary surveys (e.g., Cahalan *et al.*, 1969), and although such surveys invariably contain analysis by sex, it is rare that the data are analyzed in terms of sex differences *within different American subcultures*. Thus, we may know something of Italian-American drinking habits but we have little information about the differences and similarities (percentages who drink, where, what, with whom, etc.) in the way men and women drink. Further, the groups that have been investigated most vigorously in the past have been the Irish, the Italians, and the Jews. We are not well supplied with information about the drinking behavior and problems of blacks, Chicanos, Puerto Ricans, or—for that matter—Americans of German, Slavic, or Scandinavian origin.

Recently, several reports on black drinking behavior and problems have appeared (Robins and Guze, 1971; Sterne and Pittman, 1972). Thirty years ago, Jellinek (1942) reported that the black mortality rate from "death from alcoholism" was considerably higher than the white rate, and that the high black rate was due primarily to very high rates among black *women*. This finding was restated and reaffirmed by Bailey, Haberman, and Alksne (1965) in a study of the epidemiology of alcoholism in a residential area of New York City. In this survey, the white male-to-female ratio was 6.2 to 1, but the black ratio was 1.9 to 1. Again, Cahalan *et al.* (1969) in their survey of American drinking practices report that white and black men varied little, but that black women were quite different in drinker percentage rates than white women: a much higher proportion of abstainers and a higher proportion of heavy drinkers. Abstainer women percentages reported by Cahalan *et al.* (1969) were 51 percent black and 39 percent white, and heavy drinker percentages were 11 percent black and 4 percent white.

There are several speculations about these findings. The higher proportion of abstainers among black women may be related to low income, active membership in abstinent Protestant churches, or a particular set of historical conditions; e.g., Sterne and Pittman (1972) found in an urban study that southern rural-reared black women were less likely to be heavy drinkers than the city-bred black women. (This is generally true: Drinking rates for rural women are lower than for urban women.) The higher proportion of alcoholics or problem drinkers has also been related to more frequent head-of-household status (Bailey *et al.*, 1965) and to "particularly vulnerable occupations" (Alcohol and Health, 1971). It does seem to be true that public drinking, i.e., drinking in taverns, is more socially acceptable for women in the black community, although this generalization needs analysis in terms of age and social class differences within the black community. A study of a black urban community and its drinking practices (Sterne and Pittman, 1972) turns up some interesting information: When heavy-drinking black women are compared with more moderate-drinking black women, they turn out to be no more often heads of households, but they are more apt to drink in taverns and to drink more often for reasons of "personal escape." The heavier drinkers are less apt to be regular church attenders, less oriented toward "respectability," and more permissive toward men's drinking, and they are less often southern-rural in background. Significantly, they are more likely to have other heavy drinkers at home than are the other women and there are indications that they drink with their spouses or other adult family members, apparently a universal phenomenon for *all* women problem drinkers.

There would seem to be enough difference suggested in the role of alcohol in black and white communities to merit further study.

Therapy and Prognosis

There are a number of studies of physicians' attitudes toward alcoholism and alcoholics (e.g., Jones and Helrich, 1972) and they confirm the fact that alcoholic women tend to seek help from physicians to a greater extent than do alcoholic men. However, such reports say little about the physicians' viewpoint. The only report of this viewpoint as it pertains to alcoholic women is by Johnson (1965): The physicians ". . . believe the woman alcoholic to be a sicker person than the man alcoholic." It should be noted that less than 15 percent of the women went to the physicians with a specific complaint about alcohol problems, and while most of the physicians report their awareness of the patient's alcohol problem, many of them were loath to deal with it in a direct and open way.

We are in a circular situation: All the indices of deviance that we have indicate that the woman alcoholic is more different from nonalcoholic women than the male alcoholic is from nonalcoholic men. We are unclear about the specifics of prognosis, but in general we know that the greater the ego strengths and the better the social integration of the patient, the better the prognosis, regardless of which kind of mental or behavioral disorder we are considering. From what we know of the woman alcoholic her prognosis should indeed be poorer than that of a male alcoholic's, but what role does the defeatist attitude of the therapist play? In psychotherapy, the optimism or pessimism of the therapist can play a subtle, significant part in the outcome.

Who is the worse patient? As noted in the Introduction, there are several studies that suggest that female psychiatric services are noisier and more frantic. In a report from Italy about women alcoholics with tuberculosis (Madeddu, Lucchesi, Rivolta, and Spinola, 1969), the authors state that their prognoses are worse than male alcoholics', but at the same time they point out that in the sanatorium, the women's behavior is *more* satisfactory since they show more independence and cooperation than the men. And Garai (1970), reviewing the literature on sex differences in mental health, report that women accept being mental patients more readily. We are into complex value judgments of "good" or "bad" behavior as judged by different criteria.

When outcome of treatment or therapeutic success or failure is the measure, the results show women alcoholics as doing worse. Glatt (1961)reported almost twice as many male alcoholics as "recovered," 38.6 percent as compared with 20.8 percent of women patients. Those rated as "not improved" comprised 12.8 percent of the men and 25 percent of the women. The same kinds of figures were reported by Pemberton (1967): Successful patients comprised 20 percent of the women and 36 percent of the men; and half the men but 72 percent of the women were rated as "unsuccessful" in treatment. Both Glatt's and Pemberton's patients were British but this unfavorable comparison is also true of American patients. Curlee (1970), reporting on a group of middle-class, relatively well-integrated alcoholics, says:

> Both groups (men and women) had a few persons who were "chronic" patients, in the sense of having been in and out of treatment repeatedly. By almost any standard of chronicity, in terms of repeated treatment, more women would be included in this group.

But prognosis will, of course, vary with a complex of personal strengths and liabilities and the social–familial situation in which the woman alcoholic finds herself. One such source of variability is pointed up by Schuckit (1971), who distinguishes between women whose drinking problem is associated with an affective disorder (depression) and women who manifest "primary alcoholism," i.e., alcoholism antedating an affective disorder. He demonstrates with a three-year follow-up that the outcome is much more often favorable for the former group. Although the problem remains about the depressive symptoms, it would appear that the depressed women who lost control of their drinking, those who attempt to cope with depression by drinking, have a better prognosis than those without affective disorder. However, since "good outcome" is defined as being free of drinking problems, it is necessary to point out that good prognosis here means that the patient's drinking has been contained. This is not to minimize the achievement but to point out that *prognosis* here means less drinking, not necessarily less depression.

Other variables related to prognosis are discussed by Pemberton (1967), comparing the outcome of treatment of male and female alcoholics. He describes the better response of the male patients to "routine treatment":

> ... the males tended to support each other, to discuss their difficulties with each other and to conduct a kind of informal spontaneous group therapy ... this state of affairs was rarely found on the female side.

Pemberton also writes of "... the comparatively poor response of the widowed, separated, and divorced female group," and he describes these women as particularly dependent on their husbands for interests and contacts outside the home. If one put this together with Curlee's (1969) description of some middle-aged women alcoholics coping with an "empty nest," we see perhaps a subgroup of women alcoholics who have been overdependent on husband and children, and who cope with aging and loss by turning to alcohol. Finally, Pemberton notes about the women "successes":

> ... it was quite striking that all had succeeded in modifying the structure of their familial group and had established for themselves a personally satisfying role within it.

It is striking indeed. And it suggests how very much a study of the family interactions of alcoholic women is needed. All marriages involve ambivalences, but it is a safe guess that the constructive, positive feelings of the husband or the absence of such feelings may be the pivotal point of psychotherapy. We know that women with drinking problems tend to be very concerned about their chifldren, but we should study husband–wife and parent–child relationships within the family of the married alcoholic women, and parent–daughter and sibling relationships within the family of unmarried alcoholic women. One of the major keys to prognosis lies in this familial–social environment.

Another report of successes and failures among women alcoholics involves volunteers from Alcoholics Anonymous (Dunlap, 1961). Some of Dunlap's observations about *unremitted* women patients, based primarily on psychological test results, include: a greater need for succorance and less capacity to form dependent relationships and poorer control over aggressive and hostile impulses, which tend to appear often as depression. This would certainly fit with the observations we have made about intense needs, conflict over dependency relationships, poor impulse control, hostility, and depression. The women in Dunlap's sample who had recovered were more conforming, presumably more anxious and concerned about public opinion. They were also more "introspective." Pemberton (1967) has noted that the women who responded successfully to treatment tended to be "introverted." These terms may refer to differences in capacity for insight or simple differences in capacity for insight or simple differences in capacity to think of their problems in psychological terms.

As for *different kinds of treatments,* we have already discussed the apparent greater readiness of physicians to prescribe medication for women patients. Whether this relates to a different perception of the caretaker role as it relates to men or women patients, or whether women require more tangible evidence of the doctor's concern and interest, we do not know. The preferences and benefits of individual and group psychotherapy for men and women patients are suggested by Pemberton's (1967) observation that male patients conducted a kind of "informal, spontaneous group therapy" while women did not. Curlee (1971) studied sex difference in patients' attitudes toward alcoholism treatment and she reports that ". . . women are more responsive to individual treatment while men respond better to group activities." Curlee's men and women subjects rated lectures, contact with Alcoholics Anonymous, and counseling as most helpful, but there did appear to be an important difference in the reaction of men and women to group situations. A report on German women alcoholics suggests that they were more effectively helped with group therapy than with individual therapy (Battegay and Ladewig, 1970), but this may be because of a particular set of circumstances or a particular therapist's talents. It makes good sense that alcoholic women—fundamentally loners as adolescents, secretive about their drinking, social isolates—will be more responsive to one-to-one contact than to group therapy. Could it be that this is true of women patients in general, not only of alcoholics?

DISCUSSION

Do we really need to make the point that men and women are different and not only in biological ways? We accept the fact that there is common ground for men and women alcoholics, similarities in etiology, choice of substance, stages, effects of alcohol, social–personal side effects of drinking, response to therapeutic interventions, and so on. Common ground—*and* differences. We should know more about these differences, and for practical and theoretical reasons. Practically, we want to help women with drinking problems, to be more effective in case finding and intervention. But knowledge about alcoholic processes and behavior of alcoholic women tells us something in two other vast areas of social research: *alcoholism* itself as a form of deviant behavior and the role and status of *women,* strains and pathological manifestations.

Although there are many more women working outside the home than there was a generation ago (and we tend to forget that working outside the home was taken for granted as part of the low station of poorer women, to say nothing of slave women, since the Industrial Revolution began, and that a large proportion of women work now to supplement an income to raise living standards), the fact is that on the whole, women are at home to a far greater extent than men and the woman alcoholic is indeed more protected and covered in her drinking. She is also more despised and rejected for it. There is evidence that she is more different from nonalcoholic women than a male alcoholic is from nonalcoholic men, but we summarize these differences with the phrases "more deviant" or "poorer prognosis" without thinking through sufficiently what the implications or what the phrases mean. "Poorer prognosis," we can be sure, tends to become a self-fulfilling prophecy. We do have evidence that alcoholic women deny their problems longer and resist therapeutic intervention (e.g., conventional group therapy approaches) more. It would be foolish to attribute all the problems in working with alcoholic women to societal condemnation, as foolish as to reject her as a moral transgressor and sinner. At the same time, alcoholic women are more responsive than alcoholic men to concern about their families as a motivational base for psychotherapy and other therapies.

We know more about the middle-class, white alcoholic woman than about others and we think some of these factors are important: emotional deprivation in early life, intense needs for love and reassurance, a stormy adolescence with a characteristic pattern of submissiveness, much anxiety about dependency relationships, and a potential difficulty in impulse control. For the potential women alcoholics, marriages fail to a greater extent than is true of the general population and the sense of something missed and something missing is intensified. Conflict between the chosen feminine, passive role and aggressive, angry feelings become acute. Alcohol fills a void, acts as instrument of revenge and punishment for others, acts also as a means of self-punishment and self-abasement. If the woman is a "closet alcoholic," her social isolation deepens; if she drinks with others, her circle of acquaintances narrows down increasingly to drinking companions. Often she is married to a heavy drinker or alcoholic.

Of course, this description does not fit all white, middle-class women alcoholics. There are some who have experienced little deprivation or emotional suffering, some whose lives have been relatively easy and who drink out of sheer boredom and a search for excitement. There are others who manage their lives well enough for long periods by living vicariously through other people, often husbands or children; when loss by death or other means occurs, these women are devoid of resources, lonely, and frightened, and often alcohol serves to numb the fear and pain of aloneness.

If the women are white and poor and not from stable working-class backgrounds, they have often a myriad of problems of which problem drinking is only one. Often the alcoholic women we see in prisons have been prostitutes and their relations with men are disastrous. Relations with their mothers and other relatives and with their children are poor; they are utterly devoid of helping persons. They probably have much in common with female narcotic addicts.

Very little can be said about the black woman alcoholic. Drinking in the tavern by women is more accepted in the black community as it is in the English pub, but there are black class differences in drinking behavior and attitude that have not been studied. We know that in the whole black American community, a smaller percentage of women drink than is true among whites, but heavy drinking and alcoholism seem to occur more often among those black women who do drink than among their white counterparts.

Suggestions for Research

There are all too many questions we can ask to which we have no answer but four areas of research seem promising as places to start:

1. *The Long-Standing Problem of the Definition of Alcoholism.* Once the heavy drinking starts, women's problems are likely to intensify within the family; with men, trouble at work is more likely. Should we then use the same criteria for the definition of problem drinking and alcoholism for both men and women? Put another way, the economic ill effects of implicative and repetitive

drinking are less likely among women but the social or interpersonal ill effects are more likely. Other things being equal, do alcoholism-related diseases occur differentially among men and women and are these differences in rate, course, and prognosis? The "natural history" or order in which different symptomatic behaviors relating to alcohol appear are probably different for men and women. And how does alcoholic behavior relate to other deviant behavior syndromes?

2. *The Study of Symptom Choice.* Our potential women alcoholics have histories of deprivation, stormy adolescence, loss, and psychiatric disturbances in the family, but this hardly distinguishes them from other groups of psychiatrically disturbed women. Why is ingestion of a mind-altering substance so effective a method of coping-via-escape for them? Why alcohol? What are the differences among women who use only alcohol, women narcotic addicts, and women who use alcohol and a wide spectrum of other drugs? Availability, age, social class, ethnic group differences? What differences exist in behavior patterns and interpersonal relationships?

3. *Different Kinds of Therapeutic Interventions.* New ideas and evaluation of effectiveness are general problems in research on therapeutic intervention, but there are some cues about women alcoholics in the research literature that we should note. Group therapy may not be as effective with women alcoholics as it is with men, but women may be more responsive to a one-to-one approach. In the shortage of trained therapists, how effective would ex-alcoholic counselors, male and female, prove to be in one-to-one therapy? How effective would it be to put an intense effort into conjoint therapy, to work closely with husband, parents, siblings, and children of alcoholic women patients. The husbands tend often, as reported in the literature, to deny the problem, but often the husband is protective and concerned. How can we utilize this feeling therapeutically? Research suggests that where a marriage can be improved, prognosis is greatly improved, and perhaps major therapeutic effort needs to be bent in that direction. It is commonplace to comment on the absence of research on the husbands and lovers of alcoholic women. Perhaps we should start with the husbands who are willing to become involved in the wife's treatment, see when it works and when it fails, and study the factors in the husband–wife relationship that make for better prognosis.

4. *Crosscultural Studies.* The status and role of women, the attitude toward women's drinking and the incidence of female alcoholism needs to be researched in cross-cultural studies. The Spanish and Portuguese, the Jews, the Italians, the Chinese, and other groups have low rates of alcoholism for both men and women. Are there common threads in the status of women in low-alcoholism groups, in the clarity of female role, in the handling of dependency needs? Heavy drinking by women in these groups is relatively unusual. Why? Do we go back to crosscultural studies of primitive societies and conclude that where the feminine role is more clearly differentiated from masculine role there

is less drinking and less heavy drinking among women? Can this be generalized to modern, technological, urban society? With the sex roles becoming less differentiated, we had best think about ways to head off or minimize a potential growing problem.

REFERENCES

Adler, A., 1927, "Understanding Human Nature," Garden City Publishing Co., New York.
"Alcohol and Health," 1971, First Special Report to the U.S. Congress from the Secretary of Health, Education and Welfare.
Anastasi, A., and Foley, J. P., 1949, "Differential Psychology," Macmillan, New York.
Averbukh, I. Y., 1966, Characteristics of alcoholic psychoses in women, *Nauchne-Issled. Psikh.* 12:387 (NIMH abstract).
Bacon, S. D., 1967, The classic temperance movement of the U.S.A.: Impact today on attitudes, action and research, *Brit. J. Addiction* 62:5.
Bailey, M. B., Haberman, P. W., and Alksne, H., 1965, The epidemiology of alcoholism in an urban residential area, *Quart. J. Stud. Alc.* 26:19.
Barcha, R., Stewart, M. A., and Guze, S. B., 1968, The prevalence of alcoholism among general hospital ward patients, *Amer. J. Psychiatry* 125:681.
Bardwick, J. M., 1971, "Psychology of Women, A Study of Biocultural Conflicts," Harper and Row, New York.
Basquin, M., and Osouf, C. 1965, A study of 50 female alcoholics: sociological, clinical and etiological observations, *Rev. Alcsme* 11:173 (QJSA abstract, 1968, 29:254).
Battegay, R., and Ladewig, D., 1970, Gruppentherapie und Gruppenarbeit mit suechtigen Frauen, *Brit. J. Addiction* 65:89.
Belfer, M. L., Shader, R. I., Carroll, M., and Harmatz, J. S., 1971, Alcoholism in women, *Arch. Gen. Psychiat.* 25:540.
Berner, P., and Kryspin-Exner, K., 1965, The changes in female alcoholism in the intake area of the Vienna Clinic since 1954, *Wien. med. Wschr.* 115:860 (QJSA abstract, 1967, 28:375).
Blane, H. T., 1968, "The Personality of the Alcoholic: Guises of Dependency," Harper and Row, New York.
Blane, H. T., Barry, H. III., and Barry, H. Jr., 1971, Sex differences in birth order of alcoholics, *Brit. J. Psychiat.* 119:657.
Cahalan, D., Cisin, I. H., and Crossley, H. M., 1969, "American Drinking Practices: A National Study of Drinking Behavior and Attitudes," College and University Press, New Haven, Conn.
Carpenter, J. A., 1962, Effects of alcohol on some psychological processes, *Quart. J. Stud. Alc.* 23:274.
Cheek, F. E., 1964, A serendipitous finding: sex roles and schizophrenia, *J. Abn. Soc. Psychol.* 69:392.
Chesler, P. 1972, "Women and Madness," Doubleday and Co., New York.
Child, I. L., Barry, H. III, and Bacon, M. K., 1965, Sex differences, *in* A cross-cultural study of drinking, *Quarterly Journal of Studies on Alcohol Suppl. No. 3*, pp. 49–61.
Clinard, M. B., 1968, "Sociology of Deviant Behavior," Holt, Rinehart and Winston, New York.
Cooperstock, R., 1971, Sex differences in the use of mood-modifying drugs: an explanatory model, *J. Health and Social Behavior* 12:238.
Cramer, M. J., and Blacker, E., 1963, "Early" and "late" problem drinkers among female prisoners, *Journal of Health and Human Behavior* 4:282.

Cramer, M. J., and Blacker, E., 1966, Social class and drinking experience of female drunkenness offenders, *Journal of Health and Human Behavior* 7:276.
Curlee, J., 1967, Alcoholic women: some considerations for further research, *Bull. Menn. Clinic* 31:154.
Curlee, J., 1968, Women alcoholics, *Federal Probation* 32:16.
Curlee, J., 1969, Alcoholism and the "empty nest," *Bull. Menn. Clinic* 33:165.
Curlee, J., 1970, A comparison of male and female patients at an alcoholism treatment center, *J. of Psychology* 74:239.
Curlee, J., 1971, Sex differences in patient attitude toward alcoholism treatment, *Quart. J. Stud. Alc.* 32:643.
De Beauvoir, S., 1953, "The Second Sex," Alfred A. Knopf, New York.
Decter, M., 1972, "The New Chastity and Other Arguments Against Women's Liberation," Coward-McCann, New York.
Deutsch, H., 1944, 1945, "The Psychology of Women, Volume One, Volume Two: Motherhood," Grune and Stratton, New York.
De Lint, J. E. E. E., 1964, Alcoholism, birth rank, and parental deprivation, *Amer. J. Psychiatry* 120:1062.
Differential Drug Use Within the New York State Labor Force, 1971, New York Narcotic Addiction Control Commission.
Dollard, J., 1945, Drinking mores of the social classes, *in* "Alcohol science and society," pp. 95–104, Journal of Studies on Alcohol, Inc., New Haven, Conn.
Dunlap, N. G., 1961, Alcoholism in women: some antecedents and correlates of remission in middle-class members of Alcoholics Anonymous, Doctoral dissertation, University of Texas.
Edwards, G., Chandler, J., Hensman, C., and Pete, J., 1972, Drinking in a London suburb. III. Comparisons of drinking troubles among men and women, *in* Surveys of drinking and abstaining: Urban, suburban and national studies, *Quart. J. Stud. Alc.,* Suppl. No. 6:94–119.
Ellinwood, E. H., Smith, W. G., and Vaillant, G. E., 1966, Narcotic addiction in males and females: A comparison, *Int. J. Addictions* 1:33.
Fort, T. F., 1949, A preliminary study of social factors in the alcoholism of women, Master's thesis, Texas Christian University.
Fort, T. F., and Porterfield, A. L., 1961, Some backgrounds and types of alcoholism among women, *J. Health and Human Behavior* 2:283.
Friedan, B., 1963, "The Feminine Mystique," W. W. Norton, New York.
Garai, J. E., 1970, Sex differences in mental health, *Genet. Psychol. Monogr.* 81:123.
Glatt, M. M., 1961a, Drinking habits of English (middle class) alcoholics, *Acta Psych. Scand.* 37:88
Glatt, M. M., 1961b, Treatment results in an English mental hospital alcoholic unit, *Acta Psych. Scand.* 37:143.
Gornick, V., and Moran, B. K. (eds.), 1971, "Women in Sexist Society: Studies in Power and Powerlessness," Basic Books, New York.
Greer, G., 1971, "The Female Eunuch," McGraw-Hill, New York.
Hollingshead, A. B., and Redlich, F. C., 1958, "Social Class and Mental Illness," Wiley, New York.
Horn, J. L., and Wanberg, K. W., 1969, Symptom patterns related to excessive use of alcohol, *Quart. J. Stud. Alc.* 30:35.
Horn, J. L., and Wanberg, K. W., 1971, Females are different: some difficulties in diagnosing problems of alcohol use in women, First Annual Conference of the National Institute on Alcohol Abuse and Alcoholism, Washington, D.C.
Horney, K., 1939, "New Ways in Psychoanalysis," W. W. Norton, New York.
Howells, J. G., ed., 1972, "Modern Perspectives in Psycho-Obstetrics," Oliver and Boyd, Edinburgh.

Jacob, A. G., and Lavoie, C., 1971, A study of some characteristics of a group of women alcoholics, Conference of the North American Association of Alcoholism Programs, Hartford, Connecticut.
Jellinek, E. M., 1942, "Death from alcoholism" in the United States in 1940: A statistical analysis, *Quart. J. Stud. Alc.* 3:465.
Jesser, R., Carman, R. S., and Grossman, P. H., 1968, Expectations of need satisfaction and drinking patterns of college students, *Quart. J. Stud. Alc.* 29:101.
Johnson, M. W., 1965, Physicians' views on alcoholism with special reference to alcoholism in women, *Nebraska State Med. J.* 50:378.
Johnson, M. W., de Vries, J. C., and Houghton, M. I., 1966, The female alcoholic, *Nursing Research* 15:343–347.
Jones, M. C., 1968, Personality correlates and antecedents of drinking patterns in adult males, *J. Cons. Clin. Psychol.* 32:2.
Jones, M. C., 1971, Personality antecedents and correlates of drinking patterns in women, *J. Cons. Clin. Psychol.* 36:61.
Jones, R. W., 1972, Alcoholism among relatives of alcoholic patients, *Quart. J. Stud. Alc.*, 33:810.
Jones, R. W., and Helrich, A. R., 1972, Treatment of alcoholism by physicians in private practice, *Quart. J. Stud. Alc.* 33:117.
Kagan, J., 1964, Acquisition and significance of sex typing and sex role identity, *in* "Review of Child Development Research" (Hoffman, M. L., and Hoffman, L. W., eds.) Russell Sage Foundation, New York.
Karp, S. A., Poster, D. C., and Goodman, A., 1963, Differentiation in alcoholic women, *J. Personality* 31:386.
Kelleher, M. J., 1972, Cross-national (Anglo-Irish) differences in obsessional symptoms and traits of personality, *Psychol. Med.* 2:33.
Keller, M., and Efron, V., 1955, The prevalence of alcoholism, *Quart. J. Stud. Alc.* 16:619.
Kent, P., 1967, "An American Woman and Alcohol," Holt, Rinehart and Winston, New York.
Kinsey, B. A., 1966, "The Female Alcoholic: A Social Psychological Study," Chas. C. Thomas, Springfield, Ill.
Kinsey, B. A., 1968, Psychological factors in alcoholic women from a state hospital sample, *Amer. J. Psychiat.* 124:157.
Komarovsky, M., 1962, "Blue-Collar Marriage," Random House, New York.
Knupfer, G., 1964, Female drinking patterns, *in* Selected Papers Presented at the Fifteenth Annual Meeting of the North American Association of Alcoholism Programs, pp. 140–160, Washington, D.C.
Knupfer, G., and Room, R., 1964, Age, sex, and social class as factors in amount of drinking in a metropolitan community, *Social Problems* 12:224.
Lawrence, J. J., and Maxwell, M. A., 1962, Drinking and socio-economic status, *in* "Society, Culture and Drinking Patterns" (D. J. Pittman and C. R. Snyder, eds.), pp. 141–145, Wiley, New York.
Lengrand, J.-P., 1964, Contributions to the study of female alcoholism in the North of France, Bailleul. (QJSA abstract).
Levine, J., 1955, The sexual adjustment of alcoholics, *Quart. J. Stud. Alc.* 16:675.
Ley, H. A., Jr., 1940, The incidence of smoking and drinking among 10,000 examinees, *Proc. Life Ext. Exam.* 2:57.
Lindbeck, V., 1972, The woman alcoholic: A review of the literature, *Inter. J. Addictions* 7:567.
Lisansky, E. S., 1957, Alcoholism in women: Social and psychological concomitants. I. Social history data. *Quart. J. Stud. Alc.* 18:588.
Lisansky, E. S., 1958, The woman alcoholic. *Ann. Amer. Acad. Polit. Soc. Sci.* 315:73.
Lisansky, E. S., 1960, The etiology of alcoholism: The role of psychological predisposition, *Quart. J. Stud. Alc.* 21:314.

Lolli, G., 1953, Alcoholism in women, *Conn. Rev. on Alcoholism* 5:9.
Madeddu, A., Lucchesi, M., Rivolta, G. C., and Spinola, E., 1969, Alcoholism and tuberculosis: clinical statistical aspects, *Revue de l'alcoolisme* (Paris) 15:167 (NIAAA abstract).
Massot, Hamel, and Deliry, 1956, Alcoholism in women: statistical and psychopathological data, *J. Med. Lyon* 37:265–269.
Mayer, J., and Green, M., 1967, Group therapy of alcoholic women ex-prisoners, *Quart. J. Stud. Alc.* 28:493.
McClelland, D.C., and Watt, N. F., 1968, Sex-role alienation in schizophrenia, *J. Abn. Psychol.* 73:226.
McCord, W., and McCord, J., 1960, "Origins of Alcoholism," Stanford University Press, California.
Mead, M., 1928, "Coming of Age in Samoa," Morrow, New York.
Mead, M., 1930, "Growing Up in New Guinea," Morrow, New York.
Millett, K., 1970, "Sexual Politics," Doubleday, New York.
Mogar, R. E., Wilson, W. M., and Helm, S. T., 1970, Personality subtypes of male and female alcoholic patients, *Inter. J. Addictions* 5:99.
Myerson, D. J., 1959, Clinical observations on a group of alcoholic prisoners, with special reference to women, *Quart. J. Stud. Alc.* 20:555.
O'Donnell, J. A., Besteman, K. J., and Jones, J. P., 1966, Marital history of narcotics addicts, *Inter. J. Addictions* 2:21.
Opler, M. K., 1959, Cultural differences in mental disorders: an Italian and Irish contrast in the schizophrenias, *in* "Culture and Mental Health" (M. K. Opler, ed.), pp. 425–442, Macmillan, New York.
Palola, E. G., Dorpat, T. L., and Larson, W. R., 1962, Alcoholism and suicidal behavior, *in* "Society, Culture, and Drinking Patterns," (D. J. Pittman and C. R. Snyder, eds.) pp. 511–534, Wiley, New York.
Parker, F. B., 1972, Sex-role adjustment in women alcoholics, *Quart. J. Stud. Alc.* 33:647.
Pattison, R. M., Coe, R., and Doerr, H. O., 1973, Population variation between alcoholism treatment facilities, *Inter. J. Addictions* 8:199.
Pattison, E. M., Coe, R., and Rhodes, R. J., 1969, Evaluation of alcoholism treatment: A comparison of three facilities, *Arch. Gen. Psychiat.* 20:478.
Pemberton, D. A., 1967, A comparison of the outcome of treatment in female and male alcoholics, *Brit. J. Psychiat.* 113:367.
Podolsky, E., 1963, The woman alcoholic and premenstrual tension, *J. Amer. Med. Wom. Assn.* 18:816.
Proceedings, American Psychological Association, 1972, Honolulu, Hawaii.
Rathod, N. H. and Thomson, I. G., 1971, Women alcoholics, a clinical study, *Quart. J. Stud. Alc.* 32:45.
Riley, J. W. Jr., and Marden, C. F., 1947, The social pattern of alcoholic drinking, *Quart, J. Stud. Alc.* 8:265.
Robins, L. N., 1966, "Deviant Children Grown Up," Williams and Wilkins, Baltimore, Md.
Robins, L. N., and Guze, S. B., 1971, Drinking practices and problems in urban ghetto populations, *in* "Recent Advances in Studies of Alcoholism," National Institute on Alcohol Abuse and Alcoholism.
Robins, L. N., Bates, W. M., and O'Neal, P., 1962, Adult drinking patterns of former problem children, *in* "Society, Culture, and Drinking Patterns" (D. J. Pittman and C. R. Snyder, eds.), Wiley, New York.
Rodriguez, A. L., 1970, Attitudes toward authority of Puerto Rican lower class boys, delinquent and non-delinquent, Master's thesis, University of Puerto Rico.
Rosen, A. C., 1960, A comparative study of alcoholic and psychiatric patients with the MMPI, *Quart. J. Stud. Alc.* 21:253.

Rosenbaum, B., 1958, Married women alcoholics at the Washingtonian Hospital, *Quart. J. Stud. Alc.* 19:79.
Rosin, A. J., and Glatt, M. M., 1971, Alcohol excess in the elderly, *Quart. J. Stud. Alc.* 32:53.
Rushing, W. A., 1969, Suicide and the interaction of alcoholism (liver cirrhosis) with the social situation, *Quart. J. Stud. Alc.* 30:93.
Schmidt, W., and de Lint, J., 1969, Mortality experiences of male and female alcoholic patients, *Quart. J. Stud. Alc.* 30:112.
Schmidt, W., and de Lint, J., 1972, Causes of death of alcoholics, *Quart. J. Stud. Alc.* 33:171.
Schuckit, M. A., 1971, Depression and alcoholism in women, First Annual Conference, National Institute on Alcohol Abuse and Alcoholism.
Schuckit, M., 1972, The woman alcoholic, *Psychiatry in Medicine* 3:37.
Schuckit, M., Pitts, F. N., Reich, T., King, L. J., and Winokur, G., 1969, Alcoholism, I. Two types of alcoholism in women, *Arch. of Environmental Health* 18:301.
Sclare, A. B., 1970, The female alcoholic, *Brit. J. Addict.* 65:99.
Sears, R. R., Maccoby, E. E., and Levin, H., 1957, "Patterns of Child Rearing," Row Peterson, Evanston, Ill.
Selected Symptoms of Psychological Distress, 1970, National Center for Health Statistics, Series 11, no. 37.
Smart, R. G., 1963, The relationship between birth order and alcoholism among women, *Ont. Psychol. Ass. Quart.* 16:9.
Smart, R. G., Laforest, L., and Whitehead, P. C., 1971, The epidemiology of drug use in three Canadian cities, *Brit. J. Addict.* 66:293.
Spain, D. M., 1945, Portal cirrhosis of the liver, a review of 250 necropsies with reference to sex differences, *Amer. J. Clin. Path.* 15:215.
Srole, L., Langner, T. S., Michael, S. T., Opler, M. K., and Rennie, A. C., 1962, "Mental Health in the Metropolis, The Midtown Manhattan Study," Vol. I, McGraw-Hill, New York.
Sterne, M. W., and Pittman, D. J., 1972, "Drinking Patterns in the Ghetto," typescript, Social Science Institute, Washington University, St. Louis, Missouri.
Summerskill, W. H. J., Davidson, C. S., Dible, J. H., Mallory, G. K., Sherlock, S., Turner, M. D., and Wolfe, S. H., 1960, Cirrhosis of the liver, a study of alcoholic and nonalcoholic patients in Boston and London, *New. Engl. J. Med.* 262:1.
Thompson, C., 1941, The role of women in this culture, *Psychiatry* IV:1.
Thompson, C., 1942, Cultural pressures in the psychology of women, *Psychiatry* V:331.
Van Amberg, R. J., 1943, A study of 50 women patients hospitalized for alcohol addiction, *Dis. Nerv. System* 4:551.
Vogel, I., 1919, Effect of chronic drug abuse on function of female genitals, *Archiv fur Frauenkunde u. Konstitutionsforschung* (Wurzburg) 15:157.
Wall, J. H., 1937, A study of alcoholism in women, *Amer. J. Psychiat.* 93:943.
Wanberg, K. W., and Horn, J. L., 1970, Alcoholism symptom patterns of men and women: A comparative study, *Quart. J. Stud. Alc.* 31:40.
Wanberg, K. W., and Knapp, J., 1970, Difference in drinking symptoms and behavior of men and women alcoholics, *Brit. J. Addict.* 64:347.
Wetzel, R. D., and McClure, J. N., 1972, Suicide and the menstrual cycle: a review, *Comprehensive Psychiatry* 13:369.
Widseth, J. C., 1971, Dependent behavior and alcohol use in delinquent girls, Eastern Psychological Association, New York.
Widseth, J. C., and Mayer, J., 1971, Drinking behavior and attitudes toward alcohol in delinquent girls, *Inter. J. Addict.* 6:453.
Wilkinson, P., Santamaria, J. N., and Rankin, J. G., 1969, Epidemiology of alcoholic cirrhosis, *Australas. Ann. Med.* 18:222.

Wilkinson, P., Santamaria, J. N., Rankin, J. G., and Martin, D., 1969, Epidemiology of alcoholism: Social data and drinking patterns of a sample of Australian alcoholics, *Med. J. Aust.* 1:1020.

Wilsnack, S. C., 1972, The needs of the female drinker: Dependency, power, or what? Second Annual Conference of the National Institute on Alcohol Abuse and Alcoholism, Washington, D.C.

Wilsnack, S. C., 1973, The effects of social drinking on women's fantasy, mimeographed.

Wilsnack, S. C. 1973, Sex-role identity in female alcoholism, *J. Abn. Psychol.* 82:253–261.

Winokur, G., and Clayton, P., 1967, Family history studies. II. Sex differences and alcoholism in primary affective illness, *Brit. J. Psychiat.* 113:973.

Winokur, G., and Clayton, P. J., 1968, Family history studies. IV. Comparison of male and female alcoholics, *Quart. J. Stud. Alc.* 29:885.

Wood, H. P., and Duffy, E. L., 1964, Psychological factors in alcoholic women, *Amer. J. Psychiat.* 123:341.

Zelen, S. L., Fox, J., Gould, E., and Olson, R. W., 1966, Sex-contingent differences between male and female alcoholics, *J. Clin. Psychol.* 22:160.

Zetterman, B., 1966, Problems in an institution for female alcoholics, *Social-Med. Tidskr.* 43:326 (QJSA abstract, 1968, 29:234).

Zigler, E., and Phillips, L., 1960, Social effectiveness and symptomatic behaviors, *J. Abn. Soc. Psychol.* 61:231.

Zucker, R. A., and Barron, F. H., 1971, Do wives drive husbands to drink?, Southwestern Psychological Association, San Antonio, Texas.

CHAPTER 5

Youthful Alcohol Use, Abuse, and Alcoholism

Wallace Mandell and Harold M. Ginzburg

The Johns Hopkins University
School of Hygiene and Public Health
Baltimore, Maryland

INTRODUCTION

The fact that American teen-agers drink alcohol is obvious to the most casual observer. What has not been generally understood is that the teen-age population, for the most part, drinks in an orderly fashion and is engaging in a systematic, culturally organized induction into adulthood.

Newspaper writers usually discuss alcoholism among teen-agers. Examination of the article usually reveals that the writer is referring to the prevalence of heavy drinking, or delinquent behavior associated with any degree of use of alcohol, including occasional use. This type of writing frightens the public into periodic concern about the dangers associated with adolescent use of alcohol. There is practically no research directed toward determining whether teen-agers' physiological or psychological response to alcohol is any different from that of adults. Popular stereotypes and prejudices are often mobilized to support special interest groups' efforts to regulate alcohol use by young people. These stereotypes also confuse interpretation of the substantial data available on teen-age drinking patterns and problems associated with it.

In the present essay the very substantial literature on youthful use of alcohol has been organized around the central questions addressed by research workers in the field:

1. What do youth perceive adult use of alcohol to be?
2. What is the prevalence of the various patterns of use of alcohol among youth and has this been changing?
3. What is the influence of family, friends and other sociocultural variables in the development of the youthful drinking pattern?
4. What influence does pattern of use of alcohol have on socially unacceptable behavior?
5. What are the personal characteristics of youth who have problem drinking patterns?
6. Is there youthful addiction to alcohol?
7. Is there a relationship between use pattern and adult addiction to alcohol?

It is difficult to discuss teen-age use of alcohol because of the prejudicial language used to describe it, and because of pervasive ambivalence toward this aspect of becoming an American adult. Language used to describe teen-age drinking reflects the lack of consensus about acceptable alcohol use patterns among adults in research literature and in the popular press.

In order to make data from different studies comparable, a series of conventions have to be adopted.

The phrase "alcoholic beverages" is variously interpreted by individuals representing different cultural and class backgrounds. For instance, some groups do not consider beer to be an alcoholic beverage; others hold that if the beverage is not consumed for the purpose of producing intoxication, it need not be considered alcoholic. This is the most likely explanation of the finding in the New York State Study by Mandell *et al.* (1962) that the proportion of youth who classify themselves as users of alcohol varies with the format of the question. When teen-agers are asked about pattern of use of each beverage, many more indicate use than when the general question, "Do you use alcoholic beverages?" is posed. In the present essay, data are assembled from questions specifically asking for use of wine, beer, and whiskey unless otherwise noted.

In studying the quantity of alcohol consumed, recent research has accepted the convention of converting the alcoholic content of popular units of consumption of beverages into standard units of alcohol, either absolute alcohol, in ounces, 200 proof, or standard drink, i.e., 1 oz of 80 proof alcohol, 0.4 oz of absolute alcohol.

The following terms, representing the common patterns of use of alcohol, are used:

Infrequent user of alcohol. Someone who at least once in his lifetime has used alcohol. This use may not have continued but it may be as frequently as 11 times a year.

Occasional user of alcohol. Someone who drinks at least 12 times a year, in effect at least once a month but less than once a week on the average.

Regular user of alcohol. Someone who uses alcohol at least once a week.

Heavy drinker. The concept of heavy drinker refers to a combination of quantity and frequency of consumption. National norms have not been established for this concept among teen-agers or by age groupings. As a result, the meaning of the term varies with the author. It usually means at least seven standard drinks per week.

YOUTH'S PERCEPTION OF ADULT USE OF ALCOHOL

When interviewed, youth report that the adults with whom they interact use alcohol. Most adults in the United States have integrated alcohol use into their life-style. Fifty-seven percent use alcohol at least once a month. Thus, most youth are likely to experience alcohol as part of their parents' life-style. This is confirmed in the Chappel et al. (1953), McCluggage et al. (1956), Miller and Wahl (1956), and Mandell et al. (1962) studies, which all found more than 75 percent of high school students reporting their parents as keeping alcohol in the home and using it occasionally. Maddox reported in 1956 that students believed that a majority of adults drink at least sometimes. Youth perceive differences in the drinking patterns of adults, being sensitive to sex, ethnic, and rural–urban variations in alcohol use pattern. Their image of what adults do seems to correspond to the self-reports of adults in various surveys. This reflects accurate perception of adult behavior and that youth have accepted adults' image of themselves as alcohol users. American youth understand and accept alcohol use to be a part of the adult status and life-style.

PREVALENCE OF ALCOHOL USE AMONG YOUTH

Systematic studies of the incidence (new cases) and prevalence of youthful alcohol use are relatively recent phenomena. Since the first American study published by McCarthy and Douglass in 1948, there have been a substantial number of studies of youthful drinking.

The data from the various studies to be discussed below must be understood in the context of the methodologies that were used to gather the data. All of the studies involve the self-report of alcohol use by youth for whom such use is illegal. The data therefore contain some error contributed by deliberate misreporting, either to appear conforming to legal standards, or to exaggerate independence from these standards. There is also memory error, which varies depending on the amount of recall required of the respondents. Sometimes respondents are asked to provide general estimates of amount and frequency of use of alcohol, e.g., "How often do you usually use alcohol?" In other studies respondents are asked to give detailed reports of actual consumption for specified recent periods. The study by Mandell et al. (1962), in which several methodologies were compared, indicated that the two question formats produced considerable variation in frequency and quantity of use estimates. The more specific the question, i.e., asking for consumption of individual beverages, the greater will be the quantity and frequency of alcohol use reported. When youth are asked to report their "usual pattern" of use, they report their pattern as being more like their group's norm of behavior than is found when they report on their own quantity of use on a day-by-day basis.

Questionnaires are usually completed under conditions of anonymity intended to increase the willingness of the respondents to provide truthful information. There is no evidence that allows evaluation of the impact of this procedure.

The majority of studies report data from students who complete questionnaires while in school. Most authors overstep the limits of generalization applicable to their data by discussing youthful drinking. There are two important groups not captured by the "in school" sampling procedure that are of particular interest to the field of alcohol studies. Depending on period in history and place of study, as many as 40 percent of sixteen- and seventeen-year-olds are not in school, having legally discontinued their education. Out-of-school youth are more often from lower socioeconomic backgrounds and have a high probability of having a different use pattern. Another group with a high probability of a different alcohol use pattern were those who were absent on the day of the survey. In inner city areas between 20 percent to 30 percent of youth may be out of school on any day. This group includes many of the heaviest users of alcohol. Because of the self-report questionnaire approach, question format, and sampling limitations, most prevalence studies are likely to under-represent the extreme use patterns, particularly heavy users.

The frequent inappropriate sampling of schools to represent geographic areas, and samples of students within schools to represent the student body, will not allow using the data from most studies for precise estimation of the exact frequency or quantity of consumption of alcohol among youths.

Because most questionnaire studies ask youth to report their "usual" pattern of use, the data reflect more influence of cultural norms as perceived by youth than is desirable. They tend to report their perceptions of themselves rather than their behavior. However, the surveys do provide data on what hundreds of thousands of American youngsters report themselves to be doing. As such they are the best available source for estimating the cultural pattern of alcohol use and for discerning the cultural patterning process.

Table 1 assembles data from representative prevalence studies done in local areas between 1951 and 1973.

Among seventeen-year-old boys (the average of youth in the 12th grade) the lowest reported frequency of ever having used alcohol was 42 percent in Utah in 1951 and the highest, 81 percent, in San Mateo in 1969. The range for seventeen-year-old girls was 19 percent in Utah to 76 percent in San Mateo. The table demonstrates clearly that place and point in history are major variables related to ever having used alcohol. The factors most often used to account for the differences in findings between studies are period in history, degree of urbanization of the sample, and prevalent religious values about drinking. The data from these earlier studies will be put into perspective by an important new study still in the process of analysis.

A survey of 14,000 students in junior and senior high schools in the United States in 1974 representing a national probability sample confirmed recent regional studies' findings that between 70 percent to 90 percent of youth in school have used alcohol at some time, depending on age. The preliminary report indicated that 93 percent of 12th grade boys and 87 percent of 12th grade girls have had a drink. Among seventh graders 63 percent of the boys and 54 percent of the girls have had a drink. As might be expected, the quantity consumed at any one time also increased with age.

As indicated in the discussion of methodological problems in drinking practice studies, little is known about drinking among youth not in school. Nelson in 1968 demonstrated that the drinking pattern among school dropouts was different from those youth in school. Surveys based on in-school samples probably underestimate the frequency and quantity of drinking among youth in urban areas and may overestimate it in rural areas.

The transition from high school student status into other statuses brings with it shifts in alcohol use pattern. After high school, male youth seems to move toward the adult drinking pattern of the economic status group that they enter. A longitudinal survey in 1970 revealed that 33 percent of 1000 senior high school males drank once a week or more. During the year following graduation 44 percent drank at this frequency. The age of heaviest drinking in the youthful population was in the eighteen- to twenty-year-old period (1972 survey).

Each age cohort of the high school population seems to be adopting the drinking pattern of the cohort just ahead of them and, on the whole, seems to be moving as a group toward that level of drinking prevalent among the young

TABLE 1. Percent of High School Students Who Have Ever Used Alcoholic Beverages

Year	Region	Males							Females							Combined totals						
		N	14	15	16	17	18		N	14	15	16	17	18		N	14	15	16	17	18	All
1951	Utah	563	—	45	—	42	—		614	—	23	—	19	—		1,177	—	34	—	28	—	30
1953	Nassau Co., N.Y.																1,000	79	82	90	89	89	86
1955	Racine, Wisc.	469	—	62	76	71	79		529	47	56	57	72	77		998	51	59	65	71	78	64
1955	Metro. Kansas	629	—	56	68	63	72		578	—	33	47	55	51		1,207	—	43	59	59	63	56
1955	Nonmetro. Kansas	564	—	45	56	63	65		553	—	23	—	39	41		1,117	—	32	43	52	56	44
1962	N.Y. state	382	63	61	71	81	74		369	32	50	60	66	69		751	—	—	—	—	—	62
1964	Miss.	239	—	—	48	59	—		197	—	—	25	37	—		436	—	—	36	50	—	—
1968	7 Kentucky Counties																19,000						55
1968	San Mateo Co., Calif.	9,611	53	—	77	—	—		—	—	—	—	70	—								
1969	San Mateo, Calif.	12,337	53	—	81	—	—		—	38	—	—	76	—								
1969	Dallas	56,745	—	—	—	—	—		—	—	—	—	—	—		33,438	59	66	72	77	—	—
1973	Maryland	5,284	52	—	—	—	—		4,785	46	—	—	—	—		10,354	40	50	49	—	—	49

adults. This continuity is evident when the data from the 1972 adult survey and that from the 1974 youth survey are patched together. As demonstrated in Figure 1, there is clear continuity between the curves of youthful and adult drinking.

There has been considerable interest in estimating historical changes that may be taking place in youthful use of alcohol. Unfortunately, variations in methodology carried out at different times make invalid comparisons in estimates from national surveys. The only study that repeated the same measure of drinking for a population of students is from San Mateo County, California, from 1968 to 1973. That study (1973) showed that in recent years, there has been an increase in the proportion of students drinking at each age level. However, the most substantial shift has been among the younger age group. In 1969 52 percent of seventh grade boys had used alcohol as compared to 72 percent in 1973. In keeping with adult patterns, fewer girls than boys drank at each age, though this gap has been narrowing. Girls drank lesser amounts and less frequently than boys.

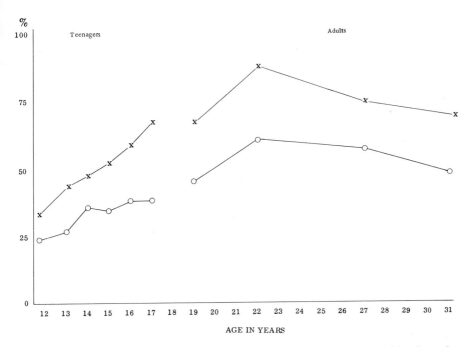

FIGURE 1. Percent of regular drinkers among youth and adults, by sex and age. Adults all use of any alcohol at least once a month, 1972. Teen-age use of beer at least once a month, 1974. Data sources: Data adapted from graphs presented in Alcohol and Health: New Knowledge, DHEW, 1974. (x) Males, (0) females.

In summary, most American youth have at least tasted alcohol by the time they graduate from high school. About 33 percent of the boys are drinking at least once a week. Once they leave high school, more of them drink at least occasionally. The age of entry into the ranks of drinkers varies with sex, urbanization of the area in which they live, and their religious affiliation. The frequency and amount of drinking in each age cohort increases through time in an orderly fashion and eventuates in a rate of frequency and quantity of alcohol continuous with adult practices. All the data support the view that there is an orderly evolution by the youthful population into adult drinking patterns. The rate of adopting the use of alcohol is modulated for particular individuals and groups by sociocultural factors: sex, socioeconomic status, locale, availability, parental and peer standards. Most youth perceive the use of alcohol as part of the young adult life pattern. The frequency and amount of alcohol use youth aspire to is more than that of their parents since older adults use alcohol less frequently than young adults. As a result, more youth are using alcohol and are probably starting somewhat younger than their parents. This discrepancy may account for some of the concern about the youthful use of alcohol.

FACTORS INFLUENCING YOUTH ALCOHOL USE PATTERN

A question of considerable interest to legislators and moralists has been establishing what factors are most influential in determining youthful use of alcohol.

Alcohol is available in most American homes. It is no surprise that most teen-agers receive their first alcohol from family members, usually in the form of beer or wine. This is consistent with reports of parental permission to drink available from some studies.

When the prevailing cultural norms are not favorable to youthful drinking, such as in rural Kansas or Utah, a larger proportion of those youth who use alcohol obtain their first alcohol outside the home.

Influence of Parents

Apparently the majority of American parents recognize the use of alcohol as a legitimate part of life for their children. Parents are mostly concerned about the age of initiation, the amount of use, and the consequences that may follow from use.

Despite prevailing laws most parents provide their youth with learning experiences with alcohol (Mandell *et al.,* 1962). When they do not, youth obtains such experiences in the company of friends. This has led to an interest in specifying the relative influence of parents and friends in determining alcohol patterns.

As Kandel (1974) pointed out, a general question exists about the patterning of youth behavior in terms of to what extent these patterns are youth's rebellion from adults, as contrasted to a manifestation of continuity with parental behavior. She noted that continuity had been demonstrated in radical political activities, religious behavior, and general life values. In fact, in most area of life parental and peer influences are generally complementary. Most studies, which have examined this question, depend on adolescent self-perception and report of parental and peer behavior. These perceptions may be molded to reduce the experience of conflict.

Kandel (1974) provided an excellent model for studies aimed at examining relative influence of parents and peers. In her study, parental use of distilled alcohol was associated with marijuana use, rising from 16 percent of children of fathers who reported no drinking to 38 percent when fathers drank distilled beverages regularly. When best friends self-reported use was examined, 15 percent of those with nonusing best friends used marijuana, as compared with 79 percent for those whose best friends were regular users. When parents' and friends' behavior diverged, the adolescents reported similarity to peers. The more involved with peers the adolescent was and the less close to mother, the more likely was the drug use pattern to correspond between peers. This study reinforced reports of greater maternal influence on youthful drug use.

The extent of adolescent independence from parents varies with the issue involved and is always relative. Kandel and Lesser (1972) believed that adolescents were responsive to pressures from members of their own generation when immediate gratifications such as alcohol use were involved. When the issues involved life goals or matters that were not salient to adolescent concern, parental generation values were more influential.

Parental practices seem to be a major determinant of drinking behavior in most studies. Mandell *et al.* (1962) pointed out that most youth had their first experience with alcohol under the supervision of family, and, in fact, during early adolescence, continued their drinking in their parent's home or in the home of a friend. When they drank in home settings, the quantity consumed was generally small and was less than when they were drinking in other settings.

When both parents drank, there was a greater likelihood that the youth would use alcohol, at least occasionally.

Several researchers have examined drinking practices of youth in relation to parental and peer norms.

Forslund and Gustafson (1970), in a study of 654 high school seniors in Albuquerque, found that parental practices had little effect on boys when there was strong contradictory peer pressure. Mothers' drinking behavior had a significant effect on sons' and daughters' behavior, while fathers' behavior had no influence on sons and some on daughters.

Alexander (1967) has examined the possibility that closeness to parents modified the impact of parentally set norms. Questionnaires were given to 1410 males in 30 high schools in eastern and Piedmont North Carolina. Sons of fathers who permitted use drank in patterns unrelated to their feelings of closeness to their fathers. Alternatively, closeness to fathers who disapproved of drinking was associated with less frequent drinking even among drinkers. When father was rejected, drinkers, even without peer support, drank more frequently. There was a slight indication that sons who felt distant from abstinent fathers, who frequently tried to "get even" with them and who had abstinent best friends, were more likely to have been drunk or passed out while using alcohol. This study supported the view that parental standards were very important in establishing youthful drinking patterns, except for a small proportion of youth for whom alcohol had taken on the meaning of personal rebellion.

In summary, a son who rejects his abstinent father, tries to "get even" with him and does not have drinking friends, is more likely to drink for psychological effect and in an uncontrolled way. It is possible to interpret this data as supporting the hypothesis that drinking friends moderate the drinking of boys who are in rebellion against their fathers. Unfortunately, the associations demonstrated in this study are low and the samples small.

Influence of Friends

Youthful drinking in the home or under parental supervision tends to be more moderate. Drinking alcoholic beverages is a social activity carried out with peers, during adolescence. The frequency and amounts consumed are related to the companions one has while drinking. When they are drinking with friends, the quantity consumed is greater.

Jessor and Jessor (1973) studied fourteen- to eighteen-year-olds in a small Colorado community who began drinking during the course of a year. Those youth who began to drink during the year had values more like the group that was already drinking. The best predictor of the change from nondrinking to drinking behavior was the teen-ager's attitudes toward drinking of his peers.

Jessor made an important point that the focus on drinking obscured the fact that other behaviors were also changing as the adolescent moved from one age status to another.

Alexander (1964) and Alexander and Campbell (1967) investigated the attitudes toward alcohol and the drinking behaviors of adolescent boys in eastern North Carolina, an area in which the dominant religious groups prescribe abstinence. One third of the sample of 1410 respondents had used alcohol in the preceding year, an extraordinarily low rate when compared to other samples. It is to be noted that respondents were required to identify themselves by name. Members of cliques could identify with few errors the drinking behaviors of other group members. Cliques tended to reject group members with deviant drinking behavior. The findings of the study can be interpreted as indicating that cliques were in part formed by individuals who were in agreement about deviant behavior and seeking social support for it.

At the other extreme from writers emphasizing parental influence is Coleman (1970), who presented the view that adolescents, because they spent most of their time segregated in schools, developed same age peer groups. These social networks of friends taught the individual how to recognize and enjoy the effects of such drugs as alcohol.

Riester and Zucker (1968) studied 143 junior and senior high school students. They found that these students divided themselves into eight natural cliques representing different social values and life-styles such as collegiate, "leathers," intellectuals. Membership in each group was related to sharp variation in drinking behavior: quantity consumed, frequency of drinking, and social occasion for drinking. Each group took on the alcohol use norms associated with that life-style as represented in older individuals.

Another perspective on youthful use emphasizes the personal motivation to drink. Alexander and Campbell (1967) reported that the majority of nondrinkers had received pressure from friends to drink. Despite this pressure, this group had not drunk. The more drinking friends the individual had, the more he reported such pressure. When the individual's parents drank and he drank, the adolescent was more attached to his group.

Sociocultural Factors and Youthful Drinking

As the data in Table 1 and Figure 1 indicate, fewer girls than boys drink at any age though this gap seems to be narrowing. Jessor and Jessor (1973) reported that religiousness and church attendance were positively related to abstinence. They interpreted church attendance as an index of involvement in the adult social control system. Nonproblem drinkers in the Mississippi study by Globetti and Windham (1967) participated more in religious activities. Maddox (1964), reviewing two nationwide surveys of American youth, concurred in this observation.

In the Johnston longitudinal study (1974), the number of young men from wealthier families who were drinking regularly increased in the year after high school by 21 percent as compared to a 5 percent increase among the poorer men. Despite this, college youth drank less than either military or working youth.

There has been only one study that looked at the drinking practices of older teen-agers. Of those teen-agers who entered the military after graduation from high school 55 percent drank regularly, while 44 percent of those who took civilian jobs and 38 percent of those who went to college drank regularly. These important data are limited in generalizability because the study did not take into account the large group of youth who did not graduate from high school.

There have been a few studies on the effect of locale of drinking among youth. Youth, after initiation into drinking at home, do more of their drinking out of the home as they grow older. The most common such site in recent times has been drinking at teen-age parties without adult supervision. Boys' drinking with friends takes place in secluded places, frequently a car. When they reach legal age, they switch to bars and do less drinking at home and in secluded spots. Girls also drink in secluded spots, but as they reach legal age, they too begin to do more of their drinking in restaurants and bars.

There has been considerable interest in the impact of the college experience on drinking practices. Straus and Bacon (1953) conducted a survey of the drinking practices of 16,000 students in 27 colleges, including a broad representation of types of education institutions and regions of the country. No data are available on the representativeness of the sample of schools. The study published in 1953 described the reluctance of college administrators in that period to adopt a clear policy toward student drinking. Stringent policies forbidding drinking were associated in college administrators' experience with resentment and rebellious student reactions. Liberal policies were associated with community and parental criticism of administrators when there were untoward consequences of alcohol use.

Students tended to perceive college administration attitudes as extreme disapproval or liberality. The more liberal the students perceived the college administration attitudes, the more users there were in the college population, but this did not influence the amount of drinking among the users. Some part of this finding may be due to self-selection on the part of students into particular colleges.

Parental sanctions against drinking were a major determinant of students' decisions to drink. However, these sanctions were correlated closely with parental drinking practices, and both sanction and parental modeling probably had effect in determining drinking patterns. All in all, the proportion of students who drank frequently and heavily was small. No more than 20 percent of the college men and 10 percent of the college women drank as frequently as once a week.

Summary

The most common reason for beginning to drink any alcoholic beverage is a holiday celebration during which such use is considered appropriate. Another large proportion of youth try alcohol because they are curious, and many because their parents offer it to them.

In the New York State Study by Mandell *et al.* (1962), the most usual setting for use of alcohol for 70 percent of high school students was their own or a friend's home. At fourteen years of age, 50 percent of the New York State boys had parental permission to drink, at least in certain circumstances. This percentage increased regularly with age, and included permission to drink away from home. The pattern was similar for girls, though the proportion with permission at each age was small.

All studies reported that there were fewer users of alcoholic beverages among children who were not given parental permission to drink. Girls apparently conformed more to parental injunction in this regard. Even among those youth who drank without parental permission, the frequency of drinking was less than among those who had permission. The number of cases in the literature is too small to rely on, but the data suggest that the small group of youth who do drink in opposition to parental wishes are more likely to drink larger quantities when they do drink.

Apparently parents who supervise their children's drinking generally have children who behave more moderately.

The use of the term "rebellion" in relationship to the general youthful use of alcohol has obscured the dynamics and complexity of the orderly learning to use alcohol in American society. There is indeed a small group of youth that is drinking alcohol in order to express defiance of parents. However, the majority of teen-agers drink with parental permission and guidance. There is a substantial group whose parents have not laid down any guidelines about the use of alcohol. These youth might more properly be described as undercontrolled rather than rebellious. To put this group with the one that is in rebellion masks the potential influence of parental control. With the data presently available, it is not possible to determine what proportion of youth would truly rebel against parental standards. The group that is truly rebelling tend to drink more heavily than their peers and tend to have problems associated with the use of alcohol. It is likely that this is one of the groups that provides the source for adult alcoholism.

Parental drinking pattern does not influence youth's report of themselves as to frequency of drinking. On the other hand, when the quantity of alcohol consumed in the preceding week is examined, the children of abstainers drink less. One of the important variables in youthful drinking seems to be whether parents keep alcohol in the home. Since parents are a significant source of alcohol, lack of supply reduces actual drinking behavior, though not self-perception by youths in terms of their pattern.

ALCOHOL USE AND SOCIALLY UNACCEPTABLE BEHAVIOR AMONG YOUTH

The use of alcohol and socially unacceptable behavior are traditionally associated with youth.

In the DHEW survey (1974) the eighteen- to twenty-year-old group had the largest proportion of persons, 27 percent, who reported themselves as having some problem in connection with drinking. The period from twenty-one to thirty-four showed a decreasing proportion who reported problems. There are several alternative explanations for the high rate of alcohol-associated problems the youthful group report. It may be that they are less experienced with alcohol and this precipitates problems that more experienced individuals do not have. Or, it may be that the youthful group has significantly greater adjustmental problems to the young adult role and alcohol use pattern represents another expression of this troubled period. Or, alcohol may be part of the attempt by late adolescents and young adults to become comfortable in this tense period.

There is considerable difficulty in interpreting research on alcohol use and socially unacceptable behavior. Often the term "problem drinking" is used to encompass this area of study. However, there is no consensus among research workers as to the definition of the concept of problem drinking even among adults. Some writers emphasize frequency of intoxication and pattern of use of alcohol such as binge drinking, thereby focusing on quantity–frequency–variability measures. Other writers focus on disruptions of normal social behavior patterns as indicated by marital, employment, or police problems associated with alcohol use.

If a standard analogous to the ones used for adult behavior were adopted, teen-age and young adult problem drinking could be defined as that which (1) produces deleterious health consequences; (2) produces poor school or work performance; (3) produces poor social relationships; and (4) produces illegal behavior.

Unfortunately, most of the studies that examine the social consequences of alcohol use are correlational in nature, rather than longitudinal and therefore have not been able to disentangle whether alcohol use is a consequence, or a cause of unacceptable behavior.

In the DHEW teen-age survey (1974) about 5 percent of 12th grade students got high at least once a week and about 23 percent had been drunk at least four times during the past year. Only 2 percent of students reported that drinking had been a problem for them. Other surveys reported from 1 percent to 15 percent of adolescents had problems associated with drinking.

Mandell *et al.* (1962) reported in 1962 that among 100 high school boys who drank at any level in a one-month period, two boys passed out, eight were unable to remember what happened, and seven drank before breakfast. Among 100 regular drinking students in a one-month period, two passed out, 49 were unable to remember what occurred while drinking, and 19 drank before breakfast. Among 100 high school girls who drank at any level, four passed out, ten were unable to remember, and two drank before breakfast. Among 100 regularly drinking girls, 14 passed out, 38 were unable to remember, and three drank before breakfast.

The relationship between school performance and problem drinking has been examined.

Mandell *et al.* (1962) found no relationship between school grades and frequency or quantity of drinking in the sample of four high schools' students, even when it was expanded to include a substantial number of school dropouts.

The Maryland Drug Abuse Administration Study in its study of tenth graders (1975) in 1973 found that current users of alcohol had a slightly lower grade point average, 2.47, compared to 2.34 for nonusers. Comparatively, in that small group of students (166) who were failing, 18 percent were using alcohol several times a week compared to 6 percent of the general student body; 15 percent of the failing students were using alcohol daily. In other words, 32.5 percent of failing students were using heavily as compared to 9 percent of the total sample.

The data on large samples indicate that, in general, use of alcohol is not associated with poor school performance. Some individuals may use alcohol in a way that interferes with school performance, but this is not common enough to influence population trends.

The relationship between alcohol use and peer social relationships has been studied. All studies reported that drinking among youth was a peer group activity. It was not surprising to find that there was no general relationship between use of alcohol and participation in formal organizations; and, in fact, that boys who drank frequently were more often leaders in high school organizations, and girls who drank more frequently belonged to fewer social organizations. This may have been because they were less effective or because they were participating in the more adult dating patterns.

Among the college students who drink, 4 percent of men and 0.7 percent of women sustained an accident or injury, and 2 percent of men had come before college authorities as a result of using alcohol. As would be expected, the students who experienced complications associated with alcohol use—missing school, loss of friends, accidents—drank more frequently and in greater quantity than their peers and were more often intoxicated. Students less than eighteen years of age had fewer complications with drinking. Among men who drank, 20 percent of Jewish students, 33 percent of Protestant and Catholic students, and 43 percent of Mormon students experienced complications. The differences were even sharper for female students. Straus and Bacon (1953) interpreted this finding to suggest that students who broke away from restrictions reacted in a more emphatic way. Among Mormons this was likely to lead to loss of friendship within the Mormon group, and this tended to be the first and most common social complication.

In the college study, 10 percent of the male students and 2 percent of the female students reported drinking surreptitiously. Nine percent of males and 1 percent of females had experienced more than one blackout. Six percent of men and 1 percent of women had become drunk alone more than once. Ten percent of men and 3 percent of women reported drinking more than once before breakfast. Seven percent of the men reported aggressive or destructive behavior had occurred more than twice in association with drinking alcohol. Five percent of male and somewhat less than 1 percent of female drinkers reported three or more of these early warning signs of alcoholism.

Some college students expressed concern about the long-range consequences of their pattern of use of alcohol. Twelve percent of men who had never been drunk expressed this concern, as compared to 27 percent of those who had been drunk five or more times. Forty percent of those men with three or four social complications associated with drinking were worried about their pattern of use of alcohol. This suggests that these problem-drinking young men recognized something about themselves that was worrisome. This self-recognition might be an early indicator of feeling of loss of control and alcoholism.

In the New York State Study by Mandell *et al.* (1962), students were asked to report on a number of behaviors of concern to society: fighting, destroying property, auto accidents, involvement with the police. At each age level, occasional and regular drinkers committed more such acts in the three months preceding the survey than students that rarely or never drank. However, they committed as many or more such acts on occasions when they were not drinking. The youth who drank often also misbehaved more. In this study the drinking did not seem to be a general cause of misbehavior. It was, however, related to the destruction of property, i.e., vandalism, which occurred more often while drinking.

As indicated in the discussion of prevalence of youthful drinking, young people are involved in a considerable amount of illegal behavior simply because it is against the law for them to obtain and use alcohol. Most of the youthful cases of alcohol-related problems coming to the attention of social agencies in the Monroe County study by Zax, Gardner, and Hart (1961) did so because of contact with the police for public intoxication. In the New York State Study 28 percent of sixteen-year-old and 73 percent of seventeen-year-old boys had been served alcohol illegally in a bar or restaurant. Almost 100 percent of sixteen- and seventeen-year-olds who ask to be served were served in bars and restaurants. Twenty percent of the youth studied in 1962 carried false proof of age.

Sterne, Pittman, and Coe examined the relative contribution of adolescents to alcohol-related offenses in St. Louis in 1960. Adolescents aged ten to twenty made up 15 percent of the city's population and contributed 11 percent of all alcohol-related offenses. They contributed 23 percent of liquor law violations, obviously related to laws that prohibit their drinking. They also contributed 19 percent of arrests for assault.

The Grand Rapids Study (by Hyman, 1968) examined 5985 drivers involved in all types of accidents and 7590 controls selected at the same location, hour, day, and month where the accident had occurred. Hyman, in his analysis of the data, found that men under eighteen were about three times as accident-vulnerable at blood alcohol content (BAC) of .01 percent to .04 percent as at 0 percent BAC and one and a half times as vulnerable at these BAC levels as men between the ages of eighteen to nineteen. However, few teen-age accidents occurred after drinking, and even fewer after heavy drinking. Thus, even though more vulnerable to the effects of alcohol, they were less likely to drive when they had been drinking even in moderate amounts.

According to Blacker, Demone, and Freeman (1965), delinquent boys used alcohol more frequently and in greater amounts than nondelinquent boys. Mandell *et al.* (1962) and Jessor, Collins, and Jessor (1972) interpreted this to mean that deviant drinking is part of more general deviant patterns of adjustment.

Jessor and Jessor (1973) reported that in a longitudinal study of high school students, problem drinkers engaged in more deviant behavior than non-problem drinkers. The deviant youth valued independence and placed less emphasis on achievement. These data supported the hypothesis put forward by Mandell *et al.* (1962) that youth who committed more unacceptable acts also drank more, and therefore, were likely to commit these acts while drinking rather than alcohol being the cause of such behavior. In fact, they suggested the hypothesis that alcohol use may be an alternative to other unacceptable behavior. The authors did not have the data necessary to control for the absolute amount of time available for unacceptable acts when drinking.

Despite lack of uniformity in the definition of the terms used, certain patterns of problems associated with youthful drinking stand out. Nearly one out of three eighteen- to twenty-year-olds perceive themselves as having experienced a problem in relation to their use of alcohol. Data from a study of highway accidents suggest that the younger group, particularly those under age eighteen, are more likely to have accidents at the same blood alcohol level as older individuals. This suggests that experience with the use of alcohol may limit some of the socially unacceptable consequences associated with it. For the youthful population at large, frequent or heavier drinking is not associated with poorer school performance or lack of success in social activities. There are, however, a group of as many as 15 percent of high school adolescents who have experienced problems associated with drinking. Students who are failing are three times as likely to be using alcohol heavily, and three times as likely to be using it several times a week. By comparison, the rates of temporary psychological dysfunction among youthful drinkers are extraordinary. Among regularly drinking boys nearly half will be unable to remember what happened while drinking at least once a month. The rate is somewhat lower for girls.

In the college group, large proportions of the drinking students experience some complications associated with drinking. Five percent of the college men and about 1 percent of the college women report at least three signs associated with problem drinking. Forty percent of these individuals with several complications from drinking were worried about their pattern of use of alcohol.

Despite these personal psychological concerns about unacceptable behavior associated with alcohol use, these high school and college age groups contribute relatively little to public problems or illegal behavior, except for violations of the law related to obtaining alcohol. For instance, though more prone to accidents while driving with even low blood-alcohol levels, they have fewer accidents than would be expected for their age group. This is apparently because they do not drive after having used alcohol.

The data do support the hypothesis that youth who behave in unacceptable or delinquent ways do drink more than other youth. But even they do not commit as much delinquent behavior while under the influence of alcohol. This has led to the hypothesis that alcohol use among youth may serve as an alternative to other unacceptable behavior.

PERSONAL CHARACTERISTICS AND ALCOHOL ABUSE AMONG YOUTH

As will be discussed below, reports of alcohol abuse among youth have been noted as long ago as biblical times. There has been very little research on the personal or psychological characteristics of individuals who develop various patterns of use.

The fragmentary evidence suggests a few plausible lines of inquiry into psychological factors that may be of importance in producing problem drinkers among adolescents.

Zucker and Van Horn (1972) examined early sibling–social structure variables among 104 boys in the sixteen to eighteen years of age range constituting a sample stratified to represent the heavy and light drinking patterns among sophomore classes of a small community's single public high school. Neither birth position nor spacing between the child and his next sibling was related to statistically significant differences in quantity–frequency of drinking or self-reported symptoms on the Park's Problem Drinking Scale. Family size was not statistically significantly related to drinking pattern. There was an interaction effect indicating that earlier children with a sibling born less than two years later might be more likely to show drinking problems, though this finding did not reach statistical significance.

Williams (1968) examined the self-evaluations of problem drinkers defined by self-report on the Park's Scale of Symptoms of Problem Drinking. His subjects were twenty-year-old volunteers from four fraternities at a New England college. He found that the 23 college problem-drinker students were similar to adult alcoholics in reporting low self-evaluations and fewer self-descriptions indicating comfort in formal social group participation. In addition, college problem drinkers also seemed less comfortable in intimate personal relationships.

In a larger study involving 83 problem drinkers at three colleges compared to 80 nonproblem drinkers, Williams (1967) compared responses to 15 variables selected from the adjective checklist by Heilbrun to indicate a trait characteristic of dependence–independence. He found clear statistically significant differences indicating greater acting independently of others, less soliciting of support, and less subordination to others among college problem drinkers.

The college problem drinkers in the Williams study seemed to be indifferent to the feelings of others and were preoccupied with themselves. They showed low endurance, low persistence in tasks undertaken, and were little involved in maintaining order in their lives. Williams believed that the problem drinkers were not well liked and may not be well socialized.

At stag drinking parties for college men, alcohol, at low levels of drinking, produced self-reports of more self-assurance and exhibitionism, more concern with heterosexuality, impulsiveness, action orientation, and less submissive and dependent behavior. At higher levels of drinking, heterosexuality and sociability were no longer prominent, and other effects became more evident. The self-reports indicated more aggression, less nurturant behavior, less heedfulness of others, more individualistic behavior, less dependability. The alcohol intensified the expression of the preexisting characteristics of the problem drinking sample.

Several researchers have examined the relationship between levels of anxiety and problem drinking. Lundin and Sawyer (1965) examined the relationship between anxiety as measured by a paper and pencil test (IPAT), and frequency and quantity of drinking among 40 undergraduate college fraternity members. The Pearson correlations between anxiety and the drinking measures ranged between .11 and .22, a small positive relationship. The small number of subjects and their homogeneity restricted the size of the correlation.

Smart (1968) used the Taylor Manifest Anxiety Scale with 123 male undergraduate business students to examine the relationship between anxiety and drinking pattern. There was no general relationship between anxiety and amount of alcohol consumed.

Blane, Hill, and Brown (1966) examined the relationship between attitudes toward temperance and irresponsible use of alcohol among 526 students grades 9–12 in a New England high school. Favorability of attitudes toward irresponsible use bore a linear positive association with alienation as measured by the Dean Alienation Scale, comprising items assessing social isolation, powerlessness, and normlessness. No relationship was found between attitudes toward temperance and self-esteem. The components of alienation, normlessness, and powerlessness contributed most importantly to the association with attitudes toward drinking.

Strassburger and Strassburger (1965) studied the relationship between attitudes toward the use of alcohol and two personality variables, social maturity and impulse expression. They found that positive attitudes toward social drinking were related to greater impulse expression. The sense of hopefulness also had been examined as a factor in problem drinking.

Globetti and Windham (1967) found 35 students out of a sample of 132 rural Mississippi high school students used alcohol. Twenty-two of these students were classified by the Straus-Bacon Index as problem drinkers. These students were from blue collar homes, where the mother worked, the family was not religious, and where family amity was low. These problem drinkers tended to be pessimistic on outlook toward the future.

Jessor, Carman, and Grossman (1968) examined expectations of need satisfaction in academic achievement and peer affection in relation to drinking behavior by means of a questionnaire administered to 88 undergraduate students at the University of Colorado, mean age 19.2 years. The group of 300 with low expectations of achievement and low expectations of affection had the highest scores on drinking measures, including drunkenness and drinking-related complications. Among female students, low expectations of achievement and affection were consistently related to drinking for personal effect. Drinking for personal effect also predicted drinking-related complications among women.

Several researchers have gathered information on the self-reported reasons for drinking among heavy and problem drinkers. Williams (1966) examined the association among three measures of self-evaluation and problem drinking as measured by the Park scale for 68 male college fraternity members.

Mandell *et al.* (1962) found that the most frequent reason given by high school students for drinking was to be sociable. Taste of the drink was the second most common reason. However, among frequently drinking boys, "getting high" was given as the reason for drinking by 15 percent, while among frequently drinking girls, 13 percent gave "feeling better" as the reason for drinking in the preceding week. Among heavy drinking boys, boredom was offered by 12 percent as the reason for drinking, and among the girls 10 percent of the heavy drinkers said they drank to forget troubles.

Sex role identification has also been examined as a factor leading to problem drinking. Zucker (1972) examined drinking history information for 20 boys, mean age 14.8 years, and 47 girls, mean age 14.7 years, for whom an obvious measure of sex role identification, the California Psychological Index (CPI), and preference for sex type movie themes used as a subtle measure of identification were available. Drinkers were more masculine on the CPI text, a vehicle allowing conscious projection of masculinity. No difference among categories of drinkers on the fantasy projection measure appeared. Zucker interpreted these data to indicate that heavier-drinking boys present a hypermasculine facade. There was a slight tendency for more frequently drinking girls to score more femininely on the CPI, $p = < .10$.

Parker (1969) administered Form A of the Terman–Miles masculinity-femininity test, called the "Attitudes–Interest Analysis Test," to 205 male undergraduate students, mean age 20.3. The Terman–Miles has two scoring systems: one standardized on normal populations, manifest masculinity; and another standardized on a population of passive male homosexuals, latent masculinity. The Manson Alcadd test was used to sort the students into three groups: abstainer–moderate, heavy drinkers, and quasi-alcoholics. The quasi-alcoholic group had higher manifest-masculinity scores than the abstainer–moderate group, $p = < .05$. The quasi-alcoholic group also had lower latent masculinity scores, indicative of a response pattern more like the homosexual standardization group, $p = < .001$. When the group was sorted into those individuals with congruent manifest and latent masculinity scores and those with incongruent scores, there was no significant difference in the drinking pattern between the two groups. Incongruence was not related to increased drinking. Despite this finding, Parker referred to incongruence as a factor in drinking pattern. A more appropriate interpretation of Parker's findings is that each measure is independently associated with alcohol use pattern.

In the Parker study the 36 subjects from homes broken before the age of seventeen provided 41 percent of the quasi-alcoholic group and only 18 percent of the abstainer–moderate drinker category. Seventy-one percent of the quasi-alcoholic group preferred their mother as a parent in contrast to 38 percent and 48 percent in the other drinking groups, $p = < .05$.

The concept that personal or psychological characteristics of young people may play an important role in their becoming heavy or problem drinkers is very attractive. Unfortunately, the research in this area is fragmentary and conclusions are based on very small and unrepresentative samples. Some hypotheses have received enough support to be worthy of further study.

By the age of college, problem drinkers have lower self-evaluations than their peers. They are less comfortable in intimate personal relationships. They act more independently of others. This independence from others does not seem to be based on greater social competence, the problem drinkers showing less endurance, persistence, and ability to maintain orderliness in their lives. Alcohol use intensifies these preexisting characteristics of the problem drinkers. At low levels of alcohol use they become more self-assured, independent, impulsive, and heterosexually oriented. At higher levels of alcohol use they become more aggressive, even more individualistic and less dependable. The effects of the alcohol are similar in nonproblem drinkers. However, the starting point for the problem drinkers is already in the direction that will be exaggerated by the alcohol.

Anxiety as a motive for use of alcohol by young problem drinkers has not been established. However, alienation defined in terms of normlessness and feelings of helplessness have been implicated. These findings are similar to those other studies that find a sense of hopelessness about achievement or obtaining affection among young problem drinkers.

The young problem drinker is described in several studies as impulsive and less orderly in organizing his life. At least in the college group, problem drinkers are more self-critical, in part reflecting their awareness that they are not living up to their ideals. Boredom and a sense of having troubles become the conscious expressions of these difficulties in adaptation and, in turn, become the triggers for drinking.

Two other hypotheses about the motivational sources for youthful problem drinking have intrigued researchers. The first of these is that problem drinking is a response to psychological conflicts in the area of sex role adjustment, namely, conflict over expression of dependence or independence and conflict over expression of masculinity or femininity.

McCord, McCord, and Gudeman (1960) noted that while adult alcoholics were rated as psychologically dependent as prealcoholics, they were counterdependent in behavior.

Lisansky (1960) believed that alcoholics had an intense dependence–independence conflict. Both Lisansky and McCord et al. saw the loss of control over alcohol and concomitant role failure during adulthood dissolving the self-image of independence. This allowed the subsequent emergence of the repressed dependency needs.

The research evidence is quite substantial in that youthful problem drinkers manifest independent behavior. It is not clear from the data available that this independence masks an unexpressed dependency. Indeed, even though dependency is expressed among adult alcoholics, it may not pre-exist but rather develop in response to later life experiences. The issue is far from closed; caution must be exerted in the face of the present evidence.

The hypotheses about conflict over expression of feminine urges are not supported by the data.

The data of the study by Zucker and Van Horn (1972) lend themselves to another interpretation. Even though the heavier-drinking boys had higher masculinity scores, they did not represent hypermasculinity in an absolute sense since the comparison of CPI scores between heavier- and lighter-drinking groups was relative. It seemed more reasonable to interpret this data to indicate that the group that was engaging in more drinking at age fifteen reported itself to be more like its appropriate older sex role group.

The data from the Parker study also did not indicate more problem drinking among those with both score patterns indicating an adjusted male pattern and homosexuality. The quasi-alcoholic group as a group had higher manifest-masculinity scores *and* higher passive homosexuality scores, suggesting that the problem-drinking group was composed of two subgroups. One subgroup was composed of individuals who had more extreme self-presentations in masculine terms and another with more extreme self-presentations characteristic of passive homosexuals.

This review of the data suggests a simple straightforward hypothesis to explain a substantial portion of youthful problem drinking. Problem drinking is an expression of psychological undercontrol of impulses, including aggression. This undercontrol interferes with the ability to achieve academic success and personal satisfaction through intimate relationships. The problem drinker has feelings of alienation and a low self-evaluation because he feels hopeless about his ability to overcome those inner forces that impede his social accomplishments and obtaining satisfaction.

JUVENILE ALCOHOLISM

There has been an increase in interest in juvenile alcoholism. This concern seems to rise and fall during different periods of history.

Healy in 1915 reported that 2.7 percent of his cases in his children's psychiatric clinic had an involvement with alcohol.

There was an increase in concern about juvenile alcoholism in medical literature during the 1930s (Rurah, 1932).

Another increase in concern became manifest in the 1940s. Teachers reported in New York in the 1940s that bright students were returning to school from lunch drowsy and dull because of wine. Lourie (1943) reports that in 1940, 121 cases in New York Children's Court were there for intoxication.

Ford (as quoted by Lourie, 1943) reported alcoholic delirium and polyneuritis in some youthful drinkers. He noted that occasional individuals became alcoholic as a result of being given wine and beer as infants.

Lourie suggested that a distinction should be made between childhood addiction to alcohol and childhood social drinking. Lourie screened 121 children at Bellevue Hospital for habitual rather than occasional intoxication. Twenty children ranging in age from five to fourteen years were determined to be habitual drunkards. He suggested that the rarity of cases in the literature is the result of physicians not asking routinely for relevant information.

Lourie ascertained that the primary reasons for these 20 children's habitual drinking were: Seven drank for escape, four drank because of identification with an alcoholic adult, four drank as part of general delinquency, three drank as part of a homosexuality adjustment pattern, and two as part of a psychosis.

Lourie reported that none of the children experienced craving for alcohol. All the children simply desired the intoxicated state. He suggested that in most children the use of alcohol is a situational response that is resolved when underlying stressful conditions are relieved. In the case of delinquent psychopathic children, use of alcohol is resumed even after prolonged interruption and is part of the delinquent life-style rather than its cause.

Malzburg (1960), in a study of New York Civil State Hospitals in 1947, reported that there were no cases of alcoholism under fifteen years of age and only two under twenty admitted to the then largest mental health system in the United States. This suggests that youthful alcoholics were not entering the mental health system.

MacKay in 1960, reporting on drinkers with problems of major significance, found 20 cases between ages thirteen to eighteen referred to an alcoholism clinic. Alcohol use in these youths interfered with schoolwork and employment. These children reported hangovers, shakiness, and blackouts. In each of these lower-class boys, drinking was an established pattern by the first year of adolescence, some drinking heavily by that time. Drinking was usually carried out in social group settings. However, the groups were loosely knit territorial associations and not effective in overcoming the drinkers' feelings of loneliness. All the boys reported their fathers to be alcoholics who abandoned the family and whom they then replaced.

MacKay described these children as hostile, impulsive with unfulfilled dependency needs, depressed, and sexually confused.

Falstein, a psychoanalyst, reported the case of a patient with alcoholic symptomatology. This fifteen-year-old boy appeared addicted and also demonstrated schizophrenic bizarre acting out. Alcohol was used to express sexual and aggressive impulses (Falstein, 1953).

Kinsey (1966) found that of 31 alcoholic women, four reported their first alcoholic symptoms to have appeared before the age of twenty.

There is only one large-scale study with data available about the prevalence of youthful alcoholism. Zax, Gardner, and Hart (1961) studied the prevalence of alcoholism in Monroe County New York in 1961 by surveying the records of all community agencies. They found 3838 individuals with a clear alcoholism problem. One per 100,000 white males and 17 per 100,000 nonwhite males under age twenty were found with a clear diagnosis of alcoholism or associated brain syndrome or more than three public intoxication arrests. It is probable that a good part of this young group had contact with the police and court as the primary point of public definition of alcohol as a problem for them. This may account, in part, for the large difference in black and white problem rates since police are more likely to arrest blacks for alcohol-related problems. However, 49 per 100,000 white males, 294 per 100,000 black males, and 6 per 100,000 white females, all under twenty, applied to an alcohol problem treatment agency, though not diagnosed as alcoholic nor having had a single public intoxication arrest. At less than twenty years of age, 19 per 100,000 males had a primary medical diagnosis of alcoholism and 349 per 100,000 had some level of alcohol-associated problems.

The literature indicates that there is indeed youthful alcoholism. However, it is not often reported nor do cases appear in the traditional mental health service system. This is probably in part because professionals do not routinely take data from youthful patients about their use of alcohol. In part it may reflect a general perception about the effects of labeling a youngster an alcoholic and the appropriateness of adult-oriented agencies for treatment. Youthful alcoholism is for the most part a response to distorting life situations and disappears when the external conditions improve. In a small number of cases the alcoholism is part of a psychopathic criminal adjustment or part of other serious psychological disturbances. In such cases the alcoholism is more resistant to treatment.

Youthful alcoholism, under age twenty, may occur at rates of 0.5 per 1000 white males, .06 per 1000 white females, 3.0 per 1000 black males, and a lower rate than .01 per 1000 black females. The differences in black and white rates may in part represent reporting differences involving police arrests. However, when voluntary applications for treatment are examined, the black male rate is many times that of white male rate. If the black difference of six times the white rate is not simply the result of reporting differences, it has important implications, both for research and service planning.

YOUTHFUL CHARACTERISTICS PREDICTIVE OF ADULT ALCOHOLISM

A number of studies provide evidence related to youthful characteristics predicting adult alcoholism.

Robins, Murphy, and Breckenridge (1968) studied drinking behavior of urban black men aged thirty to thirty-six. Of 223 interviewed, 58 percent reported that they had been, at some time, heavy drinkers, drinking at least seven drinks on one occasion every week or four or more drinks daily. Heavy drinking as adults was associated with early initiation into drinking and a pattern of truancy. Only 34 percent of high school graduates became problem drinkers at some time as compared to 73 percent of school dropouts. Being both a truant and school failure was the best predictor of later alcohol problems. Earlier drinking in high school years also predicted adult heavy drinking. "No record of delinquency" was associated with not having alcohol problems. Of high school graduates who showed no elementary school problems, 23 percent had had a drinking problem at some time, while of those who didn't graduate and had eventually become delinquent after age fifteen, 89 percent had an alcohol problem. Sons whose fathers were successful, conforming, middle-class, and present during childhood had the fewest drinking problems.

Robins, Bates, and O'Neal (1962) examined the child patient population of a St. Louis Municipal Psychiatric Clinic 30 years after they had been at that facility. The 503 white children who survived to age twenty-five with intelligence quotients above 80 were matched with 100 students selected from public school records living in the same census tract, of the same age, sex, race, and I.Q. Both groups were interviewed with regard to drinking behavior. Information was also collected from police and institutional records. Fifteen percent of the patients and 2 percent of the controls were identifiable as alcoholics as adults. The difference was even larger when examined in the male group. Twenty-one percent of the male patients were identified as adult alcoholics and 3 percent of the male controls were so identified. Eight percent of the female patients and none of the controls were identified as adult alcoholics. Antisocial behavior of the father was significantly associated with later development of alcoholism. Similarly, 45 percent of those juveniles who had appeared in court were later identified as alcoholic, as compared with 15 percent of those who had not. Roughly two thirds of the adult alcoholics and 50 percent of the problem drinkers in the study had appeared in court as children. The alcoholics also had more symptoms as children reported in their clinic records, primarily running away, theft, and assault.

McCord, McCord, and Gudeman (1960) analyzed data from the Cambridge-Somerville study, using the records of five years of observations by trained social workers, of 255 boys and their families who were participating in a project to prevent juvenile delinquency. Community records were examined. Twenty-nine of these boys were eventually identified as alcoholics as adults. They were compared with 158 boys who in adulthood were neither criminals nor alcoholics. The group eventually identified as alcoholics, as children between the ages of eleven to sixteen, did not manifest unusual social problems. About a third of the alcoholic group showed more unrestrained aggression, sadistic behavior, and hyperactivity at levels statistically significant beyond the nonalcoholic group. Their records indicated less favorable feelings toward their mothers, fewer fears, and lesser feelings of inferiority than nondeviant youngsters. Unfortunately, the comparisons presented do not allow examination of whether these characteristics are specific to alcoholics as children or to all deviant adults.

McCord *et al.* (1960) hypothesized the existence of a prealcoholic personality comprising an extreme masculine facade covering dependent feminine tendencies. They discounted the possibility that the dependency visible in many adult alcoholics is itself the result of the alcohol use pattern. They believed that the dependency must have been latent in the youthful period, and that alcohol use allowed the expression of the suppressed dependency needs. These dependency needs result from erratic satisfaction of dependency desires during childhood, and the absence of suitable male role specification because of an unsatisfying relationship with the father.

Block (1971) examined the longitudinal data that had been accumulated for two groups of normal subjects originally selected as children for developmental studies in Berkeley and Oakland, California. These individuals were approached when they were between thirty and thirty-seven years old to allow themselves to be studied again.

Because of the small sample, this study can only be seen as a source of hypotheses. By comparison, it makes available coherent pictures of youthful adjustment that could reasonably increase the probability of problem drinking later in life. The group as a whole was better educated than the general population and better placed socioeconomically. Information from the files of the subjects was rated by at least three psychologically sophisticated judges independently for the childhood, adolescent, and adult periods. The subjects were then sorted into adult personality types. As adult men, two of the five personality groups demonstrated high levels of problem drinking. Block characterized one group as "anomic extraverts." As children their middle-class family constellation involved a tense mother and somewhat withdrawn father. In junior high school they were gregarious, masculine, assertive, poised, cheerful, and conventional. They were also interested in members of the opposite sex, could express hostility directly, and were rebellious. In senior high school these boys were gregarious, assertive, and cheerful. These boys matured physically earlier. They were less controlled by their parents during adolescence and were engaged in mating behavior earlier. They were closer to, and aspired more to, their fathers' role in life, including a sense of power. On the other hand, as youths they cried or became angry easily and worried more.

The second group of males showing a high rate of problem drinking as adults are characterized by Block as "unsettled undercontrollers." In junior high school these boys were rebellious and hostile, irritable, extrapunitive, and undercontrolled. They were also self-indulgent, self-defeating, and self-pitying. During senior high school they continued to be rebellious, undercontrolled, and irritable with fluctuating moods and unconventional thoughts; they lacked a sense of meaning in life, and were more dependent on friends.

One group of women was also described by Block as having a greater potential for problems associated with alcohol. He called these women "hyperfeminine repressives." As junior high school students they were feminine, repressive, dependable, submissive, fearful, and concerned with physical appearance and dependent on their peers. During high school they had a calm manner, handled anxiety by excluding it from awareness. From middle-class families, they matured early physically, had more menstrual pain, and had friends of whom their families disapproved. They dated less, had intercourse earlier, and were unable to maintain stable interpersonal relations.

The data from these recent retrospective studies indicate that there may be characteristic adolescent personality and adaptational patterns associated with adult alcoholism. Early initiation into drinking associated with truancy and school failure form a predictive pattern among black men. Among white men, youthful adjustment problems leading to referral to a psychiatric clinic seem to be predictive of adult alcoholism. There may have been considerable similarity between the psychological adjustment of the white and black adolescents. During the period when the subjects of these studies were adolescents, there was a greater likelihood that white boys would receive referrals for professional help.

The characteristics of these boys—truancy and appearance in court for assault, theft, and running away—all clearly indicate an antisocial personality pattern. Alcoholism may be added to this pattern as part of a hedonistic lifestyle uncontrolled by social norms. Alternatively, the alcoholism may be related to more fundamental psychological adjustment problems, as suggested by the unrestrained aggression and hyperactivity found in prealcoholic boys by McCord et al. (1960).

The findings of the California studies of middle-class individuals who have problems associated with the use of alcohol, paralleled those of lower-class boys. Both groups, as adolescents, showed emotional lability, undercontrol of impulsive behavior, hostility, and physical precocity leading to early aspirations to adult status. The expression of this among middle-class boys was less likely to express itself through illegal practices.

The middle-class girls who went on to have problems also showed early physical maturity and entrance into adult behavior. However, they were more fearful, dependent, and less effective socially.

These studies highlight two hypotheses. The development of internal controls over impulses of aggression during childhood may be an important factor in the later development of alcoholism. Precocious pressure to adopt adult sex role behavior may also be associated with problems with alcohol.

JUVENILE DELINQUENCY AND TEENAGE DRINKING

There has been some research interest in the relationship between alcohol abuse, delinquency, and sociopathy.

Delinquency, by definition, involves a judicial process. Though many teen-agers perform illegal acts, if they are not caught and adjudicated they cannot be reported as delinquent. The studies relating delinquency and alcohol have all been carried out with incarcerated youngsters.

Schuckit (1963) believed it necessary to investigate the relationship between sociopathy and primary and secondary alcoholism. He defined primary alcoholics as those with no history of psychiatric disturbance prior to commencement of an excessive drinking pattern. Secondary alcoholics are those with a history of psychiatric disturbance prior to their beginning excessive drinking. Sociopathy or antisocial personality was defined as a chronic disorder beginning prior to the age of fifteen, with truancy, running away from home, criminal offenses, impulsive job changes, outbursts of rage and fighting, sexual promiscuity, a period of wanderlust, and persistent lying as significant features. The sociopathic alcoholic may have had a ten-year period of prior antisocial behavior before he was labeled as having a drinking problem.

The data demonstrating a relationship between youthful prisoner status and alcohol abuse must be viewed with caution. Drinking while attempting to carry out a crime sometimes is done to increase courage, and also may increase the chances of being caught in the act. Sociopathic prisoners may have a higher rate of alcohol abuse than the general sociopathic population.

Winokur et al. (1970) found that sociopathic alcoholic men were younger than the others at the time of onset of alcohol abuse, eighteen versus twenty-four years.

The association between delinquency and alcohol consumption in teen-agers appears to be tenuous.

Podolsky (1960) gave several alternatives for the reported relationship between sociopathy and alcoholism: (1) Many alcoholics are sociopaths; (2) there is a third factor, X, underlying both alcoholism and sociopathy; or (3) many sociopaths abuse alcohol as part of their antisocial behavior.

Blacker et al. (1965) noted in 1965 that there was little systematic study of the use of alcohol by youthful offenders. The studies done prior to that year were restricted in scope and in study-group selection.

Blacker believed that the frequent use of alcohol by teen-age delinquents may be part of the normal socialization process pattern rather than a predictor of future alcoholism. His data were suggestive that those teen-agers who could control their drinking also could control their other impulses. The drinking was symptomatic of their general difficulties in adapting to society.

His study of 500 teen-age boys referred by the courts to a state reception center in the Commonwealth of Massachusetts demonstrated that use of alcohol increased with age. Twenty-one percent of the boys twelve years and under were drinkers, as compared to 83 percent of those seventeen and older. The median age of the sample was fifteen. Pathological drinkers constituted 7 percent of the sample, with relief or "kick" drinkers accounting for 9 percent, and heavy drinkers adding another 13 percent to the sample. This distribution was not very different from that of nondelinquent boys still in high school, except for the frequency of excessive drinking. The Blacker *et al.* study failed to demonstrate that his delinquent population, as a whole, was characterized by the pathological use of alcohol. Though alcohol may not be causal, it may play a substantive role in further decreasing controls in a population that does not conform to social limits.

MacKay (1960) studied problem drinkers among 500 male delinquents. Ten percent of the sample of boys, whose average age was fifteen, were classified as problem drinkers, and 13 percent as heavy drinkers.

All problem drinkers in the delinquent group reported that their drinking was without parental approval. Three fourths of the problem drinkers drank alone. More than half of the problem drinkers had been arrested for drinking, and most had committed their criminal offenses subsequent to drinking. However, they stated that their drinking had nothing to do with their being incarcerated in the youth detention center.

MacKay (1967) studied youthful delinquents in New Hampshire and compared them to an unmatched sample of junior and senior high school students. Because of this procedure, it is not possible to attribute differences in alcohol use patterns between groups to the delinquency. However, the study is suggestive of the importance of alcohol use among delinquent youth. Fourteen percent of the delinquents and 29 percent of the students had never consumed alcoholic beverages. Only 20 percent of the delinquents, compared to 75 percent of the students, had had their first drink at home. Eighteen percent of delinquents drank the first time to become intoxicated, as compared to 2 percent of the students. Sixty percent of the delinquents and 29 percent of the students reported solitary drinking. Forty-one percent of the delinquents and 18 percent of the students reported drinking alone more than ten times. Fourteen percent of the delinquents and 2 percent of the students stated that they felt that they had a drinking problem.

On eight signs of problem drinking, the delinquents, as a group, had at least three times as many signs of problem drinking. However, the delinquents had been arrested for drinking five times as often and had been fighting while drinking six times as often. About 60 percent of the delinquents had drunk whenever they got the chance, alone, before going to a party, and had fought while drinking. Forty-six percent of the delinquents had blacked out while drinking, as compared to 15 percent of the students.

Pearce and Garrett (1970), in a study with the same methodological problems, compared the drinking behavior of delinquent youth with nondelinquent youth in Idaho and Utah in 1970. Unlike many of the samples restricted to male delinquents, females were included. The female drinking pattern, though greater than nondelinquent females, was less than that of delinquent males. Of the delinquents in this study, 31 percent had permission to drink, compared to 75 percent of students. It is not surprising that they most frequently had as a location of the first drink a friend's home, and had as sources of alcohol older friends as compared to parents. Fifty-five percent of the delinquents drank at least once a week, compared to 15 percent of the students in this low use area.

Widseth (1971) surveyed 106 delinquent girls in Massachusetts in 1971. The median age was 14.7 years with 57 percent of the girls reporting weekly consumption of alcoholic beverages. Sixteen percent were listed as abstainers with an additional 9 percent reporting that they only tasted alcohol once. Forty-four percent of the sample, or nearly 60 percent of those considered to be users of alcoholic beverages, were reported as drinking to excess. Of those who consumed alcoholic beverages more than once, 37 percent admitted to blackouts and 30 percent admitted to "passing out." Of this group, nearly 30 percent admitted to drinking alone, with 7 percent of the total drinking sample being brought to court for drinking.

The data from the studies reflect the lack of evidence of a causal association between delinquency and alcohol. The researchers indicate that more study is needed in this area, with the development of instruments that will allow more uniform reporting and a meaningful comparison of male and female delinquent and nondelinquent populations. The data are suggestive of the high levels of use of alcohol among delinquents. As discussed before, this association may reflect a problem underlying both delinquency and the alcohol use, or that delinquency produces alcohol abuse, or that alcohol abuse produces delinquency.

DISCUSSION

Regularities in drinking behavior reflect common definitions of what alcohol does for and to the individual. Each generation observes the older groups and accepts as part of its cultural heritage its ideas and language to describe and explain why and how to use alcohol. It is within this cultural pattern of use that some individuals discover personal meanings and idiosyncratic uses for alcohol.

As Maddox (1964) expressed it, responses to alcohol are part of the culturally created design for living. The teen-ager does not invent the idea of drinking, but rather learns it within a context that reflects the social background of the user. The roles teen-ager, young adult, and adult, though vaguely defined, do carry with them common expectations in terms of appropriate behavior in many areas, including the use of alcohol.

There is a pervasive ambivalence toward youthful use of alcohol in the United States. It is generally defined as illegal by organized society through governmental agencies. Yet, alcohol is provided to youth by most parents well before the legal age of use. Most youth, like most adults, classify themselves as users of alcohol. Within the concept of user, the quantity and frequency of alcohol consumed varied from once or twice a year to daily use of more than a half-pint of whiskey.

Young people are responsive to regional, ethnic, and religious values about alcohol use. This is reflected in variations in the age of first use and the frequency and quantity of consumption of alcohol.

Methodological differences in the research on prevalence of use produce variation in the reports of percentage of users at different ages. Best estimates available indicate that by age seventeen, 93 percent of boys and 87 percent of girls have had a drink of alcohol. The proportion using alcohol among school dropouts is not known at this time.

Each age cohort of the population, as it moves through the life cycle, seems to have about the same drinking pattern as the cohort that preceded it. This suggests that the youthful population is adopting in an orderly, socially controlled fashion the pattern of drinking of the young adults in the society. There has been some increase in the proportion of younger teen-agers who use alcohol in recent years. The gap between the proportion of boys and girls who use alcohol is decreasing. At this time, about one third of the boys graduating from high school are drinking at least once a week.

Youth perceive alcohol use as part of being an adult. They see it as part of the social patterns of conviviality and celebration. Youth tend to follow the drinking behavior of significant adults. Drinking is understood in terms of the benefits it provides rather than as a drug acting on the individual. Only a small proportion of youth see it as a source of relief from boredom, loneliness, or anxiety.

There is little evidence of widespread use of drinking as an expression of rebellion or hostility. In most parts of the country, parental and peer values do not conflict. Parental practices seem to be the major determinant in youthful drinking behavior. As the teen-ager becomes more involved in the process of separating from parental control in many areas of living at about sixteen, peer group standards become more important in influencing alcohol use.

Older teen-agers are involved with achieving an occupational career, leaving the home, finding love and a mate. These are not the current tasks and obligations of their adult parents. The teen-agers join together in groups that have similar concerns and ambitions. These peer groups are oriented to values related to achieving postschool adult roles. These roles are symbolized through heroes, such as athletes, scientists, or business leaders. The peer groups support their members in adopting the behaviors and life-style belonging to their heroes. Though adults may believe this behavior is premature in adolescents, it is not in general conflict with parental values. Such teen-agers are sometimes described as being in rebellion. Most studies indicate that, for the most part, they are carrying out in a comtemporary version the values, if not the actual behavior, of their parents. A very small group of youth is motivated to identify with a group chosen *primarily* because its behavior is in opposition to parental values. Such instances are more likely to reflect personal maladjustment and distorting parental–child interactions.

Teen-age use of alcohol, in general, is not the result of tensions produced by the social structure or an expression of hostility to adults. Parental drinking, permission to drink, provision of alcohol, and training of adolescents to drink all demonstrate the continuity in values and expectations between parents and children. Teen-agers' alcohol-use patterns demonstrate conformity to ethnic and religious standards, socioeconomic status, appropriate behavior, and rural–urban residence constraints. Girls, during the period covered by the studies, used alcohol in societally set gender-role appropriate ways. The nearer the teen-agers come to young adult status, the more likely they are to drink like young adults. However, young adults have always used more alcohol than the age group who are the parents of teen-agers. As a result, parents may be concerned about the discrepancy between their children's drinking behavior and their own.

In surveying the literature, it is evident that there are childhood and adolescent problems associated with the use of alcohol similar to those found among adults. These include persistent use of alcohol to levels of intoxication, interference with social performance, and physiological consequences of alcohol abuse. The prevalence of such youthful alcoholism has not been studied because there has not been a clear focusing of interest on the problem. Professionals do not ask relevant questions of individuals and populations that would allow determination of the prevalence of clinical problems associated with alcohol use. A good deal of childhood alcohol abuse seems to be situationally determined. Children living under extraordinarily stressful circumstances from which they cannot escape discover alcohol as a substance that allows them to tolerate their unhappy situations.

Heavy use of alcohol, which is associated with interference of social performance or health, is part of several life-styles in the United States. For adolescents being reared in these life-styles, heavy use of alcohol is not ab-

and is part of socialization into becoming a member of the adult group community. The group that uses alcohol because of stressful conditions gives up such use when the stress is relieved. Some adolescents live in a delinquent life-style in which alcohol use is another facet of antisocial behavior. The group with a delinquent life-style appears to continue their alcohol use on an opportunistic basis, many eventually becoming alcoholic in later life.

One group of youthful problem drinkers is composed of youngsters who are adopting a homosexual life-style that involves heavy alcohol use. Another group is composed of youngsters who are psychotic, and for whom the use of alcohol is part of attempting to cope with the inner psychological turmoil.

Some mainstream adolescents in the course of adopting adult patterns of use of alcohol are attracted to psychotropic effects that have personal significance, such as relief from feelings of isolation or feelings of inadequacy. There is not enough research available to indicate what proportion of each type goes on to be adult alcoholics, nor what proportion of the adult alcoholic group is contributed by each of these youthful groups.

There is little evidence that the general patterns of alcohol use by youth interferes with school performance, social relationships, or law-abiding behavior; this despite the frequent occurrence among teen-agers of signs that would be indicative of alcoholism among adults. Some authors believe that drinking is for some adolescents an alternative to other antisocial behavior.

More college individuals seem to worry about their alcohol use pattern than those in other career groups. Self-conscious concern about the consequences is associated with social factors. However, the proportion having problems associated with drinking is probably no more than in other groups. Between 2 percent and 6 percent of all youthful groups report very serious problems resulting from alcohol use as part of their adjustment pattern. The level of concern about alcohol-related problems may reflect social conditioning and introspectiveness rather than the actual extent of the problem.

The eighteen- to twenty-year-old group, transitioning to young adulthood, has the largest proportion of persons, 27 percent, who report themselves as having some problem in connection with drinking. The giving up of parental standards, the lack of specification of role-appropriate behavior, and the hedonistic values of this life period all contribute to cause these problems. For some individuals alcohol is used as a reenforcement to relieve the tensions of this transitional period.

The personality characteristics of youthful problem drinkers most often include evidence of impulsiveness, particularly in expression of aggression. Precocious physical development and pressure to achieve adult status early also play a part of this development. Lack of effective parental training during childhood is the cause of this lack of internal control.

A special note about the high rate of alcohol problems in black youth must be made. The rate is six times that of the white group. This, no doubt, reflects conditions of minority life and the particular difficulties this group has in entering adult male roles.

There is just beginning to be available some data that suggest that childhood and adolescent adjustment patterns are precursors of adult problems associated with the use of alcohol. Examination of clinical reports of adolescent motivation for drinking and the retrospective studies of alcoholics implicate five adolescent personality patterns. The following descriptions are offered as hypothetical types of adolescent adjustment leading to adult alcoholism.

1. *Antisocial.* an overactive, assaultive boy who truants and is an academic failure in school. This pattern reflects parental undercontrol of behavior. Alcohol is most frequently used to diminish awareness of fears.

2. *Ambitious–Impulsive.* a physically precocious boy who is motivated to assume adult and social leadership roles early, and who is also characterized by emotional lability. This pattern reflects weak internalization of control over impulses, particularly anger, because of lack of attachment to parents. Alcohol is most usually used as a part of the portrayal of manliness and involves expression of anger.

3. *Angry–Dependent.* a rebellious, hostile, irritable boy with unconventional thinking who rejects parental emotional support in favor of dependency on friends. This pattern reflects punitive overcontrol by parents. Alcohol is used to suppress fear of loss of emotional support and to allow a semblance of psychological intimacy.

4. *Hopeless.* boys and girls characterized by psychological hopelessness about achieving academic or social satisfactions. This sense of isolation may occur among teen-agers because they are members of a minority group, economically marginal, or because of personal characteristics that make them unacceptable to their parents. Alcohol is used to overcome feelings of sadness experienced as boredom and social isolation.

5. *Fearful–Dependent.* fearful, submissive girls who are unable to establish stable intimate relationships but feel under great pressure to do so. This often reflects ambivalent parental training in the expression of sexual feelings. Alcohol is used to suppress fear about loss of emotional support.

There is an urgent need to develop research both on the extent and sources of problem drinking among youth. Simple environmental hypotheses hinged to the importance of factors such as peer group standards do not hold much explanatory power for problem drinking. Simple hypotheses based on personality theory such as rebellion or hypermasculinity are also not able to explain the high rate of signs of excessive drinking among youth, nor to predict who will have social or performance problems from using alcohol. The prevalence of youthful problem drinking warrants such research.

The prediction of adult alcohol problems from youthful adjustment seems to hold some promise at this time. There are some leads from the several retrospective studies that seem to confirm the importance of factors such as impulsiveness and unexpressed anger. Research in this area also has practical potential for the development of programs for the prevention of alcoholism.

REFERENCES

Alexander, C. N., Jr., 1964, Consensus and mutual attraction in natural cliques: A study of adolescent drinkers, *Amer. J. Socio.* 69:395-403.

Alexander, C. N., Jr., 1967, Alcohol and adolescent rebellion, *Social Forces* 45:542-550.

Alexander, C. N., Jr., and Campbell, E. Q., 1967, Peer influences on adolescent drinking, *Quart. J. Stud. Alc.* 28:444-453.

Alexander, C. N., Jr., and Campbell, E. Q., 1968, Balance forces and environmental effects: Factors influencing the cohesiveness of adolescent drinking groups, *Social Forces* 46:367-374.

Blacker, E., Demone, H. W., Jr., and Freeman, H. E., 1965, Drinking behavior of delinquent boys, *Quart. J. Stud. Alc.* 26:223-237.

Blane, H. T., Hill, M. J., and Brown, E., 1966, Alienation, self-esteem and attitudes toward drinking in high school students, *Quart. J. Stud. Alc.* 27:350-354.

Block, J., and Hahn, N., 1971, "Lives Through Time," Berkeley, Bancroft Books.

Chappell, M. N., Goldberg, H. D., Ryan, J. F. X., Campbell, W. J., and Yuker, H. E., 1953, Use of alcoholic beverages among high school students, Mrs. John S. Sheppard Foundation, New York, Hofstra Research Bureau, Hofstra College, Hempstead, New York.

Coleman, J. S., 1970, Interpretation of adolescent culture, *in* J. Zubin and A. M. Freedman (eds.), "The Psychopathology of Adolescence," New York, Grune and Stratton.

Falstein, E. I., 1953, Juvenile alcoholism: A psychodynamic case study of addiction, *Orthopsychiatry* 23:530-551.

Forslund, M. A., and Gustafson, T. J., 1970, Influence of peers and parents and sex differences in drinking by high school students, *Quart. J. Stud. Alc.* 31:868-875.

Garlie, N. W., 1971, Characteristics of teen-agers with alcohol-related problems, PhD Dissertation, University of Utah.

Globetti, G., and Windham, G. O., 1967, The social adjustment of high school students and the use of beverage alcohol, *Social. Social Res.* 51:148-157.

Healy, W. A., 1915, "The Individual Delinquent." Little Brown and Co., Boston.

Hyman, M. M., 1968, Accident vulnerability and blood alcohol concentrations of drivers by demographic characteristics, *Quarterly Journal of Studies on Alcohol, Supplement 4*, pp. 34-57.

Jessor, R., Carman, R. S., and Grossman, P. H., 1968, Expectations of need satisfaction and drinking patterns of college students, *Quart. J. Stud. Alc.* 29:101-116.

Jessor, R., Collins, M. I., and Jessor, S. L., 1972, On becoming a drinker: Social-psychological aspects of an adolescent transition, *Annals New York Academy of Sciences* 197-199-213.

Jessor, R., and Jessor, S. L., 1973, Problem drinking in youth: personality, social and behavioral antecedents and correlates, Publication 144, Institute of Behavioral Sciences, University of Colorado.

Johnston, L., 1973, Drugs and American youth, Institute for Social Research, The University of Michigan, Ann Arbor, Michigan.

Johnston, L. D., Dec. 1974, Drug use during and after high school: Results of a national longitudinal study, *Amer. J. of Pub. Hlth.*, The Epidemiology of Drug Abuse, Vol. 64:29–37.

Kandel, D., 1974, Inter- and intragenerational influences on adolescent marijuana use, *Journal of Social Issues* Vol. 30, No. 2.

Kandel, D., and Lesser, G. S., 1972, "Youth in Two Worlds," Jossey-Bass, San Francisco.

Kinsey, B.A., 1966, "The Female Alcoholic: A Social Psychological Study," Charles C. Thomas, Springfield, Ill.

Lisansky, E. S., 1960, The etiology of alcoholism: The role of psychological predisposition, *Quart. J. Stud. Alc.* 21:314–343.

Lourie, R. S., 1943, Alcoholism in children, *Amer. J. Ortho. Psycho.* 13:322–338.

Lundin, R. W., and Sawyer, C. R., 1965, The relationship between test anxiety, drinking patterns and scholastic achievement in a group of undergraduate college men, *J. Gen. Psycho.* 73:143–146.

MacKay, J. R., 1960, Clinical observations on adolescent problem drinkers, *Quart. J. Stud. Alc.* 21:124–134.

MacKay, J. R., Phillips, D. L., and Bryce, F. O., 1967, Drinking behavior among teenagers: A comparison of institutionalized and non-institutionalized youth, *J. Health Soc. Behav.* 8:46–54.

McCluggage, M. M., Baur, E. J., Warriner, C. K., and Clark, C. D., 1956, Attitudes towards use of alcoholic beverages: A survey among high school students in the Wichita metropolitan area and in the non-metropolitan counties of Eastern Kansas, Mrs. John S. Sheppard Foundation, New York, Department of Sociology and Anthropology, University of Kansas, Lawrence, Kansas.

McCord, W., McCord, J., and Gudeman, J., 1960, "Origins of Alcoholism," Stanford Univ. Press, Stanford, Calif.

Maddox, G. L., 1964, High school student drinking behavior: Incidental information from two national surveys, *Quart. J. Stud. Alc.* 25:339–347.

Malzburg, B., 1960, The alcoholic psychoses: Demographic aspects at midcentury in New York State, Rutgers Center of Alcohol Studies, New Brunswick, N.J.

Mandell, W., Cooper, A., Silberstein, R. M., Novick, J., and Koloski, E., 1962, "Youthful Drinking, New York State, 1962. A report to the joint Legislative Committee on the Alcoholic Beverage Control Law of the New York State Legislature," Wakoff Research Center, Staten Island Mental Health Society, New York.

Maryland Drug Abuse Administration, March 1, 1975, L. Robert Evans, Director, Division of Statistical Research and Evaluation, Annual Report for 1974.

Miller, J. L., and Wahl, R., 1956, Attitudes of high school students toward alcoholic beverages, Bureau of Economics, Sociology and Anthropology, University of Wisconsin, Madison, Wisconsin.

Parker, F. B., 1969, Self-role strain and drinking disposition at a prealcoholic age level, *J. of Social Psychology* 78:55–61.

Pearce, J., and Garrett, H. D., 1970, A comparison of the drinking behavior of delinquent youth versus nondelinquent youth in the states of Idaho and Utah, *J. School Health* 40:131–135.

Podolsky, E., 1960, The sociopathic alcoholic, *Quart. J. Stud. Alc.* 21:292–297.

Riester, A. E., and Zucker, R. A., 1968, Adolescent social structure and drinking behavior, *Personnel and Guidance Journal* 47:304–312.

Robins, L. N., Bates, W. M., and O'Neal, P., 1962, Adult drinking patterns of former problem children, *in* Pittman, D. J., and Snyder, C. R. (eds.), "Society, Culture and Drinking Patterns," New York, Wiley, pp. 395–412.

Robins, L. N., Murphy, G. E., and Breckenridge, M. B., 1968, Drinking behavior of young urban Negro men, *Quart. J. Stud. Alc.* 29:657–684.

Rurah, J., 1932, Infantile alcoholism, *Am. J. Dis. Child.* 44:1077.

San Mateo County Department of Public Health and Welfare, 1973, Surveillance of study drug use, San Mateo County, Calif.

Schuckit, M. A., 1963, Alcoholism and sociopathy-diagnostic confusion, *Quart. J. Stud. Alc.* 34:157–164.

Smart, R. G., 1968, Alcohol consumption and anxiety in college students, *J. of General Psychology* 78:35–39.

Sterne, M. W., Pittman, D. J., and Coe, T., 1967, Teen-agers, drinking and the law: A study of arrest trends for alcohol related offenses *in* Pittman, D. (ed.), "Alcoholism," Harper and Row, New York.

Strassburger, R., and Strassburger, A., 1965, Measurement of attitudes toward alcohol and their relation to personality variables, *J. Cons. Psychol.* 29:440–445.

Straus, R., and Bacon, S. D., 1953, "Drinking in College," Yale Univ. Press, New Haven, Conn.

U.S. Department of Health, Education and Welfare, Dec. 1971, First special report to the U.S. Congress on alcohol and health, DHEW Publication No. (ADM) 74-68.

U.S. Department of Health, Education and Welfare, June 1974, New knowledge, special report to the U.S. Congress.

Widseth, J. C., and Moyer, J., 1971, Drinking behavior and attitudes toward alcohol in delinquent girls, *International Journal of Addictions* 6:453–461.

Williams, A. F., 1966, Self-concepts of college problem drinkers, 1. A comparison with alcoholics, *Quart. J. Stud. Alc.* 586–594.

Williams, A. F., 1967, Self-concepts of college problem drinkers, 2. Heilbrun need scales, *Quart. J. Stud. Alc.* 28:267–276.

Williams, A. F., 1968, Psychological needs and social drinking among college students, *Quart. J. Stud. Alc.* 29:355–363.

Winokur, G., Reich, T., Rimmer, J., and Pitts, F. N., Jr., 1970, Alcoholism: III. Diagnosis and familial psychiatric illness in 259 alcoholic probands, *Arch. Gen. Psychiatry* 23:104–111.

Zax, M., Gardner, and Hart, 1961, A survey of the prevalence of alcoholism in Monroe County, N.Y., *Quart. J. Stud. Alc.* 316–326.

Zucker, R. A., and Van Horn, H., 1972, Sibling social structure and oral behavior: Drinking and smoking in adolescence, *Quart. J. Stud. Alc.* 33:193–197.

CHAPTER 6

Family Structure and Behavior in Alcoholism: A Review of the Literature

Joan Ablon

Department of Psychiatry
University of California School of Medicine
San Francisco, California

and

The Community Mental Health Training Program
Langley Porter Neuropsychiatric Institute
San Francisco, California

> When a man meets stress, his family, willing or not, shares the anguish of his pains. He loses his job, and seeds of dissension are planted. Tensions course through the family as hardships increase; irritations chafe once smooth relationships and suppressed hostility crackles momentarily into view. The interplay within the family builds toward an emotional climax, and as the climax nears, bitter antagonisms creep from hiding and gnaw at the ties that bind the members. Often, unsuspected strengths appear to counteract antagonisms as the family stumbles toward its own tragedy and exaltation. (Hansen and Hill, 1964, p. 782)

INTRODUCTION

Theoretical Approaches to the Study of the Family

The family in its varied forms through time has constituted the primary social unit of all societies. Marriage functions as a social institution that regulates sexual mating and legitimizes children. The family characteristically has borne the chief responsibility for the nurturing and socialization of children. In keeping with its social significance, a voluminous literature of fiction and nonfiction works deals with the intricacies of specific families and of families as generic social forms.

Hill and Hansen (1960) have suggested five theoretical and conceptual frameworks for family studies: structural-functional, interactional, situational, developmental, and institutional. These frameworks are primarily sociological with a lesser emphasis on the psychological. The first three provide the conceptual context for the consideration of materials presented in this chapter on specific issues relating to family structure and behavior in alcoholism. For this reason, a general review of theory concerning the social nature of the American family and of the "problem" family initiates the discussion.

The structural-functional approach, developed largely from sociology and anthropology, views the family as a self-contained social system conposed of interacting and interdependent elements. The nature of the integration of these parts within the whole is emphasized, and consideration is given to how each element of the system relates functionally or dysfunctionally to other elements of the system and to the total system. The family as one autonomous system relates to the larger context of the overall system of society. While generic biological and human personality features are considered in relation to how human features affect the equilibrium of the system, individual personalities do not play a significant role in this approach.

The interactionist approach, developed by social psychologists and the one emphasized in this chapter, focuses on the family as a unit of interacting personalities. Each family member's respective personality, position, or status in the family, and role, i.e., certain behavior appropriate to a position within the family, has its unique significance for the functioning and character of the social unit. The individual self as a social product is emphasized in this approach; the individual continually reacting to his perception of others' attitudes toward him. The feedback effect is particularly critical in a relatively small and primary social unit such as the family. This eclectic approach has the potential for making optimum use of the available literature on the alcoholic family. Within its conceptual rubric may be combined both psychologically and socially oriented papers. Few publications combine the two viewpoints.

A chief failing of this approach as it has been implemented, and one given significant emphasis in this chapter, is that the many extrafamily social roles of the individual and the significance of these for his behavior both inside and outside of the family are not considered. Little framework is provided for the incorporation of cultural and social values that bombard the urban man exclusive of his role as a member of a family.

The situational approach, closely related to the interactionist, views the family as a social situation affecting the behavior of its members. Most of the work done in this area of research has focused on intrinsic characteristics of the family situation itself, such as the nature and tone of affectional relationships, or characteristics of family patterns, such as size and organization. External factors such as socioeconomic status, neighborhood, and health are also allowed for, although not emphasized. Only here, of all the five conceptual frameworks that have been widely used in family studies, is there room to explore ways in which specific ethnic or culturally determined characteristics affect the behavior of individuals.

The American Family

Most persons in the course of their lifetime belong to three distinct family constellations: (1) the family of orientation, in which they are reared, (2) the family of procreation, in which they function as a spouse and parent, and (3) the affinal or in-law family, relatives of their spouse. The individual as a participant and actor with certain roles in each of these families is affected in varying degrees by his participation in each family, and, accordingly, he affects the lives of the other members in each.

The urban American family form characteristically is that of the nuclear or conjugal family—a husband and wife and their unmarried children. The nuclear family is common to most countries of the world today and represents a movement away from the traditional extended family household, which may include several generations of kin. The American census defines "family" broadly: "A family consists of two or more persons living in the same household who are related to each other by blood, marriage, or adoption." For the purposes of this chapter, we refer to the two-generational family that includes parents and children, rather than, for instance, a household composed of unmarried, though related, siblings.

The American nuclear family has given over to secondary societal institutions many functions that traditionally have been associated with family life. Parsons and Bales (1955) stated that the American family has two basic functions: "First, the primary socialization of children so that they can truly become members of the society into which they have been born; second, the stabilization of the adult personalities of the population of the society" (pp. 16–17). Even these two functions are not the exclusive rights of the family. Many other social institutions and agencies are involved in their implementation. Nonetheless, crucial to the following discussion is the fact that children most frequently learn their role-taking activities in their families of orientation, and for the growing child, the adults in his household constitute models or "significant others," to use the term of the social psychologists.

An increasing concern in alcoholism research and programs is with the hidden or middle-class alcoholic who maintains low visibility in comparison with the unemployed or the "Skid Row" population. This chapter focuses on alcoholism in the American urban middle-class family. Specific parameters of that population designated as "middle-class" vary. By and large this group is characterized by fairly stable employment and family life. Occupations range from managerial and professional to clerical and sales workers. Educational achievement may end with high school or include college followed by professional training. Economic and social achievement and mobility are motivating forces in family life.

American urban life provides a highly complex and differentiated social setting for most individuals. Even without the seeking out of diversity, most urban dwellers at the very least are exposed daily to divergent forms of attitudes and behavior. Many persons routinely spend parts of their day in several social and cultural worlds. The primary socialization process has occurred within the home of orientation, and its cultural and social values characteristically form the basis of future value sets. While the most meaningful contemporary social field for the individual usually is the household wherein he lives, one may find other social fields at work or socializing with peers. These may expose him to their respective values, some of which may be in large part conflicting. No one individual incorporates all of these values into his personality structure; however, it is likely that even very divergent cultural elements that are compatible with his primary value set or that are attractive to him in certain ways because of needs or desires may be adopted.

The Problem Family

The family unit, because it is basically biological in nature, contains sufficient structural weakness to make it highly vulnerable to stressors.

> Compared with other associations in society, the average family is organizationally inferior. Its age composition is heavily weighted with dependents, and it cannot freely reject its weak members and recruit more competent teammates as do other associations. Its members receive an unearned acceptance, for there is no price for belonging. Because of its unusual age composition and its uncertain sex composition, it is intrinsically a puny work group and an awkward decision-making group. It is by no means ideally manned to withstand stress, yet the family is the bottleneck organization through which almost all troubles of modern society pass. No other institution so reflects the strains and stresses of life. The modern family experiences recurrent tension precisely because it is the great burden-carrier of the social order.
>
> The family today is not only the focal point of frustration and tension, but also the source for resolving frustrations and releasing tensions. In our society, individuals hope that their family will show great capacity of sympathy, understanding, and unlimited support, and thus act as emotional therapy for personalities bruised in the course of competitive daily living.
> (Hansen and Hill, 1964, p. 805)

How do we approach a study of "the problem family"? The labeling of a family as problematical involves a value judgment on the part of the larger community and community caregivers. While some situations may be deemed problems by agency personnel, the families in question may exhibit an unusual lifestyle or family structure because they hold values deviant to the larger society, yet be quite satisfied for their own purposes. On the other hand, many severe problem situations are hidden behind affluence and outward conformity to societal values. Specific to the concerns of this chapter, a great many problem-drinking families manage to deny or at least hide this behavior from the world. The term "closet drinker," referring to the hidden alcoholic, is being used with more regularity by community caregivers.

Following the conceptual frameworks outlined above, if the family is considered as a social system with interdependent parts consisting of interacting personalities with each having his own expected role functions, it follows that the behavior of each part sensitively affects the functioning of the others. A malfunction of one part may lead to a disequilibrium of the total system. If one party because of particular problems becomes unable to assume the customary expected behavior and activities associated with his role in the family, all other family members feel the change in balance. Because certain tasks are required for the continuing maintenance of the household, those tasks regularly borne by the malfunctioning member must be allocated to other members. For example, if the husband, because of a pattern of excessive drinking, can no longer maintain his responsibilities in the economic and authority spheres, the wife characteristically will assume these. Significant affect will accompany such transferrals of power and responsibility. The wife may react with considerable hostility, the husband with shame and guilt that precipitates more drinking. The children in this situation often are confused by role changes, and they inevitably become involved in conflict situations between the parents.

Spiegel and his associates (Spiegel, 1957; Spiegel and Kluckhohn, 1954) in a transactional approach to the study of family behavior and personality suggested that stressor events occur in a complex field of interdependent systems—the individual, the family, and the value structure. Spiegel (1957) focused on role conflict within the family. The family system is characterized by a complex and precarious interdependence that has almost an autonomy of its own.

> A constant observer of the family—or of any other persistent group process—has a somewhat contrary impression that much of what occurs in the way of behavior is not under the control of any one person or even a set of persons, but is rather the upshot of complicated processes beyond the ken of anyone involved. Something in the group process itself takes over as a steering mechanism and brings about results which no one anticipates, or wants, whether consciously or unconsciously. Or the steering mechanism may bring about a completely unexpected pleasant effect. On the basis of numerous observations, we were struck with the fact that so often what is functional for one member of the family group may be dysfunctional for the family as a

whole. The opposite also holds: What is functional for the family as a whole may have very harmful effects on one person. These phenomena take place unwittingly, not only because of the unconscious dynamics within each person, but also because of the operations of the system of relations in which the members of the family are involved. (Spiegel, 1957, p. 2)

The concept of social role—an expected behavior pattern set attached to a particular social status or social position within the family—is particularly significant for the state of family equilibrium because the role behavior of one family member cannot be considered except in relation to those of other members. When there is a discrepancy among the expectations of members concerning what the roles of others should be, conflict occurs. Spiegel emphasized the significance of the cultural value system maintained by family members and noted how crucial this system is in determining expectations of behavior. The family unit as a structural system may move from equilibrium to disequilibrium and, if it is to survive as a system, back into equilibrium.

Parad and Caplan (1960) presented a framework for studying families in crisis, suggesting that the stage for confrontation of the crisis is set by the family's "life-style." Values, roles, and communication are the chief elements characterizing the particular life-style. By values the authors mean the system of ideas, attitudes, and belief that bind the family together and make up the "culture of the family." The family functions through the patterning of roles, which provide for leadership and the division of necessary labor within the household. Communication is the network for carrying messages and transmitting information and affect among the members. When confronted by a crisis, the family develops out of the resources of its life-style "intermediate problem-solving mechanisms." Parad and Caplan stressed the importance of understanding the need-response pattern of the individual—the perception of his needs by other family members and by the culture of the family, the respect accorded to these needs, and the satisfaction of the needs in accord with the resources of the family. This scheme, like that of Spiegel, takes cognizance of the cultural values of the family, individual personality functioning, and adaptive dynamics, which may be techniques for reestablishing equilibrium or problem-solving mechanisms.

THE LITERATURE ON THE ALCOHOLIC FAMILY

Following the realization that the early stereotype of the alcoholic as a homeless derelict was an erroneous and unrealistic one, a florescence of literature dealing with family-related aspects of alcoholism appeared in the 1950s and 1960s. While the number of papers is substantial, the authors, the issues, and the orientations are limited. The most persistent subject that has been explored has been the personality and role of the wife of the alcoholic in relation to the inception and maintenance of her husband's excessive drinking patterns.

This chapter is designed to indicate the major issues covered by the literature on the alcoholic family and to summarize briefly the most salient works dealing with them. Finally, an attempt is made to tie some of the issues to the larger body of theory on family stress. To perceive alcoholism as an isolated phenomenon unrelated to other forms of family crisis and to treat it without relationship to existing theory is as myopic as were earlier attempts to study and to treat the alcoholic without regard to the networks of significant other persons, experiences, and situations in his life. Until the past several years, the literature on the alcoholic family has presented two divergent viewpoints: a psychological emphasis on specific pathological characteristics of the personalities of the two spouses, and a sociological emphasis on the total family interaction. Rarely has any reference to a larger sociocultural dimension been presented.

Bailey (1961) noted that early sampling of subjects taken largely from arrested inebriate and certain hospitalized populations tended to present a view of alcoholics as a group of "undersocialized, poorly integrated individuals, comprising homeless derelicts, chronic offenders, and the mentally ill" (p. 82). The alcoholic was regarded as a loner with few social ties. Later studies were based on middle- and upper-class nonpsychotic samples from hospitals and outpatient clinics, and alcoholics were recognized to have married to the same extent as urban males and females of comparable age, but to exhibit a significantly higher separation and divorce rate. Bacon (1945) in one of the earliest papers on alcoholism and the family argued that excessive drinking patterns are more incompatible with marriage than with any other societal institution, because these patterns effectively preclude the close interpersonal relations that constitute the essence of the marital association.

"The Wife of the Alcoholic"

Bailey noted that in the decade prior to her excellent review of the literature (Bailey, 1961), "The literature reveals a progression from the initial consideration of the wife chiefly as a part of the alcoholic patient's environment to a concern about her as a person in her own right, and finally to a current focus on the interaction between marital partners" (p. 84).

Initial clinically-oriented studies pictured the wife as a disturbed pathological personality with dependency conflicts and complex needs that directed her to choose as a husband an alcoholic or someone with a personality type susceptible to alcoholism (if it were not full-blown at time of marriage), and further to maintain the excessive drinking patterns. It was suggested that if the husband began to improve, many wives exhibited their own form of emotional disorder (Futterman, 1953; Lewis, 1954; Whalen, 1953; MacDonald, 1956; Price, 1945). A common pattern was that of a woman with great dependency needs, yet who chose a man unable to meet them. She dominated the marriage by taking over the roles he by default could not fulfill because of his drinking. She then castigated him for not being able to meet her conflicting dependency needs. Her vested psychological interests were evidenced when his achievement of sobriety commonly caused her to decompensate psychologically or physically. She could maintain her own adequacy only at his expense.

A contrasting point of view was presented in the works of Jackson (1954, 1958, 1962), who focused on how the family as a total unit adjusted to alcoholism. From her careful and systematic observation of Al-Anon Family Groups in Seattle over nearly a decade, Jackson presented in various papers a natural history of the phases of the alcoholic process for the family. Jackson organized her data on the experiences of member wives into a stable chronology of seven stages representing family behavior in the alcoholic process: (1) attempts to deny the problem, (2) attempts to eliminate the problem, (3) disorganization, (4) attempts to reorganize in spite of the problem, (5) efforts to escape the problem, (6) reorganization of part of the family, and (7) recovery and reorganization of the family (Jackson, 1954). Jackson was skeptical that current pathological characteristics of the wife could be hypothesized to constitute a set personality type and to have existed in the predrinking or premarriage period. She presented the alternative suggestion that the particular "pathological" behavior constellations exhibited by the types of wives suggested by Whalen and others might represent these women as they coped in particular ways appropriate to stresses of the respective stages.

> ... that women undergoing similar experiences of stress, within similarly unstructured situations, defined by the culture and reacted to by members of the society in such a manner as to place limits on the range of possible behavior, will emerge from this experience showing many similar neurotic personality traits. As the situation evolves some of these personality traits will also change. Change has been observed in the women studied which correlates with altered family interaction patterns. This hypothesis is supported also by observations on the behavior of individuals in other unstructured situations, in situations involving conflicting goals and loyalties, and in situations in which they were isolated from supporting group interaction. It is congruent also with the theory of reactions to increased and decreased stress. (Jackson, 1954, p. 586)

Jackson likewise noted that in her work with Al-Anon members she found very few women who decompensated when their husbands achieved sobriety. The findings of other researchers later supported this statement.

Lemert (1960), following Jackson's search for a fixed sequence of events by looking at five sample groups of wives in a California city, added a rich socioeconomic dimension to his data that is characteristically lacking in the literature. These groups were derived from: (1) city commitments to the state hospital, (2) public welfare agency clients, (3) divorce court, (4) police probation cases, and (5) Al-Anon groups. He suggested that family behavior must be seen in the larger cultural context in which it occurs. Family members may perceive and react to differing events in large part depending on socioeconomic factors. The meaning and functions of the same events (such as isolation, frequency of sexual relationships, conflict over children) for the subjects may depend on their socioeconomic class. Lemert could not find a common sequence of events in family adjustment; rather, he found events tended to cluster in general stages of early, middle, and late adjustment. While several researchers have, in passing, commented on the social class of their samples, few have emphasized this as it affects attitudes or motivation for treatment.

James and Goldman (1971) attempted to integrate Jackson's stages of family adjustment within five reasonably distinct and persistent styles of coping behavior that had been identified by Oxford and Guthrie (1968). These styles were exhibited by wives who: (1) safeguarded family interests, (2) withdrew within marriage, (3) attacked, (4) acted out, and (5) protected the alcoholic husband. Among their other findings, James and Goldman noted that the coping styles of wives in their sample changed over time as did their related problems, and all wives in the sample used more than one style of coping. While cautioning against causal theories, the authors concluded that the results of their study seem to favor Jackson's view that the wife's behavior and current coping style may be caused by the current stage of the husband's drinking, rather than the "wife's fixed personality pathologies cause the husband's pliable personality to change via alcoholism which then allows her to employ her latent coping mechanism" (James and Goldman, 1971, p. 380).

Bailey cogently summarized the issues presented above:

> Research on alcoholism and marriage has been based on one of two hypotheses: (a) that women with certain types of personality structure tend to select alcoholics or potential alcoholics as mates in order to satisfy unconscious needs of their own, and that these needs require the continued drinking of the husband; (b) that women undergoing similar experiences of stress will, as a result, manifest many neurotic traits in common. While no one denies that the spouse of the alcoholic is seriously disturbed by the time she reaches an identifiable source of help or community intervention, there is a basic question as to whether her disturbance antedates the partner's alcoholism or stems from it. Logically, the two hypotheses need not be mutually exclusive, yet their implications for treatment, for public education and for the future direction of research are very different. (Bailey, 1961, p. 90)

A few years later, Bailey (1963b) reported on new research published or in progress that presented more evidence that a conceptualization of a unitary personality for wives of alcoholics is not tenable. Even if pathology could be shown for a contemporary period, it was difficult indeed to assume the wife had chosen a man destined for alcoholism, since more recent studies indicated that a minority of alcoholics were drinking to excess at the time of marriage (Jackson, 1954; Bailey, 1965; James and Goldman, 1971). Clifford (1960) and Lemert (1960), on the other hand, stated that a majority or even all of their subjects knew about their prospective husbands' problems. Lemert found that those men who had drinking problems before marriage showed a much higher incidence and higher frequency of dependency attributes than those whose problems developed later (Lemert, 1962).

Clifford (1960) reported from a New York clinic sample that he was able to find identifiable differences in the attitudes of wives whose husbands had stopped drinking and those whose husbands were active alcoholics. Six basic attitudes toward the role of alcoholism and toward the alcoholic in the wife's life were chosen for study: (1) the effect on the family, (2) the wife's acceptance of personal responsibility, (3) her attitudes and feelings about a possible cure for her husband's alcoholism, (4) her sense of social adequacy, (5) her concept of social status, and (6) her concept of her indispensability to the welfare of the alcoholic.

Clifford found that women whose husbands had stopped drinking, his Group I, showed more concern and awareness of psychological and social damage done to the children than the women in Group II whose husbands continued to drink. The women in Group I accepted a measure of responsibility, although this varied in degree, for the husband's former drinking patterns; they searched for some device whereby they could end the husband's drinking and sought medical and psychological aid; although feeling socially inadequate in various degrees, they learned to cope with their problems and become more adequate; they were sensitive to losing social status through criticism, embarrassment, and family instability; they felt indispensable to their husbands and families. In all these respects they were in direct contrast to the wives in Group II. Clifford noted that all of the women in both groups had a history of alcoholism in their families and also a premarital awareness of their husbands' alcoholism.

Clifford's findings are dramatic in being so clear-cut, but it would be helpful to know more about his methodology. His report is written in sweeping terms that suggest unanimity on all counts in the differential groups. This contrasts with the complexities in response that almost all other researchers report.

Kogan and Jackson in the 1960s produced a series of papers investigating a variety of aspects of personality of wives of alcoholics. These aspects included emotional disturbance, role expectations, and perceptions of their spouses and themselves. Kogan et al. (1963) reported a comparison of MMPI responses of wives of alcoholics who belonged to Al-Anon and to a control group of wives who had no alcoholism problem in their families. While significantly more wives of alcoholics showed evidence of disturbance in personality function than did the comparison group, the total number of disturbed subjects was less than half on any measure, hence the investigators did not feel that generalization was warranted. They noted that the type of disturbance found was highly variable, and that no particular "personality type" could be distinguished. They concluded:

> A major research implication of these findings is that the personality of the wife of the alcoholic should be treated as an important variable rather than a constant. A related implication is the impropriety of the concept, "the wife of the alcoholic." (Kogan et al., 1963, p. 237)

Bailey et al. (1962) and Corder et al. (1964) likewise found relatively little difference indicating marked neurotic or disturbed behavior between samples of wives of alcoholics and wives of nonalcoholics. Most recently, Edwards et al. (1973) in an extensive review of the literature on "wives of alcoholics" based on articles published over some 30 years similarly concluded that no unitary personality type could be isolated, and that wives of alcoholics appear to be essentially normal persons of differing types who may exhibit personal dysfunction under specific stresses, and that the dysfunction tends to disappear when the stresses do.

Kogan and Jackson (1965b) explored both early life experience and current stress situations of their subjects, and concluded that multivariate relationships must be considered in order to understand disturbance. Women who experienced the dyad of an inadequate mother and an unhappy childhood appeared to experience more personal stress and to marry men who were alcoholic or became alcoholic.

> Women who reported an inadequate mother and childhood unhappiness were more likely to experience both personal sources of stress and symptoms of personality dysfunction and were disproportionately represented among the wives of alcoholics. Women whose childhood experience was not disturbed and who therefore could be assumed to have had an adequate feminine model and role training had a better chance of handling their marriage, even to an active alcoholic, without personality decompensation.
>
> The implication of the findings is believed to be that both early experiences and current relationships are involved in the complex interplay between life experience and personality function, and that significance cannot be assigned to specific factors as purely cause or purely effect. (Kogan and Jackson, 1965b, p. 604)

Kogan and Jackson (1965a) further tried to differentiate between types of current stress situation, categorizing some as impersonal (financial, legal) as opposed to personal (dealing with interrelationships in the family) and found that personal stress situations contributed more to current disturbance in the wives. One finding that might be noted was that there were no basic personality differences found between wives whose husbands continued to drink and those who had achieved sobriety. The authors, understandably enough in relation to many of the previously mentioned clinical studies of wives of alcoholics, wryly noted:

> Perhaps this should be interpreted as constituting evidence for the more parsimonious hypothesis that an alcoholic's recovery may turn out to be more contingent upon his own personality strengths than those of his wife. (Kogan and Jackson, 1965a, p. 436)

Bailey *et al.* (1965) in a New York sample found a pattern of "unnatural" attitudes toward drinking, i.e., excessive drinking or total abstinence in 79.2 percent of parents of women who married alcoholics, in contrast to 54.2 percent of parents of comparison groups. Clifford (1960) stated that all of the subjects in his study had a history of alcoholism in their families. Unfortunately, little data are available on this very significant aspect of the wives' early life experience.

Husbands of Women Alcoholics

Jellinek (1960) estimated that women constitute one-sixth of the alcoholic population of the United States. Belfer *et al.* (1971) quoted figures ranging from one-fifth to one-half. The latter figure would place the female alcoholic population at close to 4.5 million. Discussions with community agency personnel indeed suggest that a liberal estimate is indicated. Female alcoholics often are able to maintain low social visibility compared to their male counterparts, chiefly for economic reasons. The housewife who drinks at home may sooner or later become obvious in her habits to her family, but still may maintain a facade before her friends and the larger community.

Unfortunately, there is a dearth of material available on husbands of alcoholic women, in contrast to the substantial number of studies dealing with wives of male alcoholics. Lisansky (1957, 1958) reported on two samples of female and male alcoholics from a Connecticut prison farm and from an outpatient clinic in New York. Significant differences appeared in the comparison of women from these two samples, differences that in large part may well result from corresponding differences in socioeconomic status. The clinic sample represented a typically middle-class population, while the prison farm sample included women of a lower socioeconomic status.

Lisansky noted that a somewhat larger percentage of the women than the men had married (87 percent compared to 80 percent). The proportion of women and of men outpatients who married and who remained single was about the same as in the general population, but sizable and statistically highly significant differences occurred between alcoholics and the general population in proportions of separated and divorced. A predominance of alcoholics married other alcoholics. Relatively little data were given on husbands of female alcoholics, but among the women, one-third had had or had at the time of the study husbands who also were problem drinkers. There was a tendency for women to be married to men much older than themselves.

The prison farm women had a higher incidence of divorce, separation, and death in their families. Their problem drinking had begun at earlier ages and their marriages were characterized by more conflict and instability than those of the outpatient women. Husbands of the prison farm women were often physically abusive, unemployed, or irregularly employed, and were heavy drinkers. Some also were in jail. These women frequently supported unemployed husbands or boyfriends. Their marriages gave them very little economic or emotional security. The outpatient women, while secure in such more obvious aspects of their marriages, were conflicted in other ways. More than twice the number of outpatient women than prison farm women were raising their own children (77 percent compared to 35 percent).

Lisansky's study suggested that women's drinking more often related to and was triggered by specific life events than was the case for the males sampled. The women's problematic marriages, then, might contribute to their drinking patterns. Belfer et al. (1971) stated that they found no association between marital status and any definable subgroup of alcoholic women, but they did not deny that marital status may be an important contribution to the total picture even though not a "sufficient cause."

Fox (1956) reported from her psychiatric practice experience that compared to wives of male alcoholics, husbands of alcoholic women are less accepting of their wives' drinking patterns, and are more apt to leave "an alcoholic wife whom they feel they can no longer love." Court battles over children are frequent. Women have a greater tendency to mother and nurse a sick husband, while men are less willing to learn about alcoholism as an illness. Cultural norms allow more freedom for drinking among males, making alcoholism a less acceptable form of behavior among women. Unlike many wives, who are dependent on their husbands for subsistence, men can more easily opt for the economic independence that goes with separation.

Fox (1962) presented a typology of husbands of alcoholic women somewhat reminiscent of Whalen's typology of wives of male alcoholics, mentioning such types as the long-suffering martyr; the punishing, sadistic male; the dependent male, and others. These she differentiated from the "normal" man who "finds" himself with an alcoholic wife.

Interaction in the Alcoholic Marriage

Jackson, Bailey, and Day (1961) were for some years the lone voices calling for broader and more rigorous focus and design in research dealing with alcoholism and the family. Little research was published that examined the contemporary dynamics of the dyadic relationship between spouses or interrelationships within the total family unit. One glimpse into the nature of marital relationships in the conflicted alcoholic family was reported in a group of articles included in a symposium published in the *American Journal of Orthopsychiatry* in 1959. A major failing of previous research had been the lack of control groups and a tendency to look at each spouse as a bundle of pathological characteristics with little sensitive examination of the marital interaction. This symposium, entitled "The Interrelatedness of Alcoholism and Marital Conflict," contained three papers (by Bullock and Mudd, Ballard, and Mitchell) that reported research with couples, in some instances using both alcoholic and nonalcoholic conflicted families.

Bullock and Mudd (1959) reported a high degree of conflict in the couples observed. While the wives regarded the husbands' drinking as the main difficulty in their relationships, the husbands presented a variety of complaints about the attitudes and behavior of their wives. Conflict was of long duration and was reported to have existed in the premarital relationship in more than half the cases. Personality problems and difficult family backgrounds of both partners appeared to be important factors.

Ballard (1959) reported on MMPI results from experimental alcoholic families and control nonalcoholic families. These results are of particular interest and merit quoting in some detail.

> If the character structure of the individuals concerned be accepted as one of the factors making for marital conflict, the conflicted marriages that also involved alcoholism did not seem to constitute an altogether special case. Both the experimental and control groups contained an unusually high percentage of marriages in which one or both partners showed a characterological resemblance to clinical groups in which psychopathic, impulsive behavior emerges, or to normals who show clinical evidence of egocentricity, tactlessness, an inability to judge their own stimulus value. Of course, the two groups of husbands differed markedly in the degree to which character structure would be expected to become manifest in obviously symptomatic or deviant behavior, but even in this respect, the two groups of wives were quite similar. There were indications that a passive orientation, or actual social inadequacy, frequently characterized the husbands in this sample. There was somewhat more speculative evidence that, as a group, the wives inclined toward the complementary "nursing" or mothering sort of role but were relatively inept in its execution. Thus it would appear that, even if there were no problem of alcoholism, the marriages in the experimental group were a sort where marital conflict was likely to develop.
>
> There was nevertheless, one notable difference between the two groups of marriages. Even though the wives in the control sample appeared a bit more maladjusted than their experimental counterparts and showed a painfully sensitive, schizoid quality that was absent in their husbands, the control spouses showed a more "normal" or balanced relationship on the nonclinical

scales. This was particularly apparent in the Es, Dom, and Res Scales, where the husbands (in spite of their tendency toward passive, feminine attitudes) showed greater strength and social dominance, with a balanced sharing of social responsibility. By contrast, the wives of the experimental group (who had shown some tendency toward more masculine attitudes) seemed to have assumed some of the masculine role in terms of these traits. The control marriages seemed to offer a somewhat better base, then, for establishing a conventionally normal marital relationship.

... Although consistently appearing to be somewhat better adjusted, the wives in the experimental group nevertheless showed an unexpected degree of similarity to the alcoholic husbands. Of particular interest was their elevation on Scale 4 (Psychopathic deviate) and the failure to find differentiation on the alcoholism Scale. They differed most clearly from their husbands in their greater emphasis on repression and their notable avoidance of responses that would suggest a tendency to act out. This suggests the possibility that these wives were, within themselves, militantly defending against the kinds of behavior that emerged so obviously in their husbands.

It is now possible to speculate about the role that alcoholism may play in the marital conflict observed in these cases. From the point of view of the wife, her husband's drinking and irresponsible, impulsive behavior could offer vicarious gratification for impulses that she had sternly repressed within herself. At the same time, her militant attack on these qualities in the spouse would offer a defensive reassurance against the emergence of her own unconscious tendencies. An additional gratification may derive from the more masculine role that was "forced" on the wife by her husband's weakness. If, as this would imply, the husband's drinking provided a simultaneous and significant source of unconscious gratification and defensive reassurance for the wife, then the prime condition leading to the fixation of behavior has been met. In spite of her protestations to the contrary, the wife has an appreciable stake in maintaining the *status quo*.

The *status quo* also appeared to have some advantages from the viewpoint of the alcoholic husband. The contrast between rigid, authoritarian, moralistic attitudes and impulsive behavior of a quite contrary sort has been noted. It is possible, then, that the husband privately feels uneasy with his own deficiency of control and wants someone who will control him or, at least, reduce the feelings of uneasiness associated with his asocial behavior by punishing him for it. Theoretically, the externalization of a superego conflict not only lessens guilt but also permits greater freedom to rebel against the punitive agent. There may also be a more direct gratification in maintaining a dependent situation where the wife assumes the role of greater strength and responsibility.

Thus, both partners may be gaining a secret gratification and reassurance from the drinking behavior that is the ostensible source of marital discord. If this be true, it would be expected that the coexistence of alcoholism would make the marital conflict appreciably more resistant to change. It has already been noted that when change does begin to occur in such marriages, both partners seem to have somewhat farther to go if they are to achieve a conventionally normal marital relationship.

If, in the course of marital counseling, the husband were to give up his drinking, the implications would be somewhat different for the two partners. Both

> would, of course, lose the unconscious benefits hypothesized above; the husband, however, could experience certain compensations. Society at large, as well as the counselor immediately involved, could be expected to see this as a praiseworthy manifestation of strength on his part; such recognition could be expected to bolster his always shaky self-esteem and to assuage his incipient guilt feelings. The wife, on the other hand, would not only be deprived of hidden gratifications and reassurances but would also be in danger of losing the rationalization that frees her from any feeling of responsibility for the troubles that beset her. If the conflict continues, society is likely to be a bit more sympathetic toward her husband's position. This, presumably, is the basis for the clinical observation that when the alcoholic stops drinking, more obvious signs of maladjustment are likely to appear in the wife.
>
> Thus, it would appear that no special features need be hypothesized in the etiology of marital conflict when the husband also happens to be an alcoholic. However, in those marriages where the fixating conditions described above exist, the interaction between conflicted partners does make the whole situation less amenable to therapeutic intervention. Both the drinking and the overt conflict have unconscious meanings that tend to make them self-sustaining. (Ballard, 1959, 541-544)

Ballard's statements in part exemplify the pragmatic functional and interactionist approaches to a consideration of the marriage unit, focusing on the whole as a system of parts, interacting, with continuing feedback and reaction, rather than only on the specific individual pathology of individual partners stemming from particular childhood situations. They also suggest that the conflicted alcoholic family is not unique when compared to its nonalcoholic, conflicted counterpart in our society.

A series of papers that implicitly or explicitly regarded the interaction between spouses or the total family unit as a necessary functional context for the understanding of the individual alcoholic's drinking pattern began to appear in the 1960s.

Ewing and Fox (1968) stated that the family homeostasis is the perpetuator of drinking patterns and that this homeostasis must be changed if the drinking is to be controlled. They also noted that the wives tend to have personality dynamics similar to those of their husbands. The wife copes with her own unacceptable dependency wishes by adopting the mastery of the family when the husband's excessive drinking patterns cause him to drop responsibilities.

> By being passive, dependent, and sexually undemanding, the alcoholic implicitly encourages his wife to be protective, nurturant, and sexually unresponsive. (p. 88).
>
> Indeed, it appears that in alcoholic marriages a homeostatic mechanism is established which resists change over long periods of time. The behavior of each spouse is rigidly controlled by the other. As a result, an effort by one person to alter his typical role behavior threatens the family equilibrium and provokes renewed efforts by the spouse to maintain the status quo.
>
> The behavior of each partner serves both to express his own neurotic needs and to reinforce those of the spouse. Through drinking, the alcoholic is able,

> periodically, to give direct expression to his contained impulses. By alternating between suppression of impulses and direct expression of them, he can maintain the conflicts surrounding impulse gratification for a lifetime (p. 87). When he drinks, the alcoholic's wife feels that she has no recourse other than to respond in a maternal manner, whether this is nurturant or supportive or, at times, punitive and rejecting. This completes the other half of the interpersonal bargain. It is a form of the *quid pro quo* rule which seems to govern most, if not all enduring interpersonal relationships. The behavior of each partner serves the dual purpose of satisfying his own needs and those of the mate. Thus, any possibility of change is met with dual resistances; those operating internally and those initiated by the spouse in response to the threatened loss of a critical role relationship.
>
> Another aspect of the interpersonal bargain is that it can be kept only if it is implicit. Once the nature of the agreement is thoroughly identified it becomes difficult to maintain. There is little satisfaction in dominating someone who lets you dominate him, for then the question of who is really in control is automatically raised. The overt decision to "let" the other person behave in a particular way carries the clear implication that one can decide to do otherwise. Each person seems to feel secure only if he can force or manipulate the other into satisfying his own needs. Once this lever of manipulation is lost, other means of control must be sought. In short, the equilibrium is disturbed and another basis of homeostasis must be found. (p. 88)

In treatment, the wife's appreciation of her dependency needs and acceptance of some responsibility for the husband's drinking must accompany real changes in marital roles. Reciprocal role changes in self-defeating interactions must occur if the existing homeostasis is to be changed.

Steinglass *et al.* (1971a,b), working within an explicit systems approach, presented a specific "interactional" model of several dimensions for use in analysis of the family in alcoholism, and described the clinical application of this model in several treatment situations involving dyads of family members. The authors present the first clear statement in the literature describing the manner in which family members are involved in an ongoing alcoholic bargain that functions for the maintenance of the system, and how the untangling of the elements of this system may be useful for the therapist.

The authors suggest that family members as component parts of a system "manipulate" other members and adjust their behavior as is necessary to maintain a "complementary relationship of psychopathology, needs, strengths, cultural values, etc. within the family" (Steinglass *et al.*, 1971b, p. 405). This maneuvering is necessary for the ongoing existence of the family group as a functioning system. Drinking behavior may operate in either of two forms: (1) The drinking behavior of one member may serve as a symptom or expression of stress created by conflicts within the system. In this way, the drinking may serve, as it were, as an escape valve for such stress. Or, (2) drinking may constitute an integral part of one of the working programs within the system. In this form it might serve unconscious needs of family members in its effect on such areas as role differentiation or the distribution of power.

Davis *et al.* (1974) and Wolin *et al.* (1975) from related treatment experiments further elaborated on this interactional model. Davis *et al.* emphasized that excessive drinking patterns may indeed have certain adaptive consequences that are sufficiently reinforcing to serve as the primary factors maintaining the drinking patterns. These adaptive consequences may operate at a variety of levels: intrapsychic, intracouple, or on the family or wider social system basis. The implications for intervention are that the therapist must determine the specific manner in which drinking behavior serves the adaptive function, and secondly, once the adaptive consequences have been identified, therapy may be structured around helping the patient exhibit the adaptive behavior while sober and also to learn alternate behavior patterns.

Bowen, a chief architect of the family therapy movement, has also viewed alcoholism through family systems theory (Bowen, 1974). He outlined various alcoholic personality types characterized by certain levels of what he terms "differentiation of self"—the degree to which the person has solidly held principles by which he lives. He analyzed the alcoholic marriage in terms of the interaction between the personality characteristics of the alcoholic and the spouse, and how their level of differentiation of self mix and match. Coming from a systems theory perspective, Bowen sees family therapy as particularly appropriate for analyzing issues of individual and family function.

> How does alcoholism fit into systems concepts? From a systems viewpoint, alcoholism is one of the common human dysfunctions. As a dysfunction, it exists in the context of an imbalance in functioning in the total family system. From a theoretical viewpoint, every important family member plays a part in the dysfunction of the dysfunctional member. The theory provides a way for conceptualizing the part that each member plays. From a systems therapy viewpoint, the therapy is directed at helping the family to modify its patterns of functioning. The therapy is directed at the family member, or members, with the most resourcefulness, who have the most potential for modifying his or her own functioning. When it is possible to modify the family relationship system, the alcoholic dysfunction is alleviated, even though the dysfunctional one may not have been part of the therapy. (Bowen, 1974, p. 117).

The generic model presented by the latter authors forms a bridge between the psychological and sociological theories presented previously by explaining differing functions of excessive drinking that occur for the maintenance of the ongoing family system. The authors suggest that the specific role that alcohol plays in maintaining the family system must be carefully understood before any therapeutic intervention is planned. This systems approach indicates that the "drinking system" rather than the individual alcoholic should be considered as the appropriate basic unit of clinical research. As noted by Ewing and Fox (1968), Steinglass *et al.* (1971a,b) point up the significance of the efforts of the system to maintain itself.

> From a systems point of view, it is the protection of the functioning system itself that takes precedence over the individual concerns or needs of the members of the system. If the continuation of therapy implies a threat to the integrity or functioning of the system, then therapy will probably be rejected as an alien and dangerous force. It would seem, then, that the task of the therapist is to effect the desirable behavioral changes without appearing as an imminent threat to the ongoing system. (Steinglass *et al.*, 1971a, p. 279)

Ward and Faillace (1970) presented what they explicitly called a "systems view" of pathological drinking patterns. They described this as basically circular, self-perpetuating behavior maintained to preserve an equilibrium that involves the alcoholic, his family, his employer, and community helpers. The authors suggested that three levels of interactional systems must be considered: (1) that of the complex determinants of behavior within the individual, (2) that of the small group system of family behavior and response, and (3) that of the larger sociocultural context.

The authors discussed patterns of family behavior such as denial or punishment that the wife may employ on herself and toward the alcoholic. These patterns serve to produce homeostasis in the family and are of a complementary and circular nature. Any actions that may require behavioral changes throughout the family system may be avoided at all costs. Thus, the wife may resist change in the status quo to prevent the stress caused by disruption of the existing equilibrium in the family. This system's view explains stress and symptoms exhibited by wives of recovering or recovered alcoholics from a structural-functional approach, rather than by linking resistance to the specific individual pathology of the wife.

Ward and Faillace suggested that the actions of extrafamily community helpers in like manner may covertly serve to maintain drinking behavior because of the complementary persecution, rescue, or forgiveness needs of the helping systems.

Steiner (1971) elaborated on Berne's exposition (1964, pp. 73–81) of the transactional analysis position on alcoholism. Steiner considered alcoholism as a script or life plan for self-destruction. The individual may make an early decision—often forced and immature—that influences and predicts the rest of his life. For this decision he must have a script to follow. Steiner described three distinctive games alcoholics play: "Drunk and Proud," "Lush," and "Wino." These games are similar in their general thesis, which is an existential position exemplified by the sentence, "I'm no good and you're O.K. (ha, ha)." The actions of the alcoholic put those around him in the position of being foolish and full of blame. The games differ in their dynamics, aims, and the roles to be taken by the significant others of the alcoholic or the community caregivers who allow the game playing to continue. Roles suggested are Patsy, Dummy, Rescuer, Persecutor, Connection. The importance of each of the roles in maintaining the game is discussed and the changes that must be made in these roles to thwart the drinker's game are explained. (For an interesting parallel analysis see Al-Anon brochure, "Alcoholism: A Merry-Go-Round Named Denial," Al-Anon Family Group Headquarters, 1969. This particular piece of literature is one of their most popular pieces.) Steiner's views also appeared in an article (1969) that met with considerable conflicting responses from other clinicians.

Steiner turns only briefly to the alcoholic's family in relation to treatment but his comments are of interest. He stated that immediately following the onset of treatment (and inferred sobriety), there is usually a "honeymoon" period of increased positive interaction, which then yields to a period in which the members of the family not only expect, but almost seem to wish that the alcoholic would resume drinking. Because the wife and children had been players in his game of "alcoholic," they now feel a vacuum in their lives equal to his. Thus, they will apply pressures on him to drink. Steiner noted that treatment of a married alcoholic requires bringing about change in two or more persons, and indeed the single alcoholic may have a better prognosis. Steiner did state charitably that the positive influence of the family might overshadow this difficulty if it is worked through. In some situations he feels separation or divorce might help in the treatment process.

Gorad (1971), working within an "interpersonal–interactional" framework explored the nature of styles of communication used by alcoholics and their wives by matching a sample of alcoholic married couples with a sample of nonalcoholic control couples in an interactional game-playing situation. He found that the alcoholic couples used "one-up" messages significantly more than did nonalcoholic couples. Gorad concluded that the alcoholic uses a responsibility-avoiding style of communication when drunk or sober for gaining control, while the wife uses a more direct responsibility-accepting style of communication. Alcoholics are competitive in the sense that by using this style of communication they attempt to gain for themselves a "one-up" position in relationships. Control, then, may be sought through passive, dependent-appearing techniques as well as through action, independence, and aggression, which are the more commonly used modes. Because the alcoholic uses indirect, nonaggressive techniques, he characteristically is not seen as competitive. Gorad suggested that this may explain the frequent depiction of the alcoholic as withdrawn, dependent, and submissive, while the wife appears as dominant and controlling because of her directness. In reality they are equal in their competition for control.

Gorad found that the communication style between spouses in the alcoholic marriage is highly competitive. They did not function well in achieving joint goals. The interaction pattern was marked by rigidity of complementarity and/or an escalation of symmetry. Gorad notes:

> There has been much discussion of the issue of dominance in the literature. Most researchers have observed that the alcoholic wife is dominant over her husband . . . The present study found, in contrast, that the alcoholic's wife dominates her husband, in the sense of being "one-up" in a complementary interaction, or in control, or winning over, no more than he dominates her, and no more than normal husbands and wives dominate each other. In fact there was some suggestion, although not statistically significant, that the alcoholics in the present sample were dominant over their wives. (Gorad, 1971, p. 487)

Gorad suggested that the competitive style between spouses in the alcoholic family allows little risk-taking that might change their style to one of more mutually beneficial interaction.

Kogan and Jackson (1963), exploring role perceptions of wives of alcoholics, found that they perceived themselves as hyperfeminine, submissive, and wanting to be led and managed, more than did wives of nonalcoholics. These perceptions are contrary to the view frequently found in the literature that wives of alcoholics often have strong dominance needs that require them to select a passive dependent spouse. The portrait of the wife as dominant has been described by the spouse (Mitchell, 1959) and frequently by clinicians. Reality needs that cause the wives to overtly assume practical responsibilities within the household no doubt enter strongly into the creation of the image of the powerful wife.

Cork (1964), in a paper suggesting treatment approaches, listed and discussed characteristics of the alcoholic that most vitally affect family life. These characteristics are functionally and interactionally oriented, in that they determine the nature of ongoing relationships within the family group, the employment situation, and significant social networks. The characteristics are: (1) inability to take appropriate responsibility within the family (leading to resentment and restructuring of roles), (2) a lack of self-discipline, (3) overdependency, (4) a preoccupation with self, (5) a negative attitude toward authority, (6) a sense of inadequacy, (7) an unrealistic, immature approach, (8) limited interests, and (9) shallow or superficial ways of relating to people. Cork noted:

> Thus it may be seen that these characteristics may limit or disrupt family life more than the drinking itself. They often may be the source of early and prolonged marital discord before the drinking became really excessive. They may complicate and add to the family problems when drinking is at its height and may provide a basis for ongoing trouble when the drinking is controlled. While the drinking episode often creates a crisis or a breaking point for all family members and gives them something tangible on which to project blame for the disruption in family life, the common characteristics which I have presented, and the family's reaction and interaction to them, would seem to provide the basis for deep and more serious limitation to vital growth-producing relationships for all family members. (Cork, 1964, p. 4)

In keeping with the data on conflict described by Cork (1964) and Bullock and Mudd (1959), Bailey et al. (1965) reported from a New York study that the alcoholic couples reported more frequent disagreements and quarrels than did their nonalcoholic control groups. Such quarrels were characterized by physical or verbal abuse, walking out, silence, or moodiness. There was also less frequent settlement of these conflicts.

The Children of Alcoholics

Bacon (1945), Newall (1950), Fox (1956, 1962), Cork (1964), Jackson (1962), Bailey et al. (1965), and Chafetz et al. (1971) have discussed the destructive effects that an alcoholic parent has on the development of the children in the household. Fox's comprehensive discussion (1962) covered virtually every consideration, from the etiological factors in early life leading to alcoholism, to the complexity of interrelationships that obtain in the alcoholic family. She emphasized parent–child relations and presented a detailed consideration of how the development of the personality of the child is complicated in this problem family situation.

In an earlier paper (1956), Fox pointed out the insecurities and unpredictability that underlie some of the difficulties of the child growing up in the alcoholic household:

> It is not surprising that from 40 to 60 percent of all alcoholics come from the disturbed background of an alcoholic family. The children of alcoholics tend to be neurotic because the sense of security so necessary for the building of a strong and independent ego is rarely found in the household. The child can be utterly bewildered by the sudden shift in behavior of the alcoholic. A parent who is often affectionate, understanding and fun-loving when sober may become morose, demanding, unreasonable, touchy, noisy, and even cruel and violent when drunk. Or a naturally reserved and somewhat withdrawn parent may become sloppily sentimental and seductive in the early stages of a drinking bout, or hilariously and embarrassingly exuberant. He or she may spend money wildly and make extravagant promises which are impossible to fulfill. The frequent swing from high hopes to shattering disappointments may build up in the child such a basic distrust that all his later intimate relationships will be distorted. (Fox, 1956, p. 9)

Children of alcoholics suffer significant problems of identification and modeling. Their view and expectations of sex and family roles may be sorely distorted. They may reject the alcoholic or may take on many of his characteristics. Fighting and dissension between parents mar the children's perspective on marital relationships, and may force them to take a position vis-à-vis conflicting parents. They have particular problems in defining their own identity and self-worth. As they become sensitive to the larger social world around them, they may react with great shame and humiliation, often becoming withdrawn and isolated from their peers. Family agency personnel today stress the prevalence of drug-related problems among these children.

Bailey et al. (1965) found that children of alcoholics exhibited more negative behavioral symptoms than did the children in the control groups of nonalcoholic families. The authors listed eight behavioral symptoms that occurred frequently among the children, but noted that temper tantrums and fighting with peers at school were the symptoms that were most different in occurence between the alcoholic and comparison groups. The oldest child in the alcoholic group and in a stomach ailment control group tended to take on more household responsibilities at earlier ages than did the oldest child in the nonpathological group, a fact possibly related to the abdication of role responsibilities of the father. The children of alcoholics were more frequently identified by the New York Social Service Exchange clearance as having problems than were children in the other groups.

Chafetz et al. (1971) cited some sources suggesting that children of alcoholics fare worse in symptomatology than children of nonalcoholics, and others that find little evidence of ill health. They reported their own study of a sample of children of alcoholics in a child guidance center whom they compared to children of nonalcoholics. They did not find great differences in negative effect on health. They did, however, point out the "distinct and deleterious social consequences," for such a child with respect to potential socialization problems. "The major dissimilarities have to do with the effect of an alcohol problem in a parent on the functioning of the family as a social unit and on occurrences in the child's life which may represent impediments to becoming a socialized adult" (p. 696). Family disruption was frequent, as was a relatively high incidence of problems in school and with the law.

Bacon (1945) and Jackson (1958, 1962) have pointed out that the presence of children may intensify the problems of the alcoholic. Cognizance of failure in the responsibilities of parenthood, and related guilt, may present added pressure for excessive drinking.

Day (1961), in a succinct review article entitled "Alcoholism and the Family," listed characteristic parent–child relationships in the family of orientation that have been suggested as crucial in the etiology of male alcoholism. She discussed the effects of alcoholism in the family on children and emphasized the critical links between family process and aspects of etiology.

Factors Affecting Family Actions in Regard to the Alcoholic Problem

Bailey et al. (1962) and Jackson and Kogan (1963) focused on types of action taken by wives of alcoholics. Bailey et al. looked at three groups of wives: (1) a group who had separated from or divorced their spouses, (2) a group who were living with active alcoholics, and (3) a group whose husbands had achieved sobriety. They found that families who had separated had experienced extensive patterns of economic inadequacy and pathological behavior, with physical violence involved. Wives who stayed with an active alcoholic had experienced less stressful situations than those who had separated. There was less job loss, fewer young children, less infidelity, and less trouble with the police. In general, these wives had suffered fewer serious behavioral consequences.

Those wives whose husbands were sober at the time of the study were of higher economic status. The wives shared a concern about economic security and upward economic and social mobility. It was suggested that the anxiety over economics may have exerted a constructive force in the husband's recovery. These wives were the most isolated from friends and community professional resources. Interestingly, those women who had the best outcomes had least often been subjects of community concern.

Although the wives tested scored higher in psychophysiological disturbances than a representative sample of New York women, the differences were relatively slight except for those wives who were living with an active alcoholic. These women exhibited the highest scores on psychopathology, which may suggest that they were women who had greater basic emotional distress than those who terminate their marriages or support their husband's recovery. However, the other groups—those women who had separated, and those whose husbands were now sober—reported that they too had exhibited significantly higher symptomatology when their husbands were actively drinking. These findings support the point of view presented by Jackson (1954) that the early clinical reports might well have focused on symptomatic reaction to stress that was stage-specific and time-related rather than reflective of long-standing basic pathology.

Jackson and Kogan (1963) studied help-seeking patterns in wives who attended Al-Anon groups in Seattle. The average Al-Anon member had tried 4.5 caregivers before coming to Al-Anon. The authors found that help-seeking was a patterned activity with psychological and social correlates. The nature and effectiveness of available community resources were likewise of significance. They investigated the variety and kinds of help sought and the perceived effectiveness of these resources for the women. In this pithy article, Jackson and Kogan touched on a number of significant issues of importance to the researcher and the clinician. They found that in congruence with societal values that stress family independence and urge families to try to help themselves before seeking outside help, most indeed did initially try. However, the greater the hardship experienced by the family, the more help was sought. Lower-class families, as Lemert noted (1960), sought help earlier and more frequently than middle-class families. The former also experienced more hardship in terms of economic inadequacy, physical violence, and infidelity.

Personality disturbance of the wife was not related to the amount of help sought, but the time spent in seeking help over the duration of the active drinking problem was correlated with disturbance. The authors found that those women who sought help early (within two years of the recognition of the problem) had lower ego strength scores, whereas those who began to seek help later had proportionately higher ego strength scores. Those who consistently sought help from the clergy fell in the disturbed categories, while those who went to lawyers (more definitively bent on separation or divorce) had lower anxiety scores and higher hardship scores. There was no correlation in disturbance with those who used various counseling or medical resources.

Jackson and Kogan in this paper discussed a variety of issues surrounding conditions and circumstances of divorce. In keeping with the findings of Bailey *et al.* (1962), they noted that the divorced differ significantly in having experienced greater hardship in economics, violence, and infidelity than those who did not divorce. As also noted by Bailey *et al.,* the families of recovered alcoholics had sought significantly less help from a smaller number of professionals over a briefer period of time. Once help was sought, it was used more effectively and there was less trial and error in approaches to community resources. This might suggest greater initial family strength. In support of Jackson's early work on the stages of alcoholism for the family unit, they found that the precipitating crisis of threatened divorce may be correlated with motivating the alcoholic to stop drinking. Those women who divorced and who attempted divorces fell into categories of greater emotional health more frequently than those who had made no such attempt.

In sum, the authors suggested that a prominent characteristic of the help-seeking careers of these families was the existence of fluctuating definitions of the nature of the problem, combined with complex social and psychological factors.

TREATMENT

Families of alcoholics have sought help or "treatment" from a wide range of community caregivers—clergy, physicians, private mental health professionals of all persuasions, and various community service agencies. Emphatic feelings of failure have been mutual on the part of clients and caregivers. Clients argue that professionals do not understand their particular pernicious problem. Clinicians characteristically have felt that in their experience the wives of alcoholics are highly defensive and resistant to introspection and change, and that their problems are almost as refractory to treatment as those of their alcoholic spouses. Thus, much of the literature available on treatment approaches is not only cautious in its reporting, but commonly pervaded with pessimism. This literature deals primarily with experiments in group therapy with wives, and with family casework. A small number of writings discuss family therapy and Al-Anon groups.

Group Therapy

The early group therapy papers reported on experimental groups, describing the group process and suggesting possible alternative formats for future endeavors (which it was hoped might be more successful). By and large, traditional analytic approaches were not effective. Most clinicians stated they felt the groups had been of positive consequence for participants, despite the wives' poor attendance, initial resistance to facing their role in the alcoholic's drinking patterns, and a host of other difficulties (Burton, 1962; Cork, 1956; Ewing et al., 1961; Gliedman et al., 1956; Igersheimer, 1959; MacDonald, 1956; Pattison et al., 1965; Pixley and Stiefel, 1963). Some clinicians who ran groups for wives concurrently with their treatment endeavors with the alcoholic husbands noted that the parallel participation of wives in separate groups benefited the alcoholic in his treatment process (Ewing et al., 1961; Gliedman et al., 1956; Pattison et al., 1965; Pixley and Stiefel, 1963).

For an overview of issues and process in contrasting group modalities, the reader is directed to Pixley and Stiefel (1963) and Pattison et al. (1965). Pixley and Stiefel presented a more traditional psychoanalytic approach toward the group experience, and discussed the evolution of the group process toward an analytic model. Pattison et al. described experimental short-term group efforts in lieu of a traditional psychoanalytic approach. They emphasized certain significant socioeconomic class factors that contributed to the acceptance or rejection of the modes of treatment offered.

Meeks and Kelly (1970) reported on a project with five families in family therapy as a posthospitalization treatment program. The nonalcoholic spouse initially was seen individually while the alcoholic was hospitalized. The authors emphasized the significance of involving the total family in the treatment process. The families showed evidence of improved relationships and healthier communication between members. Likewise changes were noted in drinking behavior of the alcoholic patients.

Family Casework

Bailey shared her extensive casework experience in a paper (1963a) that cogently discussed the presenting and continuing problems of the wife of the alcoholic. Bailey's philosophical approach eclectically recognized the need for both reality-based therapy and some degree of depth analysis of the client's problems.

> Social workers must use both the psychoanalytic theories and the sociological stress theories, blending them flexibly as appropriate in each case. The strength or weakness of a person's underlying personality influences his response to stress; some persons break down more easily than others. Stress theory, however, provides a useful framework for working with the wife of an alcoholic, regardless of the configuration of her personality. No one who has worked with such a wife will deny that she is seriously upset by the time she arrives at a potential source of help and treatment. The diagnostic question is whether her personality disturbance antedates her husband's alcoholism, and indeed her marriage, or whether it is at least in part a result of the stresses inherent in living with an active alcoholic. The treatment implications of these two theories are quite different. The first point of view suggests a focus on direct treatment of the wife's underlying problems—an approach that has not been notably successful. If, on the other hand, the wife's disturbance is regarded as even partly a response to stress, the caseworker's efforts will be directed either toward modifying or removing the stress siutation or toward enabling the wife to adapt to it with less anxiety and hostility. The wife of an alcoholic often responds favorably to this approach. She becomes able to function more comfortably and may eventually motivate her husband to seek help, so that in a real sense the husband and wife recover together from alcoholism.
>
> Admittedly, the wife whose emotional problems are deep-seated will not respond to this approach. The point to be emphasized is that in the beginning phase of casework it is almost impossible to make an adequate differential diagnosis. Though the worker should certainly be alert to diagnostic clues, he may be more effective initially if he works directly with the stress situation presented by the wife and postpones judgment on the degree of underlying disturbance. (Bailey, 1963a, p. 275)

Bailey (1968) and Cohen and Krause (1971) have presented two comprehensive and thoughtful volumes that discuss salient issues and considerations in all aspects of the process of casework with the alcoholic family. Bailey's volume is a report of a training course conducted by the staff of the Alcoholism Programs of the Community Council of Greater New York. The book covers seemingly every aspect to which a trainee should attend. The section on the treatment of the wife and children outlines the fears and anxieties that wives bring and the operational processes used by the staff of the training projects. It is suggested that the treatment relationship must begin with a lengthy period of time in which the wife can ventilate her angry feelings. A didactic process then begins of interpreting the nature of alcoholism as a disease, and attempting to free the wife from her preoccupation with the drinker and with rescue operations. It is pointed out that the husband's recovery depends on his decision to help himself. The wife then must face her own role in the marriage and the continuing drinking patterns. Self-examination and work on personality problems come more slowly and secondarily in a reality-based treatment model. Problems and pitfalls in the course of treatment are discussed and illustrated with case examples, and treatment with children is touched on. Referral to other community or specialized agencies is emphasized, for supplementary aid—in addition to the services of the family agency. Joint interviewing (family therapy) is discussed briefly, as are the interactional problems that arise to cause difficulties for the therapist using this modality. The programs described here generally did not find joint interviewing helpful.

Cohen and Kraus (1971) described in detail the theoretical underpinnings and operations of an experimental program carried out in Cincinnati. The authors discussed problems of ambivalence and motivation in their subjects, and how they overcame these with great flexibility of scheduling and concern in the initial treatment period. The treatment process with wives was quite similar to that described by Bailey for the New York programs. Reality problems and anxieties about the husband's drinking were ventilated in the early period. The wife early was presented with the concept of alcoholism as a chronic disease, and her role in the maintenance of the drinking pattern was explored. Crisis situations were given special attention and supportive actions such as meeting with the wife in the home when needed were carried out. The use of other community agencies that may be of supplemental help was urged.

Al-Anon Family Groups

Al-Anon Family Groups function as a self-help, nonprofessional modality of group therapy and group education. Al-Anon Family Groups are fellowships of spouses, relatives, and friends of alcoholics who have banded together to solve the common problems encountered in living with an alcoholic. At present more than 10,000 Al-Anon Family Groups exist throughout the world, yet little is known about these groups by community caregivers, and even less by the general public.

The dynamics of the Al-Anon process work through a mixture of educational and operational principles that must be accepted and assimilated by the member if she is to change her attitudes and behavior (Ablon, 1974; Al-Anon Family Group Headquarters, 1965; 1966). Ablon has suggested that the acceptance of one basic didactic lesson and several operational principles are paramount in the Al-Anon process.

The basic didactic lesson is that alcoholism is a disease of the body and of the mind. Therefore, alcoholism is not a moral fault or a perverse whim of the alcoholic. The acceptance of the disease concept removes many burdens of hostility, guilt, and shame from the member and from her spouse if she passes this teaching on to him.

The operational principles involve changes in attitudes and behavior in the member, and are based on a revised set of the "Twelve Steps" taken from Alcoholics Anonymous. The principles revolve about the goals of loving detachment from the alcoholic, and acceptance of the fact that his drinking patterns cannot be changed by the Al-Anon member or by any person except himself. A second basic goal is the reestablishment of the self-esteem and independence of the member. She must take her own personality inventory and work on improving her own life and that of her children and her household. Ideally, the actions of the alcoholic are not discussed in meetings. Thirdly, the member learns that she cannot accomplish the above goals by herself. She must rely on a "Higher Power," to aid her. This principle constitutes the spiritual dimension of Al-Anon. In most cases, the higher power is "God" or "Jesus." For those who do not choose traditional spiritual powers, the power of the Al-Anon Group itself is the spiritual force chosen.

The group process is characterized by sharing rather than confrontation. Through the candid sharing of emotional reactions, experiences, and strategies for coping with common problems encountered in living with an alcoholic, the member is drawn into a process of self-examination and, usually, a change for the better in her own coping abilities.

Ablon (1974) has presented the fullest account of the functioning of the Al-Anon process. Bailey (1965, 1968), Jackson (various papers), and more briefly, Pattison (1965) previously presented statements on the impact of Al-Anon on its members. Jackson's rich contributions to the literature resulted from data collected through a decade of participant observation and interviewing with Al-Anon members. She did not focus on the dynamics of Al-Anon in her writings.

Bailey (1965) briefly but cogently described the process and compared attitudes of Al-Anon members with non-Al-Anon wives of alcoholics. She found that the majority of Al-Anon members defined alcoholism as a mental and physical illness; in contrast, nonmembers regarded it as a mental disturbance. Likewise, Al-Anon wives less frequently expressed moralistic attitudes toward alcoholism. Al-Anon members stated that they had found the groups of more help than any other form of aid sought. They had gained self-understanding in relation to their husbands' drinking and the likelihood of the husbands' achieving sobriety was increased by the wives' participation in Al-Anon.

Most clinicians who report on therapy with wives of alcoholics comment on their resistance to looking at their own role in the drinking patterns as a chief stumbling block in treatment of any sort. Jackson noted from her experiences in Al-Anon groups that:

> In nonthreatening situations, the wives are the first to admit their own concerns about "their sanity." Of over one hundred women who attended a discussion group at one time or another during the past six years, there was not one who failed to talk about her concerns about her own emotional health. All of the women worry about the part which their attitudes and behavior play in the persistence of the drinking and in their family's disturbances. Although no uniform personality types are discernible, they do share feelings of confusion and anxiety. Most feel ambivalent about their husbands; however, this group is composed of women who are oriented toward changing themselves and the situation rather than escaping from it. (Jackson, 1958, p. 93)

Bailey's and Jackson's writings as well as Pattison's briefer reference to Al-Anon suggest two points to be noted. Al-Anon women tend to be relatively well educated and of the middle class. They may be more motivated for a variety of reasons to expend greater efforts to salvage their marriages.

Also of significance is the fact that in a nonprofessional, nonthreatening peer group situation where all members share a common problem, certain therapeutic dynamics evolve and often work very effectively. Al-Anon's emphasis on the physical disease aspect has a strong therapeutic value for the wife, relieving her burdens of guilt and shame. In a warmly supportive milieu, wives may move nondefensively to looking at their own behavior. Bailey (1963a, 1968) suggested that professionals have moved too rapidly in trying to explore underlying problems. Bailey has made pragmatic use of Al-Anon process, and operationally her own suggested professional casework principles closely follow those of Al-Anon. Bailey (1963a, 1965, 1968) and Cohen and Kraus (1971) have pointed up the potential value of using Al-Anon as a supplement to professional treatment.

Al-Ateen groups for children of alcoholics have been discussed briefly (Bailey, 1968). Al-Ateen, too, works on a self-help principle of peers sharing common problems. The few agency personnel who have had occasion to work with children attending these groups laud their effectiveness.

SOCIOCULTURAL VARIABLES

While the importance of sociocultural factors as one significant determinant in stimulating or maintaining drinking behavior has been considered by many investigators, the role of these factors in family attitudes toward the drinker or in modes of coping with the problem drinker by his spouse and significant others has been given scant attention. Cahalan et al. (1969), for example, presented a tentative conclusion "that whether a person drinks at all is primarily a sociological and anthropological variable rather than a psychological one" (p. 200). Pittman (1967) stated, "Only by obtaining more knowledge on specific cultural attitudes toward drinking and drunkenness and the function and role of drinking in diverse cultures, will researchers begin to understand and explain pathological drinking" (p. 4).

Differences in drinking customs are recognized as existing not only between societies, but in groups such as those differentiated on the basis of age, sex, and ethnic origin. It has been suggested that in those groups where the drinking of alcoholic beverages is well established and integrated into the culture, such as among Jews, Italians, and Chinese, the rate of alcoholism will be low, (Ullman, 1958; Snyder, 1962). In contrast, the Irish Catholics, who have a high rate of alcoholism, have experienced alcohol as serving a host of secular and religious functions, and in differing segments of the Irish population, its usage has a variety of meanings (Bales, 1946, 1962). It has been noted that urban populations have higher rates of problem drinking than rural populations (Cahalan, 1970).

The great majority of the studies reviewed in this paper have made little or no comment on the sociocultural context within which their subjects exist. The attitudes and drinking patterns of their extended family, friendship circle, churchmates, and of the larger society play a great part in the attitudes of family members toward the alcoholic and in their actions in relation to him. Jackson (various papers), Lisansky (1957), Lemert (1960), Bailey (1965), and Pattison et al. (1965) have discussed the significance of the socioeconomic membership of their subjects in relation to their differential responses to questionnaires, testing, or treatment modalities. Lemert (1960) primarily emphasized the need for exploration in this area of differential perception of problems, but detailed research focusing on the specific sociocultural or socioeconomic variables in wife and family response have yet to appear. For example, Bailey et al. (1965) noted that a disproportionate number of wives of alcoholics had parents who displayed "unnatural attitudes toward drinking." Clifford (1960) stated that all of the wives in his two groups had a drinking parent. While the psychological implications of this fact in relation to modeling behavior may be apparent, further exploration into the subcultural memberships of these families would be of great interest.

OVERVIEW

The evolution of research on the family and alcoholism may be traced through several distinct phases. Initially, there appeared a series of clinical studies documenting the pathology of the spouse. This florescence of papers was followed by a series by Jackson and her associates that constituted a rebuttal to the chief thrust of the earlier studies that focused on the wife as a causal factor in the husband's drinking patterns. A primary component of this rebuttal was the exploration of various aspects of the personality of the wife through psychological tests, rather than relying on impressionistic interviews at specific crisis periods when the wife logically might be expected to be highly agitated. Wives of alcoholics were compared with "normal" nonalcoholic wives and significant differences in disturbance were not identified. The work of Jackson and Kogan presented evidence that the concept of a unitary "personality of the wife of the alcoholic" was untenable. A second and highly significant dimension of Jackson's research was sociological in its orientation, and placed emphasis on the adaptation of the total family to stresses caused by the existence of alcoholism in the household. This orientation was functional and interactional in that it drew attention to alcoholism as a total family problem and family disease.

Papers focusing on interaction between spouses began to appear in 1959, with the clearest analytical statements most recently presented by Steinglass and his colleagues, who have explicitly utilized functional systems and interactionist theories. These statements and Ward and Faillace's broader overview of alcoholism as it involves the individual, his family, and the larger community have provided the most sophisticated statements in the literature aside from the earlier foundation-building works of Jackson. Steinglass and colleagues proposed a long-awaited articulate model lending itself to significant implications for treatment. Ward and Faillace took into consideration all aspects of the drinker's environment, both as they function in regard to him, and in terms of the internal workings of the respective systems that may determine this functioning, independent of the actual reality needs of the specific alcoholic individual. Brief as this article is, it contains the directions for a great many areas of research that are essentially yet untouched.

Bailey (1961), Jackson (1962), and Day (1961) have emphasized the need for conceptually broad studies that integrate the psychological and sociological approaches. Researchers and clinicians in this field by and large have found the psychological approach to be more congenial to their training and experience and thus have accepted and focused on the personal pathology of the individual or of the two spouses, at the expense of looking to a multivariate approach combining individual personality characteristics with interpersonal interactions within the family and significant social network, and with the more diffuse and broader socioeconomic context within which the individual and the family exist. It is of interest that the multivariate approach has been suggested in recent literature in relation to the problem drinker (Cahalan, 1970; Jessor *et al.*, 1968; U.S. Department of Health, Education, and Welfare, 1971), yet largely ignored in studies of the response of his family or the significant other factors in his environment.

An equally serious failing has been the lack of studies tying in the specific alcoholic-related crisis syndrome to the existing body of theory dealing with family and community behavior in a variety of crisis situations. Researchers on the alcoholic family consistently have been parochial in their conceptualizations, seldom turning to outstanding available and relevant literature. Only a few authors such as Jackson (1958, 1962), Steinglass *et al.* (1971b), and Bowen (1974) have pointed up the similarities of family system dysfunction in alcoholism to those reported by other researchers for families in other forms of crisis.

Various papers of Hill (1958) and Hansen and Hill (1964), for example, offer a wealth of theoretical considerations for comparison. Hill (1958) classified alcoholism as an intrafamily stressful event involving "de-morale-ization" of the family, because its role patterns are disturbed. He noted that the family succeeds as a family chiefly on the basis of the adequacy of the role performance of its members. When role patterns are changed and equilibrium is disturbed, the crisis is incurred. Hill offered suggestions to the caregiver who works with families in crisis. He noted that a chief consideration is to keep the total family context in mind. In the case of families in crises of demoralization, this consideration is of particular significance. He suggested a list of family characteristics that are conducive to good adjustment to crisis. The researcher and clinician might well use these as indicators or variables of family strength or viability in an exploration of the total configuration of the alcoholic family.

Riskin (1963), in a presentation of methodology for studying family interaction, presented a similar theoretical model for examination of the family in crisis:

> The family is viewed as an ongoing system . . . It tends to maintain itself around some point of equilibrium which has been established as the family evolves. The system is a dynamic, not a static, one. There is continuous process of input into the system, and thus a continuous tendency for the system to be pushed away from the equilibrium point. The input may be in the form of an external stress such as a job change, war, or depression; or it may be an internal stress such as the birth of a new child, a death, or a biological spurt such as the onset of puberty. Over a period of time, the family develops certain repetitive, enduring techniques or patterns of interaction for maintaining its equilibrium when confronted by stress; this development tends to hold whether the stress be internal or external, acute or chronic, trivial or gross. These techniques, which are assumed to be characteristic for a given family, are regarded as homeostatic mechanisms. In the above description, the family is described somewhat analogously to an individual with his drives, defenses, character traits, and ego functions. However, these concepts differ from classical psychoanalytic ones in that they refer primarily to interpersonal organizations of behavior rather than to internal mechanisms. (Riskin, 1963, p. 344)

With few exceptions (e.g., Jackson, 1954, 1958, 1962; Ewing and Fox, 1968; Steinglass *et al.,* 1971a,b; Bowen, 1974; Ward and Faillace, 1970), researchers have not taken advantage of the rich body of theory available on the problem family.

A significant area largely ignored in the consideration of family response is the extrafamily dimension involving prevalent sociocultural attitudes toward alcoholism and the nature of the available sources of help. These may play a primary role in the development of attitudes maintained by the family and in determining what options for action they will or can take in regard to their problem. Hansen and Hill (1964) presented a classification of various family crises with the possible related community response, i.e., the attitudes and actions that community caregivers display in response to respective types of stressor situations. Again it has been primarily Jackson who has given attention to the role of the prevailing cultural attitudes toward specific alcohol-related family problems and toward the types of aid available to the family. Jackson (1962) noted that a chief complication for the family is a lack of cultural guidelines for family response to this specific problem other than a generic categorization of the excessive drinking as deviant behavior. Such a labeling process then invokes shame and stigma in family members, which contribute further to an already sensitive crisis situation.

Comparable in importance to the drinking history of the alcoholic regularly sought by caregivers might be postulated the history of help-seeking of the spouse or other family members. The area of help-seeking and community response to the needs of the family has been given relatively little consideration in regard to how this has affected the progression of the problem. Unsuccessful attempts in the help-seeking career of the spouse often contribute to the antiprofessional attitudes found among Al-Anon members as well as other wives who dismiss help-seeking endeavors with professional caregivers as futile.

The appearance in recent years of papers dealing with the alcoholic family in journals other than the *Quarterly Journal of Studies on Alcoholism* constitutes a hopeful step forward, reflecting a growing concern for the subject and an acknowledgment of its prevalence, as well as providing the opportunity for exposure of materials to professionals in a broader range of medical and caregiving fields.

THE IMPORTANCE OF SAMPLING IN THE DIRECTION OF FUTURE RESEARCH

The studies discussed in this paper generally focused on relatively small cohort samples (six to 50 persons) of women taken from outpatient clinic caseloads, Al-Anon Family Groups, family service centers, and, in a few cases, prison or probation populations. They may accordingly represent the most desperate, the most motivated, or the most coerced family members available to researchers. It is important that the fact of sample bias be noted. How representative such small samples from specialized agency or self-help groups are in relation to the millions of "hidden" casualties of alcoholic marriages is questionable.

One approach to overcome such sampling bias would be intensive, in-depth holistic studies of "normal" families in various categories of risk as indicated by survey research data (Cahalan, 1970). The prevalence of hidden alcoholic problems may be encountered in the course of such intimate exploration. Likewise, more comparative studies with conflicted, though nonalcoholic, families are needed. Still another approach would be to focus on the alcoholism component in problem families who come to the attention of caregivers with other presenting problems, and to examine how this specific element relates to other areas of marital discord. Alcoholism frequently either is presented as the only problem (which ideally when taken care of would "settle" all other problems), or is carefully hidden because it is perceived as an insignificant symptom of other deeper problems.

A chief goal of future research must be the charting of an eclectic, comprehensive approach that views alcohol-related problems as not totally unique to the class phenomenon of family crisis. Furthermore, the details of alcohol-related problems must be explored holistically and in depth, taking into account all potentially significant psychological, social, and cultural characteristics of all family members. Only such a comprehensive approach gives appropriate cognizance to the reality that alcoholism is a multifaceted affliction that crucially effects the family as a total unit and each member as an interacting individual of the collectivity.

ACKNOWLEDGMENT

The research for this paper was supported in part by the National Institutes of Mental Health, U.S.P.H.S. Grants No. MH-08375 and MH-21552.

REFERENCES

Ablon, J., 1974, Al-Anon family groups: Impetus for learning and change through the presentation of alternatives. *Amer. J. Psychother.* 28(1): 30.
Al-Anon Family Group Headquarters, Inc., 1965, "Al-Anon Faces Alcoholism," New York.
Al-Anon Family Group Headquarters, Inc., 1966, "Living with an Alcoholic," New York.
Al-Anon Family Group Headquarters, Inc., 1969, "Alcoholism: A Merry-Go-Round Named Denial," New York.
Bacon, S. D., 1945, Excessive drinking and the institution of the family, in "Alcohol, Science, and Society" (Quart. J. Studies Alc., ed.) pp. 223–238, Yale Summer School of Alcohol Studies, New Haven.
Bailey, M. B., 1961, Alcoholism and marriage: A review of research and professional literature, *Quart. J. Studies Alc.* 22(1):81.
Bailey, M. B., 1963a, The family agency's role in treating the wife of an alcoholic, *Social Casework* 44(5):273.
Bailey, M. B., 1963b, Research on alcoholism and marriage in "National Conference on Social Welfare, Columbus, Ohio" (Social Work Practice, ed.), National Conference on Social Welfare, Columbus, Ohio.
Bailey, M. B., 1965, Al-Anon family groups as an aid to wives of alcoholics, *Social Work* 10(1):68.
Bailey, M. B., 1968, "Alcoholism and Family Casework," The Community Council of Greater New York, New York.
Bailey, M. B., Haberman, P. W., and Alksne, H., 1962, Outcomes of alcoholic marriages: Endurance, termination or recovery, *Quart. J. Studies Alc.* 23(4):610.
Bailey, M. B., Haberman, P. W., and Sheinberg, J., 1965, Distinctive characteristics of the alco-

holic family, *in* "Report of the Health Research Council of the City of New York," The National Council on Alcoholism, Inc., Mimeo, New York.

Bales, R. F., 1946, Cultural differences in rates of alcoholism, *Quart. J. Studies Alc.* 6(4):480.

Bales, R. F., 1962, Attitudes toward drinking in the Irish culture, *in* "Society, Culture, and Drinking Patterns" (D. J. Pittman and C. R. Snyder, eds.) pp. 157–187, John Wiley and Sons, Inc., New York.

Ballard, R. G., 1959, The interaction between marital conflict and alcoholism as seen through MMPI's of marriage partners, *Amer. J. Orthopsychiat.* 29(3):528.

Belfer, M. L., Shader, R. I., Carroll, M., and Harmatz, J. S., 1971, Alcoholism in women, *Arch. Gen. Psychiat.* 25(6):540.

Berne, E., 1964, "Games People Play," Grove Press, New York.

Bowen, M., 1974, Alcoholism as viewed through family systems theory and psychotherapy, *Ann. N.Y. Acad. Sci.* 233:115.

Bullock, S. C., and Mudd, E. H., 1959, The interaction of alcoholic husbands and their non-alcoholic wives during counseling, *Amer. J. Orthopsychiat.* 29(3):519.

Burton, G., 1962, Group counseling with alcoholic husbands and their nonalcoholic wives, *Marriage and Family Living* 24(1):56.

Cahalan, D., 1970, "Problem Drinkers," Jossey-Bass, San Francisco.

Cahalan, D., Cisin, I., and Crossley, H. M., 1969, "American drinking practices: A national survey of behavior and attitudes," Rutgers School of Alcohol Studies, Monograph No. 6, New Brunswick, New Jersey.

Chafetz, M. E., Blane, H. T., and Hill, M. J., 1971, Children of alcoholics, *Quart. J. Studies Alc.* 32(3):687.

Clifford, B. J., 1960, A study of the wives of rehabilitated and unrehabilitated alcoholics, *Social Casework* 41(9):457.

Cohen, P. C., and Krause, M. S., 1971, "Casework with Wives of Alcoholics," Family Service Association, New York.

Corder, B. F., Hendricks, A., and Corder, R. F., 1964, An MMPI study of a group of wives of alcoholics, *Quart. J. Studies Alc.* 25(3):551.

Cork, R. M., 1956, Casework in a group setting with wives of alcoholics, *The Social Worker* 14:1.

Cork, R. M., 1964, Alcoholism and the family, A paper presented at the Alcoholism and Drug Addiction Research Foundation Annual Course, May, 1964. University of Western Ontario, Addiction Research Foundation, Province of Ontario.

Davis, D. I., Berenson, D., Steinglass, P., and Davis, S., 1974, The adaptive consequences of drinking, *Psychiatry* 37:209.

Day, B. R. D., 1961, Alcoholism and the family, *Marriage and Family Living* 23(3):253.

Edwards, P., Harvey, C., and Whitehead, P. C., 1973, Wives of alcoholics: A critical review and analysis, *Quart. J. Studies Alc.* 34:112.

Ewing, J. A., and Fox, R. E., 1968, Family therapy of alcoholism, *in* "Current Psychiatric Therapies" (J. H. Masserman, ed.), Vol. 8, pp. 86–91, Grune and Stratton, New York.

Ewing, J. A., Long, V., and Wenzel, G. G., 1961, Concurrent group psychotherapy of alcoholic patients and their wives, *International Journal of Group Psychotherapy* 11(3):329.

Fox, R., 1956, The alcoholic spouse, *in* "Neurotic Interaction in Marriage" (V. W. Eisenstein, ed.), pp. 148–168, Basic Books, New York.

Fox, R., 1962, Children in the alcoholic family, *in* "Problems in Addiction: Alcohol and Drug Addiction" (W. C. Bier, ed.) pp. 71–96, Fordham University Press, New York.

Futterman, S., 1953, Personality trends in wives of alcoholics, *Journal of Psychiatric Social Work* 23(1):37.

Gliedman, L. H., Nash, H. T., and Webb, W. L., 1956, Group psychotherapy of male alcoholics and their wives, *Diseases of the Nervous System* 17(3):90.

Gorad, S. L., 1971, Communication styles and interaction of alcoholics and their wives, *Family Process* 10(4):475.
Hansen, D. A., and Hill, R., 1964, Families under stress *in* "Handbook of Marriage and the Family" (H. T. Christensen, ed.), pp. 782–819, Rand McNally and Company, Chicago.
Hill, R., 1958, Generic features of families under stress, *Social Casework* 39(2,3):139.
Hill, R., and Hansen, D. A., 1960, The identification of conceptual frameworks utilized in family study, *Marriage and Family Living* 22:299.
Igersheimer, W. W., 1959, Group psychotherapy for nonalcoholic wives of alcoholics, *Quart. J. Studies Alc.* 20(1):77.
Jackson, J. K., 1954, The adjustment of the family to the crisis of alcoholism, *Quart. J. Studies Alc.* 15(4):562.
Jackson, J. K., 1958, Alcoholism and the family, *Annuals of the American Academy of Political and Social Science* 315:90.
Jackson, J. K., 1962, Alcoholism and the family, *in* "Society, Culture, and Drinking Patterns" (D. J. Pittman, and C. R. Snyder, eds.), pp. 472–492, Wiley, New York.
Jackson, J. K., and Kogan, K. L., 1963, The search for solutions: Help-seeking patterns of families of active and inactive alcoholics, *Quart. J. Studies Alc.* 24(3):449.
James, E., and Goldman, M., 1971, Behavior trends of wives of alcoholics, *Quart. J. Studies Alc.* 32(2):373.
Jellinek, E. M., 1960, "The Disease Concept of Alcoholism," Hillhouse Press, New Jersey.
Jessor, R., Graves, T. D., Hanson, R. C., and Jessor, S. L., 1968, "Society, Personality, and Deviant Behavior," Holt, Rinehart, and Winston, New York.
Kogan, K. L., and Jackson, J. K., 1963, Role perceptions in wives of alcoholics and of nonalcoholics, *Quart. J. Studies Alc.* 24(4):627.
Kogan, K. L., and Jackson, J. K., 1965a, Alcoholism: The fable of the noxious wife, *Mental Hygiene* 49(3):428.
Kogan, K. L., and Jackson, J. K., 1965b, Some concomitants of personal difficulties in wives of alcoholics and nonalcoholics, *Quart. J. Studies Alc.* 26(4):595.
Kogan, K. L., and Jackson, J. K., 1965c, Stress, personality, and emotional disturbance in wives of alcoholics, *Quart. J. Studies Alc.* 26(3):486.
Kogan, K. L., Fordyce, W. E., and Jackson, J. K., 1963: Personality disturbance in wives of alcoholics, *Quart. J. Studies Alc.* 24(2):227.
Lemert, E. M., 1960, The occurrence and sequence of events in adjustment of families to alcoholism, *Quart. J. Studies Alc.* 21(4):679.
Lemert, E. M., 1962, Dependency in married alcoholics, *Quart. J. Studies Alc.* 23(4):590.
Lewis, M. L., 1954, The initial contact with wives of alcoholics, *Social Casework* 35:8.
Lisansky, E. S., 1957, Alcoholism in women: Social and psychological concomitants: 1. Social history data, *Quart. J. Studies Alc.* 18(4):588.
Lisansky, E. S., 1958, The woman alcoholic, *Annals of the American Academy of Political and Social Science* 315:73.
MacDonald, D. E., 1956, Mental disorders in wives of alcoholics, *Quart. J. Studies Alc.* 17(2):282.
Meeks, D. E., and Kelly, C., 1970, Family therapy with families of recovering alcoholics, *Quart. J. Studies Alc.* 31(2):399.
Mitchell, H. E., 1959, Interpersonal perception theory applied to conflicted marriages in which alcoholism is and is not a problem, *American Journal of Orthopsychiatry* 29(3):547.
Newell, N., 1950, Alcoholism and the father image, *Quart. J. Studies Alc.* 11(1):92.
Oxford, S., and Guthrie, S., 1968, Coping behavior used by wives of alcoholics; a preliminary investigation. Abstract from International Congress Alc. Alcsm., Proc. 28th, 97.

Parad, H. J., and Caplan, G., 1960, A framework for studying families in crisis, *Social Work* 5(3):3.

Parsons, T., and Bales, R. F., 1955, "Family, Socialization and Interaction Process," The Free Press, New York.

Pattison, E. M., Courlas, P., Patti, R., Mann, B., and Mullen, D., 1965, Diagnostic therapeutic intake groups for wives of alcoholics. *Quart. J. Studies Alc.* 26(4):605.

Pittman, D. J., 1967, "Alcoholism," Harper and Row, New York.

Pittman, D. J., and Snyder, C. R., 1962, "Society, Culture, and Drinking Patterns," Wiley, New York.

Pixley, J. M., and Stiefel, J. R., 1963, Group therapy designed to meet the needs of the alcoholic wife, *Quart. J. Studies Alc.* 24(2):304.

Price, G. M., 1945, A study of the wives of 20 alcoholics, *Quart. J. Studies Alc.* 5(4):620.

Riskin, J., 1963, Methodology for studying family interaction, *Archives of General Psychiatry* 8(4):343.

Snyder, C. R., 1962, Culture and Jewish sobriety: The in-group—out-group factor, *in* "Society, Culture, and Drinking Patterns" (D. J. Pittman and C. R. Snyder, eds.), pp. 188–225, Wiley and Sons, New York.

Spiegel, J. P., 1957, The resolution of role conflict within the family, *Psychiatry* 20(1):1.

Spiegel, J. P., and Kluckhohn, F. R., 1954, "Integration and Conflict in Family Behavior," Group for the Advancement of Psychiatry, Topeka, Kansas.

Steiner, C. M., 1969, The alcoholic game, *Quart. J. Studies Alc.* 30(4):920.

Steiner, C. M., 1971, "Games Alcoholics Play," Grove Press, New York.

Steinglass, P., Weiner, S., and Mendelson, J. H., 1971a, Interactional issues as determinants of alcoholism, *Amer. J. Psychiat.* 128(3):55.

Steinglass, P., Weiner, S., and Mendelson, J. H., 1971b, A systems approach to alcoholism: A model and its clinical application, *Arch. Gen. Psychiat.* 24(5):401.

Ullman, A., 1958, Sociocultural backgrounds of alcoholism, *Ann. Amer. Acad. Polit. Soc. Sci.,* 315:48.

U.S. Department of Health, Education, and Welfare, 1971, "First Special Report to the United States Congress on Alcohol and Health," Health Services and Mental Health Administration, National Institute of Mental Health, NIAAA, Washington, D.C.

Ward, R. F., and Faillace, G. A., 1970, The alcoholic and his helpers, *Quart. J. Studies Alc.* 31(3):684.

Whalen, T., 1953, Wives of alcoholics: Four types observed in a family service agency, *Quart. J. Studies Alc.* 14(4):632.

Wolin, S., Steinglass, P., Sendroff, P., Davis, D. I., and Berenson, D., 1975, Marital interaction during experimental intoxication and the relationship to family history, *in* "Experimental Studies of Alcohol Intoxication and Withdrawal" (M. Gross, ed.), Plenum Press, New York.

CHAPTER 7

The Alcoholic Personality

Allan F. Williams

Research Department
Insurance Institute for Highway Safety
Washington, D.C.

VIEWS ON THE ROLE OF PERSONALITY FACTORS IN ALCOHOLISM

It is generally agreed that personality factors play a role in the development of alcoholism, along with sociocultural and physiological components. It has been difficult, however, to pin down just what personality characteristics are involved, and why they should lead to this particular disorder. Studies in this area have produced a wide variety of disparate findings; a review of the literature indicates that nearly every personality characteristic has been implicated in alcoholism in one study or another. Nevertheless, during the past decade substantial progress has been made in isolating personality factors involved in alcoholism. In this chapter the major studies contributing to this progress will be discussed.

One widely held view concerning the role of personality factors in alcoholism is that although such factors are important, or necessary, in the development of this disorder, more than one personality type is susceptible. "More than one" in this case usually means several, or many. Most commonly, some type of maladjustment is thought to be involved (Armstrong, 1958), but the particular form of maladjustment is seen as varying from person to person, or from culture to culture. This position derives in part from the observation that drinking can temporarily relieve a wide variety of psychological problems and is thus a possible solution for many different types of people. The view also derives from the previously mentioned situation that many different personality characteristics have been noted in alcoholics, and that alcoholism has been found in almost all diagnostic categories of psychiatric illness. Blum (1966, p. 264) concluded that "if one assumes multiple determinations instead of a unitary etiology, confusion is avoided."

An alternate view is that in all cultures there is one main personality constellation that is particularly susceptible to alcoholism. Zwerling and Rosenbaum (1959) have noted that a particular constellation of traits may be embedded in a wide variety of character structures, so that the finding of various clinical diagnoses among alcoholics does not preclude the possibility of a specific personality constellation basic to alcoholism. Similarly, they argue that "the existence of an underlying personality disorder common to alcoholics is not precluded by the observation that alcohol may be pressed into service at all levels of conflict and in association with a variety of defenses" (Zwerling and Rosenbaum, 1959, p. 627). Proponents of this school usually consider a particular personality structure to be necessary but not sufficient for the development of alcoholism, allowing for the influence of sociocultural, physiological, and other factors. Those who hold this view concerning the role of personality factors in alcoholism are talking about the "alcoholic personality," a premorbid personality structure common to all alcoholics. The main concern in this chapter is whether or not there is an alcoholic personality, and if so, what is its nature.

APPROACHES TO STUDYING PERSONALITY FACTORS IN ALCOHOLISM

Many people who deal with alcoholics have been impressed that they appear to be "different" in personality from those who do not become alcoholics, and that they have many similarities as a group. Theories and empirical research on personality factors in alcoholism have been guided by these observations. A second guiding notion is that alcohol must have some special effect on those who become alcoholics, either quantitatively or qualitatively different from effects experienced by nonalcoholics, which "explains" why they drink heavily and continue to do so despite the negative elements associated with their drinking experiences. Approaches to studying personality factors in alcoholism have followed these two lines of thought, consisting of studies of the characteristics of alcoholics and other drinkers, and studies of the emotional effect of alcohol on alcoholics and other drinkers.

Studying Alcoholics

The usual procedure in studying the role of personality factors in alcoholism has been to obtain data or to make observations on those defined as alcoholics. Almost all of the relevant studies have been based on alcoholics. Unfortunately, these studies have limited value in providing evidence pertaining to the role of personality factors in the disorder. Many of the studies have serious methodological flaws. Almost all such studies are based on the small minority of alcoholics that volunteer for treatment. Moreover, whatever alcoholic population is studied, it is unclear whether causes or effects of alcoholism are being observed.

The earliest reports of personality factors in alcoholism were not empirical studies but were based on clinical observations of a few alcoholics. Of the empirical studies that have been conducted, some have had no control groups, and many others have been inadequately controlled. For example, most studies have made comparisons between alcoholics and normal populations, although alcoholics may differ from normals only because alcoholics are more maladjusted. To understand alcoholism, rather than maladjustment in general, it is necessary to show that alcoholics differ from other maladjusted persons who do not have drinking problems. Other studies have not controlled for such variables as socioeconomic status and age, so that personality differences between alcoholics and nonalcoholics may occur but be attributable to factors other than drinking. Zucker (1968) points out that most alcoholics treated in social agencies are of low socioeconomic standing, and that it is necessary to control for this factor in comparisons with nonalcoholics. Even then, it is not possible to generalize other than to treated alcoholics of low socioeconomic standing.

The generalization problem in personality research on alcoholism is an important one. Virtually all studies of alcoholics are based on those who are being treated voluntarily, a group that may not be representative of alcoholics in general. Treated alcoholics are older than alcoholics in general, and they have admitted that they cannot control their drinking and need help. This is a crucial, psychologically meaningful admission in the life of an alcoholic, and those who take this step may have different personality characteristics from those who do not.

Most investigators who conduct personality studies of alcoholics do so in order to make inferences about etiology. However, in personality research with alcoholics, etiological statements are precluded since investigators are unable to dismiss the possibility that their findings reflect social and psychological consequences of 15 to 20 years of excessive drinking and loss of control over alcohol. Thus, even if reliable differences between alcoholics and nonalcoholics were present, it would not be known if the personality characteristics found in alcoholics existed in the premorbid state or resulted from alcoholism. The process of becoming an alcoholic is a lengthy one. The various stages comprising the typical alcoholic career have been described by Jellinek (1952) and Park (1973), and among the consequences of excessive drinking, they list grandiose behavior, loss of self-esteem, aggressive behavior, self-pity, and the like. Certainly the experiences of the alcoholic would be expected to have a marked effect on his self-image and interpersonal relationships. Since most studies are of treated alcoholics, it should also be noted that the experience of admitting a need for help and entering into a treatment relationship may additionally produce changes in the way the alcoholic sees or presents himself.

Jellinek (1952) has argued that the experience of becoming an alcoholic is likely to produce similarities in personality characteristics in this group, whereas diversity was the rule in the premorbid state. Yet, studies of alcoholics have generally failed to demonstrate consistent differences between alcoholics and nonalcoholics, even though these studies have mainly dealt with treated alcoholics. A major reason for this curious state of affairs is the methodological flaws noted earlier, plus the fact that the same instrument is rarely used to measure the same concept. Some of the methodologically adequate studies of alcoholics will be referred to later. At this point, ways in which the premorbid personality structure of the alcoholic can be investigated other than by examining alcoholics will be discussed.

Alternative Methods

A number of alternative routes to exploring personality factors that may lead to alcoholism avoid the problems inherent in studies of alcoholics. These alternative methods have received increasing attention in the last decade, and they will be given major attention in this chapter.

There are three ways in which investigators studying alcoholics have attempted to deal with the possibility that their findings may be consequences rather than antecedents of alcoholism. One technique is to study remitted alcoholics, assuming that if traits found among active alcoholics disappear, or are not found after long periods of sobriety have been achieved, these traits must be consequences of alcoholism (Machover and Puzzo, 1959). Otherwise, they are antecedents. A second technique has been to compare young and old alcoholics, assuming that the consequences of alcoholism will be less apparent in young alcoholics (Hoffman, 1970). The third technique used has been to ask parents of the alcoholic, or the alcoholic himself, what he was like when he was younger (Tähkä, 1966; Wittman, 1939). It can readily be seen that each of these techniques has limitations, and they constitute partial solutions at best.

The ideal solution is a study in which data are collected on a population in childhood, or young adulthood, before any persons in this population become alcoholics. This type of study allows a sorting out of personality characteristics that existed before and after the onset of alcoholism, and also avoids the problem of dealing only with treated alcoholics. Unfortunately, only a few such longitudinal studies have been completed that are pertinent to alcoholism.

In lieu of the rare longitudinal study, a number of investigators have undertaken studies of young men who may become alcoholics. Studies have been carried out on college students who show incipient signs of alcoholism; on delinquents with marked alcohol involvement, and on high school and college students who are heavy drinkers. These are high-risk populations, from which future alcoholics will likely be drawn. Personality characteristics found in these groups have been examined in order to determine what the premorbid personality structure of the alcoholic may be like, and personality characteristics found have been compared to those noted among alcoholics. If similar personality characteristics exist among alcoholics and young problem drinkers, this constitutes strong evidence that premorbid personality traits have been isolated. It may also be the case that there is a personality syndrome common to all heavy drinkers, including alcoholics, who are a special case of heavy drinker. Thus, one can study the premorbid alcoholic personality by examining nonalcoholic heavy drinkers, regardless of whether a significant proportion of this latter group eventually becomes alcoholics.

Studying Emotional Effects of Alcohol

Studies of the emotional effects of alcohol have undergone an evolutionary process. Until the mid-1960s the bulk of knowledge, or rather lore, concerning emotional and motivational effects of alcohol came from clinical and introspective reports. All kinds of positive emotional effects, i.e., reasons for drinking, have been ascribed to alcohol. The question is what kinds of effects on what kinds of persons are necessary to produce alcoholism. What does the alcoholic get out of drinking that is different, either quantitatively or qualitatively, from the experiences of the nonalcoholic drinker, and that spurs him to drink heavily in the first place and to continue to drink after serious problems arise? The view held by many is that there must be some powerful short-term gain experienced, in order to make sense of the continued drinking.

Much of the early writing in this area was based on clinical case studies of alcoholic patients. An assessment was made of their personal characteristics, and how their psychological needs were served by drinking. Thus, the effects of drinking were built into theories concerning why certain types of people became alcoholics. Other theorists concentrated directly on the major psychodynamic functions that appear to be served by alcohol, inferring from this analysis the types of persons most susceptible. Levy (1958) argued that the major psychological functions of alcohol are limited in number, and therefore define and limit the variety of personality structures that can use these functions as problem-solving devices via pathological drinking. The major needs that alcohol can serve, according to Levy, are masochistic and passive–infantile needs.

In terms of empirical research on emotional effects of alcohol, one technique has been simply to ask drinkers, including alcoholics, why they drink. A less direct approach was undertaken by MacAndrew and Garfinkel (1962) who had alcoholics, when in a sober condition, describe their sober, ideal, and drunk selves, using Q sorts. Other investigators have asked sober alcoholics to describe mood and other changes they think occur when they drink (Keehn, 1970; Partington, 1970; Tamerin et al., 1970).

Empirical research on the emotional effects of alcohol in which study participants actually consume alcohol is a recent phenomenon. Since the early 1960s, however, numerous studies have been carried out in which people, including alcoholics, have consumed alcohol and been tested for its emotional effects on them. In early research of this type, it was attempted to isolate the effects of alcohol per se. Kalin et al. (1965) have pointed out that this goal cannot be attained; that the results are always a product of the interaction between the experimental manipulation and the set and setting. The original investigations of emotional effects of alcohol were modeled on studies of the effects of alcohol on nonemotional behaviors such as reaction time. As Kalin et al. (1965) note, this type of experimental situation (laboratory or hospital setting, intravenous administration) is likely to be inhibiting and anxiety-arousing. Thus, comparing alcohol and placebo subjects does not isolate the effects of alcohol per se; it isolates the effects of alcohol in an anxiety-arousing atmosphere. Such an atmosphere is likely to inhibit any "positive" emotional effects that may occur under more normal drinking conditions. One can only generalize to the particular situation being studied. Thus, if the interest is in why people drink, they should be allowed to drink as they normally do, in natural settings in which a choice of drinks is available for ad lib consumption. This is the experimental paradigm for research on the emotional effects of alcohol, and several studies based on this model have been carried out. An additional concern is that participants not be aware that alcohol is the focus of the study, lest they merely produce conscious reasons for drinking. Typically, the studies are introduced as investigations of factors other than drinking behavior, such as social setting factors.

One caution should be observed in interpreting results of studies on the emotional effect of alcohol on alcoholics. Just as it is difficult to sort out antecedents and consequences when studying personality characteristics of alcoholics, so is it difficult to judge whether the effects of alcohol experienced by alcoholics now contributed to the person's becoming an alcoholic. In fact, however, most of the studies have dealt with college problem drinkers or heavy nonalcoholic drinkers, comparing emotional effects experienced by them with effects experienced by other drinkers.

On the basis of this discussion of approaches to studying the role of personality factors in alcoholism, what kinds of evidence are needed to determine whether or not there is an alcoholic personality? We need first of all evidence concerning personality characteristics that preceded the development of alcoholism. Ideally, this means a longitudinal study, and a study that represents all alcoholics rather than those of one social class or other homogeneous grouping. Secondly, we need evidence that the personality characteristics under consideration are affected by alcohol in some positive way, or evidence as to why people of this type should appreciate the effects of alcohol so much. We would like to show in controlled studies that the amount of alcohol consumed can be affected by manipulating the personality characteristic in question. Third, we need evidence that the alcoholic personality in one culture also exists in other cultures.

Earlier reviews of personality factors in alcoholism have adequately covered psychoanalytic theories of the disorder, as well as the empirical work carried out up to about 1960 (Armstrong, 1958; Blum, 1966; McCord et al., 1960; Syme, 1957; Zwerling and Rosenbaum, 1959). This material will not be covered here in any detail.

In the next sections the present evidence concerning personality factors in alcoholism is laid out, and several theories of the alcoholic personality are discussed and evaluated. The theories are evaluated in terms of the aforementioned types of evidence needed in making inferences about the alcoholic personality. Thus, studies circumventing the various problems noted in studying personality factors in alcoholism are featured. The known studies of prealcoholics and many of the studies of young heavy drinkers are reviewed first, since they form a basis for discussion of two major theories of the alcoholic personality, the dependency theory and the power theory.

STUDIES OF PREALCOHOLICS

There exist only a few studies of prealcoholic populations. One such study was carried out by McCord et al. (1960) on a predominantly lower social class population. In this study ratings were based on direct observation of the male subjects by a number of observers over a five-year period. Subjects averaged nine years of age when the first observations were made; they were last seen, on the average, in their middle teens. It was found that in their childhood, the 29 boys who became alcoholics were more likely than those who did not to be outwardly self-confident, undisturbed by abnormal fears, indifferent toward siblings, and disapproving of their mothers. They were also more likely than those who did not become alcoholics to exhibit unrestrained aggression, sadism, sexual anxiety, and activity rather than passivity.

In another longitudinal study, Robins *et al.* (1962) found that boys and girls appearing at a child guidance clinic who later became alcoholics were more likely than those who did not become alcoholics to exhibit serious antisocial behavior, as evidenced by records of juvenile court appearances. They were also more likely to have a clinic record of a variety of symptoms of antisocial behavior, such as thefts, incorrigibility, physical aggression, truancy, vagrancy, and pathological lying. Future alcoholics did not differ from those who did not become alcoholics on various neurotic symptoms.

A third longitudinal study, by Jones (1968), dealt with a middle-class, male population. Three judges rated material obtained from measurements in class, self-report inventories, observations in natural settings, and parent and teacher ratings. As junior high school students, future alcoholics, compared to those not becoming alcoholics, were rated as more undercontrolled, rebellious, overtly hostile, self-indulgent, expressive, assertive, talkative, more likely to push limits, and as less fastidious.

Loper *et al.* (1973) compared scores on the Minnesota Multiphasic Personality Inventory of 32 college freshmen males later hospitalized for alcoholism with scores of 148 male classmates. Prealcoholics scored significantly higher than their peers on the Psychopathic Deviate and Hypomania scales, but showed few signs of gross maladjustment. Compared to their peers, prealcoholics were more rebellious, nonconforming, impulsive, gregarious, and socially aggressive.

These four studies show close overall agreement in identifying prealcoholics, drawn from both middle- and lower-class populations, as active, aggressive, impulsive, and antisocial. It should be noted that except in one case, these studies dealt exclusively with males. We turn next to evidence from studies of young heavy drinkers and problem drinkers.

STUDIES OF YOUNG HEAVY DRINKERS AND PROBLEM DRINKERS

Youthful heavy drinkers and college problem drinkers constitute high risk populations from which future alcoholics are likely to be drawn. There have been a number of personality studies carried out on these populations.

A college problem-drinking scale for males developed by Park (1962) includes heavy drinking and also a number of behavioral manifestations of drinking problems such as morning drinking, becoming drunk when alone, and blackouts. Park found that problem drinkers had a low frustration threshold and a preference for immediate gratification.

Williams (1970) studied three samples of male college problem drinkers at different colleges, using the Gough and Heilbrun Adjective Check List. It was found that problem drinkers scored high on self-descriptive measures of aggression, autonomy, change (avoidance of routine), lability (inner restlessness), and low on affiliation, intraception (subjective concern for others), nurturance, self-control, deference, order, and endurance. On the basis of these descriptions, it was concluded that college problem drinkers, in comparison with nonproblem drinkers, were aggressive and independent; had a liking for the new and different and a corresponding dislike for consistency and routine; a theme of restlessness, impulsiveness, spontaneity, and action; and a lack of concern with and interest in others.

Zucker and Barron (1973) studied a male, high school population, using measures of both heavy drinking and problem drinking. A similar study of females was conducted by Zucker and Devoe (1974). The major thesis guiding this research was that problem drinking is a subtype of a broader class of antisocial or impulsive behaviors. It was found that heavy drinking and problem drinking among both males and females were highly associated with drug use, heterosexual activities, leaving the field (running away, quitting jobs, breaking dates), serious physical aggression (physical violence to humans and animals), excitement and sensation seeking (thrill seeking, joyriding, gambling, etc.), and measures of distrust, waywardness, and asocial role. In the male population, an overall 59-item measure of antisocial behavior, with alcohol-related items excluded, correlated $+0.71$ with a quantity-frequency index of drinking, and $+0.66$ with problem drinking. Among females, the antisocial behavior measure correlated $+0.57$ with quantity-frequency, and $+0.75$ with problem drinking.

The antisocial behavior theme has also been noted by Jessor et al. (1968a), who found that among high school students, problem drinking was significantly associated with a variety of indicators of deviant behavior; by Kulik et al. (1968), who found that four alcohol-related items loaded on a factor that also included skipping school, riding in a stolen car, participating in gang fights, possession of a switchblade, being late for school, and sexual activity; by Siegman (1966), who found that heavy drinking loaded highly on an antisocial behavior factor in a medical school population; by Wechsler and Thum (1973), who found that heavy drinking among high school students was associated with various antisocial behaviors, including being in fights, stealing, damaging property, and being in trouble with school authorities and police; and by Demone (1973), who found that heavy-drinking high school students manifested a wide variety of deviant actions and tended to have had both formal and informal contacts with law enforcement agencies.

Kalin (1972) undertook a series of studies of heavy-drinking college males, and found 88 items from the California Personality Inventory, the Omnibus Personality Inventory, and the Minnesota Multiphasic Personality Inventory that reliably differentiated heavy-drinking and light-drinking college students in several samples. Factor analysis of these items yielded three main factors. Heavy drinkers were characterized by assertive antisocial behavior ("As a youngster in school I used to give the teachers a lot of trouble," "I used to steal sometimes when I was a youngster"); lack of order ("I am known as a hard and steady worker," "I always like to keep my things neat and tidy and in good order"); and lively social presence ("I am a good mixer," "Other people regard me as a lively individual"). Although heavy drinkers were characterized by the lively social presence theme, items dealing with such activities as belonging to clubs and being a good leader were not related to heavy drinking. Kalin concluded that narcissistic sociability and aggressive ascendancy were the main features of the interpersonal relationships of heavy drinkers, and that their intrapsychic functioning showed impulsivity, or lack of order. Zucker and Devoe (1974) found that antisocial behavior and lack of order, but not lively social presence, were significantly related to quantity–frequency and problem drinking among high school girls.

Kalin (1967) also reported that heavy drinking by college males was associated with dangerous driving and social dancing. Zucker (1967) has noted that male heavy drinkers in a high school population dance more, date more, and drive faster than moderate or light drinkers. In another study, Kalin and Marlowe (1968) found that male heavy drinkers were more likely than moderate or light drinkers to exploit their partners in a game situation.

In other studies, Zucker (1968) reported that heavy-drinking male high school students described themselves as more masculine on the Gough femininity scale than did social drinkers. Blane and Chafetz (1971) did not replicate this finding in a population of male delinquents whose delinquency was connected with drinking, but they did find that delinquents with alcohol involvement described themselves as more independent than delinquents without alcohol involvement on a dependency situations test. Winter (1972) found heavy drinking among male college students to be part of a "stud" cluster of activities including exploitative sexual activities and vicarious power experiences such as reading *Playboy, Sports Illustrated,* or attending athletic performances.

These various studies of heavy drinkers and problem drinkers in high school and college populations have produced remarkably similar findings, as well as indicating that these groups are similar in personality to known prealcoholics. In terms of their self-descriptions and in terms of others' observations of them, prealcoholics, and youthful problem drinkers and heavy drinkers, appear to be action-oriented and exaggeratedly masculine. The major themes appearing throughout these studies are aggression, impulsivity, antisocial behavior, thrill-seeking and restlessness, marked sexual activity, and a seeming lack of concern with and for others combined with an extraverted, sociable nature. Although all but one of the studies of youthful heavy drinkers and problem drinkers were carried out on males, the one such study of females paralleled a study of males and found that personality characteristics manifested by heavy-drinking boys and girls were the same.

THE POWER AND DEPENDENCY THEORIES OF THE ALCOHOLIC PERSONALITY

The power and dependency theories of the alcoholic personality are discussed here in tandem, since the personality characterization of the prealcoholic just presented is incorporated by both theories, but the two theories differ on how this picture should be interpreted, its origins, and why people of this type should be heavy drinkers.

Description of the Power Theory

The power theory was formulated by McClelland *et al.* (1972a) on the basis of ten years of programmatic research. The theory, briefly stated, is that men who have accentuated needs for personalized power drink excessively. The aggression and the assertive thrill-seeking, antisocial activities observed in prealcoholics are indicative of this power concern. These men express power vigorously in attempting to overcome doubts about their own potency; feelings of weakness are suppressed. The concern is with personal dominance over others, with gaining power, glory, or influence without reference to others. The world is pictured as a competitive arena in which man must win out over opponents. Personalized power fantasies were found to increase at high levels of alcohol consumption, and thus drinking may be viewed as a way of feeling stronger, of accentuating the desire to dominate over opponents. It is argued that men with exaggerated needs for personalized power may well receive direct gratification from thoughts such as these at the fantasy level. Such men want power but feel weak; they drink in order to feel powerful.

The origin of heightened personal power needs is unclear. The most likely background factor is a power conflict of some sort: Men are expected to be strong, but are undercut. The power concern is thus compensatory in nature.

Description of the Dependency Theory

According to the dependency theory, the picture of heightened masculinity including aggression, antisocial behavior, and the like seen in prealcoholics is a reaction formation against underlying dependency needs. The prealcoholic has a permanently heightened desire for dependency; he is also ashamed of this need. He wants maternal care, yet wants to be free of this care, and this situation produces a dependency conflict. The strong dependency need is viewed as originating in the childhood situation, a result of behavior of the parents toward the child. Promotion of a facade of self-reliant manhood is seen as an attempt to cover up dependency needs. Drinking is regarded as a masculine activity and thus helps the person to maintain an image of independence and self-reliance. More importantly, however, drinking helps to satisfy dependency needs, by providing feelings of warmth, comfort, and omnipotence, that is, recreating the maternal care situation. The motivation for heavy drinking is to satisfy dependency needs, not to feel powerful.

The particular version of the dependency theory described here was developed by McCord *et al.* (1960). Somewhat different versions of the dependency theory of alcoholism have been presented by Bacon *et al.* (1965), Blane (1968), Knight (1937), Lisansky (1960), White (1956), and Zwerling and Rosenbaum (1959). What these alternate statements have in common is an emphasis on underlying dependency needs as basic to the development of alcoholism.

Evidence for the Power Theory

Evidence for the power theory comes from studies by McClelland and associates, as well as from other research. In a crosscultural study of heavy drinking and drunkenness (McClelland *et al.*, 1966), folk tales were used as a means of determining what people in a culture commonly think about. It was found that heavy drinking societies tended to have folk tales containing themes of physical assertiveness, interpreted as representing a power concern, and inhibition of activity (the word "not," titles of respect, fear–anxiety phrases) was mentioned less often. The same themes were found on the individual level, as manifested in imaginative stories (Thematic Apperception Test) written by working-class men (McClelland and Davis, 1972). This fantasy measure has been found to be a good indicator of what is going on in people's minds, and has been shown to be capable of predicting what actions will be taken in a wide variety of circumstances (McClelland, 1961, 1966). Men with strong power concerns and a low level of restraints in fantasy tended to be heavy drinkers. As noted earlier, the type of power concern found to be typical of heavy drinkers is personal power—seeking to increase one's power, reputation, or glory without reference to others. Personal power is to be distinguished from social power, the latter representing a concern about having more power for the good of some cause, reflected in such activities as office holding. The distinction between personal and social power was uncovered by comparing those combining an elevated need for power in general with lack of activity inhibition, who tend to be heavy drinkers, with those high in need for power and high in activity inhibition, who tend to be light drinkers.

Evidence for the Dependency Theory

The power theory represents a straightforward interpretation of the characteristics noted among prealcoholics and heavy drinkers. They act in a powerful way; they are concerned with power. The dependency theory, on the other hand, concludes the opposite: The power concern is a surface appearance shielding underlying dependency strivings. If the concern is with dependency, and not power, where is the evidence for dependency? How do we really know that underlying dependence is the main feature in alcoholism? As already noted, there is no direct evidence for dependency in prealcoholics. Whenever they describe themselves or are observed, they appear as independent rather than dependent. What kind of evidence is there for the paradoxical conclusion that they are really dependent? The evidence that exists for dependency is necessarily indirect and therefore inferential, and the interpretation of some of this evidence as indicating motivational dependency is open to challenge. Parental background factors in particular are used to infer dependency. For example, McCord et al. (1960) argue that the alternation of rejection and love found to typify mothers of prealcoholics should lead to an exaggerated concern with dependency. Bacon et al. (1965) cite lack of infant indulgence, combined with pressures toward independence and achievement, as contributing to a dependency conflict. Jones' (1968) evidence is that prealcoholics were rated as unable to function comfortably in a dependency relationship. This fact, plus the high value placed on masculine behavior, plus the ambivalence toward authority expressed in rebellious behavior, are presented as evidence for a dependency conflict. None of this evidence is direct and none of it represents conclusive evidence for dependency; in fact, McClelland (1972) argues that the parental background found to typify alcoholics could just as logically lead to a heightened personal power concern. Inability to function comfortably in a dependency relationship could also be interpreted as supporting the power theory.

Perhaps the key is to make a distinction between obvious and subtle measures of personality. If the aggressive outlook of prealcoholics is a facade, it should come through as it does on self-descriptions and observations, but evidence for underlying dependency should show up on the more subtle and indirect projective measures. Unfortunately, there is no information of this kind on prealcoholics, but there is some on young heavy drinkers. Blane and Chafetz (1971), in their study of delinquents with marked alcohol involvement, investigated both conscious dependency, as indicated by responses to a dependency situations test, and unconscious dependency, as indicated by a test of field dependence–independence. It was reported earlier that in this study delinquents with alcohol involvement were significantly more independent than delinquents without alcohol involvement on the dependency situations test. However, they also scored higher on field dependence. Blane and Chafetz concluded that this pattern provides strong support for the dependency theory of alcoholism. However, field dependence is a perceptual phenomenon that involves reliance on external cues. It has not been shown to be related to motivational dependency. McClelland (1972) notes that males are consistently more field independent crossculturally than females. He argues that the greater field dependence of alcoholics suggests doubts concerning their masculinity and a confused sense of identity that could be expected to lead to vigorous expressions of power as a means of overcoming doubts about potency. According to this argument, the field dependency finding may be viewed as supporting the power theory as much as the dependency theory.

Of course, in considering subtle measures, the folk tales and Thematic Apperception Test (TAT) used by McClelland and associates must be included. No evidence was found for dependency needs or dependency conflicts in the folk tales of heavy drinking societies, nor was there evidence for dependency in TATs written by heavy drinkers while in a sober condition. Williams (1964) also failed to find evidence for underlying dependency among college problem drinkers. In this study security-related statements of a sentence-completion test were completed twice, once for self and once for others. The null hypothesis that security, or dependence on others, is of no special concern would be reflected in an equal number of security-related completions in self and other descriptions. If a greater number of security statements are attributed to others, this is an indirect indication that the respondent himself is interested in security relationships. There were, however, no differences between problem and nonproblem drinkers on this measure of underlying dependency.

Power and Dependency Concerns in Alcoholics

Another place to look for evidence for the dependency theory is among alcoholics. The cautions mentioned earlier exist: What is observed in alcoholics may result from the experience of becoming an alcoholic, or from entering into a treatment relationship. Since the treatment situation is by nature a dependency relationship and since alcoholics are generally tested while in this setting, we might well expect to find some evidence for dependency in personality studies of alcoholics.

Instead, there is substantial evidence that alcoholics are similar to prealcoholics and to young heavy drinkers and problem drinkers in being active, aggressive, antisocial, and impulsive, behaviors that are interpretable as reflecting personal power concerns. This evidence comes from the longitudinal studies reviewed earlier for their findings concerning prealcoholics, as well as from other studies of alcoholics. Outside of clinical impressions, there is little evidence for dependency among alcoholics.

First of all, it has been found that personality scales that discriminate between alcoholics and nonalcoholics also discriminate between prealcoholics and nonprealcoholics, and between heavy- and light-drinking college students. For example, MacAndrew (1965) found that 51 items from the MMPI reliably distinguished outpatient alcoholics and nonalcoholic psychiatric outpatients. This finding has been replicated and the 51-item scale has also been found useful in identifying inpatient alcoholics, and in differentiating alcoholics from normals (Rich and Davis, 1969; Rhodes, 1969; Williams *et al.* 1971). Hoffman *et al.* (1974) reported that the MacAndrew scale significantly differentiated prealcoholics and their classmate controls. Fifteen of the 51 MacAndrew items appear on a scale of 88 items found by Kalin (1972) to discriminate between heavy- and light-drinking college students, and heavy-drinking college students scored in the same direction as alcoholics on 37 of the 51 MacAndrew items. There is also some evidence that personality scales that discriminate between heavy- and light-drinking college students also discriminate between alcoholics and nonalcoholics. Williams *et al.* (1971) administered the 88-item Kalin scale to alcoholics and nonalcoholic drinkers and found that both inpatient and outpatient alcoholics were similar in personality to heavy-drinking college students, especially on the antisocial behavior theme.

The MacAndrew items characteristic of alcoholics include the aggressive rebelliousness and impulsivity themes (MacAndrew, 1967). Of the 51 items, it appears that at least 26 are consistent with the power theory, whereas only two appear to support the dependency theory.

Aggression and impulsivity have been widely noted among alcoholics. Impulsivity, which corresponds to the concept of lack of activity inhibition, has been found in numerous studies to characterize alcoholics (Force, 1958; Halpern, 1946; Jones, 1965; Quaranta, 1947). MacAndrew and Geertsma (1963) concluded from a review of the literature that the Psychopathic Deviate scale from the Minnesota Multiphasic Personality Inventory, indicative of antisocial behavior, was the single consistent personality characteristic associated with alcoholism. Jones (1968), in her longitudinal study, found that the same qualities noted in prealcoholics were present among alcoholics, including rebelliousness, direct expression of hostility, and a tendency to "push the limits." McCord *et al.* (1960) also noted unrestrained aggressiveness among alcoholics. Ritson (1971) and Hassall and Foulds (1968) found that alcoholics scored significantly higher than nonalcoholics on measures of hostility.

Alcoholics also appear to have a style of relating to others that was found to characterize prealcoholics, and that is consistent with the personalized power syndrome. Young heavy drinkers were found to display an extraverted, exhibitionistic sociability, but they showed no evidence of the kind of social relations requiring social commitment and long-term involvement. There were indications of a lack of genuine affection or concern for others on the part of heavy drinkers and problem drinkers. There was evidence from studies by Winter (1972) and Kalin (1967) that heavy drinkers tended to exploit others, which is a direct manifestation of personalized power.

A similar pattern has consistently been found in empirical studies and clinical descriptions of alcoholics. Alcoholics tend not to belong to formal groups, which may reflect their lack of concern with socialized power (Jessor *et al.,* 1968a; McCord *et al.,* 1960). Jones (1965) found a pattern of high scores on the sociability scale of the California Personality Inventory, but a low need for affiliation as measured by the TAT. Blane (1968) noted that alcoholics are often engaging individuals, charming companions, and good storytellers, but that this sociability is superficial and rarely occurs within the context of any deeply positive emotional relationship. Force (1958) found that alcoholics were gregarious and exhibitionistic, wanting to stand out, receive fanfare, be in the public gaze. Yet they were asocial, not close in emotional contact, and favor exploitative, "selling" contacts. Shulman (1951) noted as well that alcoholics were characterized by a "salesman" type of personality, although rarely belonging to groups and being isolated. Although exceptions to these findings exist (e.g., Connor, 1962; Hoffman, 1970; Williams *et al.,* 1971), most studies and accounts of the alcoholic have noted his surface sociability combined with lack of social commitment and involvement.

The fact that most studies of alcoholics have uncovered aggression, antisocial tendencies, and other indications of power concerns raises the question of what we should expect to find in alcoholics according to the dependency theory. It turns out that theorizing differs considerably in this regard. For example, McCord *et al.* (1960) see the development of full-blown alcoholism as representing a surrender to dependency, with the facade of intense masculinity breaking down at this point and openly dependent behavior occurring. Lisansky (1960) also holds this point of view. On the other hand, Blane (1968) argues that although underlying dependence is characteristic of alcoholics, they do not necessarily act in a dependent manner. According to Blane, on the basis of his and others' clinical experience, some alcoholics appear to be dependent, some independent, and some in between. Given this situation, it is of course difficult to prove or disprove the dependency theory. One can try to rely on subtle measures, on which any underlying dependency concerns might be expected to appear. Alcoholics, like young delinquents with alcohol involvement, have been found to be characterized by perceptual, or field dependence (Witkin et al., 1959). However, as previously noted, the power theorists claim that this constitutes support for their theory as much as for the dependency theory.

There are some exceptions to the typical findings of aggressive, assertive, antisocial behavior in alcoholic populations. For example, Hoffman et al. (1974) found that alcoholics scored significantly higher than they did as prealcoholics on a scale of feminine interests. McCord et al. (1960) found evidence of overt dependence on the part of alcoholics. Twenty percent of the alcoholics were rated as dependent, compared to 6 percent of the controls. However, when a change in personality is observed from the prealcoholic to the alcoholic stage, the etiological significance of the personality dimensions found to characterize alcoholics is questionable. That is, it is not possible to conclude that underlying dependency is involved in alcoholism on the basis of the finding that alcoholics, who were rated as independent in childhood, are rated as dependent as adults. Were they always basically dependent, with the dependent behavior becoming admitted and apparent only after alcoholism is established? Or, are they really independent people who have failed in life by not being able to control their drinking, and act in a dependent manner as a result of this experience?

The same uncertainty exists in reference to various indications of oral traits that have been found occasionally in alcoholic populations (e.g., Tähkä, 1966; Wolowitz, 1964) and can be interpreted as reflecting dependency concerns. Are they causes or effects of the alcoholic experience? The latter possibility is supported by the finding that among prealcoholics observed by McCord et al. (1960), there were no differences in oral tendencies between them and the nonalcoholic group. On the other hand, Zucker and Fillmore (1968) found that the fantasies of male high school problem drinkers were more often characterized by oral themes than in the case of nonproblem drinkers, and McClelland et al. (1966) found that the folk tales of heavy-drinking societies contained oral themes significantly more often than the folk tales of light-drinking societies.

Alcoholics have also been found to have a strong attachment to their mothers, as evidenced by their agreement to the MMPI item, "The one to whom I was most attached and whom I most admired as a child was a woman" (MacAndrew, 1965). McClelland (1972) notes that this situation, which is indicative of cross-sex identity, is compatible with both the dependency and the power theories. That is, both theories argue that alcoholics have an inadequate masculine identity. The dependency theory sees underlying dependency as the reason for this situation. The power theory sees cross-sex identity, created through attachment with a strong mother figure, as leading to a defensive concern with personal power in a culture that stresses masculine prowess. However, McClelland et al. (1966) reported in their cross-cultural study that heavy drinking was not associated with folk tale themes reflecting cross-sex identity.

Two recent empirical studies also cast doubt on the suppostion that inadequate masculine identity, or more specifically, cross-sex identity, is a background factor in alcoholism. Zucker (1968) used the Gough femininity scale to measure conscious self-representation, and sex-typed preferences for books and movies as a subtle means of tapping underlying sexual identity. Although heavy-drinking high school students portrayed themselves as more masculine on the Gough measure, there was no difference on unconscious identity. Blane and Chafetz (1971) used the Gough femininity scale and the Franck test of unconscious feminine identifications with their delinquent heavy drinkers. The Franck test is a series of half-finished drawings to be completed. Sex differences have been found to exist in the way in which these drawings are completed, and these differences have been incorporated in a coding system. However, no differences were found on either measure in this study.

Effects of Alcohol on Power and Dependency Concerns

The critical data bearing on the power and dependency theories involve effects experienced under alcohol. That is, the question of whether the picture of exaggerated masculinity presented by the prealcoholic reflects a personalized power drive, or is indicative of a repressed need for dependence, is not easily resolved. However, dependency theorists have been explicit in indicating the benefits derived from alcohol by those having a dependency conflict. Alcohol is viewed as temporarily solving the dependency conflict by reducing anxiety created by this conflict, and satisfying infantile unconscious needs (Bacon et al., 1965; McCord et al., 1960; Tähkä, 1966). Alcohol is seen as producing feelings of warmth, comfort, security, acceptance, and omnipotence. As Wolowitz and Barker (1968, pp. 592-593) put it, alcohol provides "... in sum, a soft fuzziness of thinking and perceiving which yields a pleasant sense of unconcerned fulfillment, completeness and serenity."

However, no one has found empirical evidence for dependency concerns under alcohol, in studies of college heavy drinkers and problem drinkers, and adult heavy drinkers, even after very heavy drinking experiences. Nor has evidence for dependency effects been found in the few experimental studies of drinking by alcoholics (Berg, 1971; Pollack, 1965; Vanderpool, 1969). McClelland (1972, p. 279), describing his break with the dependency theory, stated, "We began to have doubts about the theory only when we found that there were no more fantasies of aroused or satisfied oral dependency needs during drinking than during nondrinking periods."

There is, on the other hand, substantial evidence from a number of sources that drinking increases feelings of aggressiveness and strength. In the McClelland studies, the typical pattern was that social power thoughts increased at low levels of alcohol consumption, and personal power fantasies predominated at higher consumption levels. This pattern tended to occur among all drinkers, but there was some evidence to suggest that personal power fantasies increased even more among heavy drinkers than they did for moderate drinkers who happened to drink heavily on that occasion (McClelland and Davis, 1972). Cutter *et al.* (1973) conducted a laboratory study in which alcoholics received either a low or high initial dose of alcohol and could choose an optional second drink ranging from 0 to 4 oz. It was found that both social power and personal power fantasies increased under alcohol, but these increases were not related to amount of alcohol consumed as in the McClelland parties. Inhibition imagery decreased under alcohol. Pretest inhibition scores predicted size of the second drink chosen, but personal power scores did not.

In other studies, Takala *et al.* (1957) found an increase in aggression in TATs written under alcohol, and Tamerin *et al.* (1970) found that alcoholics described themselves as more aggressive during intoxication. Wanner (1972) found that ethnographic accounts of drinking-bout behavior in heavy-drinking primitive societies tended to emphasize hostility, boisterousness, and exhibitionistic behavior. In a study of college problem drinkers, Williams (1968) found that at the 4-oz level, drinkers described themselves on an adjective checklist as more self-centered, self-assured, and less inhibited (exhibition); more concerned with sex (heterosexuality); more impulsive and less cautious (order); more erratic, impatient, and changeable (endurance); more resourceful, self-sufficient, and less dependent on others (succorance). At the 6-oz level and above, the heterosexuality and lack of succorance themes dropped out; exhibition, and lack of order, intraception, and endurance became more pronounced; and new themes appeared—aggression, autonomy, and decreases in nurturance, affiliation, and deference. This is a pattern that has similarities to that found by McClelland and associates. At high levels of alcohol consumption, descriptions became increasingly self-centered as social characteristics tended to drop out. Moreover, problem drinkers in a sober condition were higher on the traits that increase for everyone under alcohol, and lower on the traits that decrease under alcohol. As Williams (1968) noted, the traits characteristic of college problem drinkers appear to be unfavorable ones in terms of getting along with others. These traits become exaggerated for everyone after heavy drinking. Since being drunk brings with it a relaxation of obligations, it may well be that through drinking, problem drinkers attain a state in which they can "be themselves," that is, indulge in self-centered, aggressive behavior, without being so subject to criticism or accountability. The same process is possibly involved in the expression of personal power fantasies and behavior after heavy alcohol consumption. In any case, the available data on drinking effects support the power theory and not the dependency theory. Dependency concerns do not appear in fantasy, in self-description, or in observed behavior under alcohol, but what can be interpreted as power concerns do.

There are very few known studies in which feelings of power or dependency were manipulated, and the effect of these manipulations on subsequent drinking behavior examined. Such studies are important as a means of furnishing evidence of a direct, causal relationship between dependency, or power, and alcohol consumption. In one such study, Davis (1972) aroused feelings of power and nurturance in male drinkers in an experimental task called the Blind Man game. The results of this study are complex, but the men who were made to feel most powerful were the ones who drank the most. Men who were made to feel powerless did not drink heavily in comparison to others, and there was no support at all for the dependency theory. That is, neither those who felt deprived of nurturance nor those who were made to feel nurtured drank more than others. Higgins (1973) and Kosturn and Marlatt (1974) conducted studies of alcohol consumption of drinkers placed in what the authors suggest can be interpreted as powerless positions. In one study (Kosturn and Marlatt, 1974) male and female heavy drinkers were assigned to one of three conditions: an anger with retaliation condition; and anger with no retaliation (i.e., powerless) condition; or a no-anger, no-retaliation control condition. Subjects in the anger with no-retaliation condition drank considerably more wine than those in the anger–retaliation condition, although not significantly more than those in the control condition. Higgins (1973) compared male heavy drinkers in a social evaluation threat condition with drinkers participating in a neutral condition. Males in the evaluation threat condition were told that, as they participated in a wine-tasting task, they would be observed through a one-way mirror by female peers, and that later they would interact with these women who would subsequently be rating the subject on various dimensions of personal attractiveness. Subjects to be so evaluated, possibly interpretable as being in a powerless condition, drank nearly twice as much wine as control condition males.

Status of the Power and Dependency Theories

The dependency theory of alcoholism has enjoyed unchallenged prominence for many years. There have been numerous spokesmen for this theory, and the evidence purportedly supporting the dependency theory comes from a variety of studies. However, there has developed a tendency to accept the theory on faith, as "obviously" correct and to interpret many kinds of borderline or indirect evidence as supporting the theory. In contrast, the power theory was developed by one set of researchers working together over a period of ten years. Unlike the typical one-shot study, the power theory was built up inductively in a program of research that allowed unanswered questions to be pursued, and new questions arising during the course of the research to be addressed.

The power theory, then, represents a challenge to the dependency theory. Although the evidence is not always clear-cut, the known facts appear to fit the power theory better than they do the dependency theory. Thus, the burden of proof is on dependency theorists to demonstrate why the aggression, antisocial behavior, and other characteristics found in prealcoholics and alcoholics should be interpreted as evidence for motivational dependency, rather than taken at face value as in the power theory. It is hoped that in response to the present situation, new research efforts will be undertaken in an attempt to resolve the various issues on which the two theories differ.

It should be clear the power theory has not been validated on the basis of existing data. This theory appears to have considerable explanatory power, although McClelland and associates admit that alternative explanations are possible for some of the data and arguments presented in support of the power theory. Simmons (1972, p. 86) notes, in reference to the power theory, that ". . . the proposition that a concern about male superiority and power is one of the principal causes of alcoholism still remains a hypothesis that now needs systematic testing in a variety of interpersonal, social, and cultural settings to ascertain the extent to which it can retain and enhance its predictive power."

There are a number of unanswered questions about the power theory in its present form. For example, the origins of the power concern said to characterize those who become alcoholics are unclear. The power theorists generally take the stance that the exaggerated power concern is a compensation for felt weakness or lack of power. On the other hand, the available evidence does not support McClelland's (1972) suggestion that cross-sex identity is the main background factor leading to a compensatory power concern. And Davis' (1972) finding that men who felt powerful, rather than men who felt powerless, drank more, is also an impediment to this argument. McClelland (1972), taking note of the mixed evidence, suggested that a personalized power concern may develop in part defensively, and in part offensively. For example, he states that "a successful assertive stance toward life may help create the need for personalized power which is the drinking man's chief concern" (p. 292). But is a successful assertive stance toward life the kind of background factor that would be expected to lead to alcoholism? On a common sense basis, it is doubtful that this is the case, and clearly, additional research is needed on the origins of the personal power concern.

The second consideration related to the power theory concerns the effects of alcohol. As noted earlier, a key element in theorizing about the alcoholic personality is explanation for the symptom choice. What do future alcoholics get out of drinking that leads them to drink excessively and eventually to become alcoholics? Again, the burden of proof is on the dependency theory. The dependency theory holds that drinking is indulged in because it is regarded as a masculine activity, thus supporting the masculine facade of the prealcoholic, but that drinking in fact provides satisfaction of dependency needs. The power theory argues that drinking is regarded as a masculine activity because it actually promotes qualities of exaggerated maleness—aggression, personal power fantasies, and the like. Evidence has been found for the latter qualities, but not for dependency. White (1956) cites the example of the heavy-drinking college student who passed through a belligerent stage at drinking parties into a stage where he would lay his head on a girl's lap and weep piteously for her loving care. This kind of dependent, maternal care-seeking behavior has not been found, however, in observations of drinking bouts, and it does not appear in fantasies or self-descriptions under drinking even up to the 30-oz level.

On the other hand, none of the experimental studies of drinking has involved mixed parties. Would the effects be different when there is opportunity for interaction with women? McClelland and his associates have demonstrated that personal power thoughts increase under high levels of alcohol consumption at stag fraternity cocktail parties, and in all-male barroom drinking sessions by adult males. Yet, there is some concern that not enough typical drinking contexts have been investigated; that personal power concerns might not increase in all heavy drinking contexts, and that dependency concerns might increase in some. Additional research on effects of drinking in various contexts in which heavy drinking naturally occurs is needed in order to clarify this situation.

OTHER THEORIES OF THE ALCOHOLIC PERSONALITY

Jessor's Social Psychological Theory

Jessor and his co-workers have developed a complex social psychological theory of deviance, including excessive alcohol use, taking into account both sociocultural and personality concepts and linkages between them. (Jessor, 1964; Jessor *et al.*, 1968a). The theory is meant to apply to deviance in general, although it has the potential to account for the particular form of deviance chosen. Rotter's (1954) social learning theory of personality is the basis of conceptualizing at the individual level in the Jessor framework. The personality system is treated as being composed of a motivational instigation structure, a belief structure, and a personal control structure. At the motivational level, variables expected to be related to excessive drinking include low expectation of goal attainment (perceived opportunity) in various need areas such as achievement and recognition, and personal disjunctions, defined as the discrepancy between how much the individual values a particular need area such as achievement and the extent to which he expects to be able to attain gratifications in this area. Personal beliefs posited to be related to excessive drinking include external control (the extent to which the individual believes that what happens to him is contingent on external forces such as fate or chance, rather than being under his own control), and alienation (generalized sense of meaninglessness, helplessness, social isolation). Several of the measures at the motivational and beliefs level appear to have some relationship to powerlessness, as treated in the power theory. At the personal control level, tolerance of deviance and low religiosity are expected to be associated with excessive drinking. Drinking is viewed as adaptive behavior, as a means of attaining goals not otherwise attainable, a means of attaining alternative goals that substitute for valued goals not expected to be attainable, and through its physiological effect enabling the drinker to avoid thoughts of failure and inadequacy.

There is modest empirical support for various components of this multivariate theory. Jessor *et al.* (1968a) showed that there was generally correspondence between rates of excessive drinking of three ethnic groups in a community in Southwestern Colorado and the standing of these groups on the various explanatory variables. In surveys of high school students and adults within this community, measures of quantity-frequency and frequency of drunkenness were found to be significantly related to tolerance of deviance, and personal disjunctions in some instances, but not to expectations of goal attainment, internal control, or alienation. In another high school study (Jessor and Jessor, 1973), problem drinking among both boys and girls was significantly related to tolerance of deviance, low expectations of achievement, and to placing a greater value on independence relative to achievement. Jessor and Jessor (1973, p. 22) suggest that "problem drinking may reflect, then, both a means of coping with expected academic failure and an assertion of independence." In this high school study, problem drinking was not significantly related to alienation or religiosity.

In other studies based on the social psychological theory developed by Jessor, it was found that low expectations of need satisfaction in the areas of peer affection and academic achievement were significantly associated with various measures of problem-drinking behavior among college women, but not men (Jessor et al. (1968b). Carman (1971), however, found that among Army servicemen, low expectations for achievement, but not affection, were significantly related to excessive drinking. McClelland et al. (1972b) did not find an association between heavy drinking and personal disjunctions in working-class men.

A crosscultural study of young Italian and American Men conducted by Jessor et al. (1970) illustrates the interaction between personality and sociocultural determinants that is an important element of the Jessor theory. Among American men a linkage was established between excessive drinking and personality attributes reflecting powerlessness, frustration, and alienation: low expectations of goal attainment, alienation, and external control. In addition, a measure of drinking for personal effects featuring problem-solving or coping meanings of drinking was significantly associated both with alienation (but not low expectations or external control) and drunkenness. These relationships were not predicted to occur in the Italian samples, in view of the different socialization and institutionalization of alcohol use in Italian culture, and they did not occur. That is, in Italian culture alcohol is part of family life, diet, and religious custom, factors that make it less likely that drinking will be learned as a way of coping with frustrations and inadequacies at the personality level.

The Jessor theory, which views problem drinking as embedded in a multivariate, social psychological network is much more complex than has been described here. Given its complexity, and the inconsistent findings relating to the personality concepts in the studies reported here, a great deal of further empirical testing is needed before the theory can be evaluated adequately.

The Tension Reduction Theory

A popular theory of alcoholism is that alcohol reduces tension, and that people with acute anxiety or tension drink excessively in order to obtain relief from these symptoms. The tension reduction theory usually stands by itself, but is also a component in some expressions of the dependency theory (cf. Bacon et al., 1965) in which the posited dependency conflict is seen as generating acute anxiety that is relieved by alcohol.

Despite the popularity of the tension reduction theory, it is not well supported by empirical research. Horton (1943), in a crosscultural study, found that hunting tribes drink more heavily than settled agricultural tribes. Horton argued that hunting tribes were less economically secure than agricultural tribes, having a less adequate food supply, and that an insecure food supply produces anxiety. Drinking was seen as a means of reducing subsistence anxiety. The finding that hunting tribes drink more than agricultural tribes has been confirmed by other crosscultural studies (Field, 1962; Barry et al., 1965;

McClelland et al., 1966). However, Field (1962) questioned Horton's interpretation of this finding in terms of anxiety, and in his own study found that drinking was substantially unrelated to various measures of anxiety in a society. McClelland et al. (1966) point out that hunters may well have a more adequate food supply than agriculturalists (Lee and DeVore, 1968), a finding that breaks down the argument linking hunting and subsistence anxiety. McClelland and his co-workers argue that both Horton (1943) and Field (1962) were inferring psychological variables from crosscultural correlations without having direct measures of these states of mind. Using folk tales to get at what the cultural correlates of drinking mean psychologically, McClelland et al. (1966) found, in contrast to Horton (1943), that anxiety was expressed less often in heavy-drinking societies than in light-drinking ones.

There is little data relating to anxiety or tension in studies of prealcoholics or nonalcoholic heavy drinkers, although Williams (1966) found that male college problem drinkers had high scores on an adjective checklist measure of anxiety. There is considerable evidence that alcoholics are characterized by anxiety (e.g., Manson, 1948; Walton, 1968). However, the social isolation and various personal stresses that accompany alcoholism are very likely to foster, or aggravate, this personality characteristic. In fact, it has been found that although prealcoholics were not more anxious or tense than classmate controls, from the prealcoholic to the alcoholic stage there was a general increase in maladjustment, including increases in feelings associated with depression, distress, guilt, inadequacy, and anxiety (Kammeier et al. 1973; Hoffman et al., 1974). Rosenberg (1972) found that alcoholics were more anxious than normals, but less anxious than nonalcoholic psychiatric patients.

Evidence in regard to the hypothesis that alcohol reduces tension has come from animal research (cf. Conger, 1956) and research on human subjects. Williams (1966) found that anxiety as measured via an adjective checklist decreased significantly after the consumption of 4–6 oz of 86-proof alcohol at a fraternity cocktail party, but this trend was reversed at higher consumption levels of 8 oz and above, as anxiety rose back toward predrinking levels. Kalin et al. (1965), on the other hand, found no significant change in fear–anxiety thoughts except at levels of consumption over 9 oz of 86-proof alcohol. Higgins and Marlatt (1973) induced high and low levels of state anxiety in alcoholics and social drinkers by threatening them with either a painful or nonpainful electric shock. The amount of alcohol subsequently consumed was unrelated in both groups to the anxiety-manipulation factor. There is evidence that alcoholics often claim, when sober, that alcohol reduces their tensions (Blume and Sheppard, 1967; MacAndrew and Garfinkel, 1962; Tamerin et al., 1970). However, empirical studies have produced evidence that an increase in anxiety is experienced by alcoholics when they drink (Nathan et al., 1970; Mendelson et al., 1964; McNamee et al., 1968). Cappell and Herman (1972, p. 59), in a thorough review of the experimental evidence that alcohol reduces tension concluded that "negative, equivocal, and contradictory results are quite common if not preponderant."

ALCOHOLISM IN WOMEN

There is a much higher incidence of alcoholism among males than females, yet there are substantial numbers of women alcoholics. Most of the personality studies of alcoholics deal only with men. However, the theories generally conclude that women drink for the same reasons men do. Dependency theory proponents argue that there are fewer female than male alcoholics because women are less likely to experience dependency conflicts (McCord et al., 1960). They have more culturally approved opportunities than men to behave in a dependent manner. Thus, there is less need to resort to alcohol. The power theory argument proceeds along similar lines. Women with exaggerated power concerns are likely to become alcoholics, but since women are less likely than men to be expected to be strong and assertive, they are less likely to develop pronounced personal power concerns. They therefore have a lower incidence of alcoholism.

In contrast to these theoretical formulations, recent data point to the likelihood that women who consciously want to feel womanly, but who have doubts about their adequacy as women, are more likely to become alcoholics. Drinking is rewarding in that it makes them feel more womanly.

Evidence for this formulation comes from two studies. One study of female alcoholics (Wilsnack, 1973) indicated that alcoholics tended to value the maternal role more highly than nonalcoholics, as evidenced by their responses to interview questions. Stylistically, they were more masculine (more likely to agree to statements such as " I was sometimes sent to the principal for cutting up"; "At times I have had to be rough with people who were rude or annoying"), and they had an unconscious masculine identity as revealed by the Franck test. In addition, female alcoholics had experienced more obstetrical-gynecological problems, such as inability to conceive. This situation should raise doubts about femininity. It has often been noted in the literature that women alcoholics, to a greater extent than men, are inclined to initiate excessive drinking in response to a specific incident (Curlee, 1967; Lisansky, 1957). Twenty-four of the 26 alcoholics in the study mentioned such an incident, usually a divorce, obstetrical or gynecological problem, or children leaving the home. These are all factors that may serve to threaten a woman's sense of feminine adequacy.

The findings in the study of women alcoholics just described may in some way have resulted from alcoholism. However, a second study, on younger drinkers, tends to confirm some of these findings, and also indicates the role that drinking may play in the alcoholic process (Wilsnack, 1973). In this study young women aged twenty-one to thirty-two participated in a study of party behavior at which liquor was served. The participants wrote TAT stories in a sober condition and again at the end of these parties. Various coding systems for dependency, power, and womanliness were used in analyzing the TAT stories. The power codes were taken from the McClelland studies; the dependency codes, including resources (giving), needs for support and nurturance, and satisfactions (state of having been helped), were designed by Winter (1966). There were two coding systems for womanliness. The first was the Deprivation-Enhancement code (May, 1966). TAT stories that contain negatively toned events (deprivation) followed by positively toned events (enhancement) are interpreted as reflecting a feminine style, and this code has been shown to differentiate college men and women. The Being Orientation codes (Winter, 1966, 1969), which have been found to differentiate nursing mothers from those who had already weaned their babies, were also used as an indication of womanliness.

The data indicated that there were no changes in any of the dependency measures, or on social power under alcohol, but that drinking *reduced* personal power fantasies, and increased feelings of womanliness. In a sober condition, women who subsequently drank heavily were higher than light or moderate drinkers on personal power, and also scored in a masculine direction on the womanliness codes. This supports the position that women with an insecure feminine identification drink heavily. Like men, women who have exaggerated power concerns are heavy drinkers. But, unlike men, women high on personalized power undergo a reduction in personal power concerns under alcohol, not an increase. The implication is that they drink in order to obtain artificial feelings of womanliness. Although the effects of alcohol as related to power concerns appear to be different for women and men, they are linked in the sense that both heavy-drinking men and women use alcohol to enhance feelings in line with sex role expectations.

REFERENCES

Armstrong, J. D., 1958, The search for the alcoholic personality, *Annals of the American Academy of Political and Social Science* 315:40.

Bacon, M. K., Barry, H., III, and Child, I. L., 1965, A cross-cultural study of drinking. II. Relations to other features of culture, *Quarterly Journal of Studies on Alcohol,* Supplement No. 3:29.

Barry, H., Buchwald, C., Child, I. L., and Bacon, M. K., 1965, A cross-cultural study of drinking: IV. Comparisons with Horton ratings, *Quarterly Journal of Studies on Alcohol,* Supplement No. 3, pp. 62–77.

Berg, N. L., 1971, Effects of alcohol intoxication on self-concept: Studies of alcoholics and controls in laboratory conditions, *Quarterly Journal of Studies on Alcohol* 32:442.

Blane, H. T., 1968, "The Personality of the Alcoholic: Guises of Dependency," Harper and Row, New York.

Blane, H. T., and Chafetz, M. E., 1971, Dependency conflict and sex-role identity in drinking delinquents, *Quarterly Journal of Studies on Alcohol* 32:1025.

Blum, E. M., 1966, Psychoanalytic views of alcoholism: A review, *Quarterly Journal of Studies on Alcohol* 27:259.

Blume, S. B., and Sheppard, C., 1967, The changing effects of drinking on the changing personalities of alcoholics, *Quarterly Journal of Studies on Alcohol* 28:436.

Cappell, H., and Herman, C. P., 1972, Alcohol and tension reduction: A review, *Quarterly Journal of Studies on Alcohol* 33:33.

Carman, R. S., 1971, Expectations and socialization experiences related to drinking among U.S. servicemen, *Quarterly Journal of Studies on Alcohol* 32:1040.

Conger, J. J., 1956, Alcoholism: Theory, problem and challenge. II. Reinforcement theory and the dynamics of alcoholism, *Quarterly Journal of Studies of Alcohol* 17:296.

Connor, R. G., 1962, The self-concepts of alcoholics, *in* "Society, Culture and Drinking Patterns" (D. J. Pittman and C. R. Snyder, eds.) pp. 455–467, John Wiley, New York.

Curlee, J., 1967, Alcoholic women: Some considerations for further research, *Bulletin of the Menninger Clinic* 31:154.

Cutter, H. S. G., Key, J. C., Rothstein, E., and Jones, W. C., 1973, Alcohol, power and inhibition, *Quarterly Journal of Studies of Alcohol* 34:381.

Davis, W. N., 1972, Drinking: A search for power or nurturance? *in* "The Drinking Man: Alcohol and Human Motivation," (McClelland, D. C., Davis, W. N., Kalin, R., and Wanner, E.) pp. 198–213, Free Press, New York.

Demone, H. W., Jr., 1973, The nonuse and abuse of alcohol by the male adolescent, *in* "Proceedings of the Second Annual Conference of the National Institute on Alcohol Abuse and Alcoholism. Psychological and Social Factors in Drinking and Treatment and Treatment Evaluation" (M. E. Chafetz, ed.), pp. 24–32, DHEW Pub. No. (NIH) 74-676, U.S. Government Printing Office, Washington.

Field, P. B., 1962, A new cross-cultural study of drunkenness, *in* "Society, Culture and Drinking Patterns" (D. J. Pittman and C. R. Snyder, eds.) pp. 48–77, John Wiley, New York.

Force, R. C., 1958, Development of a covert test for the detection of alcoholism by a keying of the Kuder Preference Record, *Quarterly Journal of Studies on Alcohol* 19:72.

Halpern, F., 1946, Studies of compulsive drinkers: Psychological test results. *Quarterly Journal of Studies on Alcohol* 6:468.

Hassall, C., and Foulds, G. A., 1968, Hostility among young alcoholics, *British Journal of Addiction* 63:203.

Higgins, R. L., 1973, Manipulation of social evaluation anxiety as a determinant of alcohol consumption, Unpublished doctoral dissertation, University of Wisconsin.

Higgins, R. L., and Marlatt, G. A., 1973, Effects of anxiety arousal on the consumption of alcohol by alcoholics and social drinkers, *Journal of Consulting and Clinical Psychology* 41:426.

Hoffman, H., 1970, Personality characteristics of alcoholics in relation to age, *Psychological Reports* 27:167.

Hoffman, H., Loper, R. G., and Kammeier, M. L., 1974, Identifying future alcoholics with MMPI alcoholism scales, *Quarterly Journal of Studies on Alcohol* 35:490.

Horton, D., 1943, The functions of alcohol in primitive societies: a cross-cultural study, *Quarterly Journal of Studies on Alcohol* 4:199.

Jellinek, E. M., 1952, Phases of alcohol addiction, *Quarterly Journal of Studies on Alcohol* 13:673.

Jessor, R., 1964, Toward a social psychology of excessive alcohol use, *in* "Proceedings Research Sociologists' Conference on Alcohol Problems" (C. R. Snyder and D. R. Schweitzer, eds.) Southern Illinois University, Carbondale.

Jessor, R., and Jessor, S. L., 1973, Problem drinking in youth: Personality, social, and behavioral antecedents and correlates, *in* "Proceedings of the Second Annual Conference of the National Institute on Alcohol Abuse and Alcoholism. Psychological and Social Factors in Drinking and Treatment and Treatment Evaluation" (M. E. Chafetz, ed.), pp. 3–23, DHEW Pub. No. (NIH) 74-676, U.S. Government Printing Office, Washington.

Jessor, R., Graves, T. D., Hanson, R. C., and Jessor, S. L., 1968a, "Society, Personality and Deviant Behavior: A study of a Tri-ethnic Community," Holt, Rinehart, and Winston, New York.

Jessor, R., Carman, R. S., and Grossman, P. H., 1968b, Expectations of need satisfaction and drinking patterns of college students, *Quarterly Journal of Studies on Alcohol* 29:101.

Jessor, R., Young, H. B., Young, H. B., Young, E. B., and Tesi, G., 1970, Perceived opportunity, alienation, and drinking behavior among Italian and American youth, *Journal of Personality and Social Psychology* 15:215.

Jones, M. C., 1965, Correlates and antecedents of adult drinking patterns. Paper presented at Western Psychological Association meeting, Honolulu, June, 1965.

Jones, M. C., 1968, Personality correlates and antecedents of drinking patterns in adult males, *Journal of Consulting and Clinical Psychology* 32:2.

Kalin, R., 1967, Personality and action correlates of heavy drinking among college students, unpublished manuscript, Queens College, Kingston, Ontario.

Kalin, R., 1972, Self-descriptions of college problem drinkers, *in* "The Drinking Man: Alcohol and Human Motivation" (McClelland, D. C., Davis, W. N., Kalin, R., and Wanner, E.), pp. 217–231, Free Press, New York.

Kalin, R., and Marlowe, D., 1968, The effects of intergroup competition, personal drinking habits, and frustration on intragroup competition, *in* "Proceedings of the 76th Annual Convention, American Psychological Association," pp. 405–406, San Francisco, August, 1968.

Kalin, R., McClelland, D. C., and Kahn, M., 1965, The effects of male social drinking on fantasy, *Journal of Personality and Social Psychology* 1:441.

Kammeier, M. L., Hoffman, H., and Loper, R. G., 1973, Personality characteristics of alcoholics as college freshmen and at time of treatment, *Quarterly Journal of Studies on Alcohol* 34:390.

Keehn, J. D., 1970, Neuroticism and extraversion: Chronic alcoholics' reports on effects of drinking, *Psychological Reports* 27:767.

Knight, R. P., 1937, The psychodynamics of chronic alcoholism, *Journal of Nervous and Mental Diseases* 86:538.

Kosturn, C. F., and Marlatt, G. A., 1974, Elicitation of anger and opportunity for retaliation as determinants of alcohol consumption. Paper presented at Western Psychological Association meeting, San Francisco, April 1974.

Kulik, J. A., Stein, K. B., and Sarbin, T. R., 1968, Dimensions and problems of adolescent antisocial behavior, *Journal of Clinical and Consulting Psychology* 32:375.

Lee, R., and DeVore, I. (eds.), 1968, "Man the Hunter," Albine Publishing Company, Chicago.

Levy, R. I., 1958, The psychodynamic functions of alcohol, *Quarterly Journal of Studies on Alcohol* 19:649.

Lisansky, E. S., 1957, Alcoholism in women: Social and psychological concomitants. I. Social history data, *Quarterly Journal of Studies on Alcohol* 18:588.

Lisansky, E. S., 1960, The etiology of alcoholism: The role of psychological predisposition, *Quarterly Journal of Studies on Alcohol* 21:314.

Loper, R. G., Kammeier, M. L., and Hoffman, H., 1973, MMPI characteristics of college freshman males who later became alcoholics, *Journal of Abnormal Psychology* 82:159.

MacAndrew, C., 1965, The differentiation of male alcoholic outpatients from nonalcoholic psychiatric outpatients by means of the MMPI, *Quarterly Journal of Studies on Alcohol* 26:238.

MacAndrew, C., 1967, Self-reports of male alcoholics: A dimensional analysis of certain differences from nonalcoholic male psychiatric outpatients, *Quarterly Journal of Studies on Alcohol* 28:43.

MacAndrew, C., and Garfinkel, H., 1962, A consideration of changes attributed to intoxication as common-sense reasons for getting drunk, *Quarterly Journal of Studies on Alcohol* 23:252.

MacAndrew, C., and Geertsma, R. H., 1963, An analysis of responses of alcoholics to scale 4 of the MMPI, *Quarterly Journal of Studies on Alcohol* 24:23.

Machover, S., and Puzzo, F. S., 1959, Clinical and objective studies of personality variables in alcoholism. I. Clinical investigation of the "alcoholic personality," *Quarterly Journal of Studies on Alcohol* 20:505.

Manson, M. P., 1948, A psychometric differentiation of alcoholics from nonalcoholics, *Quarterly Journal of Studies on Alcohol* 9:175.

May, R., 1966, Sex differences in fantasy patterns, *Journal of Projective Techniques and Personality Assessment* 30:576.

McClelland, D. C., 1961, "The Achieving Society," Van Nostrand, Princeton, New Jersey.

McClelland, D. C., 1966, Longitudinal trends in the relation of thought to action, *Journal of Consulting Psychology* 30:479.

McClelland, D. C., 1972, Examining the research basis for alternative explanations of alcoholism, *in* "The Drinking Man: Alcohol and Human Motivation" (McClelland, D. C., Davis, W. N., Kalin, R., and Wanner, E.), pp. 276–315, Free Press, New York.

McClelland, D. C., and Davis, W. N., 1972, The influence of unrestrained power concerns on drinking in working-class men, *in* "The Drinking Man: Alcohol and Human Motivation" (McClelland, D. C., Davis, W. N. Kalin, R., and Wanner, E.), pp. 142–161, Free Press, New York.

McClelland, D. C., Davis, W., Wanner, E., and Kalin, R., 1966, A cross-cultural study of folktale content and drinking, *Sociometry* 29:308.

McClelland, D. C., Davis, W. N., Kalin, R., and Wanner, E., 1972a, "The Drinking Man: Alcohol and Human Motivation," Free Press, New York.

McClelland, D. C., Wanner, E., and Vanneman, R., 1972b, Drinking in the wider context of restrained and unrestrained assertive thoughts and acts, *in* "The Drinking Man: Alcohol and Human Motivation" (McClelland, D. C., Davis, W. N., Kalin, R., and Wanner, E.), pp. 162–197. Free Press, New York.

McCord, W., McCord, J., and Gudeman, J., 1960, "Origins of Alcoholism," Stanford University Press, Stanford California.

McNamee, H. B., Mello, N. K., and Mendelson, J. H., 1968, Experimental analysis of drinking patterns of alcoholics: Concurrent psychiatric observations, *American Journal of Psychiatry* 124:1063.

Mendelson, J. H., La Dou, J., and Solomon, P., 1964, Experimentally induced chronic intoxication and withdrawal in alcoholics: Pt. 3. Psychiatric findings, *Quarterly Journal of Studies on Alcohol,* Supplement No. 2, pp. 40–52.

Nathan, P. E., Titler, N. A., Lowenstein, L. M., Solomon, P., and Rossi, A. M., 1970, Behavioral analysis of chronic alcoholism, *Archives General Psychiatry* 22:419.

Park, P., 1962, Problem drinking and role deviation, *in* "Society, Culture and Drinking Patterns" (D. J. Pittman and C. Snyder, eds.), pp. 431–454, Wiley, New York.

Park, P., 1973, Developmental ordering of experiences in alcoholism, *Quarterly Journal of Studies on Alcohol* 34:473.

Partington, J. T., 1970, Dr. Jekyll and Mr. High: Multidimensional scaling of alcoholics' self-evaluations, *Journal of Abnormal Psychology* 75:131.

Pollack, D., 1965, Experimental intoxication of alcoholics and normals: Some psychological changes, Unpublished doctoral dissertation, University of California at Los Angeles.

Quaranta, J. V., 1947, Alcoholism: A study of emotional maturity and homosexuality as related factors in compulsive drinking, Unpublished doctoral dissertation, Fordham University.

Rhodes, R. J., 1969, The MacAndrew alcoholism scale: A replication, *Journal of Clinical Psychology* 25:189.

Rich, C. C., and Davis, H. G., 1969, Concurrent validity of MMPI alcoholism scales, *Journal of Clinical Psychology* 25:425.

Ritson, B., 1971, Personality and prognosis in alcoholism, *British Journal of Psychiatry* 118:79.

Robins, L. N., Bates, W. M., and O'Neal, P., 1962, Adult drinking patterns of former problem children, *in* "Society, Culture and Drinking Patterns" (D. J. Pittman and C. R. Snyder, eds.), pp. 395–412, John Wiley, New York.

Rosenberg, N., 1972, MMPI alcoholism scales, *Journal of Clinical Psychology* 28:515.

Rotter, J. B., 1954, "Social Learning and Clinical Psychology," Prentice-Hall, Englewood Cliffs, New Jersey.

Shulman, A. J., 1951, Alcohol addiction, *U. of Toronto Medical Journal* 28:219.

Siegman, A. A., 1966, Father absence during early childhood and antisocial behavior, *Journal of Abnormal Psychology* 71:71.

Simmons, O. G., 1973, Discussion: Alcohol and human motivation, *in* "Proceedings of the Second Annual Conference of the National Institute on Alcohol Abuse and Alcoholism. Psychological and Social Factors in Drinking and Treatment and Treatment Evaluation" (M. E. Chafetz, ed.), pp. 84–91, DHEW Pub. No. (NIH) 74-676, U.S. Government Printing Office, Washington.

Syme, L., 1957, Personality characteristics of the alcoholic, *Quarterly Journal of Studies on Alcohol* 18:288.

Tähkä, V., 1966, "The Alcoholic Personality," Finnish Foundation for Alcohol Studies, Helsinki, Finland.

Takala, M., Pihkynen, T. A., and Markkanen, T., 1957, "The Effects of Distilled Spirits and Brewed Beverages," Suomalaisen Kirjallisuden Kirjapaino, Helsinki.

Tamerin, J. S., Weiner, S., and Mendelson, J. H., 1970, Alcoholics' expectancies and recall of experiences during intoxication, *The American Journal of Psychiatry* 126:1697.

Vanderpool, J. A., 1969, Alcoholism and the self-concept, *Quarterly Journal of Studies on Alcohol* 30:59.

Walton, H. J., 1968, Personality as a determinant of the form of alcoholism, *British Journal of Psychiatry* 114:761.

Wanner, E., 1972, Power and inhibition: A revision of the magical potency theory, *in* "The Drinking Man: Alcohol and Human Motivation" (McClelland, D. C., Davis, W. N., Kalin, R., and Wanner, E.), pp. 73–98, Free Press, New York.

Wechsler, H., and Thum, D., 1973, Alcohol and drug use among teen-agers: a questionnaire study, *in* "Proceedings of the Second Annual Conference of the National Institute on Alcohol Abuse and Alcoholism. Psychological and Social Factors in Drinking and Treatment and Treatment Evaluation" (M. E. Chafetz, ed.), pp. 33–46, DHEW Pub. No. (NIH) 74-676, U.S. Government Printing Office, Washington.

White, R. W., 1956, "The Abnormal Personality," Ronald Press, New York.

Williams, A. F., 1964, Psychological effects of alcohol in natural party settings, Unpublished doctoral dissertation, Harvard University.

Williams, A. F., 1966, Social drinking, anxiety, and depression, *Journal of Personality and Social Psychology* 3:689.

Williams, A. F., 1968, Psychological needs and social drinking among college students, *Quarterly Journal of Studies on Alcohol* 29:355.

Williams, A. F., 1970, College problem drinkers: A personality profile, in "The Domesticated Drug: Drinking Among Collegians" (G. L. Maddox, ed.), pp. 343–360, College and University Press, New Haven, Connecticut.

Williams, A. F., McCourt, W. F., and Schneider, L., 1971, Personality self-descriptions of alcoholics and heavy drinkers, *Quarterly Journal of Studies on Alcohol* 32:310.

Wilsnack, S. C., 1973, The needs of the female drinker: Dependency, power or what? in "Proceedings of the Second Annual Conference of the National Institute on Alcohol Abuse and Alcoholism. Psychological and Social Factors in Drinking and Treatment and Treatment Evaluation" (M. E. Chafetz, ed.), pp. 65–83.

Winter, D. G., 1972, The need for power in college men: Action correlates and relationship to drinking, in "The Drinking Man: Alcohol and Human Motivation" (McClelland, D. C., Davis, W. N., Kalin, R., and Wanner, E.), pp. 99–119, Free Press, New York.

Winter, S. K., 1966, Being orientation in maternal fantasy: A content analysis of the TATS of nursing mothers. Unpublished doctoral dissertation, Harvard University.

Winter, S. K., 1969, Characteristics of fantasy while nursing, *Journal of Personality* 37:58.

Witkin, H. A., Karp, S. A., and Goodenough, D. R., 1959, Dependence in alcoholics, *Quarterly Journal of Studies on Alcohol* 20:493.

Wittman, P., 1939, Developmental characteristics and personalities of chronic alcoholics, *Journal of Abnormal and Social Psychology* 34:361.

Wolowitz, H. M., 1964, Food preferences as an index of orality, *Journal of Abnormal and Social Psychology* 69:650.

Wolowitz, H. M., and Barker, M. J., 1968, Alcoholism and oral passivity, *Quarterly Journal of Studies on Alcohol* 29:592.

Zucker, R. A., 1967, Social drinking and sentient behavior in adolescent boys-individual differences. Paper presented at Eastern Psychological Association meetings. Boston, April, 1967.

Zucker, R. A., 1968, Sex-role identity patterns and drinking behavior of adolescents, *Quarterly Journal of Studies on Alcohol* 29:868.

Zucker, R. A., and Barron, F. H., 1973, Parental behaviors associated with problem drinking and antisocial behavior among adolescent males, in "Proceedings of the First Annual Conference of the National Institute on Alcohol Abuse and Alcoholism. Research on Alcoholism: Clinical Problems and Special Populations" (M. E. Chafetz, ed.), pp. 276–296, DHEW Pub. No. (HSM) 73-9074, U.S. Government Printing Office, Washington.

Zucker, R. A., and Devoe, C. I., 1974, Life history characteristics associated with problem drinking and antisocial behavior in adolescent girls: A comparison with male findings, in "Life History Research in Psychopathology V. 4 (M. Roff, R. Wirt, and G. Winokur eds.), U. of Minnesota Press, Minneapolis.

Zucker, R. A., and Fillmore, K. M., 1968, Motivational factors and problem drinking among adolescents. Paper read at symposium on Educational Implications of Recent Adolescent Drinking Research, 28th International Congress on Alcohol and Alcoholism, Washington, September, 1968.

Zwerling, I., and Rosenbaum, M., 1959, Alcohol addiction and personality, in "American Handbook of Psychiatry" (S. Arieti, ed.), pp. 623–644, Basic Books, New York.

CHAPTER 8

Alcoholism and Mortality

Jan de Lint and Wolfgang Schmidt

Addiction Research Foundation
33 Russell Street
Toronto, Ontario, Canada

INTRODUCTION

Numerous epidemiological investigations have demonstrated quite convincingly that man's life-style in all its many facets has a profound effect on his mode of dying (Sigerist, 1943; Rubin, 1960; Polgar, 1962; Paul, 1963; Scotch, 1963). Within this context the role of alcoholism in mortality has been amply explored. However, research on this question has been somewhat uneven in character. For example, French epidemiologists have traditionally focused on the effects of chronic excessive alcohol use on mortality, whereas in North American and Scandinavian countries the literature on the sequelae of acute intoxication is extensive. Similarly, the association between alcoholism and such causes of death as suicide and cirrhosis of the liver has received far more attention than the contribution of alcoholism to death from arteriosclerotic heart disease.

This chapter describes some of the major difficulties in the epidemiology of alcoholism-related mortality. It also seeks to sketch to the reader the current state of knowledge in this area of investigation. However, it is not intended to be an exhaustive review of all the relevant prospective, retrospective, or coincidence studies. In our opinion such reviews are more useful when done within a more narrow frame of reference, e.g., on the role of alcohol consumption in death from liver cirrhosis.

THE DEFINITION OF ALCOHOLISM

One of the major difficulties in the epidemiology of alcoholism-related mortality is to define alcoholism in a useful and meaningful way. Its manifestations and consequences, such as the craving for alcohol, the inability to stop drinking at some or at many of the drinking occasions, the repetitive intake of alcohol usually in large quantities, the damage done to health or to society, are all too ambivalent to allow us to identify each and every alcoholic in any given population. How much craving, loss of control under what kind of circumstances, which quantity of alcohol consumed and in what frequency, what kind of damage to health or to society, should define alcoholism? Perhaps the elusive nature of this problem can best be illustrated by the alcohol consumption distribution curve that shows the frequency distribution of all the alcohol consumers in a population according to the quantity of alcohol each of them consumes.

For a wide variety of populations it has been shown that this distribution approximates a logarithmic normal curve as described by S. Ledermann (Ledermann, 1956; de Lint and Schmidt, 1968; Mäkelä, 1971; de Lint, 1973). For example, in a population with an average yearly consumption of 25 liters of absolute alcohol, the distribution curve would be as shown in Figure 1.

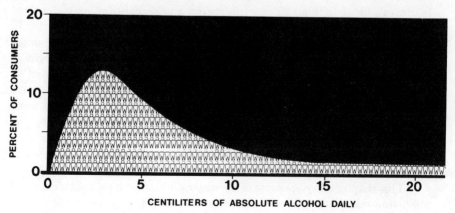

FIGURE 1. Frequency distribution of alcohol consumption ($\bar{c} = 25l$).

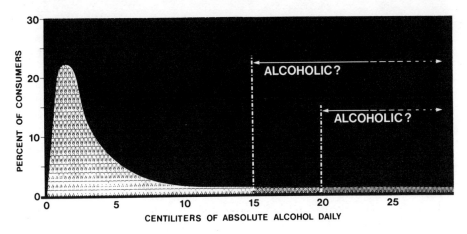

FIGURE 2. Frequency distribution of alcohol consumption ($\bar{c} = 151$).

It can be presumed that alcoholics, by almost any definition, are located somewhere in the upper ranges of the alcohol consumption distribution curves. But how to separate on the basis of average daily alcohol consumption quantities the alcoholic from the nonalcoholics? For example, is a person consuming in excess of a daily average of 15 centiliters of absolute alcohol an alcoholic or must he drink at least 20 centiliters (Figure 2) to qualify?

The results of some investigations of self-reported alcohol consumption behaviors of patients admitted to alcoholism clinics would seem to offer a solution. These findings do suggest that the condition of such patients tends to be associated with a range of alcohol consumption in excess of a daily average of about 15 centiliters of absolute alcohol (Schmidt and de Lint, 1970; de Lint, 1973). However, such a definition may not do justice to the complexity of this problem.

Thus, it can be assumed that many of the drinkers in the tail-end of the alcohol consumption distribution curve probably have sustained some damage to health as a consequence of their alcohol use, that they may exhibit loss of control during some or many of their drinking occasions, that they may experience family or job problems, have a tendency to neglect their health and proper nutrition, probably smoke heavily, and may have severe emotional difficulties. In addition, their chronic excessive alcohol consumption may well have resulted in the appearance of certain signs of physical dependence such as increased tolerance to the effects of alcohol and withdrawal symptoms.

The problem of defining drinkers in the tail-end of the distribution curve as alcoholics is further complicated by the fact that many of these behaviors and conditions are present to varying degrees. For instance, at what point can one speak of alcohol-related physical damage, or of loss of control? What is needed before it can be argued that job performance is impaired? When is a person really negligent with respect to his personal health care and nutrition? How are emotional problems defined?

Evidently, at lower levels of consumption, drinkers do exhibit fewer of the alcoholism-related behaviors and conditions and to a lesser degree than at higher levels of consumption. Conversely, drinkers of very large amounts of alcohol probably exhibit quite noticeably many alcoholism-related behaviors and conditions and would undoubtedly be identified as alcoholics by any other definition.

Because of the elusive nature of the alcoholism problem, epidemiologists working in the field of alcoholism-related mortality use a variety of operational definitions (retrospective and prospective studies) or apply indices of the magnitude of the problem (in correlational studies).

COINCIDENCE STUDIES

Typically, in these studies rates of alcohol use are compared to rates of death from causes suspected to be alcoholism-related. It has been amply documented that the overall level of alcohol use in a population is closely related to the rate of excessive use (Figure 1). It is also evident that such a rate ought to reflect the prevalence of organic damage attributable to excessive use as well as the prevalence of certain manifestations of physical dependence to some degree. However, with reference to other aspects of alcoholism, e.g., the loss of control, the emotional disturbance, the social damage and the organic damage not attributable to alcohol use per se but to habits and conditions such as cigarette smoking, exposure to cold, and neglect of health care, it should be recognized that the prevalence of these conditions may not correlate well with rates of excessive alcohol use and with overall levels of alcohol consumption. Thus, where a rate of alcohol use is employed as an index of alcoholism, it should be noted that certain aspects of the conglomerate of behaviors and conditions labeled alcoholism may be better reflected by this index than are others.

Coincidence studies can also be criticized on more specific grounds. For instance, the data employed in these studies typically comprise information routinely collected by government agencies, e.g., alcohol sales data and vital statistics. It is quite conceivable that the accuracy and/or usefulness of these data vary from jurisdiction to jurisdiction or from period to period. For instance, in countries with little illicit production an increase in the volume of alcohol beverages sold can be taken to suggest an increase in the consumption of alcoholic beverages. However, in countries with a significant illicit alcohol production, it would be more difficult to interpret a rise in the alcohol sales volume. Or, the officially reported rates of death from suicide, liver cirrhosis, and alcoholism may not be equally accurate from population to population or from period to period. Also, the observed covariance between consumption and mortality is at times difficult to interpret. For instance, in the case of the many correlations in time and space between alcohol consumption and liver cirrhosis mortality (Jolliffe and Jellinek, 1941; Brésard, 1959; Seeley, 1960; Schmidt and Bronetto, 1962; Bättig, 1967; Terris, 1967; Egoz et al., 1970; Popham, 1970; Wallgren et al., 1970; Tokuhata et al., 1971), it has been argued that variation in such factors as presence of industrial toxins, the incidence of hepatitis, the quality of diagnostic techniques may be in part or in whole responsible for these findings. On the other hand, it is difficult to conceive that all these factors show the same peculiar variation over time, space, or occupational group as do cirrhosis mortality and alcohol consumption (Terris, 1967).

One of the earliest coincidence studies was done by Bandel (1928). He observed that the trend in excess male mortality (death rate of males less death rate of females) in the forty to sixty-nine age group covaried with annual beer consumption for Bavaria 1867–1913. For Prussia, 1875–1927, he observed that male excess mortality in the forty-to-fifty age group was closest associated with variation in annual alcohol consumption (distilled spirits and wine) (Bandel, 1930). And, for Italy, 1904–1925, he noted that yearly production of wine was reflected in excess male mortality in age group forty to fifty-nine (Bandel, 1932). Ledermann (1964) extended the Italian series to the year 1958 and found that the excess male mortality in this age group and wine production continued to covary. For France, Delaporte (1941) and Ledermann (1964) compared the yearly rate of male deaths (per 10,000) and the excess male mortality in this age group to the annual consumption in liters of absolute alcohol over the period 1840–1960. Again a close relationship between these variables obtained. Ledermann discusses the difficulties pertaining to the use of male excess mortality in these analyses. He argues that undoubtedly many other differences between men and women other than drinking habits may be relevant, e.g., war, occupational hazards, celibacy, childbirth deaths, biological factors, smoking habits. However, he suggests that these factors should affect male excess mortality in many Western countries. But he observes that this excess mortality is highest in countries with high alcohol consumption. He concludes that these other factors that may perhaps explain excess male mortality should therefore be discarded.

With reference to the relationship between alcohol use and cause-specific mortality (specifically deaths from liver cirrhosis and alcoholism), many series of observations exist. Already Bandel (1930) had compared trends in the average annual consumption of distilled spirits and wine and death from alcoholism in Prussia, 1875–1927, and found considerable covariance. Ledermann (1964) showed similar data for France. In addition many temporal and spatial correlations between death from liver cirrhosis and alcohol consumption exist (Jolliffe and Jellinek, 1941; Brésard, 1959; Seeley, 1960; Schmidt and Bronetto, 1962; Bättig, 1967; Terris, 1967; Egoz et al., 1970; Popham, 1970; Wallgren et al., 1970; Tokuhata et al., 1971).

With respect to mortality patterns of different occupational groups, Bertillon (1874) reported that innkeepers and vendors of alcoholic beverages have the highest rate of death in the age group fifteen to fifty-four years and the second highest rate of death in the age group fifty-five to sixty-four (after the mine workers). Similar observations were made for Switzerland 1879–1882, England 1880–1882, France 1885–1889 and 1907–1908 (Bertillon, 1889; Huber, 1912). As can be expected specifically, death from liver cirrhosis occur much more frequently in these occupations (Febvay and Aubenque, 1957; Bronetto, 1964; Terris, 1967).

RETROSPECTIVE STUDIES

Probably the most frequently used method in the field of alcohol-related mortality and morbidity is the so-called retrospective method, where the investigator samples persons who died from a cause or are suffering from a disease suspected to be related to alcohol use and then proceeds to collect information about their life-styles preceding the death or the onset of the disease. One major difficulty with this approach concerns the reliability and validity of such information: is it really possible to obtain from a patient in a hospital or from the relatives of a deceased person good information on the quantity and frequency of his alcohol consumption over a period of several years? Another difficulty refers to the considerable variation in these investigations in their definition of excessive alcohol use and of alcoholism. For instance, in a random selection of recently published retrospective studies heavy drinkers had been defined as daily consumers in excess of 4, 5, 5.5, 6, 10, or 12.5 centiliters of absolute alcohol, respectively (Ratnoff and Patek, 1942; Brown and Campbell, 1961; Vincent and Marchetta, 1963; Zeller, 1967; Kok-Jensen, 1970; Lischner et al., 1971). Chronic alcoholics had been defined as persons who had a history of heavy alcohol intake, who had sustained severe organic damage due to drinking, who had lost control over their drinking, who fitted the Keller definition of alcoholism, or who had answered positively three questions on the Jackson Scale of Preoccupation with Alcohol (Gsell and Löffler, 1962; Palola et al., 1962; Murphy and Robins, 1967; Burch and Ansari, 1968; Viel et al., 1968).

As a consequence it is not surprising that estimates of the extent to which death or disease from a specific cause are attributable to excessive alcohol use or to "chronic alcoholism" vary considerably. For instance, in the case of recent retrospective studies of patients suffering from cirrhosis of the liver, the estimates of alcoholism involvement range from 18 percent to 89 percent (Table 1).

Similar wide variation in p values (the percentage of excessive drinkers and "alcoholics" in the sample) have been observed in patients suffering from cardiovascular diseases (Brigden, 1957; Evans, 1959; Brigden and Robinson, 1964; Massumi et al., 1965; Alexander, 1966; Tobin et al., 1967) and in persons who attempted suicide (Batchelor, 1954; Schmidt et al., 1954; Robins et al., 1957; Stengel and Cook, 1958; Palola et al., 1962; James et al., 1963; Harenko, 1967; Mayfield and Montgomery, 1972). No doubt the variation in p value found in these studies is in part attributable to variation in rates of alcoholism. However, there are many other factors to be considered.

As mentioned earlier, this variation is in part explained by differences in

TABLE 1. Estimated Percentages of Chronic Excessive Alcohol Users in Samples of Patients with a Diagnosis of Liver Cirrhosis

Investigator	Country	Year	Percentage chronic excessive alcohol users
Garceau	U.S.A	1964	89
Reid	U.S.A.	1968	84
Caroli	France	1958	83
Sepúlveda	Mexico	1956	82
Péquignot	France	1958	80
Brick	U.S.A.	1963	75
Mussini	Italy	1960	69.5
Fauvert	France	1958	65
Forshaw	England	1972	63
Rubin	U.S.A.	1962	61
Rajasuriya	India	1970	55
Gigglberger	Germany	1968	48
Gros	Germany	1955	44
Čerlek	Yugoslavia	1968	40
Creutzfeldt	Germany	1966	39
Kaeding	Germany	1955	36
Hällen	Sweden	1963	34
Stone	England	1968	33.5
Sundberg	Finland	1957	29
Chapman	Australia	1966	28
Falck	Germany	1956	26
Čerlek	Yugoslavia	1968	24
Hartmann	Germany	1956	22.5
Falck	Germany	1962	19.5
Sherlock	England	1955	18

the definition of excessive alcohol use and alcoholism employed in these investigations and by differences in the reliability of the responses between samples. In addition, many aspects of life-style and morbidity vary over time and space, and therefore one should not compare the apparent role of alcoholism in samples of diseased persons taken from quite different periods and jurisdictions. In trying to assess this role, the problem of estimating the expected frequency of alcoholism or excessive drinking should be carefully considered. If one finds that of a group of lower social class male patients suffering from pneumonia and aged between forty and sixty, 12 percent are alcoholics on the basis of life history data, does that percentage exceed the expected frequency of alcoholism for males of similar age and social class? Quite frequently little attention is paid in retrospective studies to the problem of expectancies.

Finally, many retrospective studies of patients suffering from certain diseases or exhibiting conditions suspected to be alcoholism-related exist, but few investigations have been done of the life histories of persons who have died from such causes. Most of the latter are studies of persons who committed suicide (Dahlgren, 1945; Sainsbury, 1955; Robins *et al.*, 1959; Dorpat and Ripley, 1960; Palola *et al.*, 1962; James, 1966; Žmuc, 1968). Some others examined antecedent factors in death from liver cirrhosis (Evans and Gray, 1938; Bruun *et al.*, 1960; Popper *et al.*, 1960; Lee, 1966; Kramer *et al.*, 1968; Kuller *et al.*, 1969; Sakurai, 1969).

PROSPECTIVE STUDIES

A third approach is to sample groups of alcoholics and to compare their subsequent mortality with that of the population at large. In some such studies, persons who have been treated for alcoholism are sampled; in others, persons frequently arrested for intoxication in public or persons with alcohol problems identified by some other methods are selected. These samples are then followed for varying length of time, their general and cause-specific rate of death are tabulated and usually compared with the mortality experience of the general population (alcoholics included). One difficulty with this approach is that the observations pertain to persons who have come to the attention of clinics, mental hospitals, or law enforcement agencies because of alcohol problems. However, there may still be an unknown number of persons who consume large amounts of alcohol over long periods of time who do not become identified as alcoholics in this way. Another difficulty is related to the use of vital statistics for the calculation of expected frequencies. These statistics refer to general populations that always include alcoholics. This procedure inflates the expected frequencies of those causes to which alcoholics contribute more than nonalcoholics.

One of the earliest studies of this type was reported by Gabriel (1935), who analyzed the mortality experience of 1109 alcoholics who had been treated in an alcoholism clinic attached to a mental hospital in Vienna. These patients were followed for an average of 6 to 7 years with a maximum of 13 years and a minimum of 1 year. The number of deaths in the sample exceeded expectations in the different age groups from 6 to 20 times. However, close examination of the data revealed errors in the calculation of the expected frequencies. On the basis of the information given in this report, it would seem that the observed rate of death in the sample as a whole was only about twice as high as the expectancy based on the mortality in the general population.

In the state of California, Brenner (1967) followed 1343 alcoholics for a total of 7289 man years. These patients were selected from mental hospitals, an outpatient clinic, and a prison farm. The average death rate observed in this sample was three times the expectation. In Norway, Sundby (1967) in an extensive study of 1722 male patients who were discharged from a psychiatric hospital in Oslo during the years 1925–39 with a diagnosis of alcoholism and followed for a total of 34,951 man years, observed a mortality rate that was 113 percent higher than the expectation based upon Norwegian male mortality. In Switzerland, Ciompi and Eisert (1969) reported a follow-up study concerning the mortality among 1468 alcoholics hospitalized in a psychiatric clinic in the Canton de Vaud. Eighty percent of these patients were admitted during the years 1923–1962. The remainder was treated prior to 1923. All patients were born prior to 1897. The comparison with the expectancies based on mortality rates in the general Swiss population showed a reduction of life expectancy in the sample of about 15 percent. To obtain a reduction of this magnitude requires a death rate in the sample of about twice the expected rate of death.

Kendell and Staton (1966) followed 62 alcoholics who had been assessed in the outpatient department of the Maudsley Hospital in London, England, between 1950 and 1961 but had subsequently either refused treatment or been rejected as unsuitable. The period of follow-up varied from 2 to 13 years with a mean of 6.7 years. Eleven were found to have died; this mortality rate is five times higher than the rate expected on the basis of the mortality in the general population.

Vincent and Blum (1969) reported a five-year follow-up of 128 alcoholic patients of a private sanitarium in Ontario, Canada. Of these patients 15 had died during this period. Although the ratio observed over expected number of deaths in this sample has not been given, our estimate of this ratio—based on the age distribution of the sample and the mortality rate of the Ontario population—is slightly more than two.

Schmidt and de Lint (1972) determined the rate of mortality of 6478 alcoholic patients who had been treated in the in- and outpatient facilities of the Addiction Research Foundation of Ontario, Canada, between 1951 and 1963. The follow-up period ranged from 1 to 14 years and comprised 41,145 man years. The mortality rate in the sample was 2.1 times higher than the rate in the general Ontario population.

Gillis (1969) described the mortality of 802 patients treated in the alcoholism unit of the Groote Schuur Hospital in Cape Town, South Africa, between 1959 and 1963. The follow-up appeared to cover slightly more than 3000 man years. Ninety-five patients died during the period covered, which corresponds to a ratio of observed over expected deaths of about four. In Denmark, Nørvig and Nielsen (1956) followed up 221 alcoholic patients treated in a municipal hospital in Copenhagen between 1948 and 1950. The average length of follow-up was four years. Forty-two died during this period. Undoubtedly the observed number of deaths in this study exceeded the expected number, but the data given are insufficient to calculate the extent of excess mortality. Fitzgerald *et al.* (1971) reported the mortality experience of 531 alcoholic inpatients at the Wyoming State Hospital, U.S.A. These patients were treated between 1961 and 1965 and followed up for an average of four years. Forty-one had died during this period. In view of the many cases lost from follow-up and the lack of data essential for the calculation of the ratio observed over expected deaths, it is not possible to determine excess mortality. However, it is certain that the observed number of deaths in this study exceeded the expected number considerably.

In Sweden, Dahlgren (1951) analyzed the rates of death for 10,588 men who were recorded by the Temperance Boards of Malmö and the province of Malmöhus from 1939 to 1947. The follow-up period ranged from four months to nine years. During this period 564 cases of death were observed, which is slightly less than the frequency expected on the basis of the mortality in a comparable segment of the general population. Thus, no increased mortality could be established in this sample of alcoholics.

Also in Sweden, Salum and coworkers (1972) examined the mortality of male alcoholics who had been treated in a psychiatric hospital. During the ten-year follow-up period 276 deaths occurred in this sample, which was 3.6 times higher than the expected number of 77.

Giffen *et al.* (1971) described the mortality of 343 chronic drunkenness offenders who were incarcerated for public intoxication in a Toronto jail in 1940. These men were followed up until 1961. The observed number of deaths (191) was 2.1 times higher than the number expected on the basis of the general mortality in Ontario.

Quite recently Pell and D'Alonzo (1973) concluded a five-year follow-up investigation of 899 known, suspected, or recovered alcoholics and a slightly larger number of controls who were matched by age, sex, income, and geographical location, both employed by a large U.S.A. company. The observed number of deaths among the alcoholics was 102 as against 31.2 expected. Since more alcoholics than controls were lost during the follow-up period because of layoffs or resignations (51 as against 14) the ratio 102/31.2 or 3.2 is probably an underestimate.

In England Nicholls *et al.* (1973) studied 935 patients who had been admitted to four mental hospitals in the London area and who had been given a primary or secondary diagnosis implicating abnormal drinking. The follow-up period ranged from 10 to 15 years during which 309 deaths were observed, or 2.7 times the expected number of deaths based on the general population.

In summary, all but one of the studies reviewed reported excess mortality in the samples of alcoholics. The bulk of the evidence suggests that the ratio observed over expected number of deaths ranges from two to four. In one investigation a ratio of five was found (Kendell and Staton, 1966). However, the number of cases in this instance was very small (11 deaths). Dahlgren's (1951) study fails to show any excess mortality among persons registered by the Temperance Boards. This held true also when he analyzed separately more serious cases who were committed to an institution for the treatment of alcoholism. The author suggests as an explanation for the absence of excess mortality that persons who come to the attention of Temperance Boards represent a selection of relatively healthy alcoholics, since those who suffer from chronic diseases would not be sent to such institutions. One may also consider the possibility that, as a result of the intervention and continuous control by the Temperance Boards, many in the sample became abstinent. Thus, it has been observed that alcoholics who discontinue consumption after discharge from treatment have a reduced mortality compared with those who continue excessive drinking (Gabriel, 1935).

Finally, some prospective investigations have been conducted by insurance companies to study the effect of different drinking habits on the risk of death. In all these investigations it was found that the mortality from all causes was lowest among abstainers and generally tended to increase with the regularity and volume of alcohol use (Hunter, 1932; Menge, 1950; Davies, 1965).

CAUSE-SPECIFIC MORTALITY

All the prospective investigations described above have found significantly high rates of mortality among alcoholics from cirrhosis of the liver, alcholism, pneumonia, accidents, and suicides. Many also found higher than expected rates of death from cancer of the upper digestive and respiratory tracts, heart disease, vascular lesions of the central nervous system. In addition a few reported higher than expected rates of death from tuberculosis, syphilis, peptic ulcers, and cancer of the uterus and prostate. Many of these observations are supported by the results of retrospective and coincidence studies.

Tuberculosis

For many years now, the death rate from tuberculosis has declined significantly. Most recent studies do not specifically report a high mortality from this cause. However, Sundby (1967) reported quite a considerable number of deaths from this cause, well in excess of expected values, for Norwegian alcoholics during the period 1925 to 1962, and Gabriel (1935) for Austrian alcoholics for the period 1922 to 1935. The relative absence of such deaths in more recently selected samples of alcoholics and in the general population undoubtedly reflects the introduction of highly successful therapies, e.g., the use of streptomycin (Milne, 1970; Kok-Jensen, 1972). Interestingly, many retrospective studies report a high prevalence of alcoholism among tuberculosis patients (Warnery et al., 1958; Uggla, 1960; Brown and Campbell, 1961; Lewis and Chamberlain, 1963; Pincock, 1964; Ciurdea et al., 1966; Le Coz, 1968; Milne, 1970). As mentioned earlier, such findings are difficult to interpret because of (1) the usual unreliability of self-reported life history data; (2) the variation in the definitions of alcoholism used in such studies; (3) the problem of tabulating expected values in similar age, sex, and social class populations; and (4) the likelihood that in some instances alcoholism postdates the onset of tuberculosis (Golder, 1952; Uggla, 1960; Bonfiglio and Citterio, 1963). Because of the very low rate of death from this cause in recent years, follow-up studies of much larger samples of alcoholics than are usually employed would be required to establish whether alcoholics are more apt to die from this cause.

Syphilis

The literature on the morbidity and mortality of this disease among alcoholics is quite inconsistent. In some clinical samples of alcoholics in the United States, higher than expected morbidity rates from this cause, in others lower than expected morbidity rates were observed (Stewart, 1937; Dynes, 1939; McGee, 1942; Malzberg, 1947; Barchha et al., 1968). Also in the United States, Jellinek (1942) found that the reported incidence of alcoholism as a secondary cause in deaths ascribed to syphilis was about the same as in deaths from all causes combined. Although Sundby's (1967) prospective study of Norwegian alcoholics showed a rate of death from syphilis much higher than expected, other prospective studies found no excess mortality from this cause (Gabriel, 1935; Schmidt and de Lint, 1972). In recent decades mortality from syphilis has declined considerably. As in the case of tuberculosis, it would appear that this trend is reflected in the relatively low number of deaths from this cause among recent samples of alcoholics (Gillis, 1969; Ciompi and Eisert, 1969; Tashiro and Lipscomb, 1963; Nicholls et al., 1973; Pell and D'Alonzo, 1973).

Cancers of the Upper Digestive and Respiratory Tracts

This group includes cancer of the buccal cavity; the pharynx; the larynx; the lung and the oesophagus. There is much evidence that these cancers are, to some extent, related to alcoholism. In prospective studies of alcoholic patients, very large differences between observed and expected numbers of deaths for these cancers have been reported (Gabriel, 1935; Sundby, 1967; Schmidt and de Lint, 1972; Nicholls et al., 1973; Pell and D'Alonzo, 1973). These findings have been substantiated by (1) retrospective studies of patients with such cancers that revealed a high proportion of alcoholics among them compared with the hospital populations generally (Josserand and Lejeune, 1951; Morice, 1954; Cappellano, 1960; Gsell and Löffler, 1962); and (2) coincidence studies that showed a close spatial correlation between deaths from these cancers and indices of alcoholism (Schwartz, 1966; Lasserre et al., 1967). However, current trends in alcohol use and mortality are not consistent with the apparent association between alcoholism and death from cancer of the pharynx, larynx, oesophagus, and buccal cavity. While rates of alcoholism have been increasing in Western countries for some time, mortality from these causes has not increased accordingly. For example, in Canada, between 1952 and 1969, the age-standardized rates of death from cirrhosis of the liver increased by 71 percent, while the age standardized rates of death from cancer of the larynx and oesophagus remained unchanged and from cancer of the pharynx and buccal cavity decreased by about 10 percent. On the other hand, deaths from cancer of the lung have increased greatly during this period (Vital Statistics, 1952–69). These apparent discrepancies between some of the results of spatial and temporal analyses are possibly related to the fact that the variation in rates of alcohol excess is usually larger among spatial than among temporal units.

With respect to etiology, it is not always clear to what extent the apparently high rate of death from these cancers among alcoholics, as observed in many follow-up studies, can be attributed to excessive alcohol use, to heavy smoking, or to some other cause. In the case of lung cancer, there would appear to be little doubt that it is almost entirely attributable to heavy smoking. Doll and Hill (1964), in their study of British doctors, concluded that "there is evidence of so close a relation (between the number of cigarettes smoked and mortality) that it becomes increasingly difficult to envisage any other feature correlated with smoking as being the real and underlying cause." This conclusion has since been confirmed in many prospective and retrospective studies (Report to the U.S. Surgeon General, 1971). In the case of the other cancers in this group, it would seem that the alcoholics' excessive smoking habits do not fully explain their high rate of death from these causes. In several retrospective studies, this question was specifically examined. Schwartz et al. (1962) concluded that alcohol use exerts an independent effect on the occurrence of cancers of the oral cavity, pharynx, and oesophagus. Keller and Terris (1965) came to a similar conclusion with respect to cancer of the oral cavity and pharynx. Wynder and Bross (1961) estimated that the risk of developing cancer of the oesophagus was about 25 times higher in a heavy drinker of distilled spirits than in a nondrinker when the amount of tobacco use is held constant. The results of some prospective studies are in agreement with these conclusions. Schmidt and de Lint (1972) pointed out that in their sample of alcoholic patients, for every six deaths from lung cancer four deaths from cancer of the larynx, pharynx, and oesophagus occur, whereas, in studies of smokers, this ratio is about 6:1 (Doll and Hill, 1964; Weir and Dunn, 1970). Nicholls et al. (1973) made a similar observation. They found among their alcoholics a ratio of nine deaths from lung cancer to seven deaths from cancer of the larynx, pharynx, and oesophagus instead of the expected ratio of 6:1.

In conclusion, it would appear that both heavy use of alcohol and of tobacco contribute to the high rate of death from cancer of the buccal cavity, pharynx, larynx, and oesophagus among alcoholics. With respect to their high rate of death from cancer of the lung, no evidence exists thus far to suggest that heavy alcohol use is a factor.

Cancer of the Liver

In several retrospective studies primary neoplasms of the liver have been found to occur in 10 to 30 percent of alcoholics with cirrhosis (Parker, 1957; Leevy et al., 1964; Lee, 1966). It has been suggested that in Western countries alcoholic cirrhosis is the chief factor in the etiology of this cancer (Leevy et al., 1964). The available evidence from prospective studies does not support this view. The number of deaths from this form of cancer in Sundby's (1967) Norwegian sample hardly differed from expectation, and in the Canadian sample (Schmidt and de Lint, 1972) not one death from this cause was found, although 68 deaths from cirrhosis of the liver were reported. In the other prospective studies described earlier, no specific information on deaths from this cause is given.

However, it should be noted that cancer of the liver is a relatively rare cause of death, contributing less than 2 percent of all cancer deaths in Western countries. Larger samples of alcoholics followed up over many years would be required to investigate more fully the relationship between alcoholism and this form of cancer.

Other Neoplasms

Three prospective studies have shown an increased mortality from cancer of the prostate (Sundby, 1967; Schmidt and de Lint, 1972; Pell and D'Alonzo,

1973). However, these findings were based on small numbers of deaths and a relatively low excess mortality.

International variation in rates of death from this cause does not suggest an association with alcoholism. For example, Italy and Portugal, which have very high alcohol consumption rates, have a very low mortality from cancer of the prostate; whereas Norway, a country with one of the lowest per capita alcohol consumption in the Western world, has the highest known death rates from this cause. Retrospective examination of patients suffering from this disease also has not shown a relationship with alcoholism (Wynder et al., 1971).

Excess deaths from cancer of the uterus were reported in one prospective study of alcoholics (Schmidt and de Lint, 1972). Since there were only five such deaths in the group of women alcoholics in this study, comparison with a general population must be highly unreliable. Nevertheless, it may be relevant here, that, in the etiology of this disease, sexual promiscuity (Abou-Daoud, 1967; Martin, 1967; Moghissi and Mack, 1968) and marital instability (Martin, 1967) have been indicated. It is also possible that a higher mortality from this cause results from neglect. In this disease, early treatment produces relatively high cure rates.

With reference to cancers of other sites, Sundby (1967) observed a slight excess in death from cancer of the stomach and rectum. Nicholls et al. (1973) reported 15 deaths from cancers of the stomach, intestines, rectum, and liver combined as against an expected frequency of nine, while Schmidt and de Lint (1972) found no extra mortality associated with these cancers. Generally the literature does not indicate a relationship between these causes of death and alcoholism.

Diabetes

In several prospective studies the observed mortality from diabetes exceeded the expected rate (Sundby, 1967; Schmidt and de Lint, 1972). One explanation for this excess mortality is that the cirrhotic patient suffers from an intrahepatic disturbance of sugar metabolism that may lead to diabetes. Indeed, a study of 140 patients with both cirrhosis and diabetes showed that in the vast majority of cases (103) cirrhosis had occurred first.

Another explanation is that heavy alcohol use influences the course of this disease unfavorably rather than its incidence and thus increases mortality. It would appear that the deaths from diabetes found in follow-up investigations of alcoholics are attributable to these two factors.

Alcoholism

Several investigators have observed that the ascription of death to acute and chronic alcoholism is done quite arbitrarily. Often these deaths are attributable to a variety of other conditions.

For example, in one prospective study (Schmidt and de Lint, 1972) 32 deaths from "acute" or "chronic" alcoholism were reported. In ten of these deaths pneumonia, in six coronary thrombosis, in three acute or chronic myocarditis, and in two cerebral degeneration were listed on the death certificate as contributory causes. In another study (Sundby, 1967) many of the deaths certified as due to alcoholism were in fact due to cardiomyopathy with heart failure.

Heart Diseases

Several follow-up studies of samples of alcoholics have shown that 10 to 40 percent of all excess deaths are attributable to cardiovascular diseases (Brenner, 1967; Sundby, 1967; Schmidt and de Lint, 1972; Nicholls et al., 1973; Pell and D'Alonzo, 1973). Undoubtedly some of these deaths are attributable to the cigarette smoking habit so common among alcoholics (Cartwright et al., 1959; Bates, 1965; Sundby, 1967; Dreher and Fraser, 1968; Walton, 1972; Hammond and Horn, 1958; Dorn, 1959; Doyle et al., 1964; Doll and Hill, 1964; Best et al., 1967; Friedman, 1967; Zeiner-Henriksen, 1971). A few other deaths may be attributable to alcoholic cardiomyopathy, a clinically well-described but relatively rare condition among heavy drinkers. Also of relevance here is the significantly high rate of occurrence of hypertension among alcoholics, a well-established risk factor in coronary heart disease (D'Alonzo and Pell, 1968; Leclainche and Le Gô, 1970). Indeed, in one recent prospective study it was estimated that about 25 percent of excess coronary deaths in the alcoholic sample were the result of hypertension (Pell and D'Alonzo, 1973). Whether there are other factors explaining the higher mortality from cardiovascular diseases in alcoholic samples cannot be answered at present. The underlying factor in cardiovascular diseases is usually arteriosclerosis. It is therefore of interest to relate this excess mortality in alcoholics to existing views as to the effect of excessive drinking on this degenerative vascular process. One view, based on autopsies of cirrhotic patients, is that alcoholism has a beneficial effect on this process (Cabot, 1904; Hall et al., 1953; Stare, 1961). However, this effect apparently does not extend to alcoholics without cirrhosis (Hirst et al., 1965).

Apoplexy

Several follow-up studies reported a significantly higher rate of death from cerebrovascular diseases among alcoholics (Gabriel, 1935; Sundby, 1967; Pell and D'Alonzo, 1973). In others no statistically significant difference was found (Schmidt and de Lint, 1972; Nicholls et al., 1973) and in one sample of "Skid Row" alcoholics a lower than expected rate of death from this disease occurred (Giffen et al., 1971). As was the case with coronary heart disease, arteriosclerosis usually precedes brainstrokes and in fact is mentioned frequently on the death certificate as a contributory cause (Schmidt and de Lint, 1972). These data do seem to lend further support to our earlier contention that heavy alcohol use does not prevent vascular degeneration as has sometimes been argued.

Diseases of the Respiratory Organs

The high rates of death from pneumonia frequently reported for alcoholic populations (Gabriel, 1935; Lipscomb, 1959; Sundby, 1967; Gorwitz et al., 1970; Giffen et al., 1971; Schmidt and de Lint, 1972; Nicholls et al., 1973) are difficult to interpret. In many instances a variety of other morbid conditions, e.g., heart disease, liver cirrhosis, delirium tremens, apoplexy, are present (Sundby, 1967). Indeed, the ascription of a death to pneumonia rather than to any of the other conditions appears to be somewhat arbitrary.

Many of the attributes of alcoholism are of importance in death from pneumonia. Gross intoxication, inadequate clothing, extensive exposure to cold, poor nutrition, have often been mentioned (Jellinek, 1942; Alessandri et al., 1944; Bogue, 1963; Chomet and Gach, 1967).

Deaths from other diseases of the respiratory organs such as emphysema, bronchitis, and asthma are rarely reported in prospective investigations of alcoholic samples, which is somewhat surprising in view of the known association between cigarette smoking and the occurrence of bronchitis and emphysema. One exception is Sundby's (1967) study of Norwegian alcoholics, in which 28 deaths from these causes were observed.

Gastroduodenal Ulcers

Some of the evidence available to date would seem to indicate an association between mortality from gastroduodenal ulcers and alcoholism. For instance, in the earlier-described studies of Californian, Norwegian, and Ontario alcoholics (Brenner, 1967; Sundby, 1967; Schmidt and de Lint, 1972), a somewhat elevated death rate from ulcers of stomach and duodenum was found. Jellinek (1942) observed some decades ago that in the United States "the incidence of 'alcoholism' as a secondary cause in death ascribed to ulcer of the stomach is quite high." And data pertaining to the occupational mortality in England and Wales would suggest that persons in alcohol-exposed occupations, such as proprietors and managers of hotels, tavern owners, and barmen die more often from ulcer of duodenum and stomach than persons in other occupations (Reg. Gen. Dec. Suppl., 1951).

Cirrhosis of the Liver

Probably no major cause of death has been so consistently linked to alcoholism for so many years as liver cirrhosis. Several recent reviews of the relevant literature makes it quite unnecessary to document the very significant contribution of alcoholism to mortality from this cause. It may suffice to point out that in follow-up studies the ratios of observed/expected deaths from liver cirrhosis range from eight (Sundby, 1967), ten (Brenner, 1967), 16 (Salum *et al.*, 1972), 17 (Schmidt and de Lint, 1972), to 23 (Nicholls *et al.*, 1973), and also that liver cirrhosis is frequently mentioned on the death certificate of alcoholics as a contributing cause of death (Tashiro and Lipscomb, 1963; de Lint and Schmidt, 1970; Nicholls *et al.*, 1973).

After reviewing the relevant prospective and retrospective studies, Lelbach (1973) concluded that volume and duration of alcohol consumption are the most important factors in alcoholic cirrhosis. The type of beverage alcohol consumed (e.g., beer, wine, or distilled spirits) and the pattern of drinking (e.g., periodic bouts versus daily inveterate) are apparently of little importance (Lelbach, 1974). With reference to the clinically observed higher susceptibility of female alcoholics as against male alcoholics, the evidence is still rather sketchy (Wilkinson *et al.*, 1969). One obvious reason is that most of the samples in prospective and retrospective investigations consist of males.

Accidents, Poisonings, and Violence

Accidental poisonings, motor vehicle accidents, accidental falls, accidents caused by fire and homicides contribute very significantly to the mortality of alcoholics (Dahlgren, 1951; Storby, 1953; Brenner, 1967; Sundby, 1967; Gillis, 1969; Gorwitz *et al.*, 1972). The excess deaths from these causes have been attributed to the acute effects of alcohol, to the style and conditions of life of alcoholics, and to personality traits associated with alcoholism (Graves, 1960; MacFarland and Moore, 1962; Payne and Selzer, 1962; Schmidt *et al.*, 1962; Goodwin, 1973). It is difficult to separate the relative importance of each of these factors in these deaths. Schmidt and de Lint (1972) reported that drunkenness appeared on the death certificate in 28 percent of accidental deaths as a contributory cause. They pointed out that in many other instances a state of intoxication may well have existed. For example, of deaths caused by fire many were attributable to careless smoking but the presence of alcohol was not ascertained. Also, they reported that in deaths from freezing and in traffic fatalities, alcohol was not mentioned on the death certificate, although intoxication must have played an important role in at least some of these deaths. The results of numerous retrospective investigations of such deaths would support the conclusion that the acute effect of alcohol is an important factor (Spain et al., 1951; LeRoux and Smith, 1964; Potondi et al., 1967; Waller, 1972). Studies of blood alcohol levels of persons admitted to the emergency services of general hospitals after involvement in home, transportation, or occupation accidents have shown that many were drinking heavily prior to the accident (Hindmarsh and Linde, 1934; Gjone, 1963; Haddon, 1963; Borkenstein, 1964; Birrell, 1965; Demone and Kasey, 1966; Johnsen et al., 1966; Im Obersteg and Bäumler, 1967; Kirkpatrick and Taubenhaus, 1967; Tonge, 1968; Wechsler et al., 1969; Rydberg et al., 1973). Specifically, in the case of traffic accidents, high blood alcohol levels are frequently found (Haddon et al., 1962; Tonge et al., 1962; Waller and Turkel, 1966; Bonnichsen and Åquist, 1968; Waller et al., 1969; Hossack, 1972; McBay, 1972; Waller, 1972).

Suicide

The literature on alcoholism and suicide is quite extensive. Firstly, clinicians have frequently referred to the self-destructive aspects of chronic intoxication as suicidal behavior (Menninger, 1938; Palola et al., 1962). Secondly, significant similarities in the life histories of suicidal persons and alcoholics have been noted (Gadourek, 1963). Thirdly, follow-up studies have reported high rates of suicide among alcoholics (Gabriel, 1935; Dahlgren, 1951; Nørvig and Nielsen, 1956; Lipscomb, 1959; Kessel and Grossman, 1961; Kendell and Staton, 1966; Sundby, 1967; Ciompi and Eisert, 1969; Gillis, 1969; Schmidt and de Lint, 1972; Nicholls et al., 1973), and fourthly, retrospective studies of suicides and attempted suicides have found alcoholism to be often indicated in the life history of the deceased, or of the patient (Dahlgren, 1945; Batchelor, 1954; Schmidt et al., 1954; Sainsbury, 1955; Robins et al., 1957; Stengel and Cook, 1958; Robins et al., 1959; Dorpat and Ripley, 1960; Palola et al., 1962; James et al., 1963; James, 1966; Harenko, 1967; Murphy and Robins, 1967; Žmuc, 1968; Mayfield and Montgomery, 1972). Suicides among alcoholics are usually attributed to predisposing social and personality factors (Menninger, 1938; Mowrer, 1942; Wallinga, 1949; Faris, 1961; Clinard, 1963; Rushing, 1969a; Lundquist, 1970), and to social isolation resulting from the common interpersonal disturbances in the alcoholic's way of life (Palola et al., 1962; Gadourek, 1963; Rushing, 1969b).

Relatively little attention has been given to alcohol-induced depressive states as a direct cause of suicides. Indeed, it would be extremely difficult to determine to what extent a depressed state is the psychological consequence of chronic excessive drinking or is brought about by a process of social deterioration. Some support for a direct link between heavy alcohol use and suicide may be found in the very high rates among tavern owners, restaurateurs, and hoteliers that have been observed in England (Ledermann, 1964). Evidently, the major common characteristic of these occupations is the continuous exposure to alcoholic beverages.

SOME SPECIFIC ISSUES

Many of the investigations reviewed here have demonstrated an association between, on the one hand, the conglomerate of habits and conditions of life usually labeled "alcoholism" and, on the other, death from a variety of causes. In some of these studies additional questions were raised.

Social Class

In a few prospective investigations the mortality of alcoholics of different social class was compared. Sundby (1967) found that vagrants did not differ from nonvagrants in overall mortality. However, with respect to cause-specific mortality he observed that vagrants died less often from cardiovascular diseases. Schmidt and de Lint (1970) also reported a lower rate of death from this cause among working-class alcoholics compared to middle-class alcoholics. The fact that many working-class patients and vagrants are more often engaged in physical labor whereas middle-class patients tend to have sedentary occupations may well be relevant here (Breslow and Buell, 1960; Buell and Breslow, 1960; Taylor et al., 1962; Kahn, 1963; Frank et al., 1966; Morris et al., 1968; Schmidt and de Lint, 1970).

Male–Female Mortality

It has been reported that female alcoholics are more disturbed and pass more rapidly through various stages of uncontrolled drinking than do male alcoholics (Hewitt, 1943; Berner and Solms, 1953; Lolli, 1953). However, in one prospective investigation (Schmidt and de Lint, 1969) it was found that excess mortality in male and female alcoholics is attributable to the same causes and that rate of death of female alcoholics does not exceed the male rate. Apparently the clinical impressions concerning sex differences in alcohol use and emotional state were not supported by these data. The authors suggested that the unfavorable clinical assessment of the female alcoholic may be, in part, attributable to a "double standard" on the side of the clinician, who may unknowingly compare the woman alcoholic with his ideal of womanhood rather than with the characteristics of male alcoholic patients.

Abstinence

In two prospective studies of alcoholic populations the effect of abstinence after discharge from treatment on subsequent mortality was examined (Gabriel, 1935; Pell and D'Alonzo, 1973). Not unexpectedly, it was found that the mortality of the abstinent groups was somewhat lower than of those who continued to drink. Pell and D'Alonzo (1973) suggested that "alcoholics who stop drinking retain a considerable amount of excess morbidity and mortality, possibly because: (a) certain disease processes initiated by alcohol may be irreversible; or (b) indulgence in personal habits that adversely affect health may have persisted to the same or greater degree after the alcoholics stop drinking."

Type of Beverage

Whether the chronic use of beer, wine, or distilled spirits affect mortality differently has been of considerable interest for quite some time. In many jurisdictions the use of distilled spirits is subjected to more severe restrictions (taxation and otherwise) than is the use of beer and wine on the grounds that, compared to the lower content beverages, it is more injurious to health. Indeed, Gabriel (1935) observed that the overall rate of death of distilled spirits drinkers was higher than that of drinkers of beer and wine. But he also found that such drinkers were usually in a more advanced state of their alcoholism. At any rate, in two recent reviews of the relevant experimental and clinical data on the organic pathologies attributable to alcohol consumption, it was concluded that total alcohol consumption over a drinking life rather than type of alcoholic beverage involved in such drinking is the significant variable (Wallgren and Barry, 1970; Lelbach, 1974). There may be a few exceptions. For instance, some retrospective studies of patients with myocardial disease have particularly indicated a heavy consumption of beer (Brigden and Robinson, 1964; Alexander, 1966). Similarly, studies of patients with cancers of the upper digestive tracts have frequently found distilled spirits consumption among the antecedent factors (Wu and Loucks, 1951; Wynder and Bross, 1961; Keller and Terris, 1965).

POSTSCRIPT

The characteristics of alcoholics—their drinking behavior, smoking habits, emotional state, life-style—are uniquely expressed in their mode of dying.

Thus, in death from poisonings, falls, and fire, the acute effects of alcohol use are indicated, whereas in deaths from liver cirrhosis the chronic effects are evident. In suicides the depressed state of mind of the alcoholic appears to be the more important factor, whereas the high rate of death from cancer of the lung is attributable to heavy smoking.

A more precise assessment of the contribution of each of these factors is often difficult. Thus far, the most useful information about alcoholism-related mortality has come from prospective investigations of samples of alcoholics. Undoubtedly the study of much larger samples over extended periods of time will help to clarify many of the current issues. For example, to examine the possible role of alcoholism in death from relatively rare causes requires many more man-years of observation than have thus far been obtained.

REFERENCES

Abou-Daoud, K. T., 1967, Epidemiology of cancer of the cervix uteri in Lebanese Christians and Moslems, *Cancer,* N.Y., 20:1706.

Alessandri, H., Alessandri, M., Garcia, P., Gazmuri, R., Lerner, J., Concha, E., Estevez, A., and Fritis, E., 1944, El alcoholismo como factor de enfermedad en un servicio de medicina interna, *Rev. méd. Chile* 72:199.

Alexander, C. S., 1966, Idiopathic heart disease. I. Analysis of 100 cases with special reference to chronic alcoholism, *Amer. J. Med.* 41:213.

Bandel, R., 1928, Die Trinksitte als häufigste Todesursache der Männer, *in* "Die Alkoholfrage," Heft 1/2, Verlag "Auf der Wacht," Berlin-Dahlem.

Bandel, R., 1930, Die spezifische Männersterblichkeit als Masstab des Alkoholsterblichkeit, *in* "Ergebnisse der Sozialen Hygiene und Gesundheitsfürsorge," Bd. II, pp. 424–492, Georg Thieme Verlag, Leipzig.

Bandel, R., 1932, "Nachweis der Alkoholsterblichkeit in der allgemeinen Sterbestatistik," Verlag Auf der Wacht, Berlin-Dahlem.

Barchha, R., Stewart, M. A., and Guze, S. B., 1968, The prevalence of alcoholism among general hospital ward patients, *Amer. J. Psychiat.* 125:681.

Batchelor, I. R. C., 1954, Alcoholism and attempted suicide, *J. Ment. Sci.* 100:451.

Bates, R. C., 1965, The diagnosis of alcoholism, *Appl. Therap.* 7:466.

Bättig, K., 1967, Alkoholismus: Epidemiologische Zusammenhänge und Folgen, *Natur. Rdsch.* 20:200.

Berner, P., and Solms, W., 1953, Alkoholismus bei Frauen, *Wien. Z. Nervenheilk.* 6:275.

Bertillon, J., 1874, "La Demographie Figurée de la France ou Etude Statistique de la Population Française avec Tableaux Graphiques Traduisant les Principales Conclusions: Mortalité selon l'Age, le Sexe, l'Etat Civil, etc. en chaque Département et pour la France Entière, Comparée aux Pays Etrangers," Masson, Paris.

Bertillon, J., 1889, "De la Morbidité et de la Mortalité Professionnelle," Annuaire Statistique de la ville de Paris, pp. 186–236.

Best, E. W. R., Walker, C. B., Baker, P. M., Delaquis, F. M., McGregor, J. T., and McKenzie, A. C., 1967, Summary of a Canadian study of smoking and health, *Can. Med. Assoc. J.* 96:1104.

Birrell, J. H. W., 1965, Blood alcohol levels in drunk drivers, drunk and disorderly subjects and moderate social drinkers, *Med. J. Aust.* 2:949.
Bogue, D. J., 1963, "Skid Row in American Cities," University of Chicago Press, Chicago.
Bonfiglio, G., and Citterio, C., 1963, Alcoolismo e tuberculosi, *Lav. neuropsichiat.* 32:37.
Bonnichsen, R. B., and Åquist, S., 1968, Alkoholens roll vid svenska trafikolyckor, *Alkoholfrågan* 62:202.
Borkenstein, R. F., Crowther, R. F., Shumate, R. P., Ziel, W. B., and Zylman, R., 1964, The Role of the Drinking Driver in Traffic Accidents, Department of Police Administration, Indiana University.
Brenner, B., 1967, Alcoholism and fatal accidents, *Quart. J. Stud. Alc.* 28:517.
Brésard, M., 1959, I Consommation d'alcool et mortalité par cirrhose du foie à Saint-Etienne et à Marseille. II. Consommation du tabac et alcool, *Bull. Inst. nat. Hyg.* 14:367.
Breslow, L., and Buell, P., 1960, Mortality from coronary heart disease and physical activity of work in California, *J. Chron. Dis.* 11:421.
Brick, J. B., and Palmer, E. D., 1964, One thousand cases of portal cirrhosis of the liver. Implications of esophageal varices and their management, *Arch. Intern. Med.* 113:501.
Brigden, W., 1957, Uncommon myocardial diseases. The noncoronary cardiomyopathies, *The Lancet* 2:1179, :1243.
Brigden, W., and Robinson, J., 1964, Alcoholic heart disease, *Brit. Med. J.* 2:1283.
Bronetto, J., 1964, Differential Mortality from Cirrhosis of the Liver by Occupation, Addiction Research Foundation, Mimeographed, Toronto.
Brown, K. E., and Campbell, A. H., 1961, Tobacco, alcohol and tuberculosis, *Br. J. Dis. Chest.* 55:150.
Bruun, K., Koura, E., Popham, R. E., and Seeley, J. R., 1960, Maksakirroosikuolleisuus alkoholismin levinneisyyden mittarina: tutkimuksia ns. Jellinekin kaavan soveltuvuudesta alkoholistien määrän arviontiin suomessa (Liver cirrhosis mortality as a means of measuring the prevalence of alcoholism; studies on the applicability of the Jellinek formula for the estimation of the number of alcoholics in Finland), Våkijuomakysymyksen tutkimusäätiön, Publ. No. 8/2, Helsinki.
Buell, P., and Breslow, L., 1960, Mortality from coronary heart disease in California men who work long hours, *J. Chron. Dis.* 11:615.
Burch, G. E., and Ansari, A., 1968, Chronic alcoholism and carcinoma of the pancreas, *Arch. Intern. Med.* 122:273.
Cabot, R. C., 1904, The relation of alcohol to arteriosclerosis, *J. Amer. Med. Assoc.* 43:774.
Cappellano, R., 1960, Incidência da sífilis, etilismo e tabagismo nos pacientes portadores de câncer das vias aero-digestivas superiores, *Rev. bras. Cir.* 39:518.
Caroli, J., Mainguet, P., Ricordeau, P., and Foures, A., 1959, De l'importance de la laparoscopie et de la laparascopie en couleurs dans la classification anatomo-clinique des cirrhoses du foie, cirrhoses alcoolique et nutritionelle, Rapports présentés au XXXIIe Congrès Francais de Médicine, Lausanne 1959, Masson, Paris.
Cartwright, A., Martin, F. M., and Thomson, J. G., 1959, Distribution and development of smoking habits, *The Lancet* 2:725.
Čerlek, S., 1968, Alcoholism and changes in the liver, *Alcoholism* 4:62.
Chapman, B. L., 1966, Cirrhosis of the liver, Newcastle, N. S. W.: A study of 98 cases, *Med. J. Aust.* 1:51.
Chomet, B., and Gach, B. M., 1967, Lobar pneumonia and alcoholism; an analysis of 37 cases, *Amer. J. Med. Sci.* 253:300.
Ciompi, L., and Eisert, M., 1969, Mortalité et causes de décès chez les alcooliques, *Sozial Psychiat.* 4:159.

Ciurdea, A., Marian, D., and Titei, R., 1966, Rolul unor boli asociate in patogeneza ftiziei hipercronice, *Rev. med.-chir.* (Jassy) 70:321.

Clinard, M. B., 1963, "Sociology of Deviant Behavior," Holt, Rinehart and Winston, New York.

Creutzfeldt, W., and Beck, K., 1966, Erhebungen über Atiologie, Pathogenese, Therapieerfolge und Überlebenszeit an einem unausgewählten Krankengut von 560 Patienten mit Leberzirrhose, *Dtsch. Med. Wschr.* 91:682.

Dahlgren, K. G., 1945, "On Suicide and Attempted Suicide—A Psychiatrical and Statistical Investigation," Lindstedts, Lund.

Dahlgren, K. G., 1951, On death-rates and causes of death in alcohol addicts, *Acta psychiat. neurol.* 26:297.

D'Alonzo, C. A., and Pell, S., 1968, Cardiovascular disease among problem drinkers, *J. Occup. Med.* 10:344.

Davies, K. M., 1965, The influence of alcohol on mortality, *Proc. Home Office Life Underwriters Assoc.* 46:159.

Delaporte, P. J., 1941, "Evolution de la Mortalité en Europe depuis l'Origine des Statistiques de l'Etat Civil," Statistiques générale de la France, Etudes démographiques No. 2, Imprimerie Nationale, Paris.

de Lint, J., 1976, The Epidemiology of Alcoholism: The Elusive Nature of the Problem, Estimating the Prevalence of Alcohol Use and Alcohol-Related Mortality, Current Trends and the Issue of Prevention, Proceedings of the First International Medical Conference on Alcoholism, London, September 10–15.

de Lint, J., and Schmidt, W., 1968, The distribution of alcohol consumption in Ontario, *Quart. J. Stud. Alc.* 29:968.

de Lint, J., and Schmidt, W., 1970, Mortality from liver cirrhosis and other causes in alcoholics. A follow-up study of patients with and without a history of enlarged fatty liver, *Quart. J. Stud. Alc.* 31:705.

Demone, H. W. Jr., and Kasey, E. H., 1966, Alcohol and nonmotor vehicle injuries, *Publ. Hlth. Rep.* 81:585.

Doll, R., and Hill, A. B., 1964, Mortality in relation to smoking: Ten years' observations of British doctors, *Brit. Med. J.* 1:1399, 1460.

Dorn, H. F., 1959, Tobacco consumption and mortality from cancer and other diseases, *Publ. Hlth. Rep.*, Wash. 74:581.

Dorpat, T. L., and Ripley, H. S., 1960, A study of suicide in the Seattle area, *J. Comp. Psychiat.* 6:349.

Doyle, J. T., Dawber, T. R., Kannel, W. B., Kinch, S. H., and Kahn, H. A., 1964, The relationship of cigarette smoking to coronary heart disease. The second report of the combined experience of the Albany, N.Y., and Framingham, Mass., studies, *J. Amer. Med. Assoc.* 190:886.

Dreher, K. F., and Fraser, J. G., 1968, Smoking habits of alcoholic out-patients I, II. *Int. J. Addict.* 2:259, 3:65.

Dynes, J. B., 1939, Survey of alcoholic patients admitted to the Boston Psychopathic Hospital in 1937, *New Engl. J. Med.* 220:195.

Egoz, N., Kendrick, M. A., and Mosley, J. W., 1970, Cirrhosis mortality in relation to alcohol consumption, *Gastroent.* 58:281.

Evans, W., 1959, The electrocardiogram of alcoholic cardiomyopathy, *Brit. Heart J.* 21:445.

Evans, N., and Gray, P. A., 1938, Laennec's cirrhosis. Report on 217 cases, *J. Amer. Med. Assoc.* 110:1159.

Falck, I., and Horn, G., 1957, Klinische Studien zur Leberzirrhose. IV Zur Ätiologie der Leberzirrhose. *Z. ges. inn. Med.* 12:610.

Falck, I., Heinrich, H. G., Jutzi, E., Kohler, W., Mohnicke, G., and Vetter, K., 1962, Laboratorieuntersuchungen bei Leberkrankheiten. II Mitteilung: Leberzirrhose, *Dtsch. Z. Verdau. -u. Stoffwechselkr.* 22:137.

Faris, R. E. L., 1961, "Social Disorganization," Harper and Co., New York.

Fauvert, R., Benhamou, J. P., Boivin, P., Darnis, F., and Hartmann, L., 1959, Biologie des cirrhoses alcooliques. Cirrhose alcoolique et nutritionelle, Rapports présentés au XXXIIe Congrés Francais de Médicine, Lausanne, 1959, Masson, Paris.

Febvay and Aubenque, 1957, La Mortalité par Categorie Professionnelle, Etudes statistique No. 3, I.N.S.E.E. pp. 39–44.

Fitzgerald, B. J., Pasewark, R. A., and Clark, R., 1971, Four-year follow-up of alcoholics treated at a rural state hospital, *Quart. J. Stud. Alc.* 32:636.

Forshaw, J., 1972, Alcoholic cirrhosis of the liver, *Brit. Med. J.* 4:608.

Frank, C. W., Weinblatt, E., Shapiro, S., and Sager, R. V., 1966, physical inactivity as a lethal factor in myocardial infarction among men, *Circulation* 34:1022.

Friedman, G. D., 1967, Cigarette smoking and geographic variation in coronary heart disease mortality in the United States, *J. Chron. Dis.* 20:769.

Gabriel, E., 1935, Über die Todesursachen bei Alkoholikern, *Zbl. ges. Neurol. Psychiat.* 153:385.

Gadourek, I., 1963, "Riskante gewoonten en zorg voor eigen welzijn," Wolters, Groningen.

Garceau, A. J., and the Boston Inter-Hospital Liver Group, 1964, The natural history of cirrhosis. II. The influence of alcohol and prior hepatitis on pathology and prognosis, *New Engl. J. Med.* 271:1173.

Giffen, P. J., Oki, G., and Lambert, S., 1971, The Chronic Drunkenness Offender: Ages and Causes of Death of the Chronic Drunkenness Offender Population, Chapter 13; Addiction Research Foundation, Mimeographed, Toronto.

Gigglberger, H., 1968, Zur Ätiologie der Leberzirrhose. Klinischstatistische Untersuchungen in 400 Kranken, *Acta hepat. -splenol.* 15:415.

Gillis, L. S., 1969, The mortality rate and causes of death of treated alcoholics, *S.A. Med. J.* 43:230.

Gjone, R., 1963, Kraniocerebrale trafikkskader, *T. norske Laegeforen.* 83:424.

Golder, G., 1952, Some aspects of the relationship of alcoholism and tuberculosis, *Conn. Rev. Alcsm.* 3:17.

Goodwin, D. W., 1973, Alcohol in suicides and homicides, *Quart. J. Stud. Alc.* 34:144.

Gorwitz, K., Bahn, A., Warthen, F. J., and Cooper, M., 1970, Some epidemiological data on alcoholism in Maryland. Based on admissions to psychiatric facilities, *Quart. J. Stud. Alc.* 31:423.

Graves, J. H., 1960, Suicide, homicide and psychosis. Proceedings of the 26th International Congress on Alcohol and Alcoholism, Stockholm, Aug. 1–5, 1960, Section III, p. 317.

Gros, H., 1956, Zur Ätiologie der Laennec'schen Leberzirrhose, *Medizinische* 1:686.

Gsell, V. O., and Löffler, A., 1962, Etiological factors in carcinoma of the esophagus, *Dtsch. Med. Wschr.* 87:2173.

Haddon, W., Jr., 1963, Alcohol and highway accidents. Proceedings of the Third International Conference on Alcohol and Road Traffic, pp. 3–13, BMA House, London.

Haddon, W. Jr., Valien, P., McCaroll, J. R., and Umberger, C. J., 1961, A controlled investigation of the characteristics of adult pedestrians fatally injured by motor vehicles in Manhattan, *J. Chron. Dis.* 14:655.

Hall, E. M., Olsen, A. Y., and Davies, F. E., 1953, Portal cirrhosis; clinical and pathological review of 782 cases from 16,600 necropsies, *Amer. J. Path.* 29:993.

Hällen, J., and Krook, H., 1963, Follow-up studies on an unselected ten-year material of 360 patients with liver cirrhosis in one community, *Acta med. Scand.* 173:479.

Hammond, E. C., and Horn, D., 1958, Smoking and death rates. I. Total mortality. II. Death rates by causes, *J. Amer. Med. Assoc.* 166:1159, 1294.

Harenko, A., 1967, Alkoholin osallisuus myrkytysitsemurhayrityksissä Helsingissä v. 1962–64, *Suom. Läärilehti* 22:109.

Hartmann, F., and Kottke, S., 1958, Beobachtungen zur Differentialdiagnose von Hepatitis, Cholangitis und Leberzirrhose, *Münch. Med. Wschr.* 100:705.

Hewitt, C. C., 1943, A personality study of alcohol addiction, *Quart. J. Stud. Alc.* 4:368.

Hindmarsh, J., and Linde, P., 1934, Alkoholuntersuchungen bei Unfallverletzten, *Acta chir. scand.* 75:198.

Hirst, A. E., Hadley, G. G., and Gore, J., 1965, The effect of chronic alcoholism and cirrhosis of the liver on atherosclerosis, *Amer. J. Med. Sci.* 45:143.

Hossack, D. W., 1972, Investigation of 400 people killed in road accidents, with special reference to blood alcohol levels, *Med. J. Aust.* 2:255.

Huber, M., 1912, Mortalité suivant la profession d'après les décès enregistrés en Frnce en 1907 et 1908, Bulletin de la statistique générale de la France, Fasc IV:402.

Hunter, A., 1932, Longevity and mortality as affected by the use of alcohol, *in* "Alcohol and Man" (H. A. Emerson, ed.), pp. 327–344, MacMillan, New York.

Im Obersteg, J., and Bäumler, J., 1967, Unfälle unter der Einwirkung von Arzneimitteln und Alkohol, *Schweiz. Med. Wschr.* 97:1039.

James, J. P., 1966, Blood alcohol levels following successful suicide, *Quart. J. Stud. Alc.* 27:23.

James, J. P., Scott-Orr, D. N., and Crenrow, D. H., 1963, Blood alcohol levels following attempted suicide, *Quart. J. Stud. Alc.* 24:14.

Jellinek, E. M., 1942, "Death from alcoholism" in the United States in 1940. A statistical analysis, *Quart. J. Stud. Alc.* 3:465.

Johnsen, C., Linder, B., and Lorentzon, S., 1966, Förekomst av alkoholpåverkade i ett kirurgiskt akutklientel, *Social-med. T.* 43:108.

Jolliffe, N., and Jellinek, E. M., 1941, Vitamin deficiencies and liver cirrhosis in alcoholism, Part VII. Cirrhosis of the liver, *Quart. J. Stud. Alc.* 2:544.

Josserand, A., and Lejeune, E., 1951, De l'influence aggravante de l'ethylisme sur l'évolution des epitheliomas buccpharyngés, *Lyon méd.* 184:165.

Kaeding, A., 1955, Ätiologischen Faktoren der Leberzirrhose, *Ärztl. Wschr.* 10:497.

Kahn, H. A., 1963, Relationship of reported coronary heart disease mortality and physical activity of work, *Amer. J. Publ. Hlth.* 53:1058.

Karpman, B., 1948, "The Alcoholic Woman," Linacre Press, Washington, D.C.

Keller, A. Z., and Terris, M., 1965, The association of alcohol and tobacco with cancer of the mouth and pharynx, *Amer. J. Publ. Hlth* 55:1578.

Kendell, R. E., and Staton, M. C., 1966, The fate of untreated alcoholics, *Quart. J. Stud. Alc.* 27:30.

Kessel, N., and Grossman, G., 1961, Suicide in alcoholics, *Brit. Med. J.* 2:1671.

Kirkpatrick, J. R., and Taubenhaus, L. J., 1967, Blood alcohol levels of home accident patients, *Quart. J. Stud. Alc.* 28:734.

Kok-Jensen, A., 1970, The prognosis of pulmonary TB in patients with abuse of alcohol, *Scand. J. resp. Dis.* 51:42.

Kok-Jensen, A., 1972, Pulmonary tuberculosis in well-treated alcoholics. Long-term prognosis regarding relapses compared with non-alcoholic patients, *Scand. J. resp. Dis.* 53:202.

Kramer, K., Kuller, L., and Fischer, R., 1968, The increasing mortality attributed to cirrhosis and fatty liver, in Baltimore (1957–1966), *Ann. Int. Med.* 69:273.

Kuller, L. H., Kramer, K., and Fischer, R., 1969, Changing trends in cirrhosis and fatty liver mortality, *Amer. J. Publ. Hlth.* 59:1124.

Lasserre, O., Flamant, R., Lellouch, J., and Schwartz, D., 1967, Alcool et cancer; étude de pathologie géographique portant sur les départements français, *Bull. Inst. nat. Santé* 22:53.

Le Coz, J., 1968, Alcoolisme et milieu sanatorial, *Rev. Tuberc. Pneumol.* 32:299.

Leclainche, X., and Le Gô, P. M., 1970, Alcoolisme et maladies associées, *Bull. Acad. nat. Méd.* 154:373.

Ledermann, S., 1956, "Alcool, Alcoolisme, Alcoolisation: Données Scientifiques de Caractère Physiologique, Economique et Social," Institut National d'Etudes Démographiques, Travaux et Documents, Cahier No. 29, Presses Universitaires de France, Paris.

Ledermann, S., 1964, "Alcool, Alcoolisme, Alcoolisation: Mortalité, Morbidité, Accidents du Travail," Institut National d'Etudes Démographiques, Travaux et Documents, Cahier No. 41, Presses Universitaires de France, Paris.

Lee, F. I., 1966, Cirrhosis and hepatoma in alcoholics, *Gut* 7:77.

Leevy, C. M., Gellene, R., and Ning, M., 1964, Primary liver cancer in cirrhosis of the alcoholic, *Ann. N.Y. Acad. Sci.* 114:1026.

Lelbach, W. K., 1974, Organic pathology related to volume and pattern of alcohol use, in "Research Advances in Alcohol and Drug Problems," Vol. 1 (R. J. Gibbins, Y. Israel, H. Kalant, R. E. Popham, W. Schmidt, and R. G. Smart, eds.), pp. 93–198, John Wiley and Sons, Inc., New York.

LeRoux, R. C., and Smith, L. S., 1964, Violent deaths and alcoholic intoxication, *J. Forens, Med.* 11:131.

Lewis, J. G., and Chamberlain, D. A., 1963, Alcohol consumption and smoking habits in male patients with pulmonary tuberculosis, *Brit. J. Prev. Soc. Med.* 17:149.

Lipscomb, W. R., 1959, Mortality among treated alcoholics. A three-year follow-up, *Quart. J. Stud. Alc.* 20:596.

Lischner, M. W., Alexander, J. F., and Galambos, J. T., 1971, Natural history of alcohol hepatitis. I. The acute disease, *Amer. J. Digest. Dis.* 16:481.

Lolli, G., 1953, Alcoholism in women, *Conn. Rev. Alcsm.* 5:9.

Lundquist, G. A. R., 1970, Alcohol Dependence and Depressive States. Presented at the 16th International Institute on the Prevention and Treatment of Alcoholism, Lausanne.

MacFarland, R. A., and Moore, R. C., 1962, The epidemiology of motor vehicle accidents, *J. Amer. Med. Assoc.* 180:289.

Mäkelä, K., 1971, Alkoholinkulutuksen jakautama, *Alkoholikysymys* 39:3.

Malzberg, B., 1947, A study of first admissions with general paresis to hospitals for mental disease in New York State, year ended March 31, 1945, *Psychiat. Quart.* 21:212.

Martin, C. E., 1967, Epidemiology of cancer of the cervix. II. Marital and coital factors in cervical cancer, *Amer. J. Publ. Hlth.* 57:803.

Massumi, R. A., Rios, J. C., Gooch, A. S., Nutter, D., DeVita, V. T., and Datlow, D. W., 1965, Primary myocardial disease: Report of fifty cases and review of the subject, *Circulation* 31:19.

Mayfield, D. M., and Montgomery, D., 1972, Alcoholism, alcohol intoxication and suicide attempts, *Arch. Gen. Psychiat.* 27:349.

McBay, A. J., 1972, Alcohol and highway fatalities. A study of 961 fatalities in North Carolina during the last six months of 1970, *North Carol. Med. J.* 33:769.

McGee, A. J., 1942, The incidence of syphilis in alcoholic patients; statistical study of 760 consecutive cases, *Illinois Med. J.* 81:69.

Menge, W., 1950, Mortality experience among cases involving alcohol habits, *Proc. Home Office Life Underwriters Assoc.* 31:70.

Menninger, K. A., 1938, "Man Against Himself," Harcourt, Brace, New York.

Milne, R. C., 1970, Alcoholism and tuberculosis in Victoria, *Med. J. Aust.* 2:955.

Moghissi, K. S., and Mack, H. C., 1968, Epidemiology of cervical cancer, *Amer. J. Obstet. Gynec.* 100:607.

Morice, A., 1954, De l'action d'alcool sur le développement du cancer de l'oesophage, *Bull. Acad. nat. Méd.* 138:175.

Morris, J. N., Heady, J. A., Raffel, P. A. B., Roberts, C. G., and Parks, J. W., 1968, Coronary heart disease and physical activity of work, *The Lancet* 2:1053.

Mowrer, E., 1942, "Disorganization: Personal and Social," Lippincott, Philadelphia.

Murphy, G. C., and Robins, E., 1967, Social factors in suicide, *J. Amer. Med. Assoc.* 199:303.

Mussini, H., and Chierici, C., 1960, Sulla etiopatogenesi della cirrosa epatica, *Clinica* (Bologna) 20:25.

Müting, D., Lackas, N., Reikowski, H., and Richmond, S., 1966, Cirrhosis of the liver and diabetes mellutus. A study of 140 combined cases, *Germ. Med. Monthly* 11:385.

Nicholls, P., Edwards, G., and Kyle, E., 1976, A study of alcoholics admitted to four hospitals. II. General and cause-specific mortality during follow-up, *Quart. J. Stud. Alc.*

Nørvig, J., and Nielsen, B., 1956, A follow-up study of 221 alcohol addicts in Denmark, *Quart. J. Stud. Alc.* 17:633.

Palola, E. G., Dorpat, T. L., and Larson, W. R., 1962, Alcoholism and suicidal behavior, *in* "Society, Culture and Drinking Habits" (D. J. Pittman and C. R. Snyder, eds.), pp. 511–534, John Wiley & Sons, Inc., New York.

Parker, R. G. F., 1957, The incidence of primary hepatic crcinoma in cirrhosis, *Proc. R. Soc. Med.* 50:145.

Paul, B., 1963, Anthropological perspectives on medicine and public health, *Ann. Amer. Acad. Pol. Soc. Sci.* 346:34.

Payne, C. E., and Selzer, M. L., 1962, Traffic accidents, personality and alcoholism, *J. Abdom. Surg.* 4:21.

Pell, S., and D'Alonzo, C. A., 1973, A five-year mortality study of alcoholics, *J. Occup. Med.* 15:120.

Péquignot, G., 1958, Enquête par interrogatoire sur les circonstances diététiques de la cirrhose alcooloque en France, *Bull. Inst. Nat. Hyg.* 13:719.

Pincock, T. A., 1964, Alcoholism in tuberculosis patients, *Can. Med. Assoc. J.* 91:851.

Polgar, S., 1962, Health and human behavior: Areas of interest to the social and medical sciences, *Curr. Anthrop.* 3:159.

Popham, R. E., 1970, Indirect methods of alcoholism prevalence estimation: A critical evaluation, *in* "Alcohol and Alcoholism" (R. E. Popham, ed.), pp. 678–685, University of Toronto Press, Toronto.

Popper, H., Rubin, E., Krus, S., and Schaffner, F., 1960, Postnecrotic cirrhosis in alcoholics, *Gastroent.* 39:669.

Potondi, A., Dömötör, E., and Orovecz, B., 1967, Über die Spontanverletzungen Betrunkener, *Acta Chir. Acad. Sci. Hung.* 8:337.

Rajasuriya, K., Thenabadu, P. N., and Wimalaratne, K. D. P., 1970, Aetiological factors of cirrhosis of the liver in Ceylon. A study of 206 cases, *Ceylon Med. J.* 15:80.

Ratnoff, O. D., and Patek, A. J. Jr., 1942, The natural history of Laennec's cirrhosis of the liver, *Medicine* 21:207.

Registrar-General's Decennial Supplement, England & Wales, 1951, Occupational Mortality, Vol. I, part II, Her Majesty's Stationery Office, London, 1958.

Reid, N. C. R. W., Brunt, P. W., Bias, W. B., Maddrey, W. C., Alonso, B. A., and Iber, F. L., 1968, Genetic characteristics and cirrhosis: A controlled study of 200 patients, *Brit. Med. J.* 2:463.

Report to the U.S. Surgeon General, 1971, The Health Consequences of Smoking, U.S. Department of Health, Education, and Welfare, U.S. Government Printing Office, Washington, D.C.

Robins, E., Schmidt, E., and O'Neal, P., 1957, Some interrelations of social factors and clinical diagnosis of attempted suicide; a study of 109 patients, *Amer. J. Psychiat.* 114:222.

Robins, E., Gassner, S., and Kayes, J., 1959, The communication of suicidal intent: A study of 134 consecutive cases of successful (completed) suicides, *Amer. J. Psychiat.* 115:724.

Rubin, V. (ed.), 1960, Culture, society and health, *Ann. N.Y. Acad. Sci.* 84:783.

Rubin, E., Krus, S., and Popper, H., 1962, Pathogenesis of postnecrotic cirrhosis in alcoholics, *Arch. Path.* 73:288.

Rushing, W. A., 1969a, Suicide as a possible consequence of alcoholism, in "Deviant Behavior and Social Process" (W. A. Rushing, ed.), pp. 323–327, Rand McNally, Chicago.

Rushing, W. A., 1969b, Deviance, interpersonal relations and suicide, *Hum. Relat.* 22:61.

Rydberg, U., Bjerver, K., and Goldberg, L., 1973, The alcohol factor in surgical emergency unit, *Acta Med. Leg. Soc.* 22:71.

Sainsbury, P., 1955, "Suicide in London: An Ecological Study," Chapman and Hall, London.

Sakurai, M., 1969, A histopathologic study on the effect of alcohol in cirrhosis and hepatoma of autopsy cases in Japan, *Acta Path. Jap.* 19:283.

Salum, I. (ed.), 1972, Delirium tremens and certain other acute sequels of alcohol abuse. VIII. Mortality, *Acta Psychiat.Scand.* Suppl. No. 235:86.

Schmidt, H., O'Neal, P., and Robins, E., 1954, Evaluation of suicide attempts as a guide to therapy, *J. Amer. Med. Assoc.* 155:552.

Schmidt, W., and Bronetto, J., 1962, Death from liver cirrhosis and specific alcoholic beverage consumption: An ecological study, *Amer. J. Publ. Hlth.* 52:1473.

Schmidt, W., and de Lint, J., 1969, Mortality experiences of male and female alcoholic patients, *Quart. J. Stud. Alc.* 30:112.

Schmidt, W., and de Lint, J., 1970, Social class and the mortality of clinically treated alcoholics, *Brit. J. Addict.* 64:327.

Schmidt, W., and de Lint, J., 1972, Causes of death of alcoholics, *Quart. J. Stud. Alc.* 33:171.

Schmidt, W., Smart, R. G., and Popham, R. E., 1962, The role of alcoholism in motor vehicle accidents, *Traffic Safety Res. Rev.* 6:21.

Schwartz, D., 1966, Alcool et cancer; étude de pathologie géographique, *Cancro* 19:200.

Schwartz, D., Lellouch, J., Flamant, R., and Denoix, P. F., 1962, Alcool et cancer. Résultats d'une enquête rétrospective, *Rev. Franc. Etud. Clin. Biol.* 6:590.

Scotch, N. A., 1963, Medical anthropology, *Bienn. Rev. Anthrop.* 5:30.

Seeley, J. R., 1960, Death by liver cirrhosis and the price of beverage alcohol, *Can. Med. Assoc. J.* 83:1361.

Sepúlveda, B., Rojas, E., and Landa, L., 1956, Los factores etiologicos en la cirrosis del higado, *Rev. Invest. Clin.* 8:189.

Sherlock, S., 1958, "Diseases of the Liver and Biliary System," 2nd Ed., Blackwell, London.

Sigerist, H. E., 1943, "Civilization and Disease," Cornell University Press, Ithaca, New York.

Spain, D. M., Bradess, V. A., and Eggston, A. A., 1951, Alcohol and violent death. A one-year study of consecutive cases in a representative community, *J. Amer. Med. Assoc.* 146:334.

Stare, F., 1961, Myocardial infarction in patients with portal cirrhosis, *Nutr. Rev.* 19:37.

Stenbäck, A., Achté, K. A., and Rimon, R. H., 1965, Physical disease, hypochondria and alcohol addiction in suicides committed by mental hospital patients, *Brit. J. Psychiat.* 111:933.

Stengel, E., and Cook, N. G., 1958, "Attempted Suicide," Chapman and Hall, London.

Stewart, R. A., 1937, The frequency of syphilis and neurosyphilis in chronic alcoholism, *J. Iowa St. med. Soc.* 27:291.

Stone, W. D., Islam, N. R. K., and Paton, A., 1968, The natural history of cirrhosis. Experiences with an unselected group of patients, *Quart. J. Med.* 37:119.
Storby, Å., 1953, Olycksfallsfrekvensen i ett alkoholistmaterial, *Svenska Läkartidn.* 55:2100.
Sundberg, M., and Adlercreutz, E., 1959, Zur Kenntnis der Ätiologie der Leberzirrhose in Finland, *Acta hepat.-splenol.* 6:17.
Sundby, P., 1967, "Alcoholism and Mortality," Universitetsforlaget, Oslo.
Tashiro, M., and Lipscomb, W. R., 1963, Mortality experience of alcoholics, *Quart. J. Stud. Alc.* 24:203.
Taylor, H. L., Klepetar, E., Keyes, A., Parlin, W., Blackburn, H., and Puchner, T., 1962, Death rates among physically active and sedentary employees of the railroad industry, *Amer. J. Publ. Hlth.* 52:1697.
Terris, M., 1967, Epidemiology of cirrhosis of the liver: National mortality data, *Amer. J. Publ. Hlth.* 57:2076.
Tobin, J. R., Driscoll, J. F., Lim, M. T., Sutton, G. C. Szanto, P. B., and Gunnar, R. M., 1967, Primary myocardial disease and alcoholism: The clinical manifestations and course of the disease in a selected population of patients observed for three or more years, *Circulation* 35:754.
Tokuhata, G. K., Digon, E., and Ramaswamy, K., 1971, Alcohol sales and socioeconomic factors related to cirrhosis of the liver mortality data in Pennsylvania, *HSMHA Health Reports* 86:253.
Tonge, J., 1968, Blood alcohol levels in patients attending hospital after involvement in traffic accidents, *J. Forens. Med.* 15:152.
Tonge, J., O'Reilly, M. J. J., Davison, A., and Derrick, E. H., 1964, Fatal traffic accidents in Brisbane from 1935 to 1964, *Med. J. Aust.* 2:811.
Uggla, L-G., 1960, Alcohol and Tuberculosis, Proceedings of the 26th International Congress on Alcohol and Alcoholism, Stockholm.
Viel, B., Donoso, S., and Salcedo, D., 1968, Alcoholic drinking habit and hepatic damage, *J. Chron. Dis.* 21:157.
Vincent, M. O., and Blum, D. M., 1969, A five-year follow-up study of alcoholic patients, *Rep. Alc.* 27:19.
Vincent, R. G., and Marchetta, F., 1963, The relationship of the use of tobacco and alcohol to cancer of the oral cavity, pharynx and larynx, *Amer. J. Surg.* 106:501.
Vital Statistics 1959–1962, Dominion Bureau of Statistics, Ottawa.
Waller, J. A., 1972, Non-highway injury fatalities. I. The roles of alcohol and problem drinking, drugs and medical impairment, *J. Chron. Dis.* 25:33.
Waller, J. A., and Turkel, H. W., 1966, Alcoholism and traffic deaths, *New Engl. J. Med.* 275:532.
Waller, J. A., King, E. M., Nielson, G., and Turkel, H. W., 1969, Alcohol and other factors in California highway fatalities, *J. Forens. Sci.* 14:429.
Wallgren, H., and Barry III, H., 1970, "Actions of Alcohol," Vol. II, Elsevier Publ. Co., Amsterdam.
Wallgren, H., Kosunen, A., and Nikander, S., 1970, Alkoholihaittojen kytkeytyminen juoman laatuun, Alkon Keskuslaboratorio, Seloste 8065.
Wallinga, J. V., 1949, Attempted suicide: A ten year survey, *Dis. Nerv. Syst.* 10:15.
Walton, R. G., 1972, Smoking and Alcoholism: A brief report, *Amer. J. Psychiat.* 128:1455.
Warnery, M., Voisin, R., and Corriol, R., 1958, Morbidité tuberculeuse et alcoolisme; conséquences sociales, *Rev. Tuberc.* 22:544.
Wechsler, H., Kasey, E. H., Thum, D., and Demone, H. W., Jr., 1969, Alcohol level and home accidents. A study of emergency service patients, *Publ. Hlth. Rep.* 84:1043.

Weir, J. M., and Dunn, J. E. Jr., 1970, Smoking and mortality; a prospective study, *Cancer N.Y.*, 25:105.
Wilkinson, P., Santamaria, J. N., and Rankin, J. G., 1969, Epidemiology of alcoholic cirrhosis, *Aust. Ann. Med.* 18:222.
Wynder, E. L., and Bross, J. J., 1961, A study of the etiological factors in cancer of the oesophagus, *Cancer* 14:389.
Wynder, E. L., Mabuchi, K., and Whitmore, W. F., Jr., 1971, Epidemiology of cancer of the prostate, *Cancer* 28:344.
Wu, Y. K., and Loucks, H. H., 1951, Carcinoma of esophagus or cardia of stomach; analysis of 172 cases with 81 resections, *Ann. Surg.* 134:946.
Zeiner-Henriksen, T., 1971, Cardiovascular disease symptoms in Norway. A study of prevalence and a mortality follow-up. *J. Chron. Dis.* 24:553.
Zeller, A. Z., 1967, Cirrhosis of the liver, alcoholism and heavy smoking associated with cancer of the mouth and pharynx, *Cancer* 20:1015.
Žmuc, M., 1968, Alcohol and suicide, *Alcoholism* 4:38.

CHAPTER 9

Alcohol and Unintentional Injury

Julian A. Waller
Department of Epidemiology and Environmental Health
University of Vermont
Burlington, Vermont

INJURY AS A PUBLIC HEALTH PROBLEM

Historical Perspective

Injury has three faces—that deliberately directed against others (assault, homicide), that deliberately turned against oneself (suicide), and that which is unintentional (so-called accidents). According to official mortality statistics, these three types of injury lead to almost 150,000 deaths annually in the United States, 112,000 from unintentional injury, 12,000 from homicide, and 22,000 from suicide. A more accurate estimate, however, is that as many as 50,000 deaths per year are suicides (Anonymous, 1968). An additional 50 million people annually have unintentional injuries serious enough to require medical treatment or to result in loss of a day or more of usual activities (National Safety Council, 1971a). Unintentional injury vies with heart disease as the leading medical cause of lost man-years of productivity in the United States and is the most common cause of death under age forty.

The use and abuse of alcohol is the most important human factor yet identified with severe to fatal unintentional injury, and is a major component in deliberate injury as well. This chapter, which will be limited largely to unintentional injury, describes the nature of this relationship, the types of individuals involved, and appropriate countermeasures to combat this serious health problem.

In describing the association between alcohol and injury, an effort is being made to avoid the use of the term "accident." First, to many the term has connotations of randomness, and consequently of unavoidability, which are clearly refuted by the now extensive scientific data. Second, moral connotations also abound of retribution, whether Divine or in the natural order, for sinful behavior, carelessness, thoughtlessness, or other conscious or subconscious misdeeds. As a logical consequence of such thinking, almost all emphasis on countermeasures in the past has been focused on preventing *behavior* leading to injury events, and rather little on the problem of ultimate concern, namely preventing injury itself and the death, disability, and discomfort associated with it. These moral overtones concerning injury are inextricably interwoven with those concerning the use of alcohol, and the history of injury control activities during the past half-century probably is in large part a reflection of the attitudes that promulgated Prohibition and the various post-Prohibition regulations concerning the manufacture, distribution, and use of alcohol.

A Model of Injury Events and Their Outcome

In order to understand the role of alcohol in injury events and the possible places for successful intervention, it is necessary first to understand the mechanisms by which injuries occur. Man is surrounded by and often uses for his convenience various environmental components that, through construction or use, accumulate large amounts of *physical energy*. The automobile, when moving, has kinetic energy. Stoves and matches transmit thermal energy. Many appliances utilize electrical energy, or in some cases radiation energy. Medicines, household chemicals, and alcohol itself are sources of chemical energy. Human activities such as walking or running convert calories into kinetic energy.

These energy sources serve man well, provided the energy remains under control. Such control usually requires a certain *level of performance* by the person using or potentially exposed to the energy. Performance levels, of course, vary from person to person. For any given individual they also vary from moment to moment. Some important factors, to which we will return later, that may determine performance levels are the learning process, alcohol, personality attributes, and medical impairment.

The *task demands* of dealing with the energy source also determine the extent to which the energy will remain under control. Tasks also vary from one energy source to another and even with the manipulation of a single energy source (such as driving on a narrow mountain road versus on a deserted stretch of freeway). As long as the performance level exceeds the task demands, the energy source remains under control. But if the task demands exceed performance, even if only for a moment, the energy is no longer under control and is available to damage human tissues.

The release of energy can occur in three ways. First, the task may not be excessive but performance level may be markedly decreased, as with acute alcoholic intoxication. Second, the performance may be adequate for usual tasks, but the task itself may be excessively demanding. Finally, a relative decrement of performance in combination with a relatively demanding task may bring about a situation in which task demands exceed performance levels, a pattern that is believed to be applicable to many injury events involving alcohol.

The release of energy precedes but does not in and of itself create the injury. Therefore, the events terminating in energy release are known as the pre-injury phase. In order for injury to occur, the energy must be transmitted to human tissues in sufficient quantity, at a rapid enough rate, and within a sufficiently concentrated body area to exceed injury thresholds for the particular tissues involved. The greater the energy load per units of time and body area, the greater the likelihood that injury will occur and that it will be severe. The transfer of harmful energy to the person, therefore, is known as the injury phase. (A few types of injury such as drowning and poisoning occur not because excessive energy is transferred to the person, but because the injury event interferes with normal patterns of energy transfer that control life, such as transfer of oxygen and carbon dioxide, functioning of enzyme systems, etc.)

The postinjury phase also determines the outcome of the injury event in those cases where the injury is not inherently nonsurvivable. If emergency and definitive care are adequate, a person with serious injury may have complete recovery, whereas if care is inadequate, moderate or even relatively minor injuries may have disastrous outcomes.

Two common misconceptions that this chapter will seek to correct are that alcohol is a factor only in the pre-injury phase of injury events, and that if alcohol is involved other factors are of little or no importance.

It is also worth noting that, although the above model has been used to describe unintentional injury events, it is equally applicable to deliberate events. The task, for example, may involve dealing with an inter- or intrapersonal environment instead of with a physical one. Despite the outraged cries of gun enthusiasts that "guns don't kill; people do," a bullet ejected from a gun transmits energy to tissues at many orders of magnitude greater than does a fist. And whatever the manner of injury, the presence or absence of appropriate emergency care is often a critically important determinant of outcome.

HIGHWAY CRASHES

History, and Problems of Measurement

The first fatality involving the automobile in the United States occurred in 1898. Only six years later the following editorial appeared in the *Quarterly Journal of Inebriety* (Anonymous, 1904):

> We have received a communication containing the history of twenty-five fatal accidents occurring to automobile wagons. Fifteen persons occupying these wagons were killed outright, five more died two days later, and three died a few weeks after the accident, making twenty-three persons killed. Fourteen persons were injured, some seriously. A careful inquiry showed that in nineteen of these accidents the drivers had used spirits within an hour or more of the disaster. The other six drivers were all moderate drinkers, but it was not ascertained whether they had used spirits preceding the accident. The author of this communication shows clearly that the management of automobile wagons is far more dangerous for men who drink than the driving of locomotives on steel rails. Inebriates and moderate drinkers are the most incapable of all persons to drive motor wagons. The general palsy and diminished power of control of both the reason and senses are certain to invite disaster in every attempt to guide such wagons. The precaution of railroad companies to have only total abstainers guide their engines will soon extend to the owners and drivers of these new motor wagons. The following incident illustrates this new danger: A recent race between the owners of large wagons, in which a number of gentlemen took part, was suddenly terminated by one of the owners and drivers, who persisted in using spirits. His friends deserted him, and in returning home his wagon ran off a bridge and was wrecked. With the increased popularity of these wagons, accidents of this kind will rapidly multiply, and we invite our readers to make notes of disasters of this kind.

Between 1904 and the early 1930s when the first serious research endeavors were carried out, many did "make notes of disasters of this kind." The Connecticut Commissioner of Motor Vehicles, for example, reported in 1924 that it is "time to deal harshly with the drunken operator," and in the first 11 months of that year 947 Connecticut drivers had licenses suspended and 254 were jailed for driving while intoxicated (Stoeckel, 1924). Despite the hue and cry, however, official crash statistics did not suggest that alcohol was a very important problem. In illustration, during 1932 only 29 of 378 fatal crashes in Connecticut were reported to involve alcohol (Connecticut Department of Motor Vehicles, 1933). A year later Miles (1934) commented:

> Although alcohol is directly mentioned in only 7 to 10 percent of fatal highway traffic accidents, it is the belief of informed traffic officials that one third of such accidents are at least partly chargeable to use of alcohol by the driver. Officials need a method to definitely determine whether a driver is intoxicated. . . . Determination of percentage of alcohol in the blood or urine

by biochemical means is a feasible undertaking. This method should be tried out in some representative area to secure scientific data in this controversial field.

Such quantitative testing of consecutive samples of fatalities began in a few places within the next five years (Heise, 1934; Holcomb, 1938). Nonetheless, over a quarter of a century later Haddon initiated the first of his several classical studies of alcohol and fatal crashes because he believed that statements based on police crash investigation reports indicating that as few as 1 percent of crashes involved alcohol were grossly inaccurate. His initial study, in fact, showed alcohol to be a factor in almost three out of every four fatal single vehicle crashes (Haddon and Bradess, 1959). Even today statements based only on police investigations continue to substantially underestimate the role of alcohol.

In large part this underestimation occurs because often no statement at all is made on the police report about the presence or absence of alcohol. Where statements are made, they commonly are based on impression only and, except in selected cases chosen through nonrandom sampling, they are not supported by appropriate quantitative tests to determine the blood or breath alcohol concentration (Waller, 1971a). Even when tests are done routinely, further understatement may occur because tests on children are included in the reported results, and because of inclusion of negative determinations on persons who survived six hours or longer before death and who may have entirely metabolized alcohol that was present at time of injury.

Other problems in identifying the role of alcohol in crashes have resulted from the assumption that use of alcohol by drivers is quite common. Before definitive studies were carried out, it was believed by many that alcohol was not present more often among persons involved in crashes than among persons using the roads. In order to answer this question it was necessary to test for alcohol among persons not in crashes but on the highway at times and places where crashes had occurred and to compare these data with those of persons in crashes. As will be documented in subsequent sections, these studies showed that individuals in crashes significantly more often have alcohol in their blood than do persons not in crashes. Among those with alcohol, persons in crashes have much higher blood alcohol concentrations.

It has also been assumed that impairment of *drivers* accounts for most alcohol crashes. Although this assumption is correct if statewide and nationwide data are grouped, it is important also to recognize that in urban areas impairment of pedestrians is a major factor, and in some cities it is the single most important contributor to alcohol-related highway deaths. Finally, attention has been focused on the "drinking driver" with exhortation that even one or two drinks commonly results in impairment sufficient to cause crashes. With certain exceptions, the epidemiologic data do not support these statements and, as will be shown later, the problem of alcohol in serious highway crashes is limited almost entirely to persons with blood alcohol concentrations of 50 mg/100 ml or higher.

Fatal Crashes

Overall, alcohol is estimated to be a factor in about half of all fatal highway crashes, contributing to about 27,000 highway deaths per year in the United States. However, because of the problems noted earlier of inadequate data in many parts of the United States (and most other countries as well), exact comprehensive nationwide figures are not available. Instead, the estimate is based on careful documentation of the problem over a period of many years and in many communities. As noted in the *1968 Alcohol and Highway Safety Report* to the United States Congress (U.S. Department of Transportation, 1968), "these . . . investigations have been concerned with many different types of crashes and violations and have used precise methods of chemical and physical analysis to determine blood alcohol concentrations. The results of this work on actual crashes have been overwhelmingly consistent."

Rather than examine the presence of alcohol among all highway fatalities grouped together, it is more appropriate to determine its involvement among persons whose vehicles apparently initiated the crash (i.e., fatally injured drivers in single-vehicle crashes and "responsible" drivers in two-vehicle crashes) and among fatally injured drivers who apparently were innocent victims, their crashes having been initiated by someone else. Unfortunately, data are not available about alcohol among drivers who themselves survived but who initiated crashes in which others were killed. It seems reasonable, however, to assume that they would be similar to "responsible" drivers who are fatally injured.

Table 1 shows blood alcohol concentrations among "responsible" and single-vehicle driver fatalities, drivers "not responsible" for their crashes, and drivers not involved in crashes but who were tested at times and places where fatal crashes had occurred.* These data have been chosen to represent both urban and rural experience. In sharp contrast to drivers not in crashes, most of the "responsible" and single-vehicle drivers have alcohol in their blood, and most have concentrations of 100 mg/100 ml or higher. "Nonresponsible" driver fatalities are much more similar to the uninvolved drivers than to the "responsible" drivers. Nevertheless, over 10 percent of "nonresponsible" drivers, but only 2 to 4 percent of drivers not in crashes have blood alcohol

---

* Because crash causation commonly involves the interplay of several human and environmental factors, some quite subtle, the terms referring to responsibility have been set off with quotation marks to distinguish between primary involvement in crash initiation, with which we are concerned, and legal culpability, which is not being considered.

TABLE 1. Distribution in Percent of Blood Alcohol Concentrations among Drivers Not Involved in Crashes and among Various Categories of Driver Fatalities

	Blood alcohol concentration (mg/100ml)[a]					
Category of driver	00	<50	50–99	≥100	Total %	Number of cases
All driver fatalities						
47 California counties (Nielson, 1969)	46.8	3.8	5.9	43.7	100.2	(5123)
New York City (McCarroll and Haddon, 1962)	41	\<--- 9 ---\>		50	100	(34)
Vermont (Perrine et al., 1971)	46.2	3.8	8.5	41.5	100.0	(106)
Single-vehicle fatalities						
California	34.8	3.4	5.8	55.9	99.9	(2521)
Westchester County, N.Y. (Haddon and Bradess, 1959)	26	4	13	57	100	(83)
Vermont	33	\<--- 13 ---\>		54	100	(63)
Two vehicle, "responsible"						
California	47.4	4.0	5.7	42.8	99.9	(1589)
Vermont	62	\<--- 7 ---\>		31	100	(29)
Two-vehicle, "not responsible"						
California	79.5	4.4	6.0	10.2	100.1	(685)
Vermont	71	\<--- 21 ---\>		12	99	(14)
Noninvolved drivers						
New York City	76.2	\<--- 19.8 ---\>		4.1	100.1	(217)
Vermont	85.2	7.3	5.2	2.2	99.9	(1125)

[a] Unknowns have been excluded; percents have been rounded to the nearest whole number where N is less than 100.

concentrations of 100 mg/100 ml or higher. This difference suggests that some of the "nonresponsible" drivers may have been sufficiently impaired so that they were unable to take successful evasive actions that a driver without alcohol or with less alcohol might have accomplished.

It has frequently been suggested that highway crashes could be avoided if only impaired drivers would let their passengers do the driving. This suggestion assumes both that such drivers have passengers available and that the passengers are less impaired. For a number of years, however, data on blood alcohol concentrations among passenger fatalities have suggested that at least the second of these assumptions is not correct (Heise, 1934; Nielson, 1969).

One study of crashes in which both the driver and an adult passenger were fatally injured showed that in only a quarter of such crashes did the passenger have a substantially lower blood alcohol concentration than did the driver (Waller, 1972b). Another study indicates that the first assumption also is inaccurate because "only 16% of drivers with 50 mg% or higher had a licensed unimpaired driver in the car as a passenger and about half had no passengers." (Waller et al., 1972b)

As stated earlier, impairment of pedestrians by alcohol has been an important but not commonly recognized contributor to highway fatalities. The Connecticut Department of Motor Vehicles (1933), for example, noted that during 1932 "intoxicated pedestrians caused 87 accidents and these pedestrians were the greatest sufferers.... Fifteen were killed, while deaths among intoxicated operators number 14." The classical study by Haddon et al. (1961) of pedestrian fatalities in New York City showed that 26 percent of fatally injured adult pedestrians had no alcohol in their blood at death, 32% had less than 100 mg/100 ml, and 42 percent had 100 mg/100 ml or higher. In contrast, 72 percent of noninvolved pedestrians tested at times and places of the fatalities had no alcohol, 20 percent had less than 100 mg/100 ml, and 7 percent had 100 mg/100 ml or higher.

It is much more difficult to assess who initiated a crash between a vehicle and a pedestrian than between two vehicles, and the role of alcohol is analyzed in only one such study published to date. Among fatally injured pedestrians ages fifteen to fifty-nine with no alcohol in their blood, 23 percent were considered "responsible" for the crash, whereas 61 percent of pedestrians in this age group with blood alcohol concentrations of 100 mg/100 ml or higher were felt to be "responsible" (Waller, 1972b). Comprehensive data are not available for both the fatally injured pedestrian and the surviving driver because the latter are rarely tested for alcohol. Although it is not known what proportion of the drivers had been drinking, Birrell (1971) has shown that in 38 pedestrian fatalities involving drivers impaired by alcohol the pedestrian also had been drinking in 58 percent.

On occasion, the role of alcohol has also been examined in highway crashes involving vehicles other than automobiles, and for each type of vehicle alcohol has been found to be an important contributor to such events. Thus, Birrell (1967) found alcohol in 88 percent of drivers fatally injured in collisions between automobiles and trams in Australia, while Waller and Goo (1969) demonstrated the presence of alcohol among 37 percent of fatally injured motorcyclists, 87 percent of drivers of pickup and panel trucks (Waller, 1970), and 30 percent of drivers in collisions with trains (Waller, 1968b). In Australia, where the bicycle is used as transportation for adults to a greater degree than in the United States, alcohol has been reported in the blood among 11 percent of fatally injured bicyclists (Tonge et al., 1964), and anecdotal data indicate that alcohol-impaired bicyclists at least occasionally are a problem in the United States as well (Waller, 1974).

Nonfatal Crashes

The frequency with which alcohol contributes to crashes is less well known for nonfatal crashes than for fatal ones because of legal impediments and general resistance to testing of living drivers, who commonly fear loss of insurance, and seek to avoid liability for crash damages. However, one comprehensive study has been completed comparing blood alcohol concentrations of drivers in crashes of all degrees of severity and drivers not in crashes (Borkenstein et al., 1964). These data are shown in Table 2 and clearly demonstrate, first, that persons with blood alcohol concentrations of 50 mg/100 ml or higher are overrepresented in crashes of all degrees of severity and, second, that alcohol concentrations of 110 mg/100 ml or higher are more frequently a factor in serious crashes (as defined by the investigating officer) than in minor ones.

This distinction is important because many individuals tend to make judgments about the role of alcohol in serious events based on their experience with the far more common minor events, which in this case provide an inaccurate picture of the extent of the problem. Despite the lesser frequency of alcohol in minor crashes than in severe ones, this drug is still estimated to be a contributor to at least 800,000 crashes in the United States annually (U.S. Department of Transportation, 1968).

Alcohol in the Preinjury, Injury, and Postinjury Phases

As the foregoing sections amply document, alcohol is an important contributor to the pre-injury or precrash phase. This conclusion is well sup-

TABLE 2. Distribution in Percent of Blood Alcohol Concentrations among Drivers Not in Crashes and among Drivers in Crashes of Varying Degrees of Severity[a]

Severity of crash	Blood alcohol concentration (mg/100ml)				
	00	10–49	50–109	110	Total %
Serious injury	78.0	5.7	5.0	11.3	100.0
Visible injury	80.0	6.7	6.7	6.7	100.1
No injury	83.8	6.9	4.2	5.2	100.1
All degrees of severity	83.4	6.8	4.4	5.4	100.0
Noninvolved drivers	89.0	7.8	2.6	0.6	100.0

[a] Adapted from Borkenstein et al. (1964).

ported by both laboratory and epidemiologic data. A substantial body of laboratory information exists about thresholds for impairment in performance of tasks relevant to driving. The threshold of impairment ranges from 30 mg/100 ml to 70 mg/100 ml, with occasional, light users of alcohol being more sensitive to its effects at low blood alcohol concentrations than are frequent, heavy users. However, at blood alcohol concentrations of 100 mg/100 ml or higher, virtually all persons are significantly impaired, no matter what their previous drinking patterns (Goldberg, 1943).

Laboratory studies of impairment in performing a simulated driving task show that minimal impairment can be measured at blood alcohol concentrations as low as 20–30 mg/100 ml, and that the extent of the deficit increases with each increment of alcohol in the blood (Loomis and West, 1958; Drew *et al.*, 1959). Impairment of performance is greater for complex driving tasks than for simple ones. Epidemiologic data based on actual crash experience, however, indicate that, with two possible exceptions noted later, there is no appreciable increase in risk of crashing at blood alcohol concentrations below 50 mg/100 ml (Borkenstein *et al.*, 1964; Perrine *et al.*, 1971). Above this concentration the risk rises rapidly, so that at 80 mg/100 ml the risk of crashing is four times that with no alcohol in the blood, at 100 mg/100 ml it is six to seven times, and at 150 mg/100 ml about 25 times greater than without alcohol (Borkenstein *et al.*, 1964; Perrine *et al.*, 1971).

Overwhelmingly, crashes involving alcohol occur at night, especially on weekends, when alcohol is in greatest use but when traffic may be relatively light (U.S. Department of Transportation, 1968). Zylman (1972) has asked whether persons driving with alcohol during hours of heavy traffic may be unable to cope with greater traffic demands and, therefore, may have a higher crash risk at blood alcohol concentrations not usually considered to be hazardous. His re-examination of the original Grand Rapids data shows that "after midnight even BACs of 0.07% [70 mg/100 ml] appeared in collisions less often than in the control, but during heavy daytime traffic even BACs of 0.01% to 0.04% [10-40 mg/100 ml] appeared in collisions more often than those who had nothing to drink."

Fortunately, however, as documented in the same report, relatively few persons drive with even these low blood alcohol concentrations during hours when traffic is heaviest and these low concentrations increase the over-representation in crashes only "slightly." Another reanalysis of the Grand Rapids study showed that very young and very old drivers have a modest increase in crash risk at blood alcohol concentrations under 50 mg/100 ml (Hyman, 1968).

Two other components of alcohol impairment need to be considered in the precrash phase. There is some speculation, but little definitive study as yet, that drinks with high congener content may, ounce for ounce, result in greater impairment to driving than do drinks low in congeners. Question has also been raised about the contribution to crash occurrence of impairment attributable to hangovers. Again, no hard data exist, but Glatt (1964) has noted that among alcoholics under treatment, "The danger in these alcoholic drivers was not limited to the time shortly after heavy drinking. Quite a few patients remarked spontaneously that the time they felt 'much less sure' of themselves was 'the morning after,' at a time when, as one woman alcoholic put it 'nobody could notice it in you.'"

What is the role of alcohol once a driver is threatened with a collision? As noted earlier, some drivers with alcohol may be impaired sufficiently so they are unable to take appropriate evasive actions to avoid a crash initiated by another driver. Hess (1972) has stated that recent studies of automobile and tire dynamics and feedback of information to the driver indicate that even a moment's delay in action often makes it humanly impossible for any driver, whether impaired by alcohol or entirely competent, to alter the direction of the vehicle. Thus, modern vehicle and tire design may substantially and unnecessarily exacerbate the effect of even a small amount of impairment by alcohol by unnecessarily increasing task demands.

Once the vehicle has stopped, alcohol continues to be a contributory factor to the outcome. Drivers who drown in automobiles that go into bodies of water have extraordinarily high blood alcohol concentrations, even for single-vehicle crashes, suggesting that befuddlement by alcohol rather than severity of injury may have prevented their escape (Davis and Fisk, 1964–1966). Similarly, such alcoholic confusion has resulted in deaths to persons whose cars stalled on railroad tracks as a train was approaching (Waller, 1968b).

Finally, alcohol and sequelae of chronic alcohol abuse may complicate the treatment of crash injuries. Acute alcoholic intoxication can either mask or mimic signs of injury. Routine determination should be made, therefore, of blood alcohol concentrations for all seriously injured adults under medical treatment. The outcome of other acute effects of alcohol on the immediate survival process is not well documented. Data are needed, for example, on the frequency of airway obstruction because of a combination of altered consciousness from alcohol, facial or chest injury, and the possible effect of alcohol abuse on bleeding. Both heavy smoking of cigarettes and fatty changes of the liver have been found excessively among persons in crashes with high blood alcohol concentrations (Waller et al., 1969; Waller and Thomas, 1972). These correlates of heavy alcohol use suggest that the physician called upon to treat the seriously injured may have to deal in some cases not only with the acute injury itself but also with chronic respiratory disease, altered ability of the liver to metabolize medications, and occasionally even delerium tremens.

OTHER TRANSPORTATION INJURIES

In comparison with what is known about highway crashes, very little information can currently be found about injuries involving other means of transportation. The admittedly meager data, however, do indicate that use and abuse of alcohol is a problem, and in some cases already documented to be a very serious problem, in all types of transportation mishaps in which even cursory examination has been made.

The few studies of fatal general aviation crashes (excluding commercial aviation) certainly suggest that alcohol is a frequent contributor to such events, but the extent of the problem is not yet clear. An initial study in 1964 (Harper and Albers, 1964) identified alcohol in 35 percent of 158 pilot fatalities, but it was based on examination of only a third of possible crashes. A later study involving 74 percent of potential subjects in noncommercial crashes showed 24 percent of pilots with positive blood alcohol determinations (National Transportation Safety Board, 1969). In contrast, only eight airmen had alcohol present among over 2,000 military aircraft fatalities (Davis, 1968) and none of 38 fatally injured commercial pilots had been drinking (Anonymous, 1964a).

Also relevant is the linkage of injury events involving different modes of transportation. A study of the driving records of 47 pilots involved in alcohol related general aviation crashes indicates:

> Thirty percent of the pilots had a prior record of offenses involving alcohol, principally conviction for driving while intoxicated or for public drunkenness. The 30% figure should be construed as the tip of an iceberg since motor vehicle violation records are but one source of information on a person's alcohol problem. In addition, the records of convictions of non-alcohol related offenses are, in some cases, reductions in charges of an offense involving alcohol. (Hricko, 1970)

According to a recent report of the National Transportation Safety Board (1969):

> In the case of railroad transportation accidents, we have extremely meager data on alcohol. The railroads are not required to report on this aspect of the accidents they investigate. But there are two rail accidents in which alcohol has been identified in a full investigation, and there are other recent cases in which it is suspected.

"Intoxication" is reported to be associated with less than 5 percent of deaths in commercial marine transportation and only 3 percent of fatalities in pleasure boating. However, a study by the National Transportation Safety Board (1969) of reports of "drowning by falling overboard" showed that nearly 13 percent involved persons who had been drinking. The board notes that "there is reason to believe these figures, especially for recreational boating, are the result of considerable underreporting. It would take very careful planning and very persistent study to determine the facts, because of the lack of surveillance that is characteristic of recreational boating."

The rapid increase in use of snowmobiles and, more recently, all terrain vehicles, has raised the specter of alcohol involvement in many of the frequent and serious injuries that have been occurring. One study has found that drinking occurred in 40 percent of such injury events involving adults, a frequency twice as great as for snowmobilers not injured (Waller and Lamborn, 1973). Another study identified alcohol in 27 percent of all snowmobile collisions on highways and 58 percent of such fatal collisions (Ontario Department of Transportation, 1970). In some communities winter "roadhouses" are reported to have been established that can be reached only by snowmobile.

NONTRANSPORTATION INJURY

The contribution of alcohol to nonhighway injury was noted over three thousand years ago. As early as 1500 B.C. an Egyptian papyrus carried the following warning: "Make not thyself helpless drinking in the beer shop . . . Falling down, thy limbs will be broken, and no one will give thee a hand to help thee up" (Murphy, 1972). Evidently the common misconception that "drunks never get injured because they are so relaxed when they fall" was not shared by this early writer.

More recently, Hindmarsh and Linde found in 1933 that 38 percent of 170 persons hospitalized in Stockholm for nontransportation injuries had alcohol in their blood, and 24 percent had blood alcohol concentrations of 100 mg/100 ml or higher (Miles, 1934). Nevertheless, in comparison with data on highway crashes there is extremely meager information about the frequency of alcohol involvement in injuries occurring at home, during recreation, at work, or in public places. Most of the studies have been reviewed by Demone and Kasey (1966). They state:

> Data from the reported studies tend to be scanty and scattered and varying in quality and quantity. Although some knowledge of the relation of alcohol to industrial accidents could be gained from the research, the status of such information on accidents occurring at home or in recreational areas or public places is vague. . . . Control samples were seldom built into the research design.

Subsequent work by these authors and their co-workers has provided the only study to date of nontransportation injuries of all degrees of severity in which a reasonable attempt has been made to obtain a comparison group (Wechsler *et al*, 1969). Ideally, the comparison group should be composed of persons exposed to similar circumstances of time, place, and activity but who remained in good health. Instead, the comparison group here is persons who entered an emergency room for treatment of illness rather than injury. The presumption is that illnesses are not substantially affected by use of alcohol, an assumption that we know to be incorrect in some cases. The comparison, therefore, in this study is between injuries, for which it is hypothesized that alcohol is commonly a factor, and illnesses, in which alcohol may be a factor much less often and only for specific conditions (e.g., digestive disorders and chronic respiratory conditions).

Another recent study provides data for fatalities only (Waller, 1972a). Although based on a much smaller sample, this study has the advantage of comparing the injured with persons who became acutely ill and died suddenly in similar environments to those of the injury fatalities. The author of this study believes that alcohol may be present among persons who die suddenly of illness less often than identified in this research because the illness deaths from the coroner's office do not represent a 100 percent sample of such sudden deaths in the geographic area but rather reflect some socioeconomic biases in selection. These two studies are summarized in Tables 3 and 4.

The data reveal a pattern quite consistent with that of highway crashes.

TABLE 3. Distribution in Percent of Blood Alcohol Concentrations among Persons with Nonhighway Injuries and with Illness[a]

Type of event	Blood alcohol concentration (mg/100 ml)				
	00	<50	≥50	Total %	Number of cases
Nonfatal					
Home injury	77.7	11.0	11.3	100.0	(620)
Occupational injury	84.4	10.6	4.9	99.9	(969)
Other	75.9	10.9	13.2	100.0	(808)
Noninjury—all	91.2	6.3	2.6	100.1	(2633)
	00	<100	≥100	Total %	Number of cases
Fatal					
Home injury	46	2	52	100	(52)
Recreational injury	70	15	15	100	(20)
Occupational injury	100	0	0	100	(18)
Illness at home	85	6	9	100	(162)
Illness during recreation	88	0	12	100	(8)

[a] Nonfatal data adapted from Wechsler et al. (1969); fatal data adapted from Waller, (1972a).

TABLE 4. Percent of Persons with Alcohol in Their Blood among Persons with Various Types of Nonfatal and Fatal Injuries in the Nonhighway Setting[a]

	Percent with alcohol	
Type of event	Nonfatal	Fatal
Falls	23	70
Fires	18	64
Submersion	—[b]	30
Ingestion	—	82
Cutting and piercing instrument	26	—
Collision with persons or objects	25	—
Other	11	19

[a] Nonfatal data adapted from Wechsler et al. (1969); fatal data adapted from Waller, (1972).
[b] No such category in the study.

First, except for occupational injuries, alcohol is very commonly involved in nonhighway injury events, substantially exceeding its presence among persons with acute illness. There is only modest overrepresentation of alcohol among injuries in the occupational setting. Second, as observed earlier with highway crashes, alcohol is much more often a factor in fatalities than it is in injuries of lesser severity.

The very extensive experience of the Cuyahoga County Coroner's Office (Cleveland, Ohio) has been reviewed for several recent years and indicates a pattern quite similar to that described above (Cuyahoga County Coroner, 1960-1961, 1965-1967, 1969-1970). Alcohol was found in the blood of 40 percent of persons in 851 home injury deaths, 10 percent of 208 deaths from industrial injury, 46 percent of injury deaths elsewhere in the nonhighway setting, and 10 percent of almost 21,000 deaths from "natural causes."

The above data suggest that although alcohol is involved substantially in most types of nonhighway injury events, it is not involved as often in occupational injuries. According to Observer and Maxwell (1959), problem drinkers have higher on-the-job injury rates than other workers of the same age only during their early problem drinking years, but not after they become experienced problem drinkers. Trice, as quoted by Brenner (1967), believes that five factors may account for this. The problem drinker may (1) reduce his work to a routine he knows well, (2) become especially cautious of job hazards, (3) be protected by fellow workers, (4) stay away from work while incapacitated, and (5) be assigned by supervisors to less hazardous work.

Only one attempt has been made to evaluate the relationship between blood alcohol concentration and responsibility for nontransportation injury events (Waller, 1972a). According to the author of this study:

> Many factors determine the occurrence and outcome of an injury event, and the term "responsibility" therefore is not used here to connote culpability. Persons were classified as "responsible" if they played *any* part in the initiation of the injury event. Thus, all persons who fell (unless they were pushed), drowned or died as a result of ingestion were classified as "responsible." Persons were classified as not responsible if they were injured as a result of product failure, fire that started in a neighboring apartment, or other events that clearly could not be attributed in any way to the individual who was injured.... Among 77 persons "responsible" for their deaths 57% had no alcohol, 7% had less than 100 mg% and 37% had 100 mg% or higher. In contrast 93% of 14 persons "not responsible" had no alcohol, and 7% had 100 mg% or higher.

If the alcohol experience of the above reports is applied to the entire non-highway injury experience of the United States, it would suggest that as a minimum estimate alcohol is a contributory factor in 7300–8400 fatal and 3.2 million nonfatal injuries in the home, 700 fatal and 110,000 nonfatal injuries at work, and 7400 fatal and 2.8 million nonfatal injuries in other nonhighway settings.

These estimates are arrived at by assuming contributory involvement of alcohol in 35–40 percent of 20,900 home fatalities age fifteen or older based on 1970 data (National Safety Council, 1971a), 5 percent of the 14,200 work fatalities, 40 percent of 18,400 nonhighway fatalities elsewhere age fifteen or older, 20 percent of 16 million nonfatal home injuries (20.2 million home injuries from National Health Survey data as reported by National Safety Council (1971a) and prorated to exclude persons under age fifteen), 5 percent of 2.2 million nonfatal work injuries, and 20 percent of 14 million nonfatal injuries elsewhere (prorated for age). Thus, alcohol must be identified as a very important contributor to the problem of nonhighway injury.

DELIBERATE INJURY

It is not intended in this chapter to dwell at length on the relationships between use and abuse of alcohol and occurrence of deliberate injury, whether to one's self or to others. Nevertheless, as will be discussed later, there is some evidence that the individuals who are excessively involved in alcohol-related unintentional injury events often are the same ones involved in deliberate injury. For this reason the contribution of alcohol to the deliberate events should be mentioned at least in passing.

Extensive data from the Cuyahoga County Coroner (1960–1963, 1965–1967, 1969–1970), for example, show alcohol in the blood among 29 percent of 1382 successful suicides and in 59 percent of 1263 homicides. Wolfgang (1958) has pointed out in his classical study of homicide that "it is *significant* that of these 374 cases in which alcohol was a present factor, nearly seven tenths were those involving the presence of alcohol in *both* the victim and the offender." (Italics his.) He further states, "In many cases . . . the victim is a major precipitator of the homicide and the fact that he had been drinking prior to his own death is of no small consequence in causing the offender to strike out against him." Wolfgang has also noted that homicides in which alcohol is a factor tend to be associated with more extensive violence (for example, multiple stabbing) than do those without alcohol.

ALCOHOL IN COMBINATION WITH OTHER DRUGS

Experimental evidence exists that alcohol plus other depressant drugs can produce impairment because of synergistic action that exceeds the effect of the two drugs individually (Forney and Hughes, 1968). The safety literature, therefore, is replete with warnings about the probable hazard of small amounts of alcohol in combination with other drugs. Almost no information exists, however, about the demonstrated relationship between such combinations and the occurrence of injury events. In one study of single-vehicle fatalities, 31 of 48 persons with barbiturates in their blood also had alcohol present and 27 of these had blood alcohol concentrations of 100 mg/100 ml or higher (California Department of Highway Patrol, 1967). Five out of six persons with tranquilizers had alcohol, and three had alcohol concentrations in this high range. Although these proportions are high, they are no higher than the proportions of single-vehicle fatalities with alcohol alone. Furthermore, the concentrations of alcohol observed in combination with drugs in this study were sufficiently high to explain most if not all of the impairment that led to these crashes.

These data do not and cannot indicate the frequency with which small amounts of alcohol act in synergistic combination with drugs to produce impairment that results in crashes. They do rather strongly suggest that in this particular series the amount of alcohol present was so great that any synergistic effect, if present, was probably overshadowed by the effects of the alcohol per se. The heavy drinkers who used drugs in this study were probably at greater risk of crashing because of their drinking rather than because of the other drugs consumed. Whether this pattern is the typical one in alcohol–drug combinations, or whether true synergistic action involving drugs and lower blood alcohol concentrations is more often the rule remains to be demonstrated by further study.

It has been shown that the risk of involvement in a fatal crash is greater for drivers who are heavy smokers (or have carboxyhemoglobin concentrations of 5 percent or higher) and who also have blood alcohol concentrations of 20–

99 mg/100 ml than it is either for heavy smokers or persons with these alcohol concentrations alone (Waller and Thomas, 1972). The increase in risk is only modest, and probably additive, however, rather than the result of synergism. At blood alcohol concentrations of 100 mg/100 ml or higher, the effect of alcohol appears to be sufficiently great so that it overshadows any additional effect of the smoking that might exist because of carbon monoxide or psychological reasons.

CHARACTERISTICS OF ALCOHOL USERS IN INJURY EVENTS

Historical Perspective

The first studies of alcohol and highway crashes completed during the 1930s showed that the overwhelming majority of fatalities with alcohol had blood alcohol concentrations of 100 mg/100 ml or higher (Heise, 1934). Subsequent studies have repeatedly verified and supported these data. Since such concentrations can be reached only through heavy drinking, it should have been expected that attention would be directed rather quickly to the abusive drinker as an important contributor to alcohol crashes. Instead, with only two exceptions (Canty, 1940; Selling, 1941), these epidemiologic data were virtually ignored and much more attention was paid to laboratory studies, noted earlier, indicating that some impairment can be identified at blood alcohol concentrations as low as 30 mg/100 ml.

Many of these laboratory studies were completed during the years immediately following the repeal of Prohibition when there was widespread concern about opening the floodgates of alcohol availability. It is suspected that this concern may have been an important factor in the decision to place major emphasis upon the small problem of crashes attributable to ingestion of only one or two drinks, and to completely avoid the very much larger problem of heavy use of alcohol and the persons who so abuse the drug.

Unfortunately, the public has been misled by erroneous statements about heavy drinkers. For instance, one important public education release by the insurance industry in 1957 (Association of Casualty and Surety Companies) in fact pointed out that the person who is grossly and obviously impaired by alcohol may be a much safer driver than the person with only a small amount of alcohol, a statement clearly and repeatedly refuted by the scientific facts. Even more recently the American Automobile Association (1971) published a driver education text, widely used throughout the United States, which carries the same message that the major problem in highway safety is social drinkers impaired by only small amounts of alcohol.

As will be documented on the following pages, the preponderance of scientific evidence supports the conclusion that, in the United States at least, about two thirds of drivers and pedestrians with alcohol-related mishaps on the highway are persons with drinking problems, and that the remaining third is comprised almost entirely of teen-aged experimenters with alcohol and young men who are heavy social drinkers (that is, who can still pace their drinking sufficiently to largely avoid health, social, and economic problems attributable to uncontrolled consumption). The much scantier data about characteristics of persons in nonhighway injury events involving alcohol also implicate problem drinking as the preponderant factor.

Highway Injury

In 1953 Bjerver, Goldberg, and Linde presented the first fairly comprehensive documentation of characteristics of persons in crashes involving alcohol (see Bjerver *et al.*, 1955). Although "alcohol addicts," "alcohol abusers," and "excessive drinkers" comprised 13.7 percent of men over age twenty-five in Stockholm, they comprised 32.5 percent of drivers in crashes resulting in injury and 70 percent of those with alcohol in their blood. (This study, incidentally, is virtually unknown to the many individuals who believe that the Scandinavian system of jailing drivers arrested with alcohol in their blood is highly successful because most social drinkers appear to obey the law.)

These data were soon corroborated by a study in the United States that showed not only that young airmen involved in highway crashes usually had been drinking heavily and, despite their youth, commonly had drinking problems, but also that in many cases they came from families in which one or both parents were alcoholic (Barmack and Payne, 1961a). At about the same time, three groups of researchers in the United States and Canada documented also that persons with alcoholism have twice as many crashes and traffic citations per driver (Payne and Selzer, 1962) and per unit miles traveled (Schmidt and Smart, 1959; Waller, 1965) as nonalcoholic drivers, and that their entire excess is attributable to crashes while under the influence of alcohol (Schmidt and Smart, 1959; Waller, 1968a). Furthermore, the alcoholics had ten times as many arrests for driving while intoxicated (DWI) as did the general population of drivers (Waller, 1968a).

It was not until the late 1960s, however, that more extensive information became available about characteristics of persons in the entire range of traffic experiences, from those with no previous crashes or citations, to persons with nonalcohol-related crashes or citations, and persons arrested for DWI or in alcohol crashes (Waller, 1967; Perrine *et al.*, 1971). Representative data concerning demographic characteristics of these different groups appear in Table 5.

TABLE 5. Distribution in Percent of Selected Characteristics among Drivers Not in Crashes but Stopped at Research Roadblocks, Fatally Injured Drivers with and without Alcohol, and Drivers Arrested for DWI or Nonalcohol Traffic Offenses[a]

Characteristics	Roadblock drivers with clear traffic records	Roadblock drivers, BAC < 20 mg/100 ml	Drivers with serious nonalcohol citations	Fatalities BAC <20mg/100 ml	Roadblock drivers; BAC ≥100 mg/100 ml	DWI	Fatalities BAC ≥100mg/100 ml
Male	73	79	97	77	83	98	95
Age under 25	23	28	73	35	25	22	45
Age 60 or older	13	10	0	20	0	16	5
Currently married	75	81	42	95	89	48	72
Lower occupational class	8	27	60	14	18	42	19
Usually drinks 3–4 drinks/sitting	19	15	29	0	15	27	33
Usually drinks 5 or more drinks/sitting	7	11	29	0	27	60	33
Number of cases	(63)	(969)	(40)	(49)	(24)	(50)	(44)

[a] Adapted from Perrine et al. (1971).

As this table shows, fatalities with alcohol tend to be much younger than most other groups, including drivers with alcohol who do not get into trouble and persons who are arrested for DWI. Both fatalities with alcohol and the DWIs are overwhelmingly comprised of males, and have large proportions of individuals who are not currently married and whose usual alcohol consumption is at least weekly and in medium or heavy quantity. (Medium drinkers consume three to four drinks/sitting and heavy drinkers exceed this amount.) Not noted in this table, but of some importance, is the fact that the preferred beverage of most of these individuals is beer. The DWIs are heavily weighted with persons in lower occupational classes.

Special attention is directed to drivers with serious moving violations not recorded as alcohol related. Except for their extreme youth, these persons all have the hallmarks of the DWIs, a similarity easily explained by the fact that they include many persons who probably had been drinking to some extent but who, for various reasons, were charged by the police with other offenses not labeled as involving alcohol.

Table 6 shows information from another study about the frequency of alcohol problems among these groups (Waller, 1967). Clearly, both the DWIs and drivers in alcohol crashes far exceed other groups in the frequency of previous problems, especially those involving alcohol. This conclusion is supported by other studies of both drivers and pedestrians in alcohol crashes.

Selzer et al. (1968), for example, found that 59 percent of drivers who had been drinking before involvement in fatal crashes were alcoholics, whereas only

TABLE 6. Distribution in Percent of Previous Contacts with Police and Other Community Agencies among Drivers with Clear Traffic Records, Traffic Citations with or without Alcohol, and Crashes with or without Alcohol[a]

Type of agency contact	Clear record	Nonalcohol traffic citation	Crash reported as nonalcohol	DWI	Alcohol or hit-and-run crash
Known to some community agency	19	34	39	87	76
Known to community agency for alcohol incident or problem	10	28	32	81	67
Previous arrest	15	33	34	84	67
2 or more previous alcohol arrests	3	8	14	63	50
Previous DWI	2	9	11	46	21
Previous arrest for violent behavior	5	8	8	27	36
Number of cases	(150)	(131)	(117)	(150)	(33)

[a] Adapted from Waller (1967).

3 percent of nondrinking drivers in fatal crashes were alcoholics. Among the alcoholics, 92 percent had been drinking. Furthermore, most of the alcoholics had blood alcohol concentrations of 150 mg/100 ml or higher, but most of the nonalcoholics who had been drinking had blood alcohol concentrations below 100 mg/100 ml.

Despite their similarities in many respects, several differences exist between drivers in alcohol crashes and drivers arrested for DWI. It is currently believed that most of these differences can be attributed to the fact that fewer DWIs than fatalities are under age twenty-five because young drivers commonly crash at blood alcohol concentrations under 150 mg/100 ml, at which some officers are unwilling to arrest for DWI. For administrative purposes it is important to know whether drivers arrested for DWI are likely to be the same individuals who get into alcohol crashes, or whether they are different populations. One study, therefore, attempted to compare DWIs age twenty-five or older with driver fatalities who had blood alcohol concentrations of 100 mg/100 ml or higher and who also were age twenty-five or older. According to the authors of this study, "With only a few exceptions, the data suggest that there are major similarities between DWIs and driver fatalities who had alcohol. We must conclude that, to a substantial degree, these two subgroups of high alcohol drivers were probably drawn from a single population" (Perrine et al., 1971).

Two other aspects of this problem need to be explored. First is the question whether the problem drinkers involved are by and large persons with blatant alcoholism or more commonly persons with drinking problems short of alcoholism. Although Selzer et al. (1968) on the one hand clearly identify most persons in alcohol crashes as "alcoholics," Kelleher (1971) has said that only 20 percent of persons arrested in Chicago for DWI are "alcoholics" and that 80 percent are "social drinkers." Waller (1972c), basing his conclusions on his own data and on studies by Cahalan (1970a) of the epidemiology of problem drinking, has suggested that an excessive proportion (compared to the general population) of drivers with alcohol crashes or DWI arrests are alcoholics. However, the average driver arrested for DWI does not fit the classical picture of alcoholism but more often has a prealcoholic pattern or, as Cahalan has documented, even a relatively temporary drinking problem, for which he could use professional help but which may be solved as he matures.

The second question is whether all persons with drinking problems are equally at risk of alcohol crashes, or whether some are more injury prone than others. This question is only beginning to be explored by a handful of researchers. Selzer et al. (1968) and Waller (1967) have both noted that drivers in alcohol-related crashes frequently have previous histories of assaultive behavior while under the influence of alcohol. They have suggested that heavy drinkers who tend to "act out" normally suppressed agressive tendencies after drinking are at greater risk of crashing than those without this pattern. It is possible, of course, that persons who act out tend to drive more often after drinking, and the difference in crashes, therefore, may reflect differences in exposure to risk.

Selzer *et al.* (1968) also showed that persons with paranoid thinking, suicidal thoughts or acts, and depression have greater numbers of serious crashes per driver than do persons without such patterns. Some of Selzer's data have been verified by Smart (1969), who suggests that underlying personality characteristics of the drinker may play as important a role as pharmacologic effect of alcohol in determining the extent and nature of impairment to driving. However, as noted before, most problem drinkers in crashes have such high blood alcohol concentrations that the pharmacologic action per se should not be underemphasized.

Cultural factors also must be considered. A recent study from Australia suggests that problem drinkers do not comprise a substantial proportion of Australians in alcohol crashes (Whitlock *et al.* 1971). In commenting on these data, Waller (1971b) stated:

> Patterns of social drinking vary widely from one country and one culture to another. In some cultures heavy drinking may be rather common but may not very often denote a serious problem with alcohol. In other cultures—and I believe the United States fits this pattern—heavy drinking may denote problems among a fair proportion of the individuals who consume large quantities of alcohol . . . Adequate studies of normative drinking practices should be carried out in Australia before you reach conclusions based upon the American experience.

One study in fact has shown that, unlike in the United States, a majority of Australians can be classified as usual heavy drinkers (Encel and Kotowicz, 1970). Negroes, American Indians, and Mexican-Americans have been found to be substantially overrepresented in alcohol related crashes in the United States (Freimuth *et al.*, 1958; Waller, 1967). Both heavy drinking and problem drinking are common to these racial or cultural groups, and it is not known at this time what comparative roles problem drinking and heavy social drinking play in their crashes. The following statement by Cahalan (1970b) is relevant to this question:

> Analysis of the correlates of a *typology of problem drinking* [italics his] showed that problems with obvious social consequences—problems with one's spouse or relatives, friends or neighbors, concerning one's job, or with the law or police—were most common (relative to other problems) among men under 60 (particularly in large cities), among Irish Catholics and among those of Latin-American/Caribbean origin, but extremely uncommon among Jews. These differences appear to be congruent with differences in acting-out tendencies which might be expected among men in contrast to women, among younger men (particularly in the more abrasive and alienated larger cities), and among various ethnocultural groups which differ in their styles of expressing tensions and aggressive tendencies.

Finally, attention should be directed to the special experiences of young drivers because they appear to comprise the majority of social drinkers who get into trouble with alcohol on the highway. It was noted earlier that young drivers comprise a much larger proportion of alcohol-related fatalities than of persons arrested for DWI. Two factors appear to be acting here. Drinking and heavy drinking are more common among young men than among any other age-sex group of drivers (Perrine et al., 1971). During teen years most drivers are coping with three learning curves. They are new and inexperienced drivers; they are inexperienced drinkers; and they are just learning how to combine these two activities. Furthermore, to perhaps a greater degree than at any time subsequently, they are exposed to peer pressures to conform.

It should be no surprise, therefore, that teen-aged drivers are overrepresented in crashes with alcohol and that they tend to get into trouble at lower blood alcohol concentrations than do older drivers, concentrations sufficient to increase their crash risk but sometimes not sufficiently high to convince a police officer to arrest them for DWI (Hyman, 1968; Perrine et al., 1971). Their crash experience, by and large, can be attributed to normative social drinking for their age group.

As experience is gained with driving, with drinking, and with both combined, however, there tends to be even greater experimentation because of overconfidence built upon initial successes (Schuman, Pelz, and Ehrlich, 1967). Thus, young men between the ages of twenty and twenty-five also are overrepresented in crashes. Their blood alcohol concentrations are substantially higher than those of the teen-aged driver but, again, they often get into trouble at somewhat lower concentrations than do older drivers (Perrine et al., 1971). In this age group the heaviest drinkers appear to be in the lower socioeconomic classes. Although some persons who get into trouble with alcohol at this age clearly have drinking problems or other forms of sociopathic behavior, it is suspected that most are still heavy social drinkers.

Nonhighway Injury

Only meager information is available about characteristics of persons whose injuries involve alcohol in the nonhighway setting. Individuals with known alcoholism have substantially higher death rates from nonhighway injury than have nonalcoholics (Brenner, 1967). According to one study, nonfatally injured persons with blood alcohol concentrations of 50 mg/100 ml or higher were more likely to be male, under age sixty-five, and currently unmarried (Wechsler et al., 1969). Another study found that among injury fatalities with alcohol in their blood, only 12 percent had no indicators of problem drinking, whereas 53 percent had a positive history or actual diagnosis of alcoholism,

6 percent had a fatty or cirrhotic liver, alcohol arrests or both but a negative history, and 29 percent had history unknown but at least one other indicator positive. In contrast, 78 percent of persons fatally injured without alcohol had no indicators of problem drinking, and only 5 percent had a positive history. Equally impressive, 86 percent of persons without indicators had no alcohol in their blood when they died in contrast to only 10 percent of those with a positive history or diagnosis of alcoholism (Waller, 1972a).

As noted earlier, evidence is beginning to be gathered about linkage of several types of injury events in the same population. Thus, fliers fatally injured in general aviation crashes involving alcohol commonly had DWI arrests (Hricko, 1970). A study of persons who *unintentionally* shot themselves or others showed substantial overrepresentation of previous arrests for deliberate assault, and for misuse of alcohol, and of previous highway crashes, traffic citations, and license suspensions when compared to drivers the same age and sex, most of whom own guns but who had not been involved in unintentional shootings (Waller, 1969). Considerably more research is needed on the frequency and nature of such linkages.

COUNTERMEASURES TO ALCOHOL-RELATED INJURY

Overview of Options within the Systems Approach

It was noted in the beginning of this chapter that the traditional approach to reduction of injury involving alcohol has been to try to change the actions of people who use alcohol. If the basic goal is reduction of *human morbidity and mortality,* however, rather than the prevention of *events* that might produce injury, several options are available based on the epidemiologic model described earlier. The entire range of options has been comprehensively and innovatively explicated by Haddon (1970). In this chapter six categories of countermeasures to alcohol-related injury will be examined. These are:

1. Control of availability of alcohol with respect to time, place, user, and type of drink.
2. Control of user behavior to reduce impairment or exposure to risk.
3. Early identification of alcohol abusers and their removal from hazardous situations.
4. Environmental controls aimed at the pre-injury phase.
5. Moderation of energy transfer during the injury phase.
6. Improvement of emergency and definitive care for the injured.

The first three categories are basically concerned with controlling human behavior; the latter three with intervening in other ways.

Controlling Availability of Alcohol

Relatively few hard data exist concerning the effect with respect to safety of controlling availability of alcohol, and what data do exist are limited almost entirely to highway safety. It is generally conceded that total prohibition of alcohol in the United States was un unsuccessful experiment because of massive bootlegging in response to widespread public ambivalence or even opposition to the law. Scanty data suggest that this assessment is relevant to the effect of prohibition on highway safety as well (U.S. Department of Transportation, 1968). However, short-term decreases in availability of alcohol because of strikes by distillers or beverage distributers have been associated with reduction in crash rates and in injuries from other causes as well, perhaps because the time intervals have been insufficient to permit establishment of alternative sources of supply (Anonymous, 1964b).

In many places specific legislation governs the time of sale of alcoholic beverages. The only data known concerning the safety effect of such legislation come from Australia (Royal Commission into the Sale, Supply, Disposal, or Consumption of Liquor in the State of Victoria, 1964-1965). This study showed that when the bars closed at six P.M. a peak of crashes occurred between four P.M. and eight P.M., whereas closing of the bars at ten P.M. resulted in a peak between eight P.M. and midnight. The overall 24-hour crash rate, however, was unaffected by the closing time.

It was noted earlier that presence of alcohol at times of high traffic flow may impair driving much more than at later hours. If the bars close too early, therefore, the possible gain of a reduction in total alcohol consumption per person may conceivably be offset by the exposure of drivers with alcohol to denser traffic in early or middle evening hours. Later closure, however, may have the disadvantage of permitting very heavy drinkers to reach higher blood alcohol concentrations before they drive and of bringing about serious crashes at a time when emergency services tend to be least adequate. The overall effect of these advantages and disadvantages has never been determined, however, in any real world "experiment."

Considerable argument has been raised about the safety effects of limiting alcohol to specific places. This question has come up most often when a "dry" community is adjacent to a "wet" one. One physician in a small town that, over several years, alternated between a "wet" and "dry" state has noted anecdotally that during "dry" years the crash rate increased, because residents had to travel to neighboring towns in order to consume alcohol and often drove home while impaired (Hyde, personal communication). No definitive data are available to validate his observations. It is relevant, however, that many roadhouses in the United States are so located that they can be reached only by motor vehicle and it is surprising, therefore, that there has been no real attempt to limit the licensing of bars and similar establishments only to locations that are within easy walking distance of the community or are near public transportation.

Another controversial subject has been the effect of age limits for alcohol consumption. There is rather good documentation that most persons begin to use alcohol before they are legally of age to do so, and that many of these individuals also drive after drinking (Perrine et al., 1971). It might be anticipated, therefore, that a reduction in age of permissible alcohol use, from twenty-one to eighteen for example, would have little or no effect on crash rate or on the crash involvement of young drivers. A chance to test this hypothesis has been provided by recent legislation in a number of states that has reduced the legal age to eighteen.

Data now available from several states show that there has been an increase in frequency of alcohol use among teen-agers in those areas that had previously enforced alcohol limitation by age in rigorous fashion (Douglass and Filkins, 1974). According to one report (Williams et al., 1974), this increase can be converted into about three additional deaths annually per 100,000 persons age fifteen to twenty. In those geographic areas that previously had lax enforcement of drinking laws, however, such an increase has not occurred (Douglass and Filkins, 1974).

The last form of alcohol control involves regulation of the type of drinks that may be served. Some communities permit sale of beer, but not of liquor. Since beer is an important contributor to highway crashes, even where liquor is sold, it might be expected that the availability of beer only would not result in an appreciable change, if any, in highway crashes. Again, however, data about this question do not exist.

In still other communities, alcohol is not sold by the drink; rather, the consumer must bring a bottle with him. Although studies have not been made, it is reported anecdotally that the requirement of carrying an entire bottle frequently results in heavier drinking than might occur if beverages were sold by the drink. Possession of a bottle also means that a person may carry his drinks with him for consumption in his car as well. Where he is not permitted to carry out a partly used bottle, last minute guzzling is reported to occur to avoid wasting one's purchases.

In summary, many different types of attempts have been made to control alcohol availability. Almost without exception, however, their effects on drinking practices and on various alcohol-related problems, including injury, have not been examined in any sort of systematic fashion.

Control of User Behavior to Reduce Impairment or Exposure to Risk

The history of this category of countermeasures is both long and frustrating. Much has been (and still is being) tried but little has been evaluated.

Included have been public education campaigns based on arousing fear, appeals to good citizenship or religious conscience, exhortation, and humor. With respect to alcohol and safety, perhaps two of the most important examples of this approach are the slogan, "If you drink, don't drive; if you drive, don't drink," and the admonition not to smoke in bed.

The limitation to this approach as previously and currently applied is that it has been based upon the assumption that the individuals who get into trouble with alcohol are average social drinkers who are in complete control of their drinking situations. As this chapter has already documented, however, this assumption is largely inaccurate, both because of the predominance of problem drinkers among persons who are at highest risk of injury, and because teen-age and young adult males who also are at high risk are subject perhaps more than any other group to the pressures of their peers. Neither of these problems has been considered in most of the traditional approaches to control of user behavior.

In illustration, with respect to highway safety the publicity regarding the Scandinavian approach of jailing alcohol-impaired drivers does anecdotally appear effective in preventing such driving by most social drinkers. But that group already is at low risk of alcohol crashes and, as shown by Bjerver *et al.* (1955), most persons who are involved in alcohol crashes in Stockholm are not normative social drinkers. Regarding nonhighway safety, the admonition not to smoke in bed ignores the fact that most who do smoke in bed and die as a result of the ensuing conflagration are heavy drinkers currently impaired by alcohol and therefore unlikely to heed the message, no matter how appropriate it is (Waller, 1972a).

Two other examples of attempts to control user behavior should be mentioned in further illustration of the futility of the standard patterns of establishing countermeasures without bothering to collect baseline data or to take appropriate steps to evaluate results. It is widely recommended in the United States that impaired drivers should let their passengers do the driving, as is reported anecdotally to be the pattern in Scandinavia. The limitations to such a recommendation are readily apparent in the data previously described for passengers of drivers impaired by alcohol (Waller *et al.*, 1972b).

The other example concerns programs for providing substitute transportation for persons impaired by alcohol or with suspended licenses following DWI convictions. Such programs can be quite expensive. Several have been tried among the Alcohol Safety Action Projects (ASAPs) established by the United States Department of Transportation. Most appear to have ended in failure because the number of users did not warrant the expense. One such project ended in the planning stage, however, without expenditure of funds because a survey of members of Alcoholics Anonymous and persons with previous DWI arrests showed that these individuals overwhelmingly were adverse to seeking substitute transportation when they had been drinking, or had license suspension because to do so would have created a feeling of indebtedness with which they could not or did not want to cope (Ingraham and Waller, 1971).

Can any programs succeed in controlling user behavior? If so, what factors must be considered? To date, only one successful program is known—the Lackland experiment (Barmack and Payne, 1961b). This does not mean that no other programs have achieved goals of reducing alcohol-related injury, but rather that most programs that have started with viable premises and with reasonably good countermeasure design have not included adequate evaluation to determine areas and extent of success or failure.

The Lackland experiment was based on careful characterization of the role of alcohol in highway crashes of airmen stationed at Lackland Airforce Base and of the individuals who were so involved. It became apparent that alcohol was an important factor in crashes and that the young airmen who were getting into trouble by and large were not normative social drinkers. Consequently, an intensive campaign was carried out to emphasize that "tanking up and taking off" is not evidence of manliness, but rather of sick behavior, and that persons in crashes after drinking would have psychiatric evaluation, review of their military records, and possibly dishonorable discharge.

Over the next several months, there was a marked and significant reduction in crashes among airmen stationed at the base, although a neighboring base without such a program had a rise in crash rate. Unfortunately, the countermeasure was only one of several initiated concurrently, so it is not possible to determine how much of the change was attributable to any one countermeasure or to the fact that several were combined. In addition, the military setting suggests a degree of control over personal behavior that might not be as achievable in a civilian community.

Although no definitive data are yet available, the following considerations appear to be relevant in designing programs to reduce injury by controlling user behavior:

1. There needs to be adequate characterization of the population at greatest risk of injury involving alcohol; for example, young, heavy-drinking males, or older problem drinkers.

2. Specific goals must be identified relevant to changing knowledge, attitudes, or behaviors.

3. The current status of the population to be reached with respect to the goals has to be determined. Thus, one ASAP project identified that 42 percent of heavy-drinking, teen-aged males believe they can drive safely after six beers or more, although much fewer believed they could drive safely after this number of drinks of liquor (Waller et al., 1972a). These drivers need to know that beer has more alcohol than most of them realize and that six beers are too many for

safety. In some cases, collection of such baseline data may show that the intended goal is not worth pursuing because too few people or—equally important—only persons with low risk of alcohol injury adhere to the particular knowledge, attitude, or behavior in which change is sought.

4. The most effective route for reaching the appropriate population has to be identified. For example, a public information campaign aimed at bringing about a realization that 12 ounces of beer has as much alcohol as a shot of liquor will not reach the heavy-beer-drinking, young male if it is carried on television, in the news section of the newspaper, or spoken about at meetings of the local service clubs. The message stands a better chance of reaching him if it appears at outdoor movies or auto races or on local radio, for these are the media to which such individuals are more often exposed (Waller *et al.*, 1972a).

5. The countermeasure must be culturally relevant. It is already apparent that several subcultural groups are overrepresented in alcohol-related injury events; for example, young males, lower socioeconomic groups, Negroes, American Indians, and Mexican-Americans. All appear to share a degree of alienation from the predominant American culture. It is becoming increasingly clear that efforts to control behaviors of these individuals stand little chance of success unless they are built around culturally relevant concerns *as perceived by the recipient of the message, not by the sender.*

A message about problem drinking, for example, that shows a middle-class, middle-aged man drinking martinis alone in a bar will have little impact on the lower class young man who drinks beer in his car with two or three other young men. Concern about adverse publicity in the newspaper upon being arrested for DWI may be relevant to the middle-class personnel of a safety program, but is quite unimportant to the young ghetto dweller who may be more interested in the effect of such an arrest on his insurance or on his chances for getting a job as a truckdriver. Messages in Scandinavia about alcohol-impaired driving may have much different meaning in a culture that, relative to the United States, is quite small and homogeneous, has strong roots in abstinence, and has a degree of permissiveness with respect to scrutiny by the police that would be totally unacceptable in the United States.

6. Adequate periodic or ongoing evaluation must be designed into the countermeasure program to determine areas and extent of success and of failure. In illustration, the British Road Act of 1967 established a maximum blood alcohol concentration considered legally permissible for drivers, and provided for testing of drivers by the police to determine compliance with the law. An extensive public education program preceeded and accompanied the countermeasure. Initially, there was a 30–40 percent decrease in persons with blood alcohol concentrations of 100 mg/100 ml or less and about a 25 percent decrease among persons with 150 mg/100 ml or higher, compared to baseline data. After two years of operation, there is a 22 percent decrease compared to

baseline in positive blood alcohol concentrations of 100 mg/100 ml or less by only a 5 percent decrease above that level. The program, overall, appears to be losing its effectiveness, especially among those persons who are at highest risk of injury (Havard, 1972). It is perhaps still sufficiently effective in relation to cost to warrant continuation, provided there is careful redirection based on evaluation of reasons for failure among the heavier drinkers.

Early Identification of Alcohol Abusers and Their Removal from Hazardous Situations

Impairment by alcohol can be either transient and infrequent, or constantly recurrent. This countermeasure category is aimed to both situations. One method for identification and removal of alcohol abusers is currently being applied in the field of transportation safety, and to a lesser extent in occupational safety. It involves identification of known problem drinkers and denial of drivers' licenses, or pre- and postemployment screening of truckdrivers, commercial pilots, and other employees and removal from the hazard by not hiring them, by dismissal or, all too rarely, by reassignment concurrent with appropriate rehabilitative endeavors. Once the problem drinker is identified, it is also possible to avoid the hazard of alcohol itself by use of Antabuse. A major limitation of this category of countermeasure as currently practiced is that the screening process, especially as it applies to the licensing of drivers, probably misses many, if not most problem drinkers. Furthermore, criteria are currently inadequate for determining which problem drinkers might be permitted to drive and which should not. Overall, however, especially before rehabilitation begins, evidence suggests that most problem drinkers probably are at increased risk of crash involvement (Schmidt and Smart, 1959; Waller, 1965). Another major limitation to license denial is that many individuals continue to drive, anyway, although a recent study suggests that the use of limited licenses may be a viable alternative to license revocation (Johns and Pascarella, 1971).

The second method in common use is the identification of the acutely impaired driver and his removal through DWI arrest and, in many jurisdictions, through subsequent compulsory suspension or revocation of his license. The public information and deterrent aspects of this countermeasure have already been considered above. Our concern here deals with effectiveness of the identification process, arrest patterns, court procedures, and penalties.

In the United States, in accordance with a U.S. Department of Transportation standard, most states now have a "presumptive limit" of 100 mg/100 ml (0.10 percent by weight) in which a driver is presumed to be impaired if his blood or breath alcohol concentration equals or exceeds 100 mg/100 ml. Most states presume a driver is not impaired if his blood alcohol concentration is less than 50 mg/100 ml. Between these two levels no presumption is made either way. The American Medical Association and the National Safety Council originally recommended a presumptive limit of 150 mg/100 ml in 1939; both subsequently lowered it to 100 mg/100 ml, and in 1971 the National Safety Council further reduced it to 80 mg/100 ml (American Medical Association, 1968; National Safety Council, 1971b). In New York State drivers under age twenty-one are presumed to be impaired at concentrations above 50 mg/100 ml, but older drivers are so presumed at 100 mg/100 ml (U.S. Department of Transportation, 1968).

All states also have either implied or expressed consent legislation in which a driver is automatically presumed to have given his consent to submit to a quantitative test for alcohol if a police officer has reason to believe he is impaired. (In contrast, in Sweden a driver may be tested whether or not the officer believes he is impaired.) In most communities such testing is not done until after a person is arrested for DWI, and the use of such tests for evidence has been ruled constitutional by the United States Supreme Court (U.S. Department of Transportation, 1968). A few police departments are now carrying out prearrest testing of drivers, a practice that, as of this writing, has not yet been ruled upon by the Supreme Court. Such prearrest tests, unfortunately, have sometimes been performed with inaccurate equipment specifically designed and sold for screening purposes (O'Neill and Prouty, 1972).

Despite the existence of implied consent and presumptive limit legislation, and despite the fact that at least 2 percent of drivers on the road have blood alcohol concentrations of 100 mg/100 ml or higher, there is documentation that the average police officer makes only two DWI arrests per year (Borkenstein, 1968). Why is the countermeasure working so poorly? Several formidable obstacles exist.

It is common knowledge that some individuals can consume substantial amounts of alcohol and still not appear to be impaired. In one study a physician who correctly identified a group of 60 drivers as impaired after watching them drive through a test course was able to identify only 58 percent of them as impaired using the standard clinical tests such as swaying, finger to nose, slurred speech, etc. Another physician who had not observed the driving test was able to identify as impaired only 33 percent of these subjects who were objectively impaired based on their driving performance (Coldwell, 1957). In an extensive series of police contacts with drivers, only about half of persons with blood alcohol concentrations of 100 mg/100 ml were identified by the police as under the influence of alcohol, and even at 200 mg/100 ml or higher some drivers were not identified as being impaired (Goldberg, 1951). According to one study, the average blood alcohol concentration at which police can identify drinking but not impairment is 120 mg/100 ml (Royal Commission into the

Sale, Supply, Disposal, or Consumption of Liquor in the State of Victoria, 1964–1965).

The second obstacle to effective application of the countermeasure is the frustrating system within which many police officers have to function. The arrest procedure often takes as much as two hours to complete, and the officer can expect to spend several additional hours in court. In one study most police officers reported that from one to several of the ten most recent traffic citations they issued for offenses other than DWI were to persons who in their opinion probably had blood alcohol concentrations of 100 mg/100 ml or higher. Reasons for failure to charge drivers with DWI included lack of support by the state's attorney in seeking a conviction, or by the judge in giving a meaningful sentence to those who are convicted. However, the reason given most frequently by the officers was that even if the drivers were arrested and found to have blood alcohol concentrations exceeding the presumptive limit, there would have been insufficient *additional* evidence to obtain a conviction (Waller and Worden, 1972). The absence of additional evidence has frequently been a successful defense against the presumptive limit despite the scientific evidence showing that virtually all persons are impaired in this range. For this reason it has been recommended that a limit per se be established, as is the case in Britain, so that it will not be necessary to prove impairment (U.S. Department of Transportation, 1968).

The entire judicial process, in fact, must be seen as an obstacle course to effective implementation of this countermeasure. Plea bargaining to reduce a DWI charge to that of a lesser offense is quite common, and is almost the rule for individuals who already have one DWI conviction and a second pending. Therefore, it is exceedingly rare to find individuals with more than one DWI conviction even though they have blatant alcoholism, regularly drive while impaired, and frequently get involved in alcohol crashes or get stopped because of erratic driving.

The final obstacle is the sentence imposed and its apparent success in preventing recidivism. The nature and severity of the sentence vary from community to community and even between judges within a single community. The judge commonly has little or no information at time of sentencing about the offender's background, because in most communities a presentencing investigation is not attempted (e.g., Waller and Flowers, 1972). Consequently, the fact that he is dealing with a problem drinker is not usually known by the judge and the underlying drinking problem plays little or no role in determining the nature of the sentence. Clearly, this is one extremely important area in which the current system needs to be improved, and steps are underway both at federal level and in selected communities to bring about appropriate change.

Although under the current system some drivers undoubtedly adhere to a license suspension, there is ample evidence that many, if not most, do not (Coppin and Van Oldenbeek, 1956; Ingraham and Waller, 1971). Furthermore, if they should be stopped by the police, these drivers often are not charged with driving with a suspended license. If charged and convicted, even on several occasions, the sentence for the last such offense may be more lenient than for earlier ones. The current system can only be described as Kafkaesque, ignoring the underlying drinking problem that usually exists, and not even dealing consistently with the more immediate expressions of that problem.

For several years isolated individuals have urged that DWI arrests should be used as entry points for the identification and rehabilitation of persons with drinking problems (Selzer et al., 1963; Waller, 1965). This approach is now being tried in a few communities and as of 1971 is being promoted as a joint endeavor by the U.S. Department of Transportation and the National Institute of Alcohol Abuse and Addiction. Although most of these rehabilitation activities are quite new, there is already some evidence from one of the earliest such programs suggesting that they may be effective in reducing recidivism on the highway (Crabb et al., 1971).

Environmental Controls Aimed at the Preinjury Phase

Although increasing attention is being devoted to the contribution of environmental factors, either obvious or subtle, to the initiation of injury events in general, their relation to injury events involving alcohol has only barely begun to be examined. It is clear that appropriate environmental changes, either to make tasks less demanding or to separate people from hazards, can be quite effective in reducing the frequency of injury. However, with the exception of a program, described below, aimed at reducing wrong-way driving, there has been no determination of the effect on alcohol-related injuries. It is not yet possible, therefore, to say whether cost-effectiveness in the case of alcohol-related injuries would be as great as for other types of injury events.

Some examples of excessively demanding tasks particularly relevant to impairment induced by alcohol are difficult tracking tasks such as narrow, winding roads, which exact a high toll of all drivers, but especially of impaired ones. Yet, one highway commissioner, after being told of the high crash rate on such a road that ran between a town and a roadhouse, is reported to have commented that the road should be left hazardous as a warning to people not to drink and drive (Waller et al. 1970).

Also relevant are false or inadequate cues. Examples are glass doors that may look open but actually are closed, or adjacent glass panels that look like doors but are not, or electric stoves that may give off heat without appearance of a red glow. Vehicle or other equipment controls may look identical but have divergent functions; for example, if adjacent knobs control windshield wipers and headlights, a driver can inadvertantly turn off his lights while trying to turn on the wipers. Included also are road signs that are too small, too poorly illuminated, too confusing in their messages, and too close to decision points to be handled by the driver who is either unfamiliar with the locality or otherwise impaired (Committee on Public Works, 1968a). Smith and Tamburri (1968) have documented that when even minor improvements are made in signs, signals, road geometry, etc., there may be substantial reductions in crashes.

In some instances on-and-off ramps on freeways have been so poorly marked and so close to each other that many drivers have inadvertantly driven the wrong way onto the off ramp and into oncoming traffic. One study in California showed that such mistakes were being made primarily by elderly drivers during daylight hours and by alcohol-impaired drivers at night. An extensive effort was made to develop new signs to alert people before they drove too far up the ramp. These signs proved quite effective in reducing wrong-way driving and the resultant serious crashes by both groups of drivers (Tamburri, 1969).

The requirement of protective guards on hazardous machinery has long been utilized in industry as a means of separating the person from the hazard. But such methods are only beginning to be applied in the nonindustrial setting. As already noted, alcohol is an important factor in vehicle collisions with trains; but most grade crossings are still unprotected despite ample evidence that grade crossings protected with gates have substantially lower crash rates than do unguarded crossings (Automotive Safety Foundation, 1963). Many pedestrian deaths occur because even where feasible there usually is no attempt to separate pedestrian and vehicular traffic, and the immature or grossly intoxicated pedestrian may wander onto the highway.

Spectacular reductions in head-on collisions are now being achieved through placement of median guard rails and other barriers to separate lanes of oncoming traffic. In one area, for example, there were 22 fatalities from head-on crashes before erection of a median barrier and only one during a comparable period afterward (Johnson, 1964). Since such collisions very often involve alcohol, this can be considered a successful and not overly expensive method of preventing alcohol-related injury through environmental control. (In the case of median barriers, there appears to be an increase of less serious crashes, but a decrease of more serious ones.)

Another example of such control mechanism that is largely but not entirely alcohol-specific is the automobile ignition lock currently under development. This lock would require the driver to perform a task in order to start the ignition that is probably too complicated or demanding for the seriously impaired driver. By not being able to start the car, he would thus be protected from the hazard of driving while impaired.

Moderation of Energy Transfer During the Injury Phase

Some of the best-documented contribution to injury and some of the most successful countermeasures to injury have been associated with the energy transfer phase. The documentation and the countermeasures are not specific for alcohol-related events. Moderation of energy transfer is highly relevant to alcohol-related events, however, because this category of countermeasures is aimed at reducing the frequency and severity of injury and, as already documented, alcohol injuries tend to be much more severe than those not involving alcohol. Examples of relationships and of countermeasures will be presented from both the highway and nonhighway contexts.

Within the highway context ejection from vehicles, excessive injury to face and head from impact with windshields, headers, and dashboard components, and crushing of chests of drivers by impalement on steering wheels have been shown repeatedly to be major vehicular contributors to death and disability. In fact, improper design and construction of the steering column alone is "credited" with responsibility for about 30 percent of deaths to drivers (Huelke and Gikas, 1968). Improvements in door latch mechanisms, use of passenger restraint systems, development and application of high penetration resistant windshields and of energy absorbing steering columns as countermeasures have resulted in spectacular decreases in the frequency and severity of specific injury patterns, in some cases by as much as 80 percent (Nahum and Siegal, 1968; Huelke and Sherman, 1972). New windshields have shifted the practice of surgery for facial injury from one of major emergency surgery and extensive reconstruction to somewhat less demanding procedures (Schultz, 1971). Unfortunately, there is evidence suggesting that drivers and passengers who have been drinking may be less likely to use active restraints such as seat belts than are persons without alcohol (Perrine et al., 1971). The availability of passive restraints (for example, the air bag) would avoid this problem.

Various roadside hazards involved in crashes have also been identified as important causes of unnecessary highway injury in both urban and rural settings (Committee on Public Works, 1968b). Such collisions are especially common among single-vehicle crashes that, as already noted, preponderantly involve alcohol. Studies show that in some areas between one third and one half of all highway fatalities involve collision with such unnecessary obstacles as trees bordering the road edge (in some cases these may actually be planted by highway departments for beautification), tree stumps, concrete bridge abutements, rock ledges, heavy signposts, etc. (Huelke and Gikas, 1967; Waller et al., 1970). Where modern technology has been applied, such as 30-foot clear areas along roadsides, breakaway signposts, and lamp stanchions, appropriately designed and placed energy-absorbing guard rails and similar energy attenuators, crash injury and fatality rates have dropped significantly. In fact, according to a recent federal study, "4.78 lives could be saved and 86.96 injuries could be avoided for each $1 million spent for highway safety improvement work. . . . The cost-effectiveness of the highway safety improvement work, in terms of lives saved, was shown to be about five times greater than that of regular highway construction work." The report notes that a review of 381 safety projects in California showed a 31 percent reduction in fatal crashes, 8 percent reduction in injury crashes, and 14 percent fewer crashes involving property damage only (Comptroller General of the United States, 1972). Higher standards of safety have been recommended by the Bureau of Public Roads (1968) for construction and maintenance of all roads, whether supported by federal aid or not.

Within the nonhighway context, product contribution to the severity of injury events has been identified in a minimum of 29 percent of fatalities involving alcohol, in most cases because of flammability of clothing, bedding, or chair upholstery (Waller, 1972d). The past several years have seen substantial improvement in the technology of producing flame-retardant fabrics, including cotton (which is involved in about three quarters of injuries in which clothing is ignited). These new fabrics are reasonably durable, inexpensive, and of suitable texture. Although no hard data are yet available, it is anticipated that widespread use of such fabrics, as now proposed or required by legislation in Great Britain, the United States and elsewhere, will bring about a significant decrease in flame-related deaths and injuries involving alcohol.

Improvement of Emergency and Definitive Care for the Injured

Improvement of emergency care for the injured by the United States military forces is generally credited with the major role in the substantial reduction of the military case fatality rate among persons still alive when they reached treatment facilities from 4.5 percent in World War II to less than 2 percent in the Vietnam war (Heaton, 1966). These improvements have not generally been applied in the United States, where the condition of both prehospital and emergency room emergency care is widely considered to be appalling. Studies have shown that between a fifth and a quarter of highway deaths involve persons with survivable injuries (Frey et al., 1969; Perrine et al., 1971). Among persons who do not die at the crash site but who succumb in the ambulance or the hospital, about half die of survivable injuries (Perrine et al., 1972; Gertner et al., 1972). Services are least likely to be adequate at night when injuries involving alcohol most often occur, therefore suggesting that improvement of emergency care might be most beneficial for alcohol-related injuries. However, improvement of the various segments of the civilian emergency care system is so recent a phenomenon that there has been insufficient time as yet to determine what effect such improvements actually have on survival rates in the civilian setting.

SUMMARY

Although large gaps still exist in knowledge about alcohol and injury, it is already apparent that alcohol is a major contributor to all types of serious injury events, with the possible exception of occupational injuries, and is important in less serious events as well. Data strongly suggest that the persons at highest risk of alcohol-related injury are problem drinkers. In addition, teen-aged experimenters with alcohol and young men engaged in heavy social drinking with their peers also are substantially overrepresented.

Behavioral countermeasures to alcohol-related injury commonly have been poorly conceived and inadequately evaluated. With certain important exceptions, they appear to hold only limited hope for effectiveness. Environmental countermeasures designed to reduce task demands, to separate people from hazards, and to moderate harmful energy exchanges have greater potential for reduction of morbidity and mortality. Improvement of emergency care for the injured also warrants greater attention as a useful countermeasure.

REFERENCES

American Automobile Association, 1971, "Sportsmanlike Driving," Washington, D.C.
American Medical Association, Committee on Medicolegal Problems, 1968, "Alcohol and the Impaired Driver," Chicago, Illinois.
Anonymous, 1904, Editorial *Quarterly Journal of Inebriety* 26:308 cited from *Quart. J. Stud. Alc.* Suppl. No. 4, p. vi, 1968.
Anonymous, 1964a, Flight accidents due to drinking studied by U.S., *Medical Tribune* (November 2), p. 8.
Anonymous, 1964b, Drunk driving down during liquor strike, *Medical Tribune* (August 31), p. 3.
Anonymous, 1968, Suicide prevention: NIMH wants more attention for "taboo" subject, *Science* 161:766.
Association of Casualty and Surety Companies, 1957, The influence of alcohol on traffic safety, Technical Traffic Topics No. 3 (May), New York, New York.
Automotive Safety Foundation, 1963, Traffic control and roadway elements: Their relationship to highway safety, Washington, D.C.
Barmack, J. E., and Payne, D. E., 1961a, Injury-producing private motor vehicle accidents among airmen, *Bulletin #285,* Highway Research Board, Washington, D.C.
Barmack, J. E., and Payne, D. E., 1961b, The Lackland accident countermeasure experiment, *Highway Research Board Proceedings* 40:513.
Birrell, J. H. W., 1967, A note on automobile-tram (streetcar) fatal accidents and alcohol in the city of Melbourne, *Med. J. Aust.* 2:1 (July 1).

Birrell, J. H. W., 1971, A comparison of the postmortem blood-alcohol levels of drivers and passengers compared with those of drinking drivers who kill pedestrians, *Med. J. Aust.* 2:945 (November 6).
Bjerver, K. B., Goldberg, L., and Linde, P., 1955, Blood alcohol levels in hospitalized victims of traffic accidents, *Proceedings of the Second International Conference on Alcohol and Road Traffic,* Garden City Press Co-Operative, Toronto, Canada.
Borkenstein, R. L., 1968, Technical content of state and community police traffic service, National Highway Safety Board, Washington, D.C.
Borkenstein, R. F., Crowther, R. F., Shumate, R. P., Ziel, W. B., and Zylman, R., 1964, "The Role of the Drinking Driver in Traffic Accidents," Department of Police Administration, Indiana University, Bloomington, Indiana.
Brenner, B., 1967, Alcoholism and fatal accidents, *Quart. J. Stud. Alc.* 28:517 (September).
Bureau of Public Roads, U.S. Department of Transportation, 1968, Handbook of highway safety design and operating practices, U.S. Government Printing Office, Washington, D.C.
Cahalan, D., 1970a, "Problem Drinking," Jossey-Bass, Inc., Publ., San Francisco.
Cahalan, D., 1970b, A national survey on the attributes of problem drinkers. Paper presented at 29th International Congress on Alcoholism and Drug Dependence, Sydney, Australia (February).
California Department of Highway Patrol, 1967, A report on alcohol, drugs and organic factors in fatal single vehicle traffic accidents, Sacramento, California.
Canty, A., 1940, The case study method of rehabilitating drivers, *Journal of Social Psychology* 12:271.
Coldwell, B. B., 1957, Report on impaired driving tests, Queen's Printer and Controller of Stationary, Ottawa, Canada.
Committee on Public Works, U.S. House of Representatives, 1968a, *Highway Safety Design and Operations, Freeway Signing and Related Geometrics (90-39),* Hearings before special subcommittee on the federal-aid highway program, U.S. Government Printing Office, Washington, D.C.
Committee on Public Works, U.S. House of Representatives, 1968b, *Highway Safety Design and Operations, Roadside Hazards (90-21),* Hearings before special subcommittee on the federal-aid highway program, U.S. Government Printing Office, Washington, D.C.
Comptroller General of the United States, 1972, *Problems in implementing the highway safety improvement program,* Federal Highway Administration, Department of Transportation, Report to the subcommittee on Investigations and Oversight, Committee on Public Works, House of Representatives, Washington, D.C. (May 26).
Connecticut Department of Motor Vehicles, 1933, Connecticut motor vehicle statistics for 1932, *Bulletin #91,* Research Section, Connecticut Department of Motor Vehicles, Hartford (January).
Coppin, R. S., and Van Oldenbeek, G., 1965, Driving under suspension and revocation, California Department of Motor Vehicles, Sacramento (January).
Crabb, D., Gettys, T. R., Malfetti, J. L., and Stewart, E. I., 1971, Development and Preliminary Tryout of Evaluation Measures for the Phoenix Driving-While-Intoxicated Reeducation Program, Arizona State University, Tempe.
Cuyahoga County Coroner, 1960–1963, 1965–1967, 1969–1970, "Coroner's Statistical Report, Cuyahoga County, Ohio," Cleveland.
Davis, G. L., 1968, Alcohol and military aviation fatalities, *Aerosp. Med.* 39:869.
Davis, J. H., and Fisk, A. J., 1964–1965, The Dade County, Florida study on carbon monoxide, alcohol and drugs in fatal single vehicle automobile accidents, Dade County, Florida, Coroner's Office, *Proceedings of National Association of Coroners,* pp. 197–204.

Demone, H. W., and Kasey, E. H., 1966, Alcohol and non-motor-vehicle injuries, *Pub. Health Rep.* 81:585.

Douglass, R. L., and Filkins, L. D.: 1974, The effect of lower legal drinking ages on youth crash involvement—Final Summary Report, National Highway Traffic Safety Administration, U.S. Department of Transportation, Washington, D.C.

Drew, G. C., Colquhoun, W. P., and Long, H. A., 1959, *Effect of Small Doses of Alcohol on a Skill Resembling Driving,* Medical Research Council Memorandum No. 38, Her Majesty's Stationary Office, London.

Encel, S., and Kotowicz, K., 1970, Heavy drinking and alcoholism, preliminary report, *Med. J. Aust.* 1:607 (March 21).

Forney, R. B., and Hughes, F. W., 1968, "Combined effects of alcohol and other drugs," Charles C Thomas, Publ., Springfield, Illinois.

Frey, C. F., Huelke, D. F., and Gikas, P. W., 1969, Resuscitation and survival in motor vehicle accidents, *J. Trauma* 9:292 (April).

Freimuth, H. C., Watts, S. R., and Fisher, R. S., 1958, Alcohol and highway fatalities, *J. Forensic Sci.* 3:65.

Gertner, H. R., Baker, S. P., Rutherford, R. B., and Spitz, W. U., 1972, Evaluation of the management of vehicular fatalities secondary to abdominal injury, *J. Trauma* 12:425 (May).

Glatt, M. M., 1964, Alcoholism in "impaired" and drunken driving, *Lancet* 1:161 (January 18).

Goldberg, L., 1943, Quantitative studies on alcohol tolerance in man, *Acta Physiol. Scand.* 5, Suppl. XVI.

Goldberg, L., 1951, Tolerance to alcohol in moderate and heavy drinkers and its significance to alcohol and traffic, in *Proceedings of the First International Conference on Alcohol and Road Traffic, Stockholm, 1950,* Kugelbergs Boktrycheri, Stockholm, pp. 85–106.

Haddon, W., Jr., 1970, On the escape of tigers: An ecologic note, *Amer. J. of Pub. Health* 60:2229.

Haddon, W., Jr., and Bradess, V. A., 1959, Alcohol in the single vehicle fatal accident, experience of Westchester County, New York, *J.A.M.A.* 169:1587.

Haddon, W. Jr., Valien, P., McCarroll, J. R., and Umberger, C. J., 1961, A controlled investigation of the characteristics of adult pedestrians fatally injured by motor vehicles in Manhattan, *J. Chronic Dis.* 14:655.

Harper, C. R., and Albers, W. R., 1964, Alcohol and general aviation accidents, *Aerosp. Med.* 35:462.

Havard, J., 1972, Comments made at Conference on Medical, Human and Related Factors Causing Traffic Accidents, Including Alcohol and Other Drugs, sponsored by Traffic Injury Research Foundation of Canada, Montreal, May 28–31.

Heaton, L. D., 1966, Army medical service activities in Vietnam, *Mil. Med.* 131:646.

Heise, H. A., 1934, Alcohol and automobile accidents, *J.A.M.A.* 103:739.

Hess, R., 1972, Director, University of Michigan Highway Safety Research Institute, Ann Arbor, Personal communication (March 6).

Holcomb, R. L., 1938, Alcohol in relation to traffic accidents, *J.A.M.A.* 111:1076.

Hricko, A. R., 1970, Alcohol has menacing role in flight safety as on highways, *National Underwriter,* (August 7) pp. 1, 38–39, and (August 14) pp. 10–13.

Huelke, D. F., and Gikas, P. W., 1967, Non-intersectional automobile fatalities—a problem in roadway design, Highway Research Board #152, pp. 103–118.

Huelke, D. F., and Gikas, P. W., 1968, Causes of deaths in automobile accidents, *J.A.M.A.* 203:1100.

Huelke, D. F., and Sherman, H. W., 1972, Some injury mechanisms in new car collisions, *Proceedings of Fifteenth Conference of the American Association for Automotive Medicine,* Society of Automotive Engineers, New York, New York.

Hyde, R., 1972, Vermont Department of Mental Health, Waterbury, Personal communication.
Hyman, M. M., 1968, Accident vulnerability and blood alcohol concentrations of drivers by demographic characteristics, *Quart. J. Stud. Alc.* Suppl. No. 4, pp. 34–57.
Ingraham, W. S., and Waller, J. A., 1971, Alcohol-impaired driving, license suspensions, and transportation needs during intoxication or suspension among alcoholics, CRASH Report IV-1, Project CRASH, Vermont Department of Mental Health, Waterbury (September).
Johns, T. R., and Pascarella, E. A., 1971, An assessment of the limited driving license amendment to the North Carolina statutes relating to drunk driving, Highway Safety Research Center, University of North Carolina, Chapel Hill (April).
Johnson, R. T., 1964, Effectiveness of median barriers, Traffic Department, California Department of Public Works, Sacramento (August).
Kelleher, E. J., 1971, A diagnostic evaluation of 400 drinking drivers, *Journal of Safety Research* 3:52 (June).
Loomis, T. A., and West, T. C., 1958, The influence of alcohol on automobile driving ability, *Quart. J. Stud. Alc.* 19:30.
McCarroll, J. R., and Haddon, W. Jr., 1962, A controlled study of fatal automobile accidents in New York City, *J. Chron. Dis.* 15:811.
Miles, W. R., 1934, Alcohol and motor vehicle drivers, *Proceedings of Thirteenth Annual Meeting of Highway Research Board, Washington, D.C., December 7–8, 1933, Part 1,* Division of Engineering and Industrial Research, National Research Council, Washington, D.C., pp. 362–381.
Murphy, K., 1972, Beer . . . History's Brew, *Yellowbird,* Northeast Airlines, Inc., Boston, Massachusetts, pp. 9–10.
Nahum, A., and Siegel, A. W., 1968, Statement before the U.S. Senate Committee on Commerce, Washington, D.C. (April 25).
National Safety Council, 1971a, "Accidents Facts," Chicago, Illinois.
National Safety Council, 1971b, Committee on Alcohol and Drugs, Minutes of meeting—October 28, Chicago, Illinois.
National Transportation Safety Board, 1969, Alcohol problems and transportation safety: The need for coordinated efforts, Department of Transportation, Washington, D.C. (February 20).
Nielson, R. A., 1969, "Alcohol Involvement in Fatal Motor Vehicle Accidents, California . . . 1962–1968," California Traffic Safety Foundation, San Francisco (September).
Observer, and Maxwell, M. A., 1959, A study of absenteeism, accidents and sickness payments in problem drinkers in one industry, *Quart. J. Stud. Alc.* 20:302.
O'Neill, B., and Prouty, R. W., 1972, An evaluation of some qualitative breath screening tests for alcohol, *Proceedings of Fifteenth Conference of the American Association for Automotive Medicine,* Society of Automotive Engineers, New York, New York, pp. 261–296.
Ontario Department of Transportation, 1970, Snowmobile collisions on highways and roads in Ontario, Presented at 1970 International Snowmobile Congress, Duluth, Minnesota (February 9–11).
Payne, C. E., and Selzer, M. L., 1962, Traffic accidents, personality and alcoholism; a preliminary study, *Journal of Abdominal Surgery* 4:21 (January).
Perrine, M. W., Waller, J. A., and Harris, L. S., 1971, *Alcohol and Highway Safety: Behavioral and Medical Aspects,* Final Report, Project ABETS, University of Vermont on Department of Transportation Contracts FH-11-6609 and FH-11-6899 (DOT HS-800 599), Burlington, Vermont (September).
Royal Commission into the Sale, Supply, Disposal, or Consumption of Liquor in the State of Victoria, 1964–1965, Part I, Victoria, A. C. Brooks, Government Printer, Melbourne, Australia.
Schmidt, W. S., and Smart, R. G., 1959, Alcoholics, drinking, and traffic accidents, *Quart. J. Stud. Alc.* 20:631.

Schultz, R. C., 1971, The changing character and management of soft-tissue windshield injuries, *Proceedings of the Fourteenth Annual Conference of the American Association of Automotive Medicine*, University of Michigan, Ann Arbor.

Schuman, S. H., Pelz, D. C., and Ehrlich, N. J., 1967, Young male drivers: impulse expression, accidents, and violations, *J.A.M.A.* 200:1026.

Selling, L. S., 1941, The psychopathology of the hit-and-run driver, *Amer. J. Psychiat.* 98:93.

Selzer, M. L., Payne, C. E., Gifford, J. D., and Kelly, W. L., 1963, Alcoholism, mental illness and the "drunk driver," *Amer. J. Psychiat.* 120:326.

Selzer, M. L., Rogers, J. E., and Kern, S., 1968, Fatal accidents: The role of psychopathology, social stress and acute disturbance, *Amer. J. Psychiat.* 124:1028.

Smart, R. G., 1969, Personality syndromes and the alcoholic driver, Presented at Third International Congress on Medical and Related Aspects of Motor Vehicle Accidents, New York, New York (May 31).

Smith, R. N., and Tamburri, T. N., 1968, Direct costs of California state highway accidents, Highway Research Bull. #225, Highway Research Board, Washington, D.C., pp. 9–29.

Stoeckel, R. B., 1924, Time to deal harshly with the drunken operator and the wilfully reckless, *Bulletin #14*, Connecticut Department of Motor Vehicles, Hartford (December 15).

Tamburri, T. N., 1969, Wrong-way driving accidents are reduced, Highway Research Bull. #292, Highway Research Board, Washington, D.C., pp. 24–50.

Tonge, J. I., O'Reilly, M. J. J., Davison, A., and Derrick, E. H., 1964, Fatal traffic accidents in Brisbane from 1935 to 1964, *Med. J. Aust.* 2:811 (November 21).

U.S. Department of Transportation, 1968, "1968 Alcohol and Highway Safety Report," U.S. Government Printing Office, Washington, D.C.

Waller, J. A., 1965, Chronic medical conditions and traffic safety, review of California experience, *New Engl. J. Med.* 273:1413.

Waller, J. A., 1967, Identification of problem drinking among drunken drivers, *J.A.M.A.* 200:114.

Waller, J. A., 1968a, Patterns of traffic accidents and violations related to drinking and to some medical conditions, *Quart. J. Stud. Alc.* Suppl. #4, pp. 118–137 (May).

Waller, J. A., 1968b, The role of alcohol in fatal collisions with trains, *Northwest Med.* 67:852.

Waller, J. A., 1969, Accidents and violent behavior: Are they related? *in* Mulvihill, D. J., Tumin, M. M., and Curtis, L. A., "Crimes of Violence, A Staff Report to the National Commission on the Causes and Prevention of Violence, Volume 13," Appendix 33, U.S. Government Printing Office, Washington, D.C., pp. 1525–1558.

Waller, J. A., 1970, The role of alcohol in collisions involving trucks and the fatally injured, *Arch. Environ. Health* 20:254.

Waller, J. A., 1971a, Factors associated with police evaluation of drinking in fatal highway crashes, *Journal of Safety Research* 3:35 (March).

Waller, J. A., 1971b, The drinking driver or the driving drinker? (Letter to the Editor), *Med. J. Aust.* 2:596 (September 11).

Waller, J. A., 1972a, Nonhighway injury fatalities. I. Roles of alcohol and problem drinking, drugs, and medical impairment, *J. Chronic Dis.* 25:33.

Waller, J. A., 1972b, Factors associated with alcohol and responsibility for fatal highway crashes, *Quart. J. Stud. Alc.* 33:160.

Waller, J. A., 1972c, Truths, traps and tactics concerning alcohol, other drugs and highway safety, *Calif. Med.* 116:10 (February).

Waller, J. A., 1972d, Nonhighway injury fatalities. II. Interaction of product and human factors, *J. Chronic Dis.* 25:47.

Waller, J. A., 1974, Injury in aged—Clinical and epidemiological implications, *N.Y. State J. Med.* 74:2200.

Waller, J. A., and Flowers, L., 1972, Previous police contacts, and recidivism among drivers with arrests for driving while under the influence of alcohol in Vermont—baseline data, CRASH Report V-1, Project CRASH, Vermont Department of Mental Health, Waterbury (August).

Waller, J. A., and Goo, J. T., 1969, Highway crash and citation patterns and chronic medical conditions, *Journal of Safety Research* 1:13 (March).

Waller, J. A., and Thomas, K., 1972, Carbon monoxide, smoking and fatal highway crashes, *Proceedings of Fifteenth Conference of the American Association for Automotive Medicine*, Society of Automotive Engineers, New York, New York, pp. 245-260.

Waller, J. A., and Worden, J. K., 1972, A comparison of police and male driver knowledge, attitudes, and behaviors about alcohol and highway safety, CRASH Report III-2, Project CRASH, Vermont Department of Mental Health, Waterbury (June).

Waller, J. A., King, E. M., Nielson, G., and Turkel, H. W., 1969, Alcohol and other factors in California highway fatalities, *J. Forensic Sci.*, 14:429.

Waller, J. A., Harris, L. S., Oprendek, J. J., Jr., 1970, Booby-trapped highways in the beckoning country, Associates in Community Medicine, Burlington, Vermont (September).

Waller, J. A., Worden, J. K., and Maranville, I. W., 1972a, Baseline data for public education about alcohol and highway safety in Vermont, CRASH Report I-1, Project CRASH, Vermont Department of Mental health, Waterbury (February).

Waller, J. A., Merrill, D., Flowers, L., and Maranville, I. W., 1972b, Baseline data for police enforcement against alcohol impaired driving in Vermont: Roadside survey report, CRASH Report III-1, Project CRASH, Vermont Department of Mental Health, Waterbury (March).

Waller, J. A., and Lamborn, K. R., 1973, Snowmobiling: Characteristics of Owners, Patterns of Use, and Injuries, *Proceedings of 17th Conference of the American Association for Automotive Medicine*, Oklahoma City.

Wechsler, H., Kasey, E. H., Demone, H. W., and Thum, D., 1969, Alcohol level and some home accident injuries, Final Report to U.S. Public Health Service on Grant 5 RO1 UI-00022, The Medical Foundation, Inc. (September).

Whitlock, F. A., Armstrong, J. L., Tonge, J. I., O'Reilly, M. J. J., Davison, A., Johnston, N. G. and Biltoft, R. P., 1971, The drinking driver or the driving drinker?, *Med. J. Aust.* 2:5 (July 3).

Williams, A. F., Rich, R. F., Zador, P. L., and Robertson, L. S., 1974, The legal minimum drinking age and fatal motor vehicle crashes, Insurance Institute for Highway Safety, Washington, D.C.

Wolfgang, M. E., 1958, "Patterns in Criminal Homicide," University of Pennsylvania, Philadelphia.

Zylman, R., 1972, The variability of collision involvement at low blood alcohol concentrations, The Grand Rapids curve explained, *Blutalkohol* 9:25 (January).

CHAPTER 10

Alcohol and Crimes of Violence

Kai Pernanen

Social Studies Department
Addiction Research Foundation
33 Russell Street
Toronto, Ontario

INTRODUCTION

The first problem we encounter in trying to analyze the relationship between alcohol use and crimes of violence is in the choice of a theoretical context. One possibility is to look at violent crime as a subset of deviant behavior and put it in this theoretical sociological context, if *one* such context is possible. On the other hand, we could look at it from the viewpoint of aggressive behavior in general and see violent crimes as a sample of such behavior, a sample biased in favor of extreme forms of aggressive behavior. I have chosen the latter approach, since a framework limited to sociological variables would leave out central explanatory factors that are not relevant on the aggregate sociological level of analysis. Much of the empirical and theoretical outcomes of research on deviance are applicable in the explanations, but the emphasis will be on the connections with research on aggressive behavior. The relevant sociological factors will be seen only as one set of variables in an explanatory framework that encompasses (and must encompass) research findings from several disciplines.

This chapter has been divided into two main sections. The first one reviews the evidence for a higher-than-chance association between alcohol use and alcoholism and violent crime. The second section deals with possible explanatory factors and models that would account for a positive association between the two phenomena. The task is not simple. Gaps in coverage are inevitable and idiosyncratic, selective emphases unavoidable. This is probably true of any attempt to synthesizing knowledge in a complex area, but it is perhaps more pronounced when some of the relevant fields of study are unfa miliar to the author. The reason for undertaking such a task is that, in my view, only an exhaustive approach can lead to a satisfactory explanation of a phenomenon.

I will almost exclusively deal with noninstrumental and interindividual crimes of violence. The emphasis will be on homicide, partly because it is an easily definable category of crime and thus there is the least possible definitional variation between cultures and jurisdictions. Homicides are definitely interindividual. A proportion of homicides are, however, instrumental for various reasons and thus one criterion is not optimally fulfilled.

Instrumental homicides include murders and homicides that have occurred in connection with robberies and rapes, although situational factors such as the reaction of the victim may have lead to immediate reactions in the offender that from the point of view of his original intentions were noninstrumental. (These remarks should be seen in the light of the discussion of the escalatory process below.)

However, the criterion of noninstrumentality has served to exclude some existing scattered data on the involvement of alcohol in robberies and rapes that are interpersonal but more instrumental with a higher degree of rational planning. They are thus often determined by extrasituational factors to a greater extent than most homicides and assaults.

Assaults are probably the most noninstrumental category of violent crimes. If assaultive acts are committed in connection with a robbery, they will probably be subsumed under "robberies"; if they are part of a rational attempt to kill for instrumental gains, it is likely that a large proportion will be classified under "attempted murder"; and if committed in connection with rape, they will probably be classifed as "rapes." Unfortunately, very few studies have been made on alcohol involvement in assaults, although the greater prevalence of assaults would provide data for more extensive analyses of the role of alcohol in violent behavior than is possible with homicides. The ideal study from this point of view would include assaults and noninstrumental homicides from the same jurisdiction. This is so also because it is often more or less a question of chance circumstances whether a violent act ends up as an aggravated assault or a homicide (and the nomothetic etiological factors are the same). The escalatory process of violent behavior can be cut off by outside intervention and chance circumstances often determine whether lethal weapons are available in the situation, the ambulance arrives on time, etc. (see section on the nature of the dependent variable below).

Arson and vandalism are excluded from detailed study since they are not interpersonal in nature, and explanations of these would differ in important respects from explanations of interpersonal crimes. If robbery, rape, and arson were included just because they are classified as violent crimes for nonscientific purposes, the explanatory accounting would have been extremely complex and more often misleading than not. The alcohol involvement in rape, child molesting, and arson seems to be much lower than for homicide and other purely assaultive crimes, which is another indication of the different weights of etiological factors in explaining these crimes. The definition and reporting of rape is also extremely sensitive to cultural, temporal, and jurisdictional variations. Moreover, alcohol use and concomitant factors may influence the reporting of rapes in a selective and biasing way, especially since it is estimated that a very small proportion of sexual assaults that could be classified as rapes are ever reported to the authorities. The ones that are reported are probably biased against the role of alcohol in the cases where the rape victim has been drinking. The alcohol involvement (i.e., the proportion of cases in which either offender or victim or both were "under the influence" of alcohol) has a median of between one fifth and one third in the more representative studies (Amir, 1967). In child molesting cases alcohol use by the offender has been implicated in between 20 and 30 percent of the cases (e.g., McCaghy, 1968; Nau, 1967). In arson there are very few studies of alcohol involvement available, but the indication is that alcohol is implicated in only about 10 to 20 percent of the offenders (e.g., Aleksic and Radovanovic, 1967; Gelfand, 1971). The alcohol involvement of the offender in robberies seems to be comparatively high (Shupe, 1954; Stark, 1969). The probability of being victimized in robberies is also comparatively high for individuals who have been drinking.

Suicides have sometimes been treated within the same explanatory models as homicides, presumably on the basis of common etiological factors. According to the criteria used here for delimiting an empirically and theoretically fruitful field of study, suicides will have to be excluded from our analyses. The extent to which suicides are instrumental or noninstrumental should be of central theoretical concern but is not relevant here, since the deciding factor is that suicides are not interpersonal in the sense used here. The sense in which they are violent or aggressive is not quite clear either. Classifying suicides as aggressive acts seems to hinge very much on psychoanalytic theory of the motivation of these acts. Possibly the identity criteria for using the aggregate of homicide/suicide as one dependent variable, as in many other classifications in the social sciences, are located in the connotations of everyday language. One "destroys," "does violence to" human beings (either oneself or others) or physical objects. In interpersonal violence the criteria are rather clear: a conscious willful harm to *another person*. Regarding suicide (and suicidal attempts or self-mutilation), this definition is presumably easily transformed by substitution into: a conscious willful harm to *oneself*. It seems to be a question of simple logical substitution. The easiness with which the mental substitution is made is probably explained by the fact that such substitution is unproblematic in other contexts and inferences, which in our thinking become paradigms of reasoning even outside their proper domains. It should be remembered that the etiology can be very different in complex ways, despite the ease of linguistic substitutions that guide our thinking.

Other identity criteria that have been used include the presumed motivations of, or causes "within," the offender. I have used motivational criteria of identity to set aside robberies and rapes from (most) homicides and assaults among crimes of violence. Implicit in this is the assumption that motivations are etiologically relevant and that differences in motivations would lead to different explanations. Motivational or causative criteria probably also explain some of the attempts to classify suicide and homicide together.

As with any human act, the explanation of acts classified as "violent behavior" is complicated by the fact that they can be seen as a means to a variety of end results. Thus, looking for criteria among motivational states inevitably leads to postulating a great number of motivational states (as happened with "instincts"), either conscious or subconscious; conversely, these criteria can be rendered so general, and sometimes tautological, as to become meaningless and useless. (Freud's "death wish" is a case in point.)

Besides the cluster of variables pertaining to human interaction in homicide and the common escalatory process that are both missing in suicide situations, a further indication that suicides differ in etiology from interpersonal violence is the much lesser alcohol involvement in suicide (Goodwin, 1973).

It is my feeling that although there may be common factors in explanations of suicide and homicide, such as stress situations, life histories, and certain phenomenological states, there are enough differences in the possible explanatory models so that a conglomerate of homicide–suicide would only confuse the explanations of either type of act.

REVIEW OF FINDINGS

Data on prevalences and associations of any systematic nature have to be derived from studies of police or court records.* Less extreme and more representative forms of aggressive behavior have not been documented, although attempts should be made to do so.

We must look at data on alcohol use and noninstrumental interpersonal

---

* Police records are preferable to other sources, as pointed out by Wolfgang, (1958).

violent crime as indicators (lacking better ones) of the association between alcohol use and aggressive behavior in general. This is the only way to link the information on the association between alcohol use and violent crime with general studies on effects of alcohol. I shall first look at the data available for establishing an association between homicide and (to a lesser extent) assault and the use of alcohol and, secondly, the data available for establishing an association between the two forms of violent crime and prolonged excessive alcohol use (or "alcoholism").

Association between Acute Alcohol Use and Interpersonal Noninstrumental Crimes of Violence

I will only mention in passing the studies that have been made on the association between "alcohol use," "intoxication," "being under the influence of alcohol," "drinking prior to the crime," etc., and *crime in general*. These studies may have some practical import, but their theoretical relevance is negligible, especially in the cases where alcohol-related crimes such as drunkenness are included without clear distinctions being drawn. The proportion of acute alcohol use of the "general" criminal offender shows a high and unsystematic variation between samples from various jurisdictions, providing unknown generalizability. Because the figures are theoretically almost meaningless, I shall refrain from citing any. (For a typical sample of this type of study that is based on availability, the following list will suffice: Wieser, 1964; Kinberg *et al.*, 1957; Cloninger and Guze, 1970; Ullrich, 1966; Richard, 1966).

There is a considerable amount of statistical information available on rates of crimes of violence in various countries and jurisdictions. However, the existing information on associations of alcohol use and violent crime is not systematically compiled, and it is impossible to find any reliable statistical data from culturally delimited well-defined areas. The studies that are available stem from samples of varying compositions. Few are representative of any geographical area, culture, or jurisdiction for which relevant statistical or epidemiological data would be available on other relevant variables. Some of the samples have been selected in a way that probably affects the relationship with alcohol use. The difficulties in establishing the presence of alcohol in a representative sample of offenders probably also give rise to biases. (More on these problems in the section, "The Nature of the Dependent Variable," below.)

Studies Examining Alcohol Use by Both Offender and Victim

The most systematic American treatise of the association between violent crime and presence of alcohol in the offender and victim is Wolfgang's (1958) study of 588 criminal homicides in Philadelphia between 1948 and 1952. He found that alcohol was present in both offender and victim in 43.5 percent of homicides, in the offender only in 10.9 percent and the victim only in 9.2 percent of the cases. Thus, the offender had been drinking in 54.4 percent and the victim in 52.7 percent of the homicides, and either one or both in 63.6 percent of the cases.

Mayfield (1972), reports on a study carried out in North Carolina of males who entered the prison system after being convicted of serious assaultive crimes. (Of the 307 subjects studied, the offender had been convicted of homicide in 80 percent of the cases and of a variety of felony assaults in 20 percent of the cases). Fifty-eight percent of the offenders were "definitely sober." Among the victims, 40 percent were "definitely sober" in connection with the crime and 35 percent were "definitely free of alcohol." Pittman and Handy (1964), in their study of 241 cases of *aggravated assault* in 1961, sampled from the St. Louis Metropolitan Police Department files, show that 57 offenders and 58 victims had consumed alcohol prior to the crime. They suggest that the low proportions, about 20 percent, may be due to difficulties in detecting or failure to report prior use.

In another study of 395 criminal homicides in the urban area of Chicago in 1965, it was found that "intoxicants were present, according to police records, in 53.5 percent of the homicidal scenes" (Voss and Hepburn, 1968).

A number of Finnish studies have established the prevalence of alcohol involvement of both the offender and the victim in various types of violent crime. Verkko (1951), assumes a direct causal link (via racial and physiological factors) between alcohol use and violent crime in explaining the high incidence of violent crime in Finland. He refers to a study in 1931 by a committee under the chairmanship of Bruno Salmiala. I have summarized the main findings in Table 1.

Absolute numbers are not available and it is thus not possible to get an approximate figure for alcohol involvement in homicides in general, comparable to that of Wolfgang (1958). More detailed information is, however, available for one county in Finland (Vyborg County) between 1920 and 1929 (see Table 2). The Salmiala Committee felt that these were minimum figures on the involvement of alcohol, since in many cases no evidence of the presence or absence of alcohol was available in the court records. From Table 2, we can calculate that the total alcohol involvement for murder and intentional manslaughter combined is 69.1 percent, a figure that is fairly close to Wolfgang's. An interesting difference that may have further explanatory implications is the fact that in only 34.3 percent of the offenses were both victim and offender under the influence of alcohol in the Finnish data, as compared to 43.5 percent in Wolfgang's data for Philadelphia, despite the somewhat higher general alcohol involvement in Finland.

Verkko is careful to point out that the relatively minor role that alcohol

TABLE 1. The Alcohol Involvement in Various Violent Crimes in Finland, 1904–13 and 1920–29, in Percent[a]

	Intoxicated at the time				Had drunk during the day			
	Offender		Victim		Offender		Victim	
	1904–13	1920–29	1904–13	1920–29	1904–13	1920–29	1904–13	1920–29
Murders	7.1	7.3	1.4	2.4	9.0	14.1	1.9	2.4
Intentional manslaughter, and wounding occasioning death	58.7	62.1	44.8	49.5	11.3	14.9	11.0	12.9
Wounding resulting in grievous bodily harm	63.9	66.9	36.1	38.7	9.3	13.2	8.1	13.6
Assault and battery with ensuing death or grievous bodily harm	74.1	53.4	43.1	41.7	6.9	13.5	13.8	15.9

[a] From Verkko, 1951.

plays in murders in Finland is partly definitional; there is a greater probability of a homicide being classified as manslaughter if the offender is under the influence of alcohol.

More recent data from Finland, where violence in connection with alcohol use is seen as a major problem to the extent that the rate of violent crimes (with or without evidence of alcohol involvement) is used as an indicator of "the alcohol situation," show that Salmiala's (see Verkko, 1951) figures are still approximated in Finnish data. For example, Aho (1967), in a study of the court records of 313 homicides (about 50 percent of the cases), and aggravated assaults in Helsinki between 1950 and 1965 found that 85 percent of offenders had been drinking before the crime. He classifies 69 percent as being "drunk" and 16 percent as "slightly intoxicated," the latter having in general a blood alcohol concentration or level, hereafter referred to as BAC, of less than 0.17 percent. The percentage for victims was also close to 70. Virkkunen (1974), in a study of 116 homicides in Helsinki in the period 1963–1968 found that 68.1 percent of the victims and 66.4 percent of the offenders had been under the influence of alcohol. According to Krokfors (1970), alcohol was involved in 73 percent of homicides in Finland in 1968. For aggravated assaults the figure was 68 percent and for other assaults 61 percent. He does not specify the prevalence of alcohol use separately for offender and victim.

Janowska (1970), has analyzed data on the 279 individuals sentenced for

TABLE 2. The Alcohol Involvement in Various Violent Crimes in Vyborg County, Finland, 1920–1929[a]

	Murder		Intentional manslaughter		Assaults resulting in grievous bodily harm		Assault and battery with ensuing death or grievous bodily harm
	Number	Percent	Number	Percent	Number	Percent	(Number)
Victim	0	0	74	15.3	21	8.8	3
Offender	6	10.3	109	22.5	81	33.9	7
Both	1	1.7	185	38.1	84	35.1	9
Neither	51	87.9	117	24.1	53	22.2	8
Total	58	100	485	100	239	100	27

[a] From Verkko, 1951.

homicide in Poland in 1961. She found that 64.5 percent of these, and 49 percent of the victims, were intoxicated at the time of the crime.

Studies Examining Alcohol Use by Offenders Only

Shupe (1954), has reported findings on 882 persons picked up during or immediately after the commission of a felony in Columbus, Ohio, between March, 1951, and March, 1953. Of the 30 persons arrested for murder, 25 (83 percent) had some trace of alcohol in their urine and 67 percent had a level of 0.10 percent and over (see Table 3). His data show a high percentage of involvement of alcohol also in crimes that are not by definition violent in nature. As Shupe points out, the fact that offenders who are under the influence of alcohol have a greater risk of being apprehended by the police than do sober offenders could bias the findings in the direction of too large an alcohol involvement. This should be remembered in interpreting the findings of all studies on alcohol use by crime offenders. (I will discuss this and similar biasing factors in a section below, "The Nature of the Dependent Variable.")

Macdonald (1961), has summarized the findings from ten studies that have provided figures on the proportion of homicide offenders who had been drinking prior to the crime, including the studies by Shupe and Wolfgang. The range is very wide, from 19 to 83 percent, but sizes and nature of the samples vary widely. The mode is between 50 and 60 percent, and in this interval fall all studies with a sample size of 200 and over. Assuming a random sampling procedure, there are comparatively small sampling errors, and thus a figure of 50–60 percent alcohol use by offenders in North America seems sufficiently reliable.

In addition to the studies summarized by Macdonald, geographically scattered data from widely varying types of samples show the pervasiveness of alcohol use by the offender in noninstrumental interpersonal violent crime. A study of 2,234 new arrivals to prison centers in California (1964) in 1959 showed that 60 to 63 percent of offenders in homicides and sex crimes, and other crimes of great personal risk, had been drinking before the crime. These findings were based on personal interviews. Tinklenberg (1973) has reviewed a number of additional studies of both homicides and assaults that show a high proportion of alcohol use prior to these crimes. He also cites his own data, which show that eight out of nineteen young violent offenders who had been convicted of murder, manslaughter, or assault had been under the influence of alcohol (and five had been under the influence of some other psychoactive drug). Connor (1973) reports that alcohol involvement of offenders in violent crimes seems to be more pronounced in the Soviet Union than in other countries for which data are available:

> In Gorky oblast' (province), 83 percent of all persons convicted of homicide, infliction of serious bodily injury, and rape committed their offenses in a state of intoxication; in Yaroslavl' oblast' the parallel figures for homicide and rape were 85 percent and 76 percent respectively.

According to Connor, Gertsenzon has pointed out large regional variations in the involvement of alcohol by the offender at the time of the crime in

TABLE 3. Percent of Persons Arrested in Each Crime Class Showing Various Percentages of Urine Alcohol[a]

	Cases studied	\	\	Alcohol concentration (%)	\	\	\	\
		nil	.00–.09%	.10–.19%	.20–.29%	.30–.39%	.40% plus	.10% plus
Rape	42	50	5	19	21	5	0	45
Felonious assault	64	52	5	9	20	13	2	43
Cutting	40	8	5	20	35	25	8	88
Concealed weapons	48	8	8	21	25	33	4	83
Other assaults	60	8	13	25	33	18	2	78
Murder	30	17	17	30	23	13	0	67
Shooting	33	18	3	27	33	18	0	79
Robbery	85	28	12	15	29	15	0	60
Burglary	181	29	7	24	24	14	2	64
Larceny	141	27	9	13	27	19	5	65
Auto theft	138	30	11	25	22	8	4	59
Forgery	20	40	0	20	20	20	0	60
Average total	882	27.3	8.4	20.2	25.8	15.6	2.6	

[a] From Shupe, 1954.

the Soviet Union, from a high of 80 percent to a low of 46.8 percent, the lower involvement presumably being due to cultural factors such as a predominantly Muslim population in certain regions with reservations about alcohol use in general. Husson et al. (1973) state that 69 percent of "voluntary" homicides in metropolitan France were committed under the influence of alcohol. Twenty-nine percent of aggressive assaults were committed under the influence of alcohol.

In another study 36 out of 66 "murderers" (55 percent) in the Glasgow area in Scotland were "affected by alcohol" at the time of the crime (Gillies, 1965). Gelfand (1971) attributes 50 percent ($N = 80$) of the homicides he studied among the native people in Rhodesia between 1898 and 1930 to the use of alcohol (by tradition, beer). He also studied the preliminary court documents in "murder" cases in 1967–68 and found that 63 percent of the 98 offenders had consumed alcohol prior to the crime. For assaults the percentage of alcohol involvement by offenders was surprisingly low. Thus, in 47 assault cases brought before the courts in one part of Rhodesia in 1963–67, only three were associated with alcohol.*

Additional studies on the prevalence of alcohol use prior to the commission of homicides and assaults can be found in excellent reviews by Wolfgang (1958) and Wolfgang and Strohm (1956).

The alcohol involvement of special offender samples such as offenders who have been diagnosed as mentally ill has been reported in a number of studies. These samples are probably biased against the role of alcohol as a possible causal factor, and they are worthless in comparisons with results pertaining to more randomly selected cases of violent crimes. However, the findings in these samples may be relevant for causal interpretations of possible interaction effects of alcohol use with psychiatric problems in the etiology of violent crime. For this purpose I will mention a sample of these.

Tupin et al. (1973), in a study of 50 males who had committed murder and were detained in a California *psychiatric institution,* found that 13 of them had used alcohol "heavily" at the time of the crime. This did not differ significantly, however, from a group of offenders that had been convicted for nonviolent crimes (5 out of 25). In a study by Binns et al. (1969b) of persons charged with a criminal offense who had been referred to *psychiatric examination* at a hospital in Scotland in 1965–66, 47 percent ($N = 19$) of those charged with assault had been intoxicated at the time of the crime. McKnight et al. (1964), in a study of 100 *mentally ill* homicide offenders in Ontario, Canada, found that "approximately one fifth of the patients were under the influence of various amounts of alcohol at the time of the offense; however, few, if any,

* Gelfand attributes this partly to selective reporting to the authorities, since alcohol is not usually seen as a defense in civil cases (as opposed to criminal cases under which homicides fall), and thus there may be underreporting in the records.

could be considered to have been intoxicated." Pincock (1962), on the other hand, found in a study of individuals referred for *psychiatric examination* by various courts of metropolitan Winnipeg that 31 of 42 homicidal offenders (73 percent) and 59 out of 89 offenders (66 percent) in other aggressive crimes were intoxicated coincidentally with the crime.

Studies Examining Alcohol Use by Victims Only

Wilentz and Brady (1961), in a study of violent deaths in a county in New Jersey between 1933 and 1959, found that 31 percent of the victims of homicide (54 cases out of 175) and been drinking before being killed. Of these they classified 14 percent (24 cases) as "being under the influence," i.e., having a blood alcohol level of more than 0.15 percent. Spain *et al.* (1951) studied violent deaths in Westchester County, New York. They found that seven out of the eight homicide victims in the sample had BACs ranging between 0.11 and 0.25 percent. Room (1970) reports on a study done by Hudson in North Carolina showing that in over half of all violent deaths the victim had a BAC of 0.10 or over and stating that the figure would be about 0.20 or 0.25 for homicide victims. In a study of homicide victims in the period between 1947 and 1953 in Hamilton County, Ohio, Cleveland (1955) found that out of the 225 victims (from a total to 337) who had a BAC determination, 86 percent had concentrations of at least 0.01 percent and 44 percent had concentrations of 0.15 percent and over. In a Canadian study, Tardif (1967) investigated 521 cases of homicides, assaults, rapes, and robberies in Montreal in 1964. The presence of alcohol in the victims of homicide was comparatively low, 17 percent ($N = 23$). Among the assault, rape, and robbery victims a lesser severity of injury meant a lower alcohol involvement. Victims who were hospitalized had used alcohol in 33 percent of the cases ($N = 73$), and those who were only treated but sent home had used alcohol in 26 percent of the cases ($N = 152$). Victims with light injuries had used alcohol in 16 percent ($N = 81$), and those with no injuries had used alcohol in 9 percent of the cases ($N = 155$). Bowden *et al.* (1958) report on blood and urine alcohol tests carried out on the victims of violent death in Melbourne, Australia, in 1951–56. Forty-one victims of murder were tested, and 19 of these had BACs of 0.15 percent or over. Also in Australia, Birrell (1965), in testing 47 homicide victims found 27 to have at least a trace of alcohol in the blood and 23 to have BAC levels of 0.15 percent or higher. In a Polish study Puchowski and Tulaczynski (1964) report that 38 percent of the victims of homicide (sample size unknown) had alcohol in the blood. Le Roux and Smith (1964) in a study of violent death victims in the Cape Town area found that 88 out of the 137 adult homicide victims, i.e., 64 percent, had positive blood alcohol levels and 50 percent had levels at the time of death of over 150 mg per 100 ml. Medina (1970) reports on a study by Leyton showing that in a sample of homicide victims in Santiago de Chile 62 percent of the male victims of homicides had a blood alcohol level of 0.05 percent or higher ($N = 208$). The corresponding figure for females was 35 percent ($N = 20$). Arner (1973) reports that half of homicide victims totaling about 30 among seamen on Norwegian ships between 1957 and 1964 were "more or less" intoxicated at the time of the crime.

Other Relevant Studies

In addition to the conjunctive evidence of alcohol use by the offender and victim in violent crime, I shall mention briefly other descriptive evidence that bears on the concomitance of alcohol use and aggression.

Both Wolfgang (1958) and Hopwood and Milner (1940) have noted the extremeness of the violence displayed in homicide committed in connection with alcohol use. Wolfgang objectively categorized his data of the homicides he studied to arrive at his conclusion, whereas Hopwood and Milner impressionistically note: "A striking feature of alcoholic murders is the malevolence of the crime." Stark (1969) points out the "reckless brutality" of robberies committed after drinking, which often lead to severe bodily injuries. Tardif's (1967) findings of increasing alcohol involvement of the victim with increasing seriousness of the injury in assaults, rapes, and robberies are also relevant in this context.

Anthropological studies show a disconcerting array of reactions under the influence of alcohol in different societies (Washburne, 1961; and MacAndrew and Edgerton, 1969). Child *et al.* (1965a), however, in studying literary reports on 139 societies scattered throughout the world, most of them preliterate and homogenous, found that in none of the societies reported on did the expression of hostility lessen during alcohol drinking occasions. Most of the changes were rated as slight increases in the intensity of hostility.

Since different indicators of violence and aggression were used in the above studies, some of the alternative hypotheses in accounting for the relationship between alcohol and violence have been eliminated. This is the case, e.g., for the hypothesis that the connection is due to a greater likelihood of offenders under the influence of alcohol being apprehended, which creates an inflated association between the two variables. (See the section, "The Nature of the Dependent Variable" for a discussion of these problems.)

Discussion

The data are scattered but show the pervasiveness of the association between violent crime and alcohol use in a number of different cultures, and one can presume that cultural factors or jurisdictional idiosyncracies alone cannot explain the association. However, any systematic comparison between rates from such a variety of samples is impossible and, as pointed out above, there is no way of relating these findings to other epidemiological findings or statistical information to ascertain, for example, the extent to which alcohol involvement in violent crime is related to the per capita consumption, or the consumption of specific beverages over a number of countries or states.* What is most obviously lacking is a systematic collection of data on the role of alcohol in violent crimes in samples where corresponding epidemiological data of possible causal relevance could be utilized. Especially relevant would be estimates of blood alcohol levels in both the offender and the victim at the time of the crime to get at better data on the relationship between this aspect of alcohol use and violent crime, although the practical difficulties may be insurmountable.

Studies of the alcohol involvement in violent crime should also include better information on the nature of the potential independent variable such as the type of beverage used, etc., to enable more specific epidemiological research and to provide a background for relevant experimental research on aggression.

It is enlightening to turn back to the figures in order to see what type of information would be necessary to show that the association is not due to chance, i.e., that alcohol use and violent crimes are not independent of each other. The most difficult aspect of this task lies in delimiting the universe for the statistical inferences to be drawn, i.e., establishing the population at risk, and, due to the cyclical nature of human activity, in establishing the relevant time periods for the population at risk. The ideal way of approaching the problem would be to get estimates of the proportion of time that the population at risk is intoxicated, the proportion of time that it is committing violent crimes (or aggressive acts, if we are interested in this more general question), and the proportion of time that it is committing violent crimes in an intoxicated state. Needless to say, this is an extremely simplified model, and even if unbiased empirical measurements could be obtained, one would have to take many biasing selective variables into consideration in establishing a null hypothesis.†

The samples available are of crimes of violence, however, and we do not have random time samples of human activities. Let us use Wolfgang's (1958) figures and disregard possible biasing factors such as a greater risk of being apprehended by the police for the intoxicated offender. If the null hypothesis of

---

* Wolfgang (1958) cites a comparatively old study by Fornasari di Verce, in which he found that crimes of violence "increase and decrease in direct ratio with the consumption of alcohol." The particulars of the study are unknown to the present author.
† Tinklenberg (1973) has pointed out the fact that both alcohol use and violence most frequently occur among close acquaintances. Thus, it is difficult to establish the role of alcohol in violence independently from the role of close interindividual ties in studies of an epidemiological nature. In experimental methodology, these variables could be varied independently of each other. Both Wolfgang (1958) and Amir (1967) have pointed out the difficulties in establishing a null hypothesis.

no connection between a state of intoxication and violent crime were true, we would have to accept as an empirical fact that the population at risk of becoming an offender would be intoxicated on the average 54.4 percent of their waking hours, since in his sample 54.4 percent of offenders were intoxicated at the time of the crime. Even if a very narrow definition of the population at risk is used, it seems extremely improbable that the null hypothesis would be true.

The value of this type of reasoning lies mainly in directing attention to the conditions that have to be taken into account in specifying the null hypothesis. The probabilities to be used in the null hypothesis have to be conditional probabilities and herein lies the crux of the problem. What are the conditional factors that we would have to take into account? We would probably have to take time of day as one condition that would influence the probability of both the victim and the offender being drunk. Another conditional factor or set of factors affecting the probability of being drunk would be location. The first step toward enabling a satisfactory test of (a much more complicated) null hypothesis would be to gather extensive information on the circumstances of the crime. Even the blood alcohol levels of offenders and victims do not tell us very much by themselves about the degree of association between alcohol and violent crime. This brings us to the most crucial question in the area under study.

Up to this point we have only discussed the possibility that the two events are statistically independent or related to each other. The role of alcohol in the causal explanations of the possible connection has not been touched upon yet. I will just briefly touch on the issue here and postpone a more detailed discussion until later. Even if all the necessary information were available to test the null hypothesis of no correlation between intoxication (or other aspects of alcohol use) and crimes of violence (or aggressive behavior in general) and the null hypothesis were rejected, it is not established that alcohol has causal relevance in explaining violent crime.

The existence of an association does in no way show that alcohol is a sufficient cause of violent crimes. This is so, since, in fact, few alcohol use situations result in violent crimes. Thus, we have to look for other conditional variables that together with alcohol use determine whether a situation leads to violence or not.

Before going into possible explanatory variables and causal models that would explain the correlation, I shall look at the evidence available for the proposition that alcoholism, or prolonged excessive alcohol use, and its effects, is positively associated with crimes of violence.

Association between Alcoholism and Interpersonal Noninstrumental Crimes of Violence

Again, there is a great variety of studies available in the literature relevant to establishing (at least in theory) whether there is more than a chance association between "alcoholism," "problem drinking," "heavy drinking," etc., and crime in general. Since these are of very limited usefulness for the purposes of this chapter, I shall only refer to what may be a rather randomly available sample of these (Banay, 1941–42; Kinberg et al., 1957; Smith-Moorhouse and Lynn, 1966; Edwards et al., 1972; MacKay et al., 1967; MacKay, 1963; Maule and Cooper, 1966; Edwards et al., 1971; Glatt, 1965; Bartholomew and Kelley, 1965). The proportion of alcoholics among offenders in violent crimes other than homicide and assault has also been established in a few studies. For example, Nau (1967), in West Berlin, and Grislain et al. (1968), in France, found high proportions of alcoholics in child abusers, 50 percent and 65 percent respectively. Banay (1941–42) found that approximately 28 percent of sex offenders in Sing Sing prison over a five-year period were "intemperate" users of alcohol, and in Rada's (1975) California sample of rapists ($N = 77$) 35 percent were classified as alcoholics.

Studies of Prison Populations

A large majority of studies relevant for an assessment of the association of alcoholism and violent crime use prison samples (partly because of comparatively easy accessibility, no doubt) and determine alcoholism via interviews; a much smaller number use court records and try to determine alcoholism by independent means available. Still other studies use clinical alcoholic samples and try to determine the prevalence of criminal records for these.* There would naturally be a smaller chance of finding alcoholics who have committed serious crimes, such as crimes against the person, among outpatients, since they would be more likely to be in prison. On the other hand, as Gibbens and Silberman (1970) point out, prison samples are biased in favor of offenders with long prison sentences. For prevalence estimates of alcoholism in, e.g., homicides, this fact has to be taken into account in jurisdictions where sober homicide offenders ("murderers") get longer sentences than intoxicated and more likely alcoholic offenders (in these cases there might be greater tendency to label the crime as manslaughter or to regared drunkenness as an extenuating circumstance). In this case there would be an underestimation of the prevalence of alcoholism in violent crime.

Most of the studies dealing specifically with homicides and assaults also use more or less arbitrarily selected prison populations in which indices of heavy drinking history or alcoholism are investigated, or they use available clinical populations whose police or court records are investigated. Needless to

---

* Mayfield (1972) found that the problem drinkers in his prison sample were unlikely to have been in treatment for alcoholism and "almost never voluntarily sought treatment." This indicates that the populations tapped by prison samples may be very different from the ones reflected in clinical samples of alcoholics, and results may not be comparable.

say, the criteria for problem drinking or alcoholism differ so much from study to study that it is very difficult to draw any general conclusions.*

Banay (1941–42), in his study of prisoners in Sing Sing prison mentioned above, found that alcoholic criminals (according to his definition, "delinquents whose alcoholism led to the commission of crime") accounted for 31 percent of all people who were imprisoned for assault in the fiscal year 1938–1939 (35 percent in 1939–40), and 23 percent of those imprisoned for homicide (25 percent in 1939–40). [The other crimes were committed by alcoholics as measured by Banay's criteria as follows: burglary offenders, 25 percent (33 percent in 1939–40); sex crime offenders, 22 percent (38 percent); robbers, 21 percent (19 percent); grand larceny offenders, 14 percent (19 percent); and offenders in all other crimes, 6 percent (15 percent); for 1938–39 and 1939–40, respectively.]

There seems to be considerable fluctuation over time in the proportion of inmates with alcohol problems among different types of criminals. Banay presents data for "intemperate" homicide offenders over the period 1935–1940. The proportions of these offenders out of the total homicide offender population range from 11 percent to 37 percent with a mean of approximately 23 percent (calculated by the present author). Thus, in his study the proportion of offenders with alcohol problems was higher for sex offenses (28 percent) than for homicides.

Goodwin et al. (1971) and Guze et al. (1962) report on a sample of male felons about to be released and recent parolees and probationers from Missouri penal institutions ($N = 223$). Using rather broad criteria, they arrived at a figure of 43 percent alcoholics and an additional 11 percent questionable alcoholics. Some of the criteria used for classification as an alcoholic make links with violent crime definitional to an unknown extent, as pointed out by the authors. The alcoholics were more likely to report rage reactions, feeling irritable more often, and were more likely to have had fights. Sixty-six percent of the alcoholics had one or more arrests for peace disturbance versus 17 percent of the nonalcoholic felons, and 42 percent had one or more arrests for fights compared to 10 percent for the nonalcoholics.

Mayfield (1972), in a study referred to earlier, concluded that in his sample of 307 prisoners in North Carolina convicted of serious assaultive crimes (80 percent for homicides), 36 percent were problem drinkers. These were more likely to have committed previous serious assaults than the other offenders in his sample. Ullman et al. (1957), in studying 1000 consecutive admissions to a prison in Massachusetts, found that those who had been arrested for drunkenness two or more times, or had a history of alcoholic psychosis, had 0.7 convictions for "pugnacious crimes" per individual as compared

---

* For an excellent discussion of different measures of "abnormal drinking," see Edwards et al., (1971).

to 0.4 convictions for other individuals. However, in another Massachusetts study of 95 women prisoners, it was found that there was a significantly greater likelihood for more nonviolent prisoners to be alcoholics than for violent prisoners (Climent et al., 1973). Violence was measured using objective historical data, ratings by the subject and others, and a MMPI profile. How alcoholism was measured is not mentioned. (The possibility of sex specificity in the relationship between alcoholism and violent crime cannot be ruled out.)

Gibbens and Silberman (1970), in a study of alcoholism among prisoners, found that "a history of two or more aggressive offenses (not necessarily serious ones) was much more common in alcoholics, and they included nearly all of those with a history of both aggressive and sexual offenses."

In a study carried out in New South Wales, Australia, McGeorge (1963) found that 22 out of 85 murderers and 59 of 100 offenders in assault and robbery cases were "addicted to drink." In Poland, Janowska (1970) found that 55.6 percent of the offenders in homicides had been drinking "systematically and excessively." Pittman and Wal (1968) showed that probationers to "Consultation Bureaus for Alcoholism" in Holland had a statistically highly significant tendency to have committed aggressive crimes to a greater extent than a random control sample in other probationary agencies.

A number of studies have been carried out on special offender populations that have been regarded as having proclivities toward violent behavior, and they may thus be relevant for the study of interaction effects between alcoholism and organic or psychological conditions in the etiology of violent behavior. Lanzkron (1962–63), in a study of 150 mental patients charged or indicted with murder, found that 34 percent had shown "intemperate" use of alcohol. (In only 12.7 percent of the cases could it be ascertained from the records that the accused had been "drunk" at the time of the crime.) Binns et al. (1969a) found in a study of persons charged with a criminal offense and remanded for psychiatric examination that 14 out of 107 referrals had been clinically diagnosed as alcoholics. In another similar study 19 out of 83 were thus diagnosed (Binns et al., 1969b).

The drawback with all prison samples is, as pointed out above, that they are weighted in favor of types of crime and criminals that have longer sentences. More severe crimes and, more importantly, more habitual criminals, whatever this means in terms of alcohol involvement, are overrepresented in these populations. It would be important to include length of sentence as a control variable in prison studies in general.

Studies of Other Populations

Evidence of an association between alcoholism and violent behavior tendencies (i.e., not violent crime) comes from a study of Tuason (1971), who found that 12 out of 30 violent patients hospitalized in a mental health center were "alcoholics" and six were "probable alcoholics." In reporting on the investigation of court records by the Salmiala Committee in Finland, Verkko notes that the offender in "many cases" was a confirmed alcoholic, although perhaps not under the acute influence of alcohol at the time of the crime. He does not, however, present any quantified data. Some evidence against a relationship between the two variables is to be found in Antons's (1970), study of 67 "Kurhaus" patients in which a number of different measures of aggressiveness were used, including a rating by the therapists at the Kurhaus. He found that on the objective measures of aggressiveness, the alcoholics were no more aggressive than the comparison group of 67 other Kurhaus patients. (On a self-rating scale, however, the alcoholics rate themselves as more aggressive than the comparison group.) Despite such scattered evidence against an association between alcoholism and violent behavior, the evidence is rather heavily in favor of a relationship between alcoholism and at least violent crime, although inferences here, as was the case for acute alcohol use, are hampered by the lack of control data.

Information on the prevalence of alcoholism among *victims* of homicide can be gleaned from several mortality studies. For instance, Sundby (1967) found that it was two to three times more probable for an alcoholic to die by homicide as compared to the general population. In most mortality studies of alcoholics, homicides are lumped in the same category with accidents and suicides as "violent deaths" (due to the small numbers of cases).

Discussion

The central question in assessing the association between alcoholism and violent crime (and possible causal explanations between the two) is the extent to which this typological entity explains violent behavior if the greater probability of being intoxicated at any time is controlled for, assuming that acute alcohol use is a relevant explanatory factor. At the present time, there are no reliable and generalizable data available that would permit an assessment of the relative effects of acute intoxication and the cluster of characteristics that is labeled "alcoholism" (or the factors that are causally linked to it) on the association between alcoholism and violent crime. Some indications of the extent to which intoxication plays a role in general crimes of alcoholics can be had from Banay's (1941-42) study of 200 "primary intemperate" white males. He found that 58 percent of these males had been intoxicated at the time of the crime. I have only been able to locate one study that would show the prevalence of alcohol use in connection with *violent* crimes in a sample of alcoholic offenders. Mayfield (1972) found that the problem drinkers in his sample were sober at the time of the violent crime in only 20 percent of the cases. The corresponding figure for other offenders was 40 percent.

EXPLANATORY MODELS

The problems of determining the population at risk discussed above are partly identical with the problems of explaining the association between alcohol use and alcoholism and crimes of violence. In either case one is forced to peel off layer after layer of possible biasing factors that would give rise to a spurious causal relationship between the two phenomena. The only way to test a true "direct" relationship between the two variables is through experiments but then again one is up against the problem of generalizing the findings, and this dilemma is reflected in the discussion of explanatory models below.

In the reasoning on possible causal models explaining the association between acute alcohol use and alcoholism and violent crime, I have taken for granted that the null hypothesis does not hold true.* This is based partly on the "statistical" reasoning above, but the reasons for rejecting the null hypothesis are probably based as much on "common sense" and personal knowledge of human behavior under the influence of alcohol.

In looking for explanations of the correlation between alcohol use—be it acute alcohol use or long-term excessive use—and crimes of violence, I will refrain from dividing the discussion below into fields of inquiry from which the explanatory variables have been taken. It is often difficult to assign an explanatory variable to a field. This is so because the intervening variables in an explanatory model are seldom specified, and presumably often belong to several different fields. As often is the case in explaining behavior, conditional factors in many fields of inquiry are also relevant in the explanation, and these are tacitly assumed to have some "normal" value in the phenomena explained. Thus, the fact that the assignment of explanatory or intervening variables and models to various fields (to the extent that it is attempted in the discussion below) is rather arbitrary, and reflects the state of explanations in this area of research. In ascribing the explanations to variables in different fields, one runs into difficulties that traditionally have been considered to be philosophical, such as the truth or falsehood of the positivistic doctrine of "psychophysical parallellism" or "psychophysical identity" (e.g., Polten, 1973).

Contrary to the view of some social scientists (e.g., Shoham et al., 1974), I feel that there is reason to include factors from other disciplines in the explanation of behavior, and this especially in dealing with alcohol use, which has undeniable pharmacological and psychological effects, although these may not

---

* The null hypothesis stating that a statistical positive relationship between alcoholism and crimes of violence is due merely to chance has to take into consideration the higher probability of alcohol use among alcoholic samples. Thus, in rejecting this null hypothesis it is assumed that alcoholism and causally related and concomitant phenomena, lead to a higher risk of violent behavior than expected by chance and by the higher risk of alcohol use in alcoholics.

be constant over individuals and time. I will include such variables since it is possible, and in some cases very probable, that they covary with "social" variables relevant in this context (such as social characteristics of the drinking situation).

The truly interdisciplinary nature of the problem can be seen from looking at some of the literature relevant to the problem:

1. The studies reviewed above on the proportion of offenders and victims who were under the effect of alcohol during the commission of a violent crime.

2. The few rather unspecific studies, also reviewed above, giving the proportion of "alcoholics" ("heavy drinkers," "problem drinkers," "excessive drinkers") among persons detained for violent crimes.

3. Psychological experiments on the effect of different doses and types of alcohol on the behavior of alcoholics or nonalcoholics.

4. Anthropological descriptions of predominantly primitive tribes and their behavior in alcohol use situations.

5. Studies of EEG-patterns, brain syndromes, head injuries in alcoholics and nonalcoholics or violent offenders in nonalcohol conditions and after consumption of alcohol.

6. Longitudinal studies of genetic and developmental factors, in the etiology of alcohol problems and other behavior tendencies, including proclivities toward violent behavior.

Studies of types 1 and 2 generally provide no basis for inferences as to the causal relationship between the effects of alcohol and aggressive behavior. Studies of type 3, on the other hand, are often designed to get at the "pure" effect of alcohol, making generalizations to real life situations rather questionable by leaving out relevant conditional factors, (even if we overlook the difficulties in operationalizing the concept of aggression). Type 4 studies can probably best give clues as to the normative determinants affecting the connection of alcohol use and aggressive behavior. Studies of type 5 are generally designed to study factors predisposing individuals toward violence in connection with alcohol use; and studies of type 6 give some indication whether genetic or developmental factors account for both violent behavior tendencies and a greater probability of alcohol use (and/or alcoholism).

In addition to these types of studies, there are others that are more indirectly related to the subject matter of the connection between alcohol and violence. The effects of alcohol on performance of tasks that require conceptual activity, judgment, motor skills, etc., have been studied rather extensively and these effects are relevant to the performance of any act, including violent acts. Still other studies can give rise to hypotheses regarding the proportion of violent crimes explainable by some facet of alcohol use, or can be relevant for setting up causal models in the explanation of an association between alcohol use and violence. These include the following: biological studies on the etiology of violent behavior (e.g., testosterone production, XYY chromosomal abnormalities), studies of special populations prone to violence such as temporal lobe epileptics, and general psychological studies on aggression (without alcohol use as one variable).

Explanation of the Positive Relationship between Alcohol Use and Crimes of Violence

Nature of the Independent Variable

In epidemiological studies "alcohol use prior to the crime" and other standard labels by their very nature include (in a seldom explicated form) the modal patterns and circumstances in which alcohol is used within a culture, e.g., the type of drink consumed, the selection of participants to the drinking situations, and the norms governing the behavior of these people in alcohol use situations. When the data are extracted from police or court records, the crucial operationalization of the concept is dependent on not only the criteria of the researcher, but also those of the arresting police officer and other individuals making decisions in the intermediate stages between the arrest and a possible court conviction. The testimony of the accused, the victim, and witnesses also enter in here, as well as the interpretations of the individuals doing the coding for research purposes. The same holds true for "intoxication," which probably is even less reliable in the recordings of officials and in the testimonies given. Blood alcohol level is a more exact measure, but the concepts do not overlap completely, considering the physiological and psychological factors leading to variations of subjective states and performance measures both within and between persons with the same blood alcohol concentrations (e.g., due to adaptation phenomena, see Kalant, 1961). Besides, this measure has been used very sparingly in relevant epidemiological studies of offenders in violent crime.

Experiments with Nonalcoholics. In explanations of the relationship between alcohol use and violent crime via the acute effects of alcohol, the independent variable of "alcohol use" is too vague to stand up to closer analysis. A number of different independent alcohol variables can be responsible for the correlation between violent crime and alcohol use. This has been taken into account in experimental designs (and in some interpretations of epidemiological data). In the experimental studies the most frequently used independent variable is the BAC of the subject or some other closely correlated measure such as quantity consumed per unit of body weight. Of equal *a priori,* relevance in the explanation of the relationship are some other independent variables. Some of these have been incorporated into research designs dealing with alcohol and aggressive behavior or related dependent variables. First, however, a look at a few studies using BAC as the independent variable, and the range of values used.

Hartocollis (1962) used intravenous administration of ethyl alcohol, thereby controlling for the independent effects of "suggestion" (perhaps better conceptualized as psychological or cultural "meaning" of alcohol), which could give rise to a spurious relationship between BAC and aggressive behavior. The approximate BAC level during the first hour of the experiment was 0.10 percent (1 cc absolute ethyl alcohol per kg of body weight). Bennett et al. (1969) used three different alcohol dose levels of 0.33, 0.67, and 1.00 ml (cc) of absolute alcohol per kg of body weight in the form of vodka mixed with orange juice in addition to a placebo condition. The mean BACs attained were 0.030, 0.058, and 0.086 percent respectively (measured by Breathalyzer test). Shuntich and Taylor (1972), in their experiment, which also measured alcohol effects on aggressive electrical shock settings, used a dose of 0.9 ml of 100 proof bourbon per kg of body weight as their alcohol condition, which was compared to a placebo and a control condition. They did not measure the actual BAC attained in their subjects. Hetherington and Wray (1964) gave their subjects in the alcohol condition 0.35 cc of ethyl alcohol per kg of the subject's body weight. In the nonalcohol condition they were given a placebo drink disguised as alcohol. The disguise was successful in that an approximately equal proportion of subjects in both conditions (84 and 87 out of 96) thought they had been given alcohol. Kastl (1969), in his experiment on ego functioning under alcohol, used three different alcohol doses, 0.33, 0.67, and 1.00 cc of absolute alcohol per kg body weight, in addition to the control condition where no alcohol was given. The alcoholic beverage was vodka mixed with lime juice. Katkin et al. (1970), in their experiment on the effects of the congener content of vodka and bourbon on psychomotor performance, gave their subjects an amount of 0.4 cc of absolute alcohol per kg body weight. This dose was given four times at one hour intervals. The placebo condition was tap water, and in the experimental condition alcoholic beverages were mixed with tap water. The mean BACs ranged from 0.03 percent to 0.08 percent over the different sessions. In another experiment, this one on congener effect on risk taking, Katkin et al. (1970) gave their subjects a total of 0.8 cc of absolute alcohol per kg of body weight in the form of vodka and bourbon (both with four times their normal congener content), and congener free synthetic alcohol as a control. In one condition of this experiment, the mean BAC was 0.08 percent, and in the other a falling BAC of 0.08 percent. Tarter et al. (1971), in their experiment on the effects of alcohol on "perceptual, perceptual-motor and cognitive capacities," used 95 percent commercial ethyl alcohol mixed with orange drink to attain a mean blood alcohol level of 0.08 percent. They refer to two other similar studies in which the BACs were 0.03 percent and 0.06 percent. The subjects in Bruun's (1959) experiments had relatively high blood alcohol levels ranging from a mean of 0.08 percent after two hours to 0.21 percent at the end of the six-hour experiment. The BACs were, however, based on theoretical calculations and the fact that the subjects ate a meal at two hours into the experiment means that the BACs probably did not reach this level, as Bruun points out. The subjects were allowed to choose between five different beverages, and drink as much as they wanted.

The assessment of dose levels used in animal studies is beyond the competency of this author, and consequently only short mention will be made of the type of independent variable that has been used in these studies. Weitz (1974) used three different amounts of ethanol solution in her experiment on the fighting behavior of male hooded rats, in addition to a nonalcohol condition. Raynes and Ryback (1970) used congener content as one independent variable in their experiment with Siamese fighting fish (*Betta splendens*), the other independent variable being an alcohol as compared to a nonalcohol condition. MacDonnell and Ehmer (1969) used three different dose levels 0.37, 0.75, and 1.5 grams of ethanol in cats whose brains were stimulated to produce attack behavior.

Experiments with Alcoholics. The problem inherent in the operationalization of the concepts of "alcoholism," "problem drinking," etc., are reflected in the wide range of definitions used in all types of studies dealing with this subpopulation. (For a revealing discussion of different ways of operationalizing the alcoholism concept, see Edwards *et al.*, 1971.) The extent to which the definitions and samples differ systematically between epidemiological investigations (primarily of prison samples) and clinical and experimental samples is almost impossible to ascertain. (The findings by Mayfield (1972) mentioned earlier, that a large proportion of prisoners classified as problem drinkers had not in effect sought treatment for alcoholism, points toward the possibility that experimental and clinical studies of aggression in alcoholics deal with a different population from the one tapped by epidemiological prison studies.)

Experiments with alcoholics have typically been of longer duration in order to get a representative sample of alcoholic drinking, which generally has been the intended independent variable. Docter and Bernal (1964) studied the physiological and social reactions of two male alcoholics during a drinking period of 14 days. Mendelson *et al.*, (1964) studied social interaction and other behavior of alcoholic subjects during an experimental drinking period of 24 days. Alcohol in the form of whisky was given at six occasions during the day, with four-hour intervals. The amounts were gradually increased from 6 oz to

40 oz of whisky per day. McNamee *et al.* (1968), in their study of affect, thought content, and general behavior in alcoholics, used a seven-day drinking period in which the subjects paced their own drinking of bourbon. The BACs are probably higher and the main effects of conjunctive or intervening variables correlated with prolonged drinking are probably causally relevant in this type of study.

Mayfield and Allen, (1967) in an experimental study on the effect of alcohol on the affective state of groups of alcoholic patients, severely depressed patients, and controls, used intravenously administered alcohol. The mean BACs over the three groups was approximately 0.06 percent as measured by Breathalyzer test; the range was 0.04 percent to 0.08 percent. Vannicelli (1972), in her experiment on mood and self-perception of alcoholics, attained mean BAC levels of 0.019, 0.045, 0.095 and 0.158 by giving the subjects vodka mixed with orange juice. The subjects in van der Spuy's experiment (1972) on the effects of alcohol on the mood of alcoholics received brandy in a diluted or undiluted form according to choice. Almost all the subjects drank 8 oz of brandy during a 30-minute period before testing began.* In the hypothetical 70-kg man this would mean administration of 1.3 cc per kg body weight. These experiments are relevant in determining the role of alcoholism as a conditional variable in the accounting for the link between acute alcohol use and aggressive behavior. To get at the main effects of alcoholism, we will have to consider whether alcoholics have a higher probability of aggression even outside alcohol use situations.

Discussion. The most frequently extractable independent variable in experimental studies is blood alcohol concentration. It has ranged in specificity from intravenous injections of ethyl alcohol with placebo included in the design to oral ingestion of specified amounts of absolute alcohol, generally adjusted to the weight of the subject, and, at the other end of the scale, an unspecified amount of an unspecified type of alcoholic beverage available to the subjects of a free choice basis. Another independent variable is the type of beverage; most often beer and some distilled spirit compared in their effects. Related to the latter are studies using congener content as independent variable (often with high congener concentrations not representative of any widely used beverages). Some studies carried out in more natural settings use, in addition to specific alcohol variables, the whole social drinking situation as the independent variable (at least implicitly).

In human experiments using nonalcoholics as subjects and having

---

* Assuming that hypoglycemia is relevant in the etiology of a portion of alcohol-related violence, Pawar's (1972) findings of two hypoglycemic patients who had drunk one and a half bottles of whisky and four to five pints of beer followed by another half bottle of whisky, before committing aggressive acts, are of value in assessing the representativeness of the amounts of alcohol used in experiments on aggressiveness.

aggression or related measures as dependent variables, the highest BACs used have been around 0.10. This is not representative of the BACs found in Shupe's study (1954), where 43 to 88 percent of persons arrested for various types of assault, during or immediately after commission of the act, had BACs of 0.10 percent or higher. Recalculation of Shupe's figures shows that 69 percent of offenders in assaultive crimes had BACs of 0.10 percent or higher. For homicides the percentage was 67 percent, for robbery 60 percent, and for rape 45 percent. Among offenders who had at least a trace of alcohol in their blood, the percentage of those having over 0.10 percent was 80 percent for homicides, 91 percent for purely assaultive crimes, 84 percent for robberies, and 86 percent for rapes.* The blood alcohol levels used in experimental studies are thus definitely at the lower end of those found in the offender in violent crimes, although we have to take into account the higher risk of being apprehended at high BACs. The extent to which the BACs of experimental studies are representative of the levels in less severe aggressive acts is not known. Kastl (1969) has noted the possibility that doses of 1 cc of absolute alcohol per kg of body weight may be too low for discernible effects on "regression."

Many conjunctive variables that could have had an independent main effect on the dependent variable have been eliminated with the experimental setups, but so have perhaps important conditional variables that are necessary for alcohol to increase the probability of aggression (see the section, "Conditional or Interactional Models," below, and pages 383–384, for an explanation of the different types of variables discussed here). The most rigorous explication attempt of the independent variables vaguely used in epidemiological studies or stored in common knowledge is the intravenous administration of ethyl alcohol (e.g., Hartocollis, 1962). The symbolic cues (or "suggestion") associated with alcohol use are thus eliminated, and so are possible congener effects. The next step in the line of "rigor" of the independent variable is orally given congener free alcohol, with control groups that are led to believe that they also receive alcohol (e.g., Hetherington and Wray, 1964). Next in line are experiments where congener levels are not controlled for, such as those where bourbon is given to the subjects (e.g., Shuntich and Taylor, 1972). Finally, small group experiments tend to use comparatively representative social settings as part of the methodology, and thus introduce what can best be described as a conglomerate of independent variables into the experiments (Takala *et al.*, 1957; Bruun, 1962; Kalin *et al.*, 1972; Kalin, 1972; McClelland and Davis, 1972; Boyatzis, 1974; Wilsnack, 1974). Also, in these studies drinking amounts and

---

* These figures are recalculated from Shupe's (1954). Except for purely assaultive crimes, the sample sizes are fairly small and the percentages subject to large sampling fluctuations. For homicides the base used in the recalculations is 25, for purely assaultive crimes 150, for robberies 61, and for rape 21.

types of beverage are typically available to the subjects on a free choice basis, which adds to the generalizability of the findings, but makes the explication of causally relevant independent, conjunctive, and conditional variables very difficult. Many of these small group experiments with nonalcoholics are extended in time, and this again may be a factor in explaining aggressive reactions. It has been noted that electrical shock settings tend to get higher with the passage of time in typical experiments with the "aggression machine." This is true also for nonalcohol conditions (Bennett et al., 1969; Buss, 1963). The subjects in extended small group studies may thus become more prone to interpersonal aggression due to this factor, independently of any alcohol effects. As mentioned above, in many of the much lengthier experiments with alcoholics the effects of alcohol use may be confounded with other variables correlated with the extended nature of the experiment, such as the greater probability of sleep deprivation.

The studies mentioned in this section have primarily dealt with BAC or amount drunk as independent variables. Other studies have measured the effects of type of beverage (typically beer versus some distilled beverage, Boyatzis, 1974; Takala et al., 1957) or congeners on aggression or related behavior (Katkin et al., 1970) Beside BAC, type of beverage, and congeners, there are other factors related to physiological changes after ingestion of alcohol that also have been suggested as causative in behavior changes. Kalant (1961), among others, has suggested rate of change in blood alcohol level as a potential causative factor. Tinklenberg (1973) suggests that one might expect different influences on aggressiveness depending on whether BACs are rising, stationary, or falling. (For experimental evidence of the causal significance of this factor on motor skills and cognitive skills, see Hurst and Bagley, 1972.)

Cultural factors are most relevant in determining many of the factors discussed in this section. Choice of beverage and drinking speed affect BAC levels, in addition to having other physiological effects. Other modal factors of a conditional or conjunctive nature are also determined by cultural definitions and expectations. This is true for length of drinking occasion, nutritional habits in drinking situations, the composition of the drinking group, etc. Again we find that epidemiological and statistical information of correlations between violent crime and "alcohol use" from different cultures do not specify the independent variable in this respect, although this type of data would not be hard to collect. Differences in the alcohol involvement in violent crime (and aggression generally) between different cultures could be related to these and other culturally determined factors.

Needless to say, the specification of the relevant independent variable or variables contained under the epidemiological labels has vast implications for prevention of violence connected with alcohol use. As we will see below, no systematic studies have been carried out to assess the differential causal relevance of the different independent alcohol variables discussed above and conjunctive (mainly social) variables of typical alcohol use situations.

Nature of the Dependent Variable

Although this chapter in its epidemiological review section has been limited on the whole to studies of assaults and homicides, there remain several more or less idiosyncratic factors that may have an effect on the nature of the variable that alcohol use is correlated with in these studies. The relationship may, in fact, not be with a representative sample of violent crimes, but with some more selective variable. This is true whether the data are gleaned from police or court records or by ascertaining the BACs of offenders apprehended by the police. Blum (1967) points out that "the relationship now shown between alcohol use and crime is, in fact, a relationship between being caught and being a drinker rather than in being a drinker and being a criminal." This aspect has been discussed at various points above, and by other authors (Shupe, 1954; Amir, 1967).

The proportion of unsolved crimes is a good indicator of the importance of biasing factors in apprehending the offender. In the most cited American study, Wolfgang's (1958) study on homicides, there were only 6 percent unsolved crimes in the records. This factor would thus not have much effect on Wolfgang's findings, which have been replicated in many studies, Wolfgang (1958) also surveyed 6,435 criminal homicides in 18 cities between 1948 and 1952 and found that 90 percent had been cleared by arrest. There is undoubtedly great variation between countries and jurisdictions and great variations within some of these over time. Macdonald (1961), in reviewing seven studies, found a range of between 3 and 37 percent of unsolved homicides. Other factors may cause an "internal shift" from the crime of assault (with a lesser chance of coming to the attention of the authorities) to a homicide in alcohol use situations, and thus cause an artificially high relationship between alcohol use and homicide. (On the other hand, the association with assault would weaken.) One such factor has been pointed out by Wolfgang (1958) and later by Pittman and Handy (1964) who remark on the surprising lack of studies on aggravated assault:

> "Often the line dividing aggravated assault from homicide is so thin that a factor such as the speed of an ambulance carrying the victim to the hospital will determine whether the crime will be aggravated assault or homicide."

It is probable and has been pointed out by Wolfgang (1958) that other circumstances being equal there is a smaller likelihood for an intoxicated offender, victim, or witness to an assault to phone an ambulance in time to save the victim's life.

The availability of weapons is one factor among many that will affect the escalation of aggression into interpersonal violence (Mayfield, 1972), and thus the nature of the dependent variable in current epidemiological studies. In certain subcultures it is common to bring along weapons (e.g., "Saturday night specials") for drinking occasions, especially the ones characterized as "time out" on weekends and other potentially long drinking bouts. The higher the probability of intoxication among all relevant individuals in a drinking and aggression situation, the higher the likelihood of escalation into violence, it would seem, since attempts at intervention by others after the eruption of violence will be less likely.

Even a dependent variable such as maliciousness or violent character of the crime (Wolfgang, 1958), i.e., the use of unnecessary force and excessive number of violent acts (as shown by, e.g., stab wounds) in the commission of the crime is not completely free of biasing factors of the kind discussed above. Here also, the intoxication of possible bystanders will have an influence. However, it is probably a less biased indicator of aggressive tendencies than the mere fact of homicide and was shown by Wolfgang (1958) to have a positive relationship with alcohol use by the offender.

There is another type of biasing factor of a more rational kind, which also has been noted by several authors. Alcohol use may not only be a causal factor in aggression, sexual behavior, etc., it can also, as almost anything related to human behavior can, be part of a means–ends scheme. Carpenter and Armenti (1972) mention the "planned consequences of alcohol use." Other authors, e.g., Blumer (1973), state that "drinking for courage" is not infrequent in assaultive crimes. This reasoning, however, assumes a direct causal effect of alcohol on aggressive behavior tendencies. Although "drinking for courage" may be a factor that inflates the association of alcohol with violent crime, it also assumes that alcohol effects are relevant in the causal accounting of the relationship.

Finally, there is one rational biasing factor that does not presume a causal connection, but relies more on the epidemiological association and popular assumptions of diminished mental abilities as a consequence of alcohol use. These assumptions, which are undeniably correct, are reflected in the legislation and connected with ideas about moral responsibility. The defendant (and presumed offender) in crime would, in many jurisdictions, gain special consideration if alcohol were implicated as a causal factor. Consequently, some defendants could be expected to exaggerate the role of alcohol in the commission of a crime. This is a factor mentioned by, e.g., Gelfand (1971) and Amir (1967), and has been noted particularly in connection with child abuse (McCaghy, 1968; Swanson, 1968). The offender would also, and this is important in crimes such as child abuse that are considered particularly morally repugnant, regain some of the self-esteem he may have lost by at least overplaying the causal role of alcohol in the crime. To the extent that testimony by the offender is used to determine alcohol involvement in epidemiological crime studies, this factor would lead to an exaggerated association with alcohol use.

None of the factors cited above are present in the experimental and quasi-experimental studies, which are our main resource in establishing causal relationships between alcohol use and violence.

The dependent variables used in experimental studies fall short of assaultive behavior and homicide, for obvious ethical reasons. The most common dependent variable of direct relevance to violent *behavior* is aggression as measured by electrical shock settings toward another person, who is (allegedly) frustrating the subject. Others, which have been included in the discussion below include aggressive "mood" as determined by an adjective checklist (Mayfield and Allen, 1967), feelings of aggressive "power" as determined by TAT projective stories (Kalin *et al.*, 1972; Kalin, 1972; McClelland and Davis, 1972), appreciation of humorous cartoons with an aggressive content (Hetherington and Wray, 1964). In some of the studies reviewed below, the dependent variable has been even further removed from the everyday use of the concept "aggression" but they are reviewed since they have distinct semantical, and perhaps some empirical links with aggression, and secondly, because they are seen by the investigators as linked to aggression. They could also be seen as potential intervening (and/or perhaps conditional) variables in explanatory models, and thus methodologically used as *indicators* of "aggression."

One assumption must be made to justify the relevance of the findings from these mainly experimental studies for the accounting of the association between alcohol and violent crime: there is a continuum from aggressive behavior as displayed, e.g., by electrical shocks given to another person, to the extreme forms of interpersonal violence. It is difficult to accept the existence of such a continuum unconditionally, considering the extraneous factors entering into the determination of the type of crime and whether it will escalate into a crime or just stay at the level of, e.g., verbal aggression. There may be a physiological continuum of readiness from milder forms of aggressive behavior to interpersonal violence, perhaps with some qualitative physiological threshold changes.*
However, we have to remember that we are dealing with social behavior. This

---

* Moyer (1971) has derived several different categories of aggression among animals: predatory, intermale, fear-induced, irritable, maternal, instrumental, and aggression for territorial defense. He suggests that most of these forms of aggression may have specific physiological centers and mechanisms in the central nervous system and particular endocrine bases. The different forms of aggressive behavior would be released by qualitatively different external cues. This is the basis for Moyer's categorization. The idea of one continuous variable of aggressive behavior would not be valid if Moyer is right, although for some purposes a one-dimensional scale could be useful. Moyer warns that "there is clearly no necessity for assuming that all of the kinds of aggression identifiable in the various animal species are necessarily identifiable in man."

means that social definitions of behavior and normative strictures will come in at various points in the continuum and thus make it into a nominal scale with qualitative jumps. There are also institutional means in societies to enforce this qualitative stricture, and individuals in society are socialized into these definitions and in most cases have internalized them. Thus, also on this basis, one should not fall into the trap of inferring that outside of higher values on the independent explanatory variables, e.g., blood alcohol levels, no additional factors are needed to explain violence (see section, "Direct Cause Paradigm," below).

This chapter has been limited to interpersonal noninstrumental crimes of violence, since it was felt that other types of violent crimes are etiologically different. For the same reason the studies reviewed for their explanatory significance should be limited by the same criteria. A large proportion of these studies deal with interpersonal aggression in a rather limited sense, as when there is no face-to-face contact between the two subjects and the "communication" consists of electric shocks allegedly determined in strength by the adversary. The majority of the experiments are also clearly instrumental in nature. This is true almost by definition in studies designed under the frustration–aggression paradigm. "Frustration" is operationally defined as "blocked goal-directed activity" in these studies (Buss, 1963), and the concept of goal-directedness implies that the behavior arising from frustration, i.e., the aggressive behavior, can be seen as an instrumental act designed to remove the "blocking agent," as it were.

The problems encountered in transferring the criterion of noninstrumentality from violent crime to experimental aggression are partly definitional. This is so, because extreme severity of aggression is a definitional attribute of noninstrumentality. This is to say that milder forms of aggression in response to the same situation (which had led to "noninstrumental" violent crime) could more easily have been perceived as instrumental in achieving the imputed goal of the offender. In order to get at noninstrumental aggression, experiments would have to be designed differently, and this has been tried by direct provocation, arbitrary reactions, taunting, etc. Another way out is to attack the problem on the epidemiological front and find out more about the association between alcohol use and more representative forms of aggression and the conjunctive prevalence of the two in the population at risk. We would thus not be limited to crimes of violence as the only epidemiological indicator of aggression, and we would have epidemiological knowledge of a type of aggression that could more easily be replicated in experimental settings. (A survey approach to ascertain the association and prevalence, with all its limitations (see Pernanen, 1974), seems possible). An indication of the relevance of epidemiological studies of less severe forms of aggression to the study of violent crime is the frequency of the escalation process in the etiology of violent crime (see section on escalation below). This also provides a basis for an extrapolation of the experimental findings (on "aggression") to real life situations in which violent behavior is displayed and violent crimes committed.

General Remarks on the Explanations

In reading the presentation of models that have been used to explain the connection between alcohol and violence and the many more tentative interpretations based on the scattered data that are available at this time, it should be borne in mind that the possible explanatory models are not exclusive of each other. In fact, the variables and causal connections between them may all appear in explanations of some cases of violent behavior in connection with alcohol use. It is probably true that all the variables put forth in the explanations below singly or in conjunction with others are relevant in the explanation of separate subsets of all violent behavior in our world. Some of the variables presented in the literature are rather general and thus the models derived from these do not have much explanatory value. The explanations have in general been limited to three and, in some instances, four variate cases, e.g., the independent variable, one or two conditional and/or intervening variables, and the dependent variable. The reason for this limitation is that the empirical studies relevant to the explanation of aggressive behavior and the interpretations of their findings have limited themselves to this number of variables, and that the models would be hard to conceptualize with more variables included in them. Considering the knowledge at hand, it seems unnecessary to complicate matters.

One thing should be borne in mind: There is a very important distinction between explaining individual acts of violent behavior and explaining aggregate statistics of violence. If one, for example, looks at the variations over time or geographical area or analyzes a sample of violent acts or violent offenders by using multivariate techniques, thus getting the proportion of the variation explained by each independant explanatory variable, one (a) arrives at a model that cannot strictly be called a causal one since the data are, in fact, made up of cases where different causal models of violent behavior apply and the dependent variable is the aggregate *rate* of violent behavior (in fact, the explanatory model arrived at this way may not explain any single case of violence by itself); and (b) the analysis will not detect the explanatory import of variables that have common values in all aggressive acts, because the analysis is nonexperimental and these variables are not systematically manipulated, although these common variables would enter into an explanation of each individual act of violent behavior. (Here physiological variables with threshold values come to mind.) These two types of dependent variable (aggregate rates and individual acts of violence) should be kept apart in trying to find explanations of violence.* On the individual level one should be especially skeptical of attempts at unitary explanations of all types of violence or even interpersonal noninstrumental violent crime.

"Alcohol use" and "violence" (or "aggression") will be used very inclusively and in the traditional unspecific manner in the discussion. The label "alcohol use" will potentially include all the different independent variables that have been used in experimental studies and that have been dealt with above in connection with the discussion on the nature of the independent variable. As regards the dependent variable, the focus will be on noninstrumental and interpersonal violence. The discussion will range indiscriminately over the whole presumed aggression–violence continuum.

I have decided to analyze the available data, hypotheses, interpretations of data, and theories from a more formal point of view than is usual. I have done so because of the difficulty in assigning the truly independent, or even conditional and intervening variables into any field of study, and the fact that the explanation in all probability would have to include intervening and/or conditional variables from several academic disciplines. Moreover, and at this stage of knowledge perhaps more importantly, the formal methodological designs of empirical studies in this, as in other problem areas, largely determine the substantive findings that will be arrived at and this fact can best be brought forth by this formal approach. For example, in a study that investigates the situational determinants of violence in connection with alcohol use, the contribution of predispositional or other long-term variables will be ignored and the selection of subjects for study will be determined by this. With the easy availability of college or university students as subjects or respondents and the generally small sample sizes, conditional variables of a predispositional type will be left out of consideration whether they are seen basically as biological, psychological, or sociological in nature.

In the discussion above I have talked about a number of different types of variables and relationships, e.g., "conditional" and "conjunctive" variables that are definable by their role in explanatory models. I shall attempt a rather strict definition of these formal relationships and variables that are used for analytical purposes in this chapter. The following will be dealt with: direct cause, conjunctive, conditional, interactive, and common cause relationships, and intervening variables.

---

* This is, of course, not denying that factors that explain differences in rates of violent behavior in subpopulations or in the same population, over time, also have explanatory value in individual cases of violence. It is only to say that these factors will have different explanatory values from one individual to the other, and for some, none at all; and that statistical analyses of aggregate figures on violence, by their very nature, leave out the most important variables in the causal chains leading to violence. They would, however, be extractable through experimental methods.

The *direct cause* relationship is the relationship between the independent variable and the dependent variable under the assumption that variations in the independent variable lead to variations in the dependent variable independently of variations on any other variables. This is a very strong assumption, but it has implicitly guided much of experimental research in the field of alcohol and aggression, as will be seen in the discussion of the following section.

The *conjunctive* relationship, which is a special case of a spurious relationship, is most easily explicable by symbolic representation. Let us represent alcohol by "A," violence by "V," and a third (conjunctive) variable by "Y." The epidemiological association between alcohol and violence can be represented by the following statement: $P(A \cap V) > P(A) \times P(V)$, i.e., the joint occurrence of alcohol use and violence is higher than would be the case if the two phenomena were statistically independent of each other. (The expression $P(A) \times P(V)$ represents the null hypothesis.) This association could be due to the fact that a third variable, Y, increases the probability of violent behavior and that it is statistically associated with alcohol use: $P(A \cap Y) > P(A) \times P(Y)$. Thus, conjunctive variables are the variables that are idiosyncratically clustered, often due to cultural factors, with the independent variable (alcohol use). They can give rise to a correlation between the latter and a dependent variable, e.g., violent behavior. From our point of view, the strategy would be to look for factors that are associated with alcohol use (or alcoholism) in a culture and may have independent main effects on violent behavior. The difference between this relationship and a common cause relationship is that the third factor in this case does not increase the probability of alcohol use (in addition to its effect on violent behavior).

In the *conditional* relationships, we again assume that a third variable is associated with alcohol use to a greater extent than expected by chance: $P(A \cap Y) > P(A) \times P(Y)$. In the conditional relationship, however, alcohol is a causal agent in the increased probability of violence in alcohol use situations, but only in situations in which variable Y has specified values. Alcohol in this relationship has no main effect on the probability of violence, and neither does Y. In conjunction with Y, however, alcohol use increases the probability of violence.

In the related *interactive* relationship, both alcohol use and variable Y have independent main effects on the probability of violence. These effects are, however, compounded in a nonadditive way, making the effect on the probability stronger than would be expected by the two main effects. A possible example is "pathological intoxication" in individuals with temporal lobe dysfunctions. There are probably main effects of both alcohol use and the physiological dysfunction, but these effects are exacerbated when these factors are combined.

The *common cause* relationship is too familiar a concept to need a long explanation here. Basically, what is intended by the concept in this context is that a third variable Y increases the probability of the simultaneous occurrence of both alcohol use (alcoholism), and violent behavior, thus giving rise to a spurious causal relationship between alcohol and violence.

The end result of all these relationships—the direct cause, conjunctive, conditional, interactive, and common cause relationships—is a higher-than-chance association between alcohol use (alcoholism) and violent behavior. The direct cause explanation applies to *all drinking situations* and *all individuals* taking part in these. The conjunctive, conditional, interactive, and common cause models explain the statistical association by introducing a third variable that varies independently of the other two, and that is not necessarily present in all drinking situations. Thus, it allows for the possibility that not all relevant* alcohol use increases aggressive tendencies. The association is explained by showing that it exists *in one or more subsets* of alcohol use situations (alcoholics).

The status of *intervening* variables is somewhat different from the other types of variables discussed above. All the other variables (except, of course, the dependent variable of violent behavior in the direct cause relationship) can vary independently of the independent variable (alcohol use or alcoholism). An intervening variable, on the other hand, must by its very nature have strong positive correlations with the independent variable, if the relationship between the independent and dependent variables is positive. This is so because intervening variables further specify the causal nexus without affecting the strength of the original relationship between the independent and dependent variables. (For an illustration of the role of an intervening variable in explanations of the association between alcohol use and violence, see the illustration of an explanatory model below.) The intervening variables, if they are to specify a relationship between, e.g., an independent and a dependent variable, require that the independent variable is better specified than what usually is the case. Common variables such as "sex," "social class," etc., are not well specified, even as to the system of variables to which they belong, i.e., if they are social, psychological, physiological, etc. "Sex" of drinker, for example, is not a well-specified variable, but designates an empirical conglomerate of several variables from several fields of inquiry. Introducing intervening variables requires more specificity, since the intervening variables will be different according to what aspect of being male or female is causally relevant in the context.

The causal relationships discussed above may be combined in any number of ways in causal models linking alcohol use (or alcoholism) to violence. For example, the direct cause and common cause relationships may be combined in

---

* "Relevant" here meaning, e.g., situations in which participants reach a significant blood alcohol level.

Alcohol and Crimes of Violence

the following way:

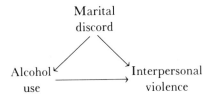

In this illustration marital discord increases the probability of both alcohol use and interpersonal violence. The alcohol use further increases this probability through its direct effects.

For the purpose of accounting for the relationship between alcohol use (or alcoholism) and violence, it will be useful to illustrate what a "complete" causal explanation of the statistical association would look like on the aggregate epidemiological level. As an illustration I will use a simple linear model to show how the explanatory share of each model in accounting for the association depends on two factors. The first of these is the *prevalence* of certain conditions, most importantly the independent variable of alcohol use and relevant conditional, interactive, and common cause variables. The second is the *strength* of the causal relationship. In the illustration below it is assumed that sleep deprivation as an intervening variable is of explanatory value, and that there is no interaction between the direct effect of alcohol use and sleep deprivation. Note that the model is not meant to show how much of the variation in the dependent variable is explained by each of the variables (as in path analysis). Instead the probabilities in brackets are probabilities that the factors will occur at any time and the probabilities beside the arrows show the conditional probability of, e.g., sleep deprivation occurring *if* alcohol use has occurred.

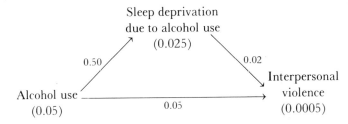

This model would have to be interpreted in the context of each separate culture or jurisdiction or any other population for which we have epidemiological data relating to the association, and most probabilities would differ from culture to culture. In this hypothetical population the probability of alcohol use occurring at any time [P(A)] is set at 0.05. The conditional probability over the total population of having significant sleep deprivation effects due to alcohol use has arbitrarily been set at 0.50: $P(A \to SD) = 0.50$. The absolute probability of anyone in the culture being deprived of sleep due to alcohol use is then $P(A) \times P(A \to SD) = 0.025$ at any time. In the illustrations the conditional probability of interpersonal violence occurring if sleep deprivation occurs is 0.02: $P(SD \to V) = 0.02$. Thus, the probability of interpersonal violence occurring due to the sleep deprivation effects of alcohol use, assuming no interaction between variables, is $0.025 \times 0.02 = 0.0005$ at any time.* The conditional probability of the effects of alcohol on interpersonal violence via other intervening variables (which have not been specified here) has been set at 0.05. Thus, the probability of interpersonal violence in alcohol use situations due to other causal relationships connected with alcohol use will be $0.05 \times 0.05 = 0.0025$, and the probability of interpersonal violence in the situations when alcohol is used will be $0.0025 + 0.0005 = 0.003$.

A corresponding model of interpersonal violence due to sleep deprivation effects pertaining to the same population in situations when alcohol is not used could be the following:

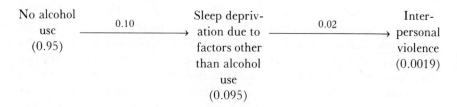

The situations when alcohol was not used will occur 95 percent of the time (since we assumed that alcohol was used 5 percent of the time). The probability of interpersonal violence due to sleep deprivation will be the same (0.02). Consequently, the probability of interpersonal violence in situations where alcohol is not used will be $0.95 \times 0.10 \times 0.02 = 0.0019$. Let us assume, for illustrative purposes, that sleep deprivation is the only causative factor in violence in nonalcohol conditions. We can now arrive at the proportion of violent acts that are connected with alcohol use, out of all violent acts committed. The probability of violent acts being committed in this population is 0.003 in alcohol use situations and 0.0019 in situations where alcohol is not used, and thus the probability of violent acts at any time is $0.003 + 0.0019 = 0.0049$. Assuming that the administrators in this population keep a record of these things (or if they are reflected in an unbiased way in their statistics of violent crime), they will find that alcohol use situations were implicated in violent acts

---

* If there were positive interaction effects between these variables, the probability of violent behavior in connection with alcohol use would be greater.

in $0.0030/0.0049 \times 100 = 61.2$ percent of the cases. (This could have been Philadelphia in the early 1950s; see Wolfgang, 1958.)*

In this type of model any other kind of intervening variable could be included, e.g., risk-taking tendencies or hypoglycemia (see below). The same type of reasoning could be applied to predispositional variables such as temporal lobe dysfunction or alcoholism. The weights given would reflect the prevalence in the population of these conditions. A complete accounting in terms of probabilities would have to take into account overlapping subpopulations and interaction factors. However, the reasoning is the same in principle: The *prevalence* of alcohol use situations (or, e.g., time spent under the influence of alcohol) and of conditional factors such as having a temporal lobe dysfunction must be known, and so much the causal *conditional probabilities* of one factor occurring if another independent factor has occurred.

This discussion on the logical and methodological aspects of explanations of the association between alcohol use and violence (or any other statistical association) may seem unnecessarily lengthy. However, an explication of the logic of explanation is needed to fit previous empirical findings into a cumulative body of knowledge and to guide research. These are the logical paradigms we use in (more verbal) explanations and they should be explicated as far as possible, to systematize knowledge and research in an efficient way, and to enable us to see connections and locate gaps in knowledge. If these aspects are not taken into account, we will still be deluged with one or two factor theories making exaggerated claims of their explanatory power, and the research carried out will be limited to bivariate analyses and "direct cause" thinking. We will then also have to be content with logical nonconcepts such as "plays a causal role" or "has explanatory potential."

It should be pointed out that a "direct cause" thinking pervades not only the research methodology of interpretations of individual studies, but the whole outlook on the proper ways of reaching valid empirical results. Instead of looking at selected samples of, e.g., males, females, mentally ill offenders, or epileptics as biased samples (e.g., Goodwin, 1973) with unknown generalizability to the general population, they should be seen as providing clues as to

---

* These models can be applied to offenders or victims or both. Assuming that sleep deprivation and other alcohol effects in the offender, the victim, or both, are of equal causal significance for interpersonal violence, we can apply the models to all alcohol involvement and compare the 61.2 percent to Wolfgang's figure of 63.6 percent. If the model is applicable only to the offenders' alcohol use, we should compare our result to Wolfgang's finding of 54.4 percent alcohol use by the offender. It should be noted that the units in our example are time spent in alcohol use situations (or under the influence of alcohol, etc.) and in interpersonal violence, whereas Wolfgang, by necessity, uses numbers of violent crime events as his units of measurement. If there is a difference between alcohol-related and other interpersonal violence in the results from these two measures, Wolfgang's measure and ours will not yield the same value for alcohol involvement. The logic of causal accounting of an epidemiological relationship can, however, most easily be illustrated by our measures.

possible conditional or interactive factors in the explanation of the association between alcohol use and violence. In an accounting of the relationship, the biasing factor should be assigned a weight reflecting the prevalence in the population at risk, and this is the only information that would be missing. Instead of the impractical and rather wasteful way of trying to get a representative sample of the population at risk, it is in fact more fruitful to select small samples of what with the present knowledge, seem to be relevant subgroups of the population and subject them to a more intensive study, taking into account the conditional nature of the findings and their prevalence in the population at risk. On the other hand, general population samples may be used in a "mapping" function to pinpoint subgroups of the population and situations in which alcohol seems to play a practically significant role in increasing the likelihood of aggressive behavior, and to arrive at prevalence estimates of these subgroups and situations. This potential has not been used at all, it seems.

The Direct Cause Paradigm

Unsystematic observations in natural settings, upon which much of our understanding of human activity is based, have not by their very nature yielded any valid information on variables that would be relevant as conditional (or interactive) factors in increasing the probability of aggression or violence in alcohol use situations. Epidemiological studies, with their limited sets of variables and their ad hoc sampling, have not provided much usable information as to conditional variables either. Clinical studies of violent offenders have yielded some limited information mainly on biological variables that could be conditional in increasing the likelihood of violence in connection with alcohol use, but even this information has been widely ignored in the experimental literature of alcohol's effects on aggression.

Consequently, although we know that alcohol use does not inevitably lead to aggression and only a very small proportion of alcohol use situations leads to violent crime, a direct cause relationship is implicit in much research on the subject. This is the case to such an extent that we can speak of a "direct cause paradigm" that has guided the methodological aspects of experimental research. Possible conditional factors have entered the studies only in the interpretative stages, when the results have not borne out the assumptions of the direct cause paradigm. It is also implicitly accepted by those adherents of the "disinhibition theory" of alcohol use, who do not care to specify any conditional factors that would determine the disinhibitory property of alcohol. (I will discuss this "direct disinhibition reasoning" in a separate section below.)

In experimental conditions designed under the influence of the direct cause paradigm, extraneous factors that are present in real life situations are controlled, but inevitably new extraneous factors are introduced with the experimental setting. Especially great care is taken to eliminate factors that may have independent main effects on the aggressive behavior (conjunctive variables), such as the cluster of variables labeled "the social setting." In trying to eliminate these conjunctive factors, many factors potentially interacting with alcohol and perhaps leading to an increase in the probability of aggression in natural settings are left out. In explaining negative findings, the tendency is then to revive these conjunctive independent variables that have been controlled (and not systematically varied) in the experiment. (It is, of course, also possible that extraneous factors introduced into the experimental situation, such as the experimental setting, and the inevitable feeling of being under observation, act as suppressors on the display of aggressive behavior.) The nature of the independent variable has varied in these experiments, but the essentially bivariate research logic determined by the paradigm has not. (This is not true to the same extent in general aggression experiments where more than one variable has often been manipulated to arrive at conditional relationships.)

The direct cause paradigm could also be called a *le cas pur* paradigm in its sampling aspect (Galtung, 1967). Since, within the paradigm, the effect of the independent variable on the dependent one is not conditional on values of other variables, the selection of subjects for experiments can be, and has been, quite arbitrary. It has been determined by economy and other practical reasons, which means that the attributes of North American university students have guided the research and explanations in the field.*

In the following, studies will be included in the discussion of the direct cause paradigm if the possibility of conditional or interactive relationships with alcohol use has not been taken into account in the design of the study, even though they may enter into the interpretation of the findings. Not one of the authors included here lacks an awareness of the relevance of conditional or interactive relationships: the inclusion is based solely on the methodology of the studies.

The most clear-cut application of the direct cause paradigm is found in animal studies, but even here the findings are not positive in any clear-cut way. The lack of conditionality in the methodology of animal studies is probably better justified than in human studies.† Experiments on the action of alcohol on aggressive behavior have been carried out on laboratory rats, mice, cats, and fish. The most direct study, leaving the least number of interpretations open as to actual conditionality of the relationship, or raising the least number of questions as to intervening variables, was carried out by MacDonnell and Ehmer (1969). The study was not intended to measure the effect of alcohol in eliciting

---

* The *le cas pur* justification seems to be implicit in much psychological research aside from any questions of availability of subjects. With university sophomores, we also have perhaps the closest possible approximation to freedom from effects of extraneous developmental factors that could be conditionally relevant in connection with all forms of behavior, including violent behavior.

† This is probably the case since symbolic and cultural aspects of alcohol use (including "suggestion") are not at play in animal samples.

aggressive behavior, but to investigate how alcohol *modified* aggressive behavior that was elicited through cortical excitation (Carpenter and Armenti, 1972). The results were somewhat equivocal; the latency of attack increased with dose, and the only aspect of the attack pattern that seemed enhanced by alcohol was the force of biting. ("Maliciousness" of a violent crime would perhaps be a far-fetched human parallel to "force of biting.")

Weitz (1974) studied rats under three different alcohol conditions and one control condition of isotonic saline. She found a clear negative correlation with increasing amounts of alcohol. Under the lower two alcohol conditions, fighting behavior was above the control condition; in the highest condition, it was below the control condition, although the difference was not statistically significant. Weitz points out, however, that there were great individual differences among the rats in the magnitude of increase in frequency of fighting behavior. This indicates that conditional factors exist even in this population.

Raynes and Ryback (1970) have reported an increase in fighting behavior among Siamese fighting fish (*Betta splendens*) when ethanol was mixed with the water in the fish tank. Bourbon and heavy congener solutions decreased the frequency of these responses compared to nonalcohol conditions. (This seems to contradict the findings by Katkin *et al.,* 1970, below, but here we naturally have to take the "phylogenetic leap" into account.) Schaaf (1971) found that fighting in rats increased with alcohol injection after electric shock. However, even a control injection of isotonic saline led to an increase.

It can be concluded that unambiguous findings as to a general main nonconditional effect of alcohol in increasing aggression cannot be found in studies of animal samples. It is significant that experimental results with animal subjects have not led to interpretations using conditional variables when the results have been negative, as has happened in interpretations in human studies. This shows that there is an implicit assumption that conditional situational and predispositional factors are not as relevant in explanations of animal behavior. Whatever the reason, there is an acceptance of the empirical findings within the direct cause paradigm in animal behavior studies, and no attempts at establishing conditional relationships.

The results in experiments on human subjects have been equally equivocal. In interpreting the results, a conditionality of the relationship between alcohol and aggressive behavior has been invoked, or conjunctive variables in natural settings blamed for the supposed association. With humans the complexities of controlling for the nature of the independent variable (especially with symbolic aspects added to the possibilities), and all possible conditional factors as well, become immense. As seen in the section on the nature of the independent variable, the BACs, when used as independent variables, have been rather standard in experiments on nonalcoholics.

Bennett *et al.* (1969) used a discrimination learning task in a laboratory setting as a pretext for eliciting electrical shocks of varying intensity from the subject toward an accomplice of the experimenter, who frustrated the subject. The subjects were 16 male graduate students who were used in a factorial design. There was only a minimal tendency toward increased aggression in alcohol conditions over the nonalcohol condition and no linear dose-specific trend. As in many general experiments without an alcohol condition, an increase in aggression (shock intensity) was noted with the progress of the experimental sessions for all conditions. In the discussion of their results Bennett et al. (1969) state the possibility that relevant variables of both predispositional (in the selection of subjects) and situational nature (in the controls for extraneous variables by imposing an experimental setting) have been eliminated. It is not quite clear, however, whether the situational cluster of variables of social setting is seen as conditional in order for alcohol use to increase the probability of aggression, or if this cluster is considered to have a main effect on aggression independently of alcohol use. If the latter is the case, the observed relationship between alcohol use and violence would be spurious. These alternatives are not exclusive of each other, of course. A third possible explanation suggested by the authors, is in the nature of the independent variable: "It seems safe to conclude that alcohol as a pharmacological agent, in the amounts used here, does not lead to aggression." The authors also see the possibility of another type of causative independent variable, the cue (or symbolic) value of alcohol use: "Alcohol could easily become a cue for behavior that would otherwise be unacceptable, or at least it might become an excuse."

Boyatzis (1974), in an experiment testing the general effects of consumption of beer and a variety of distilled spirits on aggression in a heterogenous sample of 149 males in natural settings (party settings), found that aggressive reactions increased more with spirits (self-selected gin, vodka, rum, bourbon, Scotch, and blended whiskey) than with beer, thus replicating the finding of Takala et al. (1957). He found that frequency of aggressive behavior increased with the blood alcohol level of the subjects both under the beer and distilled spirits conditions, although there was a stronger tendency with distilled spirits. He also found that over the duration of the occasion aggressive behavior increased. It is probable that this is partly a function of time passage itself and the processes related to it, since there was an increase of aggressive reactions over time also during a control condition where alcohol was not consumed (Bennett et al., 1969; and Buss, 1963). A longer duration of the social occasion could thus be a conjunctive feature of "alcohol use" with a main effect of its own on aggressive behavior in natural settings.

In an earlier study, in a comparatively structured social setting, Bruun (1962) using a quasi-experimental before–after design (Campbell and Stanley, 1963), found that aggressive reactions increased in male drinking groups with rising BAC levels (and duration of the drinking occasion). Takala et al. (1957) in a similar setting found that alcohol increased aggressive behavior in their male drinking groups. Different BACs ranging from about 0.09 percent to 0.15

percent did not significantly differ in the degree of aggression produced. They also found that beer produced less aggression than spirits with equal BACs. I will discuss these experiments further in the section on predispositional factors below. Shuntich and Taylor gave 30 college males 0.9 ml of 100 proof bourbon/kg of body weight in an experiment using the "aggression machine." Subjects in the bourbon condition delivered significantly more severe electric shocks to their presumed victims than did the subjects in the control condition.

In Shuntich and Taylor's (1972) experiment, as in the free choice experiments in natural settings, it is not possible to tell apart possible effects of the absolute alcohol consumed and the congener effects. Experiments using congener content as independent variable have generally used atypically high congener concentrations. Katkin et al. (1970) showed that risk taking, which could be an intervening variable to aggression, in a hypothetical situation, was higher in a bourbon condition than in a vodka and synthetic alcohol condition. Bourbon has about one hundred times the congener content of vodka. For the experiment the congener content of both bourbon and vodka was increased fourfold. In experiments comparing, e.g., beer and distilled spirits (Takala et al., 1957; Boyatzis, 1974) different cue values or cultural definitions and meanings of these beverages cannot be ruled out, in addition to any congener effects and the volume of liquids imbibed.

In general, it seems that the studies in which a whole social drinking situation has been the independent variable the probability of aggressive behavior has increased with drinking to a greater extent than in controlled experimental settings with a closer specification of the independent variable. Again, we have to allow for the possibility that the longer duration of these natural settings can be causally relevant. Only in the series of experiments reported by MacClelland et al. (1972) and the study by Boyatzis (1974) was it attempted to introduce a control condition with no drinking.

A great difficulty in designing experimental research lies in deciding which factors are only conjunctive and may have an independent main effect on aggression and which are also interactive or conditional to alcohol use in producing aggression. A systematic surveying and analysis of situations and individuals that are implicated in violent behavior is needed in order to find relevant conditional and interactive variables. This is where social and epidemiological research can help and where cooperation is essential.

A great drawback of the direct cause paradigm is that it tends to draw away attention from the need for a systematic variation of potential conditional factors. However, the studies reviewed here are of value as attempts to specify the nature of the independent variable. The experiments using BACs or amounts drunk have shown that there are no clear linear effects on the measures of aggression used. It should be added that this has been shown within the limited range studied. A simple linear relationship between blood alcohol concentration and risk of becoming an offender in a crime of violence or an aggressor should not be expected, even if causal relevance were established. One reason is the impairment associated with high BAC levels. Unfortunately, there are no data available on the relationship between BAC level and ability to aggress, e.g., verbally or physically. The only studies approaching this relationship in any relevant manner are the ones on the impairment caused by alcohol on some psychomotor skills, especially driving skills (e.g., Cohen et al., 1958). The level of skills needed in crimes of violence would seem to depend on many situational factors, such as the type of weapon accessible, the characteristics of the victim, etc. For this reason it is very hard to speculate about the relevance of psychomotor experiments of the type carried out to date on physical aggression and violent crimes.

The "Disinhibition" Fallacy. The experiments carried out under the direct cause paradigm (using different independent variables) have been viewed by some investigators as tests of the "disinhibiting" effects of alcohol (Bennett et al., 1969; Shuntich and Taylor, 1972). The direct cause model is in effect identical to what I shall call "the direct disinhibition model" in the discussion below. This is indicated by the interpretation of Bennett et al. (1969) of their own findings. Their experiment tells us in fact that the direct cause explanation is not true (using their specification of independent variable); it does not hold true for all individuals over all situations. The authors see the negative results as not supporting the disinhibition theory of alcohol in relation to aggressive behavior. It does, however, only refute a variant of the disinhibition theory, which states that *all individuals* (no matter what their predispositions and characteristics) have *at all times* (whatever the characteristics of the situation) aggressive inhibitions that are *always* released by alcohol. Another variant of the disinhibition theory has been analyzed by Carpenter and Armenti (1972). Before going into these two variants and a closer analysis of the disinhibition concept, it is instructive to look at the semantical labels for the idea of disinhibition that have cluttered the literature and obscured a lack of valid empirical data.

The prevalence of the disinhibition concept and more or less equivalent concepts in the explanation of the behavioral effects of alcohol is very high. Alcohol is labeled as an agent that "weakens inhibitions" (Fitzpatrick, 1974; Roebuck and Johnson, 1962), "weakens self-control" (Macdonald, 1961); "releases inhibitions" (Shuntich and Taylor, 1972); "liberates impulses and emotion which are normally under control" (Hopwood and Milner, 1940); "liberates deep features of the personality" and consequently "awakens aggressive tendencies" (Medina, 1970). It "reduces inhibitions and self-control," and leads to a "loss of inhibitory capacity and subsequent unleashing of personal predilections" (Hopwood and Milner, 1940); and it has a "disinhibiting effect" (Scott, 1968). It is known as a "disinhibiting, aggression-provoking substance" (Brill, 1970), and "as a trigger of violence" (Blumer, 1973). Its pharmacological role is described as that of "releasing aggression, removing inhibitions, etc." (Glatt, 1965).* The examples could be multiplied, but would not add much to the knowledge of the role of alcohol use in aggressive or other types of behavior alluded to as "disinhibited." It is an old concept and has been used over the centuries in discussions of the effects of alcohol use at least as far back as the time of Plato (MacAndrew and Edgerton, 1969). This should be kept in mind, if it seems that it is always legitimately used in referring to a specific model of the pharmacological actions of alcohol on the central nervous system.

A fact that should arouse one's suspicions is the general acceptance of such a concept (and purported explanatory model) by researchers and other individuals from so many diverse fields: medicine, experimental psychology, psychiatry, anthropology, alcohol epidemiology, sociology, etc. It could, of course, be seen as an indication that an explanatory disinhibition model has become so firmly established by research that it is almost universally accepted. On the other hand, knowing that this is not the case, one should ask whether all who seem to accept such a model really (comparing their backgrounds) can have the same explanatory model in mind. One possible explanation is that the disinhibition concept, which seems to be used in an explanatory function by many authors, actually is used to describe behavior that is known to occur or have occurred in a proportion of alcohol use situations, behavior that is described as being "disinhibited" or "uninhibited" in common use of language.† Everyday descriptions of behavior are common to people from all types of endeavor.

A concept of such widespread use has not been without its detractors, both in its descriptive aspect and its explanatory use. MacAndrew and Edgerton (1969) question the applicability of the concept to describe behavior in connection with alcohol use:

> How, we asked, can we square the notion that alcohol is a toxic disinhibitor with the fact that societies exist whose members' drunken comportment either (a) manifests *nothing* that can reasonably be classified as "disinhibited"; or (b) is markedly different from one socially ordered situation or circumstance to another?

---

* What will later be referred to as the formal nature of the disinhibition model is further enhanced by the fact that "alcohol," the presumed independent variable in the model, is never specified. No doubt this fact adds to the popularity of the disinhibition concept.

† The descriptive nature of the concepts used to allude to the effects of alcohol can perhaps be brought forth by asking the question, Why does "A" behave aggressively after drinking alcohol; is it because of disinhibition, loss of emotional restraint, or loss of self-restraint? How would we test these alternative hypotheses? We should perhaps not rule out the possibility that disinhibition as a descriptive concept has phenomenological connotations and these are characteristic of the experiences with alcohol use. Then, however, disinhibition is on the same logical level as "happiness," "relaxation," etc. Compare the following questions (possibly from an interview survey): "Do you feel (a) happy, (b) relaxed, (c) disinhibited, after a few drinks?"

In order to understand the widespread use of the concept in *explanations* of behavior in many fields, it must be assumed that it cannot have had much empirical content. It will be suggested below that in its explanatory use the "disinhibition" concept is not the label of a specific model that it is widely assumed to be. Its foremost use is as a type of label for certain formal properties of *definite sequences in explanation*. It is possible that this explanatory use and the descriptive use of the concept have been confused at times. I will expand on this possibility below.

To understand the concept of disinhibition in its explanatory use, we must understand that alcohol is not the only independent variable that can be characterized as a disinhibitor. In fact all possible independent variables with main, conditional, or interactive effects on behavior in the alcohol use situation can be so described. This is so because disinhibition is not an empirical aspect of an explanatory model. Instead its use is dependent on extra-explanatory factors, specifically the set of situations that one uses as a starting point in one's reasoning and one's research. If one, as usually is done, starts with a situation in which all the conditional variables have the needed values, and introduces alcohol and this causes a significant increase in aggressive behavior, it seems natural to call alcohol a disinhibitor as "releasing" or "triggering" aggression, etc. On the other hand, one could take as the starting point situations (with or without alcohol) where some other etiologically significant variable does not have the needed value. This could perhaps be a stress-inducing stimulus or aggressive behavior by another participant in a social setting. If this variable is manipulated so that there is a significant increase in aggressive behavior, this variable could just as well be called a disinhibitor.

The disinhibition concept has been interpreted as implying that there are pent-up emotions or biologically determined aggressive tendencies that are usually inhibited but released in alcohol use situations (e.g., Rada, 1975). However, in a methodological and explanatory framework these factors do not have a special logical status. They are conditional variables of a predispositional nature on par with any other predispositional variables needed in the explanation of the relationship between alcohol use and violent behavior. The same type of conditional relationship with violent behavior may exist between variables other than alcohol use. It could, for example, be the case that pent-up emotions in nonalcohol situations, in order to be causally significant, must be combined with stressful stimuli (in lieu of alcohol) for violent behavior to occur. Thus, stressful stimuli would be the disinhibitors. Consequently, even if we see the disinhibition concept as referring to a conditional model, other conditional factors besides alcohol use can be called disinhibitors, or releasors, etc., of aggression. Any of the conditional factors of a situational type in the set of variables that are relevant for the occurrence of violent behavior (in addition to alcohol) could in fact be labeled the "disinhibitor." Thus, the disinhibition model on the basis of a closer analysis ceases to be a distinct, formally identifiable explanatory model, since the concept in its explanatory interpretation does not refer to anything that is not covered by general causal considerations, except the extra-explanatory feature of which set of situations is used as a starting-point in the explanation; and consequently which variable is chosen for an analysis as to its impact on the occurrence of the dependent variable. It is to an even lesser extent *one* empirically verified or even testable model. Calling alcohol a disinhibitor does not add any explanatory power to a model. This statement only says that alcohol is one among the variables that—in conjunction with others—increases the probability of violent behavior and reveals a predilection for a special extralogical sequence of reasoning that perhaps is determined by the cultural salience of alcohol.

Situational variables, i.e., variables that can vary intra-individually and are not more or less fixed as predispositions of the individual, run the risk of being labeled "disinhibitors." In experimental situations the only variables that can be manipulated are situational. If the investigators focus on alcohol use as the main independent variable, it will by necessity be seen as the potential disinhibitor, without this having any relevance for a possible specific (physiological or any other kind of) disinhibition mechanism. Alcohol is by necessity seen as a potential disinhibitor, *because the experiment is set up in a certain way*. If stress were the main factor manipulated independently with alcohol use constant, it would be the potential disinhibitor, and we would have a disinhibition "model" of stress "explaining" aggressive behavior, without adding anything to the explanation that is not already there in the experimental setup and the empirical findings. The fact that alcohol has a traditionally well-known connection with physiological changes should not stop us from seeing that the disinhibition concept is used in this rather formal manner, and that alcohol as an independent variable just as well could be substituted by stress (which undoubtedly also induces physiological changes, but this is completely beside the point here). Alcohol could be used as a conditional variable and stress as the independent variable and thus as the potential disinhibitor. This nonempirical, and in a sense both formal and extra-explanatory, property of the disinhibition concept partly explains its wide acceptance and use by authors from a number of different disciplines.

Since Bennett *et al.* (1969) carry out their experiment under the direct cause paradigm, they also interpret the experiment as refuting a disinhibition model, which can be seen as a special case of the conditional disinhibition reasoning. In this sense, for example, stressful stimuli can also be tested under the direct cause paradigm, i.e., tested for causal effect in *all* situations for *all* individuals, and almost certainly also refuted. It would almost certainly be refuted, since only *one* experiment showing negative findings is needed to refute such a version. The conditional version of the disinhibition model is illustrated by

Carpender and Armenti (1972) in an analysis of Hetherington and Wray's (1964) findings. Carpenter and Armenti assume that a disinhibition model entails the existence of an "inhibitor." (The fact that the causal factors of this inhibition of behavior generally are not specified, as Carpenter and Armenti (1972) point out, is another indication of the formal properties of the disinhibition concept, and again explains part of its attractiveness.) Hetherington and Wray (1964) showed that the only group in their experiment, for which there was an increase in the appreciation of aggressive cartoons after ingestion of alcohol, was the group that had high scores on both need for aggression and need for social approval. Need for social approval and need for aggression (whatever this entails) are both potential conditional factors in explanatory models, since any inhibitor is on par with other conditional variables that in connection with alcohol use increase the probability of aggressive behavior. The fact that this conditional variable was singled out for the status of inhibitor probably has to do with the descriptive connotations of the concept. Here the formal use of the concept is combined with its descriptive features and thus it is used for explanation of behavior such as aggression, excessive emotionality, sexuality, i.e., behavior that is normatively regulated and that we generally would characterize as disinhibited or uninhibited. We would use it in this way, and this is an important point, even though we were not to accept a disinhibition model as an explanation for the behavior in question. This use of the disinhibition concept could be called "conditional disinhibition reasoning". The lure of this form is that the concept that labels it also is used descriptively (in common language). Thus, it may seem that one is subscribing to, and perhaps one is even led to subscribe to, a specific disinhibition model, since one accepts the description of certain forms of behavior as being disinhibited behavior. The finding by Hetherington and Wray (1964) shows that a subgroup of the population, the group of socially *inhibited* individuals who also have a high "need for aggression," exemplifies a conditional factor in the relationship between alcohol use and aggressive behavior. It should be noted that this is only a special case of the general disinhibition reasoning. It seems that a semantical shift has occurred here, aided by the descriptive connotations of the concept. This use of the disinhibition concept is incidental to its general use.

In summary, it can be said that the disinhibition concept has been used in five different ways in the reasoning on the connection between alcohol use and aggressive behavior:

1. In its descriptive use as a general label for behavior that is contrary to generally accepted social norms and values.

2. As direct disinhibition reasoning: This can be applied to any explanatory model where a threshold value is needed on an independent variable for the occurrence of an event. No conditional factors are considered causally relevant.

Thus, it can be used in the explanation of why the water in a dam starts flowing when the gates are opened. The water was inhibited in its potential flow by the dam, the inhibitor. The reason we do not call the action of the water disinhibited is that the descriptive use of the concept has anthropomorphic normative connotations.

3. The third use is a combination of (1) and (2) and it is the prevalent one in experimental testing of the disinhibitory properties of alcohol. This use explains the disinhibited behavior (descriptive concept) by the disinhibiting properties of alcohol.* This is the framework within which Bennett et al. (1969) criticized the disinhibition theory of alcohol use and which in a sense is implicit in the experimental studies using the direct cause paradigm.

4. The fourth use I have called conditional disinhibition reasoning. In this sense any situationally manipulable variable can logically be the disinhibitor, assuming specific values on other causally relevant conditional variables. The resulting values on the dependent variable cannot always be characterized as disinhibited, due to the normative connotations of the concept.

5. The fifth use again is a combination (perhaps it could be called a semantical conglomerate), this time of (1) and (4). Here the conditional variable is such that it can descriptively (in everyday language) be called an inhibitor, as inhibiting behavior and relevant values on the dependent variable can be characterized as disinhibited. This descriptive use is independent of its use as a label for the disinhibition sequence of reasoning. Hetherington and Wray's (1964) conditional variable is such that it lends itself to such an application. The observed dependent variable (aggression) can be socially described as disinhibited and the "high need for social approval" seen as an indicator of social inhibition.

In the analysis above, our use of language has been in the focus. A basic tenet has been that language is used as a part of different human activities, and that its uses cannot be understood without a reference to these activities (Wittgenstein, 1958). This is true for both everyday language and the language of the sciences. Disinhibition cannot be understood without a reference to experimental methodology and the general logic of research.

We should also note the factors that point toward a formal nature of the disinhibition concept:

1. It is accepted by authors in so many and varied disciplines that they inevitably must have very varied substantive empirical models (if any) in mind.

---

* Explanations by suitably labeled properties are one of the most primitive forms of explanation and often of a tautological nature. Experimentalists certainly do not commit this fallacy. The loose way in which the concept (as explanation) often is used makes one suspect that nonempirical treatises on the subject have fallen prey to this logical and linguistic fallacy. For a deservedly celebrated account of this type of "explanation," see Gilbert Ryle's *The Concept of Mind* (1960).

2. It is used to "explain" a wide variety of behavior: kissing, hugging, making love, fighting, and killing. This also means that the concept in its descriptive use must be very general and abstract.*

3. The causal factors in the nascent model are not specified. Such is the case, e.g., with conditional variables that could be characterized as inhibitors in conditional disinhibition reasoning.

As a consequence of this formal nature, disinhibition "theorists" will be able to say, "We were right," *whatever* the final explanatory model of the connection between alcohol use and aggression is (except if conjunctive or common cause variables explain the connection "away"). They will be able to do so as long as introducing alcohol into *any one* predetermined set of circumstances increases the likelihood of aggression.

Disinhibition as a purportedly explanatory concept is widely accepted no doubt because it hides what it purports to reveal, an empirically testable model for the explanation of human behavior. It is a largely formal pseudoexplanation that to a great extent rests on common descriptive language for its believability. It also provides a rather mechanistic and simple formula disguising a process that probably is a predominantly conceptual, symbolic activity with many conditional factors of a cultural nature.

Disinhibition theorists are correct in the realization that behavior changes in connection with alcohol use must be explained with a formal model. This will leave specification of the nature of the resulting (affectionate or aggressive) behavior largely up to situational factors. Below, in a section describing the escalation process in the etiology of violent behavior connected with alcohol use, I will outline a formal model with empirical implications of a testable nature to explain the disinhibiting property of alcohol use.

Situational and Predispositional Factors of Potential Causal Significance

The conditional and interaction models can be divided into two subcategories, the situational and the predispositional, depending on the nature of the conditional variable. This distinction is of both practical, in terms of prevention and methodological setup of experiments, and theoretical importance. It is essential to know whether the conditional variables that increase the probability of violent behavior in alcohol use situations vary between individuals, i.e., are stable characteristics of a subgroup of individuals, or if variations in the frequency of aggressive behavior in these situations can be explained mainly by variations within individuals over time.

* This is probably connected with the fact that it is essentially defined negatively as behavior that is against existing norms. Perhaps it can also be explicated as referring to *extreme* behavior without closer specification of the variable on which it is extreme. (This is true for another central concept used in describing human behavior: deviance.)

In criminology there has been a controversy over historical or genetic views of crime causation as opposed to situational views. Whereas, e.g., Sutherland and Cressey (1970) took the position of greater relevance of historical variables in the etiology of criminal behavior, others have emphasized the importance of situational factors. Among the latter, Gibbons (1971–72) speaks for the existence of a type of criminal that he labels "situational-causal": "In many cases, criminality may be a response to nothing more temporal than the provocations and attractions bound up in the immediate circumstances."

One of the clearest statements for a situational paradigm in the study of violent behavior can be found in a study by Shoham et al. (1974) on the escalation process in violence: "Biological, psychological, psychoanalytical, and sociological aspects of violence are less relevant to the explanation of violence than the actual chain of events leading up to the violent act".* There seems to be no doubt, however, that there is interaction between predispositional variables and situational ones, so that even in a situational explanation of violence the values on predispositional variables (e.g., characteristics of sampled individuals) should ideally be specified.

Here our concern is with explaining the *relationship* between alcohol use and violence. Thus, situational factors cannot be used as independent variables except as they are relevant in explaining away the relationship by using them as conjunctive to alcohol use, or as common cause factors causing both aggressive behavior and alcohol use. If this is not possible, situational factors enter explanations here only as conditional variables that bring about a greater probability of aggressive behavior in alcohol use situations than expected by a null hypothesis.

It seems probable that in many valid explanatory models, where alcohol use in some of its aspects is the independent variable, both situational and predispositional variables must be included. A specific subgroup of the population will behave aggressively only in certain types of situations. This has been acknowledged by many writers in the field, especially by experimental psychologists (e.g., Bennett et al. 1969; Kastl, 1969; and Hetherington and Wray, 1964). Indications of interactions between situational and predispositional variables in producing aggressive behavior in connection with alcohol use are not hard to find in clinical studies. For example, Bach-y-Rita et al, (1971) state that in their sample of violent patients ". . . small variations in the environment provoked massive repercussions." The task for research is to specify the subpopulations and relevant situations.

---

* Since Shoham et al. (1974) are interested in explaining violent behavior in general, it would be possible to include alcohol among the situational factors, although they do not do so. In view of the relationship between alcohol use and violent behavior, their standpoint would be stronger had they done so.

There have been no systematic attempts to manipulate both predispositional and situational variables in the same experimental studies where alcohol use has been the independent variable and aggression the dependent variable. Most studies have been carried out under the direct cause paradigm and any conditional factors, predispositional or situational, have been introduced incidentally or at the stage of interpretations. Consequently, the discussion below has to restrict itself to a discussion of possible conditional variables, without systematically attempting to put them into any relation to each other.

Situational Variables. Our definition of the set of situational factors is the following: Situational variables are variables the values of which show *intra*-individual variations over time, and only to a smaller extent relatively stable interindividual variations. The data are rather scant and a basic weakness is that few studies have been made in natural settings successfully comparing a nonalcohol condition with an alcohol condition. Consequently, it is not possible to assess the importance of the conjunctive *main* effects of relevant variables relative to their conditional (or interactive) value. Therefore, the discussion below will deal with conjunctive and conditional variables in the same subsection. Secondly, a number of variables will be reviewed that seem to be causally relevant in the accounting of the relationship as intervening variables, and thirdly, situational variables will be discussed that could explain away the association by explaining both the occurrence of the alcohol use situation (or a specific type of alcohol use situation) and the display of aggression.

1. *Conjunctive or conditional variables.* Situational variables have been invoked in the discussion of findings in studies using the direct cause paradigm. As seen above, Bennett *et al.* (1969) have discussed the causal role of the "social setting." It is not clear, however, whether they exclusively see it as the main effect variable just conjunctively related to alcohol use because of its social nature in most cultures, or whether they regard it as a conditional factor interacting with alcohol use to produce aggression. Carpenter and Armenti (1972), in reviewing the experimental studies, seem to suggest that the main effect of the milieu variables (conjunctively) explain most of the connection: "It appears that the circumstances of drinking produce greater changes in behavior than the alcohol does."

Social setting is not a variable, but a set of variables. (One sign of this is that it does not make sense to ask for a measurement of social setting or for its value). The lowest common denominator of all social settings (and perhaps what sometimes is meant by "social setting" in the literature) is the presence of other persons, whether in interaction or not. Hartocollis (1962) in his experiment on fifteen males employed as psychiatric residents at a hospital injected diluted ethyl alcohol in the amount of 1 cc per kg of body weight. He found that the subjects who were tested in groups were "more elated, boisterous, and aggressive" than the subjects tested individually. Among the latter no one showed signs of hostility. They were, on the contrary "unusually friendly to those around them." It should be noted that no control or comparison groups getting a placebo injection and subjected to both individual and group conditions were part of the research design. Because of the lack of a control group, it is not possible to determine whether interaction with or presence of other individuals in the situation had any main effect on the increased display of aggression independently of the alcohol effects, and whether there was an interactive or conditional relationship between alcohol use and the specific setting.

Besides the conglomerate of variables referred to as "social setting" (by Kastl, 1969; Bennett *et al.*, 1969; and others), not many potential conjunctively effective factors have been mentioned in the literature. Moreover, the relevant variables in the social setting have seldom been specified, and never systematically studied in studies using alcohol use as one variable. Carpenter and Armenti (1972) mention male drinking company as a possible causative conditional variable. They also suggest (in discussing the findings by Kalin *et al.*, 1972) that the social situations must be such that they have "a minimum of organization forced on them" and that in experimental situations it may be necessary to "provoke aggression by movies, harassment, personal insult, etc., before alcohol has any effect on human subjects studied in the almost isolated conditions of the individual context." In psychological experiments using the aggression machine the frustration–aggression paradigm is used in provoking aggression. Strangely enough, no attempts seem to have been made to take this paradigm into account in explaining the findings.

The possible conditional nature of frustrating or stressful stimuli have not been elaborated on (except by Boyatzis, 1974, in a short discussion), although this is what these experiments presuppose. Stressful stimuli could interact with alcohol use to produce a higher probability of aggression in alcohol as opposed to nonalcohol situations. Consequently, even in many of the experiments that have been carried out under the direct cause paradigm, this potential conditional relationship has existed. (Using the terminology of Carpenter and Armenti (1972), in their discussion of MacDonnell and Ehmer's study (1969), one could say that the experiments study the effect that alcohol has on modifying the relationship between frustration and aggression.) This limitation has not been taken into account in discussing the generalizability of positive findings and the reasons for negative findings.

What, in fact, has been tested is an interaction effect of alcohol use and frustrating or stressful stimuli versus the main effects of frustrating stimuli (in the control condition). It is, of course, impossible to test the effects of alcohol in a stimulus-free situation, since such a situation does not exist. This type of stimuli, however, is only *one* among many conditional situational variables that could be used in experiments on the effect of alcohol on aggressive behavior, and the generalizability is restricted to natural settings where frustrations exist.

There are (at least) three ways in which alcohol use situations could enter into the frustration–aggression model and explain an increased probability of violent behavior:

a. One type of explanation would assume that the relationship between alcohol intake and aggression is a spurious one. This would be the case if there were a greater probability of frustrating stimuli in a significant number of natural alcohol use situations. Thus, the increased number of frustrating stimuli would be a conjunctive variable whose independent main effect would explain the increase in aggressive behavior.

b. If the aggression threshold is lowered in alcohol use situations, quantitatively less frustration is needed to elicit aggression. The lowering of the threshold could be due to e.g., pharmacological effects of alcohol and/or the social definition of a significant number of alcohol use situations. This is a direct cause model using aggression threshold as an intervening variable, and connections with disinhibition models are obvious.

c. The perception of frustrating stimuli could be heightened not only due to quantitative changes in the aggression threshold or the number of frustrating stimuli; the perception of cues in alcohol use situations (partly due to pharmacological effects) could have changed qualitatively, so that cues that would not be interpreted in any way negatively in a sober state may be so interpreted in an intoxicated state due to a change in the conceptual model applied to the environment.*

Any combination of these three factors could, of course, be involved in explanations of subsets of violent behavior and violent crime. It should be noted that the frustration–aggression theory is not sufficient for explanation of all violence, whether in connection with alcohol use or not. It is a predominantly situational explanation, and there are numerous studies pointing toward the causal relevance of nonsituational predispositional factors. It presumes an escalatory process (never systematically studied) with little, if any, rational planning. Most experimental studies on aggression are based on this paradigm, and thus experimental evidence and explanatory models of rationally planned aggression are hard to come by.

Let us after this digression return to other possible conditional or merely

---

* A basic problem is that the situational factors may have become phenomenologically quite different in alcohol conditions as opposed to nonalcohol conditions. If we look for variables and identity criteria in the "phenomenological universe," then we do not have the same set of conditions and the situations are not comparable quantitatively.

conjunctive variables in the social setting. A conjunctive variable of possible causal import is the nutritional habits of the users of alcohol. In some cultures drinking occasions are frequently prolonged, lasting two or three days or longer (and in any culture this is true of alcoholic drinking). If sufficient nutrition is not taken during alcohol use, there will be an increased risk of hypoglycemia (e.g., Moynihan, 1965). Hypoglycemia in its turn "from whatever cause is, in many cases, associated with tendencies to hostility" (Moyer, 1971). If bad nutritional habits and alcohol use coincide, this will explain an association between alcohol use and aggression. Hypoglycemia can also be seen as an intervening variable to the extent that alcohol use increases the probability of insufficient nutrition. (See section on explanations of the association between alcoholism and violent crime.)

2. *Potential intervening variables.* If one were to revert back to 19th-century psychological categorizations, it could be said that alcohol use affects all principal psychological faculties: perception, affective state, cognition, and (as a sedative) conative functions. All of these are relevant to the occurrence of violent behavior. I have chosen to treat as intervening variables those variables that do not measure aggressive behavior directly, but have been shown to increase with alcohol use, and may be causally relevant in explaining violent behavior.

McClelland, Kalin, and co-workers (Kalin, 1972; Kalin et al., 1972; McClelland and Davis, 1972) carried out a series of studies designed to measure emotions and fantasy themes in social drinking situations. They found in the analysis of both TAT projective test results in drinking situations and folk-tale themes that increased drinking was correlated with aggressive fantasies. The social settings for drinking were stag parties and mixed parties. We can view aggressive fantasy themes as indicators of aggressive tendencies. These quasi-experiments are methodologically similar to other small group studies that have used behavior measures as dependent variables. Kalin et al. (1972) make no unwarranted claims of the generalizability of their findings outside their specific social settings and vary these along some variables. Kastl (1969) used medical students as subjects in an experimental setting. He found that alcohol ingestion had no effect on measures of aggressive impulses, and he attributes the findings of Kalin, McClelland, and co-workers to the main effects of the setting. Another possibility is, of course, that there is a conditional or interactive relationship between social setting and alcohol use that will increase the likelihood of aggressive fantasies. Wilsnack (1974) used some rather inferential measures for her dependent variables, in a partial replication of McClelland, Kalin, and co-workers' studies, with female subjects in a social setting. She interprets her findings as showing that women in alcohol use situations experience more fantasy themes concerned with feelings of womanliness, and not with power feelings. Wilsnack's explanation for the increase in such themes after drinking invokes physiological sensations as intervening variables. She suggests that sensations of "physical warmth" caused by alcohol ingestion "may be elaborated by women into feelings of emotional warmth and affection." (There is a semantical link here, but is there an empirical link?)

It thus seems that different subjects in different settings experience different types of imagery and feelings in connection with alcohol use. A pattern seems discernible, however. The first clue to the pattern is found in Kastl's (1969) study in which he also measured changes in mood with the Nowlis Mood Adjective Checklist under three different alcohol doses. Interestingly enough, there were no systematic changes over dose on any of the twelve moods studied, with the exception of one. This was the mood labeled as "happiness." A superficial phenomenological analysis of the concept of happiness indicates that it is without outer reference as opposed to, e.g., "power" feelings, "aggressive" feelings, "sexual" feelings, etc., which require a semantical reference to one or more individuals toward whom feelings of "power," etc. are directed. If we accept the causal importance of the social setting in determining the emotional and behavioral consequences of alcohol use (and there is good evidence for this in the studies that we have discussed), we can start out with this embryonic mood of happiness or well-being, and build up a possible explanatory model. The feelings of well-being could be caused by the pharmacological effects of alcohol. These feelings then could be projected into the situation to fit the cues that are salient in the situation. Pleasurable feelings in males may in an all-male situation be connected with feelings or thoughts of aggressive power and in mixed situations with concerns of sexual power and conquest. Among women, the most "natural" projection of pleasurable feelings, especially in a mixed social situation, could be with thought associations related to being womanly. A social situation will be structured according to status or power in the individuals who have a predisposition in this direction, and derive pleasurable feelings from power. In other types of situations the feelings of well-being would be "interpreted" according to the salient features of that situation, and the results of a projective test administered in a mixed male–female situation will be determined by a mental structuring experienced as pleasurable in that type of situation. (We may thus have another interpretation of why men and women drink.) In Kastl's (1969) study, the situational cues were not adequate for a structuring based on the feelings of happiness into a sexual or aggressively competitive framework, since he used an experimental setting and his measures on these variables did not show any changes over alcohol dose. In Hartocollis's (1962) experiment, the all-male group situations were structured according to the salient features of the setting and thus according to aggressive power features. In the "individual" situations, it seems that there was interaction between the subject and an attendant or nurse (full information is not available). The projection of the feeling of well-being possibly led to a quite different phenomenological structuring of the situation and a different behavior. Perhaps the structuring in a dyadic situation is more like an "intimate friend" relationship, which is a paradigm of pleasurable feelings in these situations.

There could well be great individual differences, which may depend on psychological factors, social learning, etc., in the projection and external specification of the feeling of well-being. In other words, the variables that determine what structural and other factors of a situation are experienced as salient for the feelings of well-being are determined both by idiosyncratic situational factors (such as the composition of the group) as well as more stable personal characteristics of the drinker. (The above analysis is a way of "explaining away" the specificity of feelings as a cause and effect of drinking and to get away from seeming contradictions in empirical findings. It incorporates several empirical findings into the same theoretical model.) It is possible that in some cultures all drinking will become connected with feelings of power via learning processes if drinking is carried out almost exclusively in social situations where the power structure is a salient feature.

The status of power concerns (or, more operationally, thematic physical aggression in TAT projective stories) as intervening variables has not yet been elaborated on here. The definite semantical link between "power" and "aggression" is a handy shortcut, but our attention should be directed toward establishing an *empirical* model for social settings. For this purpose a short exposition of one possible model may be useful. First, let us look at the concept of power and try to explicate its links with behavior. Max Weber's analysis of power as the probability of actualizing one's wishes and commands has been widely accepted. Using it, we could take "power concerns" to imply that the individual is concerned with having his wishes fulfilled in the situation. The somewhat different intervening variable of power feelings would probably mean that the person feels that he has power, and thus the right to expect other individuals to comply with his wishes. Either variant of the intervening variable could be used in a stochastic model of overt aggression in interpersonal situations. In a predominantly male drinking situation in a certain type of culture, a significant number of individuals will be concerned with the power aspects of the immediate situation. (The predominance of immediate situational cues can be expected because of the "here-and-now" character of perceptions under the influence of alcohol. A relatively high share of "inner cues" after alcohol ingestion will have led to a greater salience of power concerns and any other psychological states. These characteristics of alcohol effects will be discussed in the next section.) The greater the proportion of people displaying power concerns and resulting attitudes and behavior, the smaller will be the probability of compliance with anyone's wishes. The lack of compliance will be frustrating stimulus and under the frustration–aggression paradigm we can thus expect more overt aggression in the situation. This kind of sequel could then start an escalatory process culminating in violence (see next section). The point that I want to illustrate here is that we have to look at group processes in order to explain a subset of alcohol-related violence.

An intervening variable more directly related to behavior is risk taking. It is a tendency toward behavior that may lead to obnoxious stimuli as outcomes, but that also promise rewards to a greater extent than other alternatives open in the situation. For discussions of the concept of risk taking, see Cohen et al. (1958) and Katkin et al. (1970). Lemert (1967) has argued that many forms of criminal behavior show characteristics of risk-taking behavior. To the extent that crime can be seen as a subcategory of this type of behavior, and risk-taking behavior tendencies as a variable etiologically relevant in accounting for violent crime, the general literature on the association between alcohol use and crime generally (discussed briefly early in this chapter) is relevant in this context. Many suicides and suicide attempts probably also include elements of risk-taking behavior. The expression "Dutch courage" shows that the effects of alcohol use on risk-taking behavior are well established in common lore. However, the experimental studies carried out fail to show any clear-cut effects of blood alcohol level or amounts ingested (Sjoberg, 1969; Katkin et al., 1970; Hurst and Bagley, 1972; Cohen et al., 1958). There are undoubtedly differences between cultures in the extent to which drinking situations are defined as risk-taking situations (or, more generally, "time out" situations). A model *similar* to the one presented above and substituting risk-taking tendencies in social settings for power concerns as intervening variables between alcohol use and aggressive behavior could no doubt be constructed. Both of these models would then illustrate processes that on a more superficial level are labeled as a conditional relationship between alcohol use and social setting or as an interaction between the two variables. The model of the escalatory process in alcohol use situations, which will be discussed in the next section, is also an illustration of the same formal concepts.

A potential intervening variable that does not have the semantical links with aggression that power concerns and perhaps risk taking do is sleep deprivation and specifically REM-sleep deprivation. There are numerous studies showing the effects of alcohol in different amounts and the effects of withdrawal from alcohol on both general sleep deprivation and specifically REM-sleep deprivation (Knowles et al., 1968; Gresham et al., 1963; Greenberg and Pearlman, 1967; Gross and Goodenough, 1968; Yules et al., 1966; Johnson et al., 1970). The results show that increased doses of alcohol and consequent increases in BACs lead to increasing sleep deprivation, fragmentation of sleep, and a lower proportion of REM-sleep out of total sleep time both in alcoholic and nonalcoholic subjects. After prolonged alcohol use in larger doses, there is often a rebound in REM-sleep activity. Smaller doses of alcohol seem to lead to a decrease in REM initially, but if the same dose is repeated over several nights REM-sleep rebounds and then returns to normal levels. In nonalcohol experiments where sleep deprivation has been the independent variable, it has been found that deprivation of REM-sleep leads to an increase in irritability and anxiety (Gove, 1969-70). Hallucinations, delusions, and illusions sometimes reaching psychotic proportions have been noted in long periods of sleep deprivation (Tyler, 1955; Dement, 1960; Berger and Oswald, 1962; Fisher and Dement, 1963; Kollar et al., 1969). (It has been suggested by Gove (1969-70) that sleep deprivation could be an important etiological factor in the psychotic disorganization of the mentally ill.) An increase in irritability and aggression has been documented in several studies of sleep deprivation (see Gove, 1969-70), and Moyer (1971) suggests that it is one cause of "irritable aggression" in man. Bach-y-Rita et al. (1970) mention inability to sleep as a frequent etiological factor in ten men who had committed violent acts under the influence of alcohol and were diagnosed as cases of pathological intoxication. In addition to the etiological significance in delirium tremens that has been suggested by Gresham et al. (1963), among others, deprivation of REM-sleep could thus be causally relevant in (some cases of) pathological intoxication.* It is evident that alcohol use and consequent sleep deprivation, specially of REM-sleep, can lead to potent psychological and consequent behavioral disturbances. This can happen through the main effects of both alcohol intoxication and lack of REM-sleep and the interaction effects of these two conditions.

In accounting for the connection between alcohol use and violence, sleep deprivation can be regarded as a predispositional factor, if there is an interaction effect between sleep deprivation and alcohol use in causing aggressive behavior, and not merely main effects of alcohol use and sleep deprivation. Thus, sleep deprivation that occurs independently of alcohol use can be relevant for a causal accounting of the connection. Studies determining main effects and interaction effects in the causal scheme relating alcohol use, sleep deprivation, and aggression still remain to be carried out.

In cultures and subgroups of the population where large amounts of alcohol are drunk over an extended period of time, sleep patterns will become irregular and perhaps interrupted by withdrawal stages, and these actions and interactions may explain violent behavior in these circumstances. In these cul-

* Hopwood and Milner (1940) comment on the three homicide offenders in their sample of 96 violent offenders who were involved with alcohol: "Crimes committed by persons suffering from delirium tremens are usually homicidal in character, and are often connected with terrifying visual hallucinations. Although relatively rare, it is probable that homicidal crime is more frequent in delirium tremens than in any other of the common confusional states and toxic deliria."

tures we may expect a relatively strong association between alcohol use and violent crime. Finland, for example, is historically a country of low alcohol consumption, which implies that the null hypothesis of a chance association between alcohol use and violent crime is comparatively low. The association between alcohol use and violent crime in Finnish samples, however, is higher than for the United States, which has had a much higher per capita consumption and thus presumably a higher null hypothesis. The modal pattern of prolonged weekend drinking (Kuusi, 1948; Sariola, 1954; Kuusi, 1956; Pernanen, 1965) in a comparatively large subgroup of the male Finnish population may account for part of this difference through sleep deprivation effects.

3. *Situational common cause explanations.* Some factors that vary intraindividually over time can partly explain both the occurrence of the acute alcohol use situation and the increased probability of aggression in the situation. To the extent that these models explain a subset of violent crime in connection with alcohol use, the relationship would be spurious. A common cause model can incorporate the findings by McClelland and Kalin and their co-workers (1972), Wilsnack (1974), and Boyatzis (1974). The first-mentioned team, among males in drinking situations, and Wilsnack (1974), in females, found that heavier drinkers in the situation were more likely to show power concerns even *before* drinking than were lighter drinkers. It is possible that individuals who are aggressively aroused before a drinking situation (*or* have a predispositionally aggressive personality; see section on predispositional variables) would tend to drink more and, with or without the main effects of higher alcohol use, would exhibit more aggressive behavior. The correlational finding by Irgens-Jensen (1971), that nonalcoholic and alcoholic men who drink heavily consider themselves more masculine, could perhaps be fitted into this explanatory scheme. More correlational evidence is afforded by Gibbens and Silberman (1970) in their prison study. They found that heavy drinkers were more often muscular than were other prisoners. This evidence is very inferential for our purposes, however. Zucker (1968) studied high-school students on a scale measuring masculinity. He found that the heaviest drinkers among the male students were significantly more masculine than moderate drinkers. Moderate drinkers were no more masculine than nondrinkers on the scale. It should be noted that age and social class differences could explain the differences.

Macdonald (1961) suggests that consumption of alcohol and homicidal behavior both may be caused by psychological conflict. Stress situations in general may give rise to both (excessive) drinking and aggressive behavior. Correlational information exists that would fit into such a model (e.g., Linn and Stein, 1944), but by itself such information is not enough and experimental studies of drinking, aggression, and systematically varied stress are needed. It seems likely that a drinking spree and a subsequent violent act can both be determined by, e.g., marital discord, and arguments in general as stressful stimuli. The interaction effects between the aggressive arousal and alcohol use complicate any explanatory model. (In this connection it could also be asked how much alcohol-related domestic violence is due to reactions by one spouse to excessive alcohol use by the other spouse.)

4. *Escalation of aggression in alcohol use situations.* The situation that seems to have been on the mind of many a student of aggressive behavior is a stereotyped version of a barroom brawl in a working-class tavern. To this picture belongs a rapid sequence of events from an exchange of angry words to escalating relaliation, a fight, and as a possible outcome a homicide. Probably this widespread conception is influenced by westerns on television and on the screen, where saloon brawls seem to be a necessary ingredient. The etiological role of alcohol is often not made clear in these staged versions. It may be implicit, but it seems that situational factors (a gathering of men in their best fighting years) and cultural factors (the western frontier ethos and associated behavior norms) are given more prominence. The existence of expressions such as "barroom brawls" and "drunken brawls" in common parlance also indicates that, whatever the cause, there seems to be a higher risk of violence in these surroundings.

The typicality of this stereotyped situation has not been well established and the inevitability of the escalating process from an exchange of words to violence has not been documented. Wolfgang (1958), in fact, showed that the modal place of homicide was the home. Fifty-one percent of the criminal homicides in Philadelphia were committed in somebody's home. Pokorny (1965) found that 42 percent of the criminal homicides in Houston were committed in the home of either the victim or offender, and Voss and Hepburn (1968) in Chicago found that 37.6 percent of the male victims and 61.5 percent of the females were slain in the home. Mayfield (1972) found in his North Carolina sample of homicides and assaults that 46 percent of these crimes were committed in the home. Pittman and Handy (1964) note that 11.2 percent of their sample of criminal aggravated assaults in St. Louis took place in a tavern and in Pokorny's (1965) study the percentage was 13.6. In the former study the offender and victim had been drinking together prior to the crime in a majority of cases in which alcohol was involved.

Mayfield (1972) notes that 35 percent of assaults in his study took place in a drinking situation. He points out the relatively high prevalence of sudden escalatory processes resulting in violence:

> The assaults are typically sudden, impulsive acts—too frivolous of motivation to be convincingly labeled "crimes of passion." They are often a result of action and reaction between acquaintances who often do not have longstanding or deep grievance but rather a mutual state of intoxication and a readily available quick and lethal weapon.

There may be large differences between cultures and jurisdictions in the escalatory pattern. West (1968) in describing 100 homicidal offenders in Manhattan stresses the difference with English patterns:

> In the many instances of homicidal quarrels it was noteworthy how often incidents had flared up unexpectedly from trivial beginnings, sometimes from quite casual encounters between strangers in bars. In these cases the fatal outcome was invariably due to one of the participants drawing a knife or a gun, a chain of events that is less common in England.

Whatever the relevance of barroom brawls in accounting for violent crime, it has been noted that an escalatory process is present in other settings also. Gibbons (1971–72) in emphasizing situational factors in crime causation notes the typicality of this process: "Those who do engage in murder often do so within situations of marital discord or tavern fights, in which a number of provocative moves and countermoves of interactional partners culminate in acts of homicide." Pittman and Handy (1964) found in their study of aggravated assaults in St. Louis that the offender and the victim had generally been in interaction with each other before the violent act. In 181 out of 241 cases (75 percent) verbal arguments preceded the aggression: "These quarrels may range from domestic incidents to tavern disputes over who wants to sit on which bar stool. On the surface the quarrels appear to have little rationality." Other authors who have emphasized the frequency of the escalatory process include Aho (1967), Bard and Zacker (1974), Aromaa (1974), Hopwood and Milner (1940), and Washburne (1961), who provides anthropological data from several primitive cultures.

Escalation has not been studied in any detail by psychologists. In their experiments they have generally been content with studying at the most the initial cycle of the process. There are a few remarks by experimental psychologists that show an awareness of its explanatory value (e.g., Buss, 1963; Ryan, 1970; Epstein and Taylor, 1967). Shuntich and Taylor's (1972) study, in which intensities of shocks administered to the subject were gradually increased, has some bearing on escalation. Their results show a higher shock level administered to an alleged opponent over all the levels of shock that the subjects received. However, subjects in the control and placebo conditions adjusted their shock settings much more closely to the shocks given them by their "opponents."

The most detailed analysis of escalation has been undertaken by Shoham et al. (1974). The authors acknowledge that it is only a beginning of a study of situational aspects of violence. A violent act is seen as the end product of an escalating series of provocative acts, each response serving as a stimulus to the adversary. The role of the ambiguity of some acts in this series, especially the initial ones, is also built into an explanatory scheme. There is no mention of the possible effects of alcohol or other drugs on escalation into violence.

The authors emphasize the importance of the interpretation of the acts of the opponent. A remark can be interpreted in a number of ways, e.g., as an aggressive remark, as a joke, etc. Depending on the interpretation, a potentially aggressive interaction can stop soon after initiation. Epstein and Taylor (1967) also point to findings in their experimental studies that show that in "continuous aggressive competitive interaction, perception of the opponent's aggressive intent is a far more potent instigator to aggression than frustration in the form of defeat." Mayfield (1972) found that in 50 percent of the cases the victim had made an attack or a move that the offender interpreted as an impending attack immediately prior to the homicide or assault.) The role of alcohol as a possible facilitator in the escalation process has not been studied. The only systematically collected indication of the importance of an escalatory process in violent crimes in which alcohol is involved can be had from information on victim precipitation. In these instances the escalatory process contains at least one cycle of interaction. Wolfgang (1967) found that the victim had precipitated the homicide in 26 percent of the cases in his Philadelphia sample. Alcohol use was significantly more often present in the homicide situation where the victim precipitated the act of homicide (in 74 percent of the victim-precipitated cases versus 60 percent of other homicides). The victim had been drinking in 69 percent of the cases of victim-precipitation and in 47 percent of the other cases. Voss and Hepburn (1968) found that 43.9 percent of 164 criminal homicides in which alcohol was present were precipitated by the victim, whereas this was true for only 31.3 percent of the 134 cases in which alcohol was not present. Virkkunen (1974) found that in criminal homicide cases in Helsinki, Finland, "aggressive behavior and altercation" preceded the criminal act more often in cases in which alcohol had been used. In these cases the aggression sequence had been started by the victims as often as by the person who finally became the offender. An aggressive act can be conceptualized as a frustrating stimulus to the opponent in most situations. Frustration again may lead to reactions other than aggression, as Buss (1963) points out. Thus, the victim of verbal aggression may retreat, try to make a joke, try to soothe the aggressor, etc. Implicit in many of these reactions is an attempt at a redefinition of the situation to the aggressor, other participants in the situation, and perhaps to the victim of initial aggression himself. Coping devices, such as redefinitions of the situation, are learned and to a large extent culturally determined. In cultures placing great emphasis on manliness, machismo, physical prowess, etc., the use of coping devices in the face of aggression without "loss of face" (see Goffman, 1967) is probably more limited than in other cultures. The more alternative coping devices provided by the culture in situations where aggression is displayed, be they in the form of retreats or redefinitions, the less is the risk of escalation into violence of initial aggressive acts (acts interpreted as aggressive). Due to psychological effects of alcohol, it seems likely that coping devices that require an abstract conceptual command of the situation will have a smaller probability of occurring when the individual is intoxicated (see Tarter *et al.,* 1971; and Kastl, 1969). Thus, the number of coping mechanisms available probably decreases during a state of intoxication. A systematic investigation of coping mechanisms in equivalent nonalcohol and alcohol settings would shed some light on the greater likelihood of violence under the influence of alcohol. The conceptual dimension of alcohol effects could easily be investigated by experimental methods comparing the subjects' interpretations of the behavior of other individuals in alcohol use situations with those in situations where alcohol is not used. Other more specific independent variables, e.g., blood alcohol levels, should also be investigated for their effects on this dimension.

The coping mechanisms more likely to be used under the influence of alcohol will be the ones determined by the immediate situation. Jellinek and McFarland (1940), as quoted by Kastl (1969), make the following generalization in summarizing experimental results on alcohol effects: "After alcohol ingestion, associations are impoverished and follow a path of least resistance." The "path of least resistance" consists of behavior cues and stimuli that are present in the immediate situation whether these stimuli be internal or external. The statement by Jellinek and McFarland is confirmed by Kastl (1969) in his experimental measures of ideational association. Washburne (1956) suggests that a narrowing of the time dimension is "the most important factor associated with role-playing situations involving the use of alcohol." He also mentions his own observations, which indicate that the mere presence of alcohol (via the cultural definition and learned cue value, no doubt) can lead to a "decreased awareness concentrated upon the immediate situation." Washburne suggests that this probably is responsible for much of what is labeled "antisocial behavior" in concentration with alcohol use. These actions of alcohol are one aspect and/or cause of the here-and-now character of behavior under the acute influence of alcohol and in the behavior of alcoholics. (The etiological role of decreased abstracting abilities in violence among alcoholics will be discussed in a subsequent section.)

A well-known, but not sufficiently documented, feature of behavior in alcohol use situations and under the influence of alcohol is the lability of affect and behavior. It is known to observers or participants of drinking occasions that in addition to aggressive outbursts, there are displays of kissing, hugging, and backslapping in these situations (see MacAndrew and Edgerton (1969) for a thorough discussion). As mentioned earlier, the large variation in behavior exhibited in alcohol use situations probably partly explains the popularity of the disinhibition concept in explanations of drinking behavior. It is largely a formal concept and thus no specification of the resulting behavior is needed, except that it is disinhibited in a descriptive sense. Thus, all these reactions to alcohol can be incorporated into the "model." Anecdotal references to the affective lability in alcohol use situations and among alcoholics abound in the literature. Whittet (1973) mentions the unstability of the intoxicated person; he vacillates between being a "belligerent bully" and a "besotted buffoon." Hopwood and Milner (1940) make the following impressionistic observation:

> A drunken man is usually extremely unstable, whether he be a chronic or a "spree" drinker, and some slight and insignificant annoyance may produce such an exaggerated effect on the emotional tone that he may react with great violence and aggressiveness.

Aho (1967), impressionistically and in passing, and Hartocollis (1962) on the basis of observations in experimental situations, mention this fluctuation.* (One word of caution, however. If predispositional factors are relevant in the explanation of the link between alcohol use and violent and other emotionally determined behavior, we have to assume that a significant amount of intraindividual stability exists.)

A decrease in the conceptual and abstracting abilities can be used to construct a model to explain the extremeness and unpredictability of affect and behavior under the influence of alcohol. Of more relevance for this purpose, however, is the experimental evidence showing that one of the consequences of alcohol intake is a lessened ability to act upon several cues at the same time, based upon a narrowing of the perceptual field (Medina, 1970; Moskowitz and DePry, 1968). The overall findings, however, are somewhat conflicting, probably due to variations in the BACs and the measures of the dependent variable (Tarter *et al.,* 1971). Assuming that a greater number of cues are perceived in a sober state, we can deduce by statistical reasoning that a greater number of cues have to change for the interpretation of the situation (including the interpretation of the behavior of other people) to change, than in an intoxicated state. It is possible to show statistically that a combination of few elements in a universe is more randomly distributed than a combination of many. (This fact is the basis of sampling statistics.) The actions of a person can be interpreted by an intoxicated person in the same situation, in extreme, comparatively randomly determined ways, since the interpretation is determined by fewer events or cues in the situation and behavior by other individuals. Thus, for example, aggressive remarks can be interpreted as jokes and jokes as aggressive remarks to a larger extent than in a sober state. The interpretation

---

* Hartocollis, however, suggests that the extreme reactions of loquaciousness, gregariousness, aggression, and elation may be the outcome of an effort of his subjects "to cope with a generally challenging situation."

of the actions or remarks of other individuals will be determined by chance to a much larger extent than in comparative sober situations. The lack of restraint typical of many drinking occasions and the mood fluctuations (which are more extreme than in comparable sober situations) are a consequence of this. The interaction between two or more individuals who are sober is probably much less fluctuating emotionally than the interaction when one and especially if both of them are intoxicated, although the "initial interaction" and the setting are the same. This could (and should) be studied empirically by experimental and observational means.

Pastore (1952), Buss (1963), and Epstein and Taylor (1967) have showed that aggression that is seen as arbitrary, as being the result of the whim of the aggressor, elicits more aggression than aggression that can be attributed to an acceptable cause or reason. Due to the narrowing of the perceptual field, we can assume that the probability for two individuals to see *overlapping* cues as relevant in the situation will be less than in a sober situation (this is strictly statistical reasoning). Thus, the one will more likely fail to see a justification for the other person's action. Consequently, the action of the other person will seem more arbitrary and will thus evoke more aggression, which again has a higher probability of seeming arbitrary, and thus the probability of escalation into physical violence is successively increased over the comparable probabilities in a sober situation. In addition to the perception of cues, conceptual reasoning and abstracting ability are required in a justification process, in trying to "understand" the behavior of the other person, and these have been shown to deteriorate under the influence of alcohol (Tarter *et al.*, 1971). Instead of separate explanations for the actions of alcohol on aggressive and affiliative behavior we have one model that is formal in character. It can thus incorporate different types of behavior. The disinhibition "model" has at least served the function of showing us that a formal model is needed to account for behavior changes in connection with alcohol use. Instead of a mechanistic model of unalterable alcohol effects, however, we need models that account for the varying and often extreme effects of alcohol by explicating the concept of social setting. The social setting as a causally relevant conditional or interactive factor is probably very often on closer inspection a set of interactions among individuals.

In summary, it can be said that the narrowing of the perceptual field with the consequent random determination by a smaller number of cues (and possibly a concomitant preponderance of "inner" cues, or drive states), experienced as significant in the orientation to the situation and the interpretation of the behavior of others, lead to a higher likelihood of violence in drinking situations. Another causative factor is the conceptual impoverishment and decline in abstracting ability under the influence of alcohol, which decrease the likelihood of use of coping devices that go outside the immediate situation, and thus cut down the probability of alternative acts. Many other factors are probably relevant to the escalation process, many of them of conditional cultural nature, such as definition of drinking situations as "time out" or risk-taking situations; other factors are more directly related to the pharmacological effects of alcohol, such as the paresthesia induced by alcohol (Hartocollis, 1962). See Figure 1 for a summary of the above discussion. (Paresthesia would be an intervening variable in this explanatory scheme.)

Aromaa (1974) has hypothesized that individuals who are comparative strangers to each other more easily misinterpret each other's intentions, and "random behavior" thus results more easily. As Wolfgang (1958) and other investigators (e.g., Pokorny, 1965; Voss and Hepburn, 1968) have pointed out, however, most violent crime is a result of aggression between people who know each other well. Interaction characteristics and escalation cycles of strangers can thus only explain a minor portion of violent crime. Whether the individuals know each other well or not, however, it is clear that cutting down the number of possible interpretations of a situation, and the possible cues available in

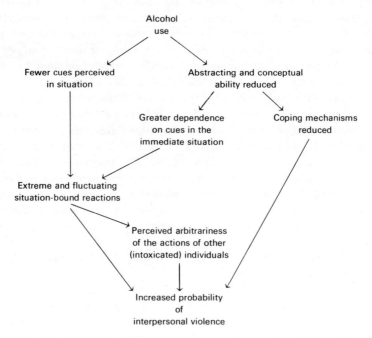

FIGURE 1. Factors in the use of alcohol leading to an increased probability of interpersonal violence.

order to decrease the probability of ambiguity through "random" selection of a few cues, will decrease the likelihood of an escalation process in this probabilistic model.* This is done in marijuana use situations by dim lights, reduced commotion, and soft music (as an inescapable common set of cues). (See, e.g., Orcutt, 1972).

We have seen above that even where coping devices are available in the culture the skills in using the ones based on more conceptual complexity have been reduced by the psychological actions of alcohol. This is all the more reason for providing within the culture unambiguous and numerically few cues to guide the interpretation of the situation and consequently the reactions of individuals when alcohol is used. If the more or less random situational cues are predominant and there are not enough culturally defined unambiguous counteracting cues present in the situation, the interpretation of which is well established via socialization (e.g., norms), the probability of aggression will increase.

It seems probable that ritualistic drinking in numerous different societies (as described in the anthropological literature by Washburne, 1961; MacAndrew and Edgerton, 1969; Heath, 1958; Madsen and Madsen, 1969; and others), with strong sanctions for breaking norms, would help to contain personalistic and situational interpretations of the situation and the behavior of other individuals. Heath's (1958) study of the Camba in Bolivia can be used here to typify the pattern of ritualized drinking. The Camba drink a beverage of very high alcohol content in amounts leading to general intoxication and even stupor at drinking occasions that can last several days. Yet, "aggression and sexual license are conspicuously absent." The answer to this puzzling pattern in the primitive societies may lie in the very ritualized sequence of drinking demanded in the culture.† MacAndrew and Edgerton (1969) discuss several other examples. The fact that many primitive tribes have built safeguards against extreme behavior in drinking situations is probably due to a greater vulnerability to disruptive behavior of small primitive societies. Even

---

* Shoham et al. (1974) have pointed out the importance of a study of factors linked to the avoidance of violence in order to arrive at a satisfactory explanation of the phenomenon. It is clear that conditional factors of a cultural nature exist, and that an explanation is unsatisfactory if it only looks at situational factors or tries to project everything onto a simple disinhibition paradigm. The existence of relevant cultural factors has been emphasized by Epstein and Taylor (1967) within the frustration–aggression model: "It is self-evident that all cultures must counteract any tendency for a simple, direct relationship to exist between frustration and aggression."

† Heath (1958) found it noteworthy that the Camba pattern of drinking did not lead to hallucinations among the participants. It is *possible* that this may have a causal connection with the consensual cues provided in the drinking situation. Perhaps ritualization in the drinking situation would also tend to decrease individual interpretations of hallucinatory perceptions possibly connected with sleep deprivation and withdrawal. This would be especially true of cultures where hallucinations are seen as having religious significance.

when aggressive displays lead to fighting, it is not uncommon that the escalation process itself is formalized (Washburne, 1961). This decreases the likelihood of randomness in the behavior cues as perceived by the opponent, and the probability of an uncontrolled escalation process.

Ritualization works in at least four ways to decrease the probability of violence in connection with alcohol use:

a. It leads to a reduction in the number of *indeterminate* cues in the situation and in the behavior of other individuals, which could lead to any number of interpretations by other participants.

b. It *provides a cognitive interpretative framework* of a *few* central cues via a cultural "meaning" of behavior that minimizes interpretations of behavior that would mainly depend on the situational cues and individual motivational states.

c. It increases the probability of "consensus" on cues seen as relevant, merely by *cutting down the number of cues* available in the situation.

d. It *provides external cues* of adequate strength to overcome a possible preponderance of inner cues and drive states. If there is a greater likelihood of mood fluctuations when the "controlling" influence of external cues is cut down, this will be reflected in the behavior.*

The risk-taking character of behavior in alcohol use situations and under the influence of alcohol, which is partly determined, no doubt, by the cultural definition of alcohol use situations as "time out" (MacAndrew and Edgerton, 1969), can probably also be explained by the narrowing of the perceptual field, and the more randomly determined behavior.

Tinklenberg (1973), in his discussion of the connection between alcohol and violence, has suggested that the model of assaultive behavior put forth by Melges and Harris (1970) applies: "Individuals with distorted temporal perspectives involving excessive focus on present here-and-now stimuli are prone to violence" (Tinklenberg, 1973). This condition would thus be due to a predispositional dysfunction in the perceptual and/or cognitive faculties of certain individuals. The reasoning above has shown that the same causal factors may be at play (intra-individually and more generally) where alcohol is used, irrespectively of any propensity of some individuals (although interaction or simple additive effects naturally may produce some high-risk subpopulations in alcohol use situations).

There is much to be gained from a dynamic situational and interactional

---

* One would probably learn a great deal from many other safeguards that have been instituted in different cultures to counteract the more or less random distribution on situational and motivational variables that could jeopardize the culturally defined "success" of a drinking situation, e.g., through violence.

approach to the problem of the role of alcohol in the etiology of violent behavior, as opposed to a mechanistic quasi-explanatory (often merely descriptive) use of the disinhibition concept, which does not take situational factors into account. (The analysis of the widely used disinhibition concept and the criticism of its vague use do not imply that the concept has not been used legitimately to refer to specific physiological (Kalant, 1961; Moyer, 1971) and psychological (e.g., Rada, 1975) models.)

It is probable that the excess of prior alcohol use by both the victim and the offender noted earlier can partly be accounted for by the escalation facilitated by the subjectively perceived arbitrariness and the objectively definable randomness of the opponent's reactions. The model suggested above explains a number of different types of disinhibited behavior and fluctuations between these in alcohol use situations, and may explain certain aspects of the escalatory process leading to violence. It should, once again, be emphasized that *interaction processes* between individuals in a drinking situation must be studied. Labeling the whole array of possible causative variables of a social nature and the group processes as "social setting" without further inquiry, will leave much of aggression and violent crime in alcohol use situations essentially unexplained. The model presented here can be seen as an illustration of the task ahead, and an attempt to arrive at parsimony in a widely scattered field.

Predispositional Variables. Most of the factors studied in connection with violent behavior in general are rather permanent variables introduced to explain violent tendencies in certain individuals. To this category belong most biological variables that have been put forth, such as general innate aggressive instincts of man as a species, temporal lobe dysfunctions, etc. Some social factors of etiological significance are also rather stable over time. This is true for the subculture variables, such as the "subculture of violence" (Wolfgang and Ferracuti, 1967). The existence of stable characteristics in the etiology of violent crime is well established by the mere fact of high recidivism rates among violent offenders (Williams, 1969; Greenland, 1971 and 1973; Walker *et al.,* 1970; Bach-y-Rita *et al.,* 1971). Predispositional factors such as temporal lobe dysfunction or violent subcultures in their main effects on aggressive behavior are not of interest to us here. They only concern us insofar as they interact with alcohol use over and above the main effects of both variables, i.e., as conditional or interactive variables.

Predispositional variables in our terminology are variables that show *inter*individual variation in a stable manner over time, with only relatively small intra-individual variation. (Values on most variables probably show both types of variation but relatively valid distinctions are still possible.) Accounting by models using predispositional variables could be called accounting by subpopulations. Before an accounting of the association between alcohol use and violent crime via a model using predisposing variables (and also the models using conditional situational variables discussed in a previous section) can be started, it must be shown that the null hypothesis is not true in the subpopulation labeled by the relevant conditional variable (e.g., alcoholics or individuals with temporal lobe dysfunctions). It has to be shown, in other words, that the probability of both using alcohol and displaying aggressive behavior at one time by chance is not sufficiently greater in these subpopulations. If there is a greater probability of alcohol use, it also has to be shown that the alcohol use does not through its main effects explain the epidemiological association. The question thus is, Is there a larger increase in the probability of violent behavior when alcohol is used by individuals with any one or more of the potential predisposing characteristics, than when other individuals without the same characteristics use alcohol? The considerations of a chance association (null hypothesis) and main effects of higher alcohol use are not relevant in properly carried out experimental work on reactions to alcohol use of, e.g., alcoholics (using nonalcoholics as controls), but these problems are the curse of epidemiological and statistical data.

For the accounting aspect of the connection between alcohol use and crimes of violence, it is also important to keep in mind the prevalence of the predisposing condition. It is evident, for example, that even if there were an interaction effect between XYY chromosomal abnormality and acute alcohol effects in aggressive behavior, this could only explain a minor part of the epidemiological relationships between alcohol use and violent crime or alcohol use and general aggressive behavior, because the prevalence of XYY chromosomal abnormality is so small (in the order of 1:700 in newborn males; see Ratcliffe *et al.,* 1970). It should be recognized, however, that many predispositional variables of a biological nature may in fact be continuously distributed in the general population, although the extreme cases turn up in clinical samples. This is suggested by Moyer (1971) in his discussion on temporal lobe dysfunctions: "Individuals manifesting inter-ictal or sub-ictal dyscontrol syndromes are on a continuum which varies from homicidal behavior to occasional "normal" irritability."

Interindividually varying but intraindividually (situationally) stable characteristics can be relevant as conjunctive variables, although they cannot covary with occurrence of or amount of alcohol use. If there is a selection of individuals to alcohol use situations, possibly due to cultural or subcultural variables, and the selective criteria are positively correlated with aggressive tendencies regardless of alcohol use, this would explain the epidemiological association by using a conjunctive variable (aggressive tendencies). However, data do not exist to lend significant support to the validity of any model of this type and it will not be discussed in any length in this section. Predisposing factors cannot be intervening variables, caused by acute alcohol use via its main effect, since intervening variables must vary with the independent variable, in this case alcohol use, which means that intervening variables must here be of the situational type. The possibility still exists that some (intervening) effects of alcohol use would interact with predisposing factors so that, e.g., sleep deprivation effects would be more extreme in individuals with temporal lobe dysfunction and lead to a greater probability of violence in this way. It would, however, take us too far into speculative detours to consider the possibility of more complicated models for which there is as yet little or no empirical data available. Predispositional intervening variables affected by alcoholic prolonged and excessive drinking will be considered in a specific section below that deals with the explanatory models accounting for the relationship between alcoholism and violent crime. Thus, in this section I will only discuss the role of predispositions mainly as conditional or interactive and briefly as common cause factors. Sometimes it is not made clear in the literature which of these two types of models are indicated, or even whether a model with the predisposition as an intervening or dependent variable and with alcoholic drinking as independent variable is invoked.

1. *Conditional or interactive variables.* Few experiments have been carried out studying the effects of alcohol on aggression in systematically selected samples of individuals whose violence-proneness has been established epidemiologically or clinically. However, experimental psychologists, even when working under the direct cause paradigm, have been aware of the limitations of their methodological assumptions. Bennett *et al.* (1969), among others, acknowledge the relevance of the characteristics of the subject population.

There are some mentions in the epidemiologically oriented literature of interaction effects between alcoholism and acute alcohol use in producing a greater probability of violent behavior than what could be expected by either variable by itself or conjointly (Hopwood and Milner, 1940; Hoff and Kryspin-Exner, 1962; Ando and Hasegawa, 1969; Bennett, 1967), but they are not substantiated with systematic empirical data. Since we do not have the information needed to establish a null hypothesis, it is impossible to tell whether these observations are due to the main effects of "alcoholism" and/or main effects of acute alcohol use.

Alcoholism of the subjects is the predispositional characteristic most often included in experimental studies that use alcohol intake as the independent variable and aggression, or a potential intervening variable such as aggressive mood changes, as the dependent variable. The relevant experimental studies have generally been designed to explain the excessive drinking of the subjects, but some of them measure variables that seem relevant to an explanation of aggressive behavior. This is due to either semantical or empirical links between the variables measured and aggression. To the extent that the links are empirical, these variables can be seen as intervening variables between alcohol use and aggression (at least in the specific subpopulations).

The labels "alcoholics" or "heavy drinkers" can most fruitfully be looked upon as a cluster of variables, some physiological or psychological and some social and subcultural. Some of the variables in this cluster may interact with acute alcohol use and the situations in which alcohol is used, to produce a higher than expected probability of aggressiveness, while others may not. I will not try to extract these in this discussion due to lack of empirical data, but instead use the general labels of "alcoholics" or "alcoholism."

In a methodologically sound study, Mayfield and Allen (1967) administered alcohol intravenously to alcoholic patients and a group of controls. The dependent variable in their study was affect as measured on the Clyde Mood Scale. The dose was rather low, the equivalent of 5.2 cl of absolute alcohol (approximately three bottles of beer). They concluded that alcohol altered several affects among which aggression was least affected. Both the preinfusion and postinfusion scores for aggressiveness were virtually identical in the three groups of subjects. Contradicting this negative finding, there is evidence from other experiments showing that alcohol effects are markedly different in alcoholics as compared to nonalcoholics, as van der Spuy (1972) points out. In reviewing the empirical literature on alcohol's effect on the mood of alcoholics, he concludes: "The alcoholic's emotional state appears to benefit considerably less from alcohol than the emotional state of the nonalcoholic." To the extent that this is the case, it could partly explain any excess clustering of violence to alcoholics. As pointed out above, however, some of the experiments on the effect of drinking on alcoholics are not comparable to experiments on nonalcoholics because the drinking is much more prolonged (for reasons of representativeness). The indication is that the increase in depression and anxiety (Mendelson et al., 1964; Nathan et al., 1970), and hostility (Nathan et al., 1970), is reported after two to four days of drinking. Sleep deprivation and other (stress) factors may be at work by that time, and this could be true also for nonalcoholics in an equivalent research setting. The amounts consumed may not be comparable either, and consequently not the blood alcohol levels. These factors could explain some of the discrepancy between Mayfield and Allen's (1967) findings and the findings from most other experiments with alcoholics, since Mayfield and Allen used a comparatively short experiment and relatively small amounts of alcohol. It could, however, be said that typical alcoholic drinking patterns increase the likelihood of aggressive reactions, but these drinking patterns would have to be considered a conjunctive independent factor and not a predispositional feature of alcoholics that would interact with alcohol use. One word of caution is in place here: There are great differences between alcoholics both situationally in their reactions to alcohol and as to more stable predisposing traits and characteristics (see, e.g., Vannicelli, 1972; Partington and Johnson, 1969) and this discussion, which treats alcoholics as one group, should not hide this fact.

There is evidence showing that alcoholism, or, more specifically, prolonged excessive use of alcohol, can give rise to stable interindividually varying characteristics, which increase the probability of violence through their main effects. Some of these characteristics will probably interact with acute alcohol use to produce an increased probability of interpersonal violence. These characteristics in their main effects on this probability will be discussed in the last section of this chapter.

Predispositional variables of a biological nature adduced as independent variables in the general explanation of violent behavior include: testosterone production (Moyer, 1971; Persky *et al.*, 1971; Eleftheriou and Scott, 1971; Williams, 1969; Hamburg, 1971), psychomotor epilepsy and temporal lobe dysfunction with abnormal EEG patterns as symptoms (Moyer, 1971; Stafford-Clark and Taylor, 1949; Mundy-Castle, 1957), history of head injury (Bach-y-Rita *et al.*, 1971; Hopwood and Milner, 1940), history of convulsions (Bach-y-Rita *et al.*, 1971), and XYY chromosomal abnormalities (e.g. Barker *et al.*, 1970). There is much less relevant data on possible biological predisposing characteristics of causal relevance to the explanation of the relationship between alcohol use and violent behavior.

Pathological intoxication is the clearest indication of an interaction mechanism of alcohol and some probably physiological predisposing variables. No representative epidemiological data exist on the proportion of cases of pathological intoxication in violent crimes. Reports of the connection with unknown representativeness of the universe of violent crime include Julius and Bohacek's (1954) study in Yugoslavia. In ten out of 19 cases of murder pathological intoxication was implicated (the sampling procedure is not known). The authors claim that the pathological reaction to alcohol was proven by test. Zakowska-Dabrowska and Strzyzewski (1969) in Poland studied a sample of 63 men who had committed crimes (of an unspecified nature) while under the influence of alcohol. They found that EEGs were abnormal prior to ingestion of alcohol in the experimental situation in 11 of the 52 cases of "simple intoxication"; after ingestion of alcohol the EEGs of an additional 23 subjects were abnormal. The EEGs were normal both before and after alcohol ingestion in the experimental situation in all the six cases where pathological intoxication had occurred at the time of the crime. Cuthbert (1970) found that in his sample of 70 murderers alcohol activated or enhanced EEG spiking in six out of seven offenders with severe temporal lobe dysfunction. Marinacci (1963) found evidence in a large sample of violent individuals that EEG patterns changed toward dysrhythmia after ingestion of alcohol and in some cases violent behavior followed. Other authors reflect the same view (Skelton, 1970; Greenblatt et al., 1944; Thompson, 1956). Bach-y-Rita et al. (1970), however, failed to replicate Marinacci's (1963) findings of epileptic discharges after alcohol ingestion in subjects who had displayed signs of pathological intoxication in the etiology of their repetitive aggressive behavior. Whatever the physiological mechanism, it is clear from the very small ingested amounts of alcohol and the dramatic displays of aggressive behavior that an interaction mechanism is at work. The behavioral effects of small amounts of alcohol over the general population and the often attested to timid behavior of the individuals in a nonalcohol state show that the main effects cannot explain the phenomenon. Moreover, the repetitiveness of the pattern of small amounts of alcohol and extremely violent behavior show the existence of interindividual differences in reactions to alcohol, and thus the applicability of interactive or conditional models. The pattern is probably not entirely determined by endogenous factors, however, as is also suggested by Bach-y-Rita et al., (1971), who noted stress as a conditional situational factor in their sample of 130 patients:

> As stress would build up, alcohol would frequently play a greater role in the patients' general daily routine, 72 reported that they had used or abused alcohol prior to their episodes.

Other claims of subpopulations with predisposing characteristics released by alcohol use that thave been made but not substantiated by empirical research include "latent schizophrenics" (Baker, 1959), individuals who are "mentally incompetent" (Pionkowski, 1965), and individuals with head injury (Hopwood and Milner, 1940). Another possible predispositional factor (which has been dealt with in some detail in a previous section) is sleep deprivation independent of alcohol use. Fasting will, in connection with alcohol use, cause hypoglycemia, which again increases the likelihood of aggressive behavior (Moyer, 1971). Variables, which sometimes are seen as causing both a greater probability of alcoholism (or acute alcohol use) and violent behavior, but often not distinguished in the discussion from their distinct and possible role as predisposing interactional factors in connection with alcohol use, include personality disorders, emotional instability, "aggressive types," etc. (These will be discussed in some detail in the next section.) The overlap of these with other labels and subpopulations discussed above is impossible to ascertain at the present time.

The experiment by Hetherington and Wray (1964) discussed previously showed that a high need for social approval combined with a high aggression need led to an increase in aggressive reactions after ingestion of alcohol, whereas this was not the case in subjects who did not have both of these attributes. The extent to which these attributes also vary intra-individually over different situations is not known, but such a variation seems likely. Similar findings were made in a study by Roebuck and Johnson (1962). They found that their sample of 40 Negro offenders who had a pattern of simultaneous "drunk and assault" charges were more often reared in homes with a rigid fundamentalist background than a comparison group of 360 other Negro offenders with a much more varied arrest pattern. The authors suggest that the rigid socialization led to a pattern where hostility could be manifested only after alcohol use.

Many predominantly social or cultural variables have been mentioned in general explanations of violent behavior or subcultures showing higher prevalences of violent behavior (with or without alcohol involvement). Accounts of drinking occasions in some cultures seem to indicate that these are characterized by anomie, normlessness. We have seen in the section on escalatory processes, however, that ritualized behavior is not uncommon even on occasions where large amounts of alcohol are consumed.

No doubt, societies, cultures, and subcultures differ as to their predisposing conditions for violence. Whether this is true also for their predispositional tendencies toward violence in connection with alcohol use, over and above the main effect of the cultural factors and of alcohol use itself, is another question. The importance of conditional, probably culturally determined, variables in the connection between alcohol use and aggression has been pointed out by a number of authors in the anthropological literature (MacAndrew and Edgerton, 1969; Washburne, 1956; Washburne 1961; Child *et al.*, 1965a). Child *et al.* note:

> Boisterousness and a combination of exaggerated sociability and hostility are typical forms of behavior in our society for someone who has consumed a large amount of alcohol. Yet in some societies these forms are not particularly conspicuous.*

One possible way in which cultural factors could produce a conditional or interactive relationship between alcohol use situations and violent behavior is through a different set of norms for alcohol use situations from those applying to nonalcohol situations. This has been suggested through the concept of "time out" in connection with alcohol use (MacAndrew and Edgerton, 1969), and

---

* It is noted among experimental psychologists that cultural factors greatly modify reactions to frustrating stimuli. The simple frustration–aggression paradigm is seen as insufficient by Epstein and Taylor (1967): "The circumstances that arouse an individual to anger, no less than those that determine whether he will act on the anger, are culturally determined, and neither bears an invariant relationship to frustration." They also state that a fruitful area for further research would be a study of conditions and individuals to find out which of these would be determined by learned social attitudes rather than by the experience of frustration.

has been documented in accounts of festivals, etc. (e.g., Listiak, 1974). Due to the methodological nature of the studies, it has not been possible to differentiate between the normative effects as opposed to the alcohol effects. Alcohol use situations could be *defined* by the culture, e.g., as risk-taking situations testing the limits of situational opportunities, and thus there would be a main effect of the cultural definition, although interaction effects are probable. Whatever the reason, allocation of responsibility is lighter in the case of aggressiveness under the influence of alcohol in many societies, which sets the stage in the culture on subculture for more license, e.g., less severe normative sanctions, and leads to the association under study. It is probable that predisposing cultural norms exist that apply to certain subpopulations and/or are sustained by certain subcultures. It could well be that the differences found by Kalin *et al.* (1972) and Wilsnack (1974) in aggressive power concerns between the sexes in drinking situations are determined by cultural definitions of drinking situations. (A model that to some extent takes this into account was suggested earlier.) Bruun (1962) has shown that personal norms applying to drinking situations tend to correlate with actual drinking behavior, and if these norms in some cultures tend toward permissiveness or perhaps positive sanctioning of aggressive behavior, an association between alcohol use and aggression will arise within the culture.

Cultural factors pervade drinking occasions and enter causal models in many ways besides their mainly conjunctive causal significance in providing typical settings for drinking, in which situational and alcohol use variables determine the outcome. One such additional way is provided by an elaboration of results arrived at by Epstein and Taylor (1967) in their psychological experiment on determinants of aggressive behavior. They found that aggression that is considered legitimate will not lead to aggression or other forms of negative reactions as often as power and aggression that are not regarded as legitimate. (This fact has also been discussed in sociological treatises of social power.) In typical Western drinking situations, at least, there probably does not exist as much basis of legitimate justification of aggression as in nonalcohol situations. This is so because drinking is done among equals (in most relevant statuses) through social selection processes, and because participation in drinking situations is often symbolically a sign of relinquishing status differences for the occasion. Thus, drinking occasions are atypical in this respect since most interactional situations are structured as to status. Epstein and Taylor (1967) found in their study that "'might makes right,' so that when a person in a position of power attempts to act aggressively it is considered to be less determined by aggressive motivation per se than when a person with lesser power exhibits the same intent." A legitimate transfer of power from other situations into the drinking situation, which is defined culturally in a quite different way from, e.g., a work situation, is probably possible only to a limited degree. If drinking situations (sometimes peripherally) include strangers, a transfer of structuring from other situations is impossible.

For purposes of accounting for the relationship between alcohol use and violent crime through conditional or interactive variables, the prevalence of the predisposing condition is crucial. Few estimates of prevalence are available for these states, and the threshold value of many predisposing states (e.g., brain dysfunction) is unknown, so that the size of the population at risk is not known.

2. *Predispositional common cause explanations.* The common cause explanations that have been suggested in the literature refer mainly to interindividually variable, but intraindividually rather stable characteristics of individuals, and they mostly refer to the association between alcoholism and (violent) crime. (These will be discussed in the next section.) Individual characteristics have, however, also been adduced in explanations of acute alcohol use and crime: "The man who gets drunk may commit crime, but this is not because drunkenness led to crime: Personality disorder or multidetermined social breakdown may be the common factor leading both to crime and to drunkenness" (Edwards *et al.,* 1971). Predispositional factors that have been adduced as common cause factors in alcoholism and violent crime are developmental or constitutional in nature: childhood experiences, affective disorders, organic brain disorders, psychopathic personality, emotional instability, "aggressive types," etc. A third possibility is the existence in the cultural matrix of social selection processes that more or less force a person who has proclivities either toward violence or excessive alcohol use into a subculture or drinking situations where heavy drinking and/or violent behavior are culturally expected, such as in a "subculture of violence" (Wolfgang, 1967). Cultural selection and subcultural expectations would thus be the determining factor in explaining the association and not alcohol use (or alcoholism).

Generally, the common cause variables adduced are the same as those found in the literature on violence proneness and alcoholism. The increased probability of acute use is not very often distinguished from "alcoholism" or "heavy drinking" as an independent variable. For this reason I will postpone the discussion to the next section and advise the reader that wherever I talk of "alcoholism" as a dependent variable in a common cause model, it could also be read as "increased probability of acute alcohol use."

Before discussing models accounting for a statistical association between alcoholism and violent behavior, however, I shall present a model suggested by Nicol *et al.* (1973), which is one of the few sufficiently specific to warrant a discussion. A similar model has been suggested by Bach-y-Rita *et al.* (1970 and 1971). The suggested model combines predisposing and situational variables with a common cause framework. Nicol *et al.* see drinking as a frequent response to stress in a subsample of violent prisoners that they studied. The probability of violence in individuals with violent tendencies is further enhanced by the pharmacological effects of "alcohol taken under these conditions," the authors suggest. Stress is also more likely to occur in these individuals, since they have difficulties in "initiating and maintaining satisfactory interpersonal relationships." The model may be schematically illustrated with the arrow diagram, below (The possible conditional or interactive relationship between stress and alcohol use in producing interpersonal violence suggested by the phrase "alcohol taken under these conditions" has not been taken into account. There is also the possibility of a causal link between stress and interpersonal violence, but it is not explicitly suggested by Nicol *et al.*):

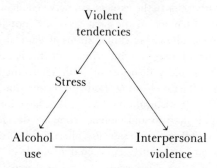

Explanation of the Positive Relationship between Alcoholism and Violent Crime

In this section I will use the label of "alcoholism' or "alcoholics" and still in the general discussion try to keep away from the possible causal relevance of labeling individuals as alcoholics and other societal reactions connected with the labeling. If labeling and other reactive processes seem to have explanatory value in accounting for the connection between alcohol use of the acute or the chronic type, such as perhaps in the rise of a criminal (or risk-taking) subculture, I will specifically mention this in the discussion. In studies of clinical alcoholic samples and their involvement in violent crime or experimental studies using clinical samples of alcoholics, the labeling aspect will probably have more explanatory value than in prison samples, which often use drinking patterns for their definitions of problem drinkers or alcoholics in prison. In prison studies that use clinical experiences and societal reactions for their definition of alcoholics, the labeling aspect will have more explanatory value.

Prolonged excessive alcohol users would be expected to have a higher probability of being under the influence of alcohol or having used alcohol immediately prior to any act. This means that the value of the null hypothesis, in testing whether the relationship between alcoholism and violent crime (or aggressive behavior in general) is higher than expected by chance, would have to be set higher in this subpopulation than in the general population. This is so because of the effects of acute alcohol use of the increased probability of aggression. If the null hypothesis thus arrived at were rejected, it would mean that prolonged excessive drinking, its conjunctive features and causal consequences (intervening variables), or common cause variables, or any combination of these would have to be used to account for the association. Acute alcohol use could also, due to the nature of drinking patterns in this population, explain away part of any excess involvement in violent crime of this population. As we have seen in the section on the nature of the independent variable, different variables in epidemiological research and statistics hidden under "alcohol use" or equivalent labels have been considered of potential explanatory value in the connection between alcohol and violence. It is possible that alcoholic drinking patterns include more of high risk features, e.g., drinking of larger amounts or beverages of high congener content (in some subpopulations of excessive drinkers). In addition to the higher probability of violence due to acute alcohol effects, there are selective factors in the epidemiological studies available in what I have called the nature of the dependent variable that can add to the overrepresentation of this type of alcohol users among violent crime offenders (the only epidemiological indicator of violent behavior available). Firstly, there is the greater risk of being apprehended by the police due to the acute effects of alcohol. Secondly, over and above this, disproportionately many alcoholic offenders are well known to the police compared to nonalcoholic offenders, which leads to a further increase in the risk of apprehension. Due to their greater risk of alcohol use, all the possible biasing factors discussed in the section on the nature of the dependent variable are relevant in any study measuring the prevalence of alcoholics in an offender population. It also seems that alcoholics have a relatively high risk of recidivism (Gibbens and Silberman, 1970), and thus perhaps a higher probability of having longer sentences that would increase their share in all crime categories in prison studies.

Although selective factors in apprehension and a greater risk of acute alcohol effects probably can explain away part of a higher-than-chance risk of violent crime in alcoholic individuals, there are a number of conjunctive features of alcoholic drinking and consequences of intake of large amounts of alcohol over long time periods (intervening variables) that seem to have explanatory value. There are also certain variables of a developmental or otherwise predisposing type that have been adduced in the explanations as common cause variables and have thus been used to explain away the seeming causal association. Prolonged excessive use of alcohol in its role as a conditional or interactive factor in causing violent behavior in connection with acute alcohol use has been discussed in the preceding section.

Direct Evidence of Main Effects of Alcoholism

I will start with the evidence for a main effect of alcoholism.* Here we assume that there is a higher probability of individuals who drink excessively to act aggressively even in nonalcohol situations or while not under the influence of alcohol than there is for nonalcoholics in equivalent situations. Evidence for a main effect is hard to come by. Jellinek (1952) mentions "marked aggressive behavior" as typical of alcoholics in the "crucial phase." There are mentions in the literature of the general irritability, aggressiveness, and feelings of hostility of alcoholics. Antons (1970), and Hassall and Foulds (1968), hqve reviewed a few of these suggestions. The latter authors replicated the findings of the two studies that they reviewed, and showed that a sample of young male alcoholics displayed more hostility than a matched control group of male hospital patients, in a projective test with items from the MMPI. On the other hand, Antons (1970) in his study did not find that alcoholic subjects were more aggressive than other Kurhaus patients on a number of measures, including a rating by treatment personnel at the institution. Mayfield and Allen (1967), in their experiment with alcoholics, severely depressed patients, and a control group of nonprofessional employees of a hospital did not find any differences in aggressiveness on the Clyde Mood Scale in the prealcohol condition of their experiment. (The uniform Kurhaus and experimental settings may have eliminated conjunctive causal factors and differences due to conditional and interactive factors present in more natural settings.)

Intervening Variables

Despite the seemingly contradictory findings of a main effect of alcoholism on aggressiveness, there are a number of effects of prolonged excessive drinking that do seem to increase the likelihood of aggressive behavior in alcoholics as compared to nonalcoholics. The factors of most immediate value in explanations based on the main effect of prolonged excessive alcohol use are the intervening variables of brain damage and especially temporal lobe dysfunction, and secondly, the formation of alcoholic subcultures. The alcoholic subcultures will be discussed later in a subsection on common cause variables, since societal reactions can be seen as a major factor in establishing an association of alcohol use and violence in these.

* "Alcoholism" will here be understood to include the long-term stable consequences of prolonged alcohol use on the physical, psychological, behavioral, and social state of the individual, and not defined exclusively as an underlying pathology or as a symptom of such a pathology.

The prevalence of epileptic brain dysfunctions caused by long-term use of alcohol in the offender population in crimes of violence has not been established, although it is generally recognized that epileptiform changes occur after prolonged excessive use of alcohol (Giove, 1964; Della Rovere and Falli, 1965; Bacher et al., 1960; Bonetti, 1962). Giove (1964) estimates that epileptic crises occur in 2 percent of alcoholics after about ten years of excessive drinking. He suggests that these attacks may be caused by the acute drinking episodes due to the sudden interruption of the depressive effects of alcohol. (This would show the interaction of this alcoholic condition with acute alcohol use.) He also mentions that (and this is more relevant in this context) cerebral lesions or atrophies caused by excessive use of alcohol may cause epileptic crises. Della Rovere and Falli (1965) also attribute part of the clinical cases of alcoholic epilepsy to irreversible change of the nervous tissue. The etiological significance of excessive alcohol use in the development of epilepsy at a comparatively late age, has also been confirmed by Bacher et al. (1960) and by Bonetti (1962) in an extensive review of the literature. Sundby (1967) suggests brain trauma as a neurological complication of excessive alcohol use responsible for epilepsy. He also found that the mortality from epilepsy was seven- to tenfold for alcoholics as compared to the general population. In the study by Wilkinson et al. (1971), 8.2 percent of the alcoholics in a large clinical sample (sample size 825) were diagnosed as having epilepsy. There is disagreement on the probability and reversibility of brain damage in alcoholic populations (e.g., Bennett, 1967; Henry, 1970). Whether it is reversible or not, however, this condition has potential explanatory value. Accepting that alcoholics have a greater probability of temporal lobe dysfunctions and of this condition leading to an increased probability of violent behavior, we could conclude that prolonged excessive drinking could explain part of any excessive violence in alcoholics even outside of alcohol use situations. In accounting for the higher prevalence of violent crime among alcoholics, however, a prevalence rate of 2 percent among alcoholics (Giove, 1964) probably is not sufficient to explain much, whereas the finding by Wilkinson et al. (1971) of 8.2 percent naturally is of higher explanatory significance. Also of relevance are the findings by Wilkinson et al. of 7.6 percent of clinical alcoholics having a chronic brain syndrome and 6.3 percent suffering from acute confusional states. (There is overlap in these percentages.) The predispositional relevance of epileptoid brain dysfunction in acute alcohol use has been dealt with in an earlier section.

One less chronic intervening factor in a possible explanatory model of increased probability of violence in alcoholics is hypoglycemia. The hypoglycemia-inducing property of alcohol has been known for at least three decades (Herman et al., 1970). There is also general agreement that alcohol does not produce hypoglycemia in individuals who have not been fasting or are not undernourished (e.g., Moynihan, 1965). The estimates of periods of fasting needed before normal subjects develop hypoglycemia range from 42 hours (Field and Williams, 1962) to 72 hours (Freinkel et al., 1962) or longer (see Moynihan, 1965). (In Volume III of this work, Hillman (1974) has reviewed the existing literature on the nutritional habits of alcoholics.) Vartia et al. (1960) in briefly reviewing the literature note that low blood sugar values are common in alcoholics even when they have not been drinking and suggest a metabolic disturbance in the carbohydrate metabolism. Hypoglycemia on its part can cause aggressive behavior (Pawar, 1972; Moyer, 1971). If more or less permanent metabolic changes have occurred in alcoholics, this would again mean that prolonged excessive drinking would have a main effect on aggressive behavior independent of acute alcoholic drinking episodes through this metabolic disturbance.*

There does not seem to be any information on the prevalence of hypoglycemia among alcoholics, although Herman et al. (1970) point out its rarity. Lacking this information, it is difficult to arrive at a population at risk, and thus to make any estimates of the relevance of hypoglycemia and malnutrition as intervening or conjunctive variables in accounting for violent behavior in connection with alcohol use and alcoholism. Insofar as hypoglycemic conditions occur comparatively frequently, it remains a task for epidemiological crosscultural research to chart the drinking patterns and relevant conjunctive patterns of nutrition in order to explain different risks of aggressive behavior in alcohol use situations, and thus explain differences in prevalences of violent behavior connected with alcohol. Variables of central concern should be amounts consumed at drinking occasions, the length of the occasions, and eating patterns in connection with the use of alcohol. Possible interaction effects of hypoglycemia with acute alcohol effects in producing aggressive behavior do not seem to be discussed in the literature.

A further conjunctive feature of alcoholic drinking (or a definitional feature, if drinking patterns are included in the definition of alcoholism; for our analysis the difference in nonconsequential), is the extension of drinking over lengthy periods of time, which induces the intervening variable of REM-sleep deprivation. This variable again can, via its main effect (and possible interactions with alcohol use), lead to near psychotic reactions and violence. It has been suggested that the effect of sleep deprivation could be heightened in combination with pathological phenomena (Oswald, 1962; Gove, 1969-70). The mental and physical health status of the population at risk in long drinking bouts is comparatively low. The effects of sleep deprivation may be profound in

---

* If alcoholics have a stronger predisposition to become hypoglycemic after ingestion of alcohol than do nonalcoholics (which Moynihan, 1965, seems to suggest), it could explain the predisposing role of alcoholism toward aggression in alcohol use situations, which was discussed in a previous section.

this population, perhaps of the same magnitude as in the experiments on subjects with histories of mental problems. These have shown extreme psychological effects.

To the extent that alcoholics are more likely to exhibit any of the characteristics discussed in connection with acute alcohol use as potentially increasing the probability of violence, over and above and acute alcohol effects, these characteristics are relevant for explanatory models of alcoholism and violent crime. This could be the case with, e.g., perceptual and cognitive attributes that were discussed in connection with a model of the escalatory process above. Gliedman (1956) and others have elaborated on the here-and-now character of alcoholic behavior, and experiments have shown a deterioration in abstract reasoning among alcoholics (e.g., Long and McLachlan, 1974). Some evidence of sudden mood fluctuations has been presented by Antons (1970) in referring to a number of other researchers. These factors all speak for a more random situational determination of behavior and a greater likelihood of escalation in alcoholics both in alcohol use situations and in other situations.

Common Cause Models

Some evidence points toward predispositional factors of a genetic or developmental nature that could increase both the probability of prolonged excessive drinking and aggressive reaction patterns. Robins *et al.* (1962), in their 30-year longitudinal study of children seen in a child guidance clinic in St. Louis, conclude:

> Not only is the occurrence of alcoholism highly related to evidences in childhood of pathology in the subjects, and their parents, but the kind of pathology related to alcoholism can best be described as antisocial rather than neurotic behavior.

The antisocial behavior pattern in childhood can in this model (also suggested by Tinklenberg, 1973) be viewed as a symptom of genetically determined violent behavior tendencies that continue into adulthood and/or as an initial behavior pattern that, through societal and interindividual reactions, is stabilized over the lifetime of the individual. The background in both alcoholism and much of criminal behavior is often one of disturbed relationships with parents and generally poor home backgrounds (Glatt, 1965).

Guze and co-workers (1962), in a study of convicted male criminals, found that 51 percent of the alcoholics reported frequent fighting leading to trouble before the age of eighteen, as compared to 32 percent of the nonalcoholics. This finding was replicated by Guze *et al.* (1968), in another study that showed that 70 percent of the alcoholics and 40 percent of nonalcoholics had an early history of excessive fighting. Assuming that alcoholic drinking patterns were not established at a very early age and thus that the fighting behavior was independent of (excessive) alcohol use, alcoholism and violent behavior can be seen as symptoms of a common developmental or genetic cause. There is further evidence from McCord and McCord's study (1962). They found that alcoholics were significantly more likely to have exhibited unrestrained aggression as boys than were the nonalcoholics in their study, 36 and 12 percent respectively. Here one should take notice of the possibility of social class differences in the backgrounds of the alcoholic and the nonalcoholic sample, which would tend to cluster both alcoholism and aggression into the same maladjustive subcultural syndrome.

Like the McCords (1962), Hagnell et al. (1973) in Sweden found an aggressive ("dangerous") subgroup of alcoholics. They suggest that the premorbid personality determines the type of alcoholic personality that emerges. De Vito et al. (1970) found that 58 percent of their sample of 300 male alcoholics belonged to an "acting-out-prone" group who drank alcohol to facilitate acting-out behavior. Wexberg (1951) presents evidence that alcoholics generally have low frustration tolerance even before their addiction. Environmental factors then determine the symptom that develops from this low tolerance. It is possible from this reasoning to see low frustration level in the dual role of causing both drinking and violent behavior in the same individuals, and as a predispositional factor that in the drinking situation, combined with the effects of alcohol, further increases the probability of aggressive behavior. Flemenbaum (1974) views affective disorders of a biological origin in the same explanatory role as Wexberg sees low frustration tolerance. (The two variables may well fit into the same explanatory model, since affective disorders are characterized by low frustration tolerance.) Affective disorders may, according to Flemenbaum, manifest themselves as alcoholism or as, e.g., antisocial acts and delinquency:

> Socio-economical class, culture, sex, age, and individual experiential factors are very likely to determine two or more "end products" or clinical manifestations of disorders with a common biological background.

Correlational evidence of an association between psychopathic personality patterns and alcoholism is evident from a number of studies (Hoff and Kryspin-Exner, 1962; Cloninger and Guze, 1970; Mader, 1972; Hagnell et al., 1973). This type of data could, however, fit any causal model.

One can, in explaining away (strictly defined) alcoholism as a causally relevant factor in violence, assume that the concomitant social factors would be influential as determinants if there is an excess of violent behavior in general in the subpopulations to which alcoholics belong. So, one would expect Skid Row type alcoholics to committ proportionately more violent crimes than the general population because the predictive factors of low income, physical deterioration, low levels of education and vocational skills, etc., characterize this subgroup of the alcoholic population. The clustering of individuals with negative attributes (from the society's point of view) into subcultures is to a great extent due to societal policies and interpersonal reactions toward these individuals. Thus, if individuals with alcohol problems and others with violent behavior tendencies (and individuals exhibiting both) enter the same subculture, one can assume an association to occur also on the individual level. This may, among other factors, occur due to the behavior norms established in such a subculture to accommodate the attributes of the members, or it may be a consequence of the greater number of frustrating stimuli encountered in such a negatively defined subculture.

In summary, we can conclude that the factors that can explain the higher prevalence of violent crime among alcoholics than among nonalcoholics are the following:

1. Alcoholics may have a greater risk of being apprehended by the police, both because of a greater risk of acute alcohol effects at any time and because of their status as alcoholics and recidivists, which generally makes them better known to the police than nonalcoholics.

2. Due to the higher risk of acute use of alcohol at any time, alcoholics are at a higher risk of displaying violent behavior, whatever the appropriate causal models of a situational nature.

3. Prolonged excessive alcohol use may be connected with predispositional attributes that increase the probability of aggressive behavior in connection with acute alcohol use. We have seen some evidence to this effect in a previous section on predisposing factors.

4. Prolonged excessive alcohol use may also give rise to predispositional changes in the individual, which outside of any alcohol use situations, increase the probability of violent behavior. An example is brain damage, especially of an epileptiform character.

5. Prolonged excessive alcohol use many be conjunctively connected with alcohol use patterns that also in nonalcoholics may give rise to states of the organism that increase the likelihood of aggressive behavior. Poor nutritional habits when drinking may lead to hypoglycemia and binge drinking may lead to REM-sleep deprivation. Both conditions by themselves increase the likelihood of violent behavior. In addition, there may be an interaction effect with alcohol use.

6. "Alcohol use" may indicate different variables for alcoholics and nonalcoholics, so that alcoholics (at least a certain subgroup of alcoholics) display alcohol use, such as use of drinks of high alcohol or congener content, that could show a comparatively strong relationship to violent behavior.

7. Prolonged excessive users of alcohol may, due to developmental or genetic factors, belong to a subpopulation that through a common cause, such as early childhood experiences or affective disorder, show a higher probability of antisocial behavior and among these, violent behavior.

8. A large proportion of excessive alcohol users are subjected to societal and interpersonal reactions. Some of these may force them into subcultures where violent behavior is condoned, expected, or technically necessary for functioning within the subculture, and in relating to the larger cultural matrix.

It has been suggested above that the greater frequency of acute alcohol use can explain part of the association between alcoholism and violent crime. The reverse is true also. The clustering of alcohol use occasions to alcoholic individuals can, through the causal effects discussed in this chapter, explain part of the statistical association between alcohol use and violent crime.

ACKNOWLEDGMENTS

The author gratefully acknowledges the invaluable assistance of Ms. Frances Tolnai throughout the entire preparation of this chapter.

REFERENCES

Aho, T., 1967, Alkoholi ja aggressiivinen käyttäytyminen, *Alkoholipolitiikka* 32(4):179.
Aleksic, Z. L., and Radovanovic, D., 1967, Arson committed by alcoholics in 1964 in the YFSR, *An. Bolnice Stojanovic* 6:298.
Amir, M., 1967, Alcohol and forcible rape, *Brit. J. Addict.* 62:219.
Ando, H., and Hasegawa, E., 1969, Drinking patterns and attitudes of alcoholics and nonalcoholics in Japan, *Quart. J. Stud. Alc.* 30:987.
Antons, K., 1970, Empirische Ergebnisse zur Aggressivität von Alkoholkranken, *Brit. J. Addict.* 65:263.
Arner, O., 1973, The role of alcohol in fatal accidents among seamen, *Brit. J. Addict.* 68:185.
Aromaa, K., 1974, Vaarallisia suhteita, *Alkoholipolitiikka* 39:70.
Bacher, F., Chanoit, P., Rouquette, J., Verdeaux, G., and Verdeaux, J., 1960, Epilepsy and alcoholism. Comparative statistical study of EEG characteristics, *Rev. Neurol.* 103:228.
Bach-y-Rita, G., and Veno, A., 1974, Habitual violence: A profile of 62 men, *Amer. J. Psychiat.* 131:9.
Bach-y-Rita, G., Lion, J. R., and Ervin, F. R., 1970, Pathological intoxication: Clinical and electroencephalographic studies, *Amer. J. Psychiat.* 127 (5):698.
Bach-y-Rita, G., Lion, J. R., Climent, C. E., and Ervin, F. R., 1971, Episodic dyscontrol: A study of 130 violent patients, *Amer. J. Psychiat.* 127(11):49.
Baker, D., Telfer, M. A., Richardson, C. E., and Clark, G. R., 1970, Chromosome errors in men with antisocial behavior, *J. Amer. Med. Assoc.* 214(5):869.
Baker, J. L., 1959, Indians, alcohol and homicide, *J. Social Ther.* 5:270 (CAAAL 9757).

Banay, R. D., 1941–42, Alcoholism and crime, *Quart. J. Stud. Alc.* 2:686.
Bard, M., and Zacker, J., 1974, Assaultiveness and alcohol use in family disputes, *Criminology* 12(3):281.
Bartholomew, A. A., and Kelley, M. F., 1965, The incidence of a criminal record in 1000 consecutive "alcoholics," *Brit. J. Crimin.* 5:143.
Bennett, A. E., 1967, Treatment of brain damage of alcoholism, *Curr. Psychiat. Therap.* 7:142.
Bennett, R. M., Buss, A. H., and Carpenter, J. A., 1969, Alcohol and human physical aggression, *Quart. J. Stud. Alc.* 30:870.
Berger, R. J., and Oswald, I., 1962, Effects of sleep deprivation on behavior, subsequent sleep, and dreaming, *J. Ment. Sci.* 108:457.
Berkowitz, L., 1970, Experimental investigations of hostility catharsis, *J. Consult. Clin. Psychol.* 35:1.
Binns, J. K., Carlisle, J. M., Nimmo, D. H., Park, R. H., and Todd, N. A. 1969a, Remanded in hospital for psychiatric examination, *Brit. J. Psychiat.* 115:1125.
Binns, J. K., Carlisle, J. M., Nimmo, D. H., Park, R. H., and Todd, N. A. 1969b, Remanded in custody for psychiatric examination, *Brit. J. Psychiat.* 115:1133.
Birrell, J. H. W., 1965, Blood alcohol levels in drunk drivers, drunk and disorderly subjects, and moderate social drinkers, *Med. J. Aust.* 2:949.
Blum, R., 1967, Mind altering drugs and dangerous drugs: Alcohol, *in* "U.S. President's Commission on Law Enforcement and Administration of Justice Task Force Report: Drunkenness", Washington D. C., U.S. Government Printing Office.
Blumer, D., 1973, Neuropsychiatric aspects of violent behavior, *in* "Proceedings of the National Symposium on Medical Sciences and the Criminal Law", pp. 73–87, Centre of Criminology, University of Toronto, Toronto.
Bonetti, U., 1962, On so-called "alcoholic epilepsy"; a review of the literature and clinicoelectroencephalographic case contribution, *Rass. Studi psichiat.* 51:477.
Bowden, K. M., Wilson, D. W., and Turner, L. K., 1958, A survey of blood alcohol testing in Victoria (1951 to 1956), *Med. J. Aust.* 45(2):13 (CAAAL 8865).
Boyatzis, R. E., 1974, The effect of alcohol consumption on the aggressive behavior of men, *Quart. J. Stud. Alc.* 35:959.
Brill, H., 1970, Drugs and aggression, *Ment. Hlth. Dig.* 2:11.
Bruun, K., 1959, "Drinking Behaviour in Small Groups", Finnish Foundation for Alcohol Studies, Vol. 9., Helsinki.
Bruun, K., 1962, The significance of roles and norms in the small group for individual behavioural changes while drinking, *in* "Society, Culture and Drinking Patterns" (D. J. Pittman and C. R. Snyder, eds.), pp. 293–303, John Wiley & Sons, Ltd., New York.
Buss, A. H., 1963, Physical aggression in relation to different frustrations, *J. Abnorm. Soc. Psychol.* 67(1):1.
California Public Health Dept., Alcoholic Rehabilitation Division, 1964, "Alcoholism and California: A Primary Analysis."
Campbell, D. T., and Stanley, J. C., 1963, "Experimental and Quasi-Experimental Designs for Research," Rand McNally & Co., Chicago.
Carpenter, J. A., and Armenti, N. P., 1972, Some effects of ethanol in human sexual and aggressive behavior, *in* "The Biology of Alcoholism" Vol. 2 (B. Kissin and H. Begleiter, eds.), pp. 509–543, Plenum Press, New York.
Child, I. L., Bacon, M. K., and Barry III, H., 1965a, Descriptive measurements of drinking customs, *Quart. J. Stud. Alc.* Suppl. No. 3:1.
Child, I. L., Barry III, H., and Bacon, M. K., 1965b, Sex differences, *Quart. J. Stud. Alc.* Suppl. No. 3:49.

Cleveland, F. P., 1955, Problems in homicide investigation IV: The relationship of alcohol to homicide, *Cincinnati J. Med.* 36:28.

Climent, C. E., Rollins, A., Ervin, F. R., and Plutchic, R., 1973, Epidemiological studies of women prisoners, *Amer. J. Psychiat.* 130:9.

Cloninger, C. R., and Guze, S. B., 1970, Psychiatric illness and female criminality: The role of sociopathy and hysteria in the antisocial woman, *Amer. J. Psychiat.* 127:303.

Cohen, J., Dearnaley, E. J., and Hansel, C. E. M., 1958, The risk taken in driving under the influence of alcohol, *Brit. Med. J.* 1:1438.

Connor, W. D., 1973, Criminal homicide, USSR/USA: Reflections on Soviet data in a comparative framework, *J. Criminal Law, Crimin. and Pol. Sci.* 64(1):111.

Cuthbert, T. M., 1970, A portfolio of murders, *Brit. J. Psychiat.* 116:1.

DellaRovere, M., and Falli, S., 1965, Contribution to the study of alcoholic epilepsy, *Lav. neuropsichiat.* 37:461.

Dement, W. C., 1960, The effect of dream deprivation, *Science* 131:1705.

DeVito, R. A., Flaherty, L. A., and Mozdzierz, G. J., 1970, Toward a psychodynamic theory of alcoholism, *Dis. Nerv. Sys.* 31:43.

Docter, R. F., and Bernal, M. E., 1964, Immediate and prolonged psychophysiological effects of sustained alcohol intake in alcoholics, *Quart. J. Stud. Alc.* 25(3):438.

Edwards, G., Hensman, C., and Peto, J., 1971, Drinking problems among recidivist prisoners, *Psychol. Med.* 1(5):388.

Edwards, G., Gattoni, F., and Hensman, C., 1972, Correlates of alcohol-dependence scores in a prison population, *Quart. J. Stud. Alc.* 33:417.

Eleftheriou, B. E., and Scott, J. P., 1971, "The Physiology of Aggression and Defeat," Plenum Press, New York.

Epstein, S., and Taylor, S. P., 1967, Instigation of aggression as a function of degree of defeat and perceived aggressive intent of the opponent, *J. Personality* 35:265.

Field, J. B., and Williams, H. E., 1962, Studies on the mechanism of ethanol-induced hypoglycemia, *J. Clin. Invest.* 41:1357.

Fisher, C., and Dement, W. C., 1963, Studies on the psychopathology of sleep and dreams, *Amer. J. Psychiat.* 119:1160.

Fitzpatrick, J. P., 1974, Drugs, alcohol and violent crime, *Addict. Dis.* 1(3):353.

Flemenbaum, A., 1974, Affective disorders and "chemical dependence": Lithium for alcohol and drug addiction? *Dis. Nerv. Sys.* 35(6):281.

Freinkel, N., Singer, D. L., Silbert, C. K., and Anderson, J. B., 1962, "Studies on the Pathogenesis and Clinical Features of "Alcoholic Hypoglycemia," Boston, Mass.

Galtung, J., 1967, "Theory and Methods of Social Research," Universitetsforlaget, Oslo.

Gelfand, M., 1971, The extent of alcohol consumption by Africans; the significance of the weapon at beer drinks, *J. Forensic Med.* 18:53.

Gibbens, T. C. N., and Silberman, M., 1970, Alcoholism among prisoners, *Psychol. Med.* 1:73.

Gibbons, D. C., 1971–72, Observations on the study of crime causation, *Amer. J. Soc.* 77(2):262.

Gillies, H., 1965, Murder in the west of Scotland, *Brit. J. Psychiat.* 111:1087.

Giove, C., 1964, Alcoholic epilepsy, *Osped. psichiat.* 32:195.

Glatt, M. M., 1965, Crime, alcohol, and alcoholism, *Howard J. Penol.* 11:274.

Gliedman, L. H., 1956, Temporal orientation and alcoholism, *Alcoholism* 3(3):11.

Goffman, E., 1967, "Interaction Ritual," Doubleday Anchor, New York.

Goodwin, D. W., 1973, Alcohol in suicide and homicide, *Quart. J. Stud. Alc.* 34:144.

Goodwin, D. W., Crane, J. B., and Guze, S. B., 1971, Felons who drink, *Quart. J. Stud. Alc.* 32:136.

Gove, W. R., 1969:70, Sleep deprivation: A cause of psychotic disorganization, *Amer. J. Sociol.* 75:782.

Greenberg, R., and Pearlman, C., 1967, Delirium tremens and dreaming, *Amer. J. Psychiat.* 124:133.
Greenblatt, M., Levin, S., and Dicori, F., 1944, The electroencephalogram associated with chronic alcoholism, alcoholic psychosis and alcoholic convulsions, *Arch. Neurol. Psychiat.* 52:290.
Greenland, C., 1971, Evaluation of violence and dangerous behavior associated with mental illness, *Sem. Psychiat.* 3(3):345.
Greenland, C., 1973, Research strategies in the evaluation of violent and dangerous behavior, *Social Worker* 41(1):4.
Gresham, S., Webb, W., and Williams, R., 1963, Alcohol and caffeine: Effect on inferred visual dreaming, *Science,* 140:1226.
Grislain, J. R., Mainard, R., Berranger, P. de, Ferron, C. de, and Brelet, G., 1968, Child abuse. Social and legal problems, *Ann. Pediat.* 15:440 (CAAAL 13588).
Gross, M. M., and Goodenough, D. R., 1968, Sleep disturbances in the acute alcoholic psychosis, in "Clinical Research in Alcoholism," (J. O. Cole, ed.), Psychiatric Research Report 24, pp. 132–147, American Psychiatric Association, Washington D.C.
Guze, S. B., Tuason, V. B., Gatfield, P. D., Stewart, M. A., and Picken, B., 1962, Psychiatric illness and crime with particular reference to alcoholism: A study of 223 criminals, *J. Nerv. Ment. Dis.* 134(6):512.
Guze, S. B., Wolfgram, E. D., McKinney, J. K., and Cantwell, D. P., 1968, Delinquency, social maladjustment, and crime: the role of alcoholism, *Dis. Nerv. Sys.* 29:238.
Hagnell, O., Nyman, E., and Tunving, K., 1973, Dangerous alcoholics, *Scand. J. Soc. Med.* 1(3):125.
Hamburg, D. A., 1971, Recent research on hormonal factors relevant to human aggressiveness, *Int. Soc. Sci. J.* 23(1):36.
Hartocollis, P., 1962, Drunkenness and suggestion: An experiment with intravenous alcohol, *Quart. J. Stud. Alc.* 23:376.
Hassall, C., and Foulds, G. A., 1968, Hostility among young alcoholics, *Brit. J. Addict.* 63:203.
Heath, D. B., 1958, Drinking patterns of the Bolivian Camba, *Quart. J. Stud. Alc.* 19:491.
Henry, C. E., 1970, The (limited) usefulness of EEG in alcoholism, *Ohio State Med. J.* 66:806.
Herman, J., Sekso, M., Trinajstic, M., Vidovic, V. and Cabrijan, T., 1970, Hypoglycemic conditions in the course of chronic alcoholism, *Alcoholism* (Zagreb) 6(2):87.
Hetherington, E. M., and Wray, N. P., 1964, Aggression, need for social approval, and humor preferences, *J. Abnorm. Soc. Psychol.* 68(6):685.
Hillman, R. W., 1974, Alcoholism and malnutrition, in "The Biology of Alcoholism" Vol. 3, (B. Kissin and H. Begleiter, eds.), Plenum Press, New York.
Hoff, H., and Kryspin-Exner, K., 1962, Persönlichkeit und Verhalten des alkoholisierten Verkehrsteilnehmers, *Blutalkohol* 1:323.
Hopwood, J. S., and Milner, K. O., 1940, Some observations on the relation of alcohol to criminal insanity, *Brit. J. Inebr.* 38(2):51.
Hurst, P. M., and Bagley, S. K., 1972, Acute adaptation to the effects of alcohol, *Quart. J. Stud. Alc.* 33:358.
Husson, F., Robert, P., and Godefroy, T., 1973, Alcoolisme et cout de crime en France dans les années 1970–1971, *Rev. Alcsme* 19(1):10.
Irgens-Jensen, O., 1971, "Problem Drinking and Personality," Universitetsforlaget, Oslo.
Janowska, H., 1970, Alkoholizam sprawców zabójstw, *Probl. Alkoholizmu* 2(185):8.
Jellinek, E. M., 1952, Phases of alcohol addiction, *Quart. J. Stud. Alc.* 13:673.
Jellinek, E. M., and McFarland, R. A., 1940, Analysis of psychological experiments on the effects of alcohol, *Quart. J. Stud. Alc.* 1:272.
Johnson, L. C., Burdick, J. A., and Smith, J., 1970, Sleep during alcohol intake and withdrawal in the chronic alcoholic, *Arch. Gen. Psychiat.* 22:406.

Julius, D., and Bohacek, N., 1954, The alcoholic psychoses in forensic psychiatry, *Neuropsichiat.* (Zagreb) 1:21 (CAAAL 7053).

Kalant, H., 1961, The pharmacology of alcohol intoxication, *Quart. J. Stud. Alc.* Suppl. No. 1:1.

Kalin, R., 1972, Social drinking in different settings, *in* "The Drinking Man" (D. C. McClelland, W. N. Davis, R. Kalin, and E. Wanner, eds.), pp. 21–44, The Free Press, New York.

Kalin, R., McClelland, D. C., and Kahn, M., 1972, The effects of male social drinking on fantasy, *in* "The Drinking Man" (D. C. McClelland, W. N. Davis, R. Kalin, and E. Wanner, eds.), pp. 3–20, The Free Press, New York.

Kastl, A. J., 1969, Changes in ego functioning under alcohol, *Quart. J. Stud. Alc.* 30:371.

Katkin, E. S., Hayes, W. N., Teger, A. I., and Pruitt, D. G., 1970, Effects of alcoholic beverages differing in congener content on psychomotor tasks and risk taking, *Quart. J. Stud. Alc.* Suppl. No. 5.

Kinberg, O., Inghe, G., and Lindberg, T., 1957, "Kriminalitet och alkoholmissbruk," Institutet för Maltdrycksforskning, Publ. No. 9, Stockholm.

Knowles, J. B., Laverty, S. G., and Kuechler, H. A., 1968, Effects of alcohol on REM sleep, *Quart. J. Stud. Alc.* 29:342.

Kollar, E. J., Pasnau, R. D., Rubin, R. T., Naitoh, P., Slater, G. G., and Kales, A., 1969, Psychological, psychophysiological, and biochemical correlates of prolonged sleep deprivation, *Amer. J. Psychiat.* 126(4):488.

Krokfors, K. J., 1970, Alkoholsituationen i Finland, *Alkoholfrågan* 64:297.

Kuusi, P., 1948, "Suomen viinapulma gallup-tutkimuksen valossa," Otava, Helsinki.

Kuusi, P., 1956, "Alcohol Sales Experiment in Rural Finland," Finnish Foundation for Alcohol Studies, Vol. 3a, Helsinki.

Lanzkron, J., 1962–63, Murder and insanity: A survey, *Amer. J. Psychiat.* 119:754.

Lemert, E. M., 1967, "Human Deviance, Social Problems, and Social Control," Prentice-Hall, Englewood Cliffs, New Jersey.

LeRoux, L. C., and Smith, L. S., 1964, Violent deaths and alcoholic intoxication, *J. Forensic Med.* 11(4):131.

Linn, L., and Stein, M. H., 1944, Acute alcoholic furor, *Med. Bull. of the North African Theatre of Operations* 2:81.

Listiak, A., 1974, "Legitimate deviance" and social class: Bar behavior during Grey Cup week, *Sociol. Focus* 7(3):13.

Long, J. A., and McLachlan, J. F. C., 1974, Abstract reasoning and perceptual-motor efficiency in alcoholics, *Quart. J. Stud. Alc.* 35:1220–1229.

MacAndrew, C., and Edgerton, R. B., 1969, "Drunken Comportment," Aldine Publ. Co., Chicago.

Macdonald, J. M., 1961, "The Murderer and His Victim," Charles C Thomas, Illinois.

MacDonnell, M. F., and Ehmer, M., 1969, Some effects of ethanol on aggressive behavior in cats, *Quart. J. Stud. Alc.* 30:312.

MacKay, J. R., 1963, Problem drinking among juvenile delinquents, *Crime and Delinquency* 9(1):29.

MacKay, J. R., Phillips, D. L., and Bryce, E. O., 1967, Drinking behavior among teen-agers: A comparison of institutionalized and non-institutionalized youth, *J. Hlth Soc. Beh.* 8:46.

Mader, R., 1972, Alkoholismus bei kriminellen Jugendlichen, *Acta Paedopsychiatr.* 39:2.

Madsen, W., and Madsen, C., 1969, The cultural structure of Mexican drinking behavior, *Quart. J. Stud. Alc.* 30:701.

Marinacci, A. A., 1963, A special type of temporal lobe (psychomotor) seizures following ingestion of alcohol, *Bull. Los Angeles Neurol. Soc.* 28(4):241.

Maule, H. G., and Cooper, J., 1966, Alcoholism and crime: A study of the drinking and criminal habits of 50 discharged prisoners, *Brit. J. Addict.* 61:201 (CAAAL 11716).

Mayfield, D., 1972, Alcoholism, alcohol intoxication and assaultive behavior, *in* "30th International Congress on Alcoholism and Drug Dependence," Amsterdam.

Mayfield, D., and Allen, D., 1967, Alcohol and affect: A psychopharmacological study, *Amer. J. Psychiat.* 123:1346.

McCaghy, C. H., 1968, Drinking and deviance disavowal; the case of child molesters, *Social Probl.* 16:43 (CAAAL 13321).

McClelland, D. C., and Davis, W. N., 1972, The influence of unrestrained power concerns on drinking in working-class men, *in* "The Drinking Man" (D. C. McClelland, W. N., Davis, R. Kalin, and E. Wanner, eds.), pp. 142–161, The Free Press, New York.

McClelland, D. C., and Wilsnack, S. C., 1972, The effects of drinking on thoughts about power and restraint, *in* "The Drinking Man" (D. C. McClelland, W. N., Davis, R. Kalin, and E. Wanner, eds.), pp. 123–141, The Free Press, New York.

McClelland, D. C., Davis, W. N., Kalin, R., and Wanner, E., (eds.) 1972, "The Drinking Man," The Free Press, New York.

McCord, W., and McCord, J., 1962, A longitudinal study of the personality of alcoholics, *in* "Society, Culture and Drinking Patterns" (D. J. Pittman and C. R. Snyder, eds.), John Wiley & Sons, Ltd., New York.

McGeorge, J., 1963, Alcohol and crime, *Med. Sci. Law* 3:27.

McKnight, C. K., Mohr, J. W., Quinsey, R. E., and Erochko, J., 1964, Mental illness and homicide, *in* "4th Research Conference on Delinquency and Criminology," Quebec Society of Criminology, Quebec.

McNamee, H. B., Mello, N. K., and Mendelson, J. H., 1968, Experimental analysis of drinking patterns in alcoholics: concurrent psychiatric observations, *Amer. J. Psychiat.* 124(8):1063.

Medina, E. L., 1970, The role of the alcoholic in accidents and violence, *in* "Alcohol and Alcoholism" (R. E. Popham, ed.), pp. 350–355, University of Toronto Press, Toronto.

Melges, F. T., and Harris, R. F., 1970, Anger and attack—a cybernetic model of violence, *in* "Violence and the Struggle for Existence" (D. N. Daniels, M. F. Gilula, and F. M. Ochberg, eds.), pp. 97–127, Little, Brown, Boston, Mass.

Mendelson, J. H., La Dou, J., and Solomon, P., 1964, Experimentally induced chronic intoxication and withdrawal in alcoholics: Part III, psychiatric findings, *Quart. J. Stud. Alc.* Suppl. No. 2:40.

Moskowitz, H., and DePry, D., 1968, Differential effect of alcohol on auditory vigilance and divided-attention tasks, *Quart. J. Stud. Alc.* 29:54.

Moyer, K. E., 1971, The physiology of aggression and the implications for aggression control, *in* "The Physiology of Hostility" (K. E. Moyer, ed.), pp. 81–101, Markham, Chicago.

Moynihan, N. H., 1965, Alcohol and blood sugar, *Alcoholism* 1(2):180.

Mundy-Castle, A. C., 1957, The EEG in 22 cases of murder or attempted murder, *Dig. Neurol. Psychiat.* 25:93.

Nathan, P. E., Titler, N. A., Lowenstein, L. M., Solomon, P., and Rossi, A. M., 1970, Behavioral analysis of chronic alcoholism, *Arch. Gen. Psychiat.* 22:419.

Nau, E., 1967, Child abuse, *Mschr. Kinderheilk.* 115:192.

Nicol, A. R., Genn, J. C., Gristwood, J., Foggitt, R. H., and Watson, J. P., 1973, The relationship of alcoholism to violent behavior resulting in long-term imprisonment, *Brit. J. Psychiat.* 123:47.

Orcutt, J. D., 1972, Toward a sociological theory of drug effects: A comparison of marijuana and alcohol, *Sociol. Soc. Res.* 56(2):242.

Oswald, I., 1962, "Sleeping and Waking," American Elsevier, New York.

Partington, J. T., and Johnson, F. G., 1969, Personality types among alcoholics, *Quart. J. Stud. Alc.* 30(1):21.

Pastore, N., 1952, The role of arbitrariness in the frustration–aggression hypothesis, *J. Abnorm. Soc. Psychol.* 47:728.
Pawar, P. B., 1972, Alcohol and hypoglycemia. *J. Alcoholism* 7:26.
Pernanen, K., 1965, Finnish Drinking Habits, Finnish Foundation for Alcohol Studies, Helsinki.
Pernanen, K., 1974, Validity of survey data on alcohol use, *in* "Research Advances in Alcohol and Drug Problems" Vol. I (R. J. Gibbins, Y. Israel, H. Kalant, R. E. Popham, W. Schmidt, and R. G. Smart, eds.), John Wiley & Sons, Inc., New York.
Persky, H., Smith, K. D., and Basu, G. K., 1971, Relation of psychologic measures of aggression and hostility to testosterone production in man, *Psychosom. Med.* 33(3):265.
Pincock, T. A., 1962, The frequency of alcoholism among self-referred persons and those referred by the courts for psychiatric examination, *Can. Med. Assoc. J.* 87:282.
Pionkowski, J., 1965, Influence of alcoholism on the delinquency of subjects with mental disorders, *Neurol. Neuroch. Psychiat. polska* 15:875 (CAAAL 11723).
Pittman, D. J., and Handy, W., 1964, Patterns in criminal aggravated assault, *J. Crim. Law Crimin. Pol. Sci.* 55:462.
Pittman, D. J., and van der Wal, H. J., 1968, Alcoholism and the legal system in the Netherlands, *Brit. J. Addict.* 63:209.
Pokorny, A. D., 1965, A comparison of homicides in two cities, *Research Reports* 56:479.
Polten, E. P., 1973, "Critique of the Psycho-Physical Identity Theory," Mouton & Co., The Hague.
Puchowski, B., and Tulaczynski, M., 1964, The toxic effect of alcohol, based on analyses of sudden deaths in Lodz, *Arch. Med. Sadowej* 14:83 (CAAAL 10692).
Rada, R. T., 1975, Alcoholism and forcible rape, *Amer. J. Psychiat.* 132:444.
Ratcliffe, S. G., Melville, M. M., Stewart, A. L., Jacobs, P. A., and Keay, A. J., 1970, Chromosome studies on 3,500 newborn male infants, *Lancet* 1:121.
Raynes, A. E., and Ryback, R. S., 1970, Effect of alcohol and congeners on aggressive response in Betta Splendens, *Quart. J. Stud. Alc.* Suppl. No. 5:130.
Richard, H., 1966, Youthful offenders and alcohol, *Youth Author. Quart.* 19(4):15.
Robins, L. N., Bates, W. M., and O'Neal, P., 1962, Adult drinking patterns of former problem children, *in* "Society, Culture and Drinking Patterns" (D. J. Pittman and C. R. Snyder, eds.), pp. 395–412, John Wiley & Sons, Inc., New York.
Roebuck, J., and Johnson, R., 1962, The Negro drinker and assaulter as a criminal type, *Crime and Delinqu.* 8(1):21 (CAAAL 10439).
Room, R., 1970, Drinking laws and drinking behavior: Some past experience, Symposium on Law and Drinking Behavior, Chapel Hill, N.C.
Ryan, E. D., 1970, The cathartic effect of vigorous motor activity on aggressive behavior, *Res. Quart.* 41:542.
Ryle, G., 1960, "The Concept of Mind," Barnes & Noble, Inc., New York.
Sariola, S., 1954, "Drinking Patterns in Finnish Lapland," Finnish Foundation for Alcohol Studies, Helsinki.
Schaaf, T. G., 1971, The effects of alcohol on pain-elicited aggression, M. S. Thesis, Eastern Washington State College.
Scott, P. D., 1968, Offenders, drunkenness and murder, *Brit. J. Addict.* 63:221.
Shoham, S. G., Ben-David, S. and Rahav, G., 1974, Interaction in violence, *Hum. Rel.* 27(5):417.
Shuntich, R. J., and Taylor, S. P., 1972, The effects of alcohol on human physical aggression, *J. Exp. Res. Personal.* (N.Y.) 6:34.
Shupe, L. M., 1954, Alcohol and crime, *J. Crim. Law. Crimin. Pol. Sci.* 44:661.
Sjöberg, L., 1969, Alcohol and gambling, *Psychopharmacol.* (Berlin) 14:284.
Skelton, W. D., 1970, Alcohol, violent behavior, and the EEG, *Southern Med. J.* 63(4):465.

Smith-Moorhouse, P. M., and Lynn, L., 1966, Drinking before detention; a survey of the population of a senior detention centre to ascertain if excessive drinking or alcoholism could be detected, *Prison Serv. J.* 5(20):29.

Spain, D. M., Bradess, V. A., and Eggston, A. A., 1951, Alcohol and violent death. A one year study of consecutive cases in a representative community, *J. Amer. Med. Assoc.* 146:334.

Stafford-Clark, D., and Taylor, F. H., 1949, Clinical and EEG studies of prisoners charged with murder. *J. Neurol. Neurosurg. Psychiat.* 12:325.

Stark, H. D., 1969, Robbery and alcohol, *in* "Raubkriminalität; Ursachen und vorbeugende Bekämpfung" (Robbery; Causes and Prevention), pp. 142–163, de Gruyter, Berlin (CAAAL 13474).

Sundby, P., 1967, "Alcoholism and Mortality," Universitetsforlaget, Oslo.

Sutherland, E. H., and Cressey, D. R., 1970, "Principles of Criminology," Lippincott, Philadelphia.

Swanson, D. W., 1968, Adult sexual abuse of children, *Dis. Nerv. Syst.* 29:667.

Takala, M., Pihkanen, T. A., and Markkanen, T., 1957, "The Effects of Distilled and Brewed Beverages," Finnish Foundation for Alcohol Studies, Helsinki.

Tardif, G., 1967, Les délits de violence à Montréal, *in* "The 5th Research Conference on Criminality and Delinquency," Quebec Society of Criminology, Quebec.

Tarter, R. E., Jones, B. M., Simpson, C. D., and Vega, A., 1971, Effects of task complexity and practice on performance during acute alcohol intoxication, *Percept. Motor Skills* 33:307.

Thompson, G. N., 1956, Legal aspects of pathological intoxication (alcoholic insanity), *J. Social Therap.* 2:182.

Tinklenberg, J., 1973, Alcohol and violence, *in* "Alcoholism: Progress in Research and Treatment" (P. G. Bourne and R. Fox, eds.), pp. 195–210, Academic Press, New York.

Tuason, V. B., 1971, The psychiatrist and the violent patient, *Dis. Nerv. Syst.* 32(11):764.

Tupin, J. P., Mahar, D., and Smith, D., 1973, Two types of violent offenders with psychosocial descriptors, *Dis. Nerv. Sys.* 33:356.

Tyler, D. B., 1955, Psychological changes during experimental sleep deprivation, *Dis Nerv. Sys.* 16:293.

Ullman, A. D., Demone, H. W., Stearns, A., and Washburne, N. F., 1957, Some social characteristics of misdemeanants, *J. Crim. Law Criminol. Pol. Sci.* 48:44 (CAAAL 8870).

Ullrich, H., 1966, Jugendkriminalität und Alkohol, *Kriminalistik* 20(5):239.

van der Spuy, H. I. J., 1972, The influence of alcohol on the mood of the alcoholic, *Brit. J. Addict.* 67:255.

Vannicelli, M. L., 1972, Mood and self-perception of alcoholics when sober and intoxicated, *Quart. J. Stud. Alc.* 33:341.

Vartia, O. K., Forsander, O. A., and Krusius, F. E., 1960, Blood sugar values in hangover, *Quart. J. Stud. Alc.* 21:957.

Verkko, V., 1951, "Homicides and Suicides in Finland, and Their Dependence on the National Character," G. E. C. Gads Forlag, Copenhagen.

Virkkunen, M., 1974, Alcohol as a factor precipitating aggression and conflict behavior leading to homicide, *Brit. J. Addict.* 69:149.

Voss, H. L., and Hepburn, J. R., 1968, Patterns in criminal homicide in Chicago, *J. Crim. Law Crimin. Pol. Sci.* 59(4):499.

Walker, N., Hammond, W., and Steer, D., 1970, Careers of violence, *in* "The Violent Offender: Reality or Illusion?" (N. Walker, W. Hammond, and D. Steer, eds.), pp. 3–35, Oxford University Press, London.

Washburne, C., 1956, Alcohol, self, and the group, *Quart. J. Stud. Alc.* 17:108.

Washburne, C., 1961, "Primitive Drinking," College and University Press, New York.

Weitz, M. K., 1974, Effects of ethanol on shock-elicited fighting behavior in rats, *Quart. J. Stud. Alc.* 35:953.

West, D. J., 1968, A note on murders in Manhattan, *Med. Sci. Law* 8:249.

Wexberg, L. E., 1951, Alcoholism as a sickness, *Quart. J. Stud. Alc.* 12:217.

Whittet, M. M., 1973, The measure of alcohol, *J. Alcoholism* 8:118.

Wieser, S., 1964, Die Persönlichkeit des Alkoholtäters, *Kriminalbiol. Gegenwartsfr.* 6:41 (CAAAL 11276).

Wilentz, W. C., and Brady, J. P., 1961, The alcohol factor in violent deaths, *Amer. Pract. Dig. Treat.* 12:829.

Wilkinson, P., Kornaczewski, A., Rankin, J. G., and Santamaria, J. W., 1971, Physical disease in alcoholism: Initial survey of 1000 patients, *Med. J. Aust.* 1:1217.

Williams, D., 1969, Neural factors related to habitual aggression, *Brain* 92:503.

Wilsnack, S. C., 1974, The effects of social drinking on women's fantasy, *J. Personal.* 42(1):43.

Wittgenstein, L., 1958, "Philosophical Investigations," Basil Blackwell, Oxford.

Wolfgang, M. E., 1958, "Patterns in Criminal Homicide," John Wiley & Sons, Inc.. New York.

Wolfgang, M. E., 1967, Victim-precipitated criminal homicide, *in* "Studies in Homicide" (M. E. Wolfgang, ed.), pp. 72–87, Harper and Row, New York.

Wolfgang, M. E., and Ferracutti, F., 1967, "The Subculture of Violence," Tavistock, London.

Wolfgang, M. E., and Strohm, R. B., 1956, The relationship between alcohol and criminal homicide, *Quart. J. Stud. Alc.* 17:411.

Yules, R. B., Freedman, D. X., and Chandler, K. A. 1966, The effect of ethyl alcohol on man's electroencephalographic sleep cycle, *EEG Clin. Neurophysiol.* 20:109.

Zakowska-Dabrowska, T., and Strzyzewski, W., 1969, Value of experimental administration of alcohol in the diagnosis of so-called pathological intoxication, *Psychiat. Polsk.* 3:565 (CAAAL 14782).

Zucker, R. A., 1968, Sex role identity patterns and drinking behavior of adolescents, *Quart. J. Stud. Alc.* 29:868.

CHAPTER 11

Alcohol Abuse and Work Organizations

Paul M. Roman
Department of Sociology
Tulane University
New Orleans, Louisiana

and
Harrison M. Trice
Department of Organizational Behavior
New York State School of Industrial and Labor Relations
Cornell University
Ithaca, New York

INTRODUCITON

With notable exceptions, the problem of alcohol abuse and potentials for its control within work organizations has received relatively little attention from researchers and practitioners in the alcoholism field. Such attention has, however, burgeoned rapidly in the past several years, largely due to significant interest and financial investment from the National Institute on Alcohol Abuse and Alcoholism (NIAAA). A set of concepts and assumptions originally articulated several decades ago has "come of age" and a large scale effort is underway to identify and provide assistance to problem drinkers within the setting of work organizations.

Much remains to be done, however, both in terms of research and in diffusing practical and acceptable programs for prevention, intervention, and treatment within work settings. In this chapter we review a range of materials dealing with alcohol abuse and the work world. We first examine the basis for attention to alcohol abuse and the job, presenting the assumptions and data that underlie this emphasis. We then outline the emergence of the subspecialty generally known as "occupational alcoholism," and assess its current status. Finally, we consider persistent problems faced in developing and successfully executing work-based strategies, which are coming to comprise a major component of the overall effort to deal with the social and personal problems accompanying alcohol abuse.

RATIONALE FOR WORK-BASED PROGRAMS

While attention to work-related aspects of alcohol abuse is not new, the question of "why the work world"? still requires ongoing examination and justification. In the field of alcohol research and treatment, the probability is low that an instant cure for the focal problem will develop to the extent that alcohol abuse disappears overnight. While there is a general mandate to deal with the problem, justification for the various strategies employed in attempting to fulfill that mandate should be an ongoing issue of concern, particularly in terms of establishing priorities. The various specialized divisions of the collectivities engaged in combating social problems are often in competition with one another for limited resources. Thus, the legitimacy of mandates for specific strategies is always open for inquiry and often calls for defense, particularly when the allocation of scarce resources is in the charge of publicly elected legislative bodies or confederations of voluntary organizations administering a pooled budget.

It is therefore appropriate to closely examine the rationale for allocating resources to the work-related aspects of alcohol abuse. In this first section we attempt to provide an objective overview of three related bases for such resource allocation: the costs of alcohol abuse to work organizations, the preventive potential of the work organization setting, and the potential etiological impact of work-based factors.

Costs of Alcohol Abuse to Work Organizations

Attention to the work world by alcoholism specialists has oftentimes been stimulated by the issue of costs, i.e., the notion that problem drinking by workers costs employers billions of dollars in terms of decreased productivity, lost time, accidents, etc. These costs are of three types: cost to employers, cost to problem drinkers, and cost to society.

If we accept the assumption that a considerable proportion of chronic problem drinkers escape detection until their disorder has progressed to a serious stage where intensive therapeutic intervention has been necessary, then it is obvious that many early-stage problem drinkers are actively engaged in work roles within industry, business, and government. Epidemiological surveys clearly indicate that large proportions of heavy drinkers are employed (Cahalan et al., 1969). Furthermore, data on treated alcoholics indicate that a very large number of them continued to be employed until the middle and late stages of their disorder (Trice, 1962; Warkov and Bacon, 1965). While we do not have adequate epidemiological data on the prevalence of problem drinking in different types and sizes of work organizations, the data from general population surveys as well as the evidence developed within several specific work organizations allow the inference that a considerable number of problem drinkers are engaged in work roles. Using such inferential evidence, prevalence rates are typically projected to be within 3 to 5 percent of the work force in any organization, with variations depending on the age, sex, and ethnic composition of the work force. From an epidemiological point of view, the inferential nature of these data cannot be overemphasized; however, they do clearly point to a substantial problem drinking population within work organizations.

Numerous studies that we have summarized elsewhere (Trice and Roman, 1972) have indicated that such problem drinkers do indeed engage in behaviors that are costly to their employers. First, the impact of regular alcohol abuse on job efficiency is clear-cut in most instances, whether such abuse is restricted to off-job hours culminating in "hangover" at work, or whether it occurs in direct conjunction with job performance. Regular alcohol use clearly interferes with efficient manual dexterity as well as accurate perception and cognition with impacts being documented while drinking is underway and for as long as 18 hours after drinking (Tichauer and Wolkenberg, 1972). The impact of alcohol abuse on job efficiency, whether through hangover or intoxication, can be seen at all job levels, although it obviously is more difficult to document at higher levels in the organizational hierarchy, particularly in the case of professional workers. Blue collar workers show lower production and increase scrap rate. Clerical and other white collar workers show lower production as well as delayed and confused decision-making. Among problem drinkers in the executive ranks, control over departmental or organizational functioning that requires sharp mental acuity is quickly impaired by problem drinking. Thus, costliness of problem drinking in terms of job efficiency and productivity is clear-cut; some degree of impact on efficiency can be expected regardless of the amount of alcohol consumed in conjunction with job performance, with the exception of an unknown proportion of anxiety-prone individuals whose productivity may be briefly enhanced by the tranquilizing effects of ethanol (Straus, 1971).

Aside from direct effects on performance, sharp costs accumulate around problem drinkers' patterns of absenteeism. We categorize absenteeism into three types: full-day, partial, and psychological. The principal disruptive feature of all three types is that the absenteeism is generally unpredictable. In the case of blue collar and other lower level workers, rates of full-day absenteeism are much higher for problem drinkers than for other workers. Such absenteeism is often acccompanied by bizarre excuses. Although the employee will rarely "project" his absenteeism in advance, patterns of absence often tend to accumulate before and after weekends, and immediately following a payday.

Among lower level workers absenteeism is often of the "partial" type. Here the worker shows an unusually high rate of tardiness, leaves the job for unexpected periods, or ends his workday before others, again using bizarre excuses for these incidents.

While absenteeism among lower level workers often provides opportunities for intervention relatively early since documentation is straightforward, absenteeism among white collar and upper level workers is less blatant and less likely to be documented on a routine basis. Workers at this level rarely stay away from the job when impaired by the effects of alcohol. Instead they may show patterns of partial absenteeism that are much more difficult to detect than in the case among lower level workers; typically looser patterns of supervision, greater privacy, and freedom of mobility at middle and upper organizational levels set the stage for such absenteeism without detection.

A prominent absenteeism pattern among upper level employees has been labeled "on the job" or psychological absenteeism, i.e., the white collar, professional, or executive employee who appears for work on time and stays throughout the working day, but who is partially intoxicated or impaired from hangover to the extent that he is unable to perform his assigned duties. Such employees go to great pains to be visibly "on the job" (Trice, 1962) although they may spend the workday lying on a couch or otherwise attempting to recover.

It is obvious that documentation of absenteeism patterns is intimately tied to the employee's success at coverup. The degree to which job performances are visible to supervision directly affects the possibility of detecting impaired job performance. Visibility includes not only the actual physical execution of work tasks, but also the extent to which appropriate work role performance is clearly defined. It can thus be deduced that the lower level employee has few alternatives but to stay away from the job when he is in an impaired state, in the hope that this excuses for the absence will prevent awareness of his drinking patterns.

As one moves up the hierarchy, however, visibility decreases on both counts: Employees have greater privacy in terms of office space and freedom of movement, and their job descriptions are much more ambiguous than those of lower level workers, allowing considerable "slack" in performance appraisal. Thus, full-day absenteeism, with its accompanying risks of problem detection, is not necessary for the upper-level employee, as long as he can manage his visibility. On-the-job absenteeism is likewise nearly impossible for the lower level worker with high visibility, whose presence on the job in impaired condition could set the stage for rapid identification.

Thus, it is obvious the problem drinkers can be very costly to their employers in terms of their job inefficiency and their patterns of absenteeism. In the case of many working problem drinkers, it is inadequate to calculate the costs of their behavior simply in terms of the loss of a day's output. Research clearly reveals that a major impact of the problem drinking employee is found in his effects on the work of others (Trice, 1965a; Maxwell, 1972). His unpredictable absences or inability to perform when present may prevent other members of the work group from functioning, this of course being contingent upon the degree of interdependence of job performances among those in the work group. The type of behavior he manifests on the job may undermine the morale of his work associates and his supervisors. In the case of upper echelon employees, much more than a day's work may be lost when an employee is impaired by alcohol use. The impact of faulty decision-making may reverberate through an organization months after it is made. Further impact on morale is found in blundering incidents of public relations that create embarrassment for organizations, manifest in loss of sales, prestige, or credibility. Clients and potential clients may be directly insulted; more importantly, their observation of alcohol-related impairment leads to a loss of confidence in the organization's abilities. Thus, without undertaking elaborate calculations of costs in terms of specific dollars and cents, it is indeed obvious that job inefficiency, absenteeism, and lowered morale can take sharp tolls on an organization's output, whether it be of the industrial, business, or governmental type.

The problem drinker's impact on the morale of his co-workers highlights one of the most basic difficulties in developing effective programs of early identification: the typical reactions of "others" to a problem drinker. On the basis of extensive research observations, we have postulated that the typical reaction to problem drinking among one's subordinates or co-workers is one of toleration, ignoring or absorbing the disruptive behaviors into the ongoing flow of activity. The basic tendency is to "reclassify" or normalize the observed deviant behavior through temporarily broadening the guidelines for acceptable behavior rather than following the idealized sequence of recognizing it as abnormal, labeling it, and attempting to intervene (Trice, 1965a).

This general tendency in social behavior—which characterizes arenas other than the work place—can be traced to a series of "pressures for normalization," some of which are related to the nature of social groups and others of which stem from mores and norms that may be unique to American society (Roman and Trice, 1971). In capsule form, these pressures include:

1. *The tendency to sustain the equilibrium of group life.* Identification of a group member as a deviant "rocks the boat" in that it draws attention away from goal attainment. If the process of recognizing and labeling an individual as deviant is successful, it may disqualify the individual from further role performance, creating the necessity of recruiting and socializing a new group member.

2. *The value of poor performers in enhancing the performance of others.* The problem drinker's absenteeism and frequently impaired performance may clearly make him the poorest producer in the group. His presence therefore makes the performance of others look relatively better than if he were absent by "setting a floor" under their performance outputs. Furthermore, his deviant activities may increase the likelihood of tolerance of degrees of deviance by others in the group that do not stand out in comparison with the actions of the problem person.

3. *Poor performers as scapegoats.* The deviant may provide the group with excuses for its shortcomings (if such exist), excuses that are acceptable both to group members and to outsiders. Removal of the deviant would eliminate this explanation for the group's problems.

4. *Fear of "boomerang labeling."* Group members do not feel confident that they can successfully label the problem drinker as a deviant, anticipating that he may in return highlight their shortcomings as a defense against their labeling attempt.

5. *Fear of attribution of responsibility.* For a range of reasons, group members may anticipate that they will be held responsible for the development of the behavioral problem if it is defined as such. In other words, if they have regularly interacted with an individual over a period of time, they may be held accountable for causing or contributing to his problem behaviors. Such fears may escalate as coverup continues; the longer the toleration, the more likely that toleration itself will be grounds for condemnation by outsiders if the existence of the problem is discovered.

6. *Exploitation of impaired performers.* In situations where the impaired performer is in competition with others for scarce rewards, the competitive advantage of the others may be enhanced by their support of his continued impairment, i.e., "giving him enough rope to hang himself."

7. *Ambiguity of problem definition.* In a context where even the professional community cannot agree on a definition of problem drinking or alcoholism, it may be indeed difficult for laymen to decide "when" a particular level of alcohol abuse constitutes a "problem" calling for intervention.

8. *Lack of diagnostic expertise.* In order to encourage humanitarian treatment of problem drinkers, a massive public educational campaign has attempted to persuade the public that "alcoholism is a treatable illness." By placing such a meaning on the consequences of excessive drinking that are observed in the community or the workplace, an individual's associates may feel incapable of defining a drinking pattern as problematic, believing that identification and intervention should be based on a diagnosis by an "expert." This is turn may result in acceptance of drinking behavior that has not been "officially" labeled as a problem.

9. *Intense norms on privacy.* Despite societal concerns over the propensity of people "not to get involved," there is a deeply embedded set of norms that emphasize individualism, privacy, and nonintervention in the affairs of others. Such norms retard translation of the meaning of alcohol abuse from a "private problem" to a group or public problem. It may indeed be the case that both formal and informal labels of "paranoid" originate from overconcern about the propriety of others' behavior.

These "pressures" can, of course, occur in varying degrees and in a range of combinations, but we would conclude that early identification of problem drinkers in the work group context is of low probability *without counterpressures to overcome the propensities to tolerate deviant behavior* until some dramatic crisis or "mess" is created. Coupled with the problem drinkers' dedicated efforts at coverup, the likelihood of early identification is further reduced.

It is noteworthy that two other types of alcohol-related behaviors frequently thought to be generators of costs for organizations have not been shown to be such by research. Problem drinking employees do not show an exceptional number of on-the-job accidents, usually due to their awareness of the affects of alcohol on their functioning and their vigilance in protecting themselves from accidents (Trice, 1959). This is, of course, another reflection of coverup behaviors, for the occurrence of accidents leads to investigations that may uncover the drinking problem. Vigilance against accidents can likewise be provided through assistance from work associates and supervisors. The problem drinking employee who works on the dangerous job may simply stay away from work when he is in an intoxicated state or suffering from severe hangover. Problem drinking employees do, however, show high rates of off-the-job accidents (Brenner, 1967) that can take their toll in organizational costs due to medical benefits, disruption of work flow, or need for the recruiting and training of replacements. The stereotype of the problem drinker as accident-prone on the job is however an incorrect generalization.

Problem drinking employees also are not characterized by high rates of turnover. Research evidence indicates that the vast majority of them stay at a steady job for the period during which their problem drinking escalates (Trice, 1962). Low rates of turnover appear to be due to the normalization of their disruptive behavior within work groups and by supervision.

Preventive Potential in the Workplace

Work organizations provide a setting for the early identification of problem drinking, together with legitimate grounds for constructive intervention. Thus, while the employer has adequate reason to attend to problem drinking because of its direct and indirect costs, the stage is set for effective intervention by the very nature of work organizations and work roles. The work setting is immediately compatible with identification efforts based on impaired job performance that lead to a confrontation that involves the offering of help to deal with the problem, with disciplinary action as the alternative.

Several assumptions underlie the preventive potential of the workplace:

1. Drinking patterns that are prodromal to chronic problem drinking and alcohol addiction become manifest in impaired job performance early in the sequence of problem development.

2. The relationship between employer and employee provides for legitimate intervention by the employer upon documentation of impaired job performance.

3. Problem drinkers and potential problem drinkers tend to be concentrated among employees with relatively heavy investments in their jobs in terms of seniority, fringe benefits, and the centrality of work in their lives.

4. Confrontation regarding impaired performance directed at relatively long-term employees tends to constitute crisis precipitation, which in turn enhances the individual's motivation to "do something" about his drinking, thereby increasing his responsiveness to treatment efforts (Smart, 1974).

The history of social concern with alcohol has been marked by the absence of settings where legitimate and effective intervention can "break up" a pattern of excessive drinking or halt the progression toward alcohol addiction (Trice, 1966). To a considerable extent, the efforts that have been undertaken have made the prevention of alcoholism a hope rather than a reality (Roman and Trice, 1974). For example, the intimate context of the family may provide the cues that a severe drinking problem has developed, but the diffuse nature of family relationships generally precludes threatening leverage that leads the problem drinker to alter his behavior. At the other extreme, exhortations by national authorities on "how to drink" and how to recognize problem drinking fail to have significant impact because of the role distance between the typical drinker and the prescriptions being made.

Work organizations, embodying bureaucratic rationality and distinctive formal properties as social systems, fall between these extremes by embodying relatively intimate role relationships in a setting characterized by legitimate use of power. The nature of work relationships ideally involves supervision of job performances, with a formalized set of provisions to deal with impaired job performance. In other words, the employer has the recognized right to intervene in instances of impaired job performance because of the implicit or explicit contract between employer and employee. Ideally, the employer and employee are not involved in a diffuse relationship where other contingencies would prevent direct attention to impaired role performance. Thus, confrontation can proceed upon documentation of impaired performance.

Turning to the assumptions that underlie our contention regarding the preventive potential of the workplace, each has been documented by a variety of research data. Problem drinking does lead to performance impairment, the detection of which is, of course, contingent upon the nature of supervision and the clarity of role performance standards. The more ambiguous the performance appraisal, the more difficult is documentation of impairment. Unlike other types of problem behaviors such as obsessive neuroses, moderate amphetamine use, or opiate addiction that often *do not* lead to impaired performance (and indeed in some instances may lead to improved performance), alcohol abuse does, however, have a direct and negative impact on cognitive and motor performance abilities (Trice and Roman, 1972; Tichauer and Wolkenberg, 1972).

Because the development of intense alcohol dependence and loss of control is a lengthy process, those who will repeatedly have alcohol-related problems are much more likely to be middle-aged workers rather than youthful workers. Younger workers do indeed have job-related problems with alcohol, but these "problems" are more likely a reflection of inexperience in "handling" alcohol as well as lower commitment to the job. The length of the dependency-development process in turn implies the likelihood of a high investment in job security among those affected so that a job impairment-related confrontation will have much greater effect than it would to a younger worker with low investment. Furthermore, it is likely that job security may lie at the center of a series of rationalizations that have served to normalize the individual's drinking problem to himself. Drinking that interferes with job performance may have been preceded by family arguments, loss of friends, and off-job accidents and injuries due to drinking, but as long as the job is secure, most rationalizations remain intact. Therefore, a confrontation that draws attention to failure on the job and offers the possibility of job jeopardy may precipitate a crisis that in itself is effective to motivate behavior change. If loss of control over drinking occurred very early in the job-life cycle when commitment and psychic investment were low, then it is less likely that a confrontation would have motivating meaning. It should be evident that this concept is in part drawn from Alcoholics Anonymous ideology about the necessity of "hitting bottom" before motivation to change one's drinking behavior occurs. In essence, the job-based confrontation is a deliberately constructed event to create a "bottom" and subsequent motivation.

A final aspect of the work world's preventive potential is the extent to which a policy calling for identification and confrontation is compatible with the organization's major goals.

Program implementation calls for attention to supervisory skills in performance appraisal, and techniques of dealing with impaired performance. Such emphases "fit" with the goals of profit-oriented organizations where adequate and effective supervision is a major component in maximizing productivity. It is therefore not necessary, at least theoretically, to train supervisors in "new" or ancillary skills in order to equip them to deal with problem drinkers. It should also be noted that if the emphasis on identifying impaired performance is sustained, then the net result for any organization should be improved supervision and productivity, regardless of the actual scope of drinking problems in the work force.

A Neglected Rationale: Etiological Factors in Work Roles

The costs of alcohol abuse to work organizations and the preventive potential of work-based identification of problem drinkers are the two principal assumptions guiding program efforts in work organizations. There is a third dimension of the rationale that has received relatively little attention: The extent to which factors in work roles may promote alcohol abuse can be identified and altered. Such a consideration is not only compatible with the overall goal *primary* prevention of alcohol abuse, but is consistent with patterns of research in industrial engineering and psychology oriented toward minimizing noxious factors in the work environment.

Research evidence to support the existence of etiological factors is sparse, and it is obvious that any such research faces all of the problems of etiological studies: locating adequate samples and legitimate comparison or control population; reliable and valid measurement of variables hypothesized to relate to etiology; and establishing the temporal sequence of measurable events while controlling for contaminating factors to the extent that causes can be separated from effects. Available research technology falls short of many of these requirements in a strict sense. Furthermore, any work-based research is immediately faced with the difficulty of gaining access to research populations, i.e., establishing rapport within a sample of organizations to gain permission for interviews, access to records, etc. Thus, we are limited largely to speculation based on related research rather than on a body of data focused directly on the environmental contributors to alcohol abuse in the work place.

Without doubt, the structure of work organizations and the combinations of demands within them that lead to the development of specific roles create numerous sources of stress for role occupants (Kahn et al., 1964; Trice and Roman, 1971). In addition to the flaws in the structuring of work roles, the very nature of organized work in a competitive society with an ever-changing technology generates role conflicts, role ambiguity, and role overloads, which in turn lead to stress among role occupants, particularly in middle and upper status levels, (French et al., 1965). The menial and repetitive nature of many blue collar roles has been found to produce poorer mental health among lower status personnel than is found among white collar workers (Kornhauser, 1965; Roman and Trice, 1972). Finally, competitive features involved in career progress and career development can generate stress that calls for coping responses among role occupants (Trice and Belasco, 1970).

A considerable bulk of research in industrial sociology and industrial social psychology has been aimed at job dissatisfaction, which may be seen as prodromal to stress or as an indicator of its presence (Vroom, 1966; Katz and Kahn, 1966; McLean, 1972). These data and speculations can easily be overgeneralized, however, and it is always important to note that the vast majority of workers at all status levels make it through their careers without psychological breakdown, and that the vast majority of workers who are alcohol users do not become problem drinkers. Nonetheless, job-related stress may lead to coping reactions that involve excessive alcohol use, and the role of work-based factors as etiological agents should not continue to be totally ignored.

Specifically, stress may generate anxieties that are relieved by alcohol use. While regular excessive alcohol use may eventually impair job performance, evidence indicates that the development of alcohol dependence is a relatively lengthy process to the extent that alcohol could be used as a coping device for job-based anxieties for some time before becoming problem drinking. Evidence from earlier research (Trice, 1962) revealed that relatively heavy off-the-job alcohol use accompanied significant career achievement in some cases before loss of control over drinking occurred, i.e., drinking was functional for career advancement over a short term. Thus, the anxieties accompanying career development may lead directly or indirectly to excessive drinking as a means of coping.

In other circumstances, stress may stem from the frustrations generated by relative or absolute failures in the workplace, and in these instances alcohol may be used as an antidepressant. McClelland and his associates (1972) found heavy alcohol use associated with needs for power, i.e., certain individuals respond positively to the effects of drinking because it provides them with a sense of power that they do not have when they are not drinking. This finding may be extended to posit alcohol use as a coping device among those who have failed according to organizational or occupational criteria and who sense a gap between achieved and aspired power and status; McClelland's data likewise may apply to the attraction to alcohol among those who are forced to remain in low status jobs throughout their work career, always subject to the power of others in the organization.

As mentioned, these etiological hypotheses are speculative for the most part, so we will not go beyond these observations regarding the possible direct relationships between work-based stress and eventual problem drinking, except to mention the potential "vicious cycle" that can develop when drinking as a coping device comes to impair job performance and lead to further anxiety of depression related to failure. There are physiological and social psychological bases for assuming the development of such a cycle (Roman, 1972).

Rather than posing specific work-based factors as direct etiological agents in problem drinking, we have found the epidemiological concept of "risk factors" to be a useful sensitizing approach to possible etiology. Generally speaking, organizational life entails risks of problem drinking to the extent that drinking is an accepted part of the rituals and routines of work. Many organizations allow or even provide for drinking at lunch; a 1972 national survey of 528 executives in large private companies revealed more than half of the respondents approved of lunchtime drinking (Roman, 1974a). More than 60 percent indicated approval of drinking during a two-day sales conference. Other than executive dining rooms, noontime drinking in officers' clubs on military bases underlines the "normality" of this behavior, particularly (and often exclusively) for upper echelon personnel. Otherwise, drinking rituals in the form of after-work tavern groups, cocktail parties, and even bars in executive offices pervade all sizes and types of work organizations. While we lack hard data, it appears that drinking has increasingly become an integral part of work life, with widespread participation that reduces the likelihood of negative reactions to individual incidents or impaired performance related to drinking; in other words, widespread participation in work-related drinking increases the difficulty of "drawing the line" in regard to problem drinking. This global risk factor likely affects middle and upper level personnel more than blue collar workers (Roman, 1974b).

Our research experience in work organizations has led us to hypothesize that certain work roles are structured so that they are more conducive to the development of drinking problems than roles that do not have these features (Roman and Trice, 1970). To some degree, the structural features of these roles may be stressful in and of themselves, but at the same time the overriding common feature of reduced visibility may afford the use of alcohol as a coping response. Other research evidence appears to document that certain social personality types are more attracted to excessive alcohol use than others (Trice, 1966; Trice and Roman, 1970; McClelland et al., 1972). It may be that when such personalities are attracted to or placed in work roles with features conducive to the excessive use of alcohol as a coping strategy, problem drinking ensues.

Briefly, the eight types of role structures posited to involve greater risks for the development of problem drinking are:

1. *Absence of clear goals.* "One-of-a-kind" jobs, jobs that involve highly technical skills, or jobs that are new in an organization all present problems of performance evaluation, making the development of any type of personal problem difficult for both the employee and his superiors to identify. Such roles may be socially isolated, representing the absence of evaluative feedback that may in itself be generative of stress.

2. *Freedom to set work hours.* These roles allow for the employee to come and go at will, frequently emphasizing the "creative" nature of the position. This situation may likewise reflect social isolation as well as absence of supervision and involvement in the informal groups that typify work life and that may be vital sources of social control.

3. *"Field" roles.* The numerous positions that involve travel and primary contacts with persons outside the organization again entail social isolation, low degrees of supervision, and absence of peer-group controls. Such jobs may generate stress through required and regular contacts with strangers on the strangers' "turf," the need to fill evening hours with some sort of activity, and the demand to meet quotas.

4. *Exploitive relationships.* In many organizations there are positions where incumbents are in competition with one another to the extent that some will benefit from another's impairment. When an individual is in a role subject to such "exploitation" by others, it is possible that his deviant behavior or other impairment would be overlooked or perhaps even encouraged by these others.

5. *Work addiction.* A pattern of compulsive involvement in work to the point that an individual literally "lives, eats, and breathes" his job may constitute a risk conducive to the development of problem drinking. The "work addict" may be held in high regard by his work associates, family, and others who benefit from his intense job involvement and accomplishments. In essence, his overactivity accumulates deposits in a "bank of conformity," providing him considerable "idiosyncrasy credit" (Hollander, 1962). Since excessive involve-

ment in work may generate considerable tension in need of release, the choice of excessive drinking in the evenings or on weekends may be a particularly attractive adaptation. Such deviance represents use of "idiosyncrasy credit" so that others tolerate the behavior as "earned," greatly reducing the likelihood of intervention by these others.

6. *Occupational obsolescence.* An increasing prominent phenomenon in a society that is highly responsive to technological change is the obsolescence of job skills acquired in a prior era. Obsolete individuals may be more prevalent in middle and upper level technical positions, reducing the likelihood of their dismissal from the organization. Removal of obsolete individuals from the "mainstream" of organizational action may not only be an inducement to drinking, but also reduces the likelihood of identification via performance appraisal since an inadequate performance is expected from such individuals.

7. *Job mobility.* Some occupational careers involve movement across different settings where the norms regarding alcohol use vary sharply. For example, in work settings such as field construction or in various overseas assignments, heavy drinking may be commonplace, accompanied by social controls that prevent adverse consequences. Risks of problem drinking may escalate when individuals move from heavy drinking-intense control situations into heavy drinking-related control settings, i.e., a potentially problematic drinking pattern is "masked" by social controls until job-related mobility into a situation lacking these controls occurs (Roman, 1970).

8. *On-the-job drinking.* A risk factor that is found with increasing frequency is drinking during work hours, typically lunchtime. Depending upon one's status, participation in such drinking may be normative, with abstainers comprising the deviant category. Regular participation in such events may be conducive to integrating alcohol use into one's everyday life, in some instances leading to escalation of consumption. Risks of problem drinking may be even more acute in occupations such as sales, where regular drinking with clients is an accepted and expected part of the job role.

There are doubtless other risk factors that would be revealed in systematic empirical research on the role assignments and career experiences of employed problem drinkers. While the etiological potential of the workplace could constitute a justification for preventive action by professionals in alcoholism, this has not been the case, for reasons discussed later in this chapter.

In summary, there are several substantial justifications for attention to alcohol abuse in the workplace and for programmatic efforts to deal with problem drinkers in such settings so that their productivity may be recovered. Such attention has finally flowered, and its development is our next topic for attention.

THE EMERGENCE OF "OCCUPATIONAL ALCOHOLISM"

The search for effective means to deal with drinking problems in American society (and to a lesser degree in other Western industrial societies) has finally come to include a commitment of resources to programs for identification, referral, and rehabilitation that are based in work organizations. This commitment has been based on two of the three sets of assumptions advanced in the foregoing sections: that the work world provides a setting for effectively dealing with problem drinking, and that these efforts can be carried out in work settings with benefits to the employer in terms of cost savings.

The resource investment in work-based programs has occurred on two levels: the development of private and governmental agencies and activities designed to promote the diffusion of program concepts across the complex of organizations of different forms and sizes that comprises the American "work world" and the development of policies and programs to deal with problem drinkers within specific organizations. Description and analysis of these two sets of resource input must be intertwined; i.e., one cannot understand the pattern of diffusion of occupational alcoholism concepts to specific organizations without looking at the agencies that have promoted this diffusion. Thus, in this section we endeavor to look at "program types" and diffusion patterns within the context of the emerging set of occupations and professions that we refer to as "occupational alcoholism."

It is important to note at the onset that we are describing the early and middle stages of a relatively unique set of concepts in the alcoholism field. Data collected early in 1974 revealed that 34 percent of major private work organizations had developed some form of program to provide assistance to problem drinking employees (Alcohol and Health Report, 1974). An intense governmental effort has been underway to promote the development of programs for both public and private sector employees, but as of early 1974, progress in the private sector is probably limited to some form of *policy development* among 25 percent of *major* employers at most, 20 percent of state governmental agencies, and an unknown but certainly very small percentage of smaller private businesses and industries. As is described further below, the federal government has developed a policy on problem drinking for its civilian work force, but the extent to which the policy has led to actual program implementation in federal agencies is far from complete.

By the same token, investments in the promotion and diffusion of a range of alcohol prevention and control efforts in private and public alcoholism organizations have been in flux since major attention to this thrust began several years ago. The ultimate priority of occupational programs within this set of efforts cannot be accurately projected. Thus, our descriptions do not comprise the final report of a successful effort to deal with alcohol abuse on a national scale, but rather an interim statement of our observations of a social movement in process, albeit a movement in which we ourselves have had considerable involvement.

Ingredients of Occupational Alcoholism Policies

Since this volume is not a "how-to-do-it" manual for practitioners, we shall not detail the steps in setting up and executing a policy and program to deal with alcohol abuse in a specific work organization. Our earlier discussion of assumptions that underlie occupational programs gives the reader a flavor of these efforts. At present, it is impossible to give a single characterization that describes all occupational alcoholism programs. At this point, however, it may be useful to outline several basic features that are common to almost all policies that have been implemented in organizations to date:

1. Nearly all policies indicate that identification of problem drinkers should occur through the identification of impaired job performance instead of through efforts to distinguish the signs and symptoms of developing alcoholism.

2. Responsibility for identification lies with individuals who have supervisory responsibilities, i.e., there is no specialist or set of specialists who monitor job performances with the goal of detecting drinking problems. It should be added that many program directors aspire toward *self-referral* of problem drinkers as the ideal mode of program operation, minimizing the need for identification agents of any sort.

3. The principal policy guideline, requiring confrontation of employees by their supervision upon documentation of impaired performance, is to be applied with equal emphasis across all levels of the organization, including middle and top management.

4. Employees with impaired performance are confronted by supervision with the evidence of declining performance, and they are informed that help will be made available to them if they cannot deal with the problem causing the poor performance. To the extent possible, supervisors avoid any mention of their suspicion of a drinking problem.

5. If an employee so confronted desires help for his problem in order to improve his job performance, he is referred to an agency outside the company specialized in appropriate counseling or treatment. This referral is usually the responsibility of an individual charged with coordinating policy implementation within the organization, i.e., supervision does not make direct referrals to outside agencies. Depending on the size of the organization, the alcoholism policy coordinator may attempt some differential diagnosis, may engage in counseling

of the employee, or may have at his disposal intramural counseling and treatment facilities.

6. It is prescribed that help-seeking activities undertaken by a problem drinker will not affect his job security or future promotional opportunities; it is help that is made available without penalty even though the employee was not performing effectively. This confidentiality is based on the premise that alcohol problems are to be treated as medical problems rather than as deviant behavior. If outside treatment is required, the typical provisions of sick leave are to apply. This is the innovative feature of these programs in many organizations in which problem drinkers had previously been identified and directly disciplined or dismissed. The help-without-penalty provision is also based on the acceptance of alcohol abuse as a health problem.

7. Employees whose performance does not improve, and who fail to take advantage to the help that is offered, may be subject to discipline after a given number of "chances," culminating in dismissal.

8. Specific attempts are made to refer problem drinking employees to counseling or treatment resources that are minimally stigmatizing and maximally appropriate for problem drinkers who are still employed and have not completely lost control of their drinking.

9. To the extent possible, health insurance policy coverage has been broadened to cover the costs of necessary treatment for the problem drinking employee by outside medical agencies.

10. Representatives of labor unions or other employee organizations are to be informed and involved at every stage in dealing with such problem drinking employees; evidence to date indicates unions have rarely thrown obstacles along the route of processing an employee under alcoholism policy guidelines.

11. In almost every instance, the costs of mounting a program and providing staffing for it are borne totally by the work organization.

12. Managements have been encouraged to adopt the concepts of occupational alcoholism on the basis that a program yields benefits greater than its costs, and that the implementation of a performance-based identification procedure "fits" with good management principles and does not require equipping supervisors with new sets of skills that may detract from their other duties. In other words, the consultation employed with management to encourage acceptance of programs is *not* usually based on the premise that the identification of problem drinkers and coverage of the costs of treatment are management's social responsibility, or that implementing such programs makes the work world part of the national "team" that is combating alcohol problems. While such sentiments may be found *after* program adoption, managerial pragmatism rather than corporate responsibility is the usual basis for encouraging adoption.

The major difference between the programs that have been developed is the breadth of their coverage of problem employees other than problem drinkers. Within the policy guidelines outlined above, programs range from those that deal only with problem drinkers to full-scale "broad-brush" or "employee assistance" programs (Wrich, 1974); i.e., provision of help to all employees with personal problems, regardless of nature and regardless of the effects on performance. The majority of program types appear to fall between these extremes. These variations are described further below, and the reasons for these differences should become clearer as we examine the organizational context within which occupational programs have been promoted and diffused.

Organizational Support for Work-Based Alcoholism and Assistance Programs

As mentioned at the beginning of this chapter, the concept that management and the worker could both benefit from job-based identification and treatment of the problem drinker is not new. By the mid-1940s, several major companies had begun identification programs. Although these programs were not widely publicized, they were apparently quite effective over the years, as indicated by reports of the proportion of identified problem drinkers who were successfully returned to their jobs as adequate performers. These efforts diffused to a few other organizations in the 1950s, but the rate of diffusion was very slow.

In almost every instance, these early programs originated in the medical departments of large companies. These departments were of adequate size to conduct most necessary counseling and treatment intramurally, avoiding the thorny issues of health insurance coverage and adequate referral resources that surround many of the new programs being developed today. To a considerable degree these pioneer efforts, based in companies such as Consolidated Edison of New York, Dupont Chemicals, Eastman Kodak, Equitable Life Assurance Society, Kemper Insurance, and others, were almost exclusively framed within a medical orientation rather than a personnel management orientation, due of course to their location in medical departments. In most instances, a physician in the medical department had become informed and interested in alcohol problems, and served as both the impetus and administrator for the programmatic efforts that followed.

A second distinctive feature was an emphasis on the identification of symptoms of alcohol abuse, with supervisors specifically trained to look for patterns of both job-related and personal behavior that indicated either on-the-job drinking or excessive drinking off the job. This is in contrast to the intense emphasis on the identification of impaired job performance that characterizes current program concepts, along with strong de-emphasis on the "diagnosis" of drinking problems by supervisory personnel.

A third feature of many of these early programs was an emphasis on "constructive confrontation" carried out by supervisory personnel. Assuming that the problem drinking employee had heavy investments in his job and its perquisites, these programs employed a slight modification of a long-standing principle of Alcoholics Anonymous: The psychological experience of "hitting bottom" is a necessary prerequisite for an individual to accept the fact that he has a drinking problem and develop a clear motivation to do something about it. Thus, constructive confrontation involves a rather strong presentation to the employee of the facts of his deteriorated performance due to drinking, *coupled* with the offer of treatment assistance. In the confrontation, a clear statement is made to the effect that failure of the employee to take action about his problem and improve his performance would lead to discipline or even dismissal. It was presumed that such job jeopardy served to "break up" the problem drinker's system of rationalizations about his drinking and its consequences, since in most instances the employees were middle-aged men whose job was the center of their identity, self-esteem, and claims to respectability. Thus, constructive confrontation or coercion is used to precipitate a crisis that parallels "hitting bottom" but it occurs at a much earlier point than a natural devolution to "bottom."

The emphasis on the usefulness of constructive confrontation has eclipsed in many current policies. A trend has appeared toward limiting the supervisor's role to the documentation of impaired performance, urging the rapid referral of the impaired performer to the company's alcoholism program coordinator, or to an outside agency. There appear to be several reasons for this. First, the absence of internal medical facilities associated with most company programs and consequent necessary reliance on community resources require referral expertise within the company that is vested in a person other than a supervisor. Second, the strong emphasis on the "job performance only" basis for referral has stemmed from the belief that the more the supervisor is involved in the confrontation, the more likely he will become entangled in diagnosing the problem underlying the deficient performance, and thereby be manipulated by the employee. Third, there are numerous treatment agencies that engage in outreach to industry in encouraging program implementation as one means of generating referrals to their agency's resources, with a consequent tendency to encourage treatment or counseling for all problem performers instead of confronting the employee and giving him the responsibility of choosing how to bring his performance back up to par. In short, while constructive confrontation was a key element in early program design, the broader programmatic efforts of today appear to be eclipsing its prominence.

A fourth feature of the early programs was a strong reliance on Alcoholics Anonymous as a referral choice for identified problem drinkers. While the breadth of treatment facilities for problem drinkers was much smaller than it is at the present (making AA the sole available resource in some instances), there is clearly an ambivalence toward the use of AA among many program administrators and policy advisors today.

All of the early programs continue to function in some form today. In some instances their founders have continued to play key roles in the diffusion of program concepts to other companies, and in the formulation of policy guidelines at the federal level. For the most part, however, it is our observation that these programs are fairly isolated from the mainstream of occupational alcoholism today.

While the early programs did not lead to rapid diffusion, the concepts of "doing something" about problem drinkers in work organizations became fairly well known among industrial physicians in the 1950s and 1960s through publications in organs such as the *Journal of Occupational Medicine* and *Industrial Medicine and Surgery*. An important element in informing the personnel management community about these program concepts (as well as creating further diffusion in the industrial community) was the 1959 publication of *The Problem Drinker on the Job,* a bulletin authored by Harrison Trice and published for widespread free distribution by the New York State School of Industrial and Labor Relations at Cornell University. This bulletin went through several printings, and the mode of distribution made it available to thousands of personnel specialists as well as industrial medical personnel.

Another early educational effort was the inclusion of work-world oriented seminars and lectures in the annual program of the Yale Summer School of Alcohol Studies, beginning in the 1940s, and continued when the school was relocated at Rutgers University. This provided for the diffusion of occupational program concepts among workers in alcoholism with other primary commitments; the extent to which these workers subsequently attempted involvements in work-world programs is questionable (Trice and Roman, 1973), but a conscientious effort was made to regularly include industrial discussions in these annual events. With the subsequent proliferation of these annual schools in various locations throughout the country (there were approximately 50 in the summer of 1975), it became routine to include discussions about work-world programs; however, with a few notable exceptions, these discussions seem to have been more ritual than substance in that many of the summer schools have not invested resources in including experienced occupational program personnel on their faculties who could deliver up-to-date information about developments in the specialty.

As mentioned, there was a gradual diffusion of new work-based programs, which accelerated in the 1960s by specific inputs by the National Council on Alcoholism and the Christopher D. Smithers Foundation, including this foundation's support of the Program on Alcoholism and Occupational Health at Cornell University, which is described further below. The major event for occupational alcoholism, indeed for the whole alcoholism field, was the formation of the National Institute on Alcohol Abuse and Alcoholism in 1970, as mandated by the 1969 congressional legislation known as the Hughes Act.

Formation of the Occupational Programs Branch of the National Institute on Alcohol Abuse and Alcoholism

Among its several provisions, the Hughes Act (Public Law 91-616) of 1970 specified the establishment of a federal agency to guide the development of National efforts to combat alcohol problems, to act as a clearinghouse for concepts and techniques in the alcohol abuse field, and to fund projects throughout the nation that indicated promise of providing insights into new strategies for dealing with alcohol problems. Thus was formed the National Institute on Alcohol Abuse and Alcoholism (NIAAA) as a part of the National Institute of Mental Health (NIMH). In 1973, NIAAA was elevated in status as one of three institutes comprising the Alcohol, Drug Abuse and Mental Health Administration (ADAMHA), the other constituents being NIMH and the National Institute of Drug Abuse.

Like most federal agencies, NIAAA consists of several divisions that in turn are comprised of branches. The Division of Special Treatment and Rehabilitation Programs includes several branches designed to deal with alcohol problems in special populations (native Americans, public inebriates, Alaskan natives, drinking drivers, women, etc.) and one of these branches is the Occupational Programs Branch, with the target "special population" of employed persons. It is here that some of the most significant decisions regarding the patterning of occupational programming have been made: The branch supervises funded activities and serves as a base for a national network of consultation activities, designed to promote the adoption of occupational program concepts among private and public employers. The branch's activities include direct consultation services, receipt and review of grant applications, state-based occupational consultation programs, a number of specialized demonstration projects that include both consultation and experimental treatment strategies for employed persons, and supervision of the alcoholism and assistance program for employees of the Public Health Service.

Most of the grant funding from the Occupational Programs Branch of NIAAA has been based on the assumption that the successful development of occupational programs throughout the nation requires the availability of expert consultants at the local level. These consultants ideally act as change agents through persuading decision-makers in the management of organizations to adopt programs; developing union cooperation in program development; providing assistance in designing programs to fit the needs of particular work organizations; and aiding in initial program implementation through providing help in supervisory training and developing viable relationships and understanding between work-based programs and appropriate community treatment and counseling resources.

The first step taken by NIAAA in developing a network of consultants involved the location in 1971 of approximately 30 persons throughout the nation who had had some prior experience in occupational alcoholism programs. These persons were retained by the branch on a consulting basis and were made available to work organizations desiring assistance, with NIAAA covering the consultants' fees and travel expenses. For about nine months, this network was used on a stop-gap basis until a state-based consultant program could be developed.

The Occupational Programs Branch then made available to each of the states and territories a three-year demonstration grant of about $50,000 per year to be administered by the state to develop their own occupational consultation activities at the state level. Grant applications were made in spring, 1972, by state authorities with designated responsibilities for alcoholism programming. NIAAA required that these grant funds be used to hire two individuals who would be trained under NIAAA guidance to function as occupational program consultants in the respective states and territories. One consultant was to be specialized in providing consultation to privately owned work organizations, whereas the other consultant was to provide consultation to state and local governments in the development of occupational programs for their employees. With three exceptions, all states and territories made application for and received these grants in 1972.

Since there were few individuals in the nation with prior experience in these activities, and since the authority to hire the potential occupational program consultants was vested in the states, the individuals originally hired for these positions represented a very wide range of prior work experience (Edwards, 1973). In order to provide a minimum set of consultation skills, a base of knowledge regarding alcohol abuse, work organizations, and unions, and to assure some standardization of role activities, the Occupational Programs Branch funded a grant to form the National Occupational Alcoholism Training Institute (NOATI) at East Carolina University to coordinate and provide training for the consultants. This consisted of a three-week training session in June, 1972, followed by three one-week training sessions at six-month intervals thereafter. NOATI received additional funding from NIAAA in 1974 to undertake training of additional occupational program consultants who had not received full exposure to the first training series. This series consisted of three one-week sessions and was completed in May, 1975.

The training design was broad in scope and initially involved the use of outside experts, many of whom were part of the initial consultant network formed by NIAAA. As solidarity developed within the group of state-based consultants, the design of training was altered in the last two sessions of the first series with a committee formed by the consultants taking charge of the training. These later sessions as well as the second series involved the sharing of successes, experiences, and problems, using training formats that minimized didactic methods and maximized small-group sharing and problem solving.

No systematic evaluation of the initial period of work performance by the state-based consultants has been completed at the time of this writing, although both authors are involved in such efforts. It is evident that numerous problems have been encountered in the inevitable role-shaping process that faces any new occupational group. From the point of view of systems analysis, it is evident that these consultants face the basic problem of integrating themselves into an existing network of policy-making, treatment, and rehabilitation efforts that were ongoing at the state level. To some extent, it appears that their role development was impeded by the ambiguous occupational identity of "consultant" (Roman, 1975a,b).

In a few states, some degree of occupational programming effort had already been developed; here it was necessary for the new consultants to integrate themselves into this network and in some instances alter the direction of the preexisting effort. Role development problems have also centered around the extent to which there has been receptivity across work organizations in the several states to the development of occupational alcoholism programs. In other words, whereas some networks of change agents are developed in response to expressed needs at the community level, a major part of the occupational program consultants' efforts was to consist of generating this felt need.

To a very large degree, the consultants felt frustrated by the absence of hard information on which to base their diffusion efforts. For example, specific estimates regarding the cost of alcohol problems to a particular work organization are extremely difficult to compute with any degree of certainty, although many consultants felt that the use of such estimates would be a primary basis for "selling" programs to cost-oriented managers of work organizations. Likewise, the consultants lacked a broad research base upon which to design specific supervisory training efforts within work organizations that may have had an interest in program development, with the exception of a single but comprehensive training study that concluded that *general* supervisory training, rather than alcohol education, was the best mode for motivating supervisors to implement program concepts (Trice and Belasco, 1969). Finally, the consultants were often frustrated in their efforts to tie a company program into appropriate treatment facilities since in many communities these facilities were either lacking or were geared to problem drinkers of the public inebriate type.

The Occupational Progams Branch of NIAAA has also supported about 40 demonstration projects throughout the nation that are designed to explore the feasibility of various types of service delivery systems for reaching problem drinkers in the employed population. One such model approach has been concerned with providing treatment services that would be used by employed persons; models utilizing this approach have been funded over the past several years in different locations throughout the country. Grant funds for these projects, which are typically based in a treatment center, are partly used to employ occupational consultants whose task is to stimulate the interest of work organizations in a given geographical area in developing programs and to subsequently assist in program development, including direct provision of supervisory training. A group of therapists especially skilled in dealing with employed persons is also supported, located in a community mental health center or similar setting. These therapists' activities are specifically geared to returning employed problem drinkers to satisfactory performance levels on their original jobs; thus, most treatment is on an outpatient basis. Several of these funded programs also encompass the care of the employees' family members who may be afflicted with a range of problems that include problem drinking.

In addition to demonstration projects oriented toward the provision of treatment services for alcohol problems appropriate for employed persons, the Occupational Programs Branch of NIAAA has also made demonstration grants for projects to test the feasibility of other models. One of these is the concept of a "consortium" arrangement whereby small businesses, otherwise incapable of supporting adequate administrative or referral personnel individually, pool their resources and use a single agent for program administration, diagnosis, referral and follow-up. Three demonstration projects are using a labor union base for developing programs within industries with organized shops in contrast to the typical management-oriented approach of most of the state-based consultants. Another project is based within a union since in this instance the union membership moves from employer to employer, and employers would not have the interest or commitment to provide identification and referral services to short-term workers. Assessment of the best means for developing program strategies within large corporate structures is a goal of another demonstration project funded by the Occupational Programs Branch. For the most part, demonstration of these model consultation and treatment efforts has been through the grants program rather than through specific contracts; thus, the interests represented in the projects have been "grass roots" in terms of groups that become motivated to submit grant proposals, in contrast to a systematically planned series of contract projects to assess the feasibility of different methodologies and modalities.

Given the scope of the employed population in the nation and the extent to which it includes the overall subpopulation at risk for developing alcohol problems, the resources allocated to the Occupational Programs Branch of NIAAA are minimal. On the other hand, the concept of identification via the workplace is relatively new and has not been comprehensively demonstrated on a nationwide basis. Furthermore, the basic emphasis in occupational alcoholism programs is for the employer and/or the labor union to assume major programmatic responsibility, tying in with local treatment resources. With such an approach, it may be that minimal federal input is appropriate.

As mentioned, evaluation of these efforts is not yet complete, and is made difficult in traditional goal-attainment terms, since no standards exist for the success of either the consultation activities or for programs operating in different types of work organizations. Data from the states indicate that over 900 private and public employers have developed written policy and procedure statements for the identification and referral of problem drinkers and other problem employees since the federal grants for consultation services were made to the states in 1972. In about half of the states, the activities of the occupational program consultants have been of adequate visibility and have been deemed adequately successful for the states to assume funding for the two original occupational program consultants. In about 10 of the states where this state-level commitment has been demonstrated, NIAAA has awarded funding for the further expansion of occupationally oriented alcoholism activities for another three-year period. These data clearly indicate that "something is happening," although any full-scale assessment of overall success is not possible at this time.

Other Organizations Involved in Occupational Alcoholism

While the Department of Health, Education, and Welfare (HEW) had had a small division dealing with alcohol problems prior to the formation of NIAAA, the new institute was the basis for a massive increase in federal involvement in the prevention, identification, and treatment of drinking problems. As is the case with any new organization, NIAAA has had to accommodate to the presence of preexisting efforts to deal with drinking problems. Given the range of NIAAA's efforts, integration and collective action with ongoing organizations of numerous types has been necessary.

The preexisting alcohol unit in HEW had had almost no involvement in occupational programming efforts, devoting most of its relatively meager resources to basic psychological and physiological research on alcohol abuse. Aside from specific work-based programs in the private sector that had been initiated over the years by internal company staff with an interest in alcohol problems, the main organizational interests in occupational programs prior to NIAAA existed in the National Council on Alcoholism, the Christopher D. Smithers Foundation, and the Program on Alcoholism and Occupational Health at the New York State School of Industrial and Labor Relations at Cornell University.

The National Council on Alcoholism (NCA), supported primarily by private donations and grants, has had occupational program specialists on its central staff for a considerable time. These specialists have done field work with organizations interested in developing occupational programs, usually through the vehicle of local conferences organized by NCA's local affiliates. NCA has had ongoing efforts to generate further interest through sponsoring conferences, dissemination of a variety of pamphlet-type publications, and staff participation in numerous summer schools of alcohol studies. In addition, the occupational specialists have worked with the nationwide network of local councils on alcoholism that are affiliated with NCA, encouraging the staff of these councils to develop contacts with work organizations in their localities and to form committees of local business and labor leaders who could further diffuse awareness of occupational programming concepts. In a few instances of some of the larger local affiliates, full-time staff members have been assigned to promoting occupational programs in their respective localities. NCA has recently received one of the largest grants ever awarded by NIAAA to fund labor and management specialists in several major cities for an intensive effort at promoting program development.

As NIAAA's occupational branch has emerged, NCA representatives have been regularly involved in program planning and development. NCA has also received some direct financial support from NIAAA to continue its preexisting occupational programming effort. NCA has maintained visibility through initiating a periodical, *The Labor-Management Alcoholism Journal,* which features articles on new developments in occupational programming; this publication is diffused widely among the various personnel working in occupational alcoholism. Finally, NCA has been more recently involved in developing union awareness about program concepts and in emphasizing the importance of union involvement in program development, particularly in terms of union policy at the national level. To this end, specialists in union–management relations have been added to the NCA staff.

The Smithers Foundation and its president, R. Brinkley Smithers, have long recognized the value of work-based programs, and the foundation has actively supported research, training, and publications specifically geared to occupational alcoholism since the late 1950s. Both of the present authors are research consultants with the foundation, with Trice carrying out a range of consultation activities with the foundation for more than 15 years. Specific activities supported by the foundation include the Program on Alcoholism and Occupational Health in the New York State School of Industrial and Labor Relations at Cornell University, headed by Trice, and begun in the mid 1960s. In addition to these activities, the Smithers Foundation published in 1962 a monograph entitled *Alcoholism in Industry,* which has had very wide distribution; the contents of this monograph were subsequently incorporated into a foundation-sponsored book, *Understanding Alcoholism.* Finally, the foundation has provided consultation support to the federal government, both in the development of the NIAAA program and in the development of the federal employee alcoholism program.

Within the support provided by the foundation, more than 30 training conferences on occupational alcoholism have been held at Cornell, with participation by management personnel, union representatives, personnel specialists, industrial physicians and nurses, and first-line supervisors. Under the aegis of the program several major research projects have been carried out at Cornell, focused on the evaluation of training in occupational alcoholism, patterns of union–management cooperation in the development of work-based programs, specific stress factors related to the development of problem drinking and emotional problems in different types of work organizations, and the assessment of drinking and drug use patterns among young workers. Cornell activities with Smithers Foundation sponsorship have also included support of graduate education that has included opportunities for research involvement in the occupational alcoholism area.

Several significant contributions to the direction of the field have come from the Cornell program. First, a small-scale research study in New York City documented the ambivalences that supervisors manifest toward alcoholic and other problem employees (Trice, 1965b). Second, the emerging strong emphasis on union involvement in occupational programs can be traced in part to research studies and conferences held at Cornell. Third, the aforementioned study of supervisory training had a direct influence on the subsequent direction of NIAAA's efforts toward the problem employee concept as a broader alternative to programs designed to deal only with employees affected by alcohol problems. By carefully evaluating reactions to different training formats, the Cornell research team revealed this broader interest to be more salient to supervisors, as well as indicating that supervisors would more directly deal with a person defined as a poor performer, regardless of cause, than a person defined as alcoholic or otherwise behaviorally impaired. Finally, the Cornell project has repeatedly sounded the theme of "prevention" in regard to occupational programs.

This effort is continuing at Cornell, and its existence has been supplemented by a similar center developed at Tulane by Paul Roman, who was formerly associated with the Cornell program.

A fourth organization that had given attention to occupational alcoholism concepts prior to the inception of NIAAA is the Alcohol and Drug Programs Association of North America. This association had had a membership that represents the range of workers in alcoholism and recently has been broadened to include those involved in the administration of drug programs. The organization holds an annual meeting oriented to the presentation of new information pertinent to the development and the execution of alcohol and drug programs. In addition, the organization maintains a central office in Washington that serves as an informal clearinghouse for the diffusion of information throughout the year, as well as a center for lobbying activities relevant to alcohol problems. Prior to NIAAA, this organization was not heavily involved with occupational programming activities, with the exception of the occasional inclusion of papers and panels on the topic added to annual meetings. During the past several years, however, a specific occupational section has been developed within the organization, and annual meetings now include a substantial number of presentations dealing with this topic. This not only provides a forum for discussion of issues related to occupational programs but also brings together North American representation other than the United States.

In sum, it appears that the integration of NIAAA and its new occupational thrust into preexisting activities has proceeded smoothly, perhaps because all of the prior activities were small in scope compared to the federal funds available to NIAAA. In our observations, there has been minimal conflict between the new and the preexisting systems. NCA has received support to continue its previously developed efforts and it has not been excluded from participating in decision-making related to NIAAA's program. It may be the case, however, that potential territorial conflict has been minimized by the granting of NIAAA funds to NCA. The Smithers Foundation provided funding for occupational alcoholism efforts when no other funding was available; its definition of funds as "seed money" has fit with subsequent developments within NIAAA.

This does not mean, of course, that the process of innovation accompanying this large-scale effort has been without problems. The next set of preexisting organizations to be considered are the state commissions and agencies designated with responsibilities for alcoholism. In our observations, some degree of conflict has occurred at the state level between the new personnel engaged in occupational programming efforts and the preexisting personnel whose programmatic activities have had other directions, particularly in providing one-to-one treatment to the individual alcoholic. While prevention-oriented activities had received lip service in most quarters, they were rarely a reality at the state level, where the orientation is usually toward developing treatment facilities to provide assistance to the highly visible alcoholic.

It should be pointed out that occupational programming concepts were not brand-new ideas at the state level, reflected in part by the inclusion of this topic at the ubiquitous "summer schools" of alcohol studies since the early 1960s that are usually operated by state alcoholism personnel. The innovation of the occupational program consultants at the state level has provided an excellent setting for a systems analysis of innovation. As is the case with many such innovations, however, the anxieties that have been present in most situations precluded research during the formative period. It is clear that in many instances occupational programming efforts, accompanied by substantial funding, have represented a threat to other alcoholism workers in the state. As we have described elsewhere, workers in the field of alcoholism have been characterized by a considerable degree of provincialism, which in some instances has made the notion of contact with work organizations difficult to accept: "Like most organizations, those in alcoholism have become partially closed systems, tending to restrict relations with organizations unlike themselves. They are 'helping' organizations whereas business and industry are profit oriented. They consist of occupations which, by their nature and traditions, are not oriented to business communities" (Trice and Roman, 1973). Thus, while alcoholism workers have traditionally been client oriented, occupational programs require them to take the perspective of management and unions in work settings, perspectives which not only require new knowledge but which they may find to conflict with traditional notions of helping the patient. Thus, the new occupational program consultants at the state level have had to accommodate themselves to preexisting perspectives and commitments within alcoholism organizations, which includes not only the state alcoholism authority but also the various treatment facilities with which they must have rapport if they are to implement programs with appropriate referral outlets.

Furthermore, in order to fulfill their consultation mission, it has been necessary for the state consultants to a large extent and for NIAAA to a lesser extent to develop rapport with the management of work organizations as well as union officials. These constitute immensely complex preexisting systems. In many instances these work-based personnel have had difficulties in accepting the consultants as consultants, instead viewing them in the social worker and treatment roles that characterize most people working in the area of alcohol problems and employed by the state alcoholism agency. This necessity for integration into a series of preexisting systems has created a context of uncertainty, making role development among the occupational program consultants even more difficult. Since most of them have entered a new occupation where the role definition itself was ambiguous, and since most of them had not had backgrounds in consulting with management and unions, identity problems have been significant. This may account for the relatively high turnover of about 60 percent that has occurred among the state-based occupational program consultants during the three years of program operation under the federal grants.

Another organization that comprises a salient element in the environment of occupational programming activity is Alcoholics Anonymous. It has long been the case that workers in the alcoholism field have been attracted to these occupations because of their own experiences in successfully coping with personal drinking problems. As an organization, AA explicitly avoids involvement in funded treatment activities, publicity campaigns, or other activities to proselytize community interest in dealing with drinking problems. This organizational stance does not preclude individual members' involvement in alcoholism activities, although conflicts between job commitments and personal AA commitments to help alcoholic people may be inevitable. The interpenetration between Alcoholics Anonymous and formal agencies dealing with drinking problems is indeed a challenging consideration for the student of organizations.

From our observations, NIAAA itself has employed relatively few personnel who are affiliates of Alcoholics Anonymous, but those that have been employed have had significant influence on decision-making and direction within this formal organization. A substantial minority of those hired at the state level to occupy the occupational program consultant positions have been AA affiliates, and this is clearly the single most common characteristic of the group of consultants. In many instances, work-based programs in both the public and private sector are manned by counseling and administrative personnel with AA backgrounds. In almost all of these instances, persons who eventually joined AA had had backgrounds in occupations other than those dealing with drinking problems. Thus, their primary orientation to drinking problems has come through socialization in AA rather than through professional or paraprofessional education.

The interpenetration of AA into the occupational programming effort seems to have had three major implications. First, there is an immediate tendency for the AA affiliate to approach programmatic activity with an enthusiasm and zeal regarding alcohol problems and the need for their alleviation. In its internal operation, AA latently places heavy emphasis on personal charisma, and these charismatic qualities are visible in many of those who have sustained leadership positions in the fellowship. To a considerable degree this enthusiasm was initially carried over into the occupational programming effort, supporting the "saving" ideology that has been very effective in creating solidarity among workers in the entire alcoholism field.

While research has shown that successful AA affiliates are likely to be extroverted types of personalities who are oriented to relationships with others (Trice, 1957; Trice and Roman, 1970a), it seems likely that this type of enthusiasm may run contrary to the orientations of management personnel who must make the decisions about starting and sustaining occupational alcoholism programs. While this positive approach may enhance effectiveness in salesmanship, one could easily anticipate "hard-nosed" managers who demand facts and figures rather than enthusiasm.

Due in part to its rapid growth and in part to the ideological tone of the preexisting alcoholism treatment collectivity, the occupational programming effort has been characterized by a populistic ideology; i.e., relatively low boundary maintenence regarding the qualifications for participation in consultation and treatment efforts. Without doubt an understanding of AA's involvement is vital in assessing both the causes and consequences of this populist thrust in occupational programming.

AA is characterized by a series of shared beliefs about the nature of alcoholism (i.e., the alcoholic has an "allergy" to ethanol to the extent that his problem drinking is predetermined), and few of these beliefs are incompatible with occupational programming efforts. Indeed, probably more than any other group in the alcoholism field, AA has a committed belief in the possibility of complete rehabilitation from drinking problems across the segments and classes of society. The occupational stability of the AA member who has achieved sobriety is a cardinal achievement, and this "comeback" is a link to the job performance criterion that is central in the evaluation of an occupational program and its participants. Likewise, the central event of crisis precipitation through supervisory confrontation in an occupational program creates a link to the crucial event of "hitting bottom" in AA experiences.

One AA belief has, however, raised some potential difficulties in occupational programming efforts, namely, the notion that it is impossible for an alcoholic to resume "normal" drinking. Given the admonition that the employer does not have the right to interfere in the employee's personal life, which is included in most policy statements, the only *legitimate* criterion of successful rehabilitation from the employer's point of view is resumption of adequate job performance. Thus, the employer cannot prescribe that the employee who has regularly drunk to excess or who has had other problems with alcohol manifest total abstinence in order to be returned to the job. While the issue of whether or not the problem drinker can ever drink normally is ambiguous, it is likely that circumstances arise where a program coordinator or counselor with an AA background requires total abstinence as a criterion of success, while the employee may contend that whatever drinking pattern he follows is his private concern as long as job performance is adequate. This difficulty is, of course, sidestepped if the employee affiliates with AA and personally adopts total abstinence as a goal. We lack systematic information on the extent to which this problem has arisen, but it represents the main instance where the interface between AA beliefs and work-based policies may create conflict.

The third major implication of AA representation in formal efforts such as occupational programming is the extent to which it may reduce AA provincialism. Although there are some programs that rely totally on AA as the treatment modality, most persons working in the area have accepted the contention that the demands of AA affiliation do not fit the personal makeup of every problem drinker. Acceptance of this limitation has in many instances led to contacts and cooperation between those with AA backgrounds and other types of treatment modalities. It has set the stage for the inclusion of AA as part of a *package* of treatment alternatives rather than using the fellowship on an all-or-nothing basis. This may indeed be a desirable outcome, although it can potentially lead to a down-playing of AA's effectiveness at the other extreme, an outcome that is enhanced by the rapidly growing vested interests in medically and psychologically oriented alcoholism treatment.

The logical implications of the effects of AA ideology on occupational programming can easily be overdrawn to the point of misrepresenting empirical reality. Many occupational program personnel are keenly aware of the potential conflicts between AA ideology and occupational programming. Statements about the limited applicability of AA affiliation in the overall population of employed alcoholics have been made frequently by occupational consultants who are members of the fellowship. Like the other consultants, they appear to identify most strongly with the management consultant role rather than with the role of recovered alcoholic functioning in an outreach capacity. This may in part be accounted for by their efforts to integrate themselves into the state alcoholism system, wherein they find their AA identity to have little salience in promoting the priority of secondary prevention and early identification. Looked at another way, since their work is primarily oriented toward developing programs rather than helping individual people deal with drinking problems, they may find that the AA identity runs contrary to the kind of role they are trying to shape for themselves within both the state alcoholism agency and the industrial/business/union community.

Finally, note should be given to two organizations that have been formed as a consequence of the growth in occupational alcoholism activities triggered by NIAAA's involvement and financial input. The first of these is the Association of Labor and Management Administrators and Consultants on Alcoholism, which now has several full-time staff members, a national office, and annual meetings for the exchange of information. This organization was developed with NIAAA encouragement, and it has been the recipient of an NIAAA grant to explore the costs and benefits of occupational programs, to develop data collection procedures for individual programs, and to initially explore standards for the certification of occupational program personnel. In its efforts to provide impetus to occupational programming efforts on a nationwide basis, it is evident that NIAAA saw the need for a "professional" organization that ould serve as a vehicle for consolidating identities and providing visibility. ALMACA has enjoyed a rather rapid growth of membership, and its coverage goes well beyond the state occupational program consultants to include program administrators and counselors in company programs and personnel in treatment agencies that have developed relationships with company programs.

For reasons that appear both complex and unclear, another organization was formed shortly after the development of ALMACA, the Occupational Program Consultants Association, which is primarily made up of state occupational program consultants, and is much smaller than ALMACA. It appears to have developed in part to deal with the problem of funding of state consultants following the expiration of the three-year NIAAA grants. Because of its small size, this organization has functioned largely on an informal basis and has not as yet publicly clarified its goals, particularly in terms of goals different from those of ALMACA. It does, however, have active leadership and a growing membership and likely will become a vector in the occupational alcoholism environment.

Development of Occupational Alcoholism and Assistance Programs for Federal Employees

For those in occupational alcoholism, one of the most significant provisions of the Hughes Act was the requirement that the federal government develop and maintain an occupational alcoholism program for its entire civilian work force. Since the United States government is the world's largest single employer, this legislative provision was seen by many as a very significant step in the diffusion of occupational programming concepts. At the present time (1976), the Federal Employee Alcoholism Program has been formally in existence for five years, and it may be informative to briefly describe how it has been implemented to deal with such a massive and dispersed work force.

The development of the program was initiated before passage of the Hughes Act. Several localized occupational alcoholism programs preceded this 1970 legislation in large federal installations where medical directors became interested in alcohol problems and designed means for identification and referral. Partly on the basis of these experiences, partly due to encouragement from those with program experience outside the federal establishment, and partly due to repeated efforts on the part of the Smithers Foundation, several conferences for high-level federal managers and personnel specialists were held in Washington during the 1960s, culminating in the formal statement of intent in 1968 by the Chairman of the Civil Service Commission to develop a policy. Program development was based in the commission, which essentially functions as a coordinating personnel department for the federal civilian work force. Thus, to some degree, the stage had already been set for implementing the mandate that was included in the 1970 Hughes Act legislation.

In July, 1971, a document (Federal Personnel Manual Letter 792-4) was issued by the Civil Service Commission and was distributed to the headquarters of all federal agencies for inclusion in the manual of personnel practices. The document outlines how program development and implementation was to proceed, indicating a series of guidelines within which federal departments and agencies were to establish their own programs for their employees. Rather than setting forth a highly specific policy, the guidelines provide for the accommodation of different programs to the diversity of working conditions found in the federal system. Furthermore, the guidelines provide flexibility in policy implementation at the level of local installations of departments and agencies so that adjustment can be made to local circumstances.

The guidelines indicate that each department and agency shall have a written policy, a central program administrator, and that each installation shall have a program coordinator to be in charge of implementing the program locally. The guidelines indicate that the coordinator shall "be allotted sufficient official time to effectively implement the agency policy . . ." In developing local programs, these program coordinators ideally operate under the guidance of the central program administrator and with the technical assistance of ten regional Occupational Health Representatives who are employed by the Civil Service Commission to guide a range of health-related activities.

The guidelines closely follow those that have been suggested for a model occupational program. Identification of problem drinkers, which is to be based on identification of impaired job performance, is the responsibility of the immediate supervisor. Thorough documentation of incidents of impaired performance is a significant concern since federal employees may make use of a multistep appeals process through the Civil Service Commission should they be subject to disciplinary action. The federal guidelines emphasize the offering of treatment assistance to the employee who has manifested impaired job performance on several occasions. Only when such help is refused (and performance is not improved) is there to be disciplinary action. According to the commission's guidelines, the supervisor who identifies impaired performance is:

> . . . to consult with the medical and/or counseling staff for advice on probable causes of the employee problem (and then) to conduct an interview . . . focusing on poor work performance, informing the employee of available counseling services in the event poor performance is caused by any personal

problem. If the employee refuses help and performance continues to be unsatisfactory, he is given a firm choice between accepting agency assistance through counseling or professional diagnosis of his problem, and cooperation in treatment if indicated, or accepting consequences provided by agency policy for unsatisfactory performance.

The guidelines include the recognition of alcoholism as an illness, and define the primary aim of the effort as rehabilitation rather than discipline. Sick leave for those requiring impatient treatment is allowed, along with coverage of treatment costs to the extent available under the federal employee health benefits plans.

Despite these rather ideal arrangements, implementation of the policy into viable programs has been a slow process. This should not be surprising, given the scope and diversity of the federal establishment. Our observations indicate several major reasons for the apparently slow diffusion of this particular occupational programming effort:

1. It appears that the policy generally occupies a low-priority position in the allocation of time and resources by federal managers. This may be due to the general absence of "enforcement" associated with the policy, in contrast to policies such as Equal Employment Opportunity. The slow development of participative support by management may also stem from the belief that drinking problems are minimal in federal service, and/or that adequate provisions for dealing with the problem drinker already exist.

2. At the same time, it is evident that the federal system is not the monolithic structure that is often presumed. From our observations in research projects currently underway, there is a great deal of flexibility in decision-making at the level of the local installation when a new policy or procedure is passed down from headquarters level *without* an indication of enforced compliance. Thus, some federal installations quickly selected a local program coordinator and provided for his training at installation expense, while others were slow in even appointing a coordinator. Similar autonomy at the local level is reported in the case of some large private corporations that have promulgated an occupational alcoholism policy from corporate headquarters, resulting in uneven policy adoption and implementation in local installations of a single major corporation.

3. A closely related impediment to program development is the fact that federal agencies as well as other public employment settings are not for profit. Thus, the "sales pitch" that projects that an occupational alcoholism program increases efficiency and cuts costs may have less meaning to a public than to a private employer. At the hypothetical extreme, documented improvement of efficiency through implementing such a program could conceivably reduce an agency's personnel budget. It may indeed be inappropriate to generalize that

goals of efficiency and improved cost-benefit ratios are held by managers in the public sector.

4. A dominant theme of federal personnel practices is protection of the employee's rights, and placement of the burden of proof of incompetence or poor performance squarely on the supervisor's shoulders. A great deal of paperwork and procedural complexity apparently accompanies any effort to "do something" about any type of problem employee. This situation is, of course, paralleled in the private sector, where union protection of members' right is extensively detailed in collective bargaining agreements and may create in both settings a basic reluctance on the part of the supervisor to take action toward a problem drinker whose performance is slipping. Fear of becoming enmeshed in procedure may cause the supervisor to waver over the problem until it becomes dramatically visible in the form of a "social mess" that can no longer be covered. Thus, procedure itself is a definite barrier to the effective implementation of employee alcoholism policies in both public and private work organizations.

5. Closely related to this reluctance is the availability of the "disability retirement" option to the federal supervisor and manager. Employees with at least 20 years of adequate service are eligible for this kind of retirement if they are physically or psychologically unable to carry out their jobs. Such options are likewise available in many private corporations. It is clear that disability retirement is often an available option for dealing with the chronic problem drinker. It should be obvious that this option may be viewed by some supervisors and managers as simpler and as "fair" (by providing a guaranteed income) to the problem person in contrast to implementing the policy guidelines. It is also noteworthy that disability retirement relieves the supervisor and manager of further coping with the problem drinker, in contrast to the need for monitoring that would follow the employee's return from treatment under typical policy guidelines.

It is conceivable that progress in the development of the federal employee program may be contingent upon the passage of time. Our field observations indicate in many installations a sincere willingness to implement the program, but weakness in facilitating ongoing *readiness* to deal with cases as they might develop. The apparent slow progress in occupational alcoholism program development in the federal government was the subject of hearings conducted by the Special Studies Subcommittee of the House Committee on Government Operations during spring, 1974.

The data collected from departments and agencies by the subcommittee staff indicated that progress in program development had been slow in most of them and that practically none of them were reaching the numbers of problem drinkers that 1970 estimates by the Comptroller General had projected. Among the recommendations stemming from the hearings and investigations was that the program be broadened to follow the employee assistance program model (discussed in the next section), that agencies should develop high level support and commitment for the program, that agencies should monitor compliance with program guidelines in field installations, that the Civil Service Commission should be more aggressive in supervising the program and increase its staff and inspection capabilities, that agencies should enlist the support of employee unions, and that negotiations with health insurance carriers should be undertaken to assure coverage for alcoholism treatment outside psychiatric clinic and hospital settings. At the time of this writing the impact of these recommendations is not clear.

While no comparative evaluative data are available, there are numerous claims of progress made in developing employee alcoholism programs for the uniformed military, these programs being outside the aegis of the Civil Service Commission. For example, each Army base is to have an Alcohol and Drug Control Officer who is usually closley tied to both on- and off-base treatment resources. The Navy has implemented "Drydock" facilities at many of its major bases for the management of early-stage alcohol problems and operates regional treatment centers for more advanced cases that require more intensive care. It appears that the Navy program has a relatively strong emphasis on Alcoholics Anonymous affiliation, while the Army program is more oriented to group therapy and individual counseling approaches.

The extent to which any of these military programs has developed successful linkages with supervisory-based identification is unclear. It also appears that there is considerable variation in the extent to which these programs provide coordinated coverage for the large civilian employee populations on various bases. (The civilian force is covered by the provisions of the Federal Employee Alcoholism Policy, while a different legislative mandate covers uniformed personnel.) Finally, observations indicate that most programmatic emphasis is toward enlisted personnel rather than officers, which parallels the typical targeting of both public and private civilian programs toward the rank-and-file rather than white-collar managerial personnel. The extent to which progress in program development in the military is a function of peacetime status is unclear, but this may well be a factor in facilitating the assignment of personnel to this activity.

PROSPECTS AND PROBLEMS FOR OCCUPATIONAL ALCOHOLISM PROGRAMS

In the preceding sections, we first outlined the theoretical and practical bases for developing strategies based in work organizations for the earlier identification of the problem drinker, and then provided a brief sociohistorical account of the development of organizational mechanisms to promote these programs in the work world. In this section we survey some of the problems that affect the future of this strategy for reaching problem drinkers and accomplishing the goal of reducing the nation's alcohol-related problems through these mechanisms.

We first examine the development of the "broad-brush" or "employee assistance" strategy, which has received considerable attention as an improved means for reaching problem drinkers in work settings. We then turn to an analysis of the potential constituencies or interest groups that are involved in the development of occupational programs, and examine the problems that these combinations of interests may present for the future of occupational programming. In the next section we examine problems inherent in the mechanisms designed to identify, refer, and return problem drinkers to productive work performance. The final portion of the discussion is focused on role of labor unions in occupational programs, program evaluation, and a projection of the needs of this specialty to effectively assure its survival.

The "Broad-Brush" or "Employee Assistance" Approach: Its Nature and Problems

A development that has come to play a significant role in the orientation of occupational programming efforts as well as being a major source of controversy is the "broad brush," "troubled employee," or "employee assistance" program design. At present, the term "employee assistance program" seems to be the one most widely used. Rather than limiting mechanisms of identification and referral to persons whose work performance problems are caused by drinking, this strategy involves identification and referral of all problem people whose job performance has shown a significant decline (Wrich, 1974).

The underlying logic of this approach centers upon the assignment to supervision of responsibility for identifying the problem drinker on the basis of his declining job performance rather than on the basis of his possible alcohol-related symptomatology. This emphasis on performance identification, in turn, has been based on the strongly held belief that a supervisory confrontation based upon presentation of alleged evidence of alcohol problems will be unsuccessful since such symptoms are ambiguous; furthermore, such a confrontation can generate intense efforts at manipulation of the supervisor–subordinate relationship on the part of the problem drinker. By contrast, attention to declining job performance is seen as the rightful prerogative and responsibility of supervision; use of this type of evidence should be much less ambiguous than evidence about the presence or absence of developing alcoholism.

Therefore, using this logic, it is obvious that persons with problems other than alcohol may experience declining job performance. The supervisor is specifically trained to avoid diagnosis. Thus, the inherent structure of the approach will lead to uncovering performance problems in addition to those caused by alcohol abuse. The rationale for providing such broad assistance has been further extended by the argument that it is in the employer's interest to identify and provide assistance to *all* problem persons with declining job performance rather than waiting until these difficulties flower into critical situations that necessitate removal of the employee. Thus, while the employee assistance approach is based on the same strategy for identification as earlier occupational alcoholism programs, those who adopt such a program accept a broader orientation toward identifying and rehabilitating all troubled people rather than an orientation toward solely rehabilitating the problem drinkers that may be in the work force.

This approach has been officially recommended by the Occupational Programs Branch of NIAAA. The institute's rationale is outlined in a widely diffused brochure, *Occupational Alcoholism: Some Problems and Some Solutions.* The historical development of this advocacy by NIAAA is somewhat difficult to unravel. As mentioned previously, a number of major corporations had developed occupational alcoholism programs prior to the emergence of NIAAA. In almost all instances these programs were designed to deal with alcohol problems and an employee assistance strategy was not evident. On the basis of observing these programs, it was alleged that their procedures involved supervisory identification based on the signs and symptoms of alcoholism rather than signs of declining job performance. This is clearly an overgeneralization, for when one examines the policy statements and procedures in these earlier programs, it is obvious that absenteeism, tardiness, and changes in performance were recommended as basic criteria for supervisors to employ in identifying the problem drinker. Rarely, if ever, did a program specifically or by inference advocate supervisory diagnosis of alcoholism, although it is certainly the case that the symptomatology of alcoholism was widely discussed in both training and program literature. One may conclude that advocates of the employee assistance strategy have attempted to strengthen the case for both the logic and the novelty of their approach through subtly posing earlier programs as being both narrow and naive.

Two sets of empirical evidence were apparently used by NIAAA to substantiate the employee assistance strategy. The first was evaluative data collected by Otto Jones, the administrator of a preexisting employee assistance program, the Kennecott Copper installations in Utah. Jones's evaluation of the program argued that the broader approach was more effective in reaching alcoholic people than an approach targeted only on alcohol problems. His data included evidence of cost effectiveness and the program procedures were presented in an attractive and persuasive manner that apparently was a major influence leading to the NIAAA policy advocacy. The particular program model at Kennecott, which relies almost exclusively on self-referral rather than supervisory identification of poor work, was not, however, the employee assistance model that NIAAA came to advocate.

The second source of evidence was data from a training evaluation study in a single organization reported by Trice and Belasco (1968). This study involved the experimental evaluation of several types of training administered to the nonacademic staff of a large university. One criterion of training success employed was supervisors' readiness to confront an alcoholic employee. Although the results of the study are fairly complex and have a number of implications, it appears that NIAAA leadership was most impressed by the finding that training oriented toward general principles of supervision rather than training oriented toward alcohol problems was more effective in creating readiness to confront a problem employee. In other words, supervisors found more salience in an approach geared to identifying performance problems than one geared to identifying alcohol problems.

Other dimensions of the rationale for the employee assistance approach that have emerged include:

1. That it increases the probability of referrals of problem drinkers by allowing a broader program label that does not include the stigma traditionally associated with alcohol problems, i.e., the supervisor need not feel that he is labeling a subordinate as an alcoholic by referring him to the program.

2. That this broader label reduces the likelihood of the employer's program being perceived as an effort to interfere with a single area of employees' lives, namely, their drinking.

3. That this in turn enhances the likelihood that programs will be defined as part of the management system in the organization rather than as a prohibitionist or social welfare effort on the part of "moral entrepreneurs."

The need for the employee assistance approach was justified on the basis that earlier programs designed to identify the employee problem drinker had been relatively unsuccessful. In point of fact, evaluative data from these earlier programs showed that they did indeed enjoy considerable success (Franco, 1965), and there is no published evidence of program failure. This raises the question as to other possible motivations that may have led to the advocacy of the employee assistance strategy (Roman, 1975c).

First, of course, is the need for any new organization either within or outside government to establish itself as credible, and to do so it should demonstrate that it is innovative. The employee assistance approach was definitely new and different; it provided NIAAA with an approach that could be viewed as its own. Had the agency chosen to advocate the earlier approach without revision, it is clear that it would have had to give both credit and leadership to those who had been instrumental in instigating these earlier programs.

It also appeared that NIAAA desired to create a broad identity for its occupational program activities that had a strong "professional management" flavor and that avoided the provincialism that has traditionally characterized the alcoholism field. By developing a program strategy that included alcohol problems among its targets but that was not limited to alcohol problems, the stage was set for socializing those who would be recruited to diffuse the program concepts with an understanding of the structure and dynamics of management and unions, rather than sustaining an orientation toward fighting the problems caused by the misuse of alcohol. There may have also been a belief that this broader orientation and emphasis on developing diffusion agents or salesmen who would see themselves and be seen as professional consultants rather than alcoholism specialists would produce greater receptivity in work organizations where an alcohol specialist would run the risk of being identified as a social worker or other "do-gooder."

Finally, a more subtle motivation may have related to the typical pattern of prior occupational alcoholism programs to be operated by physicians. The rather intense boundary maintenance that surrounds physicians' professional activities was in sharp contrast to the absence of any shared professional identity among those who were being recruited to the field of occupational programming through the newly funded federal programs. To a large extent, a flavor of populism characterized the formative period of the national occupational programming effort. This populist base was not conducive to identification or affiliation with industrial physicians. To a large extent, the employee assistance strategy is advocated as a managerial approach that parallels or is part of the organization's personnel function. By advocating that the program be placed within the personnel function rather than the medical function of a particular organization, the diffusion or consulting activities would avoid the encounter with the intense boundaries that sometimes characterize medically based activities, especially when approached by a nonprofessional or paraprofessional change agent. Furthermore, greater ease of communication might be projected since personnel functionaries parallel occupational program consultants in that their own professional identity is not clear and they are drawn from a variety of prior occupational tracks (Ritzer and Trice, 1969). Therefore, it may have been assumed that occupational programmers would feel more comfortable in their field work by dealing with those in an occupation or a semiprofession that was not altogether different from the occupationally marginal situation in which they found themselves. This general hypothesis is substantiated by the fact that most of the consultants seem to have developed personal identities as management and personnel specialists rather than as paramedical health workers.

The employee assistance strategy has elicited many concerns. For a period of time in the early development of the occupational programming field, a polarity developed between those consultants who accepted the employee assistance strategy versus those who were more in favor of a "straight alcoholism" approach. Some of the concerns about the broad-brush strategy are as follows:

1. The possibility was raised that use of the employee assistance strategy may reduce potential effectiveness of a program in reaching problem drinking employees. By generating a large number of other people who represent the entire spectrum of human problems, program staff who are handling referrals may become occupied to the point that the number of problem drinkers being reached is a low priority concern. The issue is also raised that the employee assistance strategy, in eliminating the use of the terms "alcohol" or "alcoholism," *sustains* rather than reduces stigmatization of the disorder, the reduction of which is supposed to be the goal of occupational programs, particularly those funded through NIAAA. Finally, there is a concern that the employee assistance strategy will blunt the momentum of developing efforts to reach the problem drinker, particularly since the most basic rationale for occupational programs in the first place was that the majority of the nation's problem drinkers are employed.

2. A second area of concern centers upon program effectiveness. Experience of earlier programs had indicated that early identification, confrontation, and referral for assistance were effective in bringing about behavior change and returning *the problem drinker* to productive employment. However, no such evidence exists for rehabilitating or otherwise bringing about behavior change in other types of problem people, such as emotionally disturbed persons, those with marital and domestic problems, or persons with financial and legal problems. The use of confrontation techniques with emotionally troubled persons was seen by some as particularly problematic and potentially dangerous. Awareness of this danger, on the other hand, appears to have led some to advocate eliminating confrontation altogether in program procedures, although this aspect of procedure was originally the cardinal means for creating a crisis for the problem drinker and motivating him to take action about his problem. The extent to which effective means actually exist for dealing with personal problems not related to alcohol through providing the kind of assistance that returns them to productive performance within a relatively brief period remains an open question.

3. Given the fact that the funding for occupational programming activities was initiated through NIAAA primarily by grants to state alcoholism authorities, the question has been raised as to the legitimacy of advocating the employee assistance approach from a base identified with alcohol problems. A typical response to this issue is that while an employee assistance program does generate cases with other types of problems, it is also the most successful means of reaching the problem drinker. Within this logic, the extent to which other cases are actually dealt with is a side benefit. Data from those who are operating employee assistance programs are frequently quoted to indicate that well over half of the persons identified in these programs are problem drinkers.

4. Concern has been raised about the need for professional manpower to diagnose the problems of individuals who are referred from supervision, place these individuals in the proper community treatment resources, and provide appropriate followup. The extent to which such manpower, which presumably should possess broad professional training, can be employed in a work-based program at a reasonable cost has not been widely discussed.

5. Considerable concern about the employee assistance strategy has been raised by labor organizations. There is a sense among some in organized labor that this approach constitutes a "blank check" for management to identify any and every type of problem person under the pretense of him being a problem employee, thereby infringing upon employees' individual rights. Most labor organizations tend toward accepting programs that primarily emphasize identification and assistance for problem drinkers. There is also obvious resistance by labor organizations to cooperate in these efforts when the definition of programs is centered around strategies for improving management and productivity.

6. There is an open question as to the limits to a definition of a "problem employee" from the perspective of management. It may be difficult to distinguish between persons with "behavioral–medical problems" and those who are incompetent or inappropriately placed in their current job. The extent to which the employer can sustain a commitment to provide assistance to all such persons without jeopardizing these persons' current position or their job future is indeed a thorny issue. This extension of management responsibility to all poor performers may indeed be unacceptable to cost-conscious managers.

Apparently the philosophy of the employee assistance approach defines all personal problems that affect performance within a disease framework to the extent that the individual is not responsible for them and should not be penalized for their effects on his work if he undertakes efforts to resolve the problems. Particularly in light of the open-ended definition of psychiatric problems in contemporary society, clear-cut delineation of the population of "problem employees" eligible for benefits under those programs may indeed be difficult. This highlights the unclear relationship between these programs and the preexisting specialty of industrial psychiatry, a relationship that to our knowledge has never been explored.

In some instances, the philosophy of the broad-brush program has been extended to provide assistance for dependents of the employees as well as employees themselves, with the assumption that dependents' problems may have a detrimental effect on an employee's work performance. This broader extension of management responsibility has the sound of paternalism, and there is the very real possibility that the definition of such a large potential target population might further deflect attention from the identification of employed problem drinkers. This issue highlights the matter of coverage of costs for treatment for the range of problems that may be uncovered in an employee assistance program. Such coverage is generally limited, and only in more enlightened organizations is coverage for *inpatient* treatment of alcoholism provided, with coverage of outpatient care being quite rare.

Another frequently mentioned emphasis in employee assistance programs is the encouragement of self-referrals. This is seen as a step toward primary prevention, with self-referral ideally occurring before job performance impairment is evident. The assumption has been stated that if a program established its credibility through successfully handling cases referred to it, self-referrals will increase and the need for supervisory identification of problem employees will decrease. Coupled with this contention is the assumption that self-referrals of problem drinkers will also be generated, which appears to defy most reported experiences about the need for crisis precipitation preceding behavior change on the part of most problem drinking persons, particularly those developing psychological and physiological dependence on alcohol.

The extent to which the employee assistance concept has been effectively diffused and is effectively operational remains an open question at the present time. There are many programs that have been labeled as "troubled employee" or "employee assistance" programs, but the actual internal functioning of these programs in terms of the cases identified and case outcome has not been widely documented. At the present time, it does appear that the broad-brush strategy is accepted by a substantial proportion of those engaged in consultation activities, although repeated criticism of the approach has come from others with commitments in the alcoholism field, particularly personnel affiliated with the National Council on Alcoholism as well as leadership in organized labor.

The crux of the matter is the extent to which the employee assistance approach is used as a tool to reach developing alcoholics in the work force in contrast to becoming an end in itself. It appears that the latter tendency is more evident, which carriers the risk that program satisfaction will develop on the basis of the number of problem persons seen and assisted, regardless of their problem. Without doubt, there are temptations and pressures to avoid dealing with alcohol problems directly, given the stigma of the problem and the generally low status of the alcoholism field in the overall professional health care and social welfare communities. The employee assistance approach provides an avenue toward the mental health arena, in which many program specialists may find themselves more comfortable. It does appear, however, that the value of occupational alcoholism concepts may be eclipsed unless the employee assistance strategy is clearly and openly advocated as a means for reaching the neglected population of employed problem drinkers, and is evaluated on that basis.

Constituency Support: The Beneficiaries of Occupational Programs

In this section, we shall briefly consider the interest groups involved as potential beneficiaries of occupational alcoholism and assistance programs, which reflect upon the extent to which this fledging movement has opportunities for becoming a permanent part of the overall societal effort to deal with alcohol problems.

Given the current philosophy of using federal funds as "seed money" for programs of many types, to be followed by state and local assumption of funding with combinations of local and revenue-sharing funds, it is clear that potential future support for occupational programming efforts is of considerable importance. The bulk of effort in diffusing occupational programs has been supported by federal funds to date, although these sources of support are already beginning to be withdrawn with the expiration of three-year periods of funding. These grants are rarely renewed unless substantial local funding has been generated, reflecting the above-stated federal philosophy.

One of the major ambiguities of occupational alcoholism and employee assistance programming efforts centers around the constituencies that they serve: the identity of the primary beneficiaries of occupational programs is unclear. This in turn may affect (1) the occupational and professional identities of those involved in this effort, (2) the thrust that individual programs take, and (3) the eventual constituency that becomes the advocate for continued and expanded funding of occupational programming efforts, as well as constituencies that perform as active proponents in the further diffusion and adoption of occupational programming concepts. Among the possible beneficiaries are business and industrial management, organized labor, the community, NIAAA, state and local governments, and individual problem drinkers.

The beneficiary that is the most frequently mentioned is the management of work organizations in both the public and private sectors. The early identification of problem drinkers and other problem personnel is advocated as a means of returning these individuals to productivity and avoiding the costs of hiring replacements, which entails providing the necessary training that the persons who develop problems already possess. This contention is backed by the partially substantiated notion that most persons who are identified in such programs are relatively long-term, middle-aged employees in whom the employer has considerable investment. The accompanying assumption, with greater substantiation, is that a large proportion of problem persons so identified are effectively returned to useful productivity. A closely related benefit for management is the extent to which such a program reduces costs in terms of problem employees' absenteeism, scrap rate, poor public relations, and disruption of the morale of fellow employees. These two major sets of cost-savings pose management as the primary beneficiary of occupational programs. It is crucial to point out that the evidence about tenure of problem employees and their return to successful productivity is based almost solely on data about *problem drinkers,* with minimal evidence to project these aspects of benefit across all types of cases generated in an employee assistance program.

This orientation toward managerial benefits has led to considerable resistance on the part of leadership in organized labor, particularly within the AFL-CIO. Some labor leaders regard such a program philosophy as potentially exploitive of union members. There is less but equally significant labor resistance to the concept of identification via impaired job performance. This is based on the belief that performance measurement is much clearer in the case of lower echelon employees, who are more likely members of the bargaining unit, and less likely in the case of white collar personnel, who are outside the bargaining unit. Finally, there has been considerable labor resistance to the employee assistance approach in that this strategy is seen as giving management a relatively free hand in identifying troublesome individuals and possibly bringing about their termination.

It appears that labor leadership is much more receptive to another conceptualization of benefits, namely, that an occupational alcoholism program comprises a broadening of employee health benefits. In this view, the program should provide the assurance that a problem employee will not be terminated for his drinking, but rather will be given the opportunity to seek help for his problem without jeopardizing his current job status or his job future. This is viewed independently of the program benefits that may be produced for management. As many occupational program specialists point out, the benefits projected by management and those projected by labor are not necessarily contradictory and can coexist, presumably through *joint* labor–management programs. This, however, may be overly simplistic. For example, some leaders in organized labor take the position that it is inappropriate to place limits on the efforts to rehabilitate a particular employee and that posttreatment "slips" should be reasonably tolerated as part of the rehabilitation process, which may run contrary to management's commitments. There has not yet been widespread publicity to "crucial cases" where labor has publicly protested the handling of problem employees within an occupational program. The reality and mechanics of "joint" union–management programs remain to be carefully documented and are a pressing issue at the present time.

It appears that while labor, as a collectivity, might emerge as a significant constituency for sustaining the diffusion of occupational programs, it is less likely that management will do so. Those companies that have adopted programs are supporting the activity themselves, and in some instances consider the existence of a program to be a private matter. It is conceivable that managements could band together through some form of organization to promote occupational programming concepts, but this does not appear likely. Benefits for companies are localized, and there seems to be little reason for management to work to improve the functioning of other companies with whom they have no direct beneficial relationships, and with whom they may in fact have competitive relationships. Thus, the tendency of the occupational programming movement to respond to management's needs may not enhance constituency development, whereas a greater promise of future aid probably lies with commitments and interest from organized labor.

In terms of the community as beneficiary, occupational programming efforts are novel in shifting partial responsibility onto the employer and/or labor organizations for the identification and referral of problem drinkers and other problem people. In the past, community efforts to generate earlier identification of problem drinkers have been limited to public education directed both toward the general public and toward "middlemen" such as teachers, policemen, social workers, and clergymen whose activities may bring them into contact with early stage problem drinkers. The effects of these campaigns have been questionable. It is certainly the case that assumption of a large proportion of these responsibilities by work organizations would enhance community outreach efforts. This, of course, presumes that a community accepts the goal of maximizing early identification and referral to reduce social costs of coping with those who are identified in the later stages of problem development. This benefit ties directly to the notion of "corporate social responsibility," which has been an increasing subject of public concern in regard to issues such as environmental protection.

The possibility that the employer will accept and implement a program on the bases of *both* corporate social responsibility and improved management is not assured, however. It is clear that acceptance of the social responsibility goal as primary can undermine the extent to which substantial investments in the program are made (Roman, 1975b). In other words, this goal may lead to design and use of the program for its public relations value in the community, with investments made only to the extent necessary for positive feedback from the community. It is possible, therefore, that the community's goals can alter the program's goals within a company, to the detriment of the overall purpose of reaching employed problem drinkers. The prospects here, however, may be positive. Should the assumption of responsibility by employers be viewed favorably and should the results of occupational programs visibly reduce a community's alcoholism problem, it is possible that community leadership across the nation could emerge as a constituency group for the promotion of occupational programming.

Closely related to community benefits are those desired and experienced by NIAAA. It is clear that its mission is to maximize outreach efforts to alcoholic people throughout the nation, and if the assumption is correct that a substantial proportion of developing alcoholics are located in the labor force, then occupational alcoholism programs represent a definite step toward the achievement of NIAAA's mission. The employee assistance approach does cloud this benefit to a certain extent and, as mentioned, the issue has been raised as to whether NIAAA-funded programs should extend beyond dealing with persons affected by alcohol problems.

NIAAA appears to find itself in the middle of the conflict between the perceived benefits for management and those perceived by organized labor. Observations indicate that organized labor has been very active as a specific constituency in attempting to influence the kinds of benefits associated with occupational programs that are advocated by NIAAA. To our knowledge, there have been no organized contacts between NIAAA and organizations of employers, which may in turn reflect the degree of success of the diffusion of occupational programming concepts.

Furthermore, in terms of benefits, it should be noted that as a new organization, NIAAA's visibility is enhanced by the innovative dimension of occupational programs. These programs' reliance upon local initiative and local support is consistent with the aforementioned federal emphasis upon reducing federal involvement in ongoing programmatic activities at the local level.

Another benefit associated with the occupational programs for NIAAA is the extent to which these programs further the goal of defining alcohol problems as health problems. Presumably, the diffusion of this concept to employed people through work organizations, and through specific written policies that pertain directly to employees, would comprise a large step toward persuading the general public that drinking problems are appropriately defined as health related rather than as criminal or immoral. By accomplishing this, a nationwide enhancement of the humane treatment of alcoholic persons may result. In light of this reasoning, therefore, it is a good possibility that NIAAA itself will remain a strong advocate for sustaining the occupational programming thrust.

The next set of potential beneficiaries are state governments, most of which were the recipients of three-year NIAAA occupational programming grants for support of two consultants. The mission of the state alcoholism authority and its local agencies is similar to that of NIAAA and the potential benefits stemming from occupational programs are similar to those outlined above. There are, however, two ways in which occupational programs create ambiguities for state and local governments. The first is the extent to which such government agencies can serve as an effective base for delivering services to private industry and labor organizations. Traditionally, state government's role in relation to the work world has been of a regulatory nature. This may be a detriment to the receptivity encountered by those attempting to deliver consultative and technical assistance services from this base. State and local government agencies may likewise find them in a tenuous position in terms of the potential conflict of interest between management and labor in regard to occupational programs. Advocacy of either set of interests is clearly not the role of state or local government.

The second major ambiguity of benefits to state government centers around the state's prior and ongoing investments in relation to alcohol problems. For the most part, these efforts center upon the delivery of treatment services, and, in large part, these treatment services have been geared toward the visible alcoholic person who is usually of the public inebriate type. Occupational programs call for the development of treatment resources appropriate for and acceptable to employed persons. In many instances the preexisting resources in the state, supported by state funds from various sources, may not be of this nature. Thus, occupational programs may create conflicts in terms of vested interests that have been developed around treatment of the chronic alcoholic.

Furthermore, in its idealized form, an occupational program is designed to bring about early identification in a maximum number of cases so that the need for treatment of any kind be minimized. This also may be viewed as a conflict with existing vested interests in treatment supported by state and local government. Finally, the emphasis of the employee assistance approach and its potential generation of a wide range of problem persons other than those affected by alcohol may be viewed as not only in conflict with the goals of the state and local alcoholism authorities, but as an inappropriate activity to be based within that authority. Thus, while it is clear that occupational alcoholism programs can conceivably enhance the goals of state and local alcoholism authorities, it is also clear that conflict with existing vested interests involved in treatment may be particularly sharp here.

The final set of potential beneficiaries to be considered are problem drinking persons themselves. At first blush, it appears unequivocal that occupational programs constitute an enhancement in the welfare of the alcoholic population through identifying problems at an earlier point and intervening before the consequences of excessive drinking have been personally costly.

Such an assumption may, however, require a rather rigid disease-oriented concept of alcohol problems. It assumes that all persons with the disorder, no matter what their stage of development, actually desire or can be motivated to desire help for their problems. While the strategy of constructive confrontation was originally central in occupational programming concepts and is intended to create motivation for behavior change, it is not a logical necessity that the individual accept the label of alcoholic and identify himself as part of the alcoholic population. In point of fact, there are probably a substantial number of individuals who use alcohol in such a fashion that it interferes with their job and can be identified thereby as poor performers, but who are not necessarily destined to become addicted to alcohol if intervention does not occur.

Thus, it is questionable whether early identification efforts such as occupational programs are truly tapping the potentially alcoholic population, or whether this illusion is created by the labeling processes that ensue in treatment agencies to which referrals are made subsequent to identification. Indiscriminate labeling may indeed be used as a means of providing the individual program and its management with a justification for continued existence or expansion. One of the implications of this labeling is the criterion frequently used for treatment success, i.e., the demand of total abstinence as a prerequisite for total recovery from alcohol problems (Moberg, 1974). Such a criterion may be unnecessary and inappropriately rigid in the case of early-stage problem drinkers. Thus, from both a conceptual and practical viewpoint, care should be taken in defining the characteristics of the early-identified group of problem drinkers that may be generated by an occupational program. The assumption that their inevitable progression to full-blown alcoholism is curbed by early intervention may indeed be an overgeneralization that has potentially unfortunate side effects via labeling processes.

The confusion over the foregoing set of potential beneficiaries of occupational programs is manifest in the confusion over the occupational and professional identity found among those working in the field. There is no commonality as to whether they are government agents, adjunctive aides to industrial management, or advocates for the welfare of alcoholic people. This identity confusion seems to be a barrier in the potential professionalization process that may eventually develop, particularly the kind of full professionalization including certification that is desired by the leaders of the occupational associations that have sprung up in this field.

The confusion is also reflected in the variation in organizational location of occupational programs in both public and private work organizations. In some instances the programs are found within the personnel department or function, and in other instances they are found in the medical department or function. It would appear that these two locations represent distinctively different conceptualizations of the program's role and its beneficiaries. Placement in the personnel department appears to identify the program as an explicit arm of management tied tightly into the normal disciplinary or adverse action system that has been traditionally implemented to deal with performance problems. Where a program is placed in the medical department, it is defined as part of the health benefit package accompanying employment in the organization and is fairly separated from disciplinary procedures. These departments do, of course, have different images in different companies, so generalizations of this sort should be cautious. There are numerous instances where a program is located in a position separate from either personnel or medical functions, but it would also seem inevitable that it be identified with some larger component of the organization's structure if it is to be an integral, ongoing activity.

A final example of the manifestation of the ambiguous identity of occupational programs relates directly to our concern in this section, namely, the current absence of significant public advocacy for their continued functioning among the described beneficiaries, with the exception of NIAAA. As problems of continued funding are developing, both in terms of further federal funding for consultative activities as well as funding for internal program expansion, it appears that the advocacy of the occupational programming concept is limited to those who are part of the field. In other words, no reliable constituency has developed to promote these efforts, which may place their future in a somewhat ambiguous status. This, of course, may also be an artifact of time and the possibility cannot be overlooked that a constituency may emerge as the benefits of occupational programs become more clearly documented. The relative absence of constituency development does, however, reflect upon the provincialism that has pervaded the occupational programming field so far, and its general failure to integrate itself with preexisting power structures represented by the categories of potential beneficiaries outlined in this section.

It appears that the employee assistance program approach has contributed greatly to the confusion over beneficiaries and the development of a viable constituency. By extending beyond alcohol problems and in some instances by deliberately avoiding identification with the complex of workers in the alcoholism field, the employee assistance approach has undercut potential programmatic advocacy from this sector before the concepts have diffused to the point of developing an independent constituency. The employee assistance strategy has contributed to making occupational programs "neither fish nor fowl" in terms of potential proponents outside occupational programming itself. Whether this will constitute a crippling of the "movement" is certainly unclear, but it does appear that this strategy may have been developed prematurely in terms of the crucial necessity of constituency development.

Issues in the Identification Process

The major responsibility for identifying the problem drinker lies with his immediate supervision. Identification is to be based on documented decline in job performance, and it is purported that action at this point will afford early identification of the problem drinker. In some programs, identification is to be the joint responsibility of the supervisor and the shop steward, and the importance of documentation in organized shop situations is obvious.

With supervisory identification being the pivotal point in program functioning, it is obvious that both the quality of supervision and the existence of clear-cut standards for work performance are crucial ingredients in successful identification. The problematic nature of these two precursors of identification has not received much attention in occupational alcoholism programs.

First, supervisor–subordinate relationships do not always include the clear element of monitoring job performances believed to be typically present. The ideal of readily identifying impaired job performance is most closely approximated in manufacturing settings involving assembly lines or production quotas. Quite obviously, this monitoring model cannot be generalized to the wide variety of work situations that abound in American industry, and it is evident that classic production settings directly conducive to output measurement are on the decline. The complexity of work processes has tended generally to increase, diluting the opportunity of supervision to perform its monitoring role. The monitor role remains a part of the supervisory function, but increasing complexity of work makes this function more difficult to perform. Furthermore, the wide-ranging mobility at higher occupational status levels, coupled with freedom from direct supervision of formally designated subordinates, makes the concept of "impaired performance" a rather simplistic one.

It should also be observed that in practically all "man–boss" relationships, subordinates can exercise a certain amount of power, leading through a variety of maneuvers toward "power equalization" and a subsequent ability of the subordinate to bargain and negotiate with formal supervision. This is especially true when considering the problem drinker, since he is exceptionally able at manipulation and maneuvers designed to prove that his performance is intact and that any "impairment" is normal. As a result, confrontation on the basis of declining performance may lead to the problem employee appealing to a variety of environmental conditions and excuses that he claims impact his work performance as well as that of others doing similar work.

Further complicating the use of "impaired performance" as a simplistic absolute upon which all occupational programs pivot are the built-in distortions that characterize the observation of performance process. Thus, supervision may *try* to monitor performances characterized by low visibility of both performance and productivity, i.e., the railroad brakeman. Supervision often falls prey to the "halo" effect and its opposite, i.e., professionals and higher status employees tend to perform better because of their relatively high status, while lower status personnel are correspondingly rated lower in performance quality. "Halo" effects may be particularly prominent when an employee has a superior work history that has led to upward mobility within the organization. Consequently, "impaired performance" is likely to be found among lower status employees (Warkov and Bacon, 1965).

Supervision often has available only generalized, subjective criteria of performance, forcing the use of particularized, impressionistic performance criteria. These flaws in the performance evaluation process should not, however, cast doubt on the centrality of the appraisal process in occupational programming. Judgments, interpretation, and evaluations of the performance of other persons in favorable or unfavorable terms are inherent in human interaction processes. Interaction in the workplace is infused with value judgments about the performance of fellow workers, of subordinates, and superiors. Performance evaluations are inevitably present in the perceptions of people interacting to some degree in the workplace. Distortions often blur these assessments; simplistic and ritualistic methods for extracting these judgments typically compound these flaws. In short, performance appraisal is a human process that contains the error, biases, and stereotyping present in any decision-making that people make about other people.

Its advantages for procedure in occupational programming remain clear despite these obvious flaws. Supervision does exercise authority but this authority cannot be used in an arbitrary manner. Decisions are subject to review by both the next level of supervision and often by the union. Supervisors are frequently required to clearly document their decision that a particular subordinate's performance is impaired. Combined with these pressures toward accuracy is the impersonal, production-oriented nature of the workplace. This context forces supervision to be sensitive to poor performance and legitimates supervision's awareness of the marginal performer. The persistent and accumulative impact of the problem drinker's poor job performance becomes a performance matter to which supervision is particularly sensitive. At the same time "normalization" processes and ambivalence about recognizing and taking action will typically occur (Roman and Trice, 1971). But inevitably appraisals of poor, ineffective, and impaired performance will emerge. It is the telescoping and shortening of this process that are the key goals of an occupational alcoholism or employee assistance program.

While individual performances may be difficult to document, the growing interdependence of jobs provides an auxiliary mechanism to hasten identification. These are not simply blue collar situations. Increasingly, technology has produced jobs demanding careful coordination among a group of employees. Development of client-oriented performance evaluations for use in newly emerging service occupations that involve work with people is another indication of the ubiquitous availability of performance evaluation in the work setting. When we apply these points to problem drinking employees, it seems obvious that they especially are observable through performance assessment, keeping in mind the barriers and errors described. The tendency for the impaired performance of alcoholic employees to be readily observed has been confirmed in one study (Maxwell, 1972).

To summarize, "impaired performance" exists when a supervisor or functionaries using organizational authority and standards consider it to exist; after all the deficiencies and inaccuracies of observing others are considered, ineffective performance comes into being when those persons in organizational life who are charged with "getting results" decide it exists. "Impaired performance" is what supervision *defines it to be,* and what they believe to be sustainable with their own supervision and staff personnel. While simplistic formulas cannot be offered to improve the appraisal process, the same holds true for documentation: The "evidence" will be gathered and presented in some fashion, but specification of *what* evidence will suffice across a range of work situations is likewise simplistic.

Attendance records, including both absenteeism and tardiness, comprise performance standards that are more independent of judgemental variables, but here also there are problems. First, excuses for absences and tardiness are usually items for negotiation between supervision and subordinates, and a lack of confidence in the employee's excuses leaves supervision with the undesirable and time-consuming alternative of detective work. Second, absences are contingent upon the employee's physical and social visibility on the job, such that middle and upper status employees may avoid being absent even when in an impaired condition. Again, one must accept the definition that satisfactory attendance is what supervision defines it to be.

These unresolved issues in the identification process are reflected in the enthusiasm of occupational program consultants as they attempt to proselytize the management community to the concept of "doing something" about problem drinkers and other problem employees. There has been a clear tendency for such consultants to define their primary activity as "spreading the word" in a shotgun fashion, which may not enhance their credibility to decision-makers. By presenting rather glib characterization of the supervisory process and implying that formulas for performance assessment can be established, consultants and other program advocates may inadvertently orient their prospects in management positions to rejecting the concept or to adopt programs on the basis of humanitarianism and social responsibility.

This is in no way to impugn the advocated procedure of identifying problem employees via declining performance; this is indeed the essence of the occupational alcoholism and employee assistance approaches. What is needed is a broader preparation of occupational programming personnel to truly function in change agent roles rather than in sales roles. Knowledge of organizational behavior can set the stage for "organizational diagnosis" particularly in regard to the identification process (Levinson, 1973). This in turn may produce a procedure and program for managing problem employees that is geared to a particular organization's characteristics, structure, and union atmosphere work flow. Such as approach is badly needed if viable programs are to be firmly established (Roman, 1975a).

Issues in the Referral Process

Following identification of the problem drinker or other problem employee and confronting him with documentation of his declining performance, he is offered the alternative of seeking assistance for this problem for which he will not be penalized, or the alternative of receiving the disciplinary considerations that the personnel rules provide under circumstances of repeated poor performance.

A major issue in regard to this process is the extent to which referral to treatment or counseling is made a mandatory requirement. To a considerable extent, early occupational programs emphasized the potential impact of constructive confrontation as a crisis-precipitating event that could potentially bring about an alteration of behavior in and of itself. The *minimizing of the need for treatment* was viewed as a particularly meritorious feature of this program approach, assuming that confrontation occurred at an early enough point for the individual to bring his behavior under control (Roman and Trice, 1968).

This emphasis appears to have declined sharply and even disappeared in current occupational program design. Instead, the medical model has come to prevail in the definition of all problems that cause performance decline as being in need of treatment. The "rush to treatment" emphasis appears in some instances to have even supplanted the confrontation itself. The reasons for this development seem to be (1) minimizing supervisory responsibility for and involvement in the problem itself once identification has occurred, and (2) assuring that assistance is rendered by those who are capable of making the correct diagnosis and selection of treatment or counseling regimen.

This procedure may well be effective, but it does raise several issues. First, there may be a tendency to elaborate the program structure within the work organization once this involvement in treatment becomes widespread. This contrasts with the typical contention that development of an occupational program requires the organization only to alter its procedures rather than to develop a new organizational component of significant scope.

Second, there may well be temptation on the part of the "treatment industry" to maximize the services that are necessary to return the employee to productivity. Without doubt, treatment centers survive and grow on the basis of the services that they deliver. This tendency may be more pronounced when a treatment agency itself becomes the base for setting up occupational programs in organizations.

Third, the logic behind minimizing the use of treatment services is partly based on the potential impact of labeling that inevitably occurs when persons are processed through treatment facilities. These labels not only create risks of subsequent stigmatization on the job and in the community, but can also act to alter the self-concepts of individuals, which in turn may impact on their behavior (Roman and Trice, 1968). Labels may have motivating impact when used appropriately, but widespread evidence of the casual use of labels in treatment settings, due in part to bureaucratic requirements, leads to a caution about the tendency to employ the notion of "if in doubt, treat."

Very little is known about the reaction of fellow employees to persons who have been processed through treatment and returned to their jobs. While recovery from alcoholism through Alcoholics Anonymous is believed to include effective mechanisms for minimizing stigma (Trice and Roman, 1970b), little is known about the stigmatizing consequence of other modalities for alcohol abuse or other disorders.

Role of Labor Unions in Occupational Policies and Programs

One of the main differences between what is often called "older" occupational programs and those of more recent origin is the involvement of the union. Although there were notable exceptions, earlier programs tended to be initiated by management and reacted to by unions. Recently, however, American labor unions have become much more active in the development and execution of occupational programs.

In 1958 the AFL-CIO Community Services Department expressed a strong interest to participate "in any program that attempts to treat this problem fairly" (Perlis, 1958: 536). More recently the United Auto Workers' 23rd Constitutional Convention adopted a resolution saying, in part, "The U.A.W. ... must intensify its efforts to develop cooperative programs with management . . . to help identify addicts requiring assistance and to arrange for referral to effective community treatment resources" (U.A.W., 1972: 1). These expressions of willingness to engage in cooperative and joint programming with management represent considerable change in union thinking about management-generated mental health programs in general. Often such programs, including alcoholism policies, were viewed as a guise for antiunion practices. Reinforcing this suspicion has been a tendency for occupational alcoholism programs to be applied largely to employees in the bargaining unit, excluding supervisory and professional personnel for all practical purposes. To labor unions such practices look like discrimination, which is of course accurate.

Despite these negative feelings, however, the trend toward union involvement and joint participation has continued. In 1973 the AFL-CIO's Department of Community Services, working with the Labor–Management Division of the National Council on Alcoholism devised "Suggested Guidelines for Labor–Management Agreement." These spelled out a suggested joint policy statement, the structure of a joint committee with equal membership from union and management, and specific procedures for case handling, including a section on crisis-precipitation section that advocated underscoring to the employee with declining performance "that unless his problem is identified and corrected he is subject to existing penalties for unsatisfactory job performance and attendance" (National Council on Alcoholism, 1973: 22).

Simultaneously the U.A.W. was writing into several collectively bargained contracts specific language spelling out an alcoholism policy and program. These 1973 contracts were company-wide in General Motors Corporation and in John Deere and Company, and specific to the Kenosha, Wisconsin, plant of American Motors. Each has a joint statement of policy agreed upon by labor and management regarding alcoholism and employment. They provide for the establishment of the above-mentioned joint committees, which are authorized to develop procedures for referring a problem drinker to treatment resources and to train supervisors in making referrals as well as establishing liaison with treatment facilities and evaluating their effectiveness. Other salient features of the contractual provisions are: The joint policy and program are directed toward alcohol abuse that affects an employee's job performance or attendance; he must seek treatment or improve his job performance, or both; all supervisors are responsible for assuring the employee that his security and promotional rights will not be jeopardized by diagnosis or treatment; confidentiality will be maintained; finally, if necessary, the program will not be used to avoid normal disciplinary procedures. The overriding emphasis is on the ability of line supervisors, in conjunction with shop stewards, to diagnose poor job performance, not alcoholism. They are expressly instructed not to speak of the impaired performer as "alcoholic."

A particularly important aspect of the model AFL-CIO program and the U.A.W. contract language is that they enlist the shop steward as a direct participant in the confrontation process with the supervisor, thereby bringing to bear the combined power of both the union and management for constructive confrontation (Trice and Roman, 1972). Rank and file employees are often suspicious of what management says it is doing for them; the shop steward is in an excellent position to reduce these doubts. Hopefully, these positive indications by powerful labor unions will act to convince management that unions genuinely wish to join with them in an effort to combat alcoholism. An example of such sentiments can be found in the Collective Agreement the Steel Workers' Union has with the basic steel companies (Morris, 1972: 343).

> Without detracting from the existing rights and obligations of the parties recognized in other provisions of this agreement, the company and union agree to cooperate at the plant level in encouraging employees afflicted with alcoholism to undergo a coordinated program directed to the objective of their rehabilitation.

Perhaps these kinds of concrete action will help reverse the tendency of management to fail to consult with the union when devising a program. In 1968, for example, the Industrial Conference Board reported that of 120 manufacturing firms contemplating or actually implementing occupational programs, only 17 fully consulted with their union with an additional 22 doing so "to some extent."

As encouraging as these trends are, they also generate complications. Among the most vexsome are: (1) The alcoholism policy becomes a part of the overall labor–management relations climate; (2) the grievance-arbitration machinery may become involved, placing the arbitrator as a new and potent functionary in the procedure; and (3) the development of employee assistance policies has stirred up considerable union resistance, possibly acting to dampen general union interest.

The effectiveness of any joint program obviously will be influenced by the general relations between union and company. Should be union be convinced that the company is breaking some other part of the collectively bargained agreement, it seems unlikely it will cooperate with management in the area of treatment for alcoholism. In such a hostile environment, for example, the union's reaction to a company physician who says that a shop steward has an alcohol problem will likely be to "walk the last mile" in the grievance procedure in order to fight any use of the policy (Morris, 1972). We have, however, seen instances in which the illness aspect of alcoholism has superceded adversary tendencies even through hostile relations otherwise existed. On the other hand, we have observed situations in which union cynicism about the "jointness" of a program was clearly expressed.

Arbitrators can be involved whether or not an actual alcoholism program exists. When the main points of a program are written into a contract, however, such language must be interpreted by an arbitrator should a case involving the policy be pressed by either management or the union. The record of arbitrators' actions in dealing with such cases is instructive. Although arbitrators tend to favor management's disciplinary rights where off-job behavior destroys employee efficiency, the arbitrator called for reinstatement in more than half of the published cases dealing with the discharge of problem drinkers through 1966 (Trice and Belasco, 1966). Followup of these reinstated employees, however, indicated that 75 percent were no longer employed at the same company, half of these having been discharged for a reoccurrence of the same offense.

The reasons given by arbitrators for reinstatement were similar to those in other cases of discharge, namely, "insufficient evidence," "mitigating circumstances," and lack of a "consistent policy." The largest percentage of alleged alcoholics were reinstated for "insufficient evidence." In a review of more recent arbitration awards Somers (1975) indicates "a continuation of this pattern of reasons for reinstatement." Thus, it is reasonable to estimate that in 80 to 90 percent of the cases reaching arbitration everyone loses: The employee loses his job either immediately or after a relatively short period of time, the union loses great amounts of grievance time and steward energies attempting to help the employee, and the company often loses an experienced and often valued employee. From the perspective of the alcoholic employee, reinstatement is a Pyrrhic victory. For a considerable period, he has gone to great length to convince significant others and himself that "nothing is wrong" and that he is a responsible and able worker. His fantasies and delusions are confirmed a thousand times over when he receives a formal reinstatement from a professional arbitrator. It is his supreme victory in his fight to demonstrate that he is normal. As such it probably reinforces his determination to use alcohol even more as a device to cope with life.

Thus, the involvement of the arbitrator can reduce the impact of crisis precipitation. Bound by legalistic rather than health concerns, the therapeutic value of the crisis precipitation strategy is less meaningful to him. In all probability he is not aware of the *unique* nature of alcoholism as a health problem. Compounding this ignorance is the tendency for present contract language to deal with alcohol intoxication on the job rather than with the alcohol dependency. In only a few instances do contracts deal with alcohol-related behaviors. When they do they address themselves primarily to "immediate discharge for intoxication on the job," or "discharge for proven drinking or proven drunkenness while on duty or proven under influence or possession of illegal drugs while on duty." It is this "letter of the law" that confronts the arbitrator. The behavior presented may, or may not, be alcoholism.

Arbitrators obviously need to be able to consult with alcoholism specialists in order to decide if the behavior being judged is the result of a normal employee coming to the workplace intoxicated (which happens frequently) or if it represents the developing process of a severe health problem. The use of impaired performance as the basis for occupational program policies provides a criterion free of the confusion of medical and health questions, providing maximum objectivity. All things considered, arbitrators would be well advised to (1) attempt to decide if an alcohol-related case represented problem drinking of some degree rather than social drinking that became extended into the job situation; (2) if they so decide, insist on using the opportunity to precipitate a crises by making reinstatement contingent upon entering and remaining in treatment; and (3) arbitrators who wish to make a contribution to the reduction of this severe health problem might consider a "problem-solving" approach to these types of cases rather than a judicial approach. Even though the primacy of the contract as the major governing instrument is never doubted, some mediation often creeps into the arbitration process. Alcoholism cases call for the arbitrator to introduce this education process for both parties by being less "strict constructionists" and recognizing that his responsibilities *under the contract* allow for constructive and positive innovation and imagination.

The next barrier to full union cooperation is probably more pervasive. Literally thousands of grievances never come to arbitration, and in our observations, most cases of alcoholism, fortunately, do not become embroiled in the grievance-arbitration process. Both parties, however, are constantly concerned that threatening and difficult cases will clutter the system and often go to substantial lengths to avoid them. The recent advent of the employee assistance program concept has, however, touched sensitive nerves in many union circles. Realizing full well that nonalcoholic problem persons will be surfaced by any policy based on "impaired performance," we offer the following surmises regarding why the strategy has created such resistance. First, it dramatically extends a management option about which unions have been traditionally suspicious, namely, mental health (Weiner, 1967; Holmes, 1963). If such a policy covers a large variety of behavior problems, it can easily be seen as an open-ended device that management can use to control practically any form of dissent of "trouble-making." There is a feeling that the expansion of alcoholism programs to cover any and all factors that many account for impaired performance constitutes the writing of a blank check, signed by the union.

Behind this concern lies a general uneasiness and vagueness regarding other emotional disorders. Alcohol and alcoholism are far more concrete, observable, and tangible than depression, the phobias, and the neuroses. There appears to the union to be less risk in joining management in an alcoholism program. There is, however, a justified fear that populist psychiatry may be deliberately, or with good intentions, used to unfairly harass, dominate, control, even discharge, employees unpopular with management.

There is also a more direct and defensive concern that expanded services as provided by employee assistance programs compete with services already provided by the union. In some fashion or another, most major unions provide general assistance to members through union counselors and community services representatives. Thus, "jointness" by a union with management in an employee assistance program could easily be seen as turning over the traditional role of helper for troubled union members to a program in which management plays a larger role than the union.

Furthermore, unions may see employee assistance programs as leading to many new and difficult collective bargaining and grievance problems. Once formal procedures begin, health problems would become adversary problems, delaying the appropriate rehabilitation approach. Given the right of any member to press his union to help him grieve, the union could conceivably be flooded with grievances in such a broad program, especially one written into a binding contract. There has tended to be only a partial consensus between unions and management about what that term "impaired performance" means, leading to additional possible complications. In sum, many unions may well believe that employee assistance programs will compound and confound the grievance machinery operating within the labor contract.

Like management, labor comes in many local forms and structures. No simplistic policy or general reaction can be said to characterize the union reaction to employee assistance programs. We have observed instances of a union's complete acceptance and participation in joint employee assistance programs. We have observed the traditional "wait-and-see" cautions, and we have observed open, irrational hostility to such programs. In many ways, the conflict is a semantic and symbolic one. Perhaps the best approach for both parties is for vigorous and full implementation of their alcoholism policies on the basis of identifying declining performance and adaptation to whatever other problem behaviors emerge. Quite assuredly these will be many.

Evaluation of Occupational Alcoholism Programs

Evaluation of occupational programming emerges not only from the pressure of funding agencies to account for expenditures of federal and state monies involved in stimulating public and private sector programs. It comes also from concerns and questions about the major strategy in occupational programs, namely, that these programs produce substantial results in terms of altering the behavior of alcohol-dependent persons. The relative absence of evaluative studies is striking when contrasted with current interest and increased efforts.

In terms of evaluative efforts, two basic designs have appeared. First, and most frequent, have been the historical, time-sequence studies to evaluate programs. In this strategy data on job behavior are collected for a uniform period prior to the intervention of the program and its strategies and for a uniform period following the intervention. Comparisons are then made between how program participants behave before they experienced the program with how they perform on the same behavioral criteria after going through the program. A second and more recent design, is to attempt some experimental pattern, using some type of comparison group in an effort to demonstrate the program's effectiveness. The strategy here is to assess whether those who experience the program's intervention improve performance "in a desirable direction" in contrast to those who do not. Thus, persons who voluntarily entered a rehabilitation program are compared with those who were "coerced" to undergo rehabilitation. This second type of evaluation is more complex and obviously may be flawed in methodology.

Common to both these approaches is the problem to criteria or the "yardsticks" of behavioral or attitude change that determine if an intervention has produced desirable results. These measures range from the number of days of sickness absences to overall global criteria made up of improvement in job status and location as well as salary and working conditions. Other criteria that have been used include job maintenance, supervisory improvement rating scales, medically assessed recovery scales, off- and on-the job accidents, promotions, formal leaves of absences, grievances, and disciplinary actions as indexes into possible improvement.

An early evaluation of the historical-time series type without control groups was reported by Franco (1960). Using job maintenance as one criterion, he reports that 72 percent of those who stayed in treatment for one year maintained their jobs. After four years, 51 percent had maintained their jobs. Those who refused treatment also have a high level of job maintenance (61 percent). Using supervisor ratings of work performance, sick absenteeism, and medical assessment he also uses a recovery scale to evaluate improvement over time. The scale permitted a decision concerning degrees of recovery both before and after treatment or confrontation without treatment. Of those who were described as maintaining their jobs over the four-year follow-up, 60 percent were reported to be rehabilitated and 30 percent to be improved, with only 3 percent unimproved. Franco (1965) later updated his evaluation. Overall, using all his indexes, he concludes that his company program has rehabilitated "about half" of those recognized as alcoholics. At the same time he concludes that certain employees, about 10 percent, can continue to work while still drinking. Consequently his overall figure runs close to 60 percent successes when the criterion of job maintenance is applied.

A more recent study of the same type (Hilker, 1972) included data on job performance, sick absences, promotions, sobriety, and on-the-job, off-the-job accidents five years before intervention (treatment or rehabilitation of some type in the company medical department) and for five years after. Dramatic differences are reported. Fifty-seven percent had not used alcohol for one year or more. Another 15 percent were performing satisfactorily on the job even though still drinking, making a total of 72 percent. Job efficiency of those who went through the program also changed substantially in a favorable direction. For both male and female employees who underwent the rehabilitation program, approximately 10 percent were rated by supervision as "good" for the five-year period prior to rehabilitation contrasted with 46 percent with this rating in the five-year period after experiencing the program. Additionally the reduced costs of days of sickness disability absences for the 306 employees on whom data could be secured indicated an estimated saving of $459,000, plus savings in reduced use of hospital insurance plans. More dramatic results emerged when on-the-job accidents requiring medical treatment showed a drop over the study period from 57 to 11.

Another study of effectiveness is a "one-sided" time series follow-up (Kammer and Dupong, 1969). Data were collected on job maintenance as the outcome criterion only after the program intervention. No history of job performance for program participants was available before they entered the program; consequently no time-series contrasts could be made. Job maintenance was, unfortunately, not measured by performance ratings but by "status at follow-up": either terminated because of alcoholism or terminated for reasons other than alcoholism. Within this strategy, 80 percent of the 300 employees who came to the program either kept their jobs or left employment for other than alcohol reasons. Approximately eight to ten other reports have some elements of the historical, time-series strategy (cf. Clyne, 1965).

No systematic effort to randomly assign alcoholic employees to occupational programs, on the one hand, and to a "leave them alone" group, on the other, has been attempted. Relatively simplistic comparisons between problem drinking employees who have fully experienced an occupational program and problem drinking employees who have experienced other varieties of intervention have, however, been made. Smart (1974) compared employed alcoholics who were mandatory referrals ("coerced") and those who were voluntary referrals to the same treatment center. An overall improvement rating scale composed of sobriety, work, family and social behavior, financial, residence, and incarceration scales was used to assess posttreatment changes. Patients who entered treatment feeling highly coerced by their company's program did not have better overall or drinking improvement rates than those who felt less coerced. Mandatory patients did, however, show significant improvements in supervisor ratings of dependability, productivity, overall job performance, drinking on the job, hangover, lateness, and absenteeism.

A somewhat comparable effort has been reported by Moberg (1974). Comparing employed persons entering an inpatient treatment facility from companies with an operative occupational program with employed patients from other companies who had not incorporated the occupational alcoholism program, he found a higher percentage of the former to be in earlier stages of alcoholism. In treatment terms, however, he reports no significant differences between the two groups in terms of abstinence, or changes in "need satisfaction" after a nine-month follow-up period.

Another reported study compared 24 employees known to have alcohol- or drug-related problems, but who chose not to enter a rehabilitation program, with 117 employees who actively participated in a referral program to outside treatment facilities. Lost man hours, sickness and accident benefits, accidents, formal leaves of absence, grievances, and disciplinary actions made up the criteria on which comparisons were made. Behaviors of both groups of employees on these indexes were observed one year before the participation data for both groups and for one year after that date, whenever possible. Although the report is unclear whether the comparison group shows less improvement on the criteria, it does indicate that for both groups there was a very significant drop in wages lost for individuals before program involvement ($463,520) contrasted to $237,176 after program involvement. "The hours lost also reflect a high loss to management, not only in dollars reflected in lost production but in additional monies paid out to replacement employees. The figures on accidents, grievances, and disciplinary actions also reflect a substantial recovery in company expenditures after program involvement" Alander and Campbell, 1973).

On balance, both types of evaluative efforts are realistic, beginning efforts. All, however, suffer serious flaws. The time-series studies have no way to control for intervening variables. Many of those who experienced the program's intervention might well have improved if merely left alone. Furthermore, it is quite possible for a program to "stack the deck" in its favor by selecting those who give most promise of succeeding. The "comparison group" studies suffer from the higher degree of illegitimacy of their comparisons. Thus, "voluntary" admissions probably represent highly motivated employed persons who probably have been, in some fashion pressured by friends, family, or physician to seek treatment. "Coerced" patients appear to have been confronted by their problems at work as the means to motivate them. In sum, the comparison appears to be between those obviously possessing high motivation reinforced by significant others and those largely coerced by the work community. It is a genuine tribute to the effectiveness of occupational alcoholism programs that in this relatively unfair comparison they still did as well as their better-motivated comparison group. Furthermore, no effort was made in any of these "comparison groups" studies to equate the two groups either by sampling techniques or by matching methods. No effort was made to determine if the strategy of constructive confrontation genuinely took place in a relatively uniform fashion among those "coerced."

Finally, we turn to the popular evaluative notion of "penetration rates" (U.S. Civil Service Commission, 1975). To determine this rate, an organization must assume a rate of alcoholism among its employees, 6 percent, for example. This means that an organization with 1000 employees would experience 60 alcoholic employees (over what period of time is unclear). If 15 employees in a given year were constructively confronted, then the penetration rate for that year would be 25 percent of the population "at risk." In this fashion an evaluation of a specific organization's program can be accomplished. Turfboer (1959) estimated that 6 percent or approximately 400 of a company's personnel were problem drinkers. Over an eighteen-month period, 160 employees were seen and given some form of clinic treatment. Thus, over the year and a half, a penetration rate of 40 percent is accomplished, or about 28 percent penetration per year into the problem drinking population.

Unfortunately, this strategy suffers from the lack of a rationale for selecting 3 percent, 4 percent, 5 percent, 6 percent, even 7 percent as the assumed figure. Other deficiencies include the influence of organization size. At the same time it is an excellent example of the "indigenous model" of evaluation (Trice and Roman, 1974). This model takes as its starting point the indigenous efforts of an action group to evaluate itself and then attempts to aid that group in refining and improving its own technique. Thus, it seems reasonable, as a first refinement, to suggest rationales for selecting between the estimated prevalence percentages. For example, it seems reasonable to select the lowest estimates (2 or 3 percent) if an organization's work force is predominately female (75 percent) with a majority (75 percent) of its personnel in the twenty-six to thirty-nine age brackets and located in a large urban center (over 100,000 population). If an organization has a high proportion of Jewish and Italian ethnic groups in its work force, the likelihood is that its alcoholic population would be low. Penetration rates could be improved so that a program will not be made to look overly effective or ineffective because of an unwarranted prevalence estimate.

Conclusion: Needed Research

With the large number of personnel and heavy investment of financial resources in occupational alcoholism, it may be appropriate to regard this set of ideology and activities as a federally funded social movement at this point in time. It is not without parallel, for many major social programs that have had centralized support from the federal government have been mounted rapidly and diffused throughout the nation without a thorough and carefully constructed research base. The impetus for such programs, and the reason for rapid development, has often been a congressional mandate accompanied by substantial fiscal appropriations. Thus, those who are charged with program execution are usually pressured to move rapidly. With new programs, a set of experienced professionals are rarely available to undertake program execution, calling for either the development of new occupations or the role expansion of existing occupations. Sociological understanding of the course and patterns of such movements is poorly developed.

The development of community mental health programs on a nationwide basis, beginning with a congressional mandate in 1963, may parallel the contemporary "alcoholism movement." This overall program, which heavily emphasized local participation and decision-making, proceeded in a rather unstructured way and clearly lacked a research base for many of its assumptions (Roman and Trice, 1974). Some new occupations have been developed, such as the indigenous nonprofessional health workers, and existing occupations expanded, such as a considerable degree of direct therapeutic involvement by social workers and the creation of a new psychiatric specialty of community psychiatry. It seems obvious that the speed with which this program developed was related to the need to use the appropriated funds. At the same time, an intense ideological commitment developed among many in the movement. New ways of handling psychiatric illness outside of traditional hospital settings were seen as comprising "psychiatry's third revolution." Many have viewed these new strategies as bringing about a dramatic reduction in the suffering and isolation of the traditional mental patient. Now, after the community mental health movement has completed a decade of activity, dramatic confrontations are underway as the fiscal responsibility for sustaining these programs is shifted to local governments.

There are many similarities between these developments and the patterns that can be seen in the occupational alcoholism movement. The Hughes Act made funds immediately available and early decisions were made to allocate a considerable proportion of NIAAA's budget to occupational programs. While many of the assumptions that formed the basis for program development make a good deal of common sense, there is a paucity of sound research evidence to back up many of the assumptions. Very few persons who are truly specialized in occupational alcoholism were available to execute the effort, and the plans to diffuse occupational programs on a national scope clearly required the development of a new occupation or set of occupations. These parallels with the community mental health movement are also demonstrated by the clear-cut pressure to "move ahead" and use the appropriated funds. A further parallel is found in the fact that the programs funded by the occupational branch of

NIAAA are defined as "demonstration" activities. This clearly foreshadows the eventual shifting of both fiscal and policy responsibility to the local level. The scope of funding for community mental health programs has been, of course, much greater than that of the occupational alcoholism effort. This is turn produces potential visibility and provides for greater flexibility in the eventual shift to local responsibility.

As researchers, we have a basic concern with the data base upon which occupational alcoholism programs are being developed. In summary form, basic "data gaps" exist in the following areas:

1. *Epidemiology of problem drinking in work organizations.* Accurate estimates of the extent to which problem drinking can be expected to be found in various types of work organizations is a fundamental item of information. It is particularly essential in persuading management people to develop programs. Such a data base, if it could be developed, would set the stage for planning and resource allocation within particular work organizations. As is well known, sound epidemiological research constitutes a major problem for the entire field of alcohol research. The major formulations that have been developed, such as the Jellinek estimation formula, are based upon late-stage cases of alcoholism. By contrast, occupational programs are specifically oriented toward early identification. Accurate measurement of drinking behaviors to the extent that problem estimates can be developed is always faced by the problem of data reliability. The collection of data on drinking behavior in work organizations appears to particularly highlight these validity problems. Employees are especially reluctant to provide statements about their drinking behavior if they feel in any way that such information will filter back to management. Furthermore, any epidemiological efforts are hindered by the diversity of work organizations, since generalization from one type of company of a certain size and with a certain set of activities may be completely inaccurate. Substantive research does appear possible, however, with the development of projections based on the experiences of a range of different types and sizes of organizations that have had active occupational alcoholism programs for a period of time. Assuming access to an adequate set of records, it appears likely that useful epidemiological data could be developed from such a set of sources. Nonetheless, no specific research of this nature has yet been undertaken.

2. *The consultation process.* With several hundred occupational consultants throughout the country attempting to develop programs within different types of work organizations, basic research on successes and failures in this consultation process is badly needed. While there is substantial anecdotal evidence on why work organizations do *not* initiate occupational alcoholism programs, we lack a research understanding of the crucial factors that result in

program adoption by an organization. Following the model of research diffusion and adoption that has been extensively developed in sociology (Rogers and Shoemaker, 1972), it would appear fruitful to conduct research directed at isolating the pivotal factors that lead to the decision of program adoption. Likewise, systematic information on program rejection would be necessary for useful data to understand the consultation process. With the vast number of such consultations apparently underway, such research is clearly feasible, although it is likely that the newly trained consultants might be resistant to such a study. Such research has not yet been undertaken, despite its pivotal importance for the success of the national program.

3. *The process of identification.* Nearly a decade ago one of us (Trice, 1965a; 1965b) collected data on the factors that played a role in the identification of employees subsequently diagnosed as alcoholic, neurotic, and psychotic. This research has been used widely to demonstrate the various barriers that prevent identification. Further studies of this type are, however, badly needed. The reported research effort was conducted in only a single organization that had a well-developed occupational program. Research directed at the identification process in a variety of different organizations that have active programs would provide a test for some of these earlier assumptions, as well as badly needed information on identification via impaired performance in contrast to identification via symptoms. While a large amount of anecdotal information has accumulated from program directors, it is obvious that an outside researcher with no vested interests in program success could generate illuminating data on the events that precede identification, the reaction of supervisors and work associates to these events, the resistances of the problem employee to identification, and the manner in which the referral processes subsequent to identification proceed.

4. *Processing through treatment resources.* One of the principal goals of occupational programs is to offer treatment assistance to problem drinking employees to the extent that they may be returned to the job as productive employees as soon as possible. When programs are first initiated in work organizations, it is reported that "hard-core" chronic cases of alcoholism surface first, oftentimes requiring inpatient treatment that is of a long-term nature and does not result in full rehabilitation. Once such cases have been handled, program coordinators aspire to reaching early stage problems that require minimal treatment on an outpatient basis to bring the drinking problem under control. Research on the patterning of these referral processes and the extent to which different treatment modalities are successful in bringing about rehabilitation is badly needed.

5. *Posttreatment job performance.* While numerous internal evaluations have been reported that indicate that alcoholism programs lead to successful

rehabilitation in a majority of instances (Kamner and Dupong, 1969; Franco, 1960; Hilker, 1972), we lack substantive data on the impact on the career of an employee who is identified and referred for help. The extent to which there is recidivism among these cases is alone a very important piece of information. There have been some anectodal comments to the effect that posttreatment job experiences include stigmatization that affects opportunities for advancement. Thus, to understand fully the nature of occupational programs and their effects, data on the posttreatment job placement of problem drinkers, the reaction of supervisors and work associates to such treated individuals, and the subsequent career development of treated employees appear to be of primary importance. We know of no such research underway at the present time.

6. *Program diffusion within organizations.* While most persons working in this area, including the present authors, often refer to "programs" in a rather glib fashion, we lack an operational definition of a "program." It is obvious that there are different levels of investment and commitment that work organizations can make to occupational alcoholism programs. Furthermore, the extent to which program concepts and actual supervisory participation in program implementation diffuses through an organization is variable. Research is needed on how a program is brought into supervisory awareness as well as data on the reactions of supervisors and other employees to program concepts. The existence of a program coordinator and a written policy statement tells us little about the extent to which the program is accepted with the organization. Furthermore, we need to know the extent to which supervisors actually participate in the referral process versus referral being initiated by the direct and indirect efforts of the program coordinator. Thus, further pursuit of this research question is necessary not only to understand internal diffusion processes and the extent of program participation, but also to construct a typology of the range of activities included under the rubric of "occupational alcoholism programs."

7. *Program development in smaller organizations.* The program concepts and strategies of development that currently dominate the occupational alcoholism scene are geared to large-scale organizations. A considerable proportion of the American work force is either self-employed or employed in small work organizations, ranging from a few employees to several hundred workers. In most of these small organizations, personnel management is carried out in a highly informal fashion and is allocated few, if any, resources. Likewise profit and loss have greater visibility than in the giant corporation. The extent to which these smaller organizations could be feasibly involved in occupational alcoholism programs is currently being explored in several projects where a number of small businesses have pooled resources to provide for program coordination and access to treatment resources. Given the extent to

which small business and self-employment represent American workers, a considerable degree of further research attention to program development in these types of organizations is called for.

This provides only some of the highlights of the research needs in the occupational alcoholism area. To repeat our earlier point, program design and development has proceeded without a firm research base. The situation is not unlike that which has occurred in the launching of other program efforts based on innovative concepts; if and when such needed research is completed, however, it will be of interest to note the extent to which programming efforts are subsequently modified. Since the effort has had to operate without this research base during the major phase of its development, it is likely that many assumptions will become embedded and vested interests developed, making research-based change difficult.

ACKNOWLEDGMENTS

During the preparation of this manuscript, the authors received partial support from U.S. Public Health Service Grants Nos. AA-00494 and AA-01504 to Tulane University and Grant No. AA-00493 to Cornell University. Both authors also have received partial support for their activities from the Christopher D. Smithers Foundation, which is gratefully acknowledged.

REFERENCES

Alander, R., and Campbell, T., 1973, *One organization's approach: An evaluative study of an alcohol and drug recovery program,* Oldsmobile Division, General Motors Corp., East Lansing, Michigan.
Brenner, B., 1967, Alcoholism and fatal accidents, *Quart. J. Stud. Alc.* 28:517–527.
Cahalan, D., Cisin, I., and Crossley, H., 1969, "American Drinking Practices," College and University Press, New Haven.
Clyne, R. M., 1965, Detection and rehabilitation of the problem drinker in industry, *J. of Occup. Med.* 9:265–268.
Edwards, D., 1973, *Final Evaluation Report: A National Training Program Designed to Modify Service Delivery in Occupational Systems,* National Occupational Alcoholism Training Institute, Greenville, N.C.
Franco, S. C., 1960, A company program for problem drinking: A ten years follow-up, *J. Occup. Med.* 2:157–162.
Franco, S. C., 1965, Alcoholism in industry. Mimeographed lecture, annual meeting, Maryland Industrial Physicians Association, Baltimore.
French, J. R. P., et al., 1965, *The Work Load of University Professors,* University of Michigan Survey Research Center, Ann Arbor.

Hilker, Robert R. J., 1972, A company-sponsored alcoholic rehabilitation program, *J. Occup. Med.* 14:769–772.
Hollander, E. P., 1962, Conformity, status and idiosyncrasy credit, *in* "Current Perspectives in Social Psychology" (E. Hollander and R. Hunt, eds), Oxford University Press, New York.
Holmes, D., 1963, *Report of the New York City Alcoholism Project,* National Council on Alcoholism, New York.
Kahn, R. L., Snoek, W., Quinn, R., Wolfe, D., and Rosenthal, R., 1964, "Organizational Stress," John Wiley, New York.
Kammer, M. E., and Dupong, W. G., 1969, Alcohol problems: Study by an industrial medical department, *New York State J. of Med.* 88:3105–3110.
Katz, D., and Kahn, R. L., 1966, "The Social Psychology of Organizations," John Wiley, New York.
Kornhauser, A., 1965, "The Mental Health of the Industrial Worker," John Wiley, New York.
Levinson, H., 1973, "Organizational Diagnosis," Harvard University Press, Cambridge, Mass.
Maxwell, M. A., 1972, Alcoholic employees: Behavior changes and occupational alcoholism programs, *Alcoholism* 8:174–180.
McClelland, D., Davis, W., Kalin, R., and Wanner, E., 1972, "The Drinking Man," The Free Press, New York.
McLean, A. A. (ed.), 1972, "Mental Health in Work Organizations," Rand McNally, Chicago.
Moberg, P., 1974, Followup study of persons referred for inpatient treatment from an industrial alcoholism program, Paper presented to the Occupational Section of the Alcohol and Drug Problems Association, San Francisco, Calif.
Morris, J., 1972, The unions look at alcohol and drug dependency, *Internat. Labor Rev.* 106:335–346.
National Council on Alcoholism, 1973, Suggested guidelines for labor management agreements, *Labor–Management Alcoholism Newsletter* 2:18–22.
National Institute on Alcohol Abuse and Alcoholism, 1974, *Alcohol and Health: Second Report to the Congress by NIAAA,* Government Printing Office, Washington.
Perlis, L., 1958, Labor's viewpoint on alcoholism in industry, *Indust. Med. Surg.* 27:535–538.
Ritzer, G., and Trice, H., 1969, "An Occupation in Conflict: The Personnel Manager," Cornell University, Ithaca, N.Y.
Rogers, E. M., and Shoemaker, E., 1972, "Communication of Innovations," The Free Press, New York.
Roman, P., 1970, The future professor, *in* "The Domesticated Drug" (G. Maddox, ed.), College and University Press, New Haven, Connecticut.
Roman, P., 1972, Sleep deprivation, drug use and psychiatric disorders, *Amer. J. Sociol.* 77:907–911.
Roman, P., 1974a, Executives and problem drinking employees, *in* "Alcoholism: A Multilevel Problem" (M. Chafetz, ed.), Government Printing Office, Washington.
Roman, P., 1974b, Settings for successful devaince, *in* "Deviant Behavior: Occupational and Organizational Bases" (C. Bryant, ed.), Rand McNally, Chicago.
Roman, P., 1975a, The misplaced concept of consultant in occupational alcoholism programming, Tulane University Project on Monitoring and Evaluation of Occupational Alcoholism Programming, New Orleans.
Roman, P., 1975b, Spirits at work: needed strategies in occupational programming in *Proceedings of the Fifth National Conference of NIAAA* (M. Chafetz, ed.), Government Printing Office, Washington.
Roman, P., 1975c, Secondary prevention of alcoholism: problems and prospects in occupational programming, *J. Drug Issues* 5:327–343.

Roman, P., and Trice, H., 1968, The sick role, labeling theory and the deviant drinker, *Int. J. Soc. Psychiat.* 12:245–251.
Roman, P., and Trice, H., 1970, The development of deviant drinking behavior, *Arch. Environ. Health* 20:424–435.
Roman, P., and Trice, H., 1971, Normalization: A neglected dimension of labeling theory, Paper presented to American Sociological Association, Denver.
Roman, P., and Trice, H., 1972, Psychiatric impairment among "middle Americans," *Soc. Psychiat.* 7:157–166.
Roman, P., and Trice, H., 1974, Strategies of preventive psychiatry, in "Sociological Perspectives on Community Mental Health" (P. Roman and H. Trice, eds.), F. A. Davis, Philadelphia.
Sadler, M., and Horst, J., 1972, Company/union programs for alcoholics, *Harvard Bus. Rev.* 50:22–41.
Smart, R., 1974, Employed alcoholics treated voluntarily and under constructive coercion: A follow-up study, *Quart. J. Stud. Alc.* 35:196–209.
Somers, G., 1975, Alcohol and just cause for discharge, paper delivered to the National Academy of Arbitrators, San Juan, Puerto Rico.
Straus, R., 1971, Alcohol and alcoholism, in "Contemporary Social Problems," (R. Merton and R. Nisbet, eds.) Harcourt Brace Jovanovich, New York.
Tichauer, E., and Wolkenberg, R., 1972, *Delayed Effects of Acute Alcoholic Intoxication on Occupational Safety and Health,* New York University Institute of Rehabilitation Medicine, New York.
Trice, H., 1957, A study of the process of affiliation with Alcoholics Anonymous, *Quart. J. Stud. Alc.* 18:34–37.
Trice, H., 1959, Work accidents and the problem drinker, *I.L.R. Research* 3:18–32.
Trice, H., 1962, The job behavior of problem drinkers, in *Society, Culture, and Drinking Patterns* (D. Pittman and C. Snyder, eds.), John Wiley, New York.
Trice, H., 1965a, Reactions of supervisors to emotionally disturbed employees, *J. Occup. Med.* 7:177–188.
Trice, H., 1965b, Alcoholic employees: A comparison with psychotic, neurotic and normal personnel, *J. Occup. Med.* 7:94–99.
Trice, H., 1966, "Alcoholism in America," McGraw-Hill, New York.
Trice, H., and Belasco, J., 1966, *Emotional Health and Employer Responsibility,* Bulletin #57, New York State School of Industrial and Labor Relations, Cornell Univ., Ithaca.
Trice, H., and Belasco, J., 1968, Supervisory training about alcoholic and other problem employees, *Quart. J. Stud. Alc.* 29:382–398.
Trice, H., and Belasco, J., 1970, The aging collegian: drinking pathologies among executive and professional alumni, in "The Domesticated Drug" (G. Maddox, ed.), College and University Press, New Haven, Connecticut.
Trice, H., and Roman P., 1970a, Sociopsychological predictors of affiliation with Alcoholics Anonymous, *Soc. Psychiat.* 5:51–59.
Trice, H., and Roman, P., 1970b, Delabeling, relabeling and Alcoholics Anonymous, *Soc. Probs.* 17:468–480.
Trice, H., and Roman, P., 1971, Occupational risk factors in mental health and the impact of role change experience, in "Compensation in Psychiatric Disability and Rehabilitation" (J. Leedy, ed.), C. C. Thomas, Springfield, Ill.
Trice, H., and Roman, P., 1972, "Spirits and Demons at Work: Alcohol and Other Drugs on the Job," Cornell University, Ithaca, N.Y.
Trice, H., and Roman, P., 1973, Alcoholism and the worker, in "Alcoholism: Progress in Research and Treatment, (R. Fox and P. Bourne, eds.), Academic Press, New York.
Trice, H., and Roman, P., 1974, Dilemmas of evaluation in community mental health, in "Socio-

logical Perspectives on Community Mental Health" (P. Roman and H. Trice, eds.), F. A. Davis, Philadelphia.

Turfboer, R., 1959, The effects of in-plant rehabilitation of alcoholics, *The Medical Bulletin of Standard Oil of New Jersey,* 19:108–128.

United Automobile Workers of America, 1972, Resolutions Committee Report #2, 23rd U.A.W. constitutional convention, Miami Beach, Florida.

U.S. Civil Service Commission 1975, *Internal Evaluation of Agency Alcoholism and Drug Abuse Programs,* Bulletin #792-15, Washington, D.C.

Vroom, V. 1966, "Work and Motivation," John Wiley, New York.

Warkov, S., and Bacon, S., 1965, Social correlates of industrial problem drinking, *Quart. J. Stud. Alc.,* 26:58–71.

Weiner, H., 1967, Labor-management relations and mental health, *in* "To Work Is Human," (A. A. McLean, ed.), Macmillan, New York.

Wrich, J., 1974, "The Employee Assistance Program," The Hazleden Foundation, Center City, Minn.

Chapter 12

Education and the Prevention of Alcoholism

Howard T. Blane
Division of Specialized Professional Development
University of Pittsburgh
Pittsburgh, Pennsylvania

INTRODUCTION

Prevention and control of alcohol problems have received increasing attention from the scientific and professional communities in recent years, especially in the United States. Beginning with the initial report of the Cooperative Commission on the Study of Alcoholism (Plaut, 1967), and followed by the Commission's report on prevention (Wilkinson, 1970), this trend has continued and accelerated since the establishment of the National Institute on Alcohol Abuse and Alcoholism (NIAAA) in 1971. Education has been advanced as a major tool of prevention in the United States, although in other countries education may assume a less prominent role. In this chapter, attention is directed to (1) a description and discussion of major models of prevention; (2) an account of the principal techniques of prevention; (3) detailed discussion of alcohol education as a critical technique of prevention; and (4) an analysis of some key issues in preventive and educational action.

The literature on prevention and control of alcohol problems is extensive, amorphous, scattered, hard to locate, and of extremely uneven quality. It is, therefore, hardly surprising that it has never been systematically or comprehensively reviewed, although particular aspects of prevention have been considered. Since this chapter is intended as a close look at developments over the past ten years rather than an exhaustive examination of the literature, the interested reader may wish to refer to Kelly's (1964) historical account of control programs in the United States, Jellinek's (1963) review of programs in other countries, a report by the Scottish Home and Health Department (1970) synthesizing literature relevant to primary and secondary prevention, and Warburton's (1968) study of the effects of Prohibition in the United States. Partial reviews may also be found in Popham *et al.* (1971), Room (1971), Whitehead (1972a; 1972b), and Wilkinson (1970). With regard to alcohol education, the early work on endeavors in school settings in the United States was reviewed in an influential and comprehensive study by Roe (1943), with later reviews by McCarthy (1952), Monroe and Stewart (1959), Freeman and Scott (1966), and the Scottish Home and Health Department (1970). Also, a recent review of literature on drinking behavior in childhood and adolescence (Stacey and Davies, 1970) contains implications for alcohol education programs. With regard to public information and education programs, whether conducted through mass media, local groups, or other means, the very small body of literature that does exist has not been reviewed.

CONTEMPORARY MODELS OF PREVENTION

There are three major models of prevention, termed here the social science model, the distribution of consumption model, and the proscriptive model. The first two draw more or less on classical public health principles of agent–host–environment relationships and distinctions among primary, secondary, and tertiary prevention. The relative emphasis has shifted considerably in the past 20 years so that now the focus is heavily on environmental manipulation and primary prevention. Each model, as well as the traditional public health approach, will be discussed.

Social Science

The social science model is less a unitary model of prevention than a series of interrelated proposals derived from research and clinical applications by anthropologists, sociologists, psychologists, and psychiatrists. Some proposals have a strong conceptual foundation while others do not, and some versions of the same proposal are highly conceptualized while others appear to spring fullborn from a common wisdom that requires no explanation. Four distinct aspects of the social science model may be arbitrarily distinguished: (1) normative structure, (2) integrated drinking, (3) socialization, and (4) level of mental health and social conditions. Two things should be noted. First, there is no standard nomenclature within prevention to refer to the submodels, but in my labels I have tried to stay as close as possible to the sense of the original. Second, in a given work presented as a unit, an author may cover more than one submodel, so that the classification scheme used here may, for the sake of clarity, do violence to the integrity of a program.

With the exception of the level-of-mental-health submodel, the submodels find their empirical basis in observations by sociologists, anthropologists, and epidemiologists that drinking practices vary along national, ethnic, and/or religious divisions, and that these variations are related to differences in mores, customs, beliefs, values, and sanctions about alcohol and its use by individuals in a society. The first three submodels are based on observation of such differences within a society, noting their relationship to differential rates of alcohol problems, and on examination of differences among individual societies to discover cultural, social, and psychological factors that systematically separate groups with high or low problem rates. Sociocultural patterns of drinking and alcohol problems are reviewed in earlier chapters of this volume. The interested reader may also wish to refer to accounts by Bacon and Jones (1968), Bales (1946), Blacker (1966), and Wilkinson (1970). The first two submodels, normative structure and integrated drinking, and by implication the third, socialization, are referred to collectively by Whitehead (1972a; 1972b) as the "sociocultural model" of prevention, and by Room (1971) as the "inoculation hypothesis," because they carry with them the implication of introducing young people to responsible drinking in order to "immunize" them from drinking irresponsibly.

Normative Structure

From this essentially sociological perspective, variations in severity of drinking problems across socioculturally defined groups are viewed as a function of differences in the quality and structure of the norms a society holds about drinking. According to Blacker (1966), societies that proscribe excessive drinking, prescribe moderate drinking, and show a high degree of group and individual consensus about the rules governing drinking, will have a low rate of problem drinking. As Williams *et al.* (1968) point out, this formulation fails to take nondrinking into account, except perhaps to proscribe it by default in societies where the majority of the population consists of drinkers. This suggests that nondrinking norms be permissive (guidelines for educational programs formulated by NIAAA stress "the right to be abstinent"). Thus, in a society where official statements and individuals are in agreement that drunkenness is bad, a little bit of drinking is good, and that it is all right for a person either to drink or not to drink, the alcohol problem rate, however measured, will be low. The general implication of the model is that societies can prevent or substantially reduce alcohol problems by manipulating the normative structure. Several programs and specific proposals have been based on one or more aspects of the normative structure submodel.

Chafetz (1967; 1973) argues for a change in attitude toward alcohol in complex industrialized societies that would include "the development of responsible drinking behavior" and "a negative sanctioning of the intoxicated state." As one step toward establishing proscriptive norms for excessive drinking, he assumes that the frequency of occurrence of episodes of intoxication may be used as a sign of early problem drinking, and specifically proposes that four episodes of intoxication per year be used as one criterion of problem drinking. Plaut (1967; 1972) proposes "a clarification of the distinctions between socially acceptable and socially unacceptable types of drinking behavior", i.e., between prescribed and proscribed norms, which will evolve from and be reinforced by social definitions and group sanctions. As he puts it, groups should "be responsible for the drinking behaviors of their members and invoke appropriate sanctions when drinking deviates significantly from agreed-upon standards." In a similar vein, a clear distinction between drunkenness and drinking *per se* is said to be needed, with "strong taboos on the state of heavy intoxication" (Wilkinson, 1970) and labeling of drunkenness as "immature and stupid behaviour" (Whitehead, 1972b). A more positive tone is adopted by the Task Force on Alcohol and Health in its recommendation for the creation of "a National climate that encourages responsible attitudes toward drinking for those who choose to drink; that is, using alcohol in a way which does not harm oneself or society" (U.S. Department of Health, Education, and Welfare, 1971).

The thrust of these proposals is to create a set of ideal norms that proscribe excessive drinking but prescribe moderate drinking, while attempting to allow room for the acceptability of nondrinking. Before one considers the forbidding problems in bringing about the adoption of this set of ideal norms and translating them into behavioral norms, one is faced with problems of definition of excessive drinking, irresponsible drinking, intoxication, and drunkenness, on the one hand, and responsible drinking, moderate drinking, and socially acceptable drinking, on the other. The Task Force on Alcohol and Health (U.S. Department of Health, Education, and Welfare, 1971) and Chafetz (1967) have taken steps in this direction, but the criteria of the one are overprecisely arbitrary, and those of the other vague, but humanistic.

Integrated Drinking

The term "integrated drinking" has been employed in two different ways to refer (1) to unambiguous agreement at all levels within a society about the values, norms, and practices that regulate the use of alcohol in that society (Ullman, 1958), and (2) to the subordination of the use of alcohol in a society to other activities, particularly family, religious, and recreational pursuits, rather than as a prime organizing principle of a social activity (Plaut, 1967; Whitehead, 1972a). Here we will be concerned with integrated drinking in the second sense, noting only that integrated drinking in its first usage is the cement that binds normative structure together. In countries and groups with low prevalence of alcohol problems, the use of alcohol may be commonplace but is merely one aspect of a more prominent activity. Two frequently cited examples are Jews and Italians. Among Jews, especially Orthodox Jews, the use of alcohol is a necessary, integral part of religious observance, and drinking is rarely done for its own sake. Among Italians, alcohol, particularly wine, is viewed as food and as a necessary part of the dinner table and of celebrations, such as weddings, christenings, and holidays. While wine consumption is viewed positively, nondrinking is not negatively sanctioned (evidence for this latter point is not convincing). Groups with high problem rates, on the contrary, value drinking for its own sake and effects, and define social occasions by alcohol use; the Irish, perhaps erroneously, are often given as an example. Chapter 2 of this volume contains summaries of Jewish, Irish, and Italian drinking practices, which may also be found in Bacon and Jones (1968), Plaut (1967), and Wilkinson (1970); comprehensive analyses of each group may be found in Bales (1962) and Stivers (1971) for the Irish, Snyder (1958) for the Jews, and Lolli et al. (1958) for the Italians.

The preventive implications drawn from the inverse relationship between integrated drinking and alcohol problem rates are that educational efforts and legislative control measures should be directed toward enhancing the integration of drinking in the family and other relevant social situations while conversely discouraging unintegrated drinking patterns and occasions. Many of the specific recommendations made by Wilkinson (1970) in a comprehensive program of prevention proceed along these lines. He suggests, among other things, that drinking be promoted in "situations of restraint" and "on occasions when drinking itself, being only one of several integrated activities, does not become an overwhelming focus of the group's attention." By "situations of restraint" he means those where informal controls make immoderate drinking out of the question. To enhance drinking in the family (a "situation of restraint"), for example, he recommends "removal of all legal-age controls on drinking at home" and that "private family activities provide alcohol for young people in controlled circumstances, without obliging them to drink." To facilitate drinking integrated with other activities, he proposes that retail licenses be issued "to integrate new drinking places with varied leisure activities," such as bowling and sports events. Interestingly, little mention is made of mechanisms specifically designed to discourage unintegrated drinking, such as that proposed by Lemert (1962) to the effect that various civic and professional groups can play an effective role in the informal social control of unintegrated drinking. Plaut (1972) makes the assumption that when drinking "is incidental to other activities, the rules of the game in relation to those other activities generally operate as a constraining force on the drinking behavior." He therefore wishes to discourage drinking that occurs for its own sake and to encourage drinking that is integrated with other activities, and along with Chafetz (1967) with Whitehead (1972a; 1972b) he also recommends drinking with meals and in family settings. Again, mechanisms for discouraging unintegrated drinking are not spelled out.

Socialization

Socialization in its most general sense refers to those psychological and social processes whereby individuals from early infancy onward become fully functioning and accepted members of their society; learning is postulated as the basic process underlying socialization. In the present context, socialization is used to refer to the processes and mechanisms, as well as their correlates, involved in the development of a youngster's attitudes and behavior in regard to the use of alcohol, and their progression into adulthood. While a focus on socialization derives to some extent from the crosscultural and crossnational research referred to earlier, it gains its major impetus from the conception, common to most social and behavioral scientists, that the use of alcohol, like any other behavior, is learned; and more specifically from research on adolescent drinking behavior that shows that there is considerable continuity between parent–child drinking behaviors, i.e., light-drinking parents tend to have light-drinking children, and so on (Bacon and Jones, 1968; Stacey and Davies, 1970; Braucht et al., 1973). Clearly, socialization may be the vehicle by which "good" or "bad" behaviors are carried from one generation to another and so perpetuated and reinforced. Those concerned with prevention of alcohol problems use concepts borrowed from or implied by socialization for the purpose of furthering preventive goals, such as integrated drinking or clearcut norms about the use of alcohol discussed above. Interestingly, however, little has been said about attempts to abort socialization chains that facilitate irresponsible drinking.

Most of the preventive proposals proceeding from a social science perspective invoke the family and the schools as the major agents of socialization, although there is some difference of opinion within and among various authors. Chafetz (1967) expresses the opinion that the family is no longer the center for the development of social responsibility, and that, for better or worse, schools and the mass media have usurped this function; therefore, these institutions have the highest potential for socialization of responsible drinking. Nevertheless, he also lays a major portion of the socialization burden on parents, suggesting that they (if drinkers) introduce their children to alcohol in an integrated fashion. Plaut (1967) clearly sees the schools as the more important socializing agent and devotes considerable effort to the characteristics of effective alcohol education programs (see section of this chapter entitled "Education in School Systems"). He, nevertheless, holds the view that while the family should be the primary setting for learning social behavior, it is unable to assume its proper role in regard to drinking because of pervasive ambivalence about alcohol. Therefore, he proposes that parents should be helped to become more effective agents of socialization in this area through adult alcohol education efforts. Wilkinson (1970) apparently finds no difference between the relative socializing capabilities of school and family, and would rely on both. It is striking that in none of these proposals is the socializing effect of the church ever seriously invoked, despite the repeated observation not only that drinking practices vary along religious lines but that degree of religious participation is inversely associated with immoderate drinking (Cahalan *et al.*, 1969). The potential of the church for socialization of responsible drinking, particularly now when most major Protestant denominations accept the idea of drinking in moderation (Burnett, 1973; Plaut, 1967), is little explored in these preventive proposals.

Level of Mental Health and Social Conditions

Plaut (1967; 1972) has developed a program of prevention of alcohol problems divided into specific and nonspecific approaches. Nonspecific prevention refers "... to activities or intervention directed at other problems, but which may also help to reduce rates of problem drinking." Included in nonspecific prevention are recommendations for measures to improve the level of general mental health and to create a "better society." An increased level of general mental health will mean a heightened capacity for coping with psychological stress, thus reducing the need to develop maladaptive behaviors such as problem drinking. As means of raising the level of mental health, Plaut cites approaches to improving the quality of family life, helping people to cope effectively with crises, and increasing understanding of the role of emotion and interpersonal relations in human behavior. With regard to social conditions, Plaut finds it reasonable to assume that the creation of a better and more humane society will be accompanied by a lower rate of alcohol problems. While this reasonableness is open to challenge, few would question the essential humanitarianism of his proposals: reduction of poverty, deprivation, inhumanity of man to man, injustice, alienation, discrimination; increasing sense of community and belongingness; insuring equal access to the opportunity structure; and establishing preventive and curative medical services for all (Plaut, 1967; 1972).

Among advocates of a social science approach to the prevention of alcohol problems, Plaut stands out for recommending indirect preventive means. Virtually all other programs proposed to prevent alcohol problems adopt an alcohol-specific approach, thus narrowing the programs' perspective and divorcing them from consideration of other social realities in which alcohol problems may be embedded. Plaut's view, leaving aside the merit of his specific suggestions, places alcohol problems and their prevention in a broad social context. It is an important reminder that answers to alcohol problems are not to be found solely in the study of alcohol and its use, but are intimately and inextricably linked to the human condition. At the level of social action, Plaut's philosophy raises the question of whether preventive actions should be categorical or, rather, integrated and coordinated with more general health-productive activities. Plaut in fact recommends both, but if one follows the logic underlying nonspecific prevention, other possibilities suggest themselves. For instance, one might find education about alcohol embedded in programs aimed at helping people become self-potentiating members of a complex technological society: programs that focus on feelings, interpersonal relations, family life, and ways of effectively coping with crises (all cited by Plaut as examples), in which the instructor would be prepared to deal with alcohol, drugs, sex, or any other topic that came up during the course of exploring how people get along with others. Such programs would have quite different implications for teacher preparation, content, curriculum, and organization than the commonly suggested alcohol education unit even when seen as part of an integrated health-education sequence. In sum, Plaut's nonspecific approach, which has much in common with that of Sanford (1972), is an important part of the literature on prevention of alcohol problems.

Distribution of Consumption

The distribution of consumption model, so termed by Whitehead (1972a), is as neat and tidy as the social science approach is diffuse and vague. Based largely on investigations undertaken over the past 15 years by members of the staff of the Addiction Research Foundation (ARF) in Toronto, its major preventive thrust is to manipulate the relative price of alcohol upward to a point where per capita consumption of absolute alcohol falls, resulting in a reduction of alcoholism rates.

The rationale and empirical support that lay behind the development of this approach have been described in detail in Chapter 7 and will not be repeated here. The ARF group concludes that the form of the distribution of consumption of alcohol is constant across populations and is such that excessive, or alcoholic, consumption varies with the level of general consumption. Further, of measures that countries have employed to control consumption, only relative price (per unit cost of alcohol as a proportion of average annual disposable income) is negatively related to general consumption and cirrhosis mortality. Other governmental measures, including state versus private control of distribution and sale of alcohol; hours of sale; and number, dispersion, and diversity of outlets, are unrelated to general consumption levels. It is therefore inferred that changes in relative price alone can "cause" changes in level of consumption, which in turn can effect changes in excessive consumption and its correlates, cirrhosis and alcohol mortality. Thus, manipulation of relative price becomes the prime means of preventing alcoholism.

Specific proposals for manipulation of relative price include (1) setting the price of alcoholic beverages so that the price per unit of absolute alcohol is the same regardless of the type of beverage; (2) making the price of alcohol a constant proportion of average annual disposable income, with adjustments according to annual changes in income; and (3) gradually increasing the relative price of alcohol over a period of years. The mechanism of control would be governmental control by taxation. Education is considered to be secondary to manipulation of relative price and designed to inform "the public of the hazards associated with heavy drinking" and to help people see price increases as a valuable public health measure, rather than just another rise in the cost of living (Popham et al., 1971).

The empirical evidence offered in support of the distribution of consumption model is incontrovertible. The relationship between cirrhosis mortality rates and high chronic alcohol intake is beyond question. It is clear that the greater the per capita consumption of alcohol, the higher the alcoholism rate as measured by physical indices. It is similarly evident that per capita consumption rises and falls with variation in relative price. Nevertheless, the findings have been challenged on technical grounds and the interpretation of them has been criticized.

The work of Popham et al. (1971) demonstrates that consumption of alcohol among drinkers approximates the lognormal distribution for the many populations they studied. They assume that the proportion of a population falling above a critical "alcoholic" level of consumption (usually defined as 15 centiliters of absolute alcohol per day) varies directly with the average of the distribution. Thus, a country with a low average per capita consumption will have a lower proportion of the population falling above the critical level than will a country with a higher average per capita consumption. This assumption of direct covariation between mean consumption and alcoholic consumption is crucial to the distribution of consumption model. That the assumption is, strictly speaking, not tenable has been shown by O'Neill and Wells (1971) who demonstrated that lognormal distributions with different proportions falling above specified points of distributions may have the same mean. Alternatively, distributions with the same proportions falling above a specified level may have different means. Although covariation may not be theoretically defensible (Room, 1973), Ekholm (1972) notes that there are empirical reasons to suppose that a general tendency to covariation does exist. Despite technical considerations about the properties of the lognormal distribution, it does appear that higher per capita consumption is related to higher proportions of excessive drinkers.

Whether general consumption causes more excessive drinkers, however, or vice versa, is an open question (Room, 1971). Grosswiler (1972), for example, independently draws quite different conclusions from those of the ARF group. Estimating that a small percentage (12 percent) of the drinking population of the United States is responsible for the bulk of the alcohol consumed (over two thirds), he argues that there are two economic demands, that of heavy drinkers who drink large amounts but are in the minority and that of moderate drinkers who are in the majority but who consume a small total amount. If price increases are used to control problem drinking, he predicts that reduction in consumption is likely to occur in the moderate rather than the heavy drinking group. In support he cites findings by Wanberg and Horn (1971) that alcoholics with marginal median incomes nevertheless spend from 10–33 percent of their incomes on alcohol. Grosswiler further speculates that (1) problem drinkers' difficulties would increase as income is diverted to alcohol and away from other goods needed by their families, and (2) nonproblem drinkers would sacrifice pleasure to the extent that they could not afford increased costs. The ARF group recognizes that there may be other interpretations of the covariation of consumption averages and alcoholism prevalence; for example, an increase in consumption level might be due to a higher rate of alcoholism (de Lint, 1973). This interpretation is discarded on the assumption that the form of the lognormal distribution is invariant (de Lint, 1973), which, as we have seen earlier, is not necessarily the case.

Whitehead (1972a) has made the point that the distribution of consumption model tends to define alcoholism exclusively in terms of physical pathology, usually alcoholism and/or liver cirrhosis mortality, and thus more narrowly than the social science approach, which includes physical pathologies and alcoholism as only one aspect of alcohol problems (Plaut, 1967) or of problem drinking (Grosswiler, 1972; Wilkinson, 1970). De Lint and Schmidt (1971), in referring to Finnish "explosive" drinking, which is accompanied by violent behavior and is apparently in part a consequence of high relative price, make the explicit point that "the problem of occasional intoxication should not be confused with alcoholism." Focus on one aspect of alcohol use weakens the distribution of consumption approach because in suggesting a cure it tends to ignore possible side effects on other aspects of alcohol use and on other social phenomena. At least one study (Grosswiler, 1972) shows unexpected relationships between per capita consumption and social problems demonstrated in other investigations to be linked to use of alcohol. Grosswiler found negative correlations between consumption and highway deaths and between consumption and murder, while consumption and larceny were positively correlated (for 50 states in the United States, 1968). However, as Grosswiler correctly observes, there are too many possible intervening variables to support an hypothesis that alcohol causes larceny or causes reduced homicide or highway fatalities.

The dangers of attributing causal significance to correlational data without complete investigation of competing interpretations and search for intervening variables are further complicated when single rather than multiple causation is invoked. At the very least, as Room (1973) states, the lognormal distribution is "susceptible to a large variety of competing and overlapping potential 'explanations'. These explanations carry highly divergent implications for policy, but the work of elucidating the competing models and theories and turning them into hypotheses subject to empirical testing has not even begun." Many hold that effective explanations will of necessity be complex. Skog (1973), after reanalyzing much of the reported data on consumption levels, says:

> The existing data are not sufficient to prove that a change in average consumption is the instrument of choice in the fight against alcoholism. The causal relationships relevant to the problem of alcoholism are much too complicated for that to be the case.

This thought is echoed by Edwards (1973): "It is unlikely that any unifactorial preventive policy will be widely effective" in light of the multivariate nature and determination of alcohol problems.

In addition to concerns about basing a program on unifactorial interpretations of correlations, the program of the ARF may be questioned, as mentioned above, because of its possibly deleterious side effects. Grosswiler (1972) has speculated that price increases may reduce the already marginal financial condition that characterizes many alcoholics. Whitehead (1972a; 1972b) wonders whether increased prices for alcohol might not make large-scale illicit production of alcohol profitable, a not unlikely possibility, considering Seeley's (1960) estimate that prices in Ontario would need to be tripled in order to cause a significant decrease in average consumption. Along with raising the specter of the association between crime and Prohibition in the United States, Whitehead suggests that illegal sales and associated clandestine behavior would heighten the ambivalent mystique that surrounds use of alcohol in many Western societies, thus laying the foundations for future social problems with alcohol. Another possible side effect is an increase in home production of alcoholic beverages, which occurred to some extent during Prohibition and is a distinct trend currently in the United States in the absence of significant price increases. Implicit in the Toronto group's specific proposals is the gradual as opposed to precipitous raising of prices, which might tend to offset increases in home or illegal production of alcohol.

As de Lint and Schmidt (1971) are aware, high prices may be related to increases in occasional excessive drinking, with its attendant problems. They suggest that this may occur in Ireland and also is found in Finnish "explosive" drinking. However, since they view the social costs of physical pathologies associated with chronic high alcohol intake as greater than those related to asocial or antisocial behavior associated with occasional high intake, they see reduction of the former as beneficial, even if the means of accomplishing this results in an increase of the latter. Assessing relative social costs is a tricky business. Aggravated assault, for example, including its homicidal extreme, has been shown to be significantly related to the use, not necessarily chronic, of alcohol (Wolfgang and Strohm, 1956). The evidence on periodic dyscontrol, while not entirely clear-cut, suggests a similar phenomenon (Maletsky, 1973). Some automobile accidents may occur because motor skills and judgment are disrupted by acute intoxication in nonchronic drinkers or by drinking among youthful drivers who may be unpracticed in drinking, driving, or both (Hurst, 1972). Limiting availability of alcohol through raising price would not appear to reduce, and might even increase, these kinds of social costs.

Partly in recognition of these issues, Whitehead (1972a; 1972b) has attempted to combine aspects of the consumption and social science models in ways that maintain the preventive thrust of each. Focusing primarily on integrated drinking, he outlines three fundamental approaches: (1) increasing integration of drinking practices without modifying per capita consumption; (2) reducing per capita consumption without changing integration of drinking practices; and (3) simultaneously increasing integration of drinking practices and reducing per capita consumption. For the first approach he suggests legal changes that would permit parents to serve alcohol to children at home and in restaurants. He predicts that average consumption would not be affected because moderate use of alcohol in the home would hopefully be offset by decreased use outside the home. Rather than direct price controls to effect the second approach, he suggests reduction of alcohol content in alcoholic beverages and of the size of standard containers, which would lead to lower average consumption without risking the dangers of illegal production and sale of alcohol. To implement the third approach, he suggests the elimination of all advertising of alcoholic beverages because it tends to block integrated drinking by emphasizing sex, power, and social mobility in relation to alcohol. The absence of advertising, he holds, might reduce the incidence of alcohol use as well as exposure to stimuli urging one to drink; the combined effect would be to lower average consumption. It might be noted in this connection that a British ban on advertising of cigarettes on television had no discernible effect on the prevalence of cigarette smoking (Zacune and Hensman, 1971).

Whatever the merits of Whitehead's specific proposals, his analysis indicates that there are ways of affecting consumption levels that do not run the risk of creating new problems or augmenting old ones and that direct themselves to social health as well as physical health components of alcohol problems. The analysis, of course, presupposes that the distribution of consumption model is indeed valid, not only empirically, as the evidence seems to indicate, but theoretically as a causal explanatory sequence. It is here that questions remain.

Proscriptive

The proscriptive model sees any use of alcohol as a problem. The ideal norm for individuals is nondrinking, and production, distribution, and sale of alcohol should be prohibited. Recognizing that the social and human forces that gave rise to this model are extremely complex and have a centuries-old history, it may not be too great an oversimplification to say that today the basic impetus for the model may be found in the moral precepts, among Western countries at least, of major Protestant denominations. It may seem out of place in a scientific work to discuss a "model" that is ascientific and to which references in the scientific periodical literature are rare, to say the least. It is included here because of its past and current sociopolitical impact on actual alcohol problem control programs in a number of countries, because in making recommendations based on rational inquiry most scientific professionals fail to take the proscriptive model into account, and finally because any viable preventive program will depend on the support and active cooperation of the model's adherents, who reflect sentiments widely held within a population.

The swing of the cultural pendulum in recent decades has lessened the effect of the proscriptive model on control of alcohol consumption, as witnessed in continuing liberalization of liquor laws in the United States, Canada, and many European countries. As one aspect of this process, and perhaps partly in reaction to it, there has been a noticeable trend toward the liberalization of norms governing the use of alcohol by many of the Protestant churches (Burnett, 1973; Plaut, 1967). They have officially begun to change the model by permitting the use of alcohol within proscriptive precepts of moderation. What effect this change in policy will have on group behavioral norms is not yet clear, though it is certain that liberalization has been in fact a recognition that the previous ideal proscriptive norm was widely violated.

An apparently simultaneous trend is the incorporation of scientific findings to support the proscriptive model. The logic of the distribution of consumption model, taken in uncritical isolation from other social realities, ultimately would argue for prohibition of alcoholic beverages, and the Temperance movement is not insensitive to this implication (de Lint, 1973). Research findings that may be interpreted as indicating disruptive physical, psychological, or social effects, temporary or permanent, from ingestion of small quantities of alcohol are used as scientific support for abstinence. Advocates of abstinence, writing about education, which is one of their major preventive tools, argue for an accurate scientific informational approach (Geraty, 1971; Hill, 1971a; 1971b; Ivy, 1971a; 1971b) presented so as to "persuade" students to "use the facts rightly" (Ivy, 1971a). In Sweden, where the Temperance movement has long been officially included in the state's comprehensive and well-coordinated alcohol education program, careful attention is paid to the informational accuracy of the effects of alcohol on the human body and other aspects of alcohol use (Jellinek, 1963). The prospects for success of proponents of the traditional proscriptive model for prohibition of alcoholic beverages have been diminished by the opposing tenor of the times as well as by unpleasant memories of earlier experiments with it. Recent shifts in church policy suggest that the proscriptive model will move closer in the direction of the social science model. To the extent that the proscriptive model is rooted in religious institutions and that the social science model has ignored those institutions in its proposals, this movement would appear to offer the social science approach a powerful force for socialization of integrated drinking and changes in ideal and behavioral drinking norms.

Traditional Public Health

The classic public health model describes and analyzes the temporal–spatial topography of a disorder and its agent, host, and environmental interrelationships, using epidemiological and/or ecological techniques of investigation. From the results obtained by this mapping over time and space, proposals for the control and prevention of the disorder may emerge. The success of this model in confining, controlling, and in many instances preventing infectious diseases made it natural to attempt to extend it to chronic diseases and then to socially problematic behaviors that could be defined as disease entities. The 1950s and early 1960s saw widespread application of the public health model to nonspecific mental disorders, juvenile delinquency, suicidal behavior, drug addiction, and alcoholism.

Gordon (1956; 1958) and Plunkett and Gordon (1960) made a systematic attempt to apply the infectious disease model to alcoholism. They proposed a biologic gradient of alcoholism that sequentially distinguishes susceptible cases from those with inapparent or latent infection and from those with the clinical disease. Gordon (1958) defined alcoholism as "the use of alcohol as a beverage, irrespective of amount, place, periodicity, or practice," and subdivided "alcoholics" into social or customary drinkers, symptomatic drinkers, and addictive or advanced drinkers. This formulation has an analogical advantage; for instance, by suggesting that inapparent infection (social drinking) may exert an immunizing effect, whereas susceptibles (abstainers) may have less immunity when they become "infected." The breadth of their definition, which Plunkett and Gordon found conceptually necessary for full epidemiological study because narrower definitions "preclude the finding of all but the extreme cases," may be contrasted with the endpoint criterion of mortality rates adopted by the ARF group. At present, however, the analysis by Gordon and his associates within a classic framework has little currency, partly because its emphasis on analogies with communicable diseases does not strike a responsive chord among today's social scientists.

Through the 1960s, many concerned with prevention spoke in terms of levels of prevention, distinguishing among primary, secondary, and tertiary prevention (Blane, 1968a; 1968b; Chafetz and Demone, 1962; Cross, 1967; Cumming, 1963; Jellinek, 1963; Philp, 1967; Scottish Home and Health Department, 1970). Primary prevention consists of removing or preventing the causes of a disorder, or of increasing the number of those who are immune or resistant in the population. Secondary prevention is the arresting of a disorder through early treatment before it becomes fully developed. Tertiary prevention signifies treatment of the full-blown condition in order to prevent chronic or permanent disability or to effect cure (Cumming, 1963).

The usefulness of a levels conceptualization in prevention has been frequently and variously challenged on pragmatic, empirical, and theoretical grounds (Blane, 1968b; Cross, 1967; Gordon, 1956; 1958; McGavran, 1963; Philp, 1967; Plaut, 1963, 1967, 1972; Plunkett and Gordon, 1960). Tertiary prevention, identified wholly with treatment, has been criticized first of all because treatment and prevention are logically separate domains, so that tertiary prevention *qua* prevention is a contradiction in terms. Furthermore, as Gordon (1958) categorically states: "No mass disease has ever been adequately controlled by attempt to treat the affected individual," and goes on to say: "A program based on treatment of the exaggerated illness is temporizing and with no great promise of productive result; it is good clinical medicine but poor public health." Also, tertiary prevention does nothing to reduce incidence (Blane, 1968b), although it is conceivable that long-term decreases (over one or more generations) might accrue because of interruption of sociogenetic links. Finally, in terms of effects on preventive programs, it matters little whether one calls rehabilitation tertiary prevention or not, since treatment procedures and activities are in either case unlikely to bear operational relationship to prevention.

Secondary prevention has been criticized on similar bases (McGavran, 1963; Plaut, 1967). Referring to infectious diseases McGavran (1963) has said:

> Contrary to our beliefs generally, there is no evidence that control or eradication of any disease has been accomplished by the approach, procedures, techniques, and activities directed at early diagnosis and treatment of disease in individuals.

This point of view fails to take into account two factors: (1) the possibility of partial control over communicable diseases like syphilis and gonorrhea given casefinding oriented toward early detection; and (2) the prevention of social problems may not follow in all aspects the preventive model found effective for infectious diseases.

From a pragmatic point of view, it has been suggested that the bulk of resources available for prevention be channeled to secondary prevention because too little is known to mount an effective primary prevention program (Cross, 1967; Philp, 1967). Following the presentation of broad guidelines for a primary preventive program, Cross (1967) indicates that effective techniques of implementation are not presently available, and therefore calls for secondary preventive programs, pointing without citation to the success of the development of techniques and services in the area. Philp (1967), in a similar vein, says:

> There is no primary prevention which has been known and shown to be effective. It is true that there are some clues as to what might be attempted ... but these clues are essentially untested and at this point in time have not been incorporated into any specific program or any significant demonstration.

As an alternative, secondary prevention is offered:

> The concept of early detection, early intervention, and early treatment which has proved effective in other public health problems is also effective in approaching the alcoholism problem. The earlier the detection and the earlier the intervention, the better the result. Evidence continues to accumulate to support this statement.

Two observations may be made: (1) since the time that Philp wrote the above, several specific primary preventive programs have appeared, although they have not been tested and demonstrated; and (2) no solid evidence exists for *or* against secondary preventive measures despite positive claims by Cross and Philp and disavowals by McGavran (1963) and Plaut (1967).

Current positions about prevention of alcohol problems tend to have moved away from a levels conceptualization, some dealing exclusively with operations and activities that may be subsumed under primary prevention, others invoking secondary prevention measures though still featuring a predominant emphasis on primary prevention. Tertiary prevention is hardly mentioned. The ARF group speaks simply of prevention and employs it solely in its primary sense, as, with minor exceptions, does Wilkinson (1970). Plaut (1963; 1967) in his earlier work dealt exclusively in terms of primary prevention (Plaut, 1963), or simply prevention (Plaut, 1967), but distinguished between its specific and nonspecific forms. In a recent publication (Plaut, 1972), he maintains this distinction, but for the first time introduces secondary preventive measures. The National Institute on Alcohol Abuse and Alcoholism, while shifting toward primary prevention, a development that can be traced from the Secretary's First Report to the U.S. Congress on Alcohol and Health (U.S. Department of Health, Education, and Welfare, 1971) through a recent description of its prevention program (National Institute of Mental Health, 1973), continues to maintain a secondary prevention component. The inclusion of secondary preventive measures in a predominantly primary prevention program has been suggested by Blane (1968a), who reasoned that such elements are necessary until conclusive evidence is available about the utility of programs for high-risk groups and early detection; their omission might alter a possibly catalytic effect.

Current approaches to prevention of alcohol problems may be viewed through the public health model, and are, indeed, to one extent or another, products of it. Yet, it appears that the model does not possess the potentialities it once seemed to for preventing social problems, including those associated with alcohol. Of current models, distribution of consumption probably reflects most closely the practices and principles of public health. None, however, show much concern with the biological gradient of disease or the intricacies of agent–host–environment relationships. The remaining heritage is evident in aspects of epidemiological method, growing emphasis on primary prevention and therefore populations rather than individuals, and a focus largely on environmental manipulation.

TECHNIQUES OF PREVENTION

The principal classes of techniques of prevention of alcohol problems include education through the school system; information and education of the public at large, including advertising; manipulation of substance, person, and environmental factors affecting consumption patterns; and singling out for special attention subpopulations having characteristics that make them especially suitable targets for preventive work. These techniques are simply tools of social control available to agents of public policy, and say nothing about the philosophy, aims, and direction of any particular preventive policy. In discussing each class of techniques, available research or evaluative evidence about its effectiveness is summarized, with the understanding that the little research done in the area has often been exploratory, poorly conceived and executed, or of limited generalizability. Coverage of the recommendation or application of preventive techniques is reasonably thorough for the United States and Canada; experience in European countries is drawn upon illustratively when it deals with demonstration or evaluation of a method or with the introduction of a novel technique. Except for education in the schools, which is treated separately in the section that follows, each class of techniques is discussed in this section.

Public Information and Education

Public information and education are more or less explicit components of all prevention programs. In countries where prevention programs are centralized in the government, access to multimedia channels is controlled and results in a "united front" approach. In many countries, notably in Western Europe, the United States, and Canada, public information and education programs may emanate from several sources, including the government, public interest groups (e.g., the National Council on Alcoholism; the National Safety Council), church groups, and business (e.g., insurance companies), including the alcoholic beverage producers themselves. Within government, different agencies may present differing points of view, as has occurred, for instance, between the National Highway Traffic Safety Administration (NHTSA) and NIAAA in the United States. Further distinction may be made between public interest messages, whose basic intent is to encourage rational, socially responsible decision-making, and advertising, whose goal is to maintain or increase profits. In practice, this distinction is often blurred because public interest messages have come to rely increasingly on Madison Avenue technqiues, while private advertising, in the United States at least, has shown some salutary attempts at including moderating morals in its messages.

Wilkinson (1970) suggests several public education and related commercial promotion measures. In the area of public education, he recommends that federal and state agencies initiate a variety of programs to promote moderate drinking practices, respect for nondrinkers, and discussion of drinking problems; in addition to the usual media, classrooms, state driving tests, and contact with business managements are included. Under advertising and promotion, he suggests that regulations be revised to make it more possible to stress moderate and safe drinking in advertising, including the promotion of low-proof beverages, particularly table wines. Much of the content of

Wilkinson's proposals may be found in the multimedia campaign sponsored by NIAAA (Chafetz, 1973). Wilkinson's proposals also follow fairly closely on those put forth by Plaut (1967) in a more general way:

> Widespread public discussion about drinking practices will enable people to obtain a perspective on their own drinking behavior and that of others through a detached and objective examination of current practices and beliefs relating to alcohol.

In addition to employing the usual media, he suggests the use of health and welfare associations, youth organizations, Parent–Teacher Associations, and church groups. The National Institute on Alcohol Abuse and Alcoholism is currently utilizing the Junior Chambers of Commerce (Jaycees) for a cooperative nationwide educational program. The initial thrust is "toward widespread education of the Jaycee membership about the need for local and state educational programs on responsible drinking" (National Institute of Mental Health, 1973).

The distribution-of-consumption proponents view public information and education programs as a vital means of gaining support for price increases and countering the "formidable political and emotional obstacles" the introduction of price rises for alcohol would encounter. The aims of the programs would be to generate public awareness of the hazards of high consumption levels and acceptance of the preventive value of price control (Popham *et al.*, 1971). Several countries—Norway, Switzerland, Finland, Russia, Sweden—have strong state-supported antialcohol public information and education programs (Connor, 1972; Jellinek, 1963). In Switzerland, in particular, the explicit aim is to reduce alcohol consumption. The campaign in France, too, has as its goal reduction of consumption, but through moderation. Among slogans it has used are: "Never more than one litre of wine a day" and "No alcoholic beverages outside of meals." In some countries, the Netherlands and many southern European nations, for example, there is no state-supported public information or educational effort.

A recently instituted service in the United States is NIAAA's National Clearinghouse for Alcohol Information, which serves as a resource for the availability of all types of information in all forms of media about alcohol, its use and its abuse. The Clearinghouse not only serves the professional and caretaking community, but is available and relatively accessible to community groups and individuals who desire to establish public education and information programs at state or local levels.

The effect of public information and education programs on alteration of drinking patterns is indeterminate, although it is generally felt to be slight (Connor, 1972; Jellinek, 1963). Research evidence is almost totally lacking; all that exists are bits and pieces of information that are no more than suggestive. Thus, the fact that France was the only major Western nation to reduce its apparent per capita consumption in a recent ten-year period (Efron *et al.,* 1972) might be taken by some as an indication that its campaign for moderation in drinking was effective. On the other hand, the increase in proportions of drinkers in recent decades in countries where abstinence or temperance sentiments pervade information and education campaigns (Lindgren, 1973; Wallace, 1972) might be taken to demonstrate that these campaigns are self-defeating. Such statistics are perhaps more conservatively interpreted within the context of generally increasing consumption levels and increasing numbers of drinkers in virtually *all* Western nations, regardless of official stances about prevention.

Drinking–driving campaigns mounted in the United States during the 1950s and 1960s are generally thought to have had little effect on lowering the relative frequency of driving while intoxicated or accidents related to the use of alcohol. These campaigns had a common theme: to make drinking and driving mutually exclusive activities. Creating an atmosphere of the dire consequences that may follow drinking and driving, the message either graphically portrayed these consequences or left them to the viewer's imagination. In retrospect, the theme had low credibility, since people know or have experienced that drinking and driving frequently occur together without an accident. Despite low credibility and apparent lack of effect on target behaviors, messages around this theme have been shown to have persistent recall value. A survey conducted in 1973 by Harris and Associates (1973) showed that 20 percent of the respondents recalled verbatim messages that had not been used for several years (e.g., "If you drink don't drive, if you drive don't drink—the life you save may be your own"). Recent campaigns emphasize limiting oneself to no more than two drinks if one is going to drive and isolating the drunken driver as a deviant in need of professional help; no data are yet available on the effects of these campaigns, although Room (1972c) predicts they will if anything increase alcohol problems. The more general NIAAA campaign, referred to above, has been partially evaluated by Harris and Associates (1973). The evaluation considered the amount of television exposure of NIAAA advertisements, the distinguishing characteristics of viewers who recalled them, and changes in attitudes that might be attributed to exposure and recall. Exposure and viewer awareness were considered to be satisfactory, comparing "well with major paid advertising campaigns." Results for changes in attitudes showed some shifts in the desired direction, but for the most part there was no meaningful change. Also, due to several sources of uncontrolled variation, it would be difficult to attribute attudinal findings to the advertisements.

Haskins (1969), in a review of research on the effects of mass communication on traffic safety in general and drinking–driving in particular, concludes that "... there has been relatively little research of any kind on the effects of the numerous campaigns that have been conducted and, due to inadequate research design, one cannot be confident of the results obtained in most published studies." The major flaws he identified include absence of control groups, *post hoc* statistical analysis, unrealistic laboratory environments, and overreliance on verbal measures. In order to avoid these inadequacies in evaluations of overall campaign effects, Haskins (1970) states that four basic requirements must be met: naturalistic communications conditions, unobtrusive and valid measurement, accurate execution of the total communications and research design, and a clear relationship between cause and effect.

Clues about effectiveness may also be sought in results of preventive campaigns in other areas; for example, smoking. While some desired alterations in smoking behavior have occurred, causal linkages to public information efforts are difficult to determine. Campaigns in the United States appear to have had little effect on the incidence of new youthful smokers, although the health-risk orientation of the campaign does appear to have had an effect on middle-aged and older smokers, apparent in a rise in the incidence of exsmokers (United States Department of Health, Education, and Welfare, 1970). In Britain, on the other hand, the incidence of exsmokers also rose after publication of the Royal College of Physicians' report on smoking and health; the effect was, however, short-lived and the incidence of exsmokers has since been in a gradual decline (Zacune and Hensman, 1971), a finding that may be attributed to lower health concern in Britain than the United States, differences in campaign continuity, or other factors. Direct tests of the effect of antismoking information programs are rarely done; in one of the few studies available Cartwright *et al.* (1960) found that a community smoking information and education campaign in Britain changed attitudes but not behavior. It is generally conceded that antismoking campaigns have enjoyed limited success (United States Department of Health, Education, and Welfare, 1970; Zacune and Hensman, 1971).

A factor complicating evaluation in countries, such as the United States, where programs may be initiated by any interest group is the simultaneous presence of several programs that have divergent aims, objectives, and practices. The effects on public opinion and behavior of contradictory or partly conflicting multimedia messages from different sources within the government and from nongovernment sources are largely indeterminate because evaluation has not been attempted. Evaluation in this area would in any case prove exceedingly difficult. It may, however, be hypothesized that pluralistic messages tend to reinforce pluralistic values and behavior. Preventive alcohol policies that aim to heighten normative consensus would thus tend to be subverted by contradictory messages from other sources. This would argue for the development of an alcohol policy that takes into account exigencies of all relevant interest groups, rather than one that continues past policies of lack of coordination among groups. A similar situation, proceeding without coordination or guidance from the results of evaluation, characterizes drug information and education activities in the United States, on a very large financial scale (Ford Foundation, 1972). It may additionally be noted that in 1973 a moratorium was declared by the government on the production of new drug information materials.

The potential benefits of applying social science theory and methods to an understanding of the conditions under which public information and education messages are most effective in attaining social goals have been little explored. Leventhal (1964) conducted a heuristic analysis of the effects of alcohol advertising on drinking behavior, drawing upon theory and research on voting behavior in lieu of available data on alcohol advertising per se to support his arguments. The impact of his work is not that it says a great deal about effects of alcohol advertising, but that it documents the kinds of intervening variables and the relationships among these variables that must be considered before one can begin to draw conclusions about the relative effects of alcohol information and education programs. McGuire's (1969) review of the experimental social psychology of attitude change provides a similar conceptual frame of reference. Another example of the application of social science methods to evaluation of public information campaigns is the work of the panel brought together by the National Academy of Sciences–National Research Council to determine effective ways of assessing drug information programs (Rittenhouse, 1973). The panel recommended combining techniques successfully used to evaluate advertising effectiveness with procedures used in social science research. These latter include random assignment, control groups, pretesting, and identification of target groups and behavioral goals.

Public information and education campaigns are necessary elements of any preventive plan. To gain their ultimate objective with an acceptable degree of success, they must be thoroughly and rigorously pretested before mass distribution. During the campaign, monitoring should proceed routinely and with the same care as that given to the pretest. In view of the general failure of population-wide campaigns to change health behaviors significantly, it may be that financial and other resources allocated for their support might more profitably be used to develop limited programs, based on pilot research, with defined target populations and goals specific to those populations. It is encouraging to note that there is some activity along these lines. Worden et al. (1973b), conducting an action and research program on alcohol and highway safety, have undertaken a variety of pretests prior to the institution of a public education program in Vermont. In Canada, a prospective study of the impact of a month-long drinking–driving campaign in an "experimental" and "control" community has been recently reported (Information Canada, 1973). Short-term hypothesized changes were obtained for attitudes and informational level; observed decreases in the percentage of drivers with blood-alcohol levels above

0.08 percent were small and subject to bias, but were sufficiently encouraging to merit further investigation.

Manipulation of Substance, Person, and Environmental Factors

Manipulation of substance, person, and environmental factors to affect consumption patterns follows the classic public health model of control and represents the key techniques of primary prevention, whether the aim is to lower overall consumption levels, as in the distribution of consumption model, or to adjust consumption patterns that result in problems, as in the social science model. Actual proposals follow the general preventive trend of emphasizing environmental manipulation, but there are a number of suggestions that center directly on substance (agent) and person (host) factors.

Substance Factors

Plaut (1972) lists the hypothetical possibility that alcohol could be altered so as to reduce its habit-forming properties, perhaps by adding some substance that would change its manner of absorption or rate of metabolism. Reviewing the research literature on antagonists, Wallgren and Barry (1970) conclude that central nervous system stimulants are the only substances of promise, but are dangerous in therapeutic use because of their convulsant properties. Disulfiram (Antabuse) is an antagonist of sorts, but it has not been successful in controlling drinking problems because its use is voluntary, must be regular, and can be dangerous. Experience with disulfiram and other substances proposed to control alcohol problems shows that while single-factor solutions appeal in theory, they infrequently work in practice. Less esoteric substance manipulations include lowering the alcohol content of beverages, promoting weaker over stronger beverages, making substitute wines with no alcoholic content, and recommending the use of substances that have the same cue properties to fun and conviviality that characterize alcohol but that do not have alcohol's pharmacological properties (Jellinek, 1963; Plaut, 1967, 1972; Whitehead, 1972a, 1972b; Wilkinson, 1970).

Person Factors

Age, sex, and prior drinking behavior are person-variables that have usually been manipulated in attempts to control use of alcoholic beverages. Most societies have rules governing drinking according to age and, to a lesser extent, sex. Wilkinson (1970) has catalogued the interstate inconsistencies of such regulations in the United States. In some Scandinavian countries, persons who have had officially noticed trouble with alcohol are restricted with regard to the purchase of alcohol; this also occurs informally in bars and liquor stores in other localities by the individual actions of bartenders and salespersons who decide when a patron has had "enough" and is to be refused further service. Both Wilkinson (1970) and Plaut (1967; 1972) recommend either removing or lowering age restrictions against the sale of alcoholic beverages on grounds that such action would subvert the development of "forbidden fruit" images and fantasies about alcohol and its use, would be consistent with actual practice, and would recognize that adulthood begins before the age of twenty-one. Further, they contend that in localities where such action has been taken, there is no evidence to show an increase in problems related to alcohol. Since these proposals were made, many states have indeed lowered the age limit; the effects however, have not yet been studied.

Environmental Factors

The manipulation of environmental factors is the largest arena of proposed action. Numerous measures have been taken or proposed and include: provision by colleges for safe drinking facilities on campus; revision of tax scales to increase the financial attractiveness of purchasing low-strength beverages; revision of tax scales to make every beverage equally expensive per unit of absolute alcohol; elimination of restrictions on decor and interior of on-premise drinking places; development of model neighborhood taverns; removal of prohibitions against Sunday drinking; integration of drinking with varied leisure activities; permission for grocery stores to sell alcoholic beverages and liquor stores to sell food; provision of training in alcohol control, drinking, and alcohol problems for retailers; establishment of penalties for selling alcohol or extending credit to intoxicated persons; coordination of alcohol control and health promotion functions of governmental agencies; control of hours of sale; manipulation of numbers and types of outlets; planning for the presence of nondrinking drivers at social and recreational activities where alcohol is served; promotion of activities that are functionally equivalent or preferable to the use of alcohol; abolition of all liquor advertising; display of natural drinking situations in advertising; and removal of information on labels that indicates amount of alcohol content.

This list of measures has been adapted from the ARF group, Plaut (1967; 1972), Whitehead (1972a; 1972b), and Wilkinson (1970), as well as from measures used in other nations. It does not pretend to cover all measures, but to give an idea of the rather bewildering, often contradictory variety of proposals suggested or adopted, and a sense of the continued emotionalism, to use Plaut's term, that surrounds presumably rational discussions about prevention of alcohol problems.

One way of viewing these proposals is to examine what is known about their effectiveness. While the literature on the subject is scientifically and methodologically weak, and while controversy abounds, some conclusions may be drawn, especially about environmental manipulations that affect availability and accessibility of alcoholic beverages in space and time. Strict systems of legal and organizational control appear to be related to low per capita consumption but to a high incidence of visible behavior problems. This has been reported by Christie (1965) for the Scandinavian countries and for the United States in studies undertaken under the auspices of the New York Moreland Commission (Bacon, 1965). Room's (1971) analysis suggests that other factors, such as urbanization, may be more decisive in explaining variation in problem indices than strictness of control systems themselves. In the United States, Prohibition reduced alcohol consumption (Warburton, 1968) and cirrhosis mortality precipitously (Terris, 1967) while bringing about new and unanticipated problems. This brings into focus other attributes of environmental manipulation measures; namely, the uncertainty of their ultimate preventive effects (Kuusi, 1957) and the fact that unintended effects may occur (Room, 1971) even though immediate intended effects can be predicted. A fairly conclusive finding is that consumption levels by type of beverage (spirits, wine, beer) can be readily influenced by differential pricing policies (Grosswiler, 1972; Jellinek, 1963; Kuusi, 1957; Room, 1971; Wilkinson, 1970), as the instances of Belgium, Denmark, Norway, Sweden, and Switzerland demonstrate. This finding, combined with the known power of commercial advertising in influencing choices of brand for a product, has implications for preventive programs that aim to alter drinking patterns by emphasizing beverages with low alcoholic content. Such factors as number, type, and location of drinking places; on-premise and off-premise drinking; hours of sale; private license or state monopoly; and local option rules, all of which have been used to control drinking and problems associated with it, have not been demonstrated to have specific preventive effects (de Lint and Schmidt, 1971; Popham, 1962; Popham et al., 1971), except perhaps for hours of sale in Britain (Terris, 1967). More detailed information on legal controls may be found in Chapter 13.

The relative absence of research in this area and the poor quality of much of what has been reported have often been deplored, yet the effects noted above seem relatively well established. Perhaps the most interesting aspect of these findings is that while they reveal undoubted effects on drinking behavior, similar effects on alcohol problems are not evident. To the contrary, the general tenor in the field indicates if anything that alcohol problems are on the rise in industrialized societies. This is the import of data that show rising consumption levels in most countries in Europe as well as America, and of reports about such disparate localities as Russia (Connor, 1972), Japan (Yamamuro, 1973), and Hong Kong (Singer, 1972). One wonders whether preventive activity, which does many things but does not always prevent, is analogous to treatment activity with respect to alcoholism: new miracle treatments have been discovered periodically for decades, yet there is still no definitive or even empirically preferable treatment.

As Terris (1967) has suggested, an anomalously salutary picture may obtain for the control system of Great Britain. It therefore merits closer examination. In the United Kingdom, the consumption of alcohol is regulated by both price and sales controls. The former, which makes alcohol more expensive through taxation, is not intentionally directed toward reduction of alcohol problems, but presumably lowers consumption levels, especially in lower socioeconomic class strata. The major feature of sales control is the limitation of hours of sale, which are generally from 11:00 A.M. to 3:00 P.M., and from 5:30 P.M. to 10:30 P.M. (Zacune and Hensman, 1971). Since the institution of this basic system during World War I—it has had minor changes over the years with recent liberalization of conditions of scale—Great Britain has witnessed dramatic decreases in arrests for drunkenness (Jellinek, 1963) and in liver cirrhosis mortality (Terris, 1967). Terris attributes the lower cirrhosis death rate to reduced per capita levels of consumption, which he sees as causally related to the sales-price control system. That other unknown variables are operating is suggested by the observation that while the per capita consumption in England and Wales is in the midrange when compared to other countries, its cirrhosis mortality rate (along with that of Ireland) is lowest (de Lint and Schmidt, 1971). The United States, for example, with a per capita consumption in the same range had a cirrhosis death rate five times that for England and Wales, while Canada with virtually the same consumption had a death rate two-and-one-half times that for England and Wales. In other words, giving complete credit to the formal control system does not explain Britain's relatively low alcohol problem rate. Perhaps, as Wilkinson (1970) has suggested, the measures introduced of necessity in World War I "catalyzed and reinforced a growing concern about drunkenness," which continued to be expressed after the war until about 1950, when a reversal in consumption level and cirrhosis rates appeared. Jellinek (1963) also suggests that the reduction in drunkenness arrests cannot be attributed solely to the control measures, because the reduction occurred during a period when decreases in consumption and drunkenness were taking place in most European countries. Nevertheless, the very fact that moderate levels of consumption are combined with relatively low incidences of physical and visible social problems related to alcohol use in the United Kingdom, and possibly Ireland, suggests that study of these countries may have more general preventive implications.

Secondary Prevention

Reliance on secondary preventive measures has been recommended many times (Blane, 1968a, 1968b; Chafetz, 1967; Cross, 1967; Cumming, 1963;

National Institute of Mental Health, 1973; Philp, 1967; Plaut, 1972; Ritson, 1972; Wilkinson, 1970), yet it is difficult to locate descriptions of actual programs that meet definitions of secondary prevention. In this section, general recommendations that have been made will be listed, and known programs of secondary prevention will be described.

The two most commonly proposed populations for the introduction of secondary preventive measures are early problem drinkers and groups whose members run a high actuarial risk of becoming problem drinkers. Chafetz (1967) has presented a set of behavioral guidelines to be used for early case finding, some of which have been criticized on the basis that they are indicative of late-stage rather than early-stage alcohol problems (Block, 1967). Ritson (1972) states that there are a number of clear signs that clinicians and others may use in the early identification of young problem drinkers, including recurrent minor illnesses and accidents, regular loss of work, and frequent hangovers at work. Early signs of developing problems with alcohol are well-known, but are not routinely taught to future health professionals. Further, there is a reluctance on the part of professionals and others to use observation of these signs as a basis for intervention. Other than attempts to identify problem-drinking workers in industry (Trice and Roman, 1972), no large-scale program has been established to identify early cases systematically, although Plaut (1972) has spoken of the possibility of employing mass-screening techniques if knowledge were available about the psychological, social, or physiological precursors to drinking problems.

Focus on high-risk groups has been discussed by Blane (1968a; 1968b) who has suggested as possible target groups children of alcoholics, delinquents, recurrent absentees, drinking drivers, and collegians who suffer accidental injury. Research by Maddox (1966) has shown an association between heavy drinking and injury among college students, while Chafetz et al. (1971) found more social–behavioral difficulties in a group of children of alcoholics than in a matched group of children of nonalcoholics. High proportions of drinking delinquents show distinct evidence of having problems with the use of alcohol (Blane and Chafetz, 1971).

Interventive measures suggested for members of high-risk groups include treatment, special educational programs, disciplinary action, and imposition of legal sanctions. Wilkinson (1970) has suggested probationary "learning sessions" for youthful drinking offenders. Focus on drinking drivers and traffic safety has high priority now, and three of the measures proposed by Wilkinson (1970) in this connection have been widely adopted, a phenomenon that began to occur prior to publication of his book. These include passage of implied consent laws, "learning sessions" for convicted driving-while-intoxicated (DWI) offenders, and suspension of offenders' driving licenses. A fourth suggestion, establishment of court-attached clinics, has been less widely adopted (Crabb et al., 1971).

Most secondary prevention programs that have actually been put into operation deal with drinking and driving. Under the impetus of a high-priority program on alcohol and traffic safety by NHTSA, a number of programs for DWI offenders have been initiated throughout the United States. These may be considered to fall within the purview of secondary prevention, even though they contain a proportion of persons with late-stage alcohol problems. For the most part, these programs have contented themselves with educational approaches and imposition of legal sanctions, using existing treatment resources for alcoholic persons secondarily as indicated. This stems largely from the policy of the NHTSA not to become involved in the provision of direct health and welfare services. There is considerable uniformity in educational approaches for DWI offenders from community to community, most being based on some variant of the one originally developed in Phoenix, Arizona (Stewart and Malfetti, 1970). In Vermont, an active research program is underway to develop alcohol content in high-school driver education courses (Worden *et al.*, 1973a). Some programs have the unusual distinction in the alcohol prevention field of having been subjected to relatively rigorous evaluation (Blumenthal and Ross, 1973). The results are not encouraging, since no evidence was found that the legal sanctions or interventive measures employed were superior to the traditional fine in changing behavior in the desired direction. Ross (1973), in an exhaustive review of the evidence on the effects of the British drinking–driving program instituted in 1967, concludes that early successes of the program failed to be maintained over time because initial subjective certainty of punishment on the public's part, which acted as a deterrent, was not reinforced by actual enforcement on the part of the police and the legal system.

A demonstration program for drinking delinquents, conducted at the Massachusetts General Hospital in Boston (Blane, 1968a, 1968b; Blane and Chafetz, 1971), grew out of the need to assess secondary prevention approaches. The choice of adolescent male delinquents as a target group was based on the assumption that they run a higher-than-average risk of developing severe drinking problems in adulthood. The delinquents were assessed with a wide variety of social, psychological, and behavioral instruments and then randomly assigned to treatment and no-treatment groups. Treatment was individually designed for each youth in the treatment group and proceeded from a mental health orientation that viewed the individual as functioning within psychological and social space. Like the evaluation of the DWI project above, the results of the delinquency project were negative when discrepancy scores between pretreatment and posttreatment measures were analyzed (Blane, 1971b). This held true at social, psychological, and behavioral levels of analysis. Although individual cases of remarkable, beneficial change could be noted within the treatment group, statistical analysis revealed no differences between treatment versus no-treatment groups.

The studies referred to above are important because they are rare instances of research in the field of prevention of alcohol problems and because their findings are discouragingly negative. The latter all too clearly highlights the need for controlled pilot demonstrations, the results of which may be used as one element to inform the development of policy. This may help to avoid costly, but unwise policy decisions about widespread adoption of programs.

EDUCATION IN SCHOOL SYSTEMS

Basic Assumptions

Health education is founded on values that are for the most part widely and uncritically accepted. It is commonly believed that health education is effective in reaching its aims of promoting health and defeating ill health. It is further taken for granted that one of the proper functions of a public educational system is to instruct children about health with a view to instilling health-oriented values and behavior. This is not readily subject to empirical verification, but reflects the values and beliefs our society holds about the functions of education.

The beliefs and assumptions that characterize health education appear typical of alcohol education, too. However, in recent years several reviews and research reports have appeared that suggest that there is a need to reassess the implicit foundations upon which alcohol education rests. It can no longer be assumed that alcohol education is good simply because it exists. While the question of effectiveness is major, there are also other lesser but related questions. Is there a priority among socializing agents with regard to dominance in shaping drinking behavior that may make the schools' contributions redundant? Is the school the best setting to socialize drinking attitudes and behavior? If so, should the agents be teachers or specialized personnel?

While definitive work in the socialization of drinking behavior remains to be done, currently available evidence points to the centrality of the family in shaping drinking behavior (Bacon and Jones, 1968; Braucht et al., 1973). Religious affiliation and participation (Bacon and Jones, 1968; Cahalan et al., 1969) are related to differences in adult drinking behavior; this and other evidence (Snyder, 1958) suggest that religious training may also have an important socializing effect. Peer-learning is another possible source of socialization, although the evidence here is contradictory (Bacon and Jones, 1968; Forslund and Gustafson, 1970). Evidence for the socializing impact of the school is limited, and though adult drinking behavior varies with educational level, this is in all likelihood a function of social class rather than of schooling itself. If the family does exert the major influence on the child's developing view of and relationship to alcohol, with the church and peers occupying an intermediate position, and the schools exerting the least influence, it may be the soundest policy to allocate resources in such ways as to take full advantage of the socializing agents that have proved themselves. Or, it may be argued that the schools have not been doing the job they could have done if they had had the resources. Edwards (1970) implies that current socialization agents are sufficient when he says alcohol education may be unnecessary in view of the fact that most people grow up to become responsible drinkers. The point is that alcohol education for children and young people does not necessarily have to take place in the schools.

The suitability of the school as a viable setting for alcohol education and of teachers as the providers of that education may be questioned. In areas involving styles of behavior, teachers tend to have low credibility in the eyes of students, and often the relationship between teachers and students is less tutorial than compliant but essentially adversary in nature. When teachers speak of the necessity to control behaviors that have strong elements of attraction to young people, or where discrepancies exist between teacher description and student observation, credibility decreases along with message effectiveness. With regard to the authoritarian nature of the teacher–student relationship, Davies and Stacey (1972) say that "since young people who drink heavily . . . appear to have negative or hostile attitudes toward authority in general and school teachers in particular, it is doubtful if school teachers are the ideal people to undertake the task of alcohol education."

The question of the effectiveness of alcohol education cannot be answered in our present state of knowledge. As with other aspects of preventing alcohol problems, the available literature on evaluation of effectiveness and other relevant background topics is sparse. It would appear that alcohol education has proceeded without serious concern or discussion about the tenability of the assumptions that underly it. A disjointed body of knowledge and literature about alcohol education has been created, much of it impressionistic and rooted in faith, some of it reflecting sound, insightful reasoning, but little of it based on solid evidence or representing a cumulative record of information and experience in the area. It is this literature that will be selectively drawn upon in the following sections.

Philosophies of Alcohol Education

The philosophies that underlie alcohol education programs have been developed in considerable detail for proponents of the social science and proscriptive models of prevention, and the general outlines of the distribution of consumption model may be surmised. The social science model tends to dominate alcohol education in the United States and Canada and in such European countries as Great Britain and Germany, while the proscriptive model predominates in the official alcohol education programs of most Scandinavian countries, Switzerland, and Russia.

Social Science Model

In the United States, and to a lesser extent in Canada and Great Britain, there is general agreement about a broad philosophy of alcohol education among those who have written on the subject (Blane, 1971a; Chafetz, 1973; Davies and Stacey, 1971, 1972; Fraser, 1967; Freeman and Scott, 1966; Globetti, 1971; Hayman, 1966; Jahoda and Cramond, 1972; McCarthy, 1952; Mullin, 1968; Pasciutti, 1962, 1963; Plaut, 1967; Robinson, 1969; Russell, 1965, 1969; Scottish Home and Health Department, 1970; Unterberger and DiCicco, 1968; Weir, 1967; Williams *et al.*, 1968), although this philosophy is not always reflected in textbooks used in health education courses (Anderson, 1968; Grout, 1968; Kogan, 1970). The essence of this philosophy is that young people require information about alcohol sufficient to enable them to make intelligent decisions about their own behavior with regard to it. Information is broadly conceived to include personality, emotional, and social dimensions as well as more factual material on what alcoholic beverages are, how they are made, and the effects of alcohol on the human body. There is also strong emphasis on creating attitudes conducive to healthy use or nonuse of alcohol, a movement away from emphasizing alcoholic problems, and an avoidance of fear messages.

These components of the social science philosophy of alcohol education are reflected in statements such as these: "Alcohol education accepts the fact people will drink and focuses on what kind of drinker they will be. It is believed that alcohol education can modify the attitudes and expectations that people bring to drinking" (Fraser, 1967); "The objective of alcohol education programs for teen-agers is to develop or strengthen positive accepting attitudes toward abstaining and moderate social drinking and negative attitudes toward excessive drinking" (Williams *et al.*, 1968); "The schools have the responsibility of dispelling myths and altering sociocultural attitudes about alcohol and alcoholism by providing scientific data and fostering healthy attitudes about drinking" (Mullin, 1968); and "Alcohol education for teen-agers should be directed toward adolescent drinking itself, rather than toward adult drinking problems and their remedy, with emphasis on a reduction of ambivalence and of disparity between attitudes and behavior" (Freeman and Scott, 1966). As with other philosophies of alcohol education, a major underlying assumption is that education will ultimately be related to a decrease in the incidence of alcohol-related problems (Fraser, 1967; Williams *et al.*, 1968).

Proscriptive Model

Alcohol education influenced by the philosophy of the proscriptive model was prominent in the United States as recently as the 1940s (Roe, 1943), and state laws regarding alcohol education that were enacted following the repeal of Prohibition made teaching about the evils of alcohol mandatory. While many of these laws are still on the books, their influence on alcohol education is much less today than it used to be. Temperance groups still maintain active lobbies and achieve occasional successes, as, for example, when the Florida State Department of Education recently adopted a work sponsored by a temperance organization as its official alcohol education text, but their impact is on the decline. This is not the case in several European countries where temperance groups have semiofficial status and make major contributions to alcohol education campaigns. While every attempt is made to ensure that information presented is scientifically accurate, the temperance group materials and approach there have been likened to those available in the United States 25–30 years ago (Jellinek, 1963).

In the United States, the philosophy of education espoused by adherents of the proscriptive model closely follows that described above for the social science model; i.e., young people require information to help them make intelligent personal decisions about the use of alcohol. One critical difference, however, obtains: the belief that factual education about alcohol is incomplete without a moral foundation. The integration of moral and factual alcohol education will lead to one correct decision: not to drink. Ivy (1971b) relates the "moral basis of abstinence" to the "scientific approach . . . to alcohol education" that "must not only collect, organize, and present the facts regarding the effect of alcohol on man and society, but . . . must also ascertain from these facts whether it is right or wrong, beneficial or deleterious, for man to use alcohol as a beverage. Furthermore, each one of us should work out for himself the best way he can present the facts so as to persuade others to use the facts rightly." Hill (1971a), while abjuring the use of fear-producing teaching materials, misinformation, and misleading classroom demonstrations in abstinence-oriented alcohol education, and castigating teaching that has "little regard for contemporary scientific investigation and the evaluation of concomitant problems," also stresses the moral–ethical foundations of alcohol education, embedding them in a set of what he presents as the goals of education in general (Hill, 1971a; 1971b). Unlike Ivy, who directly stresses *correct* outcomes of alcohol education, Hill argues that students have the right to make their own decisions; the educator, however, must help students accept that responsibility for their decisions and "teach young people to do systematic thinking in order to arrive at right conclusions" (Hill, 1971a; 1971b). Thus, as we have seen earlier, American temperance groups make good use of the latest thinking and research about alcohol in their battles against it.

In many European countries where temperance movements are strong, the means employed to achieve the goal of abstinence differ from those recommended by American temperance groups, and include reliance on authoritarian statements, one-sided presentation of information, and fear appeals. After an extensive review of educational materials, Olkinuora (1971) concludes that Finnish temperance education in the public schools emphasizes the effects of alcoholism as if they were the same as the effects of social drinking, and he decries the dictatorial teaching methods employed. In the Soviet Union, anti-alcohol propaganda and education use fear messages and appeals to patriotism to attack drunkenness in particular, and also emphasize the dangers involved in drinking itself (Connor, 1972).

Distribution of Consumption Model

For the distribution of consumption model, alcohol education is clearly a secondary supportive measure to ensure public acceptance of price increases of alcohol. It is definitely not viewed by the ARF group, as it is in both the other models, as a major weapon in the struggle to prevent alcohol problems. The basic objectives of education within this framework are to gain public recognition of the dangers of heavy consumption, to demonstrate the preventive utility of increased relative prices for alcohol, and thereby to generate popular support for a proposal that its authors recognize as being manifestly unpopular.

Beyond general statements, the ARF group has not elaborated upon its educational proposals. Presumably, however, specific content would highlight such physical consequences of prolonged, heavy drinking as cirrhosis and chronic brain syndromes. This might be combined with popularized presentations of the association between level of per capita consumption and alcohol mortality rates. While the ARF group has not discussed target populations for education, it may be presumed that schoolchildren and young people would be among them. To the extent the program did not address itself directly to alcohol and its individual use, it might well be conducted within civics, economics, or history classes, and might carry with it feelings quite different from those engendered in educational activities proceeding from either of the other models. On the other hand, if the program focused on the physical liabilities of heavy consumption, it might run the risk that allegedly accrues to fear messages or it might have low credibility for most drinkers.

In any event, we are in the realm of speculation and the Canadian group's proposals are still on the drawing boards; because of this, and also because of the relative decline in influence of the proscriptive model, the following sections will concentrate heavily on alcohol education as it is seen from the perspective of the social science model.

Materials

Even cursory examination of available materials—books, pamphlets, films, slide presentations, posters, curriculum guides—reveals a wealth of up-to-date, technically polished, and factually accurate materials. The old complaint that they are not available, that their accuracy is questionable or that their presentation is old-fashioned no longer holds, although one or more of these certainly obtained as little as 20–30 years ago. Roe's (1943) comprehensive study of alcohol education in the late 1930s and early 1940s showed that most educational materials had been strongly influenced by temperance groups and evidenced little respect for objective presentation of information. Later reviews (Freeman and Scott, 1966; Monroe and Stewart, 1959) that reported similar trends may have been overly influenced by Roe's work and the ethos that grew out of it; also, being less comprehensive than Roe, these reviewers may have missed newer materials that were beginning to appear. Actually, it appears that in the late 1950s and through the 1960s up to the present, there has been an increasing flow of materials that meet the criticisms first voiced by Roe. General sources for materials include Finn and Platt (1972), Mental Health Materials Center (1973), and National Clearinghouse for Alcoholic Information (1972).

Most state departments of education have prepared model curriculum guides for health education and, at the secondary level at least, these nearly always contain sections on alcohol and its problems. The tremendous interest in drug education in the last five to ten years has stimulated the production of alcohol education materials as well. More recent impetus may be found in the programs of NIAAA and NHTSA's high-priority alcohol and traffic safety program, both of which got underway in 1970. State curricular guides are periodically updated, depending on the state, but heightened federal activity in the alcohol abuse area appears to have increased the frequency with which updating is undertaken. An example of federal activity may be found in the publication of a multivolume curriculum manual on alcohol and alcohol safety for elementary, junior high, and senior high school levels (Finn and Platt, 1972), the preparation of which was supported by the National Highway Traffic Safety Administration. These manuals will undoubtedly influence the preparation of materials at the state level.

Most health education textbooks at high school, collegiate, and graduate levels contain at least some material on the use and abuse of alcohol. Many devote substantial space to the question and generally follow an approach consistent with the social science model, although there is often a moralistic tone. For example, in one text alcohol is considered under the general heading of "The Personality in Trouble" (Otto et al., 1971), and in another under "Dangerous Habits" (Miller et al., 1971). Commonly, too, textbooks dwell at length on alcohol problems at the expense of exploring drinking as a developmentally ordered and socially and psychologically regulated behavior that may have health consequences (Hein et al., 1970; Miller et al., 1971; Otto et al., 1971), although occasionally there is more balance in the presentation of material on drinking and the problems that result from it (Guild et al., 1969).

A common assumption about alcohol education is that a paucity of materials exists (Richards, 1971; Robinson, 1969; Unterberger and DiCicco, 1968), but this does not appear to be the case. Well-presented, factual, and imaginatively conceived materials are readily available. The correct question has to do not with availability of materials but with whether and how they are used. Although evidence is virtually lacking, the general impression is that alcohol education has never had much of an active place in the total education picture, and that curriculum guides recommended by state education authorities are little used locally.

Target Groups

The current trend is toward inclusion of all school levels as target groups: elementary, junior high or middle, and senior high (U.S. Department of Health, Education, and Welfare, 1971). This represents a gradual lowering over the years of the age level at which young people are thought to be appropriate targets, in terms of need, readiness, and outcome, for alcohol education. The early emphasis was almost exclusively on high school students. With the accumulation of research findings that showed that the largest proportion of adolescents start to drink between fourteen and fifteen years of age, junior high school students increasingly were included as another suitable target group. This research has been summarized by Bacon and Jones (1968) and reviewed by Stacey and Davies (1970).

More recently, the growing belief that the family as a primary unit of socialization is decisive in the formation of drinking attitudes and behaviors has suggested that the school should get involved in alcohol education as early as possible if its efforts are to be effective. This belief is supported by impressions that young children are more exposed to and aware of alcohol, at least contextually, than is usually supposed, although there has been little scientific evidence to back up these impressions, other than some mildly suggestive and subjective data reported by Byler et al. (1969). It was against this background that the recent move toward the development of K-12 (kindergarten through grade 12) curriculum guides for alcohol education was initiated.

Interestingly, research support for this development has appeared after the fact, although it will undoubtedly accelerate the trend, in the form of an elegant research program conducted by Jahoda and Cramond (1972) with six-to-ten-year-old schoolchildren in Glasgow. Their conclusions are of major significance and merit direct quotation:

> The results of the study clearly support the expectation that children begin to learn about alcohol early in life, even before primary school. By age 6 a majority recognize the behavioral manifestations of drunkenness, and many are capable of identifying some alcoholic drinks by smell alone; they also perceive people in different social roles to like alcohol in different degrees. By the age of 8 most children have attained a mastery of the concept of alcohol, and in general, the rate at which children acquire a broad understanding in this sphere was greater than had been anticipated.

Furthermore, Jahoda and Cramond found that negative attitudes toward alcohol and drinking increase with age, a finding they interpret as meaning that children learn to disapprove of alcohol from people in such institutions as church and school. These findings clearly offer support for the extension of alcohol education to the elementary school level.

Curricular Placement

The placement of alcohol education in the school curriculum is a topic that has generated considerable discussion. In most published curriculum guides and health education textbooks, alcohol is treated either as a separate unit, as a component of a substance abuse unit that also includes drugs and tobacco, or as a component of a mental health unit. In practice, alcohol is often treated in single school-wide meetings, arranged by the school administration, that consist of a lecture or other presentation by an outside "expert." Alcohol education using either of these forms, the isolated short-term unit or the "one-shot" presentation, has been criticized on grounds that it fails to engage students, is arbitrary, and neglects individual student readiness and interest (Freeman and Scott, 1966; Globetti, 1971; Sinacore, 1971; Unterberger and DiCicco, 1968).

An alternative is to incorporate content about alcohol wherever it seems natural and relevant in the curriculum: in social studies, English, biology, chemistry, etc. (Blane, 1971a; Ferrier, 1964; National Institute of Mental Health, 1973; Pasciutti, 1963; Plaut, 1967; Sinacore, 1971; U.S. Department of Health, Education, and Welfare, 1971). By making the study of alcohol a natural part of the curriculum, the student would acquire knowledge of alcohol's physical properties, physiological and behavioral effects, legal status, historical relation to politics, role in the economy, variegated use over space and time, and place in literature and the arts. Treating alcohol as a matter-of-fact part of the learning process would tend to reduce the "emotionalism" (Plaut, 1967) that surrounds alcohol, thereby heightening students' readiness to examine their anticipations and behavior with regard to the use of alcohol. Further, alcohol education would not be concentrated as a focal point in the health education sequence, thereby diminishing the singled-out aspect to which many students respond negatively (Plaut, 1967).

Another alternative may be found in suggestions that proceed from the conception that young people today are growing up to become members of a rapidly changing, complex technological society characterized by conflicting values and information, dramatic shifts in space and time, and increasing social problems. In order to prepare young people not only to cope but also to potentiate themselves, "courses" may be designed that operate on group process principles and in which students may feel free to explore their feelings and to gain knowledge about areas of development and issues of interest as they become important to them. Typical topics would include alcohol and drinking, as well as such subjects as sex, drugs, identity, relation to authority, and work. Such an approach to alcohol education is implied in Plaut's (1967; 1972) recommendation of raising the level of mental health and is tentatively discussed by Blane (1971a), Chafetz (1973), Merry (1973), Sinacore (1971), and included in the preventive programs of the National Institute on Alcohol Abuse and Alcoholism (National Institute of Mental Health, 1973). Aside from these examples, one finds little reference to this approach in the alcohol education literature, although it enjoys some currency among mental health educators (Long, 1971).

Curriculum placement is an important consideration in alcohol education because it is generally thought that placement affects outcome, although no research is available to support this assumption. Placement is also important because it has implications for the nature of teacher preparation, a topic that can be more completely explored after discussing methods of instruction. Current thinking appears to favor a separate alcohol education effort, perhaps combined with education about other substances, but with alcohol content also dispersed throughout the curriculum.

Methods of Instruction

There is fairly general agreement that the simple presentation of information by traditional methods, as, for example, lectures accompanied by question-and-answer periods, discussion, and/or demonstrations, is not effective in attaining alcohol education goals. This also holds for the use of fear messages in presentation of information. Conversely, there is equally strong agreement that small-group discussion and interaction, led by a teacher experienced in group dynamics and in guiding group discussions, is the most effective means of attaining educational goals in this area.

Despite this consensus, there seem to be no published reports of studies investigating differences among instructional methods in alcohol education, although the results of a study conducted by Williams *et al.* (1968) have a direct bearing on a small-group discussion method. Using a small-group discussion method with junior high school students, with a nonalcohol education control group, they found positive short-term changes in attitudes and knowledge level, with attitude change decaying more than information over time. One of the most interesting findings was that self-reports of drinking behavior one year later showed that in comparison with the control subjects more of the alcohol education than control subjects drank, but proportionately fewer of them reported numerous episodes of intoxication. The study by Williams and his group stands out as much for its uniqueness as for the care with which it was done.

It is unlikely that an either/or approach to choice of method of instruction will be ultimately effective. In certain communities with certain target groups and goals, an informational technique will be the method of choice. Under other conditions, other methods will be more effective. Similarly with fear messages: Under certain conditions and for certain targets, fear messages may be potent influencers of attitude change (Higbee, 1969; Leventhal, 1965, 1970; McGuire, 1969). Actual choice of method or technique will always be the product of the confluence of many factors, but thus far the potentially valuable input that may be derived from research and evaluation has been ignored. Further, the applicability of findings from research on attitude change by social psychologists has been virtually untapped (Edwards, 1970; Stacey and Davies, 1970).

Teacher Preparation

It is generally thought that systematic efforts to prepare teachers for alcohol education are virtually nonexistent in schools of education and that most teacher preparation is conducted in programs not associated with universities (Blane, 1971a; Ferrier, 1964; Richards, 1971; Scottish Home and Health Department, 1970). According to Ferrier (1964), there are a number of summer programs for in-service teachers and school administrators, as well as in-service training opportunities in several states through extension courses and adult classes in various cities and communities. The national impact of these endeavors appears to have been minimal. Further, such programs often are created out of the interest of individuals and have little institutionally supported life of their own; thus, they die when their originator leaves. Courses in health education and driver education for teachers in training rarely devote more than one or two class sessions to alcohol. In order to document the nature and extent of teacher preparation for alcohol education, NIAAA has recently proposed to conduct a survey of the area (National Institute of Mental Health, 1973).

Due to the recent interest in drugs, most undergraduate colleges now offer courses dealing with drugs; these frequently include segments on alcohol. The courses, sometimes offered on a multidisciplinary basis and sometimes emanating from psychology, sociology, biology, or pharmacology departments, are extremely popular with students. To the extent that they are taken by prospective teachers, these courses provide an unplanned program for instruction and development of attitudes about alcohol that may be useful in their later careers.

While educational activities for teachers or teachers-to-be seem actually to be rather hit-or-miss affairs, there has been no dearth of suggestions about how the situation should be improved. Recommendations about teacher preparation in alcohol education focus on aspects such as the nature of its target groups, its placement in the curriculum, preferred methods of instruction, and an appraisal of developments and characteristics of the teacher preparation system. The most general suggestions include well-intentioned pleas simply to devote some time in teacher preparation curricula to alcohol education. Those who view alcohol education as being integrated into the entire curriculum from kindergarten onward suggest specific content about alcohol for both elementary and secondary teacher preparation programs. Nearly all argue that alcohol education should be mandatory for students who plan to become high school teachers.

While nearly everyone interested in alcohol education deplores the informational, lecture method of instruction, relatively few have considered that alternative methods relying heavily on psychological and group dynamics require special preparation on the part of the teachers. Emphasis is placed on teacher attributes—such as warmth, openness, interest, honesty, ability to relate to students, ability to stimulate discussion, and feeling comfortable with one's own attitudes about alcohol—with the implicit assumption that possession of all or some of these traits is all that is necessary (along with some information about alcohol) for the teacher to conduct small-group discussions that examine alcohol and its use in relation to individual personality differences. Unterberger and DiCicco (1968), on the other hand, point out that teacher preparation for the small-group method is time-consuming, exacting, and costly; they recommend teaching teachers by means of eight to ten small-group sessions where group dynamics are learned by participation:

> Our experience has demonstrated that this amount of time is required for adults to clarify their feelings about their own drinking and about teen-age drinking, and to determine goals and methods in their approach to young people.

Blane (1971a) suggests systematic development of health education counselors who might serve school districts or areas and who would have special preparation in counseling as well as in health education. This notion is supported from a Scottish perspective by Davies and Stacey (1971) who consider the disadvantages of having teachers handle alcohol education, favoring the training of "teacher counselors" or "specialist teachers"; in any event, whoever is responsible for actual instruction should be a "detached neutral" (Jahoda and Cramond, 1972).

Issues in Implementation

There are many issues that have to be taken into account by anyone seriously involved in the business of alcohol education. Several of these have been touched upon in the preceding sections, including the need for alcohol education, definition of goals and objectives, nature of materials, teaching methods, type of instructor, teacher training, and the general failure of social scientists to become engaged in research and evaluation activities that could provide policy-makers and administrators with answers to some of these issues. All the issues are important and must be considered, but there is one overriding issue that alcohol educators have generally neglected: the problem of implementation on a mass basis. The educational establishment, from its teacher preparation component through its various administrative levels to the teacher in the classroom, is at best indifferent to alcohol education. There are exceptions, but they have had little discernible, long-lasting effect on alcohol education.

The most concerted systematic effort in recent times to implement alcohol education nationally is to be found in a campaign conducted by the National Institute of Mental Health between 1959 and 1962. The campaign consisted of a series of short conferences, supported by federal technical assistance project funds, held throughout the country. Each conference was developed locally, with the assistance and consultation of federal staff, under sponsorship of state authorities responsible for activities related to alcohol use and abuse. The organization of the conferences consisted of lectures presented by three or four resource persons to a group of 50 or so participants, invited because of prior interest or their role in the community. Following large-group presentations, smaller discussion groups led by resource persons were held. Finally, small-group reports were made at a plenary session, and an attempt was made to arrive at a consensus about an operational community-level alcohol education program. The goal of these conferences was to bring together for a period of time people in a community with interest in and/or positional relevance to the development of alcohol education endeavors in order to begin to develop a plan for alcohol education that would be viable within the context of the conditions that prevailed locally. Proceedings of each conference were prepared and published. Conferences and workshops were held in many states during the three-year period, including, for instance, Alabama, Georgia, Kansas, Minnesota, Mississippi, New Jersey, North Carolina, North Dakota, Rhode Island, and Vermont.

No formal evaluation of the effect of the campaign was ever undertaken, but discussions with involved staff members at the National Institute of Mental Health suggest that outcomes were not long-lived. Typically, conference participants were highly motivated and ready for action by the end of a workshop, but after they returned to their regular jobs and responsibilities their ardor gradually waned. Thus, it would appear that mechanisms for continuing local involvement and implementation were not contained in the campaign. Further, participants were often chosen not because of their demonstrated record in program implementation or of their potential leverage within the educational establishment, but because they worked in a school or an alcoholism agency. The general conclusion of those involved as organizers or participants in these conferences is that their ultimate influence on alcohol education was minimal. Since 1962, there has been no concerted effort to mount a nationwide alcohol education program, although the National Institute on Alcohol Abuse and Alcoholism (National Institute of Mental Health, 1973) describes major plans in the area.

Among factors requiring consideration in implementation are recognition that the educational establishment is on the whole indifferent toward alcohol education, knowledge about power distribution in school systems and chains of influence in the educational system, and recognition of the role played by local school boards as they represent the community. With regard to the first, normal institutional resistance to change has been justified in the particular instance of education on grounds that universities and colleges preparing teachers, as well as school administrators, are constantly bombarded by requests from special interest groups for inclusion of categorical programs or curricular content. As the argument goes, if educators responded to these requests there would soon be no time in the university curriculum to train teachers *qua* teachers or for teachers in public schools to devote to their primary responsibility of teaching academic subjects. This rationale, difficult to overthrow in practice, has little logic to recommend it, because granting a request for the addition of one new program does not mean that all requests must be granted. New programs have been added; the most recent example in health education is the widespread introduction of drug education into public school curricula. When a program is turned down it means it is not wanted, and this appears to be the case for alcohol education.

Secondly, distribution of power in the schools and chains of influence in the educational system have major implications for the implementation of programs, which require institutional support if they are to be maintained. Most school systems are run by a relatively self-contained group of career educators (teachers, supervisors, principals and other administrators) who are the major effectors of curricular policy and change at the line level (Blane, 1971a). New teachers, with low power and high turnover, do not effect changes in policy or major attitudinal shifts in public schools, although they may have high impact as individuals on their classes. Career educators, as a group, are responsive to two primary sources of influence: the profession and the local community. Professional influence on career educators tends to be stable over time and perceived by them to be positive. The stance and attitude adopted by state boards of education, national and regional professional organizations, the federal Office of Education, and private and special educational associations serve as role model determinants for gaining professional identity and as routes for achieving status and prestige within the profession and community.

The influence of the community, thirdly, stems from many local sources (parents, reputational leaders, political figures), but appears in its most distilled and direct form in the local school board. Unlike professional influence, it tends to be variable and situational, and may be perceived by educators as a threat. Nevertheless, the general impression is that programs desired by educational careerists will receive local approval, within obvious financial limits, as long as the program is not an inflammatory issue within the community. Alcohol education is not such an issue, although in initiating alcohol education in local schools it would certainly be unwise to neglect to involve key figures, organizations, and institutions (Mullin, 1968; Russell, 1969).

One strategy of implementation implied by the above factors is to concentrate efforts on career educators through the organizations and media to which they respond. This means enlisting the interest, cooperation, and support of the leaders of national, state, and regional educational organizations, with financial support as necessary. It would appear that maximum influence and leverage for implementing alcohol education on a widespread scale might be obtained through systematic utilization of journals, newsletters, meetings, and other means of communication to create a positive image of alcohol education as a necessary and prestigeful activity. Careful analysis of the chains of influence within the educational system would also be necessary in order to exert the most leverage in the shortest period of time on a nationwide basis (Blane, 1971a).

Specific suggestions have included the assignment of specialists in alcohol education to the U.S. Office of Education and state departments of education to work with local school authorities and teacher-training institutions to develop in-service training programs (Plaut, 1967); the establishment of an incentive program for teachers to become involved in alcohol education, perhaps coordinated with the Training of Training Teachers program (Blane, 1971a); and the development of a public service campaign directed specifically toward educators to sensitize them to attitudinal factors in alcohol use, by use of material inserted in professional organs and other media of the teaching profession (Blane, 1971a).

In addition to public school education, attention might be directed to education within churches. The attitude of the Roman Catholic Church toward the use of alcohol has traditionally been *laissez-faire*. However, concern about alcohol problems among Catholic clergymen has been expressed within the Church. It may be that Catholic, like Protestant, groups are now receptive to supporting alcohol education that is consistent with a social science model approach. The importance of religious groups in forming responsible drinking behavior has been repeatedly demonstrated; the church is often mentioned by those interested in alcohol education as an invaluable ally; yet, systematic attempts to make churches working partners in alcohol education are totally lacking.

ISSUES IN CONTEMPORARY APPROACHES TO PREVENTION

Untested Assumptions and Interpretations

The recommendation or introduction of preventive measures has nearly always been based on assumptions that have unknown empirical validity and/or on interpretations of research information that are tenuous or fail to take into account alternative explanations. This is to be expected within the framework of the proscriptive model, which fundamentally is not concerned with questions that have relative and probabilistic answers but with the categorical imperatives of good and evil. Adherents of this model have, nevertheless, become adept at interpreting research findings in a manner congenial to their beliefs, without considering other possible interpretations.

The distribution of consumption model, on the other hand, was developed and elaborated in accordance with established scientific principles. Its proponents have tested its hypotheses from varying vantage points, have followed promising leads with care and imagination, and have attempted to take into account criticisms made in the literature. The positive relationship between consumption levels and cirrhosis mortality is established almost beyond doubt, although it has been and probably will continue to be challenged. Similarly, evidence for the negative association between consumption levels and relative price of alcohol is convincingly impressive. These findings have been interpreted by the ARF as showing that variations in relative price "cause" variations in consumption levels, which, in turn, "cause" variations in cirrhosis mortality rates. It has been pointed out that this interpretation is but one alternative, and might therefore be considered an insufficient basis from which to recommend a major preventive program, certainly lacking a pilot study.

The social science model, unlike the distribution of consumption model, lacks a unified empirical basis for its premises, but it too has been prone to overinterpret research results. There has been no concerted attempt to form an integrated research foundation from which its programs may be derived. Much of the crossnational work from which the components of normative structure, integrated drinking, and socialization were developed does not meet acceptable canons of research practice. It was in fact hypothesis-development research. Unfortunately, the hypotheses, once developed, were not systematically pursued, with the result that the research is now outdated, and in some instances evidence has accumulated to bring earlier findings into question (Room, 1972b). Independent of the quality of research, adherents of a social science approach have been prone to engage in extended, almost speculative interpretation. In these instances it is difficult to separate interpretation from application. For example, because socialization of drinking among Jews is accompanied by low incidence and prevalence of alcohol problems, it has in essence been suggested that components of the Jewish experience with alcohol—developmentally and normatively—form the basis for norms about drinking in the United States. This broad-ranging view fails utterly to take into account the ritual, religious, and therefore deeply symbolic qualities that pervade Jewish drinking, and the essential foreignness of these qualities in American life.

Basing programs on untested assumptions, tenuous interpretations, overgeneralization, and sheer belief is unlikely to lead to attainment of desired objectives. This kind of situation predominates for most programs in most countries. Possible exceptions include the long-standing research program on alcohol control by the Finnish Foundation for Alcohol Studies; recent work in Scotland (Davies and Stacey, 1972; Jahoda and Cramond, 1972) where recommendations for alcohol education were based on carefully conducted research, analysis of the literature, attention to the psychology of attitude change, and recognition of social and educational realities in Scotland; and the research per se of the Addiction Research Foundation in Canada.

Research and Evaluation

As is evident from reading almost any section of this chapter, a great need exists for the conduct of sound research and evaluation in the area of prevention. Even the most obvious task of conducting trial demonstrations of preventive measures has been infrequently undertaken. Room (1971), in a critical review of the literature on legal controls, could cite only the study of Kuusi (1957) as meeting the criteria of a controlled field experiment, although other investigations have been reported (Buckner, 1967; Törnudd, 1968). In the education area, the report of Williams *et al.* (1968) is a hallmark simply by virtue of being the only published study to evaluate the effects of an approach to alcohol education, although Weir's study (1967) is also relevant in this connection. In the drinking–driving area are the recent prospective investigations by Blumenthal and Ross (1973) on the effects of court-imposed sanctions on subsequent behavior of persons convicted for driving under the influence of alcohol. Other studies include the controlled, short-term evaluation of a Canadian alcohol and traffic safety campaign (Information Canada, 1973) and the preliminary work being conducted on Vermont's proposed public education program on drinking and driving (Worden *et al.*, 1973a; 1973b). In addition to this handful of studies of "trial balloons" are attempts to evaluate programs after they have been mounted on a broad scale, presumably to help decide whether they should be continued (Crabb *et al.*, 1971; Harris and Associates, 1973). Finally, there is a group of after-the-fact studies that attempt to assess the effects of social or legal changes over time or at a given point in time (for example, Christie, 1965; Room, 1973; Ross, 1973; Seeley, 1960; Terris, 1967; Warburton, 1968). In the meantime, legal and social changes are being initiated on a national basis and at considerable cost. In the United States, for instance, many of the recommendations made by the Cooperative Commission on the Study of Alcoholism (Plaut, 1967; Wilkinson, 1970) have been adopted by federal and state governments without trial demonstrations. A substantial number of states have lowered the legal age for drinking, and NIAAA has initiated a nationwide information–education campaign through the mass media; neither has been informed by the results of a prior social experiment.

From the point of view of social action, evaluation of demonstration projects has a high priority because its results may inform and guide social policy. Room (1972a), in discussing the effects of the lack of "hard policy-relevant knowledge" in the alcohol field, argues for the initiation of "a wide variety of small-scale controlled experiments on all aspects of prevention ... before money is committed on a long-term basis to large-scale programs." However, as Rivlin (1974) points out, social experiments run many risks, including those "which arise from the conflict between the desire to obtain valid, reliable results and the equally urgent desire to obtain results quickly and at a low cost ... Politicians rarely get excited enough about a problem to finance an experiment until they are nearly ready to make the decision. Then they want immediate results. But a 'quick and dirty' experiment may be worse than none." If care is not taken to make demonstration projects as sound, sensible, and useful as possible, their results will not be accurate guides in decision-making and their use will decay or become nominal.

In addition to direct tests of demonstrations is the whole area of conceptual and research endeavor necessary for the sound construction of programs and their careful evaluation. This may be termed background research and covers a wide variety of content and activities. As Braucht (1973) has pointed out in regard to the impossibility of drawing valid conclusions from evaluations of drug education, effective evaluation is multifaceted and depends on delineation within target groups, program components, and definition and measurement of desired objectives. While research in such areas is important, it is generally lacking, except for limited work on educational objectives (Carroll, 1965; Milgram, 1969), receptivity and perceived need for education (Globetti, 1971; Globetti et al., 1969; Milgram, 1969; Pomeroy and Windham, 1966; Wyatt, 1972), and measurement of objectives (Williams et al., 1968).

Background research in economic, legal, and governmental factors as they relate to the development and structuring of preventive programs is in its infancy (Driver, 1970; Grosswiler, 1972), but what has been done indicates the high potential for prevention of research that attacks alcohol use and problems with the conceptual and methodological tools of diverse disciplines. Driver's (1970) study of a state beverage control commission reveals the lack of awareness on the part of control authorities about the relationship between beverage control and alcohol problems as well as their failure to implement mechanisms available in the legal system to reduce alcohol problems. Grosswiler's (1972) economic analysis of the social costs of problem drinking leads him to the conclusions that price elasticity calculations are not a suitable basis for preventive policy and that when subjective utility of consumers is taken into consideration, increasing relative price may produce negative results; i.e., light and moderate drinkers will drink less, but heavy drinkers will continue to drink the same quantity, thus lowering their remaining income and not reducing prevalence of alcohol problems.

Since most preventive programs depend heavily on the effectiveness of communications in molding or changing attitudes and behavior, research in the social psychology of attitude change should receive high priority. As Edwards (1970) has remarked, "Educational methods have often paid only scant attention to what is known of the psychology of attitude change." Fear messages, for example, are deplored by the majority of those involved in public information and school education programs; yet, it is well established experimentally that increases in fear generally increase persuasion, with a preponderance of studies demonstrating a positive relationship between intensity of fear arousal and amount of attitude change (Higbee, 1969; Leventhal, 1965; 1970; McGuire, 1969). Effectiveness of fear communications in changing attitudes is a function of many conditions, including variables such as characteristics of the source of the message (prestige, credibility, age), nature of the message (intensity, credibility, complexity, clarity), and characteristics of the recipient (intelligence, coping style, anxiety level). Clearly, attitude change involves processes far more complex than the assertions of many alcohol educators would suggest. Study of these processes can have important applications to the structuring of preventive measures that are essentially communicative and aim toward changes in attitudes and behavior. Neglect of the psychology of attitude change in alcohol prevention is noticeable in reviews listing the health concerns—smoking, tetanus, safe driving, mental health, roundworms—in which the effects of communications on attitudes have been studied (Higbee, 1969; Leventhal, 1965). In these reviews, no mention is made of studies in which the target attitude or behavior involved alcohol.

Research that assesses needs and readiness for the introduction of a specific preventive measure or program, interprets findings in light of existing social realities, makes good use of existing literature, and develops alternative strategies for arriving at objectives is all too rare. In this connection, the aforementioned studies by Davies and Stacey (1972) on teen-agers and Jahoda and Cramond (1972) on children, both conducted under the auspices of Scottish Home and Health Department, are exemplary. The implications and recommendations of both studies are coordinated with the empirical results of investigations the authors conducted with samples of Scottish schoolchildren, high school and college youths, and form the empirical and conceptual basis for an alcohol education program tuned to local conditions.

Development of Realistic Preventive Policy

Recommendations about prevention of alcohol problems are sometimes visionary and so carry within them the seeds of their own defeat in terms of viable implementation. Threats to keep alcohol away from a person smack too much of Prohibition, increasing relative price is political suicide, teaching youngsters to drink is immoral and irresponsible; many proposed measures are of a "not here, not now" variety. The mere enunciation of objectives and means is not enough. Analysis of the paths to implementation of goals and measures as well as analysis of the roadblocks that hinder that attainment must be undertaken. Some paths may be immediately traveled and some roadblocks may be immediately removed, but for others the time will not be right or a great deal of preparation, including trial runs and background research, will be necessary before they can be pursued. The careful thinking-through and analysis of social, economic, political, and other realities that this process involves seem rarely to have been undertaken, although there is some indication that such a process may be now occurring in the United States under the aegis of NIAAA.

Development of effective social policy in the United States has been hampered in the past by the long-standing conflict between interest groups representing prohibitionist and antiprohibitionist ideologies. Plaut (1967), in the influential report of the Cooperative Commission on the Study of Alcoholism, argues convincingly that the "wet" versus "dry" controversy has become an increasingly empty issue as the temperance groups have lost strength and a drinking ethic has become predominant. "The weakening of the Temperance Movement, and a more general acceptance of social drinking contribute to a climate favorable to the eventual development of a reasonable American alcohol policy" (Plaut, 1967). One nevertheless continues to hear the "wet–dry" controversy invoked as a major factor to be considered in developing social policy. Although it may be a consideration, and even a major factor in some local circumstances, its main import now is largely historical.

The demise of the "wet–dry" conflict has served to bring into focus other issues, concerns, and conflicts that must be taken into consideration in attempting to formulate an integrated alcohol policy. Positions and attitudes that grow out of the vested interests of the helping professions, the alcohol industry, and alcohol control agencies sometimes result in stances that block communication and even recognition that other positions exist, thus effectively obstructing the formulation of coordinated, mutually accepted policies. The role of the alcohol industry and its governmental regulation in this regard has been described by Wilkinson (1970). As will be seen, the industry itself is more amenable to engaging in discussions of the roles it can play in prevention than are the governmental agencies that have been established to regulate the industry and to control the public's alcohol consumption.

The alcohol control system in the United States consists of the federal Alcohol and Tobacco Tax Division of the Treasury Department, the alcoholic beverage control boards or commissions of each state, and the Joint Commission of the States to Study Alcoholic Beverage Laws, a national body representing the state boards and commissions. Alcohol control itself refers to the establishment, administration, and enforcement of rules that regulate the conditions of manufacture, distribution, sale, and consumption of alcohol. Alcohol control bodies view their general functions as law enforcement, generation of income through taxes, and, when the state operates a monopoly and therefore buys from the alcohol industry, as business. Since the basic enabling legislation for alcohol control in most states was drafted at the time Prohibition was repealed, it often contains a temperance element that may be adopted as policy by the administration of an alcohol control agency, but that may conflict with the tax-generating ethic held by many agencies. Most states have regulations that might, if enforced, have preventive implications, although none have control regulations that were originally conceived as preventive in orientation (except to reflect lingering Prohibitionistic ideology). Driver (1970), for example, in a study of control in New Jersey, reports that "the violation most closely related to alcohol problems, serving apparently intoxicated persons, is poorly enforced." If New Jersey is typical, alcohol control agencies have little awareness of the relationship between their operations and alcohol problems. Although state control bodies have considerable leeway in setting policy, none have defined their mission so as to include coordination of their activities with those of other state agencies involved in amelioration of alcohol problems. The relationship of state control bodies with the alcohol industry, on the other hand, is characterized by a complex network of associations and exchange of personnel (Wilkinson, 1970). Alcohol control agencies are sensitive to movements by the industry, particularly in monopoly states; to shifts in alcohol control and tax legislation; and to public pressure having to do with their image with regard to preferential treatment of applications for licensure, due process of alleged violators, etc. (Driver, 1970).

The alcohol control system has a vital role to play in any preventive program, yet it is ideologically geared in directions that do not permit it to recognize its potential. Occasional attempts have been made to bring representatives of alcohol control and ameliorative agencies together, but these have led to no tangible results. Nevertheless, any integrated social policy on alcohol and prevention of alcohol problems will have to rely on formal controls to some extent and the alcohol control system must of necessity become involved. A major step in this direction could occur at the national level through working agreements between the Departments of the Treasury and of Health, Education, and Welfare with regard to their respective alcohol components.

As for the alcohol industry, it still recalls Prohibition, and its trade associations (for example, Distilled Spirits Council of the United States, Inc., United States Brewers Association, Wine Institute) as well as individual producers are conscious of the image they project. Their goal is to maintain and to increase profit margins, and at the same time appear to be against and even to separate themselves from the problem use of their products. This makes them receptive to some preventive measures but not to others. Measures that threaten to decrease the total level of consumption without compensation are strongly resisted. Thus, programs based on a level-of-consumption model have little chance of receiving their support. On the other hand, measures that do not affect consumption or that may increase it are supported. The social science model's advocacy of lowering age requirements for legal purchase of alcohol and teaching people how to drink responsibly would fit into this category. Recommendations for focusing on low-alcohol content beverages, however, are resisted by whiskey producers but supported by the vintners and brewers. The point is that the alcohol industry can be involved in prevention programs that do not threaten the economic integrity of the industry. Because of its unique history, it may be more prepared to participate in policy planning and implementation than other industries where product use is connected to ill health (for example, the tobacco and automotive industries).

The helping professions collectively maintain positions that also must be taken into account in the development of alcohol policy. Their task is the amelioration of alcohol problems through treatment, and secondarily, prevention. The numerous facilities and local groups dealing with alcohol problems in the United States are represented by one or more of the national associations or give allegiance to a funding state or federal agency. Individual practitioners often seek representation through their national professional association. Alcoholic persons themselves work independently or are loosely represented by Alcoholics Anonymous, the major self-help group. At the federal level, NIAAA has general responsibility for alcohol programs, while NHTSA's Office of Alcohol Countermeasures assumes primary responsibility for alcohol use as it relates to traffic safety. At the state level, most states have an ameliorative program under the department of mental health or public health, or in a separate coordinating commission. The major national associations are the National Council on Alcoholism and the Alcohol and Drug Problems Association of North America.

A preventive theme in the tradition of the social science model has been sounded by NIAAA but has not been fully reciprocated by other ameliorative groups and their representatives. Nearly everyone in the helping professions gives due respect to the term "prevention," but for most obeisance is more nominal than substantive. In general, individuals in the helping professions know little about preventing alcohol problems. By virtue of training and experience they are oriented toward individuals, families, and small groups rather than populations or high-risk subpopulations whose members currently show no discernible difficulties with alcohol. Further, their task, treatment of alcoholic persons, is already fully time-consuming. Within much of the helping professions, there is a belief that prevention will not work, that the treatment need has priority, and an undercurrent of concern that if prevention becomes central, treatment endeavors will lose their base of support. Helping associations, such as the National Council on Alcoholism, were created at a time when community responses to alcohol problems were typically punitive and moralistic and when services for alcoholic persons were virtually nonexistent. The thrust of these associations and their members was, and continues to be, more on alcoholism and the plight of alcoholics, and less on drinking and alcohol. This view tends to be supported by former alcoholics and Alcoholics Anonymous.

The need for support from the helping professions in the development of social policy on alcohol is complicated by the negative aura that surrounds prevention. In general, prevention does not capture the imagination, and implies an element of control that is distasteful to many. Litigation concerning infringement of civil liberties is common when public health measures are introduced on a population-wide basis (e.g., fluoridation of water supplies). The control of personal decisions about one's life, such as drinking, smoking, or living itself, is considered by many to be no business for the state to be in. Prevention is a process of conservation rather than growth, again inimical to American values. Finally, prevention is helping in an indirect rather than a direct sense. This may be a reason why clinicians on the average are more inclined toward secondary than primary prevention. Planners of preventive policy must take pains to counter these pejorative views of prevention so as to insure fullest involvement by the helping professions.

There are, of course, many in the helping professions and social sciences who are actively engaged in prevention. It is to be hoped that their internecine struggles will be resolved or put aside in the interest of developing an integrated preventive policy. Polarization of positions with regard to goals of eradication versus reduction, or techniques of primary versus secondary prevention, or a sole focus on prevention to the exclusion of treatment, are to be avoided. In our current state of knowledge, extreme positions are indefensible. While eradication may be a rational goal in preventing infectious diseases, it is highly unlikely to be so in the control of social problems, including those associated with the use of alcohol. Sound policy will therefore contain provisions for both primary *and* secondary modes of prevention, provided that specific measures are adequately field-tested and subjected to continued, long-term evaluation. The same may be said for treatment: as long as alcohol problems exist and the likelihood of their eradication is small, there should be adequate facilities for their alleviation.

The issue of balance between formal and informal controls is also an important consideration in arriving at an effective alcohol policy. There are many who abjure the application of severe formal controls, on grounds that such controls do not work or constitute an infringement on civil liberties. Commonly cited are the instances of Prohibition and the ready availability of alcohol to teen-agers who are below the legal age limit for its purchase. It is sometimes argued that informal controls are the primary immediate determinants of behavior with regard to alcohol, and that formal controls often merely codify and legitimize behaviors that are already widespread and informally approved. The introduction of formal controls to change behavior that is widespread, as in the instance of Prohibition, is doomed to failure. When formal and informal systems of control are in opposition, behavior will usually violate the formal rather than the informal system. Further, it has been shown that alcohol control legislation that restricts the availability of alcohol may lead to increased level of violations of the law (Bacon, 1965), may have less-than-expected effects on target behavior (Bacon and Jones, 1968; Room, 1971), and may have unanticipated side effects (Grosswiler, 1972; Room, 1971; Whitehead, 1972a). Another consideration involves assessing the costs of administering a control system against the benefits derived (Cisin, 1971), which may be thought of in purely financial terms or in a social cost-benefit framework that combines social with financial factors (Grosswiler, 1972). In addition, social system organizational and process factors in the application of formal controls must be taken into account if effective enforcement is to occur (Cisin, 1971). Even in areas that have broad public support, such as control of drunk driving, enforcement is haphazard and desired effects minimal (Blumenthal and Ross, 1973; Ross, 1973; Zylman, 1970). Effective enforcement means more work for police officers. They, along with prosecutors and judges, are often ambivalent about the severity of penalties, and defendants who can afford to buy expert legal counsel tend not to be convicted.

It would, of course, be absurd to conclude that formal control systems do not affect behavior or, further, do not change behavior in ways that reflect the intent of the law. Effective alcohol policy should attempt to keep formal control as minimal as possible and with clear-cut objectives. As many have observed, specific control measures over the conditions of sale in the United States have no discernible objectives, but seem merely to be control for control's sake (Bacon, 1965; Cisin, 1971; Plaut, 1967; Wilkinson, 1970). Guiding principles for developing alcohol policy are to align formal and informal control systems where it is possible, not to introduce formal controls where informal controls are already effective, and not to introduce formal controls that run counter to widely accepted behaviors.

The development of public policy in any area presupposes that the area is publicly and politically salient. However, Beauchamp (1973) has pointed out that the alcoholism issue "has attracted a very limited base of public recognition and support." His analysis of the reasons for this is also germane to the formation of a preventive public policy about alcohol. He finds that "the wide legitimacy of social drinking prohibits any public representation of alcoholism that implicates the substance of alcohol or actual drinking practices." He also indicates that no public personality has emerged to catalyze public opinion (he notes the exception of Senator Hughes), and no professional specialty has grown up around the alcoholism issue to claim support for it. Further, the low visibility of much deviant drinking reduces prospects of public endorsement by legitimate groups and rewards to community groups for attacking the problem. Beauchamp is pessimistic about the future of recent advances at the federal level, including the establishment of NIAAA and the programs emanating from NHTSA; however, he neglects some of the advantages that accrue to low visibility of a public issue, notably that legislation on such an issue may often be accomplished more efficiently and be less compromised than on a "hot potato" issue. Understanding the nature and potential influence of the broad factors critical to the adoption of public policy is just as important as an appreciation of the constituencies and other forces that are operative in the alcohol field, if an integrated and viable alcohol policy is to be established.

SUMMARY

Three models of prevention may be identified: the social science model, the distribution of consumption model, and the proscriptive model. The first two, in particular, are rooted in concepts of the public health approach that has proved valuable in the prevention and control of many communicable diseases. Concepts of levels of prevention and agent–host–environment relationships have left their mark on prevention of alcohol problems. However, it appears questionable that direct application of concepts and techniques effective in preventing infectious diseases will prove equally effective in preventing social problems. Preventive programs based on a social science approach involving sharpening the normative structure about drinking and promoting integrated drinking through socialization in the broadest sense appear to predominate currently, especially in the United States. Programs proposed by proponents of the other models are oriented toward controlling consumption by making alcohol less available or unavailable; these programs tend to go against prevailing sociopolitical trends in most countries.

Numerous techniques of control and prevention have been used or proposed. They include a variety of public information and education campaigns; education about alcohol in school systems; manipulation of substance, person, and environmental factors; and secondary prevention programs involving early detection of and work with high-risk populations. Education in the school systems is a key preventive technique. A wide variety of sophisticated and accurate teaching materials are available for alcohol education programs from kindergarten through high school levels. While it appears that most alcohol education is not now integrated with the rest of the curriculum, there is a trend in the United States toward broad integration of alcohol-related materials in the general as well as the health-education curriculum. The use of small-group discussion methods rather than typical classroom lectures for alcohol education is favored. This implies that the training of special counselor-teachers skilled in small-group methods will be necessary.

Prevention of alcohol problems has remained largely uninformed because of the paucity of research that meets acceptable standards. This has affected both the development of alcohol policy and the implementation of effective programs. The low visibility of alcohol problems, the fact that they have had few charismatic champions, and the high social acceptance of drinking have seriously hampered the emergence of strong public support for development of an alcohol policy. Factors such as these must be taken into account in deliberations about the establishment of policy that aims toward the reduction of problems associated with the use of alcohol.

ACKNOWLEDGMENTS

The preparation of this chapter was supported in part by grants AA00491 and MH13179 from the National Institute on Alcohol Abuse and Alcoholism.

REFERENCES

Anderson, C. L., 1968, "School Health Practice," 4th Ed., Mosby, St. Louis.
Bacon, M., and Jones, M. B., 1968, "Teen-Age Drinking," Crowell, New York.
Bacon, S., 1965, American experiences in legislation and control dealing with the use of beverage alcohol, in "The Legal Issues in Alcoholism and Alcohol Usage," pp. 123–141, Boston University Law-Medicine Institute, Boston.
Bales, R. F., 1946, Cultural differences in rates of alcoholism, *Quarterly Journal of Studies on Alcohol* 6:480.
Bales, R. F., 1962, Attitudes toward drinking in the Irish culture, in "Society, Culture, and Drinking Practices" (D. J. Pittman and C. R. Synder, eds.), pp. 157–187, Wiley, New York.
Beauchamp, D. E., 1973, "Precarious Politics: Alcoholism and Public Policy," Doctoral Dissertation, Johns Hopkins University.
Blacker, E., 1966, Sociocultural factors in alcoholism, *International Psychiatry Clinics* 3(2):51.
Blane, H. T., 1968a, "The Personality of the Alcoholic, Guises of Dependency," Harper and Row, New York.
Blane, H. T., 1968b, Trends in the prevention of alcoholism, *Psychiatric Research Report* 24:1.
Blane, H. T., 1971a, Alcohol Education and Traffic Safety, Paper presented at NIMH-NHTSA Planning Meeting on a K-12 Curriculum in Alcohol Abuse and Traffic Safety (February 23–24, 1971), Washington, D.C.
Blane, H. T., 1971b, Unpublished data, University of Pittsburgh.
Blane, H. T., and Chafetz, M. E., 1971, Dependency conflict and sex-role identity in drinking delinquents, *Quarterly Journal of Studies on Alcohol* 32:1025.
Block, M. A., 1967, Alcoholism prevention and reality; comment on the article by M. E. Chafetz, *Quarterly Journal of Studies on Alcohol* 28:551.
Blumenthal, M., and Ross, H. L., 1973, "Two Experimental Studies of Traffic Law. Volume 1. The Effect of Legal Sanctions on DUI Offenders," DOT Publ. No. HS-800, 825, United States National Highway Traffic Safety Administration, Washington, D.C.
Braucht, G. N., 1973, A Psychosocial Typology of Adolescent Alcohol and Drug Users, Paper presented at the Third Annual Alcoholism Conference of the National Institute on Alcohol Abuse and Alcoholism (June, 1973), Washington, D.C.
Braucht, G. N., Brakarsh, D., Follingstad, D., and Berry, K. L., 1973, Deviant drug use in adolescence: A review of psychosocial correlates, *Psychological Bulletin* 79:92.
Buckner, D. R., 1967, "The Influence of Residence Hall Alcoholic Beverage and Study Hour Regulations on Student Behavior," Doctoral Dissertation, American University.
Burnett, I. B., 1973, "Methodism and Alcohol: Recommendations for a Beverage Alcohol Policy

Based on the Everchanging Historic Disciplinal Positions of American Methodism," Doctoral Dissertation, School of Theology at Claremont.

Byler, R., Lewis, G., and Totman, R., 1969, "Teach Us What We Want to Know," Mental Health Materials Center, New York.

Cahalan, D., Cisin, I. H., and Crossley, H. M., 1969, "American Drinking Practices," Rutgers Center of Alcohol Studies, New Brunswick, New Jersey.

Carroll, C. R., 1965, "Application of the Taxonomy of Educational Objectives to Alcohol Education," Doctoral Dissertation, Ohio State University.

Cartwright, A., Martin, F. M., and Thomson, J. E., 1960, Efficacy of an anti-smoking campaign, *Lancet* 1960:327.

Chafetz, M. E., 1967, Alcoholism prevention and reality, *Quarterly Journal of Studies on Alcohol* 28:345.

Chafetz, M. E., 1973, Problems of reaching youth, *Journal of School Health* 43:40.

Chafetz, M. E., and Demone, H. W., Jr., 1962, "Alcoholism and Society," Oxford University Press, New York.

Chafetz, M. E., Blane, H. T., and Hill, M. J., 1971, Children of alcoholics; observations in a child guidance clinic, *Quarterly Journal of Studies on Alcohol* 32:687.

Christie, N., 1965, Scandinavian experience in legislation and control, *in* "The Legal Issues in Alcoholism and Alcohol Usage," pp. 101–122, Boston University Law-Medicine Institute, Boston.

Cisin, I. H., 1971, Formal and informal controls over drinking practices, *in* "Law and Drinking Behavior" (J. A. Ewing and B. A. Rouse, eds.), pp. 17–28, Center for Alcohol Studies, University of North Carolina, Chapel Hill (Mimeographed).

Connor, W. D., 1972, "Deviance in Soviet Society; Crime, Delinquency, and Alcoholism," Columbia University, New York.

Crabb, D., Gettys, T. R., Malfetti, J. L., and Stewart, E. I., 1971, "Development and Preliminary Tryout of Evaluation Procedures for the Phoenix Driving-While-Intoxicated Reeducation Program," Arizona State University, Tempe, Arizona.

Cross, J. N., 1967, Epidemiologic studies and control programs in alcoholism. I. Public health approach to alcoholism control, *American Journal of Public Health* 57:955.

Cumming, E., 1963, Pathways to prevention, *in* "Key Issues in the Prevention of Alcoholism," pp. 11–25, Department of Health, Harrisburg, Pennsylvania.

Davies, J., and Stacey, B., 1971, Alcohol and health education in schools, *Health Bulletin* 29(1):1.

Davies, J., and Stacey, B., 1972, "Teen-agers and Alcohol; A Developmental Study in Glasgow," Vol. II, Her Majesty's Stationery Office, London.

de Lint, J., 1973, The validity of the theory that the distribution of alcohol consumption approximates a logarithmic normal curve of the type proposed by Sully Lederman: A brief note, *Drinking and Drug Practices Surveyor* 1973(7):15.

de Lint, J., and Schmidt, N., 1971, Consumption averages and alcoholism prevalence: A brief review of epidemiological investigations, *British Journal of Addiction* 66:97.

Driver, R. J., 1970, "The Control of Beverage Alcohol in New Jersey: Goal Orientation in a Governmental Organization," Doctoral Dissertation, Rutgers University.

Edwards, G., 1970, Place of treatment professions in society's response to chemical abuse, *British Medical Journal* 2:195.

Edwards, G., 1973, Epidemiology applied to alcoholism: A review and examination of purposes, *Quarterly Journal of Studies on Alcohol* 34:28.

Efron, V., Keller, M., and Gurioli, C., 1972, "Statistics on Consumption of Alcohol and on Alcoholism," Rutgers Center of Alcohol Studies, New Brunswick, New Jersey.

Ekholm, A., 1972, The lognormal distribution of blood alcohol concentrations in drivers, *Quarterly Journal of Studies on Alcohol* 33:508.

Ferrier, W. K., 1964, Alcohol education in the public school curriculum, *in* "Alcohol Education for Classroom and Community" (R. G. McCarthy, ed.), pp. 48–66, McGraw-Hill, New York.

Finn, F., and Platt, J., 1972, "Alcohol and Alcohol Safety," 6 vols., Publ. No. DOT HS 800 709, U.S. Government Printing Office, Washington, D.C.

Ford Foundation, 1972, "Dealing with Drug Abuse," Praeger, New York.

Forslund, M. A., and Gustafson, T. J., 1970, Influence of peers and parents and sex differences in drinking by high-school students, *Quarterly Journal of Studies on Alcohol* 31:868.

Fraser, F., 1967, Drinking and mass education, *Canadian Mental Health* 15:30.

Freeman, H. E., and Scott, J. F., 1966, A critical review of alcohol education for adolescents, *Community Mental Health Journal* 2:222.

Geraty, T. S., 1971, The role of the educator in the field of prevention, *in* "Toward Prevention; Scientific Studies on Alcohol and Alcoholism," pp. 213–216, Narcotics Education, Washington, D.C.

Globetti, G., 1971, Alcohol education in the school, *Journal of Drug Education* 1:241.

Globetti, G., Pomeroy, G. S., and Bennett, W. H., 1969, Attitudes toward Alcohol Education, So.-An. No. 14 (August, 1969), Mississippi State University, State College, Mississippi, p. 33.

Gordon, J. E., 1956, The epidemiology of alcoholism, *in* "Alcoholism as a Medical Problem" (H. D. Kruse, ed.), Hoeber, New York.

Gordon, J. E., 1958, The epidemiology of alcoholism, *New York State Journal of Medicine* 58:1911.

Grosswiler, R. A., 1972, "The Economic Dimensions of Problem Drinking in the United States Related to Prevention and Treatment Programs," Doctoral Dissertation, University of Colorado.

Grout, R. E., 1968, "Health Teaching in Schools," 5th Ed., Saunders, Philadelphia.

Guild, W. R., Fuisz, R. E., and Bojar, S., 1969, "The Science of Health," Prentice-Hall, Englewood Cliffs, N.J.

Harris and Associates, 1973, Public Awareness of a NIAAA Advertising Campaign and Public Attitudes Toward Drinking and Alcohol Abuse, Phase Two: Spring, 1973, Study No. 2318 (May, 1973) p. 83.

Haskins, J. B., 1969, Effects of safety communication campaigns: a review of the research evidence, *Journal of Safety Research* 1:58.

Haskins, J. B., 1970, Evaluative research on the effects of mass communication safety campaigns: A methodological critique, *Journal of Safety Research* 2:86.

Hayman, M., 1966, "Alcoholism: Mechanism and Management," C. C. Thomas, Springfield, Illinois.

Hein, F. V., Farnsworth, D. L., and Richardson, C. E., 1970, "Living; Health, Behavior, and Environment," 5th Ed., Scott, Foresman, Glenview, Illinois.

Higbee, K. L., 1969, Fifteen years of fear-arousal: Research on threat appeals: 1953–1968, *Psychological Bulletin* 72:426.

Hill, H. H., 1971a, Education as a means of prevention, *in* "Toward Prevention; Scientific Studies on Alcohol and Alcoholism," pp. 217–225, Narcotics Education, Washington, D.C.

Hill, H. H., 1971b, Philosophies of alcohol education, *in* "Toward Prevention; Scientific Studies on Alcohol and Alcoholism," pp. 227–231, Narcotics Education, Washington, D.C.

Hurst, P. M., 1972, Blood alcohol and highway crashes: A selective review of epidemiological findings. Paper presented at Alcohol and Traffic Safety Conference, TRANSPO 72 (October, 1972) p. 32.

Information Canada, 1973, The Edmonton Study; The Impact of a Drinking Driving Campaign, Catalogue No. T46-273, Information Canada, Ottawa, p. 60.

Ivy, A. C., 1971a, Citizenship and community responsibility for the prevention of alcoholism, *in* "Toward Prevention; Scientific Studies on Alcohol and Alcoholism," pp. 207–211, Narcotics Education, Washington, D.C.

Ivy, A. C., 1971b, The philosophical background for the prevention of alcoholism, *in* "Toward Prevention; Scientific Studies on Alcohol and Alcoholism," pp. 201–206, Narcotics Education, Washington, D.C.

Jahoda, G., and Cramond, J., 1972, "Children and Alcohol; A Developmental Study in Glasgow," Vol. I, Her Majesty's Stationery Office, London.

Jellinek, E. M., 1963, Government Programs on Alcoholism, A Review of Activities in Some Foreign Countries, Report Series Memorandum No. 6, Mental Health Division, Department of National Health and Welfare, Ottawa, Canada (April, 1963) p. 90.

Kelly, N., 1964, Social and legal programs of control, *in* "Alcohol Education for Classroom and Community" (R. McCarthy, ed.), pp. 11–31, McGraw-Hill, New York.

Kogan, B. A., 1970, "Health: Man in a Changing Environment," Harcourt, Brace and World, New York.

Kuusi, P., 1957, "Alcohol Sales Experiment in Rural Finland," Finnish Foundation for Alcohol Studies, Helsinki.

Lemert, E. M., 1962, Alcohol, values, and social control, *in* "Society, Culture, and Drinking Patterns" (D. J. Pittman and C. R. Snyder, eds.), pp. 553–571, Wiley, New York.

Leventhal, H., 1964, An analysis of the influence of alcoholic beverage advertising on drinking customs, *in* "Alcohol Education for Classroom and Community" (R. McCarthy, ed.), pp. 267–297, McGraw-Hill, New York.

Leventhal, H., 1965, Fear communications in the acceptance of preventive health practices, *Bulletin of the New York Academy of Medicine* 41:1144.

Leventhal, H., 1970, Findings and theory in the study of fear communications, *in* "Advances in Social Psychology" (L. Berkowitz, ed.) Vol. 5, pp. 119–186, Academic Press, New York.

Lindgren, A., 1973, Some results from an international series of drinking surveys, *Drinking and Drug Practices Surveyor* 1973(8):34.

Lolli, G., Serianni, E., Golder, G. M., and Luzzatto-Fegiz, P., 1958, "Alcohol in Italian Culture," Free Press, Glencoe, Illinois.

Long, B. E., 1971, An approach for mental health education, *in* "Orthopsychiatry and Education" (E. M. Bower, ed.), pp. 133–150, Wayne State University Press, Detroit.

McCarthy, R. G., 1952, Alcoholism, 1941–1951: A survey of activities in research, education and therapy. VII. Activities of state departments of education concerning instruction about alcohol. *Quarterly Journal of Studies on Alcohol* 13:496.

McGavran, E. G., 1963, Facing reality in public health, *in* "Key Issues in the Prevention of Alcoholism," pp. 55–61, Department of Health, Harrisburg, Pennsylvania.

McGuire, W. J., 1969, The nature of attitudes and attitude change, *in* "The Handbook of Social Psychology," 2nd Ed., Vol. III (G. Lindzey and E. Aronson, eds.), pp. 136–314, Addison-Wesley, Reading, Mass.

Maddox, G. L., 1966, Teen-agers and alcohol: Recent research, *Annals of the New York Academy of Science* 33:856.

Maletzky, B. M., 1973, The episodic dyscontrol syndrome, *Diseases of the Nervous System* 34:178.

Mental Health Materials Center, 1973, "Selective Guide to Materials for Mental Health and Family Life Evaluation," Perennial Education, Northfield, Illinois.

Merry, J., 1973, Causes and prevention of alcohol abuse, *Lancet* 1965: 421.

Milgram, G. G., 1969, "Teenage Drinking Behavior and Alcohol Education in High School Perceived by Selected Reference Groups," Doctoral Dissertation, Rutgers University.

Miller, B. F., Rosenberg, E. B., and Stackowski, B. L., 1971, "Investigating Your Health," Houghton Mifflin, Boston.

Monroe, M. E., and Stewart, J., 1959, "Alcohol Education for the Layman: A Bibliography," Rutgers University Press, New Brunswick, N.J.

Mullin, L. S., 1968, Alcohol education: The school's responsibility, *Journal of School Health* 38:518.
National Clearinghouse for Alcohol Information, 1972, "Curriculum Guides for Primary and Secondary Levels of Alcohol Education," Rockville, Maryland.
National Institute of Mental Health, 1973, Responsible Drinking . . . and Prevention of Alcohol Abuse, Undated pamphlet, Rockville, Maryland, p. 15.
Olkinuora, H., 1971, Content Analysis of Alcohol Information in Education Plans and Textbooks of Public Schools, Research Report No. 30, Department of Education, University of Jyväskylä. English abstract in *Quarterly Journal of Studies on Alcohol,* 1972, 33:903.
O'Neill, B., and Wells, W. T., 1971, Blood alcohol levels in drivers not involved in accidents, *Quarterly Journal of Studies on Alcohol* 32:798.
Otto, J. H., Julian, C. J., and Tether, J. E., 1971, "Modern Health," Holt, Rinehart and Winston, New York.
Pasciutti, J. J., 1962, Educational approach, *in* "Problems in Addiction: Alcoholism and Narcotics" (W. C. Bier, ed.), pp. 138–150, Fordham University Press, New York.
Pasciutti, J. J., 1963, Current emphasis in instruction about alcohol, *in* "Interpreting Current Knowledge about Alcohol and Alcoholism to a College Community," pp. 62–73, New York State Department of Mental Hygiene, Albany.
Philp, J. R., 1967, Epidemiologic studies and control programs in alcoholism. IV. Discussion, *American Journal of Public Health* 57:971.
Plaut, T. F. A., 1963, Translating concepts into action, *in* "Key Issues in the Prevention of Alcoholism," pp. 62–72, Department of Health, Harrisburg, Pennsylvania.
Plaut, T. F. A., 1967, "Alcohol Problems, A Report to the Nation," Oxford, New York.
Plaut, T. F. A., 1972, Prevention of alcoholism, *in* "Handbook of Community Mental Health" (S. E. Golann and C. Eisdorfer, eds.), pp. 421–438, Appleton-Century-Crofts, New York.
Plunkett, R. J., and Gordon, J. E., 1960, "Epidemiology and Mental Illness," Basic Books, New York.
Pomeroy, G. S., and Windham, G. O., 1966, Attitudes of Selected Adult Groups toward Alcohol Education, So.-An. No. 4 (August, 1966), Mississippi State University, State College, Mississippi, p. 18.
Popham, R. E., 1962, The urban tavern; some preliminary remarks, *Addictions* 9:16–28 (Abstract, 1963, *Quarterly Journal of Studies on Alcohol* 24:761–762).
Popham, R. E., Schmidt, W., and de Lint, J., 1971, The Prevention of Alcoholism: Epidemiological Studies of the Effects of Government Control Measures, Project J 100, Substudy 2-2, 4, 10–71, Addiction Research Foundation, Toronto, Canada, p. 28.
Richards, W. T., 1971, (no title), Paper presented at NIMH-NHTSA Planning Meeting on a K-12 Curriculum in Alcohol Abuse and Traffic Safety (February 23–24, 1971), Washington, D.C.
Ritson, E. B., 1972, Drinking among young people, *in* "Notes on Alcohol and Alcoholism" (S. Carvana, ed.), Section XI, pp. 1–4, Medical Council on Alcoholism, Edsall, London.
Rittenhouse, J. D., 1973, Assessing Drug Information Programs, Paper presented at the Annual Meeting of the American Association for Public Opinion Research (May 20, 1973), Ashville, North Carolina.
Rivlin, A. M., 1974, Social experiments: Promise and problems, *Science* 183(4120):35.
Robinson, R. R., 1969, The prospect of adequate education about alcohol and alcoholism, *Journal of Alcohol Education* 14(2):1.
Roe, A., 1943, "A Survey of Alcohol Education in Elementary and High Schools in the United States," *Quarterly Journal of Studies on Alcohol,* New Haven.
Room, R., 1971, Drinking laws and drinking behavior: Some past experience, *in* "Law and Drinking Behavior" (J. A. Ewing and B. A. Rouse, eds.), pp. 29–108, Center for Alcohol Studies, University of North Carolina, Chapel Hill (Mimeographed).

Room, R., 1972a, Notes on alcohol policies in the light of general-population studies, *Drinking and Drug Practices Surveyor* 1972(6):10.
Room, R., 1972b, Some propositions on the analysis of crosscultural data on alcohol, *Drinking and Drug Practices Surveyor* 1972(6):2.
Room, R., 1972c, Strategies of prevention and alcohol opinion campaigns, *Drinking and Drug Practices Surveyor* 1972(6):16.
Room, R., 1973, Notes on the implications of the lognormal curve, *Drinking and Drug Practices Surveyor* 1973(7):18.
Ross, H. L., 1973, Law, science, and accidents: The British Road Safety Act of 1967, *Journal of Legal Studies* 2:1.
Russell, R. D., 1965, What do you mean-alcohol education? *Journal of School Health* 35:351.
Russell, R. D., 1969, Education about alcohol . . . for real American youth, *Journal of Alcohol Education* 14(3):1.
Sanford, N., 1972, Is the concept of prevention necessary or useful? *in* "Handbook of Community Mental Health" (S. E. Golann and C. Eisdorfer, eds.), pp. 461–471, Appleton-Century-Crofts, New York.
Scottish Home and Health Department, 1970, Health Education and Alcohol, Scottish Health Service Studies No. 14, p. 41.
Seeley, J. R., 1960, Death by liver cirrhosis and the price of beverage alcohol, *Canadian Medical Association Journal* 83:1361.
Sinacore, J. S., 1971, Alcohol Education, Paper presented at NIMH-NHTSA Planning Meeting on a K-12 Curriculum in Alcohol Abuse and Traffic Safety (February 23–24, 1971), Washington, D.C.
Singer, K., 1972, Drinking patterns and alcoholism in the Chinese, *British Journal of Addiction* 67:3.
Skog, O-J., 1973, Less alcohol-fewer alcoholics? *Drinking and Drug Practices Surveyor* 1973(7):7.
Snyder, C. R., 1958, "Alcohol and the Jews," Free Press, Glencoe, Ill.
Stacey, B., and Davies, J., 1970, Drinking behaviour in childhood and adolescence: An evaluative review, *British Journal of Addiction* 65:203.
Stewart, E. I., and Malfetti, J. L., 1970, "Rehabilitation of the Drunken Driver; a Corrective Course in Phoenix, Arizona for Persons Convicted of Driving under the Influence of Alcohol," Columbia University Teachers College, New York.
Stivers, R. A., 1971, "The Bachelor Group Ethic and Irish Drinking," Doctoral Dissertation, Southern Illinois University.
Terris, M., 1967, Epidemiology of cirrhosis of the liver: National mortality data, *American Journal of Public Health* 57:2076.
Törnudd, P., 1968, The preventive effect of fines for drunkenness: A controlled experiment, *Scandinavian Studies in Criminology* 2:109.
Trice, H. M., and Roman, P. M., 1972, "Spirits and Demons at Work: Alcohol and Other Drugs on the Job," Cornell University, Ithaca, New York.
Ullman, A. D., 1958, Sociocultural backgrounds of alcoholism, *Annals of the American Academy of Political and Social Science* 315:48.
United States Department of Health, Education, and Welfare, 1970, "Changes in Cigarette Smoking Habits Between 1955 and 1966, "U.S. Government Printing Office, Washington, D.C.
United States Department of Health, Education, and Welfare, 1971, "Alcohol and Health," DHEW Publication No. (HSM) 72-9099, Washington, D.C.
Unterberger, H., and DiCicco, L., 1968, Alcohol education reevaluated, *Bulletin of the National Association of Secondary School Principals* 52:15.
Wallace, J. G., 1972, Drinkers and abstainers in Norway; a national study, *Quarterly Journal of Studies on Alcohol,* Supplement No. 6:129.

Wallgren, H., and Barry, H., III, 1970, "Actions of Alcohol, Vol. II, Chronic and Clinical Aspects," Elsevier, Amsterdam.
Wanberg, K. W., and Horn, J. L., 1971, "A Descriptive Analysis of Symptom Patterns Related to the Excessive Use of Alcohol," Fort Logan Mental Health Center, Denver, Colo.
Warburton, C., 1968, "The Economic Results of Prohibition," AMS Press, New York (reprint of 1932 edition).
Weir, W. R., 1967, "A Program of Alcohol Education and Counseling for High School Students with and without a Family Alcohol Problem," Doctoral Dissertation, University of North Dakota.
Whitehead, P. C., 1972a, The Prevention of Alcoholism: An Analysis of Two Approaches, Paper presented to the Canadian Sociology and Anthropology Association (May, 1972).
Whitehead, P. C., 1972b, Toward a New Programmatic Approach to the Prevention of Alcoholism: A Reconciliation of the SocioCultural and Distribution of Consumption Approaches, Paper presented to the 30th International Congress on Alcoholism and Drug Dependence, Amsterdam (September, 1972).
Wilkinson, R., 1970, "The Prevention of Drinking Problems," Oxford, New York.
Williams, A. F., DiCicco, L., and Unterberger, H., 1968, Philosophy and evaluation of an alcohol education program, *Quarterly Journal of Studies on Alcohol* 29:685.
Wolfgang, M. E., and Strohm, R. B., 1956, The relationship between alcohol and criminal homicide, *Quarterly Journal of Studies on Alcohol* 17:411.
Worden, J. K., Riley, T. J., and Waller, J. A., 1973a, The Development and Evaluation of the Vermont Driver Education Program in Alcohol Safety, CRASH Report I-3 (June, 1973), Waterbury, Vermont, p. 41.
Worden, J. K., Waller, J. A., Riley, T. J., and Flowers, L., 1973b, Pre-Campaign Data for Public Education about Alcohol and Highway Safety in Vermont, CRASH Report I-2 (February, 1973), Waterbury, Vermont, p. 96.
Wyatt, P. D., 1972, "Alcohol Education: An Exploratory Study of Teacher Opinions and Drinking Practices," Doctoral Dissertation, University of the Pacific.
Yamamuro, B., 1973, Alcoholism in Tokyo, *Quarterly Journal of Studies on Alcohol* 34:950.
Zacune, J., and Hensman, C., 1971, "Drugs Alcohol and Tobacco in Britain," Heinemann Medical Books, London.
Zylman, R., 1970, Are drinking-driving laws enforced?, *Police Chief* 37:48.

CHAPTER 13

The Effects of Legal Restraint on Drinking

Robert E. Popham, Wolfgang Schmidt, and Jan de Lint

Addiction Research Foundation
33 Russell Street
Toronto M5S-2S1
Ontario, Canada

INTRODUCTION

The purpose of this review is to assess the evidence bearing on the effects of legal measures believed to have some primary preventive value with respect to the incidence of alcohol problems. Excluded from consideration are laws concerned solely with such special segments of the drinking population as incorrigible alcoholics, public inebriates, and impaired drivers. It is true that these laws are sometimes seen to have a possible deterrent effect on the appearance of new cases (Bruun, 1970). However, with the exception of certain legislation directed against impaired driving, the primary effect if any is more likely to be on recidivism rates. In contrast, our focus is on legislation and derivative measures aimed at the whole population of consumers (or potential consumers) of alcohol, and intended to prevent the occurrence of alcohol problems through regulation of the amount or character of alcohol consumption.

Legislative enactments dealing with one or another aspect of alcohol use are probably as old as written laws. For example, the famous Code of Hammurabi, formulated some four thousand years ago, contained four articles on the topic (Harper, 1904). But it was not until comparatively recent times that the principal concern in alcohol legislation came to be with the prevalence of alcohol problems. At first, in both the Old World and the New, the usual objectives were to prevent fraudulent practices on the part of sellers, assure availability, and secure revenue for the state (Catlin, 1931; Krout, 1925). Scattered attempts, ostensibly to cope with intemperance, and including the total prohibition of public drinking places, are found in Classical and Medieval legislation (King, 1947). In England during the thirteenth and fourteenth centuries, taverns were increasingly a target of restrictive legislation, efforts being made to reduce their numbers, hours of sale, and the amount of time patrons spent in them; and in 1551 what appears to have been the first licensing system was introduced by Edward VI with the control of excessive use in view (Shadwell, 1915). However, it was in the late eighteenth and nineteenth centuries, with the rise of industrialism, that the prevention of drunkenness and its consequences became the overriding issue, and the volume and complexity of alcohol legislation reached a peak (Askwith, 1928; King, 1974; Krout, 1925). In North American and some other areas, these efforts culuminated during World War I in total prohibition. This was repealed in most jurisdictions in the 1920s and 1930s, and followed by the adoption of the diverse measures and systems of control that prevail today.

The literature concerned with the effects of the legislative approach to the prevention of alcohol problems is vast. In addition to the assessments of historians, physicians, clergymen, jurists, journalists, and others, there are many official government reports. For example, in England there have been inquiries roughly once every 20 years since the Select Committee published its findings in 1834. In Canada and the United States there are reports of federally instigated inquiries, and of countless provincial or state bodies concerned with the merits and demerits of the control systems in their jurisdictions. While some of this literature contains data worthy of further analysis, most of it contributes little of value to the present review. Typically, the conclusions are based on the personal tastes or beliefs of the author, on *ex cathedra* arguments, or on the weight of opinion of persons with little or no direct knowledge of the matters at issue.* And even in cases where objective evidence of change is provided, one is usually left unsure, as Room (1971) remarked, whether the law was the cause, or itself a product of prior changes in public sentiment. In short, scientifically acceptable attempts to evaluate the effects of particular control measures are seldom encountered.

---

* To cite a characteristic instance, the authors of "The Pub and the People" (Mass Observation, 1943) noted that a recent Royal Commission on Licensing (1932), ostensibly setting out to conduct an impartial study of taverns, only interviewed persons who rarely or never patronized such establishments; not a single regular patron was asked to give evidence, nor were direct observational studies undertaken.

This dearth of scientific studies may seem surprising, especially when one contemplates the very great expansion of research interest in the alcohol field during the last 30 years. Room (1971) has offered some plausible reasons which include: (1) lack of sufficient specificity in the stated objectives of most measures; (2) the formidable complexity of the factors that may be involved in the production of any changes observed to follow the introduction of a new law; and (3) the fact that those who enact legislation normally are not influenced by a desire to develop rational policy through appropriate testing of alternatives, but by the probable reactions of their constituents. However, it is likely that another reason, also cited by Room (1971) and by Mäkelä (1972), is of greater importance, notably the rise of the "disease concept" of alcoholism. The view that normal drinkers and alcoholics comprise two quite separate groups within the population, which this concept has meant for many workers, has rendered meaningless or at least of low priority the contemplation of measures intended to affect the prevalence of alcoholism through the general regulation of alcohol consumption. The drinking of the alcoholic came to be seen as independent of other drinking: a symptom of pathological factors peculiar to him, and therefore, not amenable to change by measures that would affect the normal drinker. As a consequence, scientists understandably concentrated their attention mainly on the alcoholic per se in an effort to discover the distinctive causal factors (Armstrong, 1958; Keller, 1972).

We shall return to this and related issues in the section on Models of Prevention. Suffice to say for the present, that a rapidly growing body of evidence casts serious doubt on the validity of so narrow a concept of alcoholism, and indicates that the overall level of consumption in a population may well play a crucial role in its prevalence (de Lint and Schmidt, 1971). The result in some quarters has been a renewal of interest in the possible preventive value of legal measures and the appearance of a number of relevant studies.

In the following sections, the evidence is reviewed respecting the effects or lack of effects of the principal control measures which have been employed in recent times and most of which are now in use. However, whether or not the law will be, or should be applied to preventive ends depends on many factors besides the objective effectiveness of particular measures. Accordingly, in the final sections some of these factors are examined, as well as current models of prevention in relation to the role of legal restraint on drinking.

CONTROL OF OUTLET FREQUENCY

Perhaps no single control measure has been more frequently and widely employed over the centuries than the regulation of the number of places in which alcoholic beverages may be purchased (King, 1947; Krout, 1925;

Shadwell, 1915). Outlets for on-premise consumption have been particular targets, and the immediate aim most commonly has been to reduce their frequency. While from time to time those who influenced the legislators had objectives other than the prevention of insobriety (Lee, 1944; Lemert, 1962; Odegard, 1928; Popham, 1962), the latter has been typically the stated justification. The underlying assumption, often made explicit by temperance writers, is simply that the more opportunities there are for people to drink, the more they will be tempted to do so and the more drunkenness there will be. However, during the post-World War II years in particular, the opinion of such interested groups as the clergy and of the public at large has come to be divided on the issue (Canada Facts, 1946; Wolch, 1957). Now there are those who, by implication at least, favor increases in outlet frequency, still with the promotion of sobriety in mind. The notion of the "forbidden fruit" is cited, and it is argued that if alcohol were everywhere available, man would not desire it so much and would therefore drink in a moderate and civilized manner. As might be expected, this view has received considerable support from the alcoholic beverage industries.

On both sides, similar types of statistical data are often used to support the arguments for or against changes in outlet frequency. These data include arrests or convictions for drunkenness, alcohol sales figures, and alcohol-related mortality and morbidity statistics. The validity of such data as indices of the prevalence of alcohol problems has been questioned many times, most recently, for example, by Walsh and Walsh (1973). Thus, official statistics of drunkenness may depend not only on the actual prevalence of drunkenness but to varying degrees on prevailing legal sanctions, police instructions, the manner in which drunkenness is defined, the extent to which refuges from police vigilance are available, and local attitudes toward insobriety. In addition, rates generally are not separated into first offenders and repeaters; they refer simply to the number of arrests or convictions rather than to different individuals. Official sales figures, among other deficiencies, do not reflect the legal consumption of homemade wines and beers, or the consumption of illicitly produced and distributed beverages. An upward trend in per capita sales may reflect the addition of new reporting jurisdictions or an increase in the proportion of users rather than a true increase in average consumption. Alcohol-related hospital admission and vital statistical data may be subject to numerous extraneous influences such as attitudes of physicians, incomplete reporting, and trends in diagnosis and treatment.*

There is no question but that these potential sources of error need to be kept in mind when the effects of changes in control measures are under

---

* For detailed commentaries on the artifacts that may affect variation in alcohol statistics, see Bruun et al. (1960), Jellinek (1947), Popham and Schmidt (1958), and with particular reference to legal statistics of drunkenness, Ahlström-Laakso (1971).

scrutiny. At the same time, it is well to guard against the tendency to overestimate their importance, particularly when the implications of an apparent change in rates are not consistent with one's presuppositions. Thus, it is easy to explain away a drop in alcohol sales following stricter control measures by appeal to an artifact (such as illicit consumption) whose contribution may be negligible in reality. In any case, several types of alcohol statistics, especially sales and mortality data, have been shown to be valid indicators of the magnitude of alcohol problems in an area (Jellinek, 1947; Ledermann, 1956, 1964; Popham, 1970; Schmidt and de Lint, 1970). Furthermore, very considerable regional and temporal differences in these indicators have been found, and it is difficult to believe that much of this variation is attributable to errors in the indices rather than to real differences in the prevalence of alcohol problems (de Lint and Schmidt, 1971). Accordingly, it is of interest to review attempts to determine if, in fact, there is a relationship between the indicators mentioned and outlet frequency.

Popham *et al.* (1976) reported that, among the provinces and certain larger cities in Canada, higher rates of arrest or conviction for drunkenness tended to be found where there were fewest public drinking places per unit of population. In addition, trends through time in Ontario suggested a similarly inverse relationship (Popham, 1962; Popham and Schmidt, 1958). During the latter years of the nineteenth century, when the frequency of outlets for on-premise consumption achieved its all-time high, rates of conviction for drunkenness were comparatively low. Following this, the outlet rate fell steadily until the introduction of Prohibition in 1916. On the other hand, the conviction rate rose to a peak in 1912–14, then fell during Prohibition to reach a low point in the early Depression years. After the reintroduction of licensed drinking places in 1934, the outlet rate remained fairly static until recent years. But during the same period, drunkenness conviction rates rose steadily to achieve a level markedly higher than in any previous period. Per capita alcohol sales and liver cirrhosis mortality rates, long considered one of the best indicators of the prevalence of alcoholism (Jolliffe and Jellinek, 1941; Expert Committee on Mental Health, 1951; Popham, 1970), showed roughly similar trends over the years for which figures were available.

Mass Observation (1943) also reported an apparently negative association between drunkenness charges and outlet rates in English Data. And Popham *et al.*, (1976) ran a linear correlation analysis on two other English series: figures for 84 county boroughs and for 52 counties, excluding boroughs. The coefficients of correlation between convictions for drunkenness and on-premise licenses per 10,000 of population were negative but small (−0.18 and −0.19, respectively), and not significantly different from zero. Equally extensive legal statistics on drunkenness have not been studied for the United States. However, coefficients of correlation between tavern rates and per capita alcohol sales, and tavern rates and alcoholism prevalence estimates for 49 states proved readily attributable to chance (Popham *et al.*, 1976). With respect to other types of outlet in the United States, Entine (1963) concluded that limiting the number of package stores did not reduce off-premise consumption. On the other hand, Simon (1966b) found per capita sales to be related positively to the frequency of such stores, but felt on further analysis that this variable was more likely to be dependent on sales than the reverse.

The tendency toward a negative relationship between outlet frequency and drunkenness rates may be due to one or more of a number of factors. It may be that where a large number of outlets are tolerated, a relatively more liberal attitude toward drinking and drunkenness prevails so that there are fewer arrests (Popham, 1962). Or, as the authors of "The Pub and the People" (Mass Observation, 1943) noted, drunkenness rates are usually higher in urban areas and vary with the business cycle, rising and falling with the bank rate and other measures of prosperity. Outlet rates, on the contrary, are controlled by a licensing authority whose objectives may lead to fewer outlets where or when drunkenness is considered to be prevalent.

The most thorough analysis of this issue, of which we are aware, is that of Ahlström-Laakso (1971). In her study, the point of departure was the fact that the rate of arrest for drunkenness was markedly higher in Helsinki than in Copenhagen. On the other hand, the level of alcohol consumption and the frequency of public drinking places was very much higher in the latter city. Among the many possible explanations, the author showed that differences in the control system were especially important. Thus, for example, fewer taverns in Helsinki, and in particular of those catering to drinkers of the lower social strata, meant fewer places to become drunk unobserved by the police, and a greater likelihood that heavy drinking would occur in parks and other public areas. Differences in behavior when intoxicated, in the probability of intoxication on any given drinking occasion, and in enforcement vigor were also considered to contribute significantly. However, it is quite possible that such factors are more likely to be among the primary determinants of variation among different countries than of regional or temporal variation within the same country. Through time in Finland, for example, the trend in arrests for drunkenness appeared to follow rather closely the trends in other indicators of the level of alcohol consumption and alcohol problems (Bruun *et al.*, 1960).

Summing up the evidence, it would seem clear that in the populations examined, variations in indicators of the prevalence of inebriety are not dependent on outlet frequency. It is important to emphasize, however, that the variations considered ranged from situations where outlets were ubiquitous to those where some customers may have been mildly inconvenienced. That under the latter circumstances, there may be little or no effect on consumption, especially in a world of high speed transportation, is suggested by the figures shown below. These data relate to two rural municipalities in Ontario of about the same size. The only alcoholic beverage store in the district was located in one of the two communities so that residents of the other had to travel several miles to make a purchase. It can be seen that this circumstance did not prevent a slightly higher patronage by them (Table 1).

Finally, some attention should be given to studies bearing on the effects of control measures that create or alter situations of extremely low accessibility: that is to say, situations substantially different from those reviewed in the foregoing paragraphs. The classic instance on one side of the question is Prohibition when, in several countries, the frequency of legal outlets was reduced virtually to zero. There can be little doubt that during the first few years of Prohibition in Canada (Popham, 1956), Finland (Bruun et al., 1960), and the United States (Jolliffe and Jellinek, 1941; Warburton, 1932) all indicators of alcohol consumption and alcohol problems reached the lowest level yet achieved in any period for which there are relevant data. It is also clear that in later years—say, roughly 1923-1933 in the United States—as an illegal trade became well established and the speakeasy and other clandestine outlets made their appearance, consumption increased substantially (Warburton, 1932).

It may be that during Prohibition there was an increase in poison deaths as a consequence of toxic impurities in poor quality beverages, or because of the use of toxic substitutes. Such effects have been alleged to buttress the contention that Prohibition was a complete failure (see, for example, citations in Room, 1971). We are not aware, however, of any scientifically acceptable study that has demonstrated an excess mortality from these causes of epidemiologically significant proportions. And the studies already noted would suggest that, even in the later years of Prohibition in Canada and the United States, the level of alcohol consumption and the prevalence of alcohol-related health problems were significantly lower than before or since. A rationally based argument against Prohibition therefore must seek other grounds for rejection, and many such have been well documented (e.g., Asbury, 1950; Feldman, 1930; Lemert, 1962; Mäkelä, 1972; Warburton, 1932).

TABLE 1. Buyers of Alcoholic Beverages in Two Rural Municipalities in Ontario[a]

Municipality	Buyers per 1000 adults	Frequent buyers per 1000 adults[b]
With store	247	42
Without store	267	50

[a] From de Lint and Schmidt (1966).
[b] Buyers who purchased alcoholic beverages 4 or more times during the one-month study period.

At the opposite extreme is the effect of a change in control policy that renders alcoholic beverages readily accessible in areas previously isolated from a legal supply through geographic conditions. For example, in parts of northern Canada, the nearest outlet to some communities may take an expensive day or more to reach. Under these circumstances, it might be hypothesized that the introduction of outlets close by would have an appreciable effect on consumption. The results of Kuusi's alcohol policy experiment in rural Finland would seem to support this hypothesis (Kuusi, 1957). In the study, stores for the sale of beer and wine were opened in selected market towns that had previously been "dry" for many years. The drinking habits of the affected population and of a control population were studied in depth prior to the change, and monitored in detail afterward. An increase in overall consumption could be attributed to the new outlets. At the same time there was no evidence of a change in the frequency of intoxication, and indications that the increased consumption was partly offset by a decline in the consumption of illicit alcohol.* Very similar results have since been obtained by Amundsen (1965), who studied the consequences of the first introduction of alcoholic beverage stores to isolated "dry" areas in Norway.

Mäkelä (1972) recently reported a more dramatic effect in Finland following a very considerable and rapid rise in number of outlets, many of which were established in previously "dry" areas. Thus, in 1969 medium-strength beer was released for unrestricted retail distribution. Shortly thereafter, apparent alcohol consumption in the country increased by 48 percent. Beer accounted for most of the increase. Although effects on other indicators of the prevalence of alcohol problems have not been reported as yet, Mäkelä (1971) showed that such an increase tended to be spread over the drinking population in a manner that inevitably brought about an increase in the proportion of heavy consumers.

REGULATION OF TYPE AND LOCATION OF OUTLETS

A related area of control, which has been a subject of much debate and legislative or other governmental action over the years, concerns the character and distribution of outlets to be permitted. Usually, the focus has been on one or another aspect of such questions as: In what type of setting and under what conditions may different classes of alcoholic beverage be sold for on- or off-

---

* The study stands as a model for those wishing to undertake tests of control measures in the alcohol field or, for that matter, of any proposed change in social policy. In the report, the formidable methodological and practical problems encountered are dealt with at length, as well as the relation of the findings to the intricacies of developing an acceptable control policy.

premise consumption? What entertainment or recreational facilities may be provided in public drinking places? With special reference to the latter, may outlets be located in or near certain categories of public building, commercial establishments, or institutions such as schools and churches?

An almost bewildering array of regulatory measures have been attempted at one time or another within this area of control. Some—for example, the prohibition of treating or round-buying—have failed if only because they proved unenforceable. Some—for example, the requirement that alcoholic beverages could only be sold for on-premise consumption if accompanied by food (which led to the "reusable sandwich"—were easily circumvented and objects of ridicule. Some were dropped in response to negative reactions from the public. Many still survive, and none, so far as we are aware, have ever been adequately assessed as to their effectiveness for the purpose intended.

Many earlier writers (and a few in recent times) considered the facilities and physical features of the American saloon or the English public-house to be among the seducers of the working man to a life of insobriety. For them the overriding goals have been the abolition of these places, and the development of nonalcoholic alternatives in the community (e.g., Calkins, 1901; Levy, 1951). Others have been more moderate and have advocated only a number of restrictions on the operation of public drinking places, and the establishment of counter attractions such as free garden plots, libraries, museums, parks, and a variety of other recreational facilities (e.g., Select Committee, 1834). As Lemert (1962) has suggested, the implicit assumption would seem to have been that drinking occurred in response to deficiencies in community life, and that if these were eliminated the need for alcohol would disappear.

Perhaps it is obvious that this assumption has not entirely withstood the test of time. Certainly the use of alcohol is still very much in evidence although most of the gaps in social life to which its attractiveness was commonly attributed have long since been filled. It is probably true, on the other hand, that the importance of public drinking places—at least in North America—has declined since the turn of the nineteenth century. And this may be due to the growth of competing attractions as well as to overall improvements in the standard of living (and therefore of the home as a drinking place). Nonetheless, the on-premise outlet has remained a significant element in the recreational, if not emotional, life of a great many people, and a substantial portion of all drinking takes place in it (Cavan, 1966; Mass Observation, 1943; Popham, 1962).

A number of writers have focussed on the physical improvement of on-premise outlets, rather than on their abolition, in the belief that structural and aesthetic features were important determinants of sociability and sobriety. The models that have most captured the attention of these scholars are the inns and taverns of the Elizabethan period and of the following two centuries. Maskell's work is typical in this regard (1927). It was written in the hope that "the sympathy of some readers may be enlisted in the movement to rescue the public house from obloquy and neglect" (p. vii). It was in fact a plea for the renaissance of the taverns and tavern life eulogized by literary figures from Shakespeare to Dickens and celebrated by poets from Chaucer to Longfellow. In the introduction it is submitted that "decadence began in England when the kindly landlord gave way to the brewer's manager and the inn became a public-house where men go just to drink in sordid and demoralizing surroundings" (p. x). In short, public drinking places of both the past and the present are viewed selectively, with the result that the former are associated with the picturesque, the healthy, and the desirable, and the latter with the ugly, the sordid, and the undesirable.

This school of thought apparently has had some influence on government control policies; at least in England. For example, in 1916 the government acquired the entire alcoholic beverage trade, including more than 100 public houses, in and around the City of Carlisle. This became widely known as the "Carlisle Experiment." The justification was the necessity of strict control to prevent insobriety among local munitions workers. Among the measures introduced to achieve this end, considerable emphasis was placed on a program to remodel or renovate many of the public-houses (Askwith, 1928; Shadwell, 1923). A more recent example is the "trust-house" concept of government control over the restoration and maintenance of the original appearance of historic drinking places. The point of view is also sometimes implicit in arguments to support the "tied-house" system in England. Thus, it is contended that the large brewery owners of public-houses are more likely to have the capital and motivation to make their outlets attractive than are many small entrepreneurs (Oliver, 1947).

Until comparatively recent years, just the opposite philosophy seems to have dominated control policies respecting the operation, appearance, and facilities of both on- and off-premise outlets in most of Canada, a numer of American jurisdictions, and some other areas. This has been especially evident where a substantial "dry" sentiment continued to prevail following the repeal of Prohibition. The Province of Ontario is a typical instance. Prohibition was repealed in much of the province in 1927, and a limited number of government-owned and -managed package stores were opened. These were generally located well off the main shopping streets, and the beverages offered for sale were never displayed or readily accessible to customers for examination. In 1934 outlets for on-premise consumption were reintroduced in the form of "beverage rooms," that is, establishments licensed to sell beer only. They were constructed so that the activity within could not be observed from the outside. Most were exceedingly plain both as to furnishings (usually restricted to simple tables and chairs) and decoration. No professional entertainment, game, singing, or dancing was permitted. Typically, there was a room for men only, and one for women only or women with escorts. A customer was not permitted to carry his drink from one table to another, to drink while standing, to buy more than one drink at a time, or to buy on credit. There were many other rules but the foregoing will serve to convey the relevant picture. Clearly, the objective was to minimize the attractions of the public drinking place, and the likelihood of prolonged and lively social interaction occurring among its patrons.

Since World War II, public sentiment in Ontario has increasingly favored a more permissive policy with respect to drinking, and this has been reflected in several changes in legislation and derivative regulations. One notable change in 1947 led to a diversification in types of outlet for on-premise consumption. Beverage rooms continued to operate under the same restraints as before, but now "cocktail and dining lounges" were permitted in the province as well. Essentially, this meant a slight increase in outlet rate, the sale of wines and distilled liquors by the glass with or without meals, and higher standards of decor for the establishments licensed to sell such beverages. Most beverage rooms catered to a working-class clientele; the new establishments were intended to attract a middle-class patronage. Music, professional entertainment, and dancing were allowed in the lounges, and no segregation of patrons by sex was imposed.

In an effort to assess the effect of the 1947 legislation, Popham et al., (1976) analyzed trends in alcohol statistics for eight-year periods before and after the change. By way of control data, the trends were compared with those in the adjacent Province of Manitoba. There no significant changes in policy had occurred over the 16 years, and the regulations and types of outlet were much the same as prevailed in Ontario prior to 1947. It was found that, with the exception of alcoholism (in effect, liver cirrhosis mortality) rates, the percentage increases were much greater in both provinces before than after 1947. From 1947 to 1954, alcohol sales increased slightly more in Ontario (19 percent) than in Manitoba (14 percent), but drunkenness conviction and alcoholism rates showed greater increases in Manitoba.

A criticism which may be levied against this study is that it did not take account of a possible differential change in the proportion of users of alcohol in the two provinces; the rates were all based on either the drinking age or adult populations. However, the rather scanty survey data available for the period (Popham and Schmidt, 1958) do not suggest changes of a magnitude that would significantly affect the results reported. Bryant (1954) examined the consequences of a similar change in the State of Washington: the introduction of "liquor by the drink." He concluded that there was no evidence that increases in consumption or in alcohol-related offenses could be attributed to the change in control policy. And in Saskatchewan, Dewar and Sommer (1962) conducted a before-and-after study of a small community in which a "beer parlor" for men only was replaced by a beer and wine tavern catering to both sexes. They could discern few effects. After the change, more drinking occurred away from home but there was no apparent alteration in the overall consumption level.

It should be stated at once that exclusive dependence on studies such as these is seldom likely to permit a definitive conclusion respecting the effects of a change in legal restraint. In the first place, when a change applies to a whole state or province, let alone a nation, it will be difficult if not impossible to find an adequately matched control population. Secondly, it is rare to find instances when a single specific change has occurred. Usually, several changes are introduced simultaneously, as in the case of the 1947 outlet diversification in Ontario. Under these circumstances, if an effect is identified it will not be known to which changes it should be attributed and to which not. This greatly reduces the value of the results insofar as rational policy development is concerned.

Thus, it is important that other approaches to assessment be undertaken as well. It seems to us that one of the most promising—for the area of regulation under consideration—would be careful observation and interview studies of behavior in beverage outlets before and after changes. Good models of the essentially ethnographic approach involved are provided by the tavern studies of Cavan (1966) and Mass Observation (1943). Ideally, such investigations would be done with the cooperation of the local licensing authority so that contemplated changes might be introduced in a manner that facilitated controlled examination. Another approach, which has received far too little attention in the alcohol field, is direct experimental manipulation using small groups in a simulated or, where practicable, real tavern setting. Bruun's study of a group of regular drinkers, conducted in one of the rooms of a licensed outlet in Helsinki, is an excellent example of the potentialities of the approach (Bruun, 1959).

In the past 25 years in Ontario, as in many other jurisdictions of the Western world, there have been numerous changes in alcohol control policy, mainly in the direction of relaxation of restrictions. Since the mid-1950s television has been allowed in beverage rooms; in the past few years, games and other recreational facilities have been permitted in establishments with lounge licenses; some "stand-up" bars have been allowed; displays of beverages have been introduced to package outlets; self-service package stores have been established in main shopping areas; and licenses have been issued for on-premise outlets in locations not previously contemplated; for example, museums and other public buildings, theaters, office complexes, large department stores, and sidewalk cafés.

It may be that, taken individually, these and other changes in type and location of outlet permitted have little effect on general consumption and attendant problems. But this has not yet been substantiated with even a reasonable degree of certainty. It is clear that an overall shift in alcohol control policy is well underway in many countries. In the past, a strong influence was exerted by those who considered that changes that increased the attractiveness of beverage outlets, or their visibility to the public would lead to increased alcohol use with undesirable consequences. Today it is more common to hear that by improving the appearance and facilities of outlets, and introducing them into all areas of everyday life, the mysticism associated with alcohol will be reduced; it will come to be regarded as no more remarkable than any other consumer product and therefore will be used moderately. The essential point is that, whichever view is argued, the appeal to date has had to be to theory rather than fact, or to the weight of prevailing sentiment rather than to evidence systematically gathered and objectively analyzed.

CONTROL OF HOURS AND DAYS OF SALE

The opening hours of alcoholic beverage outlets, especially those for on-premise consumption, have been almost as common a target of regulatory measure through the ages as their frequency (King, 1947; Shadwell, 1915, 1923). Usually, the legislation has provided for a reduction in opening times, and the stated intent has been to combat the problems of insobriety. In recent years, however, and in keeping with trends in other areas of control, new regulations have sometimes extended both the hours and days of sale. In Ontario, for example, the hours of dining lounges were extended from the usual midnight closing to 2:00 A.M., and such establishments were permitted to sell alcoholic beverages with meals on Sunday, traditionally a general closing day. In addition, the 6:30 to 8:00 P.M. closing of beverage rooms was not imposed on premises with lounge or dining lounge licenses when these were allowed in 1947.

Despite the long-standing belief in the efficacy of regulating opening hours, there have been exceedingly few attempts to put the matter to test. Popham (1962) found an apparent correlation between the opening hours of beverage rooms in Toronto and the hourly pattern of arrests for drunkenness exhibited between 8:00 A.M. Monday and 8:00 A.M. the following Sunday. However, when arrests were plotted for the period 8:00 A.M. Sunday to 8:00 A.M. Monday morning, during which time all beverage outlets were closed, an almost identical pattern emerged. This might be taken to indicate that the hours of sale reflected the drinking pattern of at least one segment of the community rather than the reverse. On the other hand, the opening hours that prevailed at the time of the study had been in force for many years, and originally may have shaped the characteristic circadian pattern observed. Indeed, the results of a recent study in Victoria, Australia, would suggest this as the more probable relationship.

For many years in much of Australia a 6:00 P.M. closing was enforced with respect to on-premise outlets. One of the original objectives was to reduce the amount of drinking by working-class males and assure their arrival home at a reasonable hour. As Room (1971) noted, the measure had an unforeseen side effect, namely, the "six o'clock swill." In other words, patrons seemed to consume substantially more than they otherwise would have, to make the most of the short period between leaving work and the closing of the outlet. During the mid-1960s in Victoria, the closing time was extended to 10:00 P.M. Raymond (1969) has reported an enlightening analysis of the effects of this extension on motor vehicle accidents. She found no change in the overall total of personal injury accidents, or in the proportion of the total occurring in the new extra hours (plus the hour immediately after closing). There was, however, a marked change in the hourly pattern. After the extension of closing time, the previous accident peak between 6:00 and 7:00 P.M. altogether disappeared, and was replaced by a new peak between 10:00 and 11:00 P.M. On Sunday when all outlets were closed this accident pattern was not present. Raymond's arguments are convincing that the accidents that comprised the shift in peaks did in fact involve patrons of on-premise outlets. From the results of this study, one would conclude that changes in closing hours can have a significant effect on pattern of consumption. But the total consumption or, at least, the frequency with which patrons consume impairing amounts may remain unaffected.

It would seem clear, especially in view of Raymond's findings, that the effects of changes in hours and days of sale deserve further research. Jurisdictions should be examined in which substantial changes have occurred, and for which trustworthy indicators of drinking and alochol problems are available for extended periods before and after. Certainly in the United Kingdom, where closing hours have often been manipulated in an effort to control alcohol-related problems, there is some conviction that this is an effective measure. For example, Shadwell (1923) contended that measures such as shorter hours "have proved really efficacious, while others—particularly state ownership and control, the reduction of licensed houses, alteration of premises, disinterested management, and supply of food—have failed to exert any perceptible influence on sobriety and public order" (p. 150).

LIMITATION OF DRINKING AGE

Most jurisdictions have sought through legislation to limit or prevent the use of alcohol by persons below a specified age. At present, this age limit differs substantially from one area to another, and does not necessarily correspond to other legal age limits; for example, those pertaining to marriage, driving, military service, or voting. In some cases, the proscription is comprehensive: It is illegal for a person under the stated age to purchase or possess alcoholic beverages, or for any one of age to supply these to such a person. In others, no attempt is made to outlaw drinking in the home; only consumption in public drinking places is prohibited. In certain European countries there are two age limits: one (lower) at which the use of beer and wine is legal, and another (higher) at which distilled beverages are permitted. We are unaware of instances where the minimum age is greater than twenty-one or less than fourteen years. Eighteen and twenty-one have been the most common age limits in North America with a trend toward the former in recent years.

As in the case of other restrictions discussed, there are divergent opinions respecting the desirability and effects of age limits, and very little objective data upon which to base a rational judgment. Some consider that the law should seek to postpone the introduction to alcohol as long as possible, and that to reduce or dispense with an age limit is to encourage the early development of alcohol-related health problems. Others point to evidence of extensive underage use and argue that, since the limit is out of step with actual behavior, it serves only to generate disrespect for the law. Still another view is that age limits (especially when high) encourage clandestine drinking by the young, and reduce the likelihood that healthy attitudes toward alcohol use will be learned. A lower limit or none at all, it is felt, would open the door for parents to introduce alcohol to their children in ways that would foster the evolution of moderate, guilt-free drinking practices. A detailed examination of the arguments for and against this and opposing positions has been provided by Wilkinson (1970).

Given some evidence of illicit alcohol use by young people, current views would suggest that lowering an age limit will produce: (1) an increase in consumption and alcohol problems, at least in the age group affected, by no longer exercising a restraining influence; (2) no change because it will simply legalize the status quo; or (3) a decrease in both consumption and alcohol problems by removing guilt or anxiety associated with secretive drinking. Bruun and Hauge (1963) sought to determine through surveys the amount and extent of drinking among male teen-agers in the capitals of four Scandinavian countries. There were substantial differences among the four in control measures aimed at young drinkers, including age limits, penalties for infractions, and intensity of enforcement. Yet, only minor differences in consumption and extent of use were found and not, in the investigators' view, of the magnitude to be expected if the prevailing controls were exerting appreciable effects. However, this cannot be considered an acceptable test of the influence of age limits for at least two reasons: (1) because of the variety of potentially relevant cultural and other differences that were inevitably uncontrolled; and (2) because estimates of alcohol consumption based on verbal reports are of doubtful validity, usually falling very short of actual intake (Pernanen, 1973).

By way of a preliminary examination of the effects of a change in age limit, Schmidt (1972), and Schmidt and Kornaczewski (1973a,b) studied trends in alcohol sales and motor vehicle accidents before and after a reduction in the minimum age in Ontario from twenty-one to eighteeen years. For some years prior to the change (which became effective on July 31, 1971), and in the seven months immediately preceding it, the proportion of alcoholic beverage sales for on-premise consumption had been declining moderately but steadily. During the first five months after the change, a notable alteration occurred in this trend as a consequence of increased sales, especially of beer, in public drinking places. The increase was reflected in total sales, and therefore, was not due primarily to a shift in setting from off- to on-premise consumption. Schmidt (1972) concluded that the reduction in age limit had led to a substantial rise in the consumption level of the eighteen to twenty age group, who apparently chose the more visible way to exercise their new legal status. There was also a distinct increase in the involvement of the age group in alcohol-related motor vehicle accidents in the five months following the change (Schmidt and Kornaczewski, 1973a). Nor does this seem to have been due to an alteration in enforcement practice. Clearly, the increase in alcohol-related accidents constituted a significant *addition* to the total expected accident involvement of the age group (Schmidt and Kornaczewski, 1973b).

The approach taken in these studies would appear to offer a comparatively simple and inexpensive way to obtain indications of the consequences of a change in age limit. Similar assessments should be undertaken elsewhere to increase confidence that the results were not solely a reflection of conditions peculiar to Ontario. This should not be difficult. The documentary statistical data employed are quite readily available in many jurisdictions, and usually provide a reasonably accurate reflection of the prevalence of the relevant behavior. A shortcoming that is more difficult to avoid is that effects observed in a short period immediately after a change may be transitory. Here an effort could be made to follow trends in sales and accident rates over a longer period to determine whether or not the new levels persist. Changes in the age distribution of clinic populations of alcoholics might also be monitored to obtain indications of long-term health consequences. However, it has to be recognized that when trend studies are extended in this manner, the difficulty of controlling for extraneous influences increases, and confidence in the conclusions decreases in proportion to the length of time covered. In any case, it is clear that the issue of legal age limitation requires further study. The findings of the Ontario investigations cast doubt on the generality of the contention that age limits do not exercise a restraining influence, or that their reduction may lead to a decline in alcohol problems.

PRICE CONTROL

As suggested at the outset of this review, measures affecting the price of alcoholic beverages have been among the earliest forms of control imposed by law. However, the original objectives were to prevent overcharging rather than overdrinking, and most particularly, to secure revenue for government. It was not until the rise of substantial Temperance agitation, especially in the late eighteenth and throughout the nineteenth centuries, that taxation was often justified as a means to control insobriety and protect the health, morals, and stability of society. As one parliamentarian put it in 1792:

> It is of very little consequence to the morals of the people (if they will get drunk) what they get drunk with: it is, however, the duty of the legislature, as much as in them lies, to make the means of intoxication as difficult to come by as they possibly can: This can only be done by laying duties as high as the article will bear.*

During the early years of the present century in the United Kingdom and certain European countries, heavy taxation was accepted by many as one of the most effective ways to combat insobriety (Nielsen and Strömgren, 1969; Reuss, 1959; Shadwell, 1923). In North American areas, following the repeal of Prohibition, similar beliefs appear to have influenced government control policies. To give one local example, in the Third Report of the Liquor Control Board of Ontario for the year 1929, it was stated that "beyond all doubt consumption increases with low prices, and decreases with high prices." And with reference to decreased drunkenness in the United Kingdom:

> While propaganda has been carried on against drunkenness, the best-informed workers and observers are generally of the opinion that the higher cost of spirits coupled with industrial conditions and unemployment largely contribute to the decreased consumption (p. 10).

Today there are probably no jurisdictions, where alcoholic beverages are legally sold, that do not impose some tax upon them. This has continued to be seen as a lucrative source of revenue for the state. Occasionally, the influence of economic conditions on amount consumed is noted. For example, the chief statistician of the Finnish State Alcohol Monopoly remarked: "The connection between liquor consumption and economic conditions is so plain that the volume of alcohol sales has even been used in scientific economic research as a meter to register economic fluctuations" (Harvola, 1956, p. 26). But taxation is now rarely justified as a protective measure, and the suggestion that it might be effective in this regard is apt to encounter strong resistance and sometimes

---

* The Chancellor of the Exchequer responding in the Irish parliament to opposition holding that the brewers should not be taxed to the same degree as the distillers. Cited in Walsh and Walsh (1970, p. 115).

hostility. This change in attitude is consistent with the trend, previously alluded to, toward greater permissiveness in the control area and rising levels of public acceptance of drinking. Indeed, in many jurisdictions tax increases have lagged behind income levels so that, in effect, the economic accessibility of beverage alcohol has been steadily increased (Popham et al., 1976).

Since, from a purely economic standpoint, price level would seem a likely, if not obvious determinant of the amount of alcohol consumed, it may seem surprising that it has not received more attention from contemporary students of alcohol problems. There would appear to be three principal reasons for this. The first is a reflection of liberal attitudes now widely encountered in the Western world, toward individual rights and the control of behavior. Presumably, as Mäkelä has intimated, scientists are not immune to such trends in their cultures, and therefore have tended to denigrate a priori the legal control of alcohol use, and particularly measures that may be perceived to involve economic class differences in their effects. The tendency instead has been to seek preventive measures consistent with the thrust toward self-determination in this and other areas of social concern.

A second reason for neglect, fostered to a significant degree by the etiological views of Alcoholics Anonymous, has been the rise of the "disease concept" of alcoholism (Jellinek, 1960a). As noted earlier, this has led to a restricted view of the problem and the belief that the factors that may influence the demand of most people for alcohol are largely irrelevant to the demand of the alcoholic. Research effort, accordingly, has been directed mainly to a search for determinants peculiar to him.*

The third reason appears to reflect a persistence of some of the thinking characteristic of the Temperance movement both in North America and Europe. This is the assumption, discussed at length elsewhere (Popham et al., 1976), that the demand for alcoholic beverages is qualitatively different from that for other commodities, and consequently not subject to the same factors as affect the demand for other consumer goods. For example, it is doubtful if anyone would expect an increase in the number of gas stations to cause an increase or decrease in the amount of gasoline purchased (barring cases of extreme inaccessibility). The number of gas purveyors is readily perceived to be responsive to demand rather than the reverse. On the other hand, one would expect sales volume to be affected by price. But in our view, the special

---

* For an enlightening statement of the ramifications of the "disease concept," see Powell vs. Texas, Supreme Court of the United States, No. 405–October Term, 1967. In this much debated test case, a man found guilty of public intoxication appealed on the grounds that he was "afflicted with the disease of chronic alcoholism," and therefore had not appeared drunk in public of his own volition. A revealing sidelight of his prior appeal at the county level emerged when it was noted that he had taken one drink before appearing in court. Questioned as to why he had not consumed more and was able to attend in a sober condition, the appellant admitted he had not had the price of a second drink.

attitudes and beliefs that have prevailed in the case of alcoholic beverages have tended to distract research attention from this commonplace economic variable, and have stimulated some of the doubts that have been expressed about the evidence of price effects and their preventive implications.

The evidence of an association between the price of beverage alcohol, apparent consumption, and the prevalence of alcohol-related health problems is certainly persuasive if not compelling. Despite some methodological differences and shortcomings, econometric studies of data for Australia, Canada, England, Finland, Ireland, Poland, Sweden, and the United States have consistently shown price to be a significant predictor of the demand for alcohol (Lau, 1975; Nyberg, 1967; Tolkan, 1969). With respect to the association between demand levels and the prevalence of alcohol problems, the evidence is equally convincing. Many years ago, Jolliffe and Jellinek (1941) demonstrated that the death rate from liver cirrhosis in the United States varied closely with per capita alcohol sales. Since then, several studies have established the validity of liver cirrhosis mortality as an index of alcoholism prevalence (Popham, 1970; Schmidt and de Lint, 1970); and a substantial body of epidemiological data has accumulated to confirm the close association between its rate of occurrence and the level of apparent alcohol consumption (Ledermann, 1964; de Lint and Schmidt, 1971; Popham, 1970).

In 1960, Seeley reported the results of a regression analysis of temporal data for Ontario in which the interrelations of the three variables—price, sales, and liver cirrhosis mortality—were examined simultaneously. He took the ratio of price to disposable income (in effect, a measure of an economically average individual's ability to buy alcohol) to be the relevant economic variable. A strong inverse relationship was found between this index of accessibility and both per capita alcohol sales and liver cirrhosis death rates. Seeley's findings were confirmed by the present authors, who extended his time series and also examined regional and temporal variation in many different North American and European jurisdictions (Bronetto, 1960–63; Popham *et al.*, 1976).

Both Lau (1975) and Simon (1966a) have pointed out that the demand for alcohol is generally more sensitive to changes in income level than in price, and the former has objected to the use of an index that combines the two and thereby obscures their separate effects. This criticism is justified in that it is essential to establish whether or not price independently contributed to the covariation reported; it is, after all, price that best lends itself to manipulation for control purposes. However, Lau's own studies of Canadian data (1973a; 1973b), and those of others recently reviewed by him (1975) have demonstrated clearly a significant price effect with income held constant. The "relative price" index employed in Seeley's and subsequent work remains of value from the standpoint of policy development and implementation. Thus, it draws attention to the fact that increases in taxation may not have the intended effects on economic accessibility unless concurrent changes in income level are taken into account when the amount of increase is determined.

We do not wish to convey the impression that, in our opinion, economic accessibility fully accounts for the observed variation in indices of consumption and alcoholism. It seems to be a powerful determinant, statistically speaking, and one with clear theoretical implications for the development of a health-oriented control policy. However, there are some noteworthy exceptions to the generality of the associations reported (Alcoholic Beverage Study Committee, 1973; Popham, 1970; Popham et al., 1976), and these deserve careful study in the future. In addition, there is another factor that also appears to exert a strong effect on the indicators mentioned, notably, level of acceptance of drinking. Thus, such indices as proportion of total abstainers, responses to attitude surveys on drinking, and voting behavior on alcohol control issues have been demonstrated to vary with alcoholism rates (Jellinek, 1960a; Seeley, 1962).

It is probable that the two factors are not entirely independent: Where there is a high level of acceptance, there is apt to be a high degree of economic accessibility, and conversely, low accessibility may be associated with low acceptance. However, it is clear that this is not always the case, and acceptance levels may help to explain some of the departures from expectation referred to above. For example, until very recently rates of apparent consumption and liver cirrhosis mortality in The Netherlands were among the lowest on an international basis, despite a high level of economic accessibility (de Lint and Schmidt, 1971). This may well have been due to the strong disapproval of excess in all areas of behavior that prevails especially among the Calvinist segments of the population (Gadourek, 1963; Jellinek, 1960a). On the other hand, there is substantial evidence that the Republic of Ireland is a high acceptance area (Bales, 1946). Nevertheless, consumption and alcoholism levels are estimated to be very low (Blaney, 1967; Lynn and Hampson, 1970). This is in accord with the high price of beverage alcohol relative to the average income level of the Irish (Walsh and Walsh, 1970).

An objection quite often heard concerns the validity of liver cirrhosis mortality as an index of alcoholism prevalence. Specifically, it is argued that a chronic disease will not respond so rapidly to transitory changes in the etiological factor. But, as Terris (1967) has pointed out, in the case of liver cirrhosis the "phenomenon is consistent with the clinical course of the disease. In many cases the cirrhotic process can be halted and decompensation prevented by avoiding further use of alcohol. Conversely, resumption of heavy alcohol use after a period of abstinence can decompensate a previously injured liver in a relatively short period of time" (p. 2078). However, in practice, a slight lag in the mortality trend would be expected, since advanced cases are likely to die prematurely even in the absence of drinking. Such a lag was clearly evident in the data for Ontario (Popham et al., 1976).

A criticism more frequently encountered is that the evidence is correlational, and therefore does not indicate a cause–effect relationship. Indeed, it has been suggested by Room (1971) that variations in liver cirrhosis mortality and per capita sales may both reflect changes in the consumption of excessive drinkers. His argument is to the effect that a substantial portion of all alcohol consumed is contributed by such drinkers. Accordingly, if for some reason there were a drop in their prevalence, there would inevitably be a corresponding decrease in alcohol sales and presumably also in the mortality rate.* However, Room did not take into account the strong association with relative price that could not be explained satisfactorily by this line of reasoning. To attempt to do so would involve making one or other of two highly doubtful assumptions: (1) that the demand of alcoholics for alcohol was more price and income elastic than that of other drinkers; or (2) that changes in the number of alcoholics influence income and taxation levels in such a way as to reduce economic accessibility when their prevalence is low and increase it when their prevalence is high.

As to the more general issue of possible spurious correlation, this is often likely to be a theoretical shortcoming of the type of epidemiological work with which we have been concerned. Definitive experiments will seldom be possible for practical, political, and ethical reasons. For example, the case for a cause–effect relationship between cigarette consumption and the prevalence of pulmonary disease must rest on essentially circumstantial evidence rather than experimental proof.† However, if the associations are consistently encountered in both regional and temporal series for different populations, if the range of variation in the indices is substantial, if trends through time have been in both directions, and if the character of the relationship is in accord with expectation based on established knowledge, then the onus of proof shifts to those who

---

* Since, as Room noted, up to 50 percent of total consumption may be attributable to excessive drinkers (although the modal figure is probably substantially less), there is bound to be some spurious correlation between the two variables. But if the causal chain were in the direction implied by him, one would expect percent changes in consumption to be generally less than those in liver cirrhosis mortality. For example, if alcoholics contributed 50 percent of all consumption and their prevalence fell by 20 percent, then a 10 percent drop in per capita consumption would be expected. Exactly the opposite is the case: On the average, a given percent change in per capita sales is accompanied by a substantially smaller change in the liver cirrhosis mortality rate (Seeley, 1960).

† This is not to suggest that no relevant experimentation whatever is possible, or that source of data other than those commonly exploited by the epidemiologist may not provide highly pertinent additional evidence. For example, that a long history of heavy alcohol consumption is a common cause of liver cirrhosis is well established on clinical grounds (Lelbach, 1974). And recently an experimental study was reported in which the cost of alcohol available to a small sample of hospitalized alcoholics was manipulated: Both the amount of work required to earn a drink and the price per drink (which increased if subjects sought to buy more than two per hour) were varied. It was found that increasing the cost could be employed both to reduce the total amount consumed over a given period, and to alter the temporal pattern or spacing of consumption (Bigelow and Liebson, 1972).

contend that cause-and-effect is not involved. In our view this applies reasonably well to the present status of the evidence bearing on the role of economic accessibility as a determinant of rates of alcohol consumption and alcoholism.

Elsewhere (Popham *et al.*, 1976), we have endeavored to outline the principal elements of a price policy that would be consistent with this evidence. We concluded that it would require:

1. A price structure such that the cost of a given quantity of absolute alcohol was the same for the cheapest source of alcohol in each class of beverage: beer, wine, and spirits. (Prices would then be scaled upward in each class, depending upon the producers' values for different brands and other relevant commercial considerations.);

2. Adjustment of prices as often as required to maintain a constant relationship between the cost of beverage alcohol so established and the disposable income level of the population; and

3. The ultimate establishment of a relative price level that would maximize the preventive value of this control measure.

Our assumption was that the optimum level might differ from one jurisdiction to another, and result from a careful weighing of the costs of too high a price level—for example, the possible rise of an extensive illicit trade and loss of some of the social and economic benefits derived from alcohol use—against the public health costs of too low a price. In theory, implementation of the first two elements would serve to stabilize accessibility and thereby prevent, or at least retard, further increases in rates of consumption and alcoholism.

One of the strong criticisms levied against such a policy has been that it would penalize the poor and leave the rich unaffected. Thus, Terris (1967) has shown that in the United Kingdom, where alcohol taxes are high, liver cirrhosis mortality tends to be concentrated in the upper classes; in contrast, the death rate from this cause is highest among working-class drinkers in the United States. However, it is also relevant that the overall prevalence of alcoholism in the United States is probably at least four times that in the United Kingdom (Expert Committee on Mental Health, 1951; Jellinek, 1947; Popham, 1970).

As we noted in our original paper on the topic (Popham *et al.*, 1976), objections of this character raise questions of critical importance, but ones that fall into the realm of ethics and political philosophy rather than of science.* One of these is: Should legal controls ever be employed to protect public health? If the answer is yes, then it is a matter of how many will be helped and

---

* For an excellent discussion of the role of science and the scientist in the resolution of such issues, see Kalant and Kalant (1971).

not of class discrimination, and of weighing the foreseeable costs (economic and otherwise) of a particular proposal against the costs of retention of the status quo. At present, the price control policy outlined above is not likely to be viewed in this light in jurisdictions where acceptance of drinking and accessibility are comparatively high. Therefore, it would be politically difficult to implement in these areas. The policy would first have to be seen as a valid protective measure, and this would require a substantial increase in awareness of the public health consequences of rising consumption levels.

DIFFERENTIAL TAXATION

There is an old proverb, apparently of central European origin, that asserts: "Getting drunk on beer makes a man vulgar; getting drunk on spirits makes him dangerous; and getting drunk on wine makes him charming." Such traditional beliefs that alcoholic beverages differ in their effects on behavior, by implication for reasons other than of amount of alcohol involved, are widely encountered, with variations in detail from one culture to another. Probably the most influential, with respect to control measures, has been the belief that beer is a drink of moderation and that the problems of alcohol can be attributed mainly to the use of spirituous liquors. It is a view promulgated by the brewers* and opposed by the distillers; it has often been embodied in reports and briefs to government on the liquor control question; a few contemporary students of alcohol problems have cited evidence in its favor (e.g., Isaksson, 1957; Wallgren, 1960); and it has been the explicit justification in many jurisdictions for the imposition of substantially higher taxes on distilled beverages than on any other class (Jellinek, 1963). Clearly if the belief were valid, a policy of differential taxation would be preferable to the policy of uniform price control discussed in the previous section. It would not penalize the "moderate drinker," or by definition, the consumer of a beverage shown to be relatively harmless.

Currently, the studies most frequently cited in support of arguments for a tax differential favoring beer have been those of Goldberg and his associates on relative intoxication potential (Isaksson, 1957). Recently, these were more or less replicated by a Canadian group with similar results (Alcoholic Beverage Study Committee, 1973). The experiments demonstrated that a higher peak blood alcohol level, and correspondingly greater psychophysical impairment were achieved after the ingestion of spirits than after ingestion of the same quantity of alcohol in beer. It is safe to assert that the validity of this difference,

---

* For a sophisticated recent example, see the report of the Alcoholic Beverage Study Committee (1973), which was prepared under the auspices of the Brewers' Association of Canada.

under the experimental conditions imposed, has been generally accepted by workers in the field for some years. Indeed, it is in accord with an established pharmacological principle respecting the effect of the concentration of alcohol ingested on rate of absorption. In the present context, the relevant questions are: Is the finding likely to be applicable to actual human drinking habits? Is any meaningful difference that may exist reflected in the type or prevalence of alcohol problems encountered?

As to the first question, the experimental conditions would seem to have been far from realistic analogues of most human drinking practices. A single very large dose of beverage alcohol had to be consumed in a very short period of time, and no further drinking was permitted. In addition, drinking took place after a protracted fast (usually at least 12 hours). It is also noteworthy that the largest difference was obtained when the distilled beverage was taken undiluted, a mode of consumption encountered in Scandinavian and Slavic countries but comparatively rare elsewhere. Kalant *et al.* (1975) recently completed a comparison of the effects of beer, wine, and spirits in which subjects achieved substantial blood alcohol levels through progressive drinking. The minimum time since last eating was 45 minutes. Under these more realistic circumstances, differences among the blood alcohol curves for the three classes of beverage virtually disappeared; nor were any significant differences in impairment found on the performance tests employed.

In the same experiment, each subject was asked to rate his own sense of impairment on a specially designed scale. A tendency toward overestimation of impairment appeared when intoxication was based on spirits, and underestimation when based on beer. It is tempting to hypothesize that this was a reflection of the beliefs held by subjects about differences in the two types of beverage. One might also speculate as to the effects of such self-ratings on the type of behavior ordinarily exhibited when intoxicated. In short, would an expected effect operate as a self-fulfilling prophecy? Evidently it did not in the case of the performance measured by Kalant *et al.*. On the other hand, Takala *et al.* (1957), in an experiment conducted in Finland, found that subjects were more prone to violence when the same blood alcohol level was reached through drinking spirits than when reached through beer. In Finnish culture there is a well-known pattern of explosive intoxication associated with the concentrated consumption of spirits (Kuusi, 1948; Sariola, 1956). Accordingly, it may be that their results were a function of a difference anticipated by their subjects. If so, this would not make the difference any less real, but it would tend to shift the etiological focus to variables in the area of culture and personality rather than of pharmacological action.

A related speculation concerns the possible relation of subjective assessments of impairment to risk-taking behavior. If impairment resulting from spirits consumption tends to be overestimated, is there a greater likelihood that high risk situations will be avoided after such consumption, or that more caution will be exercised in an effort to compensate for impairment? Is the reverse the case with beer if impairment is underestimated? We are not aware of any research that has been directly concerned with these questions. However, their potential importance is highlighted by the results of a careful study of the role of alcohol in traffic accidents, conducted by Borkenstein and his colleagues (1964). The findings also bear on the second major issue posed above, namely, the social significance of any difference in effect between spirits and beer.

In the Borkenstein study, the blood alcohol levels of a large sample of drivers in accidents were determined, and of a control sample of drivers stopped at random within the constraints of an appropriate statistical design. Information was obtained through interviews as to the class of beverage typically consumed by the drivers. The results confirmed earlier work in that a higher overall proportion of drinking drivers was found in the accident group, and the probability of accident involvement increased rapidly with higher blood alcohol levels. But in addition, significantly more of the accident drivers (64 percent) than of the control drivers (58 percent) reported beer to be their beverage of choice. Conversely, fewer in the accident (30 percent) than in the control group (36 percent) reported spirits to be their customary beverage. It was further shown that the proportion of the control group who described themselves as beer drinkers increased greatly with higher blood levels. Among those not found to have been drinking at the time of the test, the ratio was 1.5 beer drinkers to one spirits drinker; among those with blood alcohol levels of .08 percent or higher, the ratio was 4:1.

These results certainly do not provide grounds for complacency among those who contend that beer is a harmless drink of moderation. This is not to say, however, that the study is entirely conclusive. Although the likelihood of a discrepancy in most cases seems small, the possibility cannot be altogether discounted that a driver's beverage of choice was not the beverage that produced the blood alcohol level detected. Also, it may be argued that the reports of beverage preferences were influenced by a belief that beer is less impairing than spirits. An accident driver, in particular, might feel that reporting beer rather than spirits consumption would be less damaging to his position. The investigators made every reasonable effort to divorce themselves from the activities of the police, and to reassure drivers that information given would have no bearing on official investigations. The drivers were questioned about their drinking habits in general, rather than about their consumption immediately preceding the blood alcohol test. Nevertheless, these shortcomings, together with the potential importance of Borkenstein's results, indicate the desirability of further research focused specifically on the contribution of the different types of beverage to impaired driving.

Turning to the evidence of chronic effects, crossnational epidemiological studies have failed to find convincing evidence to indict any one class of beverage over another (de Lint and Bronetto, 1966; de Lint and Schmidt, 1971). With respect to distilled beverages, when total apparent alcohol consumption was held constant—by pairing a number of countries with nearly identical sales rates—the percent contribution of spirits to the total showed no relationship to variations in liver cirrhosis mortality (Popham et al., 1976). It is noteworthy that for the United States, Schmidt and Bronetto (1962) showed that the best predictor of the liver cirrhosis death rate was per capita wine sales. The reason appeared to be that a larger proportion of all wine sold than of spirits or beer was consumed by alcoholics. The most likely explanation for the attraction of wine to the latter was its relatively low cost. This interpretation was supported by the results of a direct study of alcohol buying in Ontario (de Lint, 1962; de Lint et al., 1967), where certain fortified wines also constituted the cheapest source of beverage alcohol (on a cost per ounce of pure alcohol basis).* The vast majority of consumers of such wines were found to be chronic drunkenness offenders, Skid Row alcoholics, and other relatively impoverished pathological drinkers. A similar finding has recently been reported for Finland (Mäkelä, 1971). Inexpensive wines tended to be the favored beverage of heavy drinkers in the lower economic strata of the population.

So far as beer is concerned, examination of the clinical literature reveals that it has been implicated in alcoholism and its consequences about as frequently as spirits and wine. For example, in Australia (Wilkinson et al., 1969), Czechoslovakia (Skala, 1967), and parts of southern Germany (Ledermann, 1964; Lelbach, 1967), beer was reported to be the beverage of choice of a majority of alcoholics; and in clinical populations studied in Ontario, it was commonly the principal beverage used by patients (de Lint, 1964; Schmidt and Popham, 1965). There is also substantial evidence to indicate a causal connection between heavy beer consumption and various ailments, but particularly, liver cirrhosis and myocardial disease (e.g., Alexander, 1966; Anonymous, 1968; Frank et al., 1967; Lelbach, 1967; McDermott et al., 1966; Sjöberg, 1969). Wallgren (1970) and Lelbach (1974) have reviewed the experimental and clinical data on organic pathologies attributable to alcohol consumption, and agree in the conclusion that type of beverage is irrelevant; total alcohol consumed and the duration of heavy drinking are the significant variables. As to alcoholism, the conclusion would seem inescapable that a dependence can be sustained as well with beer or wine as with spirits. Indeed, the beverage preferences of alcoholics seem to reflect simply the preferences characteristic of the populations in which they live, with a distinct tendency

---

* In Ontario as of May, 1973, the cheapest alcohol obtainable through a distilled beverage cost 55 percent more than the cheapest in wine. The cost of alcohol in beer fell exactly midway between these two.

toward the cheaper sources of alcohol (Devrient and Lolli, 1962; Lolli *et al.*, 1958, 1960; Parreiras *et al.*, 1956; Sadoun and Lolli, 1962; Terry *et al.*, 1957).

In summary, the available evidence, pertaining to both acute and chronic effects, offers little to justify the view that beer is a comparatively harmless beverage of moderation while spirituous liquor is a comparatively harmful beverage of excess. At the same time, if the stomach is empty and there is determination to become as drunk as possible in as short a time as possible, a distilled beverage is to be preferred. For the same reason, the theoretical risks of severe alcoholic states such as delirium tremens or acute alcohol poisoning may be greater in the case of spirits. However, in practice, these risks have not been convincingly shown to materialize as significant social problems.

If a sound rationale for differential taxation does not reside in a differential liability to generate alcohol problems, it may yet exist in quite another direction. In his analysis of Canadian data, Lau (1973a; 1973b) found that, unlike wine or spirits, the demand for beer was largely price inelastic. Walsh and Walsh (1970) came to a similar conclusion in their study of consumption trends in the Republic of Ireland. Thus, with income held constant, it would seem that when the price of beer increases, the tendency is to maintain the same consumption, presumably at the expense of something else. Walsh and Walsh (1970) pointed out that, in the case of low income families, a sharp increase in the price of beer might lead to the neglect of necessities, nutritional or otherwise. The possibility of such an effect has to be weighed against the likelihood that differential control will ultimately result in the substitution of beer for the more costly sources of alcohol, and thereby negate the preventive effect sought. In this regard, the experience of Denmark and Belgium are of special interest.

Prior to the First World War in both countries, cheap spirituous liquors were readily available and widely favored. Levels of apparent consumption were high, and problems of insobriety and alcoholism seem to have been frequent. In Denmark an extremely heavy tax was levied on distilled beverages in 1918, rendering these largely inaccessible to much of the population. For example, the price of Danish vodka (*aquavit*) increased twelvefold between 1917 and 1918; the price of beer remained almost unchanged. In Belgium the Vandervelde Act of 1919 limited the quantity of spirits that could be purchased at any one time to a minimum of two liters. The cost of so large an amount drastically reduced the accessibility of the working classes to such beverages.

There can be little doubt that in both Denmark (Nielson and Strömgren, 1969) and Belgium (Reuss, 1959), these measures led initially to a sharp and very substantial decrease in the consumption of distilled spirits, resulting in a decline in the total volume of alcohol consumption. It is equally clear, however, that the lasting change was in the pattern rather than the level of consumption.

The decline in use of spirits was eventually replaced by a rise in the consumption of beer. In Denmark overall alcohol sales per drinker—based largely on beer—are now the highest in Scandinavia; Belgium came to be one of the leading beer-consuming nations of the world; and in both countries, a substantial prevalence of alcoholism was again achieved (Popham et al., 1976).

The uniform price control policy, discussed in the previous section, took into account the likelihood of substitution. Its objective, however, is far more modest—and in our view, realistic—than that of the drastic measures just described. Its essential aim is to prevent further increases in consumption rather than to reduce an existing level. Since the policy ties price to income, once implemented, the proportion of a drinker's budget required to maintain his consumption level would not change on the average. Under these circumstances, and provided initial price restructuring were gradual, the probability of generating a significant shift in the purchasing patterns of low income drinkers from essentials to beer would be at a minimum. If, on the other hand, the desire were to *reduce* the overall consumption level, then the possible consequences flowing from the price inelasticity of beer would need to be carefully weighed against the potential health benefits of increases in the relative price level.

THE MONOPOLY SYSTEM OF CONTROL

It has been widely believed that when the alcoholic beverage trade is in the hands of private enterprise, competition inevitably leads to practices that stimulate greater consumption and, consequently, a higher prevalence of alcohol problems. Proponents of this view have, therefore, favored a system whereby the state or its official agency maintains a monopoly of the whole or a significant part of the trade. One of the earliest examples of such a monopoly was the Gothenburg System initiated in Sweden in 1865 in the city from which its name derived. Under a special statute, a company was established that controlled all licensing of outlets for the sale of distilled beverages. Through this company, the municipal government in effect maintained a monopoly of the local retail trade, the net profits from which were paid over to the city treasurer (Marcus, 1946).

Essentially as an alternative to a growing sentiment in favor of Prohibition, the Gothenburg System was displaced in 1971 by the far more comprehensive Bratt System (named so for its inventor). This was a statewide monopoly of the trade in all alcoholic beverages except beer. A feature of the system, which occasioned worldwide interest among Temperance workers, was the attempt to control individual consumption through rationing. To purchase distilled beverages it was necessary to possess a special permit or pass-book. The amount allowed an individual per month (up to a maximum set by law) was specified in the permit and determined by the issuing authority. The decision of the latter was based on an evaluation of the applicant's needs and habits; such characteristics were taken into account as age, drinking history, sex, and marital status. Women, for example, almost invariably were allowed less than men, and if married were not granted a permit. A record of alcoholism or arrests for drunkenness were other reasons for rejection. A permit was not required for wine but, in theory at least, amounts purchased by each individual were recorded, and if these exceeded limits judged to be reasonable, further sales would be curtailed or refused. In later years, with mounting evidence of circumventions and shortcomings, the effectiveness of the system was increasingly called into question. It was finally abolished in 1955 and replaced by a much less restrictive form of monopoly control (Elmer, 1957).

An attempt was made to assess the effect of this change by von Euler but, unfortunately, support was withdrawn before the study could be completed (Boalt and von Euler, 1959). His preliminary data suggested a transitory effect on consumption level and alcohol-related offenses. There was a rapid initial increase in total alcohol sales per capita, mainly due to increased use of spirits by both sexes. The frequency of drunkenness arrests and impaired driving also peaked in 1955 and 1956. Thereafter the rates declined, and by 1958, apparent consumption was lower than in 1954. However, the conclusion that the more liberal controls prevailing after 1955 were as effective as the Bratt System seems unwarranted. One reason is that increases were evident for a few years prior to the change, and quite possibly in anticipation of it, reflected a decline in the vigor of enforcement. Another is that substantial tax increases were imposed in subsequent years. These would be expected to offset any upward trends that might otherwise have occurred.

There have been other instances of strict monopoly control systems, including the Carlisle Experiment, noted in a previous section; and in some cases, direct monitoring of individual consumption, combined with the exclusion from purchasing of identified problem drinkers has been attempted, more or less on the Swedish model. The permit and interdiction system, which prevailed in Ontario for many years after Prohibition, is an example. Usually these forms of control have been abandoned because they proved impractical in large urban populations, and gave rise to extensive public criticism and resentment. In only one case of which we are aware was an effort made to rigorously evaluate the effectiveness of a form of individual control comparable in its objectives to the Bratt System. This was Lanu's study of a buyer surveillance program launched in Finland shortly after the conclusion of World War II (Lanu, 1956).

As in Sweden, the instrument of control was a purchase permit. All holders were registered and their purchases of alcoholic beverages were recorded in this document. The State Alcohol Monopoly maintained a file of alcohol "misusers" comprising persons considered to have made abnormally large purchases over a period of time. In addition, the names of persons who had received treatment for alcoholism were included, as well as those reported by the police to have been found intoxicated or charged with alcohol-related offenses. Sanctions were then applied that ranged from warnings, through home visits and detailed interrogation, to suspension of purchasing privileges for periods up to a year. Lanu conducted intensive interviews with a large sample of "misusers" to determine whether or not there had been changes in behavior attributable to these sanctions. He compared the results with those for a control sample of permit holders—matched for age, sex, social class, and other relevant variables—who had never been listed as "misusers." He was unable to find evidence that favorable changes in drinking behavior had occurred more often among those exposed to sanctions, and concluded that any changes that were manifested could be attributed to other factors. Lanu also reported that only about a quarter of all persons arrested for drunkenness were holders of permits, and that, on the other hand, most "normal" purchasers viewed buyer surveillance as an unnecessary nuisance. These problems ultimately led to the abolition of the program by the State Monopoly.

Following the repeal of Prohibition in North American jurisdictions, all provinces of Canada and about a third of the states of the United States adopted a government monopoly system. In the remaining states, the alcoholic beverage trade was given over to private enterprise with control exercised through licensing. Jellinek (1947) analyzed trends within the states under each system for the period 1930 to 1945. He could find no evidence of an effect on "rate of inebriety," and with respect to trends in apparent consumption, noted that "the monopoly system did not prevent fairly large increases; nor did the license system lack small increases" (p. 16). The present authors examined differences between the two groups of states for the year 1964 (Popham et al., 1976). The rates of liver cirrhosis mortality, total alcohol sales, and sales by type of beverage were all slightly higher in the license group. However, only the difference for wine sales was statistically significant, and this was entirely due to the presence of California, the leading wine-producing area of the country. An index of urbanism also proved slightly larger for the license states, and it was suggested that this might account for the tendency toward higher rates in the group. Degree of urbanism is well known to be positively associated with variations in indices of alcohol use (Jellinek, 1947; Seeley, 1962), although its influence may depend partly on the higher income levels prevailing in more urban areas (Popham, 1956).

We must emphasize that to contrast monopoly and license states is not to compare "control" with "no control," nor even to compare areas where competition for private profit is largely absent with those where it is rampant. There are many important differences in the administration of the monopoly system among the states that have adopted it (Barker, 1957; Landis, 1948; McCarthy and Douglass, 1959; Simon, 1966b). In many cases, only the wholesale distribution and package retailing of distilled beverages are monopolized; domestic producers, beer and wine stores, and outlets for on-premise consumption are often privately owned. Among both license and monopoly states there are differences in the extent to which emphasis is placed on the revenue-producing rather than the problem-controlling functions of the system. (In recent years, the former seems more frequently to have become the dominant interest.) In short, a more meaningful assessment than any reported to date would disregard the official classification, and base a comparison on relevant differences in operation independently validated.

It certainly may be that where the monopoly is rather comprehensive, as in some Scandinavian countries, and the primary objective is control, alcohol problems are less prevalent than they otherwise would be (Christie, 1965). But then the question becomes whether or not the particular measures that prove effective could not be applied equally well under a licensing system. A state monopoly would seem to reduce both the number and the potential political influence of those with a vested economic interest in higher levels of alcohol consumption. This will depend, of course, on the extent to which the trade is, in fact, monopolized. It would also seem that the implementation and enforcement of certain types of control measure, and the monitoring of their effects are facilitated by a centralized system. However, purely from the researcher's standpoint—and in most jurisdictions very much in theory—the greatest benefits of a monopoly system lie in the possibilities for collaborative studies of the effects of different control measures, and in the greater likelihood that research results will be taken into account in the development of policy. That this is not an idle fantasy is best demonstrated by the nature and influence of Finnish work in the area (mentioned at several points in this review), all of which has been carried out under the aegis of the State Alcohol Monopoly.

MODELS OF PREVENTION

In the course of this review, attention has been drawn to the divergent beliefs that have prevailed with respect to the role of legal measures in the control of alcohol problems. Currently, in this regard, three schools of thought or "models of prevention" may be distinguished. These are as follows:

1. *The Bimodal Model.* The heavy alcohol consumption of problem drinkers and alcoholics is considered to be symptomatic of some disorder peculiar to them. Therefore, by implication, the distribution of consumption in a population will be bimodal in character, and factors that may cause a change in the consumption level of normal drinkers will have little or no effect on that of pathological drinkers. It is stressed that the problem lies in drunkenness, not drinking, and in alcoholism, not alcohol; and that since the causes of the problem are unknown, no method of primary prevention exists. Accordingly, legal measures intended to restrain overall consumption are held to be misdirected, ineffective, and unjustified. Room (1971) has noted that this essentially null hypothesis is favored by spokesmen for the distilled beverage industry, who may feel that they cannot afford skepticism on the point. However, as Mäkelä (1972) pointed out, the view has not been confined to those with a vested economic interest in it. Such scholars as Bales (1946) and Ullman (1958) have also denied that the level of alcohol consumption in a group has anything to do with the rate of alcoholism or insobriety. Indeed, "bimodal thinking" is widely encountered among students of alcohol problems, and in our view, is largely attributable to the influence of the "disease concept" of alcoholism (Jellinek, 1960a; Popham et al., 1976).

2. *The Integration Model.* It is believed that to eradicate or even substantially reduce the prevalence of alcoholism would require fundamental changes in the social and mental health status of the population. However, it is also believed that more problems result from the mysticism associated with alcohol, and the guilt attendant upon its use than would occur if drinking were an integrated part of everyday life. In this respect, the model is closely akin to the older "forbidden fruit" concept. Thus, it is argued that young people should be introduced to alcoholic beverages at an early age so that they may learn to drink moderately, and come to regard the activity as of no greater significance than eating. Restrictive control measures are seen both as reinforcers of an unhealthy ambivalence toward drinking, and as impediments to the adoption of healthy drinking styles. With reference to the latter, the practices of certain European countries, for example, the use of wine with meals in France and Italy, are held to be conducive to moderation. It is felt that such customs should be promoted through education, and where required, through appropriate relaxation of legal restrictions. The model is especially popular among North American students, and received its most articulate expression in the recommendations of the Cooperative Commission on the Study of Alcoholism (Wilkinson, 1970). Parenthetically, Mäkelä (1972) has made the astute observation that its proponents themselves reflect the ambivalence toward drinking that they seek to combat. On the one hand, they tend to uphold the middle-class ideal of moderation, and decry the uncontrolled drinking traditionally associated with the working classes. But at the same time, the puri-

tanism of the Temperance movement is eschewed, and the restrictive system of control that is advocated, opposed.

3. *The Single Distribution Model.* Proponents argue that the frequency distribution of alcohol consumption per drinker in any given population is continuous, unimodal, and positively skewed. It is also contended that the same theoretical curve (of lognormal type) adequately describes the distribution in quite different populations: The variance is constant and only the mean differs from one group to another. Under these circumstances, the relative frequency of high level consumers (e.g., those consuming at or above the level at which most alcoholics drink) depends upon the mean per-drinker consumption in a population, and factors that alter the latter may be expected to alter the former. Since the same distribution is found to obtain in populations that differ greatly in attitudes toward drinking, beverage preferences, drinking customs, and in the educational and legal measures employed to combat the problems of alcohol, it is concluded that there is as yet no way to modify the prevalence of heavy consumers without altering the average consumption of other drinkers. Regional and temporal variations in mean consumption are believed to be caused mainly by differences in the economic accessibility of beverage alcohol and in the level of acceptance of drinking. It is argued that relaxation of legal restrictions to encourage the adoption of new, allegedly healthier drinking styles, serves only to reinforce higher levels of acceptance, and to add to, rather than replace existing drinking habits. It is felt that upward trends in the prevalence of alcohol problems could be prevented through appropriate taxation measures, restraint on the liberalization of existing restrictions, and vigorous education to raise awareness of the social health consequences of a rising consumption level. Recently, this model has been attributed to the present authors (Mäkelä, 1972; Room, 1971), who have certainly propounded it (see, for example, de Lint and Schmidt, 1968, 1971; Popham *et al.,* 1976). It has in common with older Temperance thinking the emphasis on the control of insobriety and alcoholism through the control of overall consumption. However, its roots do not lie in moral or religious preconceptions but in the empirical work of Ledermann (1956, 1964) and Jellinek (1947, 1960a, 1960b), especially the former. It is beyond the scope of this review to attempt an examination of all the arguments that have been, or might be advanced for and against each of these models. Nor can we review such recent variants of one or another as that proposed by Whitehead (1972), or implied by the recommendations of the Study Committee of the Canadian Brewers Association (Alcoholic Beverage Study Committee, 1973). There are, however, three central issues that deserve special attention: (1) the relation between problems of intoxication and problems of alcoholism; (2) the distribution of alcohol consumption and the concept of alcoholism; and (3) the effect of moderating legal restrictions to promote new drinking habits.

Room (1971) considered a principal difference between the integration and single distribution models to be one of focus: Adherents to the former emphasize the "social disruptions of explosive but often intermittent drinking associated with 'dry' environments"; the latter emphasize "the long-term medical complications of chronic excessive drinking, associated with 'wet' environments" (p. 21). While there is justification for this characterization, it is only one aspect of the issue. In fact, much of the research underlying the bimodal and integration models either did not have available at the time, or failed to take account of adequate epidemiological data. One result was the fallacious assumption that the occurrence of highly visible intoxication on many or most drinking occasions reflected a high prevalence of insobriety. Conversely, the apparent rarity of drunkenness in an atmosphere of frequent drinking was taken as evidence of moderate overall use and a low prevalence of alcohol problems.

The study of Irish drinking is a case in point. The prevalence of insobriety—both acute and chronic—is almost certainly quite low on an international scale (Blaney, 1967; Lynn and Hampson, 1970). But attention has been diverted from the broad picture by the observation that gross intoxication is often the result on a drinking occasion (Bales, 1946), and may have deleterious consequences (Walsh and Walsh, 1973). On the other hand, the frequency of drinking occasions is very low compared, for example, with France and Italy. In these countries, where the use of alcohol is said to exemplify "civilized drinking," frequent use throughout the day is the norm, and drinking is a fully accepted part of most activities. There, not only is the rate of alcoholism high—in the case of France, the highest known—but also the rate of "acute" problems such as alcohol-related industrial and traffic accidents (Bonfiglio, 1963; Ledermann, 1956, 1964).

Also revealing in this connection are the results of a study by Mäkelä (1971) in which the pattern and amount of consumption of middle-class and working-class drinkers were compared. He found, in accord with the common assumption, that the frequency of drinking occasions was greater, and the average intake per occasion smaller in the middle class than in the working-class group. However, he also found that intoxicating quantities were ingested by both groups with equal frequency, and concluded that the "temperate ritual drinking of the middle class does not supersede the bouts [of the] working class but only enters the picture as an additional feature" (p. 16).

Our intention is not to minimize the importance of problems of intoxication—or, as Jellinek (1960a) put it, of "occasional excess"—as against the health consequences of prolonged heavy use. Certainly the former can give rise to great social concern. The Finnish pattern of concentrated consumption associated with violent behavior is a striking instance (Kuusi, 1948, 1957; Sariola, 1956). We wish simply to stress that a high prevalence of "chronic" alcohol problems is very unlikely to occur in the absence of a high prevalence of "acute" problems; nor, for that matter, is the reverse likely to be found. The two types of problem overlap to a considerable degree. For example, Mäkelä (1971) found that 56 percent of all consumption for the purpose of becoming intoxicated was attributable to the 10 percent of consumers (comprising most of the alcoholics) at the highest per capita consumption level. And in Ontario, Schmidt *et al.* (1962) estimated that at least 28 percent of alcohol-impaired drivers in accidents were alcoholics in the clinical sense of the term.

It is true that the character of intoxicated behavior may differ from one culture to another, and probably also the degree of intoxication typically achieved (Jellinek, 1960a). But in all countries, alcoholism involves a long history of frequent impairment, and this inevitably means frequent problems of personal incapacity and social disruption (Jellinek, 1960b). Therefore, it is all but inconceivable that measures that successfully controlled the prevalence of alcoholism would fail to exert a controlling influence on the frequency of intoxication and attendant problems.

At the same time, a difference in the *number* of alcoholics should not be confused with a difference in the *character* of the problem. Where per capita consumption and alcoholism prevalence are low—say, as a consequence of restrictive measures, a substantial "dry" sentiment, or both—alcohol problems are apt to be seen as both numerous and serious. This is partly an observational artifact. To put it bluntly, in a sober world the drunk obtrudes himself upon the attention. But it is also partly due to a difference in the character or composition of the alcoholic population. In relatively low consumption countries, or among subcultural groups having strong taboos against drinking, the proportion of extremely alienated, socially unstable, and psychiatrically very ill persons among those who do become alcoholics tends to be greater than in the alcoholic population of high consumption, high prevalence groups (Jellinek, 1960a; Popham, 1959). The confusion of high social and psychological vulnerability with a high prevalence is evident in the stress placed by the integration theorists on observations of alcoholism in such strongly disapproving groups as Ascetic Protestants (Mäkelä, 1972). On the other hand, prevalence is erroneously assumed to be low in populations where drinking is so commonplace that the visibility of the highly deteriorated alcoholic is obscured.

Another error often encountered in arguments favoring a more liberal approach to the control of drinking has been the confusion of extent or frequency of use with volume of intake. For example, Mäkelä (1972) drew attention to the influence that studies of Jewish and other ethnic groups have had on the thinking of the integration school. Thus, much is made of Jewish sobriety as evidence that, despite extensive drinking, extremely few alcohol problems occur. And Room (1971), in his critical commentary on the single distribution model, doubted its validity because there were groups in the United States with a high proportion of weekly users but a low proportion of heavy users, and other groups in which just the reverse was the case. However, the critical variable is not extent of use—that is, the proportion who report that they are moderate or frequent users rather than abstainers—but the average intake of alcohol per user on an annual basis. The latter may be quite low, as it undoubtedly is in most American Jewish groups, even though nearly everyone takes a drink on certain occasions. Then, from the standpoint of the single distribution model, a small proportion of heavy users would be predicted.

This leads to the second major issue noted above, namely the relation between average consumption and problem consumption. It might be argued that the existence of a connection between the two was established by the evidence reviewed in the section on price control. However, this evidence is not entirely convincing, partly because of the interpretive difficulties inherent in correlational data, and partly because of heavy dependence on an index of alcoholism (liver cirrhosis mortality) that theoretically might be subject to independent sources of error. So far as we are aware, Ledermann (1956) was the first to determine directly the statistical properties of the distribution of alcohol consumption. Employing mainly documentary data of a reliable character, he found the empirical distributions for Finland, France, Sweden, and one sample in the United States to be lognormal in form. He then developed a single theoretical curve with constant variance that adequately described the data for these areas.

Following a prodigious amount of work with about a quarter of a million alcoholic-beverage sales slips, containing the buyers' names and addresses, de Lint and Schmidt (1968) demonstrated that the Ledermann curve fully described the distribution of purchasing in Ontario. Smart and Schmidt (1970) found the distribution of blood alcohol levels in a large sample of nonaccident drivers in Michigan to conform reasonably well to the curve. Its application to Finnish data was reaffirmed by Mäkelä (1971), and Skog (1971) considered it to give a satisfactory description of Norwegian consumption data. There has been some debate, on mathematical grounds, as to the precision of fit, and the conclusions that it is theoretically permissible to draw from the curve (Ekholm, 1972; O'Neill and Wells, 1971; Skog, 1971). This has been partly due to a misunderstanding of the particular equation involved. In any case, there does not appear to be any doubt as to the essential characteristics of the distribution: continuity, unimodality, and marked positive skewness.

In view of the similarity of the distribution in a variety of jurisdictions, and other considerations that have been mentioned, the present authors concluded that "for practical purposes, the essential character of the distribution is unalterable and the prevalence of consumption by those we label "alcoholic" is inextricably linked to general consumption" (Popham et al., 1976, p. 14). Confidence in this conclusion was further increased by the finding that estimates of alcoholism prevalence, derived through an application of the curve, were in close agreement with those based on mortality data and a case-finding survey (Schmidt and de Lint, 1970). For the purpose, alcoholics were defined as those persons whose average intake on an annual basis was the equivalent of 15 centiliters of pure alcohol daily or more. This level of consumption was chosen on the basis of the reported intake of clinical alcoholics in Ontario (Schmidt and Popham, 1965), France (Péquignot, 1958), West Germany (Lelbach, 1966), and Australia (Wilkinson et al., 1969). It is important to emphasize that no justification for the selection of a particular level is provided by the distribution itself; statistically, the selection is arbitrary. However, once chosen, the proportion of persons who consume at or above the level depends only on the mean per-drinker consumption of the population.

The question now arises as to the relation between these results and the "disease concept" of alcoholism. Few experienced clinicians would doubt that most persons now labeled "alcoholic" are ill, certainly by virtue of the consequences of long-term heavy consumption, and often also because of preexisting physical, mental, or social problems. But there is no compelling evidence of a *unique* predisposing factor, or an irreversible change due to chronic intake, which renders the individual permanently incapable of controlling his alcohol consumption, and insensitive to environmental contingencies. The search for specific etiological factors has been eminently unsuccessful in both the psychological (Lisansky, 1960) and the physiological (Lester, 1966) areas. "Oddities" there are (Keller, 1972), but the many ways in which alcoholics are found to differ from nonalcoholics are either of no etiological significance, or are nonspecific; that is, found also in other deviant and psychiatric populations. In short, as Jellinek (1960b) once remarked, about all that alcoholics have in common are drinking and damage.

On the other side, the finding that alcohol consumption is distributed unimodally is consistent with the view that the alcoholic suffers from a nonspecific behavioral disorder. This, in turn, is in accord with the results of recent experimental research that suggests that the reversibility of physical dependence on alcohol lies in the realm of learning (Gibbins et al., 1971; Kalant et al., 1971; LeBlanc et al., 1969, 1973). Indeed, Bacon (1973) from a sociologist's standpoint, and Keehn (1969a; 1969b) from a psychologist's, have found it possible to formulate persuasive models of alcoholism without appeal to underlying factors in either the physical or psychic constitution. The notion of permanent loss of control has been challenged on clinical grounds (Davies, 1962; Pattison et al., 1968); and the response of alcoholics to newer treatment modalities that are entirely concerned with the manipulation of environmental reinforcers has been by no means discouraging (e.g., Hunt and Azrin, 1973).

So far as the single distribution model is concerned, the position has been very well summarized by Edwards (1971):

> The reasons why a person drinks abnormally are connected both with his personality and with his environment; his drinking will in fact result from an interaction of the two, so that, for example, an anxious person living where alcohol is cheap and attitudes to drinking are permissive will be more likely to become an alcoholic than an anxious person who finds alcohol difficult to obtain and attitudes disapproving. (p. 424)

It remains to add that the principal aim of a preventive measure is to affect the incidence (new cases) rather than the prevalence of a condition. Schmidt and de Lint (1969, 1972) showed that the death rate of clinical alcoholics from all causes was more than double that to be expected in the absence of alcoholism. Therefore, natural attrition alone would rapidly diminish prevalence if the inflow of new cases were reduced. It is with respect to the latter that appropriate control measures might be expected to have their most significant effects.

Room (1971) pointed out that a weakness of the single distribution model was its dependence on observations of regional variation. He rightly contended that direct evidence was required to demonstrate that a change in a control measure, which altered the mean consumption of a population, in fact caused a shift in the whole distribution. Ekholm (1972) similarly concluded that the relevant distribution should be shown to give an accurate description of consumption data both before and after such a change. The importance of future research in this direction cannot be denied. However, the task is formidable, and rendered more so by the disappearance from most jurisdictions of purchase recording systems that require the names and addresses of purchasers. The best hope seems to lie in the improvement of verbal survey techniques to the point where much more detailed and accurate consumption data are secured than has been usual in the past (Pernanen, 1974).

In any event, Mäkelä's (1971, 1972) study of the effects of a change in control policy in Finland, though not a mathematically ideal test, provides strong support for the theory. His findings also have an important bearing on the third issue mentioned at the outset: the consequences of altering legal restrictions to foster new drinking habits.

In the section on control of outlet frequency, reference was made to the relaxation of restrictions on the sale of medium beer in Finland. Mäkelä examined the pattern and amount of alcohol consumption before and after this change. The new policy was introduced by the State Monopoly with the clear expectation that it would encourage moderate drinking. However, Mäkelä found the effect to be additive rather than substitutive.* Sales of beer increased

---

* The only substitutive effect noted by Mäkelä (1972) related to the older price policy of the State Monopoly, which favored wine; also in the belief that moderate use would result. The principal effect was a shift in the beverage preference of price-conscious heavy consumers to the cheaper source of intoxication thereby provided.

substantially, and accounted for a large part of the 48 percent increase in total alcohol sales during the first year. But subsequently, the sales of stronger beverages showed increases at the same rates as before the change. Nor did the overall increase simply reflect a more equitable distribution of consumption: "All the consumer groups increased their ingestion of alcohol roughly in relation to their previous consumption level" (Mäkelä, 1972, p. 14). This inevitably meant an increase in the number of persons drinking at hazardous levels (for example, at or above the 15 cl level previously discussed). Mäkelä further showed that the increase in total consumption was by no means solely attributable to new drinking occasions involving moderate intake (such as drinking a little beer with meals in addition to the traditional weekly binge). The frequency of heavy drinking occasions also increased after the change.

To summarize, the evidence may have to be classed as presumptive, but it nonetheless heavily favors Mäkelä's (1972, p. 21) conclusion over the largely unsubstantiated positions on the bimodal and integration schools: "Liberalizing the system leads to an increase in consumption, and an increase in consumption adds to the complications." At the same time, complete rejection of the integration model on this basis is unwarranted. The stand on liberalization of control systems derives from a much broader, long-range goal that envisions fundamental changes in attitudes and behavior. Should the latter come about, the harmful effects currently observed might not occur. However, some writers have expressed understandable skepticism as to the likelihood of early achievement (Bättig, 1967; Room, 1971). The goal is not a little reminiscent of the traditional aim of preventive medicine: to rid the world of disease. No doubt utopic ends are worthy of pursuit, but it is questionable whether or not this is a sound justification in the interim to promote measures detrimental to public health, or to deny those of potential value. As Edwards (1971, p. 424) succinctly put it:

> Since we are not able to manipulate personality and produce a race with no neurosis, the only realistic method of exerting a benign influence on the prevalence of alcohol addiction is by control of the environmental conditions of drinking, and it is the availability element that remains the prime candidate for control.

There would appear to be agreement among several students of law and drinking behavior on at least three points: (1) Highly restrictive controls on accessibility lead to lower consumption and fewer alcohol problems; (2) such controls are unlikely to be implemented in the absence of substantial public support; and (3) such controls are apt to involve costs that eventually will be perceived to outweigh their benefits (Bruun, 1970; Christie, 1965; Lemert, 1962; Mäkelä, 1972; Room, 1971; Shadwell, 1923). These costs—depending on the nature and degree of suppression—include, for example, resentment of the system by nonproblem drinkers; increased use of toxic substitutes by at least some determined heavy users; the rise of an illicit trade with attendant criminality; the expense and difficulties of enforcement; loss of revenue by producers, sellers, and government; loss of respect for the law through widespread evasions; and loss or substantial reduction of personal pleasures and social benefits derived from alcohol use.

It is regrettable that the contemplation of these and other consequences of drastic measures seems to lead to a rejection of legal restraint in any form. At present, in North America and elsewhere the pendulum is evidently swinging to the other extreme. This may well carry equally high costs, although differently derived. Thus, in France, which probably has the fewest restrictions on accessibility, and certainly has the highest known consumption level of any country, it is estimated that over 40 percent of the total health bill is attributable to the treatment of alcohol-related diseases, and about 50 percent of all hospital beds are occupied by patients suffering from these conditions (Brésard, 1969). This is to say nothing of other types of cost inevitably associated with the frequent impairment of some 9 percent of the adult population (Ledermann, 1956, 1964; de Lint and Schmidt, 1971).

Unrealistic expectations in the past of the ability of the law to solve alcohol problems are doubtless partly responsible for present loss of confidence in its preventive value. The failure (i.e., adverse consequences) of Prohibition and other highly restrictive legislation is still often cited to buttress arguments against legal measures with far more modest aims. We would hope that this review has indicated the potential importance of legal restraint, and hence, of research to increase our understanding of the effects of diverse control measures. In this way, a sound basis ultimately can be provided for the development of policies that maximize the preventive potentiality of the law and minimize its costs.

As to current models of prevention, the available evidence supports the principal tenets of the single distribution model, contradicts those of the bimodal, and casts serious doubt on the contention of the integration theorists that problem drinking can be reduced through the relaxation of restrictions on accessibility. The single distribution model seems to hold most promise as a means to prevent upward trends in jurisdictions still substantially removed from the French level of apparent saturation. The model does not involve drastic measures, and its aim is correspondingly modest. The main impediment to its application in many jurisdictions may be, as Archibald (1971) recently suggested, that the moderate consumer quite understandably does not see himself as in any way involved in alcohol-related health problems. Presumably, therefore, to render full implementation politically feasible would require vigorous education to alter awareness of the health consequences of rising consumption, and indicate the need for suitable legal controls as protective measures. Given the importance now attached to matters of public health in the Western world, there is no reason to think this an unrealistic objective.

ACKNOWLEDGMENTS

We are very grateful to Oriana Kalant and Robin Room for their critical reading of the manuscript, and attempts to protect us from bias. If they did not entirely succeed in the latter, the fault is certainly ours.

REFERENCES

Ahlström-Laakso, S., 1971, Arrests for drunkenness—two capital cities compared, *Scandinavian Studies in Criminology* 3:89–105.
Alcoholic Beverage Study Committee, 1973, "Beer, Wine and Spirits: Beverage Differences and Public Policy in Canada," Brewers Association of Canada, Ottawa.
Alexander, C. S., 1966, Idiopathic heart disease. I. Analysis of 100 cases with special reference to chronic alcoholism, *Amer. J. Med.* 41:213–234.
Amundsen, A., 1965, Hva skjer når et nytt vinutsalg åpnes? Statens Institutt for Alkoholforskning, Oslo [Ms. on file].
Anonymous, 1968, Heart disease and beer drinking, *Pub. Hlth. Reports* 83:998.
Archibald, H. D., 1971, Alcohol and drugs: Government responsibility, in 29th International Congress on Alcoholism and Drug Dependence (L. Kiloh and D. S. Bell, eds.), pp. 238–256, Butterworths, Melbourne.
Armstrong, J. D., 1958, The search for the alcoholic personality, in "Understanding Alcoholism" (S. D. Bacon, ed.), Vol. 315, pp. 40–47, Annals of the American Academy of Political and Social Science, Philadelphia.
Asbury, H., 1950, "The Great Illusion: An Informal History of Prohibition," Doubleday, New York.
Askwith, G. R., 1928, "British Taverns: Their History and Laws," Routledge and Kegan Paul, London.
Bacon, S. D., 1973, The process of addiction to alcohol, *Quart. J. Stud. Alc.* 34:1–27.
Bales, R. F., 1946, Cultural differences in rates of alcoholism, *Quart. J. Stud. Alc.* 6:480–499.
Bättig, K., 1967, Alkoholismus: Epidemiologische Zusammenhänge und Folgen, *Natur. W. Rdsch.* 20:200–204.
Barker, T. W., 1957, The States in the liquor business, *Quart. J. Stud. Alc.* 18:492–502.
Bigelow, G., and Liebson, I., 1972, Cost factors controlling alcoholic drinking, *The Psychological Record* 22:305–314.
Blaney, R., 1967, The prevalence of alcoholism in Northern Ireland, *Ulster Med. J.* 36:33–43.
Boalt, G., and von Euler, R., 1959, "Alkoholproblem," Bohusläns Grafiska, Uddevalla.
Bonfiglio, G., 1963, Alcoholism in Italy, *Brit. J. Addictions* 59:3–10.
Borkenstein, R. F., Crowther, R. F., Shumate, R. P., Ziel, W. B., and Zylman, R., 1964, The role of the drinking driver in traffic accidents, report of the Department of Police Administration, Indiana University, Bloomington, Indiana.
Brésard, M., 1969, Alcoolisme et conscience collective, *Revue d'Alcoolisme* 15(2):81–96.

Bronetto, J., 1960-63, Alcohol Price, Alcohol Consumption and Death by Liver Cirrhosis, Substudies 1-8-60 to 1.7-8-63, Addiction Research Foundation, Toronto.

Bruun, K., 1959, Drinking Behaviour in Small Groups, Finnish Foundation for Alcohol Studies, Publication No. 9, Helsinki.

Bruun, K., 1970, Legislation and alcoholism, in "Alcohol and Alcoholism" (R. E. Popham, ed.), pp. 356-359, U. of Toronto Press, Toronto.

Bruun, K., and Hauge, R., 1963, Drinking Habits among Northern Youth: A Cross-National Study of Male Teenage Drinking in the Northern Capitals, Finnish Foundation for Alcohol Studies, Publication No. 12, Helsinki.

Bruun, K., Koura, E., Popham, R. E., and Seeley, J. R., 1960, Liver Cirrhosis Mortality as a Means to Measure the Prevalence of Alcoholism, The Finnish Foundation for Alcohol Studies, Publication No. 8, Part 2, Helsinki.

Bryant, C. W., 1954, Effects of sale of liquor by the drink in the State of Washington, *Quart. J. Stud. Alc.* 15:320-324.

Calkins, R., 1901, "Substitutes for the Saloon," Houghton Mifflin, Boston.

Canada Facts, 1946, Report of Results of a Study of Attitudes of the Ontario Public Towards the Distribution of Alcoholic Beverages, Canada Facts Ltd., Toronto.

Catlin, G., 1931, "Liquor Control," Holt, Rinehart and Winston, New York.

Cavan, S., 1966, "Liquor License: An Ethnography of Bar Behavior," Aldine, Chicago.

Christie, N., 1965, Scandinavian experience in legislation and control, in National Conference on Legal Issues in Alcoholism and Alcohol Usage, pp. 101-122, Boston U. Law-Medicine Institute, Boston.

Davies, D. L., 1962, Normal drinking in recovered alcohol addicts. *Quart. J. Stud. Alc.* 23:94-104.

de Lint, J., 1962, Pathological Wine Consumption in Toronto: A Study of Its Definition for Sociological Analysis, M. A. thesis, U. of Toronto (Sociology), Toronto.

de Lint, J., 1964, The Distribution of Alcohol Consumption in Male and Female Alcoholic Samples, Addiction Research Foundation, Substudy 16-10-64, Toronto.

de Lint, J., and Bronetto, J., 1966, Konsumtion av destillerade alkoholdrycker och Dödligheten i levercirrhos, *Alkoholpolitik* 29:62-64.

de Lint, J., and Schmidt, W., 1966, Unpublished Data from a Study of Alcohol Buying in Ontario, Addiction Research Foundation, Toronto.

de Lint, J., and Schmidt, W., 1968, The distribution of alcohol consumption in Ontario, *Quart. J. Stud. Alc.* 29:968-973.

de Lint, J., and Schmidt, W., 1971, The epidemiology of alcoholism, in "Biological Basis of Alcoholism" (Y. Israel and J. Mardones, eds.), pp. 423-442, Wiley, New York.

de Lint, J., Schmidt, W., and Jorge, F., 1967, Statistics of Alcohol Buying in Toronto, Addiction Research Foundation, Toronto (Ms. on file).

Devrient, P., and Lolli, G., 1962, Choice of alcoholic beverage among 240 alcoholics in Switzerland, *Quart. J. Stud. Alc.* 23:459-467.

Dewar, R., and Sommer, R., 1962, The Consumption of Alcohol in a Saskatchewan Community Before and After the Opening of a New Liquor Outlet, Bureau on Alcoholism, Regina [Mimeographed].

Edwards, G., 1971, Public health implications of liquor control, *The Lancet* (August 21, 1971) pp. 424-425.

Ekholm, A., 1972, The lognormal distribution of blood alcohol concentrations in drivers, *Quart. J. Stud. Alc.* 33:508-512.

Elmer, A., 1957, The change of temperance policy in Sweden, *Brit. J. Addictions* 54:55-58.

Entine, A. D., 1963, The Relationship between the Number of Sales Outlets and the Consumption of Alcoholic Beverages in New York and Other States, New York State Moreland Commission of the Alcoholic Beverage Control Law, Study Paper No. 2, Albany.

Expert Committee on Mental Health, 1951, Report on the First Session of the Alcoholism Subcommittee, World Health Organization Technical Report, Series, No. 42, Geneva.

Feldman, H., 1930, "Prohibition: Its Economic and Industrial Aspects," Appleton, New York.

Frank, H., Heil, W., and Leodolter, I., 1967, Leber und Bierkonsum; Vergleichende Untersuchungen an 450 Arbeitern, *Munch. med. Wschr.* 109:892–897.

Gadourek, I., 1963, "Riskante gewoonten en zorg voor eigen welzijn," Wolters, Groningen.

Gibbins, R. J., Kalant, H., LeBlanc, A. E., and Clark, J. W., 1971, The effects of chronic administration of ethanol on startle thresholds in rats, *Psychopharmacologia* 19:95–104.

Harper, R. F., 1904, "The Code of Hammurabi, King of Babylon," U. of Chicago Press, Chicago.

Harvola, V., 1956, [English summary of lead article], *Alkoholpolitik* 19:26.

Hunt, G. M., and Azrin, N. H., 1973, A community-reinforcement approach to alcoholism, *Behav. Res. and Therapy* 11:91–104.

Isaksson, D., 1957, Förslag till progressivt Beskattningssystem för alkohol-haltiga Drycker, baserat på Dryckernas relativa Intoxikationseffekt, *Alkoholfrågan* 51:72–80.

Jellinek, E. M., 1947, Recent trends in alcoholism and in alcohol consumption, *Quart. J. Stud. Alc.* 8:1–42.

Jellinek, E. M., 1960a "The Disease Concept of Alcoholism," Hillhouse, New Haven.

Jellinek, E. M., 1960b, Alcoholism, a genus and some of its species, *Can. Med. Assoc. J.* 83:1341–1345.

Jellinek, E. M., 1963, Government Programs on Alcoholism; A Review of the Activities in Some Foreign Countries, Dept. National Health and Welfare, Mental Health Division Report Series Memo No. 6, Ottawa.

Jolliffe, N., and Jellinek, E. M., 1941, Vitamin deficiencies in alcoholism. Part VII. Cirrhosis of the liver *Quart. J. Stud. Alc.* 2:544–583.

Kalant, H., and Kalant, O. J., 1971, "Drugs, Society and Personal Choice," General Publishing, Toronto.

Kalant, H., LeBlanc, A. E., Wilson, A., and Homalis, S., 1975, Sensorymotor and physiological effects of various alcoholic beverages, *Can. Med. Assoc. J.* 112:953–958.

Kalant, H., LeBlanc, A. E., and Gibbins, R. J., 1971, Tolerance to, and dependence on, some non-opiate psychotropic drugs, *Pharmacol. Reviews* 23:135–191.

Keehn, J. D., 1969a, Psychological paradigms of dependence, *Internat. J. Addictions* 4:499–506.

Keehn, J. D., 1969b, Translating behavioral research into practical terms for alcoholism, *The Canadian Psychologist* 10:438–446.

Keller, M., 1972, The oddities of alcoholics, *Quart. J. Stud. Alc.* 33:1147–1148.

King, F. A., 1947, "Beer Has a History," Hutchinson's London.

Krout, J. A., 1925, "The Origins of Prohibition," Knopf, New York.

Kuusi, P., 1948, "Suomen Viinapulma: Gallup-Tutkimuksen Valossa," Otava, Helsinki.

Kuusi, P., 1957, Alcohol Sales Experiment in Rural Finland, The Finnish Foundation for Alcohol Studies, Publication No. 3, Helsinki.

Landis, B. Y., 1948, Economic aspects of state alcoholic beverage monopoly enterprises, 1937–1946, *Quart. J. Stud. Alc.* 9:259–269.

Lanu, K. E., 1956, Control of Deviating Drinking Behavior: An Experimental Study of the Effect of Formal Control over Drinking Behavior, The Finnish Foundation for Alcohol Studies, Publication No. 2, Helsinki.

Lau, H-H., 1973a, Time Series Regression Analysis of Per Adult Consumption of Alcoholic Beverages I. Canada 1949–1969, Addiction Research Foundation, Substudy No. 542, Toronto.

Lau, H-H., 1973b, Time Series Regression Analysis of Per Adult Consumption of Alcoholic Beverages II. Eight Canadian Provinces 1933–1969, Addiction Research Foundation, Substudy No. 543, Toronto.

Lau, H-H., 1975, Cost of alcoholic beverages as a determinant of alcohol consumption, in "Research Advances in Alcohol and Drug Problems" (R. J. Gibbins, Y. Israel, H. Kalant, R. E. Popham, W. Schmidt, and R. G. Smart, eds.), Vol. 2, Wiley, New York.

LeBlanc, A. E., Kalant, H., Gibbins, R. J., and Berman, N. D., 1969, Acquisition and loss of tolerance to ethanol in the rat, *J. Pharmacol. Exp. Therap.* 168:244–250.

LeBlanc, A. E., Gibbins, R. J., and Kalant, H., 1973, Behavioral augmentation of tolerance to ethanol in the rat, *Psychopharmacologia* 30:117–122.

Ledermann, S. C., 1956, Alcool, Alcoolisme, Alcoolisation; Données Scientifiques de Caractère Physiologique, Économique et Social, Inst. Nat. Études Démog., Trav. et Doc., Cah. No. 29, Paris.

Ledermann, S., 1964, Alcool, Alcoolisme, Alcoolisation; Mortalité, Morbidité, Accidents du Travail, Inst. Nat. Études Démog., Trav. et Doc., Cah. No. 41, Paris.

Lee, A. M., 1944, Techniques of social reform: An analysis of the new prohibition drive, *Amer. Soc. Rev.* 9:65–77.

Lelbach, W. K., 1966, Leberschäden bei chronischen Alkoholismus, *Acta Hep.-Splenol.* 13:321–349.

Lelbach, W. K., 1967, Zur leberschädigenden Wirkung verschiedener Alkoholika, *Dtsch. med. Wschr.* 92:233–238.

Lelbach, W. K., 1974, Organic pathology related to volume and pattern of alcohol use, in "Research Advances in Alcohol and Drug Problems" (R. J. Gibbins, Y. Israel, H. Kalant, R. E. Popham, W. Schmidt, and R. G. Smart, eds.), Vol. 1, Wiley, New York.

Lemert, E. M., 1962, Alcohol, values, and social control, in "Society, Culture, and Drinking Patterns" (D. J. Pittman and C. R. Snyder, eds.), pp. 553–571, Wiley, New York.

Lester, D., 1966, Self-selection of alcohol by animals, human variation and the etiology of alcoholism: A critical review, *Quart. J. Stud. Alc.* 27:395–438.

Levy, H., 1951, "Drink: An Economic and Social Study," Routledge and Kegan Paul, London.

Lisansky, E., 1960, The etiology of alcoholism: The role of psychological predisposition, *Quart. J. Stud. Alc.* 21:314–343.

Lolli, G., Golder, G. M., Serianni, E., Bonfiglio, G., and Balboni, C., 1958, Choice of alcoholic beverage among 178 alcoholics in Italy, *Quart. J. Stud. Alc.* 19:303–308.

Lolli, G., Schesler, E., and Golder, G. M., 1960, Choice of alcoholic beverage among 105 alcoholics in New York, *Quart. J. Stud. Alc.* 21:475–482.

Lynn, R., and Hampson, S., 1970, Alcoholism and alcohol consumption in Ireland, *J. Irish Med. Assoc.* 63:39–42.

Mäkelä, K., 1971, Concentration of alcohol consumption, *Scandinavian Studies in Criminology* 3:77–88.

Mäkelä, K., 1972, Consumption Level and Cultural Drinking Patterns as Determinants of Alcohol Problems: Paper presented at the Thirtieth International Congress on Alcoholism and Drug Dependence, Amsterdam.

Marcus, M., 1946, "The Liquor Control System in Sweden," Norstedt and Söner, Stockholm.

Maskell, H. P., 1927, "The Taverns of Old England," Allan, London.

Mass Observation, 1943, "The Pub and the People," Gollancz, London.

McCarthy, R. G., and Douglass, E. M., 1959, Systems of legal control, in "Drinking and Intoxication" (R. G. McCarthy, ed.), pp. 429–435, Free Press, Glencoe.

McDermott, P. H., Delaney, R. L., Egan, J. D., and Sullivan, J. F., 1966, Myocardosis and cardiac failure in men, *J. Amer. Med. Assoc.* 198:253–256.

Nielsen, J., and Strömgren, E., 1969, Über die Abhängigkeit des Alkoholkonsums und der Alkoholkrankheiten vom Preis alkoholischer Getränke, *Akt. Fragen Psychiat. Neurol.* 9:165–170.

Nyberg, A., 1967, Alkoholijuomien Kulutus ja Hinnat, The Finnish Foundation for Alcohol Studies, Publication No. 15, Helsinki.

Odegard, P. H., 1928, "Pressure Politics: The Story of the Anti-Saloon League," Columbia U. Press, New York.

Oliver, B., 1947, "The Renaissance of the English Public House," Faber, London.

O'Neill, B., and Wells, W. T., 1971, Blood alcohol levels in drivers not involved in accidents and the lognormal distribution, *Quart. J. Stud. Alc.* 32:798–803.

Parreiras, D., Lolli, G., and Golder, G. M., 1956, Choice of alcoholic beverage among 500 alcoholics in Brazil, *Quart. J. Stud. Alc.* 17:629–632.

Pattison, E. M., Headley, E. B., Gleser, G. C., and Gottschalk, L. A., 1968, Abstinence and normal drinking: An assessment of changes in drinking patterns in alcoholics after treatment, *Quart. J. Stud. Alc.* 29:610–633.

Péquignot, G., 1958, Enquête par interrogatoire sur les circonstances diététiques de la cirrhose alcoolique en France, *Bull. Inst. Nat. Hyg.* 13:719–739.

Pernanen, K., 1974, Validity of survey data on alcohol use, *in* "Research Advances in Alcohol and Drug Problems" (R. J. Gibbins, Y. Israel, H. Kalant, R. E. Popham, W. Schmidt, and R. G. Smart, eds.), Vol. 1, Wiley, New York.

Popham, R. E., 1956, The Jellinek alcoholism estimation formula and its application to Canadian data, *Quart. J. Stud. Alc.* 17:559–593.

Popham, R. E., 1959, Some social and cultural aspects of alcoholism, *Can. Psychiat. Assoc. J.* 4:222–229.

Popham, R. E., 1962, The urban tavern: Some preliminary remarks, *Addictions* 9(2):16–28.

Popham, R. E., 1970, Indirect methods of alcoholism prevalence estimation: a critical evaluation, *in* "Alcohol and Alcoholism" (R. E. Popham, ed.), pp. 294–306, U. of Toronto Press, Toronto.

Popham, R. E., and Schmidt, W., 1958, "Statistics of Alcohol Use and Alcoholism in Canada 1871–1956," U. of Toronto Press, Toronto.

Popham, R. E., Schmidt, W., and de Lint, J. E. E., 1976, The prevention of hazardous drinking: implications for research on the effects of government control measures, *in* "Drinking" (J. A. Ewing and B. A. Rouse, eds.), Nelson-Hall, Chicago.

Raymond, A., 1969, Ten o'clock closing—the effect of the change in hotel bar closing time on road accidents in the metropolitan area of Victoria, *Australian Road Research* 3(10):3–17.

Reuss, C., 1959, History of Beer Consumption in Belgium 1900–1957, Inst. Rech. Econ. Soc. Univ. Louvain, Louvain [Ms. on file].

Room, R., 1971, The Effects of Drinking Laws on Drinking Behavior, Paper presented at the Annual Meeting of the Society for the Study of Social Problems, Denver, Colorado.

Royal Commission on Licensing (England and Wales), 1932, Report, H. M. Stationary Office, London.

Sadoun, R., and Lolli, G., 1962, Choice and alcoholic beverage among 120 alcoholics in France, *Quart. J. Stud. Alc.* 23:449–458.

Sariola, S., 1956, Drinking Patterns in Finnish Lapland, Finnish Foundation for Alcohol Studies, Publication No. 1, Helsinki.

Schmidt, W., 1972, A Note on the Effect of Lowering the Legal Drinking Age on the Consumption of Alcoholic Beverages, Addiction Research Foundation, Substudy No. 525, Toronto.

Schmidt, W., and Bronetto, J., 1962, Death from liver cirrhosis and specific alcoholic beverage consumption: An ecological study, *Amer. J. Pub. Hlth.* 52:1473–1482.

Schmidt, W., and de Lint, J., 1969, Mortality experience of male and female alcoholic patients, *Quart. J. Stud. Alc.* 30:112–118.

Schmidt, W., and de Lint, J., 1970, Estimating the prevalence of alcoholism from alcohol consumption and mortality data, *Quart. J. Stud. Alc.* 31:957–964.

Schmidt, W., and de Lint, J., 1972, Causes of death of alcoholics, *Quart. J. Stud. Alc.* 33:171–185.

Schmidt, W., and Kornaczewski, A., 1973a, A Note on the Effect of Lowering the Legal Drinking Age on Alcohol Related Motor Vehicle Accidents, Addiction Research Foundation, Substudy No. 552, Toronto.

Schmidt, W., and Kornaczewski, A., 1973b, A Further Note on the Effect of Lowering the Legal Drinking Age on Alcohol Related Motor Vehicle Accidents, Addiction Research Foundation, Substudy No. 558, Toronto.

Schmidt, W., and Popham, R. E., 1965, Unpublished Data on an Out-Patient Clinical Population in Toronto, Addiction Research Foundation, Toronto.

Schmidt, W., Smart, R. G., and Popham, R. E., 1962, The role of alcoholism in motor vehicle accidents, *Traffic Safety Research Review* 6(4):21–27.

Seeley, J. R., 1960, Death by liver cirrhosis and the price of beverage alcohol, *Can. Med. Assoc. J.* 83:1361–1366.

Seeley, J. R., 1962, The ecology of alcoholism: A beginning, *in* "Society, Culture, and Drinking Patterns" (D. J. Pittman and C. R. Snyder, eds.), pp. 330–344, Wiley, New York.

Select Committee Appointed to Inquire into the Extent, Causes and Consequences of the Prevailing Vice of Intoxication among the Labouring Classes, 1834, Report, Home Office Library, London.

Shadwell, A., 1915, "Drink, Temperance and Legislation," Longmans Green, New York.

Shadwell, A., 1923, "Drink in 1914–1922: A Lesson in Control," Longmans Green, London.

Simon, J. L., 1966a, The price elasticity of liquor in the U.S. and a simple method of determination, *Econometrics* 34:193–205.

Simon, J. L., 1966b, The economic effects of state monopoly of packaged-liquor retailing, *J. Political Economy* 74:188–194.

Sjöberg, C., 1969, Olutalkoholismi lisääntyy ruotsissa, *Alkoholikysymys* 3:81–84.

Skala, J., 1967, Some Characteristic Signs of Alcoholism in Czechoslovakia, Selected Papers Presented at the 12th International Institute on the Prevention and Treatment of Alcoholism, International Council on Alcohol and Alcoholism, Vol. 1, pp. 21–34, Lausanne.

Skog, O-J., 1971, Alkoholkonsumets Fordeling i Befolkningen, Statens Institutt for Alkoholforskning, Oslo [Ms. on file].

Smart, R. G., and Schmidt, W., 1970, Blood alcohol levels in drivers not involved in accidents, *Quart. J. Stud. Alc.* 31:968–971.

Takala, M., Pihkanen, T. A., and Markkanen, T., 1957, The Effects of Distilled and Brewed Beverages, Finnish Foundation for Alcohol Studies, Publication No. 4, Helsinki.

Terris, M., 1967, Epidemiology of cirrhosis of the liver: National mortality data, *Amer. J. Pub. Hlth.* 57:2076–2088.

Terry, J., Lolli, G., and Golder, G. M., 1957, Choice of alcoholic beverage among 531 alcoholics in California, *Quart. J. Stud. Alc.* 18:417–428.

Tolkan, M., 1969, Polityka cen a wzrost spozycia alkoholu, *Problemy Alkoholizmu* 4(1):7–10.

Ullman, A. D., 1958, Sociocultural backgrounds of alcoholism, *in* "Understanding Alcoholism" (S. D. Bacon, ed.), Vol. 315, pp. 48–54, Annals of the American Academy of Political and Social Science, Philadelphia.

Wallgren, H., 1960, Alkoholism och alkoholförbrukning, *Alkoholpolitik* 23:146–149.

Wallgren, H., 1970, On the Relationship of the Consumption of Alcoholic Beverages to the Genesis of Alcoholic Disorders, Alkon Keskuslaboratorio, Report 7378, Helsinki.

Walsh, B. M., and Walsh, D., 1970, Economic aspects of alcohol consumption in the Republic of Ireland, *Econ. Soc. Review* (Ireland) 2:115–138.

Walsh, B. M., and Walsh, D., 1973, Validity of indices of alcoholism, *Brit. J. Prev. Soc. Med.* 27:18–26.

Warburton, C., 1932, The Economic Results of Prohibition, Studies in History, Economics and Public Law No. 379, Columbia U. Press, New York.

Whitehead, P. C., 1972, The Prevention of Alcoholism: An Analysis of Two Approaches, Paper Presented at the Annual Meeting of the Canadian Sociology and Anthropology Association, Montreal, Quebec.

Wilkinson, P., Santamaria, J. N., Rankin, J. G., and Martin, D., 1969, Epidemiology of alcoholism: Social data and drinking patterns of a sample of Australian alcoholics, *Med. J. Australia* 1:1020–1025.

Wilkinson, R., 1970, "The Prevention of Drinking Problems," Oxford, New York.

Wolch, C., 1957, How Ontario clergy look at alcoholism, *Alcoholism Research 4(4):1–7.*

Index

Absenteeism, psychological, 448
Abstinence, 295
  moral basis for, 550
  mortality of group, 295
Accidents on highway, see Highway
Acculturation pressures, 21, 47
Achievement, pressure toward, 28
Addiction Research Foundation of Ontario, Toronto, 283, 526, 562
Adolescent
  adjustment leading to adult alcoholism, 201
  ambitious–impulsive, 201
  angry–dependent, 201
  antisocial, 201
  fearful–dependent, 201
  hopeless, 201
Affective disorder, 427-434
AFL–CIO Community Services Department, 500-502
Age and drinking, legislation of, 592-594
Aggravated assault
  lack of studies on, 377
Aggression–frustration model, 403
  indicator of, 379
  most dependent variable, 379
  in party setting, 391
  severity, 380

Aggressive behavior, 391, 397, 402
  Clyde Mood Scale, 430
  fantasy theme, 404
  machine, 376, 392
  mood, 379
  personality, 409
Aggressor, whim of, 415
Ainu of Japan, 8, 12
Al-Anon Family Group, 232-234
  basic lesson, 232
  for children, 234
  dynamics of, 232
  operational principles, 232
  spiritual dimension, 233
  "twelve steps", 232
  women of, 234
Alcohol
  aboriginal use of, 13, 15
  abuse
    among youth, 184-189
    among youthful prisoners, 195
    medical problems of, 42
    occupational, 445-517
      cost to employee, 446
        to employer, 446, 447
        to society, 446
      effect on others, 449
      work organization and, 445-517

Alcohol (cont'd)
  abuser, 337-340
    early identification, 337-340
    removal from hazardous situations, 337-340
  adult use as perceived by youth, 169
  advertising and drinking behavior, 540
  affective behavior, 413, 414
  affective liability, 414
  altering human consciousness, 37
  American values and attitudes, 77-83
  anthropological perspectives on the social biology of, 37-76
  and anthropology, physical, 42-44
  in archeology and history, 40-42
  assaultive behavior and, 328
    associated problems, 180
    violent crimes, 364
  attitudes in America, 77-83
  in blood of drunk drivers, see Blood alcohol level
  in blood of suicides, 322
  consumption,
    behavior, 277
    distribution curve, 276
    effect of age limit, 333
    frequency distribution, 276, 277
    Lederman's distribution curve (1956), 614
  contributing to nonhighway injury, 322
  coping mechanisms of, 413
  crashes on highways in the U.S.A., 315
  crimes of violence and, 351-444
    explanatory models, 369-444
    review of findings, 354-368
  and culture, classical, in the Mediterranean, 41
  conference on, 44
  curse of driving class, 117
  dependency conflict, 260
  as a disinhibitor, 396
    properties, 399
    theory of disinhibition, 388
  and drivers in crashes, 328
  and drugs, 323-324
  and Drug Control Officer, U.S. Army, 481
  and Drug Program Association of North America, 472
  education, philosophies of models, 548-552

Alcohol (cont'd)
  effects, 82
    on anxiety, 21
    on emotion, 247-249
    of limiting alcohol, 332
  in escalation process, 412
  escape, 47
  fatalities due to, 327
  and Health Report (1974), 459
  healthful use of, 42, 43
  heavy use of, 199
  and Highway Safety Report to U.S. Congress (1968), 312
  in history and archeology, 40-42
  homosexual lifestyle, 200
  impairment in drivers, 311, 316
    in pedestrians, 311, 314
  -induced depressive states as cause of suicide, 293
  industry, 567
  infrequent user of, 169
  and injury, 308
    police investigation of, 311
    unintentional, 307-349
  and linguistics, 42
  medicinal use, 1, 42
  in Mexico, 41
  on North American frontier, 41
  nonhighway injury, 322
  occasional user, 169
  as part of parents' lifestyle, 169
  patterns of use, 3
  in post-injury phase, 315-317
  in pre-injury phase, 315-317
  reduction of emotional tension, 18, 266, 267
  references in early writings, 41
  regular user, 169
  Safety Action Project, 334
  and smoking, 323
  social customs, 3
  sociocultural anthropology, 44-48
  specific issues, 294-295
  statistics, artifacts of, 582
  as stress—coping device, 456
  subcultures, 430
  threshold of impairment, 316
  in urine of arrested persons, 359
  use, 382, 429
    aggression, escalation of, 410
    chronic, 281

Index

Alcohol (cont'd)
  use (cont'd)
    coincidence studies, 278-280
      criticism, 278
      early studies, 279
      in England, 280
      in France, 280
      in Switzerland, 280
    crimes of violence, 351-444
      in Poland, 358
    different sexes and, 14-17
    independent variable, 371
    in industry, 371
    prior to crime, 371
    program for controlling behavior, 335-337
      for controlling exposure to risk, 333-337
      for controlling impairment, 333-337
      for reducing injury, 335-337
    studies, prospective, 282-285
      retrospective, 280-282
    by victims, 361
    among youth, 169-174
      age, 171
      American study (1948), 169
      availability, 174
      college experience, 178
      effect of locale, 178
      frequency, 171
      historical changes, 173
      inappropriate sampling, 170
      influence of friends, 176-177
      parental influence, 174, 175
      percentages, 172
      religion, influence of, 177
      school questionnaire, 170
      sex differences, 177
      socially unacceptable behavior, 180-189
      sociocultural factors, 177
    youthful abuse, 167-204
Alcoholics Anonymous, 105, 106, 157, 464, 474-476
Alcoholic beverages, 16, 90, 168
  African considerations, 42
  availability, 16
  Latin American considerations, 42
  legislation governing, 332
  quantity consumed, 168
  standard units of, 168

Alcoholic beverages (cont'd)
  symbolism of, 42
Alcoholic Beverage Study Committee (1970), 598, 601
Alcoholic family
  adjustments, 213
  attachments, 259
  excessive drinking, 211
  help seeking, 238
  literature on, 210-229
  problems due to, 227-229
  seven stages, 212
  treatment, 229-234
Alcoholic marriage
  communication style, 224
  competitive style, 225
  family members in, 221
  family therapy, 222
  increase of quarrels, 226
  interaction in, 217-226
Alcoholic personality, 243-274
  in highway crashes, 325
  power and dependency, 253-264
  theories, 264-268
Alcoholic, wife of, 211-216, 235
  action taken by, 227
  awareness, 214
  dependency needs, 212
  dominance needs, 225
  help-seeking pattern, 228
  history of alcoholics in family, 216
  pathological personality, 211
  personality of, 213-215
  stress relationships, 215
  typology of, 217
Alcoholic women
  adolescent, 131-132, 146
    dependency relationship, 132
  attitudes toward, 118, 155
  behavior, 148-149
  black, 154, 159
  enhanced womanly feeling, 268, 269
  heterogeneous group, 139
  husbands of, 146, 216-217
  as lesbians, 149
  power concerns, 268
  unconscious masculine identity, 268
  white middle class, 158
Alcoholics
  aggression and impulsivity, 257, 259
  aggressive subgroups of, 434

Alcoholics (cont'd)
  antisocial behavior pattern, 433
  cardiomyopathy, 290
  career, 246
  in child abuse, 365
  children of, 226-227
  chronic, defined, 280
  compare young and old, 246
  deaths from accidents, 292-293
    from poisonings, 292-293
  deaths from violence, 292-293
  underlying dependency, 258
  destructive effects, 226
  difficulties for growth, 226
  drinking history of, 238
  fighting, excessive, 433
  high measure of hostility, 258
  hypertension among, 290
  identification problems, 226
  institutionalized, 83
  labeling of, 422, 428
  masculine identity is weak, 259
  middle class, 294
  mood of, 374
  negative behavior symptoms, 227
  populations of, 259
  power and dependency concerns, 256-260
  primary, 195
  remitted, study of, 246
  secondary, 195
  self-perception of, 374
  single, 224
  sociability, 258
  social class, 294
  study of, 245-246
  suicide, high rate of, 293
  temper tantrum, 227
  test for emotional effect, 248
  theory of origin, 248
  violent crimes by, 351-444
  wife of, 138
Alcoholism, 46, 81, 107, 290
  absenteeism, 448
  adult, predicted by youthful characteristics, 191-194, 201
    truancy, 192
  anxiety and, 18, 267
  aspects of, 278
  and behavior, 205-242
  in black women, 154

Alcoholism (cont'd)
  and cancer, 287-289
  in Chinese, rare, 45
  community caregivers, 229
  and crime, violent, 351-444
  definition, 159, 276-278, 430, 533
  as deviant behavior, 158
  diagnostic categories, 244
  as disease, 596, 615
  early stages of study, 283
  education, 519-578
  failure feelings, 229
  family structure, 205-242
    interaction, 142
  frigidity in women, 142
  gynecological problems, 141
  homicide victims, 368
  juvenile, 189-191, 199
    ambivalence toward, 198
    in black youth, 201
    lifestyle influence, 190
    study of, in Monroe County, New York, 190
  literature review, 205-242
  longitudinal study, 247
  menopausal depression, 142
  a merry-go-round named denial, 224
  models for studying, 104-109
    bad habit model, 105
    crime model, 105
    disease model, 104-107
    limitations, 106-109
    physiological disease submodel, 106
    social problems model, 104-107
    vice model, 104-107
  mortality, 275-305
  among Navaho men, 6
  obstetrical problems, 141
  occupational, 445-517
    cost of alcohol abuse, 446-452
    dismissal, 461
    etiological factors, 454-458
    health insurance, 461
    help without penalty, 461
    hitting bottom, 463
    identification program, 462
    policies, 460-462
    preventive potential, 452-454
    work-based programs, 446-458
  personality factors, 244-249
    methods for studying, 246-247

Index

Alcoholism (cont'd)
  personality research, 245
  prevention, 519-578
  Programs of the Community Council of Greater New York, 231
  in Prussia, 280
  psychoanalytic theories, 249
  psychopathic criminal adjustment, 191
  rate of, 87, 325
    among Chinese, 45, 234
    among Italians, 234
    among Jews, 234
  in relatives, 130
  self-destruction, 223
  social behavior, 449
  suicide, 293, 294
  treatment, 157, 229-234
  ulcers, 291
  underlying dependency, 254
  variables, type of, 382-383 283
    common cause relationship, 383
    conditional relationship, 383
    conjunctive relationship, 383
    direct cause relationship, 383
    interactive relationship, 383
    intervening variables, 384
  in women, 117-166, 268-269
    black women, 154
Aleut, 15
Ambivalence, 49
American family, 207-208
  nuclear, 207
  urban, 207
    middle class, 208
American Prohibition, 78, 80
  teetotalism, 78
American state of values and attitudes, 80-83
American urban life, 208
American women
  social drinking of, 124-128
    sanctions, 125-126
"Anatomy is destiny", 119
Anomic extrovert, 193
Anomie theory, 106
Antabuse, 337, 541
Anthropology, subfields, 38, 39, 44-48
Antisocial personality patterns, 194, 251, 433
Anxiety, 18, 19, 48, 49, 188

Apoplexy, 290-291
Arbitrator and the problem drinker, 503
  crisis precipitation, 503
  discharge, 503
  reinstatement, 503
ARF, see Addiction Research Foundation
Army Alcohol and Drug Control Officer, 481
Arson, 353
Assault, 352
  aggravated, 356, 530
  and alcohol, 328
  model of, 418
Association of Labor and Management Administrators and Consultants on Alcoholism, 476
Aviation crashes
  drunk pilot, 318, 331
  fatal, 318, 331
Azande, drinking customs of, 4

BAC, see Blood alcohol level
Baganda, 3, 5
Banana wine, 4
Barroom brawl, 410, 411
BBC, see British Broadcasting Corporation
Beer, 90, 601, 604, 605
  drinking, 79
  in Nigeria, 6
  parlor, 589
  procurement, 8
Behavior, dependent
  in adulthood, 23
    aggregate rate of, 381
    definition, 380
    path of least resistance, 413
    violent, 354, 423
Behavior, social
  boomerang labeling, 450
  diagnostic expertise, 450
  exploitation, 450
  fear of attribution of responsibility, 450
  group life attribution, 450
  normalization pressures in the U.S.A., 450
  poor performance, 450
  privacy, 450
  problem definition, 450
  scapegoat, 450
Being orientation codes, 269

Beverage, alcoholic, *see also* Beer, Whiskey, Wine, etc.
  home production, 530
  legislation of age limit for consumption, 333
    place of sale, 332
    time of sale, 332
    type of drink served, 333
  reduction in alcohol content, 530
Beverage room, *see also* Tavern
  in Ontario, Canada, 588-589
  opening hour, 591
  six-o-clock swill in Australia, 592
Beverage type, *see* Beer, Whiskey, Wine, etc.
Bicyclist, fatally injured, 314
Black ghetto, drinking customs, 4
Blind man game, 262
Blood alcohol level, 183, 320, 371-376, 393
  in drivers, 313, 315, 321, 603
Bolivia, 46
Borkenstein study of blood alcohol levels in drivers, 603
Brain damage, 430
Bratt system of alcohol monopoly (1971), 606
Breath test, 338
British Broadcasting Corporation (BBC) study, 12, 20, 22, 24, 25
  crosscultural, 20

Cactus, fermented liquor from, 8
California, 283
  prison, new arrivals at, 359
  psychiatric institutions, murders committed in, 360
California Personality Inventory, 252, 258
California Psychological Index, 187
California Psychological Inventory, 143
Camba, Bolivia, 417
Cambridge—Somerville study, 192
Cancer and alcoholism, 287-289
  of esophagus, 288
  of liver, 288
  of lung, 288
  of prostate, 288, 289
  of rectum, 289
  of respiratory tract, 287-288
  and smoking, 288
  of stomach, 289
  of upper digestive tract, 287-288
  of uterus, 289

Cantonese, drinking pattern, 4
Carlisle experiment, 588, 607
Case study method, 7
Casual relationship, strength of, 385
Catholics, 88, 99, 101
Cerebrovascular disease, death rates, 290
Chicago, homicides in, 356
Chicha, 10
Child
  abuse, 378
  training, and drunkenness, 20, 21
    variables of, 20, 21
Childhood experiences, 427
Chinese, drinking pattern, 4, 45, 234
Christopher D. Smithers Foundation, 465, 470
Chromosome abnormality, 420, 423
Cigarette smoking, 111
  advertising ban on TV, 531
  and cancer, 288
Classified Abstracts Archive of the Alcohol Literature, 40
Cleveland, Ohio, alcohol use by victims, 361
Clyde Mood Scale, 422
Cocktail lounge, 589
Code of Hamurabi, 580
College students, drinking patterns, 181, 182, 200
  problem drinkers, 185-188
Columbus, Ohio, felonies, 358
Community caregivers, 229
Conditional variables, *see* Variables
Congeners, 376
Conjunctive variables, *see* Variables
Cooperative Commission on the Study of Alcoholism (1967), 92, 519, 563, 566
Copenhagen, Denmark, arrests for drunkenness, 584
Crashes involving alcohol, 316
  and drowning in car, 317
  and hangover, 317
Crime of violence, 351-444
  and alcohol use, 355-364, 377
  homicide, 410
Crosscultural studies, 11-29, 48
  correlations, 47-48
  methods, 11
  research, 30
Cross-sex identity, 260
Cultural factors, 376
Cuna, drinking patterns, 10

Index

Dean Alienation Scale, 186
Delinquency, definition of, 195
Denmark, 284, see also Copenhagen
Dependence
 field, 256
 -independence conflict, 23, 188
 variables of, 25
Dependency
 behavior, 22, 23, 48
 conflict hypothesis, 22
  multiple regression analysis, 26
  test of, 24
  variables, 26
 theory
  description, 254
  evidence, 254-256
   little, 257
  open to challenge, 255
  status, 262-264
Dependent behavior in adulthood, see Behavior
Depression, 147-148
Deprivation—enhancement code, 269
Descriptive studies, 3-7
Deviant drinking, see Drinking behavior, deviant
Diabetes, 289
 mortality from, 289
Differential association process, 106
Direct cause thinking, 387-388, 392
Disease model of alcoholism, 109
Disinhibition theory, 388, 393
 acceptance, 394
 concepts, 395, 397-399
 conditional, 398
 fallacy, 392
 use of, 395
Distilled spirits, 79, 90
Distilled Spirits Council of the U.S.A., Inc., 567
Distribution of consumption model, 526-531, 551
 price effect on, 527-529
Disulfiram, see Antabuse
Divorce, 229
Domestic stress, 133, 135
Drinker
 cultural factor, 329
 heavy, 409
 personality, 329
Drinking
 in aboriginal times, 13

Drinking (cont'd)
 age, 88-89
 in Asia, 41
 association with school performance, 184
 behavior in blacks, 153, 154
 correlates, 90-91
 delinquent girls, 153
 determinants, 17-29
 dependency—conflict hypothesis, 22
 development, 522
 deviant, 183
 ethnic background, 151-153
 ethnoscience, methods of, 6
 in Europe, 41
 folklore of, 42
 integrated, 522-524
 of Irish, 523
 of Italians, 523
 of Jews, 523
 at lunch, 456
 motivation for power, 24, 48
 motive for drinking, 18
 patterns of change, 103
 personality types, 457
 psychological explanations for, 17
 psychological state of dependence, 21
 in relation to aggression, 49
  to race, 151-153
  to religion, 151-153
 social status, 89
 socialization, 524-525
  church, 525
  family, 524
  school, 524
 sociocultural variables, 234-235
 subcultural variations, 83
 transcultural variables, 11-14
 belligerence associated with, 94-96
 comportment, learned, 43
 correlation with drinking problems, 86-90
 for courage, 378
 customs
  aboriginal, 13
  in Aleutian community, 5
  among Andean Indians, 6
  of the Azande, 4
  of black ghetto dwellers, 4
  among Bolivian Camba, 6
  among Bulawayo, 6
  in Central America, 5

Drinking (cont'd)
  customs (cont'd)
    comparative studies, 11
    among Eskimos of the Mackenzie River delta, 5
    in French culture, 4
    in Hawaiian plantation communities, 5
    among Indians, 5-6
      of Ecuadorean sierra, 5
    intensive case study analysis, 7
    of Irish, 4, 9
    of Iroquois, 6
    of Italians, 4, 9
    of Jews, 4, 9
    of Lovedu in Africa, 8
    of Machupe Indians, 6
    in Mexico, 6
    among Mohave community, 5
    among Navaho Indians, 5, 9, 10
    in Northern Brazil, 6
    among Ogalala Sioux Indians, 5
    in Polynesian societies, 5
    among Salish Indians, 5
    scientific interest in, 2
    on Society Islands, 6
    among Standing Rock Sioux Indians, 6
    in Tepepan, 5
    among tribal peoples, 3
    among White Mountain Apache Indians, 6
  and driving, 546
    campaigns in the U.S.A., 538
  and economy, 20
  effects of, 127-128
  heavy, 104
  legal restraints, 579-625
    days of sale, 591-592
    differential taxation, 601-606
    drinking age, 592-594
    hours of sale, 591-592
    location of outlet, 586-591
    monopoly system, 606-609
    outlet frequency, 581-586
    outlet type, 586-591
    price control, 595-601
  motivation for, 18
    reduction of anxiety, 18, 19, 48, 49
    patterns, 44-46
    for personal effect, 186
    for personal reasons, 4
    practices, 5, 84
      correlates, 87

Drinking (cont'd)
  problems among men, 21-59, 96-99
    among poor, 97
    among women, 127-128
  problems, factors of, 99-103
    environmental factors, 102
    multiple correlations of 51 variables, 102
    typology, 100-101, 108, 329
  promiscuity, 145
  reasons for, 17
  reduction of anxiety, 18, 19, 48, 49
  response to stress, 428
  ritualistic, 417
  social reasons, 4, 82
  spree, 83
  in Sweden, 284
  in Switzerland, 283
  time-out situations, 416
  weekend, prolonged, 409
  among women, 136-138
    in "closet", 136-138
    in public, 136-138
    with spouse, 138-139
    womanliness, search for, 128
    womanly feelings enhanced, 127
  at work, 456-458
  in Yugoslavia, 423
Drinks, see Beverage, alcoholic
Driver
  and alcohol, 324-328
  contact with police, 327
  drowning in automobile, 317
  stalled at railroad track, 317
  stopped at roadblocks, 326
  young, 330
Driving while intoxicated (DWI), 325-328
  arrests, 328
  entry point for identification of alcoholic, 340
  lack of evidence, 339
  plea bargaining, 339
  sentence imposed, 339
Drowning, 317, 318
  in car, 317
  falling overboard, 318
Drug use, 150-151
  by women alcoholics, 150-151
"Dutch courage", 407
DWI, see Driving while intoxicated

Index

Early life experience, 129-133
    disruption in, 130-131
East Indians, 45
Economy and degree of drinking, 20
Education on alcoholism
    curricular placement, 554-555
    implementation, 558-561
    materials, 552-553
    methods of instruction, 555-556
    philosophies, 548-552
        disruption of consumption model, 551
        proscriptive model, 550
        social science model, 549
    in schools, 547-561
    target groups for, 553-554
    teacher preparation, 556-558
Education, public
    fear-based, 334
    on highway safety, 334
Egoistic dominance, 29
Egypt, ancient, 41
Electroencephalogram, normal, 423
Emergency care, appalling, 343
Employee assistance, 482-489
    benefits, 489-496
    "broad brush", 482
    community as beneficiary, 491
    confusion, 494-495
    constituency support, 489-496
    definition of a problem employee, 487
    federal "seed" money, 489
    industrial physician, role of, 485
    industrial psychiatry, 488
    labeling, 494
    labor resistance to, 490, 500-505
    program effectiveness, 486
    self-referral, 488
Employment stress, 133, 135
England, 283, 285, 294, *see also* Tavern in
Epidemiology as a discipline, 109
    and behavioral science perspectives, 104-112
Epilepsy, 431
    and brain dysfunction, 431
    psychomotor-, 423
Episcopalians, 88
Escalation, analysis of, 411
    model of, 433
Eskimo on Baffin Island, 5
Ethnic background, *see* Ethnocultural group
Ethnocultural group, 86-90
Ethnography, 6

Family
    adaptation to stress of alcoholism, 235
    behavior in alcoholism, 235-238
    behavior in cultural context, 213
    case work, 230-232
    conflict, 208
    in crisis, 237
    equilibrium, 210
    homeostasis, 223
    lifestyle, 210
    problems, alcohol related, 238
    role during conflict within, 208
    as a social environment, 157
    as a social system, 208
    study of, theoretical approach, 205-207
    success, 237
    system dysfunction in alcoholism, 236
    system theory, 222
    therapy, 222, 232
    as a unit is vulnerable to stress, 208
Fasting, 424
Fat metabolism, 149
Fear message, 556, 564
Fear reduction due to alcohol, 17-19
Federal Employee Alcoholism Policy, 481
Federal Employee Alcoholism Program, 477-481
    guidelines, 478-479
    rehabilitation, not discipline, 479
    slow diffusion of program, 479
Feelings of failure, 229
Female alcoholism, sharp rise in, 125
    in France, 147
    in Quebec, 147
    in Scotland, 147
    in United States, 147
Feminine identification, 269
Feminine mystique, 119
Finland
    beer consumption, 586
    court records, 368
    drinking pattern, 584, 612, 616
    explosive drinking, 528
    prohibition, 585
    violent behavior, 528
    violent crime in, 356-358
Finnish Foundation for Alcohol Studies, 562

Finnish State Alcohol Monopoly, 595
Firewater myth, 43, 46
Flammability of items, 343
  of bedding, 343
  of clothing, 343
Folktales, 24, 42, 256
  psychological meanings, 267
Food supply, insecure, 21
Forest Potawatomi Indians, 5
France
  drinking pattern in, 4, 612
  female alcoholism in, 147
  homicide, voluntary, 360
Franck test, unconscious feminine identification, 260
Freud's view of women, 119
Frontier drinking patterns, 78
Frustration, 412
  aggression, 144
  defined, 380
  tolerance, low, 434

Gastroduodenal ulcer, 291-292
  mortality, 291
Gothenburg system of monopoly (1865), 606
Gough Femininity Scale, 252, 260
Gough and Heilbrun Adjective Check List, 251
Greeks in Southeastern United States, 14
Group discussion, 556, 557
  ineffective, 559
Group therapy, 230
  psychoanalytical approach, 230
  short term efforts, 230
  wife's participation, 230
Guidance, 111-112
Guilt, 110
Gypsy, 45

Hamurabi, code of, 41, 580
Happiness as concept, 405
Head injury, 424
Heart disease, 290
  cigarette smoking deaths, 290
Heavy drinkers, 17, 86, 169
  college males as, 252
  defined, 280
  labels, 422
  motivation of, 254
  smoking in bed, 334

Heavy drinkers (*cont'd*)
  social presence theme, 252
  young, 250
Helping profession and alcoholism, 568, 569
Helsinki, arrests for drunkenness, 584
High risk group, identification, 545
Highway and Alcohol Safety Report (1968), 312
Highway injury, 325-330
  and alcohol, 325
  and crashes, 183, 309-317
  fatal, 312-315
  first studies (1934), 324
  history, 309-312
  nonfatal, 315
Highway Safety Campaign, 334
Homicide, 412
  and alcohol, 322, 323, 365
  criminal in Chicago, 356
  instrumental, 352
  unsolved, 377
Hoover, Herbert, 79
Hostility while drinking, 16
Hughes Act, 465, 477, 510
Hyperfeminine repressives, 194
Hypertension among alcoholics, 290
Hypoglycemia, 404, 431

Identification, 496-499
  absenteeism, 498, 506
  attendance record, 498
  diagnose the job performance, not alcoholism, 501
  "halo" effect, 497
  performance evaluation, 497, 506
    client-oriented, 498
    impaired, 498, 499
  problem drinkers, 496
  by supervisor, 496, 506
  tardiness, 498
Idiosyncratic factors, 377
Infant behavior, 22
  treatment, 22
Impairment of driver, 311
  of pedestrian, 311, 314
Impairment
  due to hangover, 317
  due to simulated drinking task, 316
  threshold of, 316
Injury
  against oneself, 307

Injury (cont'd)
    against others, 407
    alcohol-related, 324-331
        history, 324-325
        option, 331
    countermeasures to alcohol-related, 331-344
        controlling availability of alcohol, 332-333
        controlling user behavior, 333-337
        early identification of alcohol abuser, 337-340
        environmental controls aimed at the pre-injury phase, 340-341
        improvement of emergency care, 343-344
        moderation of energy transfer, 342-343
        removal from hazardous occupation, 337-340
        systems approach, 331
    deliberate, 322-323
        homicide, 322-323
        suicide, 322
    door latch mechanisms and, 342
    energy transfer in, 342-343
    events, model of, 308-309
    history of, 307-308
    mechanism of occurrence, 308
    nonhighway, 330-331
    phase
        alcohol user in, 324-331
        collisions in
            with abutment, 342
            deliberate, 322-323
            with signpost, 342
            with steering column, 342
            with tree, 342
            with windshield, 342
    as a public health problem, 307-309
    release of energy, 309
    shooting, accidental, 331
    unintentional, 307
Inoculation hypothesis, 521
Insecurity of subsistence, 19
    positive association with measures of insobriety, 19
Integrated drinking, 12-14
    "integrated drinking" factor, 12
Integration model of prevention, 610-611
    forbidden fruit concept, 610
Intervention, single point, 109

Intoxication, 126-127, see also Alcoholism, Drinking
    social attitudes toward women, 126
Ireland, 87
Irish, 101
    drinking patterns, 4, 9, 523, 612
Italians, 123, 124
    drinking patterns, 4, 9, 153, 234, 523

Jackson Scale of Preoccupation with Alcohol, 280
Japan, 45
Jellinek's classical nosology, 107
Jessor's social psychological theory, 264-266
    theory of deviance, 264-266
Jews, 77, 88, 99, 101, 124, 234
    and drinking, 4, 9, 41, 46, 523, 562, 613, 614
Jivaro Indians, 8
Junior Chamber of Commerce, 537
Juvenile alcoholism, 189-191
    Monroe County, New York, study of, 190
Juvenile delinquency
    and teenage drinking, 195-197

Kinship features, 21
Kurhaus patients, 368

Labeling theory, 106
Labor union
    and alcoholism, 500-505
    arbitrator, 503
    barrier to cooperation, 504
    constructive confrontation, 502
Lackland Airforce Base experiment, 335
Latin America, 87
Lepcha, 8
Liver cirrhosis, 110
    death rate in Canada, 287
    in Holland, 598
    in the U.S.A., 279, 281, 597
    linked to alcoholism, 292
    and per capita sale of alcoholic beverages, 599
    and upper class, 600
Loss of face, 412
Lunch-time drinking, 456

Male—female mortality, 294-295
Manson—Alcadd test, 187

Marihuana, 81
Marital conflict, 136
Marital discord, 385
Marriage and children, 145
Maryland Drug Abuse Administration study, 181
Masculine power hypothesis, 24
　support for, 25
Massachusetts study of women prisoners, 367
　alcoholism in prison, 366
McClelland-formulated power theory, 253
Melbourne, Australia, alcohol use by victims, 361
Mental disorder, 121, 149
　epidemiology, 121
　relation to social class, 123
Mental Health Materials Center (1973), 552
Mental patients, study of, 367
Mentally incompetent, 424
Mescalero Apache, 5
Micronesia, 14
Middle-age identity crisis, 136
Milieu variables, 401
Minnesota Multiphasic Personality Inventory, 250, 252, 257
Missouri penal institutions, 366
Monopoly system of control of alcohol by state, 606-609
　Bratt system (1971), 606
　Carlisle experiment, 607
　Gothenburg system in Sweden (1865), 606
　purchase permit, 608
Monroe County, New York, study of juvenile alcoholism, 190
Montreal, alcohol use by victims, 361
Mortality
　cause-specific, 285-294
　in Groote Schuur Hospital, Cape Town, South Africa, 284
　male—female, 294-295
　and suicide, 150
Multicollinearity, 27
Multiple regression analysis, 26, 27
Murder, attempted, 352
　in Columbus, Ohio, 358
　in Glasgow, Scotland, 360
Muslims, 45

National Clearinghouse for Alcohol Information, 537-552
National Commission on Marihuana and Drug Abuse report, first, 80
National Council on Alcoholism 465, 470, 488, 536, 568
National Highway Traffic Safety Administration, 536, 552
National Institute of Alcohol Abuse and Alcoholism (NIAAA), 81, 84, 445, 465, 559
　consultant network, 466
　demonstration projects, 468
　Division of Special Treatment and Rehabilitation Programs, 465
　national surveys, 84
　Occupational Programs Branch, 465-469
　training conferences, 471
National Institute of Mental Health, 558
National Occupational Alcoholism Training Institute (NOATI) at East Carolina University, 466
National Safety Council, 536
Navaho Indians, 5, 6, 43
Neoplasm, 288-289
New Jersey, alcohol use by victims, 361
New York Moreland Commission, 543
New York State Study, 182
NIAAA, *see* National Institute on Alcohol Abuse and Alcoholism
Nontransportation injury, 319-322
　at home, 320
　occupational, 320, 321
　recreational, 320
North Carolina, alcohol use by victims, 361
North Carolina, alcoholics in prisons, 366
Norway, alcoholics, 283, 286
Nowlis Mood Adjective Checklist, 405
Nurturance, diffusion of, 28

Oakland Growth Study, 143
Occupational alcoholism: Some Problems and Some Solutions, 483
Occupational alcoholism, gaps exist in
　consultation, 511
　epidemiology, 511
　identification, 512
　post-treatment job performance, 512
　program diffusion, 513
　treatment, 512
Occupational Program Consultants Association, 477
Omaha, 15

Index

Omnibus Personality Inventory, 252
Ontario, homicide offenders, 360

Papago, 7
Paradigm of pleasurable feelings, 406
Parents
  drinking pattern, 179
  influence on youthful drinking, 90, 174-176, 179
  loss of, 130
  psychiatric problems, 130
Park's Problem Drinking Scale, 185
Park's Scale of Symptoms, 185
Pathological drinking patterns, systems view of, 223
Pathological intoxication, 423
Pedestrian
  and alcohol, 325
  fatalities, 314
    in New York City, 314
Penetration rate concept, 509
Penis envy, 119
Permissiveness, 125
Personality
  antisocial pattern, 194, 251, 433
  characteristics of drinker, 329
  cultural factors, 329
  research on alcoholism, 245
Physical anthropologists, 39
Pilot of aircraft, screening of, 337
Pneumonia, death rate in alcoholics, 291
Poland, 423
  alcohol addiction, 367
  homicide offenders, 358
Polynesia, 47
Population at risk
  establishing, 363, 364
  problems of determining, 369
Portal cirrhosis, 149
Postinjury phase, 309
Potawatomi Indians, 5
Power concerns, 260-262, 406
Power and dependency concerns, effects of alcohol, 260-262
Power fantasies, 261
  personalized increase with alcohol level, 253
Power syndrome, 258
Power theory, 255, 263
  available data, 262
  description, 253

Power theory (*cont'd*)
  evidence for, 254-255
  by McClelland, 253
  status of, 262-264
Prealcoholic
  evidence for dependency, 255
  exaggerated masculinity, 253, 260
  personality, 193, 328
  study of, 249-250
  youthful problem drinkers, 253
Predispositional factors, 427
  affective disorders, 427
  brain disorder, 427
  childhood experiences, 427
  emotional instability, 427
  psychopathic personality, 427
Preinjury phase, 309, 340-341
  automobile ignition lock, 341
  electric stove burner, 340
  freeway ramp, 341
  glass door, 340
  guard rail, 341
  railway crossing, 341
Premenstrual discomfort, 141
Premenstrual tension, 140
Prevention of alcoholism, 519-578, 609-619
  models for
    bimodal, 610
    distribution of consumption, 526-531, 561
    primary, 454, 533-535
    proscriptive, 531-532, 550, 561
    public health, 532-535, 551
    secondary, 533-534, 544-547
    single distribution, 611-619
    single distribution model, 611-619
      Finnish drinking, 612-616
      French drinking, 612
      Irish drinking, 612
      Jewish drinking, 613-614
      weakness of, 616
  social science, 520-526, 549, 561
    integrated drinking, 522-524
    mental health, 525-526
    normative structure, 521-522
    social conditions, 525-526
    socialization, 524-525
  techniques for, 535-547
    environment, 542-544
    public education, 536-541
    public information, 536-541

Prevention of alcoholism (*cont'd*)
  techniques for (*cont'd*)
    person, 541-542
    substance, 541
    tertiary, 533, 534
Preventive policy on alcoholism, development of, 565-571
Price control and drinking, 595-601
Primitive societies, drinking patterns, 2
Prison farm women, 217
  husbands of, 217
Problem drinker, 103, 104
  confrontation at work, 460, 463
  identification of, 460, 482
  impaired job performance, 460, 482
  referral, 460
  studies of young, 250-253
  The Problem Drinker on the Job, 464
Problem drinking, 79, 180, 265
  and anxiety level, 185
  in college, 185
  definition, 92
  labeling, 110
  scale for college males by Park (1962), 251
  school performance, 181
  surveys, 91
  typology, 329
  women
    characteristics, 127
    longitudinal studies, 131
Problems of alcoholism with
  behavior, 133
  binge drinking, 93, 95
  definition, 92-96
  family, 208-210
  financial, 94, 95
  frequent intoxication, 92, 95
  friends, 93, 95
  health, 94
  job, 93, 95
  with law, 94, 95
  neighbors, 93, 95
  psychological dependence, 93, 95
  relatives, 93, 95
  spouse, 93, 95
  symptomatic drinking, 93, 95
Prognosis of alcoholism, 155-158
Program on Alcoholism and Occupational Health at Cornell University, 465, 470, 471

Prohibition in
  Canada, 585
  Finland, 585
  U.S.A., 46, 110, 520, 529, 569, 585, 608
Proscriptive model of alcohol prevention, 531-532
  moral precepts, 531
  temperance movement, 532
Protestant denominations, 91, 99, 101
  conservative, 89
  fundamentalist, 88
Psychiatry, 510
  community health programs, 510, 511
  third revolution, 510
Psychoanalytic ideas, 118
Psychodynamics, 139-149
  sexual role, 140-147
  variability, 139-140
Psychological distress, 121
Psychopathic deviate
  hypomania scale, 250
Psychopathology
  research, 121
  sex differences, 120-124
Psychophysical identity, 369
Public education campaigns, 334
Public health model of alcohol prevention, 532-535
  analogy with infectious disease, 532
Puritan, 78
  ethics, 41

Racial tolerance to alcohol, 43
Rape, 127, 352
Rapid eye movement, *see* REM
Rapist, alcoholic, 365, 375
Rat behavior, 375
  increased fighting after alcohol injection, 390
Referral process, 499-500
  rush-to-treatment stigma, 499, 500
Rehabilitation, 111
REM-sleep
  activity, 408
  deprivation, 407, 408, 432
  rebound, 408
Respiratory organ, diseases of, 290
Rhodesia, homocides in, 360
Risk-taking
  factor, 418
  situation, 407, 416, 426

Index

Ritualization decreases violence, 418
Robbery, 352
  alcoholic homicide, 375
  committed after drinking, 362
  reckless brutality, 362
Role structure and problem drinking, 457
Rotter's social learning theory, 265
Rutger's Center for Alcoholic Studies, 40

Sain Louis, Missouri
  metropolitan police department, 356
  municipal psychiatric clinic, 192
Saké, 9
Santiago de Chile, alcohol use by victims, 362
Schizophrenia
  characteristics, 122
  ethnic background, 123
  English, 123
  Irish, 123
  Italian, 123
  latent, 424
  sex differences, 122-123
Scotland, 147, 360, *see also* Glasgow
Self-identification, 110
Sexual activity and alcohol, 127
  adjustment, 144
  frigidity, 144
Sexual differences, 118-120
  differential psychology, 118
  empirical measurements, test-laboratory, 118
  interchangeable jobs, 120
Sexual differentiation pattern
  drinking conforms to, 16
Sexual offenders, 365
Sexual role
  abortion, 140
  adjustment, 188
  alienation, 122
  biological phenomena, 140
  hysterectomy, 140
  identification, 187
  identity, 142, 144
  infertility, 140
  menopausal depression, 140
  miscarriage, 140
  premenstrual difficulties, 140
  reversal of, 122
Siamese fighting fish, 373, 390
Sing Sing prison, alcoholics in, 366
Situational values, 401, 409

Sleep deprivation, *see also* REM
  with alcohol, 385
  without alcohol, 386, 421
Smith, Al, 79
Smokers, heavy
  in bed, 334
  as drivers, 323
Snowmobile, 319
Social class, 151-153
  status, 129
Social control, contributions of, 109-112
  guidance, 111-112
  prevention, early education, 110
  rehabilitation, 111
  self-identification, 110
  therapy, 111-112
Social deviance disorder, 122
Social drinking situation, 404
Social position, two-factor index, 96
Social science model of prevention of alcoholism, 520-526
  inoculation hypothesis, 521
  mental health submodel, 521
  sociocultural model, 521
Social setting
  empirical model for, 406
  as a variable, 401
Society, integration of, 49
Sociocultural anthropology, 38, 39, 44-48
Sociopathy and alcohol, 195
South America, drinking patterns in, 14
Soviet Union, violent crime and alcohol, 359, 360
Spree drinking, 83
Straus–Bacon index, 186
Stress, 428
  domestic, 133, 135
Substitute transportation for drunks, 334
Suicide, 282, 293-294, 353
  rates for men and women, 150
  social isolation, 293
Superfemininity, 128, 132
Sweden, drinking pattern, 284
Switzerland, drinking pattern, 283
Syphilis among alcoholics, 286
  in the U.S.A., 286

Task Force on Alcohol and Health, 522
Tavern
  in England, 580

Tavern (cont'd)
  first licensing system (1551), 580
  frequency and drunkenness, 583-584
  regulation of location, 586-591
    of number, 581-586
    of type, 586-591
  "reusable" sandwich, 587
  round-buying outlawed, 587
  studies on, 590
Taxation
  of beer, 601
  of drinking, 595
  of spirits, 601
  of wine, 601
Taylor Manifest Anxiety Scale, 186
Temperance Boards of Malmö, Sweden, 284, 285
Temperance Movement, 3, 78, 117, 532, 550, 551, 566
  wet—dry conflict, 566
Temporal lobe, 423
  dysfunction, 430
Tension reduction theory, 266-268
Terman—Miles masculinity—feminity test, 143, 187
Testosterone production, 423
Thematic Apperception Test, 254, 256, 258
Therapy, 111-112, 155-158
  physician's attitude, 155
  reality-based, 230
  response to, 156
  treatment, 157
  women response, 157
Tied-house system in England, 588
Time-out concept, 425
Time series, 506-508
Traffic accidents, 293
Transportation injury, 318-319
  in boating, 319
  in flying, 318
  in snowmobile, 319
Tribal societies, alcohol use, 1-36
Truckdriver, screening of, 337
Tuberculosis and alcoholism, 286
"Twelve steps" of Alcoholics Anonymous, 232
Two-factor index of social position, 96

United States of America
  alcohol-related problems, 91-104
  Brewers Association, 567

United States of America (cont'd)
  drinking behavior and problems, 77-115
  drinking practices, 83-91
  drinking problems, 77-115
  normalization pressures, 450
  "Unsettled undercontrollers," 193

Vagrants, 294
Vandalism, 353
Variables
  conditional, 401, 421
  conjunctive, 401
  potential intervening, 404
  predispositional, 419
Violence
  crimes of, 351-444
  escalation process, 400
  models of, 369-444
  proneness, 421
  unitary explanation, 382
Vodka, 372, 374
  in Denmark, 605

Weapon
  availability, 377
  "Saturday night special," 378
Westchester County, New York, alcohol use by victims, 361
Wet—dry controversy, 79
Wine, 1, 90
Wine Institute (U.S.A.), 567
Winnipeg, Manitoba, homicide offenders, 361
Women
  alcoholism in, 117-166, 268-269
    complications of, 129, 149-151
    concealment, 128
    drinking patterns, 129, 133-139
    early experience, 129-133
    psychodynamics, 129, 139-149
    research reports on, 128
    social differences, 129, 151-155
    treatment, 129, 155-158
  Christian Temperance Union, 78, 117
  Freud's view of, 119
  liberation movement, 117
  patterns of drinking and alcoholism, 133-139
    age of onset, 134
    stress, 135-136
    telescoping: later onset, 134-135
  on prison farm, 217

Women (*cont'd*)
  psychology encompassing "anatomy is destiny," 119
    cultural pressures, 119
  social drinking, 128
    increased womanly feelings, 128
  work role and alcoholism, 126
Working class patients, 294
Workplace
  and alcoholism, occupational, 452-454

Workplace (*cont'd*)
  job dissatisfaction, 455
  impairment, 453
  stress, 455

Yale (University) Summer School of Alcohol Studies, 464
Youngster, psychotic, 200
Yugoslavia, drinking patterns, 423